Lecture Notes in Electrical Engineering

Volume 833

The book series *Lecture Notes in Electrical Engineering* (LNEE) publishes the latest developments in Electrical Engineering - quickly, informally and in high quality. While original research reported in proceedings and monographs has traditionally formed the core of LNEE, we also encourage authors to submit books devoted to supporting student education and professional training in the various fields and applications areas of electrical engineering. The series cover classical and emerging topics concerning:

- Communication Engineering, Information Theory and Networks
- Electronics Engineering and Microelectronics
- Signal, Image and Speech Processing
- Wireless and Mobile Communication
- Circuits and Systems
- Energy Systems, Power Electronics and Electrical Machines
- Electro-optical Engineering
- Instrumentation Engineering
- Avionics Engineering
- Control Systems
- Internet-of-Things and Cybersecurity
- Biomedical Devices, MEMS and NEMS

For general information about this book series, comments or suggestions, please contact leontina.dicecco@springer.com.

To submit a proposal or request further information, please contact the Publishing Editor in your country:

China

Jasmine Dou, Editor (jasmine.dou@springer.com)

India, Japan, Rest of Asia

Swati Meherishi, Editorial Director (Swati.Meherishi@springer.com)

Southeast Asia, Australia, New Zealand

Ramesh Nath Premnath, Editor (ramesh.premnath@springernature.com)

USA, Canada:

Michael Luby, Senior Editor (michael.luby@springer.com)

All other Countries:

Leontina Di Cecco, Senior Editor (leontina.dicecco@springer.com)

**** This series is indexed by EI Compendex and Scopus databases. ****

More information about this series at https://link.springer.com/bookseries/7818

Shu-Chuan Chu · Jerry Chun-Wei Lin ·
Jianpo Li · Jeng-Shyang Pan
Editors

Genetic and Evolutionary Computing

Proceedings of the Fourteenth International Conference on Genetic and Evolutionary Computing, October 21–23, 2021, Jilin, China

Springer

Editors
Shu-Chuan Chu
College of Computer Science
and Engineering
Shandong University of Science
and Technology
Qingdao, Shandong, China

Jianpo Li
Northeast Electric Power University
Jilin City, Jilin, China

Jerry Chun-Wei Lin
Western Norway University
of Applied Sciences
Bergen, Norway

Jeng-Shyang Pan
College of Computer Science
and Engineering
Shandong University of Science
and Technology
Qingdao, Shandong, China

ISSN 1876-1100 ISSN 1876-1119 (electronic)
Lecture Notes in Electrical Engineering
ISBN 978-981-16-8429-6 ISBN 978-981-16-8430-2 (eBook)
https://doi.org/10.1007/978-981-16-8430-2

This Springer imprint is published by the registered company Springer Nature Singapore Pte Ltd.
The registered company address is: 152 Beach Road, #21-01/04 Gateway East, Singapore 189721, Singapore

Conference Organization

Conference Founder

Jeng-Shyang Pan	Shandong University of Science and Technology, China

Honorary Chairs

Jeng-Shyang Pan	Shandong University of Science and Technology, China
Leon S. L. Wang	National University of Kaohsiung, Taiwan
Yao Qingguo	Fujian University of Technology, China

Advisory Committee Chairs

Tzung-Pei Hong	National University of Kaohsiung, Taiwan
Josh Han-Chieh Chao	National Dong Hwa University, Taiwan

General Chairs

Jerry Chun-Wei Lin	Western Norway University of Applied Sciences, Norway
Shu-Chuan Chu	Flinders University, Australia
Liang Yonquan	Shandong University of Science and Technology, China

Program Chairs

I-Hsin Ting	National University of Kaohsiung, Taiwan
Philippe Fournier-Viger	Harbin Institute of Technology (Shenzhen), China

Jimmy Ming-Tai Wu	Shandong University of Science and Technology, China
Hong-Bo Liu	Dalian Maritime University, China

Local Organization Chairs

Weimin Zhang	Shandong University of Science and Technology, China
Chien-Ming Chen	Shandong University of Science and Technology, China

Electronic Media Chairs

Tsu-Yang Wu	Shandong University of Science and Technology, China
Thi-Thi Zin	University of Miyazaki, Japan
Matin Pirouz Nia	California State University, Fresno, USA

Invited Session Chairs

Peng Yanjun	Shandong University of Science and Technology, China
Voznak Miroslav	VSB-Technical University of Ostrava, Czech Republic

Publication Chair

Zhongying Zhao	Shandong University of Science and Technology, China

Finance Chair

Jui-Fang Chang	National Kaohsiung University of Sciences and Technology, Taiwan

Program Committee Members

Yongtang Bao	Shandong University of Science and Technology, China
Binge Cui	Shandong University of Science and Technology, China
Hsing-Chung Chen	Asia University, Taiwan
Chun-Hao Chen	National Taipei University of Technology, Taiwan

Chien-Ming Chen	Shandong University of Science and Technology, China
Youcef Djenouri	SINTEF Digital, Norway
Hamido Fujita	Iwate Prefectural University, Japan
Vicente Garcia Diaz	University of Oviedo, Spain
Philippe Fournier-Viger	Shenzhen University, China
Tzung-Pei Hong	National University of Kaohsiung, Taiwan
Jean Lai	Hong Kong Baptist University, Hong Kong
Jerry Chun-Wei Lin	Western Norway University of Applied Sciences, Norway
Wei-Cheng Lin	National Kaohsiung University of Science and Technology, Taiwan
Irfan Mehmood	Bradford university, UK
Weijian Ni	Shandong University of Science and Technology, China
Matin Pirouz Nia	California State University, USA
Seungmin Rho	Sejong University, South Korea
Ja-Hwung Su	National University of Kaohsiung, Taiwan
Wei Song	North China University of Technology, China
I-Hsing Ting	National University of Kaohsiung, Taiwan
Pei-Wei Tsai	Swinburne University of Technology, Australia
Cun-Wei Tsai	National Sun-Yat Sen University, Taiwan
Raylin Tso	National Chengchi University, Taiwan
Bay Vo	Ho Chi Minh City University of Technology, Vietnam
Miroslav Voznák	VSB-Technical University of Ostrava, Czech Republic
Chen-Shu Wang	National Taipei University of Technology, Taiwan
Mu-En Wu	National Taipei University of Technology, Taiwan
Wei-min Chen	Yili Vocational and Technical College, China
Wen-quan Zeng	Guangdong University of Science & Technology, China
Tsu-Yang Wu	Shandong University of Science and Technology, China
Jimmy Ming-Tai Wu	Shandong University of Science and Technology, China
Cheng-Wei Wu	National Ilan University, Taiwan
Chen Xin	Shandong University of Science and Technology, China
Lu Yan	Shandong University of Science and Technology, China
Ji Zhang	University of Southern Queensland, Australia
Thi-Thi Zin	University of Miyazaki, Japan

Zhongying Zhao	Shandong University of Science and Technology, China
Jianli Zhao	Shandong University of Science and Technology, China
Weimin Zheng	Shandong University of Science and Technology, China

Contents

Neural Network and Deep Learning

NLP and Clustering

Nature Inspired Constrained Optimization

Adaptive Droop Control Strategy for Island Microgrid Based on Improved Particle Swarm Optimization Algorithm

Xiangbiao Leng[1(✉)], Hao Zheng[2], Taiping Yuan[1], Yubin Lai[1], Fei Peng[1], Haixiang Yu[1], and Jingfeng Ou[3]

[1] China Southern Power Grid Energy Development Research Institute CSG, Guangzhou 510700, China

[2] Jining Power Supply Company, State Grid Shandong Electric Power Company, Jining 272000, China

[3] Guangdong Chengyu Engineer Consultation and Supervision Co., Ltd., Foshan 528200, China

Abstract. The isolated island microgrid with multiple distributed power sources operating in parallel cannot ensure that the voltage information is equal everywhere due to the difference in line impedance. As a result, its droop control may cause some problems such as unbalanced power distribution and bus voltage fluctuation. For this, a method to improve particle swarm optimization (IPSO) using fuzzy rule system to optimize droop control is proposed. Firstly, the principle and shortcomings of traditional droop control are analyzed. Then, in order to reach optimization of droop factor, a particle swarm (PSO) algorithm with fuzzy rule system is proposed, which can dynamically adjust the learning factor and inertia weight of the particle swarm algorithm, and effectively improve the convergence ability and search speed of the algorithm. The experiment results show that the proposed IPSO algorithm can maintain the real-time stability of bus voltage and microgrid frequency under complex operating conditions, efficaciously improve the accuracy of power balance distribution, and enhance the dynamic performance and stability of islanded microgrid.

Keywords: Microgrid · PSO algorithm · Fuzzy rule-based systems · Droop control · Power distribution

1 Introduction

As a small power system, the microgrid integrates distributed generations, energy storing equipment, energy transformers, protection devices and loads, and has two operating modes: islanding and grid-connected [1]. The control mode of microgrid is classified as master-slave control and equivalence control [2]. Compared with the master-slave control, the equivalence control mode simulates the external characteristics of the grid-connected operation of the synchronous motor. Under this control mode, the droop control of the distributed generations (DG) can automatically participate in the distribution of the output power, which is convenience for realizing the plug-and-play of the distributed generations [3, 4].

S.-C. Chu et al. (Eds.): ICGEC 2021, LNEE 833, pp. 3–13, 2022.
https://doi.org/10.1007/978-981-16-8430-2_1

The droop controller of the alternating current (AC) microgrid mainly includes two parts: the inner loop that uses the droop curve to generate the voltage reference quantity, and the outer voltage-current loop that tracks the reference value generated by the inner loop [5]. Whereas, the existence of line impedance and local load makes the voltage and frequency information of each DG not equal everywhere, which will lead to insufficient power allocation accuracy and bus voltage drop [6]. At present, artificial intelligence control algorithm plays an increasingly significant role in solving complex problems in power grid. Artificial intelligence control algorithm has incomparable advantages. Artificial intelligence control algorithms have been the primary way to study and settle electrical power system problems in control, modeling, operation and analysis [7].

Therefore, this paper proposed the IPSO algorithm by using the fuzzy rule-based system to optimize the droop control. The function of introducing fuzzy rule system is to dynamically adjust the learning factor and inertia weight of particle swarm algorithm, and improve the global search ability and local search ability of the algorithm. The simulation experiment results show that the proposed IPSO algorithm can sustain real-time voltage and frequency stability under different operating conditions, effectively improve the power allocation accuracy and the dynamic performance and stability of the microgrid.

2 Traditional Droop Control

The topology of the AC microgrid droop controller is similar to equivalent circuit of synchronous generator. The topology of the simplified multi-distributed power microgrid in Fig. 1 is obtained [8].

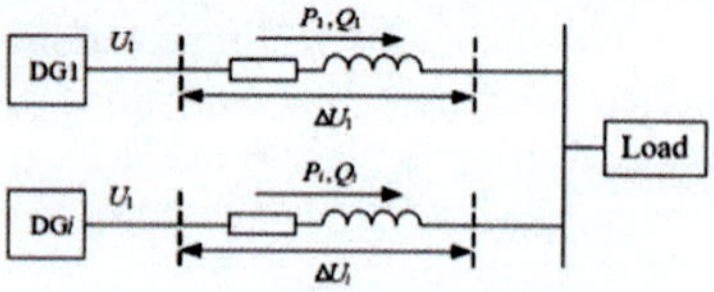

Fig. 1. Simplified structure of microgrid.

As illustrated in the picture above, the DG is connected to the microgrid through inverter and LC filter respectively and is connected to the AC bus through line impedance to supply power to the load [9]. Since $R << X$, the resistance part of the line impedance does not need to be calculated. The power output P_i and Q_i of the i-th DG are:

$$P_i = \frac{U_i U_{PCC} \sin \delta_i}{X_i} \quad (1)$$

$$Q_i = \frac{U_i U_{PCC} \cos \delta_i - U_{PCC}^2}{X_i} \quad (2)$$

In the formulas, U_i is the amplitude of the i-th DG output voltage, U_{PCC} is the amplitude of the AC microgrid bus voltage, X_i is the line reactance from the i-th DG to the AC bus, and δ_i is the phase of the i-th DG output voltage Phase difference with PCC voltage.

According to formula (1) and (2), the variation factors of active power P and reactive power Q of the DGs are voltage phase difference δ and voltage magnitude U. Thus the droop characteristics of microgrid are illustrated in the picture blow.

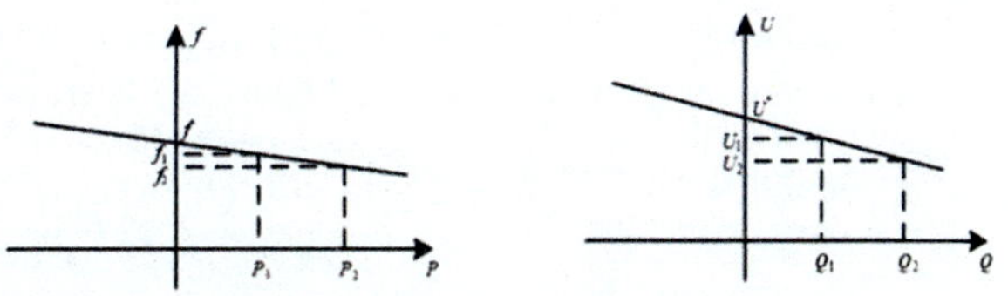

(a) P-f Droop characteristics. (b) Q-U Droop characteristics

Fig. 2. Droop characteristics.

From Fig. 2, we can get that the droop characteristic equation of the DG_i are

$$f_i = f^* - k_p P_i \tag{3}$$

$$U_i = U^* - k_q Q_i, \tag{4}$$

Where k_p is the frequency droop factor in the microgrid droop control, and k_q is the voltage droop factor in the microgrid droop control, f^* is the reference frequency value of the microgrid system, U^* is the reference voltage value of the microgrid system.

Therefore, the traditional droop control principle is illustrated in Fig. 3 [10].

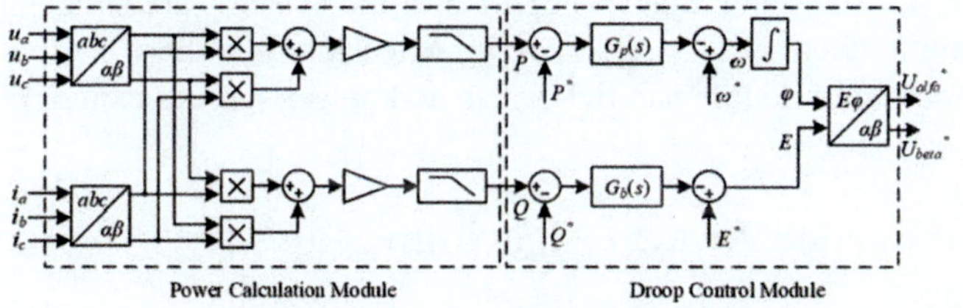

Fig. 3. Traditional droop control schematic

In the microgrid, the reactive power of the DG can 't be accurately distributed by droop control according to the local variable output voltage. Hereby, we should need to optimize the distribution of the DG reactive power output [8].

In the simplified microgrid topology diagram, the following formulas can be obtained:

$$U_i = U_{PCC} + \Delta U_i \tag{5}$$

$$\Delta U_i = \frac{X_i Q_i + R_i P_i}{U^*} \tag{6}$$

Among them, ΔU is the voltage drop of the i-th DG line impedance. As mentioned earlier, because of $R << X$, the above formula can be simplified to:

$$\Delta U_i = \frac{X_i}{U^*} Q_i \tag{7}$$

According to the simplified microgrid topology diagram in Fig. 1 and Eqs. (5) to (7), the reactive power transmission characteristic curves of each DG is shown as follows:

$$U_1 = U_{PCC} + \frac{X_1}{U^*} Q_1 = U_{PCC} + k_1 Q_1 \tag{8}$$

$$U_2 = U_{PCC} + \frac{X_2}{U^*} Q_2 = U_{PCC} + k_2 Q_2 \tag{9}$$

Among them, k_1 and k_2 are the impedance factors of the DG1 and DG2 exit line. Combined with the droop characteristic diagram shown in formula (4), the reactive power distribution relationship of double parallel distributed power sources is illustrated in the picture blow.

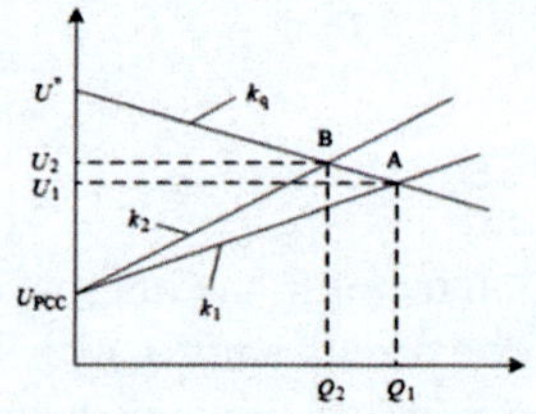

Fig. 4. Reactive power distribution relationship

When $X_1 \neq X_2$, the stable operation points of the systems are situated at A and B in Fig. 4. Assuming that there is $\Delta Q_1 = \Delta Q_2$, when $X_1 < X_2$, there is $Q_1 > Q_2$, and the reactive power output of double parallel distributed power sources cannot be accurately distributed.

3 Improved Particle Swarm Algorithm

3.1 Basic Particle Swarm Algorithm

The PSO algorithm has the characters of structure simplification and good optimization effect, and can search for appropriate parameters without establishing an accurate mathematical model [12]. The initial parameters of PSO are quantitative random particles. In the PSO running process, the particles update the parameters of individual extremum and global extremum through each iteration in order to find the optimum result [13].

In PSO algorithm, the particle velocity and position update formula are as follows:

$$v_{id} = \omega \times v_{id} + c_1 r_1 (p_{id} - x_{id}) + c_2 r_2 \left(p_{gd} - x_{id}\right) \tag{10}$$

$$x_{id} = x_{id} + v_{id} \tag{11}$$

Among them, c_1 and c_2 are study rates of the PSO algorithm; r_1 and r_2 are symmetrical random numbers in [0,1]; ω is the adaptive weight value.

3.2 Introduction of Fuzzy Rule System

Despite of simple structure of the basic PSO algorithm, when optimizing complex non-linear microgrids based on droop control strategy, the basic PSO algorithm has the shortcomings of insufficient global search and detection ability. Therefore, a series of improvements to the basic PSO are required.

This thesis introduces a fuzzy rule-based system on the basis of the proposed method in the reference [14] to enhance the performance of the PSO algorithm, and dynamically adjusts the learning factor c_1 and c_2 of the particle swarm algorithm through fuzzy inference and the inertia weight ω. Among them, the iteration coefficient λ of each particle is the ratio of the current iteration number to the maximum iteration number, and the iteration coefficient λ of each particle is between 0 and 1. The calculation formula is:

$$\lambda = \frac{t}{t_{max}} \tag{12}$$

Among them, t is the particle current iteration number, and t_{max} is the maximum iteration number of the particle. During the initial operation of the algorithm, λ approaches 0. With the change of the iteration number, λ gradually increases and approximates 1.

The particle similarity γ is the reciprocal of the square root of the average value of the optimum result of all particle distances in the whole world, and its value is between 0 and 1. The calculation formula is:

$$r = \frac{1}{\sqrt{\frac{1}{m}\sum_{i=1}^{m}\sqrt{\sum_{j=1}^{d}\left(x_{ij}(t-1) - gbest_i(t-1)\right)^2} + 1}} \tag{13}$$

Among them, m is the total particle number, d is the search space dimension, x_{ij} is the position of the i-th particle in the j-th dimension during the previous iteration, and $gbest_j$ is the j-th dimension of the optimum result found in the previous iteration. When the distance between the particle and the optimum result in the algorithm is larger, the average value is also larger, and the particle similarity is smaller, until it is 0; otherwise, the particle similarity is larger, until it is 1.

The fitness error ε of a particle is the difference between the average of the ratio of the fitness value of the global optimal particle to the fitness value of all particles and 1, and its value is between 0 and 1. The calculation formula is:

$$\varepsilon = 1 - \frac{1}{m}\sum_{i=1}^{m}\frac{f_{gbest}}{fit(i)} \tag{14}$$

Among them, f_{gbest} is the fitness value of the algorithm's global optimal particle, and $fit(i)$ is the adaptation degree of the i-th particle in the IPSO algorithm. When the algorithm has a good convergence effect, the particles approach the global optimum result, the ratio of f_{gbest} and $fit(i)$ becomes larger, approaching 1, and the fitness error approaches 0. Conversely, the difference between f_{gbest} and $fit(i)$ The ratio becomes smaller and approaches 0, and the fitness error approaches 1.

In the fuzzy rule-based system, the inputs are the iteration coefficient λ, the particle similarity γ and the particle fitness error ε, and the output is the learning factor c_1, c_2

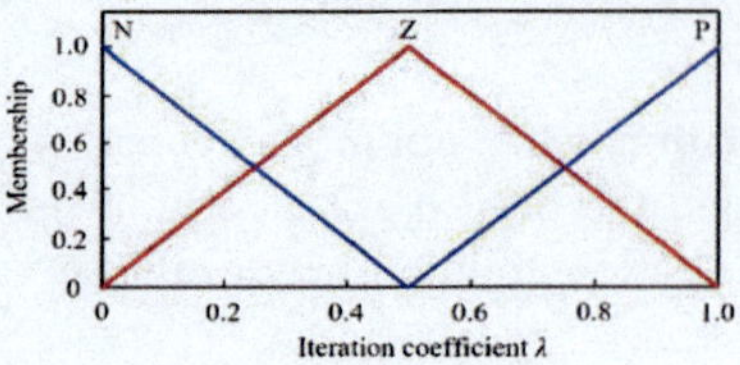

(a) Iteration coefficient λ membership degree

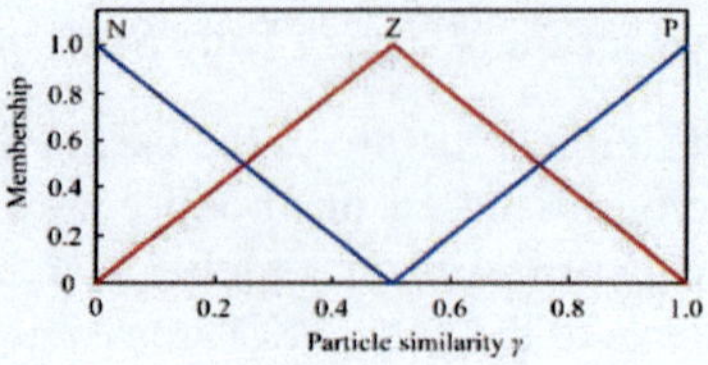

(b) Particle similarity γ membership degree

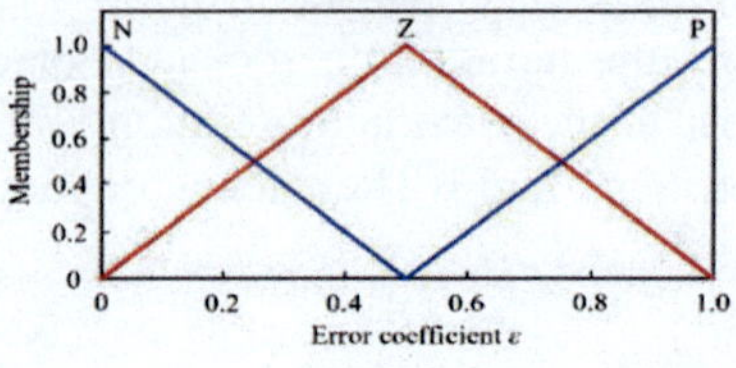

(c) Error coefficient ε membership degree

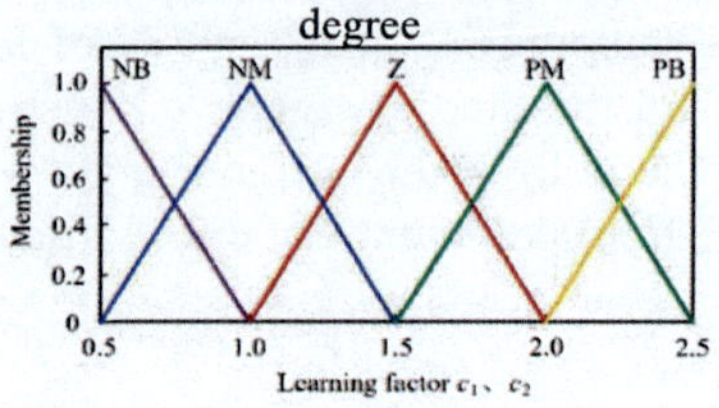

(d) Learning factors c_1 and c_2 membership degree

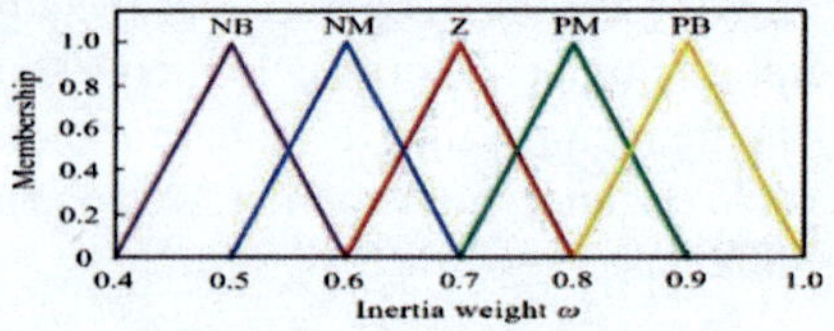

(e) Inertial weight ω membership degree

Fig. 5. Fuzzy rule system membership degree

and the inertia weight ω of the PSO algorithm, and there are 27 rules. For one thing, this article sets the delimited region of iteration coefficient λ, particle similarity γ and particle fitness error ε to [0,1], the delimited region of ω to [0.4,1], and the delimited region of c_1 and c_2 to [0.5,2.5]. Afterwards, this article set the number of input fuzzy divisions to 3. The number of output fuzzy divisions is 5. The fuzzy inference system adopts the Mamdani type system and the centroid method of gravity for the defuzzification [10]. Otherwise, because of the excellent performance and good application result, the triangular membership function is applied to the IPSO algorithm proposed in the thesis.

The application of fuzzy rule-based system must meet two requirements as follow. In the early stage of the IPSO operation, the large distance between the particle and the optimum result and the small value of the particle iteration coefficient λ, the particle similarity γ, and the particle fitness error ε can make the IPSO algorithm obtain the better exploration ability. In the late stage of the IPSO operation, the small distance between the particles and the optimum result, and the large value of the particle iteration coefficient λ, the particle similarity γ and the particle fitness error ε can make the IPSO algorithm obtain the local searching capability.

3.3 Fitness Function

In this paper, an island microgrid with double parallel distributed power sources is the research object, considering the power balanced distribution, bus voltage fluctuations and frequency oscillations, a control method using an IPSO algorithm to find the optimal droop factor is proposed.

The fitness function optimized for the IPSO algorithm is:

$$F = \Delta P + \Delta Q + \Delta U + \Delta f \tag{15}$$

In the above formula, ΔP can be expressed as:

$$\Delta P = |P_1 - P_2| \tag{16}$$

$$\Delta Q = |Q_1 - Q_2| \tag{17}$$

Among them, P_1 and P_2 are the active power output, and Q_1 and Q_2 are the reactive power output.

Similarly, ΔU can be expressed as:

$$\Delta U = \Delta U_1 - \Delta U_2 = |U_1 - U_0| + |U_2 - U_0| \tag{18}$$

Among them, ΔU and ΔU_2 are the output voltage deviations, U_1 and U_2 are the output voltages, and U_0 is the system rated voltage.

Similarly, Δf can be expressed as

$$\Delta f = \Delta f_1 - \Delta f_2 = |f_1 - f_0| + |f_2 - f_0| \tag{19}$$

Among them, Δf_1 and Δf_2 are the frequency deviations, f_1 and f_2 are the frequencies and f_0 is the power frequency of 50 Hz.

4 Simulation Analysis

4.1 System Control Structure and Parameter Design

There is the IPSO algorithm present in the second section of this article in application to the droop control of the island microgrid as demonstrated in Fig. 6. Among them, DG i is the i-th distributed power supply put into operation in the system, and droop control gives control signals for inverter i. The IPSO optimization module and the droop control part achieve the intercommunication of the microgrid system: when the microgrid is running, the measured system deviation information is input into the IPSO part, and the IPSO part receives the input signal to optimize the particles, and at the same time fuzzy rules the system Through the calculated particle information, the parameters of IPSO can be dynamically adjusted to speed up the convergence speed of the algorithm, and then update its own speed and position. After completing the final optimization to get the optimum result of the particle, the obtained two values d_p and d_q will be used as the input of the droop control strategy to get the new droop factor, and then the reference voltage and reference current are obtained to generate inverter control signal through the voltage and current loop.

The frequency band of the current and voltage loop are set to 200 Hz and 26.6 Hz, and their parameters are set to: $k_{pc} = 0.3$, $k_{ic} = 6$ and $k_{pv} = 4$ and $k_{iv} = 82.8$.

The parameter settings of the microgrid system are demonstrated in the table blow.

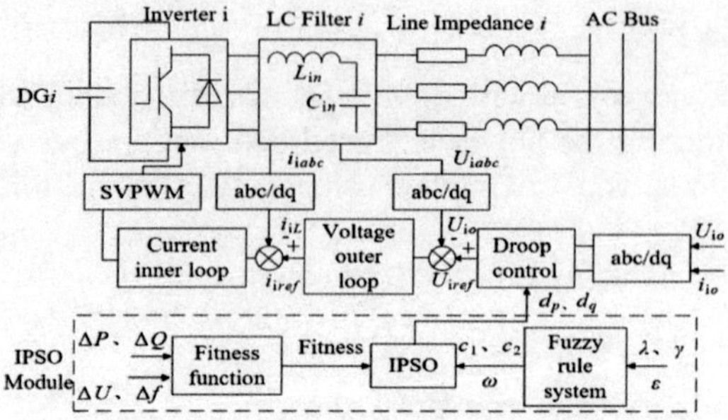

Fig. 6. Microgrid droop control structure diagram based on IPSO

Table 1. Parameter setup of simulation experiment

Parameter	Value
DC voltage source/V	700
Initial active power droop factor	2×10^{-4}
Initial reactive power droop factor	1×10^{-5}
Inverter resistance/Ω	0.3
System frequency/Hz	50
Filter inductance/mH	3
Simulation time/s	3
Initial load power/VA	2000 + j2000
Filter capacitor/μ_F	15
Effective value of bus voltage/V	220
DG1 line impedance/Ω	$0.642 + j2.64 \times 10^{-4}$
DG2 line impedance/Ω	$0.321 + j1.32 \times 10^{-4}$

4.2 Simulation Results

In the thesis, an island microgrid with double parallel distributed power source is used as the experimental subject. In the simulation experiment, DG1 and DG2 have the same capacity, the power distribution ratio of DG1 and DG2 is set to 3:2 [15]. The other parameters are illustrated in the above table. The initial active and reactive power are set to 2000W and 2000var. The load active and reactive power are 1000W and 1000var in 1s, but the reactive power become 500var in 2s. The simulation result under original droop control is demonstrated in the pictures.

In accordance with the first section of this article, the active power corresponding to the DG output can be accurately allocated according to the droop curve, as shown in Fig. 7(a), the overall distribution accuracy of the distributed power source active power output is high, and its fluctuation is small, but due to the inherent inflexibility of the droop factor, the active power output by DG1 and DG2 still has a small amplitude difference,

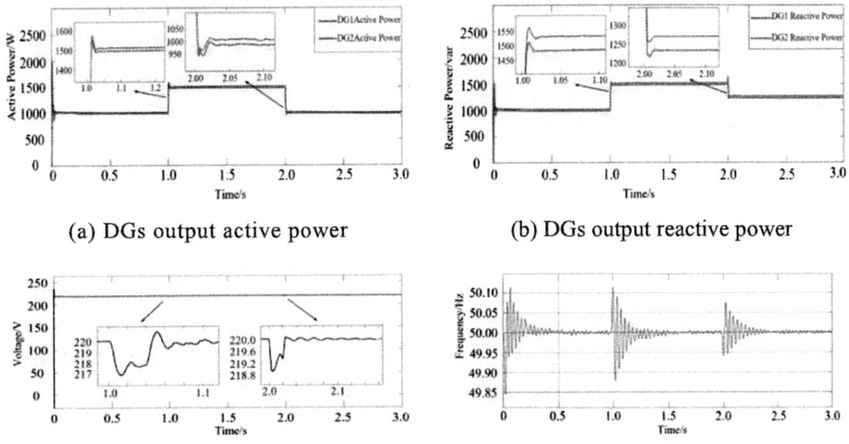

(a) DGs output active power (b) DGs output reactive power

(c) Effective value of Bus voltage (d) System frequency

Fig. 7. Simulation results of original droop control

with a power difference of 20 W. On the contrary, the reactive power cannot be accurately distributed through the original control strategy. As shown in Fig. 7(b), because of the different line impedance, the reactive power output by the two power supplies has a large difference, which is 50W, and the fluctuation range is large, and the reactive output of the two sources is unstable. Therefore, the reactive distribution under the traditional control needs to be optimized urgently. In addition, as shown in Fig. 7(c) and (d), as the load varies in the traditional control, the system frequency and bus voltage will have large fluctuation in value, and the duration will be longer, and the bus voltage fluctuation time will continue 0.1s, the fluctuation size reaches 3V, the system frequency fluctuation time lasts 0.5 s, and the fluctuation range is 0.1 Hz.

Figure 8 gives the experiment results of the IPSO optimized droop control present in this thesis. We could learn that the active and reactive output by DG1 and DG2 can be distributed in a high-precision and balanced manner in the above picture, and the output power is relatively stable with little fluctuation, realizing the coordinated output between the two power sources. In Fig. 8(c) and (d), under the improvement of the algorithm present in the thesis, when the load changes, the bus voltage and frequency are significantly improved in comparison with the traditional control strategy. The voltage fluctuation is controlled within 1V, and the frequency fluctuation is within 0.05 Hz, the bus voltage recovery time is also shortened from 0.1 s to 0.05 s, and the system frequency fluctuation time is shortened from 0.5 s to 0.2 s, which meets the requirements of system operation.

To sum up, the improved control strategy present in the thesis has great advantages in power allocation, voltage and frequency stability, and can enhance the performance of system operation.

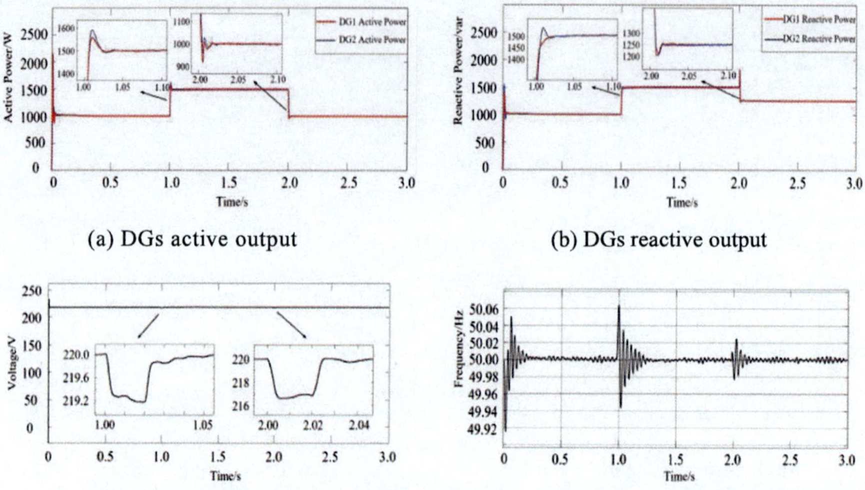

(a) DGs active output

(b) DGs reactive output

(c) Effective value of bus voltage

(d) System frequency

Fig. 8. Adaptive droop control simulation results based on IPSO

5 Conclusion

This thesis presents a power distribution control strategy for parallel multi-distributed power source microgrid based on an IPSO algorithm. The introduction of fuzzy rule system can make the PSO algorithm obtain better droop factor and enhance the convergence and optimization ability of the original algorithm. Then the IPSO algorithm is used to find the optimum droop factor. The simulation effects demonstrate that the present strategy realizes the intercommunication between the PSO part and the control part, and the optimalization and allocation of power under the condition of unequal impedance, and achieves the non-difference power control, and effectively enhances the performance and stability in all aspects of the microgrid.

References

1. Liu, Z., Wang, X., Zhuo, R., et al.: Flexible network planning of autonomy microgrid. IET Renew. Power Gener. **12**(16), 1931–1940 (2018)
2. Shahparasti, M., Mohamadian, M., Baboli, P.T., et al.: Toward power quality management in hybrid AC-DC microgrid using LTC-L utility interactive inverter: load voltage-grid current tradeoff. IEEE Trans. Smart Grid **8**(2), 857–867 (2017)
3. Wang, C.S., Li, W., Wang, Y.F., Meng, Z., Yang, L.: DC Bus voltage fluctuation classification and restraint methods review for DC microgrid. Proc. CSEE **37**(1), 84–97 (2017)
4. Yan, L.F., Liu, J., Shi, M.X., Chen, X., Wen, J.Y.: Adaptive power allocation strategy based on fuzzy logic algorithm for hybrid energy storage system in DC microgrid. Proc. CSEE **39**(9), 2658–2670 (2019)
5. Yong, J., Xu, X., Zeng, L.Q., Li, L.L.: A review of low voltage DC power distribution system. Proc. CSEE **33**(7), 42–52 (2013)

6. Mi, Y., Wu, Y.W., Zhu, Y.Z., Fu, Y., Wang, C.S.: Coordinated control for autonomous DC microgrid with dynamic load power sharing. Power Syst. Technol. **41**(2), 440–447 (2017)
7. Xu, Q.W., Hu, X.L., Wang, P., et al.: A decentralized dynamic power sharing strategy for hybrid energy storage system in autonomous DC microgrid. IEEE Trans. Ind. Electron. **64**(7), 5930–5941 (2017)
8. Zhu, S.S., Wang, F., Guo, H., Wang, Q.F., Gao, Y.X.: Overview of droop control in DC microgrid. Proc. CSEE **38**(1), 72–84 (2018)
9. Zhang, L., Zheng, H., Hu, Q.G., Su, B., Lyu, L.: An adaptive droop control strategy for Islanded microgrid based on improved particle swarm optimization. IEEE Access **8**, 3579–3593 (2020)
10. Xia, Y.H., Wei, W., Peng, Y.G., et al.: Decentralized coordination control for parallel bidirectional power converters in a grid-connected DC microgrid. IEEE Trans. Smart Grid **9**(6), 6850–6861 (2018)
11. Cai, C.C., Cheng, S.S., Deng, Z.X., Jiang, B., He, W.G., Ma, J.X.: DC Microgrid equivalent modeling based on fuzzy-RBF artificial neural network. Power Syst. Technol. **40**(11), 3446–3452 (2016)
12. Khorsandi, A., Ashourloo, M., Mokhtari, H.: A decentralized control method for a low-voltage DC microgrid. IEEE Trans. Energy Conversion **29**(4), 793–801 (2014)
13. Huang, Z.H., Dang, D.S., Gao, C.C., Wang, L.: Grey verhulst power load forecasting method based on background value optimization. IOP Conf. Ser. Earth Environ. Sci. **218**(1) (2019)
14. Yang, J., Jin, X.M., Wu, X.Z., Jiang, C.: An improved load current sharing control method in DC microgrids. Proc. CSEE **36**(01), 59–67 (2016)
15. Zhang, L., Zheng, H., Wan, T., Shi, D.H., Lyu, L., Cai, G.W.: An integrated control algorithm of power distribution for islanded microgrid based on improved virtual synchronous generator. IET Renew. Power Gener. **15**(12), 2674–2685 (2021)

Coordinated Optimization Control Technology of Clean Energy Consumption Based on Spatiotemporal Characteristics of Controllable Load

Weichun Ge[1], Shunjiang Wang[1], Qingbo Yin[2](✉), Feng Sun[1], Peng Jin[1], and Shengbin Chi[3]

[1] State Grid Liao Ning Electric Power Co. Ltd., Shenyang, China
[2] College of Electrical Engineering, Northeast Electric Power University, Ji Lin, China
[3] State Grid Shandong Electric Power Company Zibo Power Supply Company, Zibo, China

Abstract. As controllable load, electric vehicles (EVs) have great adjustable ability. This paper mainly analyzes how to charge EVs orderly according to the obtained state information, in order to shrink the power fluctuation and peak-to-valley gap of power grid. This paper analyzes the spacio-temporal characteristics of the travel of EV owners, and sets up the spacio-temporal model of the travel of EV owners based on the travel chain. The state information of the EV in the spacio-temporal model is obtained through the PLC communication system between the EV and the charging pile. Firstly, this paper analyzes the spacio-temporal characteristics of EV travel on the basis of the travel chain; Secondly, the Monte Carlo technique is applied to count the charging power of a certain number of EVs during disorderly charging; Finally, considering the basic load of residential area, genetic algorithm is introduced to shorten the charging time of EVs, the simulation experiment is carried out with the peak valley difference as the optimization objective.

Keywords: Electric vehicle · Space-time characteristics · Genetic algorithm · Orderly charging

1 Introduction

By 2020, there will be more than 5 million EVs in China. It can be predicted that with the popularity of EVs in next years, large-scale EVs connected to the power grid charging will have great influence on power system planning and operation. Among them, one of the important impacts is that large-scale EV charging will bring a new round of load growth, especially in the peak period of EV charging, which will farther increase the peak-valley gap of power grid load. Accordingly, there is an urgent need to study and lead the charging behavior of EVs to enhance the economy and reliability of power grid operation.

In [1], a space-time forecasting model of EV charging load for user travel simulation on the condition of the traffic network restrictions is proposed. Reference [2] present

S.-C. Chu et al. (Eds.): ICGEC 2021, LNEE 833, pp. 14–24, 2022.
https://doi.org/10.1007/978-981-16-8430-2_2

a probabilistic model for nodal charging requirement on the basis of the space-time dynamics of moving EVs, and set up the space-time model of moving EV loads according to random trip chain and Markov decision process (MDP). Considering the user's acceptance range of battery state of charge change, reference [3] sets up the travel and charging model of EV. In [4], the author comprehensively considers the stochastic behaviors of EVs, for example charging time, charging place and driving mileage. In view of the shortcomings of the existing behavior analysis model of urban traffic travel chain, this paper analyzes the basic concept of urban traffic travel chain and the possible path of simple travel chain [5]. Reference [6] present a new methodology for calculating the different charging stations load according to basis of trip chain. Based on nhts data set, the travel chain of EV is decoupled and reconstructed, and the spacio-temporal model of EV travel chain is established [7].

2 Space Time Characteristics of EV (Electric Vehicle)

2.1 Spatial Characteristics Modeling of EV Travel

The travel destination of EV owners is closely related to the charging power requirement of EVs. The travel chain can simulate the random law of the EV travel. Travel chain is a time-series series of EV travel that describes that the owner of EV starts from the starting point, passes through multiple intermediate destinations, and finally returns to the starting point [5].

Travel chain refers to the whole process of starting from the original place, going through several trips, and finally returning to the original place. By studying the daily travel chain data of users, the daily travel characteristics of vehicles can be mined. Assuming that the travel habits of traditional vehicle users can be completely consistent when the travel tools are replaced by EVs, the probability of travel characteristic parameters can be calculated by investigating the daily travel chain data of traditional vehicle users, and then the daily travel parameter state of EV users can be predicted.

In accordance with the different driving destinations of EVs, the location in the space can be divided into home (H), workplace (W) and leisure related (L). According to literature [2], the average length of travel chain of EV is 3, and the diagram of travel chain is shown in Fig. 1. It mainly consists of the following three parts: first, start in the morning, from the dwelling district to the working district, then go to the leisure area for entertainment activities after work in the afternoon, such as shopping, dinner or other purposes, and finally drive back to the residential area. Although not all of the EVs travel in line with the above rules, but the vast majority of EVs in line with the above travel rules. Therefore, when considering the spatial randomness modeling of EVs, this paper mainly considers the proposed three-stage journey.

The distance of each journey of EV not only affects the driving time, but also determines the *SOC* value of EV. Therefore, it is necessary to estimate the distance of travel chain. The starting and ending points of the journey decide the travel distance of EVs.

The distribution of the statistical results of EV travel distance of NHTS in 2017 adopts the roulette selection method to select the travel distance of EVs.

The travel distance d is illustrated in the formula below:

$$d = max(d_2) + a_2 * (max(d_1) - max(d_2)) \tag{1}$$

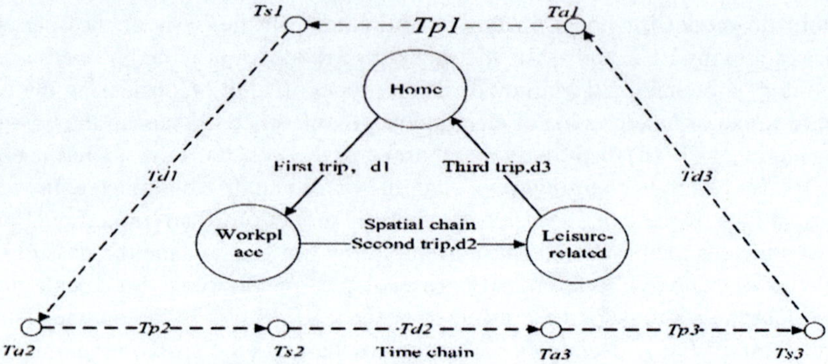

Fig. 1. Schematic diagram of travel chain

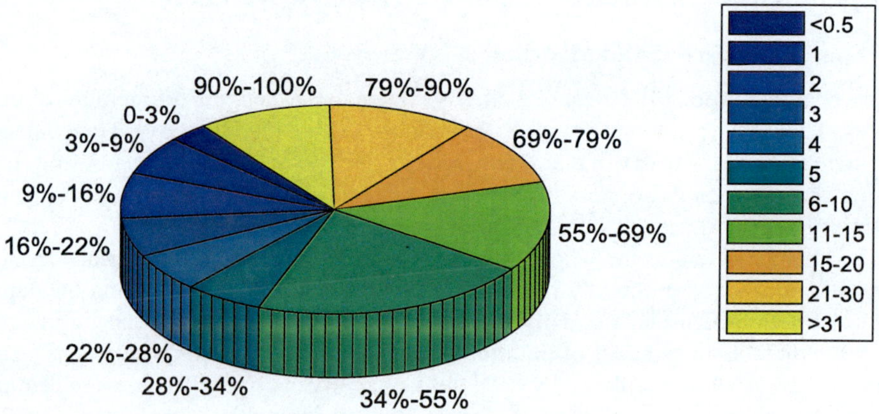

Fig. 2. Travel distance ratio of EVs

2.2 Time Characteristics Modeling of EV Travel

The travel time of EV includes the departure time t_s, driving time t_d, parking time t_p and arrival time t_a.

According to reference [1], the first trip time of EV, that is the time when EV starts from residential area, obeys the normal distribution law:

$$f(t_s, \mu_{t_s}, \sigma_{t_s}) = \frac{1}{\sqrt{2\pi\sigma_{t_s}^2}} \cdot e^{-\frac{(t_s - \mu_{t_s})}{2\sigma_{t_s}^2}} \tag{2}$$

Where, t_s is the departure time of the EV from the residential area, $\mu_{\mathrm{t_s}}$ and $\sigma_{\mathrm{t_s}}$ are the mean value and standard deviation of t_s, in this paper, $\mu_{\mathrm{t_s}} = 7$, $\sigma_{\mathrm{t_s}} = 1.7$.

Suppose that the whole EV is running at a uniform speed v the driving time t_d is mainly related to the travel distance d and the speed v of the EV. The travel distance d

can be obtained from the previous section. According to the literature [2], v is generally 28.5 km/h,

$$t_{\mathrm{d}} = \frac{d}{v} \tag{3}$$

On the premise of determining the travel time ts and driving time t_d of EV Therefore, the time t_a when the EV arrives at the workplace can be obtained from the following equation,

$$t_{\mathrm{a}} = t_s + t_d = t_s + \frac{d}{v} \tag{4}$$

By default, the EV will stay in the workplace after driving to the workplace, and the EV will enter the next journey after work. Set the dwell time of the EV in this process as t_p.

According to the research of [2], the parking time in different parking places follows the normal distribution,

$$f(t_{\mathrm{p}}, \mu_{t_p}, \sigma_{t_p}) = \frac{1}{\sqrt{2\pi\sigma_{t_p}^2}} \cdot e^{-\frac{(t_p - \mu_{tp})}{2\sigma_{tp}^2}} \tag{5}$$

When the parking place is the work area, $\mu_{tp} = 10$, $\sigma_{tp} = 1.67$; when the parking place is the leisure area, $\mu_{tp} = 1.5$, $\sigma_{tp} = 0.5$, the arrival time t_a and the stay time t_p can be obtained through the above process, so the next departure time $t_{s,2}$ of the EV can be obtained, as shown in formula (6).

$$\mathrm{t}_{s,2} = t_a + t_p \tag{6}$$

The low range and long charging time of EVs are the crucial causes for the "mileage anxiety" of EV owners. At the same time, overcharge and overdischarge of EVs will shorten the service life of the battery, so the owners of EVs usually neither fully charge the battery nor completely consume the battery energy. In this paper, we set a psychologically acceptable maximum SOC (SOC_{max}) and a psychologically acceptable minimum SOC (SOC_{min}) for EV owners. The default value of SOC_{max} is the original SOC value of the EV when the EV travels for the first time. When the SOC value of the EV is less than SOC_{min} after the end of the journey, or the SOC value of the EV in the next trip is not enough to complete the journey, the EV will charge when it reaches the destination.

According to references [3] and [4], the initial value of SOC_{min} and SOC_{max} of EV charging can be expressed by probability density function, and SOC_{min} obeys normal distribution,

$$f(SOC_{\mathrm{min}}) = \frac{1}{\sigma\sqrt{2\pi}} \cdot e^{-\frac{(SOC_{\mathrm{min}} - \mu)}{2\sigma^2}} \tag{7}$$

In the above formula, μ is the mean, and σ is the standard deviation of SOC_{min}.

The probability density estimation of SOC_{max} is,

$$f(SOC_{max}) = \begin{cases} 4.352 \cdot SOC_{max}^{3.352}, 0 < SOC_{max} \le 1 \\ 0, \quad otherwise \end{cases} \tag{8}$$

In order to avoid the problem that the randomly generated SOC_{min} value is larger than SOC_{max} value or the difference between them is small, it is stipulated that the minimum difference between SOC_{max} and SOC_{min} should be 20% or more, so as to ensure the normal travel of EVs, i.e.

$$SOC_{max} - SOC_{min} \geq 20\% \tag{9}$$

The SOC_a value of EV at the end of the journey depends on the distance d of the trip, the power consumption per kilometer (0.1678 KWh/km), and the SOC_S value of the last trip.

$$SOC_{\mathrm{a}} = \mathrm{SOC_s} - \frac{\omega \cdot d}{E_v} \tag{10}$$

Where, E_V is the capacity of EV battery, taking 16 KW·h.

When the SOC value of the EV after the end of the travel period is lower than SOC_{min} or the SOC value of the EV during the driving process of the next segment will not meet the driving requirements, the EV shall be charged until it is charged to SOC_{max}.

Therefore, the charging time t_c needed by the EV can be estimated,

$$t_c = \frac{SOC_{\mathrm{max}} - SOC_a}{\eta \cdot P_c} \tag{11}$$

Where η is the EV battery charging efficiency, and in this paper, η is 0.85, P_C is the EV battery charging power.

The conventional charging mode is adopted, and the charging power is 3.3 KW and 7 KW.

2.3 Monte Carlo Algorithm

By using Monte Carlo method, a 24-h day is divided into 96 segments (that is, each segment is 15 min), and the total charging load of M EVs in the i-th segment is shown in formula (12),

$$P_{\mathrm{i}} = \sum\nolimits_{m=1}^{M} P_{m,i}, i = 1, 2, ..., 96 \tag{12}$$

2.4 Mathematical Model of Orderly Charging for EVs

To avoid the phenomenon of peak on the peak, we need to set up the relevant mathematical models under the prerequisite condition of satisfying the charging requirement of EV possessors. The 24 h were divided into 96 time periods for optimal control, in which 15 min was a time period. The optimized time calculation point is the end of the time period. We can assume that the charging power of the EV connected in a period is calculated at this time calculation point. This paper mainly considers the secure and steady operation of the power grid, and presents the orderly charging control of EVs to find the optimum solution for shrinking the peak valley difference of the load on the grid side.

In the example of the load optimization of residential area, assuming that the valley value of load curve of EV charging into residential area grid is min (*P(t)*), and the peak value of load curve is max (*P(t)*), then the optimization objective is:

$$f = \min\{\max(P(t)) - \min(P(t))\} \tag{13}$$

The constraints need to satisfy the charging requirement of EV possessors and the power constraints of power grid:

The power constraint is the peak load constraint:

$$P(\mathrm{t}) \leq P_{\mathrm{max}} \tag{14}$$

In the above formula, *P(t)* is the total power grid load, and P_{max} is the maximum allowable peak load of the local power grid.

State of charge constraints:

$$\mathrm{S}_{\mathrm{i,t}} \leq \mathrm{S}_{\mathrm{i,max}} \tag{15}$$

In the above formula, S_i and *t* refer to the *SOC* value of the i-th EV at time *t*, and S_i and Max are the maximum SOC value of the i-th EV, which can be obtained from formula (8).

The state continuity constraint of EV can be expressed by formula (16)

$$\mathrm{S}_{\mathrm{i,t+1}} = S_{i,t} + \eta P_c \backslash E_{\mathrm{v}} \tag{16}$$

Where η is the charging efficiency, in this paper, 0.85, E_V is the battery capacity, 16 KWh, P_C is the charging power.

3 Case Analysis

In the example of a residential area, the orderly charging under different number of EVs is studied by using genetic algorithm. The daily load curve of the residential area is illustrated in the picture below.

The parameters of genetic algorithm are set as follows: the maximum evolution algebra *G*, the population size, the crossover probability P_C and the mutation probability P_M are 800, 100, 0.8 and 0.05 respectively [8]. When the number of EVs is 500 and 1000 respectively, the total charging load is optimized.

3.1 Case 1 When the Number of EVs in the Community is 500

According to the time to residential area, *SOC* status, and time to leave residential area generated in 2.1 trip chain as the input parameters of simulation, when the number of EVs in the residential area is 500, the load curve obtained is shown in Fig. 4.

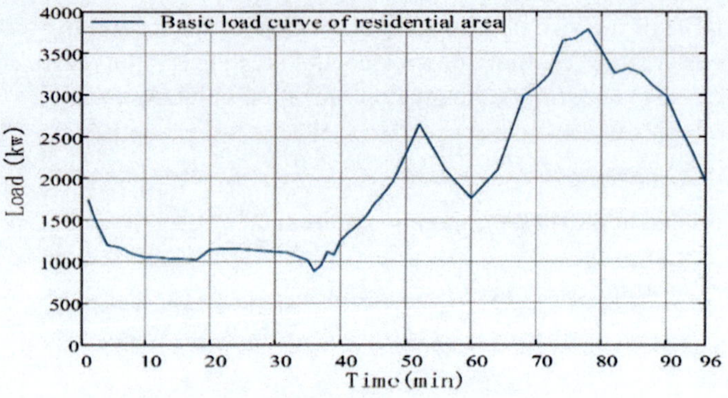

Fig. 3. Basic curve of the residential area load

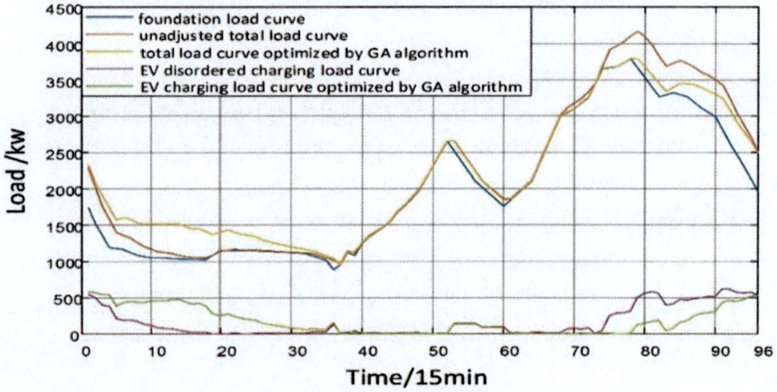

Fig. 4. Load curve of 500EVs

According to the analysis of purple curve and brown red curve in Fig. 4 above, the charging peak in disorderly charging period is concentrated between the 76th point and the 88th point, that is, during 19:00–22:00 p.m., which will cause the increase of load curve, increase the daily peak load, cause the phenomenon of "peak on peak" and threaten the secure and steady operation of power grid. The charging load curve optimized by genetic algorithm is shown in the green curve in Fig. 4. It can delay the charging time of EVs to the first point to the 28th point, that is, between 0:00–7:00 of the next day, on the premise of meeting the charging demand of EV owners, which can significantly reduce the peak load of EVs in disorderly charging. Figure 5 shows the distribution of charging parameters of charging start time, charging end time and the last travel distance before returning to the residential area before and after optimization.

The black points in Figs. 4, 5, 6 and 7 are the parameters optimized by genetic algorithm, and the green points are the parameters before optimization. Figures 4, 5, 6 and 7 illustrate intuitively that after optimization by genetic algorithm, the charging time of EV is delayed from 19:00–22:00 in the evening to 0:00–7:00 in the next day.

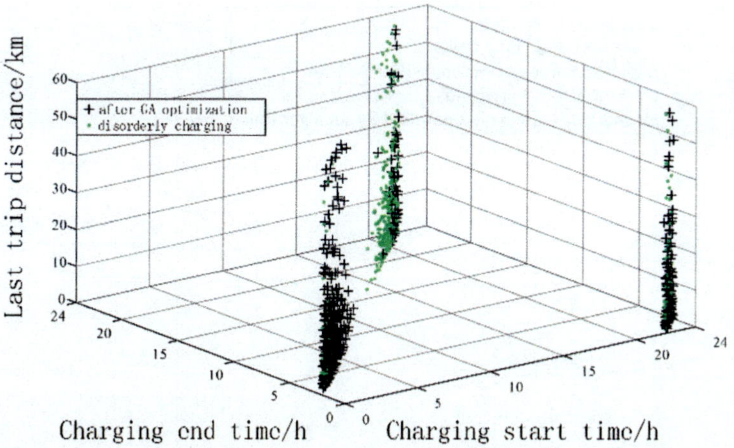

Fig. 5. Distribution of charging parameters when the number of EVs is 500

Table 1 lists the peak-to-valley gap of power grid load and the peak-to-valley gap of basic load of 500 EVs in the residential area under disordered charging and orderly charging.

Table 1. Peak-to-valley gap and peak load when the number of EVs is 500

Parameters	Base load curve	Total load without optimization	Total load optimized by GA algorithm
Maximum (KW)	3791	4169	3801
Minimum (KW)	887	973	963
Peak valley difference	2904	3196	2838

It can be concluded from the data in Table 1 that when the EV is charged disorderly, the peak-to-valley gap of load increases by 292 KW compared with the peak-to-valley gap of basic load; after orderly charging, the peak valley difference of load decreases by 66 KW compared with the peak-to-valley gap of basic load, and the effect is obvious. After the verification of simulation experiments, when the number of EVs is 500, the genetic algorithm is used to optimize the charging time of EVs, which can significantly avoid the problem that the peak-to-valley gap becomes larger during the charging period of EVs.

3.2 Case 2 When the Number of EVs in the Community is 1000

When the number of EVs is 1000, the load curve is illustrated in the picture.

The simulation results show that the charging time of most EVs is arranged between the first point and the 28th point, namely 0:00–7:00 on the next day. At 0:00–7:00, the

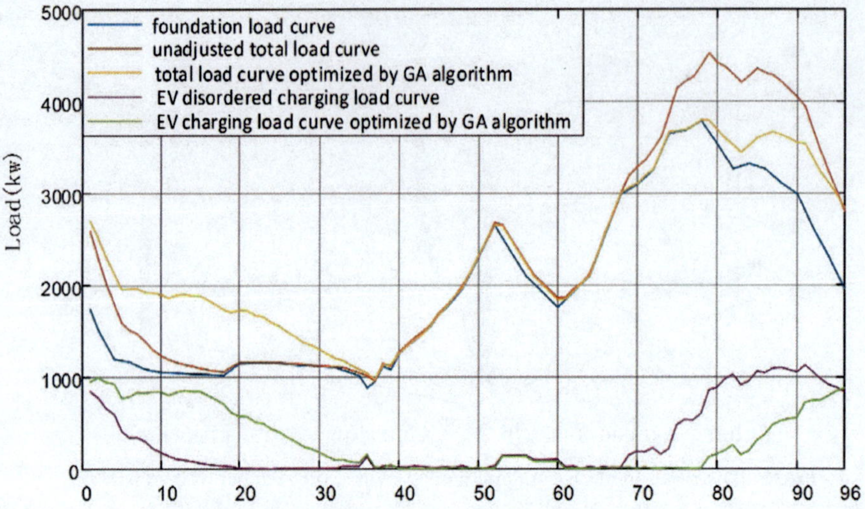

Fig. 6. Load curve of 1000 eV in residential area

basic load of the community is lower, which can possibly arrange the starting charging time of EVs, significantly reduce the peak load of the EVs in disorderly charging, and ensure the secure and steady operation of the power grid. The distribution of charging parameters of start time, end time and last travel distance before and after the EV optimization is illustrated in the picture below.

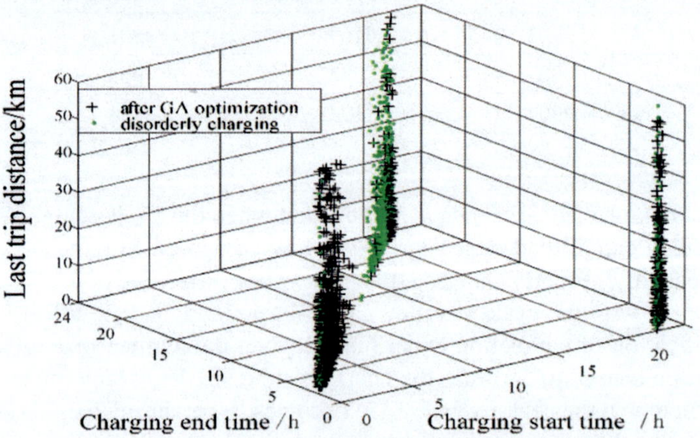

Fig. 7. Distribution diagram of charging parameters

The black points in Fig. 7 are the parameters of the EV after optimization by genetic algorithm, and the green points are the parameters of the EV before optimization. The below picture shows that the charging time of the EV after optimization by genetic algorithm is delayed from 18:00–23:00 in the evening to 0:00–7:00 in the next day.

Table 2. Peak valley difference and load peak

Parameters	Base load curve	Total load without optimization	Total load optimized by GA algorithm
Maximum (KW)	3791	4523	3811
Minimum (KW)	887	976	987
Peak valley difference	2904	3547	2824

It can be concluded from the data in Table 2 that when the EV is charging in disorder, the difference between peak and valley of load is increased by 643 KW compared with that of the basic load; the peak-to-valley gap of EVs with orderly charging optimized by GA algorithm is reduced by 80KW compared with that of the basic load. The simulation effects demonstrates that when the number of EVs is 1000, the genetic algorithm is used to optimize the charging time of EVs, which can significantly shrink the problem that the peak valley difference becomes larger during the charging of EVs.

4 Conclusion

The orderly charging strategy of EVs based on genetic algorithm proposed in this paper can extend the charging time of EVs, effectively reduce the peak valley difference of residential load during the charging period of EVs, so as to ensure the secure and steady operation of power grid. For two cases, in compare with the residential area load in disorderly charging, the simulation results show that, on the premise of knowing the vehicle state information, the genetic algorithm can extend the charging time of EVs, and can effectively shrink the peak-to-valley gap of residential area load during the charging period of EVs, so as to make the power grid safe and stable operation.

References

1. Yu., L.H., Bin, D.Z., Dan, C.L., et al.: The model and evaluation of electric vehicle charging load based on travel simulation. Power Syst. Autom. **43**(21), 88–96 (2019)
2. Tang, D., Wang, P.: Probabilistic modeling of nodal charging demand based on spatial-temporal dynamics of moving electric vehicles. IEEE Trans. Smart Grid **7**(2), 627–636 (2016)
3. Fan, M.X., Ying, L., Hao, W., et al.: Research on demand of charging point based on random simulation of electric vehicle travel. Acta Electrotechnics Sinica **32**(S2), 190–202 (2017)
4. Smart, J., Schey, S.: Battery EV driving and charging behavior observed early in the EV project. SAE Int. J. Alternative Powertrains **1**(1), 27–33 (2012)
5. Zhou, J.Z., Zhang, D.Y.: Behavior model of urban traffic trip chain under multi-modal traffic network. J. South China Univ. Technol. (NATURAL SCIENCE EDITION) **42**(2), 125–131 (2014)

6. Gong, L., Cao Wm Zhao J.: Load modeling method for EV charging stations based on trip chain. In: 2017 IEEE Conference on Energy Internet and Energy System Integration.Beijing, IEEE (2017)
7. Yue, L.: Research on Electric Vehicle Charging Strategy and Charging Station Planning Based on Time-Space Characteristics. University of Science and Technology of China (2019)
8. Tang, Y.C., Xu, Y.: Application of fuzzy Naive Bayes and a real-valued genetic algorithm in identification of fuzzy model. Inf. Sci. **169**(3), 205–226 (2004)

A Beta Distribution Based Optimization Algorithm and Its Application in Power Load Forecasting

Jiao Wang[1], Shu-Chuan Chu[2,3], Shi-Jian Liu[1], and Jeng-Shyang Pan[1,2](✉)

[1] School of Computer Science and Mathematics, Fujian University of Technology, Fuzhou, China
wj@smail.fjut.edu.cn
[2] College of Computer Science and Engineering, Shandong University of Science and Technology, Qingdao, China
[3] College of Science and Engineering, Flinders University, 1284 South Road, Clovelly Park, SA 5042, Australia

Abstract. Population-based evolution algorithms imitate lots of swarm-based behavior of creatures and even natural phenomena. In the entire evolutionary process, randomness and diversity make solution set convergence to the global optima quickly, and at the same time, make them jump out local optimums. However, this search mode based on the solution pool needs much memory space to store real solutions. The blindness of search will inevitably increase the time-consuming. Solving these disadvantages has been a hot topic in research, which is the goal of our work. In this paper, we propose a novel optimization algorithm based on Beta distribution. A known probability distribution, simulated by Beta distribution, transforms the solution set into a sampling sample. The specimen keeps updating toward the direction we expect, according to the characteristics of the probability distribution. This model reduces the memory space occupied by candidate solutions and also reduces the complexity of the algorithm. Finally, we analyzed and verified the performance of the proposed algorithm through numerical experiments, especially when dealing with large-dimensional problems. And its performance at power load forecasting has also proved its advantages in handling big data.

Keywords: Beta distribution · Probability · Weather station selection · Load forecasting

1 Introduction

In the field of intelligent computing, neural network technology is rapidly developing [1–3], but evolutionary computation, especially particle swarm optimization (PSO) [4, 5], still occupies an important position. And various population-based heuristic algorithms have emerged one after another, such as ant colony optimization (ACO) [6], cat swarm optimization (CSO) [7], bat algorithm (BA) [8], monkey king evolutionary

S.-C. Chu et al. (Eds.): ICGEC 2021, LNEE 833, pp. 25–35, 2022.
https://doi.org/10.1007/978-981-16-8430-2_3

(MKE) algorithm [9], genetic algorithms (GA) [10, 11], differential evolution (DE) [12], fish migration optimization (FMO) [13] and quasi-affine transformation evolutionary (QUATRE) algorithm [14, 15]. Their core technology is the same as that of PSO. First, initialize the candidate solution pool, and then evolve the loop, which is: (1) evaluate the fitness of each candidate in the solution pool, and generate the current optimal solution accordingly (2) randomly generates trial solutions based on fitness information of candidates, and (3) comprehensively consideration of candidate solutions and trial solutions in a certain way to update the candidate solution pool. From the perspective of the group level, it has a direction. However, what follows is a series of problems related to hardware performance, such as insufficient memory space, a large amount of calculation, and significant time consumption caused by group behavior.

Estimation of distribution algorithms (EDAs) [16–21] belongs to another type of evolution algorithm, and it has a close relation to a probability distribution. It has the most critical difference from PSO in the model update iteration. EDAs mainly estimate the distribution of problem variables. And then sample from the population according to the probability distribution to update the next generation of individuals, without mutation and cross operation. However, EDAs still have population characteristics, which means that the problems encountered by PSO still exists in EDAs. So, in this paper, a novel Beta Distribution based optimization Algorithm (BDA) is proposed that suppose an existing population conforms to a probability distribution. An individual sampled from it according to the same distribution. In other words, this virtual individual represents the continuous iterative evolution of the population, and its variables conform to this probability distribution.

2 Related Works

2.1 Mathematical Model of EDAs

EDAs are a type of swarm-based heuristic algorithm, and that each individual in the group is a candidate solution for the problem to be solved. Compared with the traditional evolutionary algorithm, its evolution is different in that estimating and sampling the joint probability distribution is the main operation to guide the evolutionary. In this section, the Gaussian model is taken as an example to further expanded. For simplicity, assume that the N-dimensional variables, which compose the individual, are independent of each other and conform to a Gaussian distribution, as shown in Eqs. (1) and (2).

$$X_i^g = N\left(\mu_i^g, \sigma_i^g\right) \tag{1}$$

$$f_{N\left(\mu_i^g, \sigma_i^g\right)}(x_i) = \left(1/\sqrt{2\pi}\sigma_i^g\right)e^{-(x_i-\mu_i^g)^2/2(\sigma_i^g)^2} \tag{2}$$

where $i = 1, 2, \cdots, n$. g represents the number of current iterations. Mean μ and standard deviation σ in point x_i.

In each generation, to select the current optimal solution, the tournament rule is introduced. From the perspective of individual variables in dimension, comparing corresponding values from two individuals randomly generated, and retaining the better.

Then an offspring group is obtained through a loop, used to produce μ and σ of x_i^{g+1}. With the increase of iterations, the population gradually evolves to the optimal direction under the guidance of μ and σ.

2.2 Beta Distribution

In probability and statistics, Beta distribution [22–24] is a set of continuous probability distributions defined at (0, 1). a and b are adjustable parameters in Beta distribution, utilized to control the shape of the distribution curve, that is, simulate various probability distributions.

Algorithm 1 Pseudo-code of BDA

Input: MaxIteration, r_1, μ, σ

Output: *Gbest*

```
Initialize vector C
While (t < MaxIteration)
   Generate a single sample X
   Calculate the mean value of X in (μ − σ, μ + σ)
   Evaluate fitness of X
   Update the current global optimal Gbest and μ^{g+1}
   If mod(function calls, g_1)==0
      C generated by the uniform distribution
   Else
      C generated by the Gaussian distribution
   End If
   If mod(function calls, g_2)==0
      σ = σ*r_1
   End If
   t = t + 1
End While
```

The mathematical expression revealed in Eq. (3) indicates the probability of occurrence in the range of $(-\infty, x)$, which is considered sampling probability $\boldsymbol{C}$. The horizontal axis of the cumulative distribution of Beta distribution is in (0, 1), so it is regarded as the occurrence probability of $\boldsymbol{C}$.

$$C = \int_{-\infty}^{x} f(s)\mathrm{d}(s) \tag{3}$$

3 The Proposed Novel Optimization

Gaussian distribution is everywhere [25, 26]. And many mathematical statistics methods are based on Gaussian distribution, such as t-distribution, F-distribution, and Chi-square distribution. Therefore, the Gaussian model is used to be the primary foundation of BDA. Determines the peak position of the Gaussian distribution through μ, which means the central area of the distribution. And σ controls the range of Gaussian distribution, which means the domain of data distribution is proportional to the value of σ. Accordingly, μ charges population moving direction, and σ limits the search range in BDA.

In the iterative process, in the form of a single sample that reflects the probability distribution of population, simulating the population evolution. Variables for the single representation sampled from independent Gaussian distributions. Then, what kind of distribution should the sampling probability of each dimension conform to, which area of the Gaussian distribution the sample value comes from, can the sample not lose generality and reflect the diversity of the population. In BDA, we use the Beta distribution to simulate the uniform distribution and the Gaussian distribution, using an alternating sampling strategy.

Algorithm 2 Pseudo-code of BDA-based weather station select method

Input: MaxIteration, g_1, g_2, μ, σ, r_1
Output: Z
1: Initialize **y**
2: **While** (t < MaxIteration)
3: Generate **X**
4: Generate **Z**
5: Fit model with the average temperature of the weather station combination, and assess forecast errors
6: **Update** $\boldsymbol{\mu}$
7: $\mu = \text{mean}(\mathbf{X}(\mathbf{Z}))$
8: **Update y**
9: **If** mod(Current iteration number, g_1) == 0
10: **y** accords with uniform distribution
11: **Else**
12: **y** conforms to Gaussian distribution
13: **End If**
14: **Update** $\boldsymbol{\sigma}$
15: $\sigma = r_1 * \sigma$
16: t = t + 1
17: **End While**

Algorithm 1 is the pseudo-code of the running process of BDA. Specifically, first of all, the Beta distribution simulates the Gaussian distribution, randomly generating the vector ***C***, which is performed by line 1. Then takes sample **X** from the inverse mapping of the cumulative distribution according to vector ***C***, which is performed by line 3. Updates the current global optima and takes the mean value of **X** in $(\mu - \sigma, \mu + \sigma)$ to update μ, guiding the virtual group to move to the optimal solution when the sample vector **X** is dominant, which are performed by lines 4 to 6. During the iteration process, to increase the diversity of representation and avoid gradually falling into the local optimum, the vector C is alternately generated by the Gaussian distribution and the uniform distribution at a certain number of iterations g_1, which are performed by lines 7 to 11. After a fixed number of iterations g_2, the σ decreases through the linear equation, limiting the search range gradually, which is performed by lines 12 to 14. Through experiments, the coefficient r_1 of the linear equation is 0.9. That not only accelerates convergence but also prevents the search space from converging too quickly and falling into local extremes.

In this paper, a novel weather station selection method based on BDA is proposed. This method reduces the amount of calculation and saves lots of time. Algorithm 2 is its pseudo code. In detail, the decision variables of the vector **y** respectively correspond to the probabilities of the weather station is selected. **X** is a randomly generated sample with the same population distribution based on **y**. **Z** is a binary vector, and it represents a combination of weather stations used to fit the predictive model. In the loop, the prediction error continuously evaluate. Thereby **y** and **Z** are updated until the stop conditions reach.

4 Experimental Analysis and Application

4.1 Numerical Experiment

In this section, 20 representative test functions verify the performance of BDA for large-dimensional problems [27–29]. Table 1 describes test functions $f_1 - f_{12}$ in detail. Unimodal functions contain $f_1 - f_4$. $f_7 - f_{12}$ are multi-peak functions that the number of their local minima is proportional to the number of decision variables. Composition functions are black-box tests, which come from $cec_{23} - cec_{30}$ in CEC2014 [30] that the equation is not displayed. Besides, f_5 is a step function with only a discrete minimum, and f_6 is a quartic function with noise. Four algorithms with different technical characteristics are comparison algorithms. They are PSO, GA [31], DE [32], and compact PSO [33].

Out of fairness, each optimal solution is obtained independently under the same experimental environment. Each algorithm runs independently ten times, and the fitness functions are invoked 50,000 times in each run. According to the definition of CEC2014, the dimension of the test function $cec_{23} - cec_{30}$ is 100.

Figure 1 shows that BDA has good convergence performance. The horizontal axis represents the number of times the test function has called. And the vertical axis represents the average optimal fitness value of test functions running independently ten times. Due to space limitations, we only exhibit the convergence curves of f_1, f_4, f_9, f_{10}, cec_{23}, and cec_{24}. They represent by a, b, c, d, e, and f, respectively. At the same time, Table 2 also clearly shows the advantages of BDA in search accuracy. Bold font indicates which algorithm obtains better results in the current test function. Overall, the BDA has superiorities, dealing with large-dimensional numerical problems, which meets our expectations.

4.2 Application in Load Forecasting

In this section, to verify the processing capacity in large-scale data, long-term power load forecasting in a large geographic area will be used as an application example [34]. In a predestined geographic zone, the fluctuation of power load is affected by many factors.

A large number of studies have shown that electricity consumption has a strong correlation with temperature. Therefore, how to combine weather stations, which are distributed in different geographical locations, to forecast power load has become a research goal that cannot ignore. In 2015, Hong et al. first proposed a weather station selection method [35] for power load forecasting in a predestined geographic area.

The weather station selection method selects and combines the weather stations distributed in the target zone [36] from all possibilities to predict the power load. In other words, it is that how many and which weather stations prophesy the power load in the target area. In 2020, Santiago et al. proposed a new GA-based weather station selection method [37] that successfully solved the problem.

However, this method needs to first find the optimal solution corresponding to the combination of different numbers of weather stations through computing large-scale training data, and then use a large amount of test data for cross-validation, and finally get the best combination of weather stations. This will consume a lot of time.

However, the BDA-based weather station selection method makes full use of the advantages of the meta-heuristic algorithm in terms of randomness and diversity of solutions, and at the same time, reduces the amount of calculation and saves lots of time. Our scheme does not need to find the best weather station combination corresponding to different numbers of weather stations but directly finds the final result. Therefore, there is no need for cross-validation, which further improves the efficiency. For this experiment, the data comes from GEFCom2014 [38]. The actual data from 2007–2009 used as the training set, and from 2010 as a test set.

Tao's Vanilla model as a predictive model and MAPE as the evaluation standard. Table 3 shows the prediction accuracy and time-consuming of two weather station selection methods based on different evolutionary algorithms. With similar prediction accuracy, the BDA-based method is more time-saving. That further proves BDA has advantaged processing large-scale data.

Table 1. The detail of test functions $f_1 - f_{12}$.

Functions	Dim	Best
$f_1(x) = \sum_{i=1}^{n} x_i^2$	1000	0
$f_2(x) = \sum_{i=1}^{n} (\sum_{j=1}^{i} x_j)^2$	1000	0
$f_3(x) = \max_i\{\lvert x_i \rvert, 1 \le i \le n\}$	1000	0
$f_4(x) = \sum_{i=1}^{n-1} \left[100(x_{i+1} - x_i^2)^2 + (x_i - 1)^2\right]$	1000	0
$f_5(x) = \sum_{i=1}^{n} (\lfloor x_i + 0.5 \rfloor)^2$	1000	0
$f_6(x) = \sum_{i=1}^{n} ix_i^4 + random[0, 1)$	1000	0
$f_7(x) = \sum_{i=1}^{n} -x_i \sin(\sqrt{\lvert x_i \rvert})$	1000	−12569.5
$f_8(x) = \sum_{i=1}^{n} \left[x_i^2 - 10\cos(2\pi x_i) + 10\right]$	1000	0
$f_9(x) = -20\exp\left(-0.2\sqrt{\frac{1}{n}\sum_{i=1}^{n} x_i^2}\right) - \exp\left(\frac{1}{n}\sum_{i=1}^{n}\cos 2\pi x_i\right) + 20 + e$	1000	0
$f_{10}(x) = \frac{1}{4000}\sum_{i=1}^{n} x_i^2 - \prod_{i=1}^{n}\cos\left(\frac{x_i}{\sqrt{i}}\right) + 1$	1000	0
$f_{11}(x) = \frac{\pi}{n}\left\{10\sin^2(\pi y_i) + \sum_{i=1}^{n-1}(y_i - 1)^2\left[1 + 10\sin^2(\pi y_{i+1})\right] + (y_n - 1)^2\right\} + \sum_{i=1}^{n} u(x_i, 10, 100, 4) y_i = 1 + \frac{1}{4}(x_i + 1)$ $u(x_i, a, k, m) = \begin{cases} k(x_i - a)^m, x_i > a \\ 0, -a \le x_i \le a \\ k(-x_i - a)^m, x_i < -a \end{cases}$	1000	0
$f_{12}(x) = 0.1\{\sin^2(3\pi x_1) + \sum_{i=1}^{n}(x_i - 1)^2[1 + \sin^2(3\pi x_i + 1)] + (x_n - 1)^2[1 + \sin^2(2\pi x_n)]\}$ $+ \sum_{i=1}^{n} u(x_i, 5, 100, 4)$	1000	0

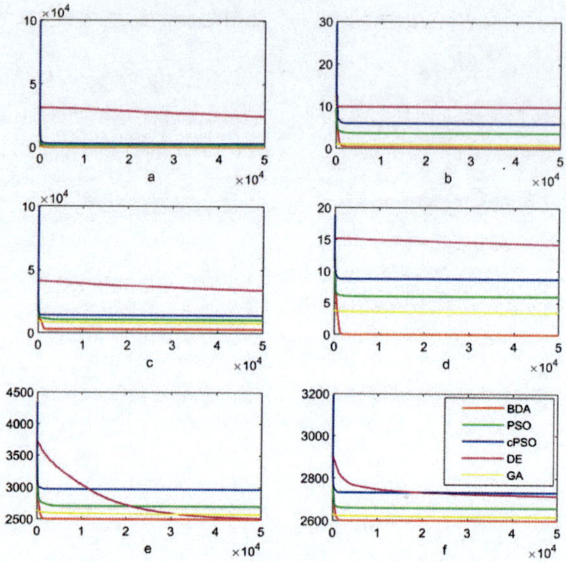

Fig. 1. Convergence curve of test functions.

Table 2. Comparison of average best values of numerical functions.

F	CPSO	DE	GA	BDA	PSO
f_1	3.0217E+03	2.4838E+04	2.3705E+02	**1.4586E-137**	1.6535E+03
f_2	3.0729E+07	3.6361E+05	5.0854E+07	**1.6999E-80**	3.2443E+04
f_3	6.0617E+00	9.9554E+00	9.7020E-01	**4.2370E-01**	3.7531E+00
f_4	9.3604E+04	1.8210E+08	1.4744E+04	**1.1412E+03**	7.1127E+05
f_5	1.4584E+04	2.4820E+04	8.8987E+02	**0.0000E+00**	1.6491E+03
f_6	1.9558E+07	7.6022E+08	5.3903E+04	**3.3370E-01**	3.3587E+06
f_7	– 5.0725E+155	– 1.7369E+03	– 4.4390E+02	**–3.9453E+03**	– 1.2268E+03
f_8	1.4073E+04	3.3957E+04	7.9449E+03	**2.9263E+03**	1.0488E+04
f_9	8.7561E+00	1.4263E+01	3.4918E+00	**1.4744E-14**	6.0477E+00
f_{10}	2.0134E+00	7.2343E+00	4.3200E-01	**1.4000E-02**	1.3749E+00
f_{11}	5.5792E+00	3.1727E+01	3.4917E+00	**4.1886E-09**	2.5011E+00
f_{12}	2.7723E+05	5.3632E+06	3.4697E+01	**1.3498E-32**	3.6334E+02
cec_{23}	2.9695E+03	2.5146E+03	2.5776E+03	**2.5000E+03**	2.7054E+03
cec_{24}	2.7335E+03	2.7165E+03	2.6200E+03	**2.6010E+03**	2.6608E+03

(*continued*)

Table 2. (*continued*)

F	CPSO	DE	GA	BDA	PSO
cec_{25}	2.7219E+03	2.7120E+03	2.7028E+03	**2.7000E+03**	2.7084E+03
cec_{26}	2.8025E+03	2.8034E+03	**2.8000E+03**	**2.8000E+03**	2.8002E+03
cec_{27}	7.4614E+03	3.5833E+04	3.2840E+03	**2.0000E+03**	5.7919E+03
cec_{28}	3.7156E+04	3.5056E+04	3.5226E+03	**3.0000E+03**	7.0351E+03
cec_{29}	2.4793E+06	4.1748E+07	2.1841E+08	**3.1000E+03**	5.6607E+08
cec_{30}	2.2211E+08	1.0931E+07	3.7796E+07	**3.2000E+03**	1.0549E+08

Table 3. Comparison of different weather station selection methods.

Method	In-sample		Out-sample
	Precision	Time(s)	Precision
BDA-based	5.2147	12418.6222	4.4961
GA-based	5.0227	25095.0435	4.3156

5 Conclusion

Although the traditional population-based EAs and EDAs have different evolutionary strategies, they still cannot avoid the adverse effects on computing resources and time-consuming issues brought by the enormous candidate solution set. Therefore, this article proposes a new type of EDA. The population evolution is reconstituted by that of a virtual sample based on the joint probability distribution. The diversity of solutions will be affected to a certain extent. But the BDA has less complexity. And it also has outstanding advantages in reducing the input of computing resources and improving the actual effect. These verify from the results of numerical experiments and application examples. At the same time, this also provides some exploration for dealing with large-dimensional or large-scale problems.

Acknowledgments. This work is supported by the National Natural Science Foundations of China (No. 61872085), the Scientific Research Project of Fujian Education Department (JK2017029, JAT190069), and the Scientific Research and Development Foundation of Fujian University of Technology (GY-Z18181, GY-Z20068, XF-X19017).

References

1. Chan, Z.M., Lau, C.Y., Thang, K.F.: Visual speech recognition of lips images using convolutional neural network in VGG-M model. J. Inf. Hiding Multimed. Signal Process. **11**(3), 116–125 (2020)

2. Wu, C.C., Wu, S.L.: Research and implementation of road traffic sign identification system. J. Network Intell. **4**(2), 47–57 (2019)
3. Wang, Y., Wu, H., Ma, H., Ma, Q., Ding, Q.: Speech denoising method based on improved least squares GAN. J. Network Intell. **5**(3), 113–121 (2020)
4. Zhan, Z., Zhang, J., Li, Y., Chung, H.S.: Adaptive particle swarm optimization. IEEE Trans. Syst. Man Cybern. Part B (Cybernetics) **39**(6), 1362–1381 (2009)
5. Chang, J.F., Chu, S.C., Roddick, J., Pan, J.S.: A parallel particle swarm optimization algorithm with communication strategies. J. Inf. Sci. Eng. **21**, 809–818 (2005)
6. Dorigo, M., Birattari, M., Stutzle, T.: Ant colony optimization. IEEE Comput. Intell. Mag. **1**(4), 28–39 (2006)
7. Kong, L., Pan, J.S., Tsai, P.W., Vaclav, S., Ho, J.H.: A balanced power consumption algorithm based on enhanced parallel cat swarm optimization for wireless sensor network. Int. J. Distrib. Sensor Networks **2015**(3), 729680 (2015)
8. Dao, T.-K., Chu, S.-C., Pan, J.-S., Nguyen, T.-T., Ngo, T.-G., Nguyen, T.-D., Tran, H.-T.: An improved bat algorithm based on hybrid with ant lion optimizer. In: Pan, J.-S., Lin, J.-W., Liang, Y., Chu, S.-C. (eds.) ICGEC 2019. AISC, vol. 1107, pp. 50–60. Springer, Singapore (2020). https://doi.org/10.1007/978-981-15-3308-2_6
9. Meng, Z., Pan, J.S.: Monkey king evolution. Knowl. Based Syst. **97**, 144–157 (2016)
10. Alsaleh, D., Larabi-Marie-Sainte, S.: Arabic text classification using convolutional neural network and genetic algorithms. IEEE Access **9**, 91670–91685 (2021)
11. Wang, J., Pan, B.W., Wang, Q.R., Ding, Q.: A chaotic key expansion algorithm based on genetic algorithm. J. Inf. Hiding Multimed. Signal Process. **10**(2), 289–299 (2019)
12. Das, S., Suganthan, P.N.: Differential evolution: a survey of the state-of-the-art. IEEE Trans. Evol. Comput. **15**(1), 4–31 (2011)
13. Pan, J.S., Hu, P., Chu, S.C.: Binary fish migration optimization for solving unit commitment. Energy **226**, 120329 (2021)
14. Meng, Z., Pan, J.S., Xu, H.: Quasi-affine transformation evolutionary (QUATRE) algorithm. Knowl. Based Syst. **109**, 104–121 (2016)
15. Meng, Z., Pan, J.S.: Quasi-affine transformation evolution with external archive (QUATRE-EAR): an enhanced structure for differential evolution. Knowl. Based Syst. **155**, 35–53 (2018)
16. Bajer, L.: Estimation of distribution algorithms: a recent approach to evolutionary computation. In: WDS'10 Proceedings of Contributed Papers, Part I, pp. 48–53 (2010)
17. Bengoetxea, E., Larrañaga, P., Bloch, I., Perchant, A.: Estimation of distribution algorithms: a new evolutionary computation approach for graph matching problems. In: Figueiredo, M., Zerubia, J., Jain, A.K. (eds.) EMMCVPR 2001. LNCS, vol. 2134, pp. 454–469. Springer, Heidelberg (2001). https://doi.org/10.1007/3-540-44745-8_30
18. Hauschild, M., Pelikan, M.: An introduction and survey of estimation of distribution algorithms. Swarm Evol. Comput. **1**(3), 111–128 (2011)
19. Morshedian, A., Razmara, J., Lotfi, S.: A novel approach for protein structure prediction based on an estimation of distribution algorithm. Soft. Comput. **23**(13), 4777–4788 (2019). https://doi.org/10.1007/s00500-018-3130-0
20. Wang, X., Xing, K., Li, X., Luo, J.: An estimation of distribution algorithm for scheduling problem of flexible manufacturing systems using petri nets. Appl. Math. Model. **55**, 776–788 (2018)
21. Liu, R.-D., Zhan, Z.-H., Chen, W.-N., Yu, Z., Zhang, J.: Estimation of distribution algorithm for autonomous underwater vehicles path planning. In: Huang, T., Lv, J., Sun, C., Tuzikov, A.V. (eds.) ISNN 2018. LNCS, vol. 10878, pp. 647–655. Springer, Cham (2018). https://doi.org/10.1007/978-3-319-92537-0_74
22. Ermolova, N.Y., Tirkkonen, O.: Using beta distributions for modeling distances in random finite networks. IEEE Commun. Lett. **20**(2), 308–311 (2016)

23. Ramos, P.L., Louzada, F.: Bayesian reference analysis for the generalized gamma distribution. IEEE Commun. Lett. **22**(9), 1950–1953 (2018)
24. Gómez-Déniz, E., Sarabia, J.M.: A family of generalised beta distributions: properties and applications. Ann. Data Sci. **5**(3), 401–420 (2018)
25. Krithikadatta, J.: Normal distribution. J. Conservative Dent. **17**(1), 96–97 (2014)
26. Dhargupta, S., Chakraborty, A., Ghosal, S.K., Saha, S., Sarkar, R.: Fuzzy edge detection based steganography using modified Gaussian distribution. Multimed. Tools Appl. **78**(13), 17589–17606 (2019). https://doi.org/10.1007/s11042-018-7123-x
27. Yao, X., Liu, Y., Lin, G.: Evolutionary programming made faster. IEEE Trans. Evol. Comput. **3**(2), 82–102 (1999)
28. Digalakis, J.G., Margaritis, K.G.: On benchmarking functions for genetic algorithms. Int. J. Comput. Math. **77**(4), 481–506 (2001)
29. Mirjalili, S., Mirjalili, S.M., Lewis, A.: Grey wolf optimizer. Adv. Eng. Softw. **69**, 46–61 (2014)
30. Liang, J., Qu, B.Y., Suganthan, P.N.: Problem definitions and evaluation criteria for the CEC 2014 special session and competition on single objective real-parameter numerical optimization. Zhengzhou University, Zhengzhou China, Computational Intelligence Laboratory (2013)
31. Li, J., Li, L.: A hybrid genetic algorithm based on information entropy and game theory. IEEE Access **8**, 36602–36611 (2020)
32. Thangaraj, R., Pant, M., Bouvry, P., Abraham, A.: Solving stochastic programming problems using modified differential evolution algorithms. Logic J. IGPL **20**(4), 732–746 (2012)
33. Neri, F., Mininno, E., Iacca, G.: Compact particle swarm optimization. Inf. Sci. **239**, 96–121 (2013)
34. Hong, T.: Long-term spatial load forecasting using human-machine co-construct intelligence framework. Dissertation, North Carolina State University (2008)
35. Hong, T., Wang, P., White, L.: Weather station selection for electric load forecasting. Int. J. Forecast. **31**(2), 286–295 (2015)
36. Hong, T., Pinson, P., Fan, S.: Global energy forecasting competition 2012. Int. J. Forecast. **30**(2), 357–363 (2014)
37. Moreno-Carbonell, S., Sánchez-Úbeda, E.F., Muñoz, A.: Rethinking weather station selection for electric load forecasting using genetic algorithms. Int. J. Forecast. **36**(2), 695–712 (2020)
38. Hong, T., Pinson, P., Fan, S., Zareipour, H., Troccoli, A., Hyndman, R.J.: Probabilistic energy forecasting: global energy forecasting competition 2014 and beyond. Int. J. Forecast. **32**(3), 896–913 (2016)

Optimization of Li-ion Battery Charge and Discharge Model Based on Genetic Algorithm

Xiang Li[1], Wen Yang[1], Huan Wang[1], Xiaopeng Chang[1], Rongxin Dang[2], and Dejun Wang[1](✉)

[1] Key Laboratory of Materials Design and Quantum Simulation, College of Science, Changchun University, Changchun 130022, People's Republic of China
wangdj@ccu.edu.cn

[2] Key Laboratory of Advanced Structural Materials, Ministry of Education & Advanced Institute of Materials Science, Changchun University of Technology, Changchun 130012, People's Republic of China

Abstract. In this paper, a strategy using genetic algorithm to optimize the charging process of lithium-ion batteries is proposed, which is based on the coupling of computer optimization algorithm and battery charging model. Visual Studio 2010 is used for programming, and COMSOL Multiphysics 5.5 is used to establish an electrochemical model of the battery to simulate various processes in the battery to verify the reliability and accuracy of the design algorithm. The genetic algorithm can be used to find the suitable charging current, and the optimization can be picked in a large number of databases by using the optimal strategy to reduce the time complexity. By comparing the battery's discharge curve and SOC result, it is found that the battery using genetic algorithm and intelligent optimization is better than that using the traditional constant current and constant voltage charging algorithm in charging time, cycle life, charging efficiency and other performance indicators.

Keywords: Genetic algorithm · Intelligent optimization · Lithium ion battery

1 Introduction

With the rapid development of science and technology, portable electronic devices have become necessities of people's lives today. For example, mobile phones, computers and various power supply systems all have a huge demand for portable batteries. Lithium-ion batteries have the advantages of long cycle life, recyclability and low self-discharge rate. In recent years, environmental and energy issues have become increasingly prominent, reducing battery costs and improving the safety and stability of lithium-ion batteries have become one of the problems that people need to solve urgently.

In recent years, the research of lithium battery charging and discharging optimization algorithm mainly focused on the traditional CC/CV method, multi-level charging optimization algorithm based on particle swarm optimization, charging optimization

S.-C. Chu et al. (Eds.): ICGEC 2021, LNEE 833, pp. 36–46, 2022.
https://doi.org/10.1007/978-981-16-8430-2_4

algorithm based on dynamic programming, and charging optimization method based on genetic algorithm.Some researches reduce the capacity loss and thermal runaway by optimizing the charging current, and only a few researches start from multi-objective optimization.There are some studies to optimize charge and discharge based on battery temperature and electrochemical mechanism.However, battery performance parameters and charging current are inseparable from SOC [1].

The battery operating system generally consists of a battery and a battery management system.

The quality of battery charging plays an important role in the operating efficiency of the battery management system, and the charging efficiency of a battery also has a great impact on battery performance and cycle life. Therefore, a good algorithm is needed to improve and optimize the battery charging process, and a model or system can be formed. It is proposed to use genetic algorithm and program design to optimize the battery charging process. The original intention of designing the optimization algorithm is to use a large number of data clusters in the database, and at the beginning of the optimization data, a set of initial parent nodes will be randomly generated. For each parent object on the tree, use the method of crossover and mutation to calculate the corresponding chromosome as the child node after this iteration. After collecting these child nodes and parent nodes, the scores of these filtered nodes are calculated according to the multi-objective function. Next, in the next iteration, the candidate node with a higher score is selected as the parent object. After the final iteration is completed, the optimal solution can be determined. The proposed algorithm is implemented in C language, and the parameters in the battery are obtained according to the least squares estimation algorithm. COMSOL software realizes the simulation of the battery, the purpose is to verify the accuracy and reliability of the algorithm.

The rest of this paper is arranged as follows. In the second section, various optimization algorithms are discussed and analyzed. The third section introduces and discusses the charging process optimized by genetic algorithm. The fourth section discusses the program design and implementation algorithm. The fifth section introduces and discusses the superiority of the simulation battery model verification algorithm. Finally, The conclusion is drawn in the sixth section.

2 Optimization Algorithm

2.1 Constant Current Constant Voltage Optimization Algorithm

The constant current constant voltage (CC/CV) charging method [2] has been well applied because of its simplicity and ease of use, and this method is widely used in lithium-ion battery charging. Its charging process is to apply a constant current to the battery until the battery voltage reaches the preset maximum charging voltage, keep the charging voltage at the constant maximum charging voltage, and the charging current will drop rapidly. When the charging current reaches the set minimum current value, the charging will stop. However, CC/CV charging also has some problems. The temperature and open circuit voltage of the battery will affect the charging process. Excessive temperature will affect the service life of the battery. Open circuit voltage lower than the minimum cut-off voltage and higher than the maximum cut-off voltage will affect the stability and cycle life of the battery. Therefore, protective measures must be taken to protect the battery when using the CC/CV method to charge the battery. Using the loop structure and condition structure, according to the preset boundary conditions, the main function is given, including all data sets and corresponding variables. The judgment condition is that the preset condition function is also embedded in the main function. Figure 1 shows the flow chart of the CC/CV charging process based on the optimized algorithm. Firstly, retrieve the initial conditions of the battery [3], whether the battery temperature and open circuit voltage are within the normal operating range. If the open circuit voltage is less than the cut-off voltage V_{cutoff}, the battery is charged in TC mode with small current until the battery voltage reaches the cut-off voltage, where 0.3C represents the charging current and C represents the nominal capacity of the battery. As long as the battery voltage exceeds the cut-off voltage, CC mode will be used to charge the battery. The size of the charging current is selected according to the specifications of the battery used, and the current range can be selected from 0.4C to 3.5c.When the voltage of the battery reaches a constant current applied to charge the battery until the voltage reaches the preset maximum charging voltage $V_{pre\text{-}set}$, the charging state is switched to CV charging mode, assuming that the battery is charged at a voltage of 4.2V and the charging current decreases accordingly. End the cycle with $t_{max} \leq t_{pre\text{-}set}$ or minimum current $I_{min} < 0.1C$ as the stop condition. The traditional CC/CV charging method is improved by adding selection conditions for efficient search in big data.

2.2 Multistage Current Charging Optimization Algorithm

Multistage current charging method is a deformation of CC/CV charging method, but this charging method also has defects, so the ant colony intelligent optimization method is proposed to optimize this method [4], and the ant colony system is used to optimize the current distribution, so that the optimal charging current of each charging stage can be determined. The ant colony optimization algorithm is realized by computer programming. Figure 2 shows the flow chart of the ant colony optimization algorithm. The computer must write a program to realize the algorithm, and the algorithm flow chart is essential. After this algorithm improves the multistage current charging method, the steps of the algorithm are simpler than the traditional CC/CV charging algorithm steps, and the charging efficiency is higher.

3 Charging Process Optimized by Genetic Algorithm

3.1 Genetic Algorithm

Genetic algorithm (GA) is a technique for random search and optimization in computational mathematics. It was originally developed by imitating the process of biological evolution [5], these phenomena include principle based crossover, selection and mutation; The optimal solution can be searched out and its child nodes are generated. To optimize the charging process with GA is to use GA technology to extract battery parameters. In the process of extracting the parameters, it is to extract the charging parameters involved in the battery charging process, search for the optimal parameters, and achieve the purpose of optimizing the charging.

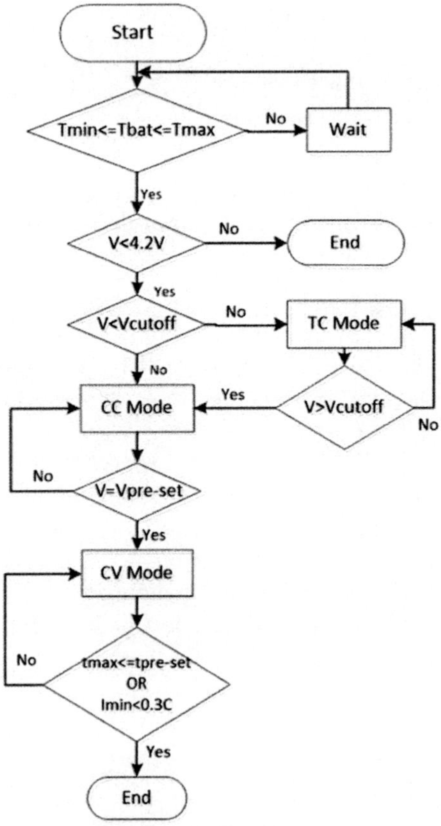

Fig. 1. Flow chart based on optimized CC/CV charging algorithm.

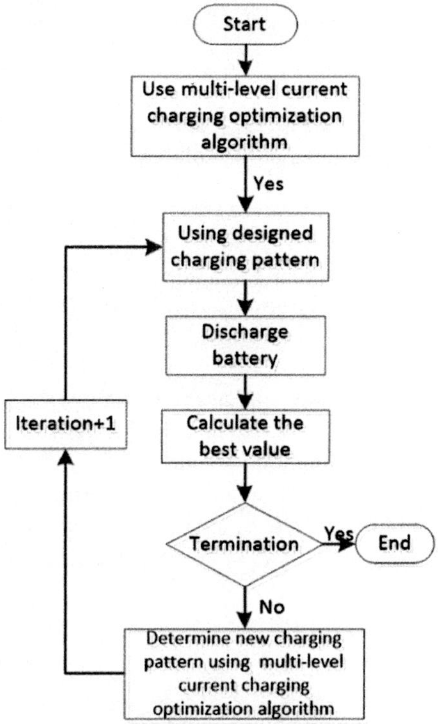

Fig. 2. Charging flowchart of ant colony optimization algorithm.

3.2 Optimization of Charging Parameters by Genetic Algorithm

In order to optimize the charging process with GA technology, the first step is to use the algorithm to search for the optimal parameters in a large number of parameter sets. GA uses standard mutation and crossover operators to calculate and automatically assign random values [6].

1) Parameter specifications: The charge or discharge rate of lithium battery depends on the coefficient of polynomial a_1–a_{21} GA optimization, set a size Sn to 300 × 20 matrix. Set the length of the string to 15. The mutation probability P_m is 0.45, and the crossover probability P_c is 0.75.
2) Generate a library of initial solutions: Firstly, discretize the entire library of problem-solving methods, and use a set of binary strings $a_{mi,1} \cdots a_{mi,n}$ to define variables. Each string is a set of strings. is a free combination of different ways $a_{m1,n} \cdots a_{mi,n}$;.Let Sn be the set of random solutions of the charging coefficient

of lithium battery [7], and Sn is an m × n matrix, Given in (1).

$$S_n = \begin{pmatrix} a_{m1,1} & \cdots & a_{m1,n} \\ \vdots & \ddots & \vdots \\ a_{mi,1} & \cdots & a_{mi,n} \end{pmatrix} \tag{1}$$

Where a and n represent the n-th coefficient of the i-th solution set, and the n-th coefficient of the i-th string in the solution set is 15 digits.

$$a_{mi,n} = (x_1 x_2 \cdots x_m) \tag{2}$$

Among them, x_1 and x_m represent the lowest and highest bits of the coefficient. Different genetic operators are used to calculate the string [8], and the binary number will be converted into an integer.

3) Adjust fitness function: In order to improve the quality of the method library, a fitness function F(x) can be established to improve the quality of the method library. The fitness function is determined by each solution method [9] according to its fitness. The fitness function is the maximum when F(x) = 1; the corresponding fitness is calculated, the optimal solution list is updated, and the lowest fitness is deleted. If we want to make f (x) larger, then f (x) will be smaller, expressed as

$$F(X) = \frac{1}{f(x) + 1} \tag{3}$$

4) Screening process: It is to save the calculated value of high fitness in the child nodes of the tree, and filter the optimal solution according to the maximum solution of F(x).The larger the value of F (x), the worse the fitness of the solution set. The fitness value of the next iteration can be calculated from the corresponding mathematical functions of (4) to (6). Where F_{avg} is the average value of the fitness function, E_{co} is the expected count value, and P_{es} is the probability of finding each string.

$$F_{avg} = \frac{\sum f(x)}{S_n} \tag{4}$$

$$E_{co} = \frac{F(x)}{F_{avg}} \tag{5}$$

$$P_{es} = \frac{E_{co}}{S_n} \tag{6}$$

5) Reproduction process: In the process of reproduction, some poor f (x) values will be discarded and the best value with high fitness will be selected. A mating set is formed, and the selected optimal value is added to the mating set in the form of probability. In this way, a random number between 0 and 1 can be generated to provide the next generation of string selection.

6) Crossover operator calculation: The crossover operator calculates the selected solution library. In the process of cross calculation [10], a new character string is generated for the character string exchange information in the pairing set. Select several parent strings for cross calculation to get the subclass string. When the probability of the selected parent node and parent node $P_{co} \leq 0.8$, the cross calculation must be started, otherwise the string will be copied to the next generation [11]. To illustrate the crossover process, two parent nodes P_1 and P_2 are found in the mating set. The ninth is the intersection. As shown in Fig. 3, two descendants Q_1 and Q_2 are generated by exchanging information between the character strings in the queue string. When two genes reproduce, one progeny gene can be produced. If there is a suitable intersection, then when the parent string is combined, it will form a good sub-category string. But this process will also produce many inappropriate child nodes.

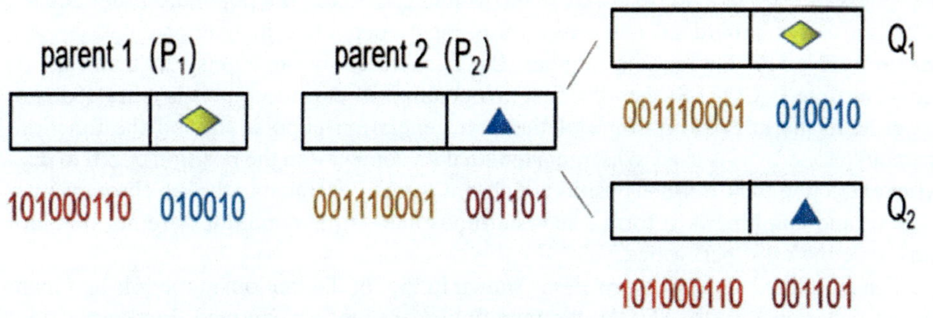

Fig. 3. Parent node P_1 and parent node P_2 produce new children Q_1 and Q_2.

7) Mutation operator calculation: During the mutation process, the parent class will select a specific subclass, and change 0 to 1 to change one or more characters in the string. If the number selected by the system is less than Pm, the system will reselect the string to generate a good subclass character The mutation calculation is performed by string; otherwise, the system will not change the mutation position of the next generation. As shown in Fig. 4, the mutation process of the offspring.

4 Programming

In order to realize the optimization of the algorithm for battery charging, when designing the program, a structure is first defined, that is, a chromosome is defined to generate a random number between 0 and 1, and then the fitness of each individual is calculated to find the best in the population Individuals select crossover individuals and crossover points to perform crossover calculations. In order to realize the optimization of the algorithm for battery charging, when designing the program, a structure is first defined, that

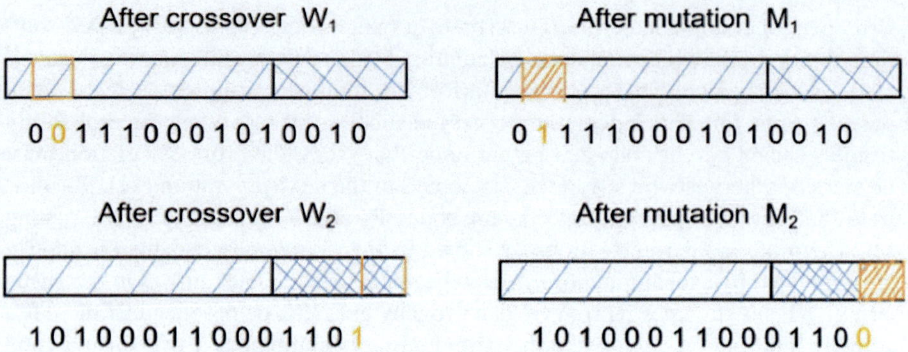

Fig. 4. Variation process of the nth generation.

is [12], a chromosome is defined to generate a random number between 0 and 1, and then the fitness of each individual is calculated to find the best in the population Individuals select crossover individuals and crossover points to perform crossover calculations. As shown in Fig. 5, if the random number (0–1) generated by randd() is less than the set crossover rate P_CORSS, then the first pos position of chromosome i is extracted and assigned to the kth chromosome of the next generation population, and the function memcpy(&a, &b, n) is used to assign n length data copies from the position of &b to &a. After extracting the LEN-pos position of chromosome j, assign it to the kth chromosome of next, and combine it to form a new chromosome. If the conditions are not met, no cross calculation is performed.

The mutation calculation process is shown in Fig. 6. If a random number is less than the set mutation P_MUTATION, the mutation operation is performed. Find a position from the middle of the chromosome, perform mutation, bitwise XOR [13], and assign a value function. The same is 0 but the difference is 1. Here, the value at the pos position is set to 0. Finally, the value of the entire optimized next generation is assigned to the current population cur for optimization in the next cycle.

5 Simulation

In order to verify the optimization of charging and discharging parameters, COMSOL multiphysics 5.5 multi physical field coupling simulation software was used to study the charging and discharging process of the battery. By substituting the optimized discharge rate parameters of the algorithm into the model [14], the performance changes during the charging and discharging process of the battery are studied. As shown in Fig. 7, the discharge curve of the battery is shown. When the discharge rate is 0.1C, 1C, 2C and 4C, the capacity is 18.6 Ah/m^2 at 0.1C and 13.7 Ah/m^2 at 4C, indicating that the capacity loss increases with the increase of discharge rate. It also shows that the discharge rate is optimized through the algorithm, and the battery capacity can be maintained above 12 Ah/m^2.

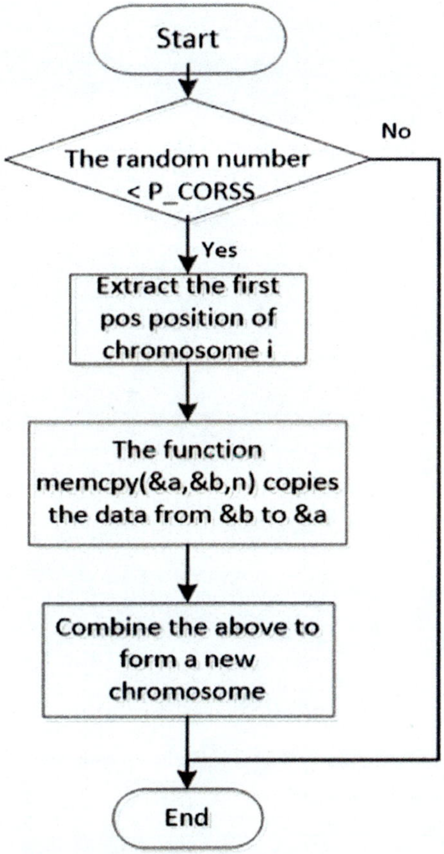

Fig. 5. Algorithm flow chart of cross calculation.

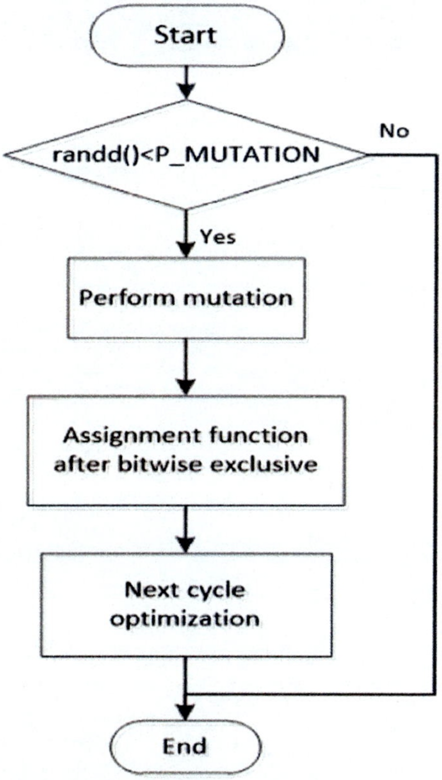

Fig. 6. Algorithm flow chart of mutation calculation.

Figure 8 shows the voltage and current curve of the battery during charging and discharging. The battery first discharged for 2000 s, then stopped for 300 s, and the circuit remained interrupted. Recharge for 2000 s, and finally to the end of charging. It can be seen from the figure that due to the relaxation from voltage to rest, the voltage has a rising stage [15], which is caused by ohmic polarization and concentration polarization. The discharge rate and charge-discharge time are selected according to the algorithm. It is proved that the simulation of voltage and current has good convergence.

The best state of charge of the battery is optimized by the algorithm, and the change curve of electrode SOC can be obtained. As shown in Fig. 9, after discharging and charging on the negative electrode, the SOC of the battery finally tends to be stable, reaching 0.56. Compared with the set standard SOC of 1, the utilization rate of the battery reaches more than half, which improves the utilization rate of the battery better. At the same time, when the positive SOC tends to be stable, the SOC value above 0 indicates that the lithium ion is not exhausted. The optimization of parameters prevents

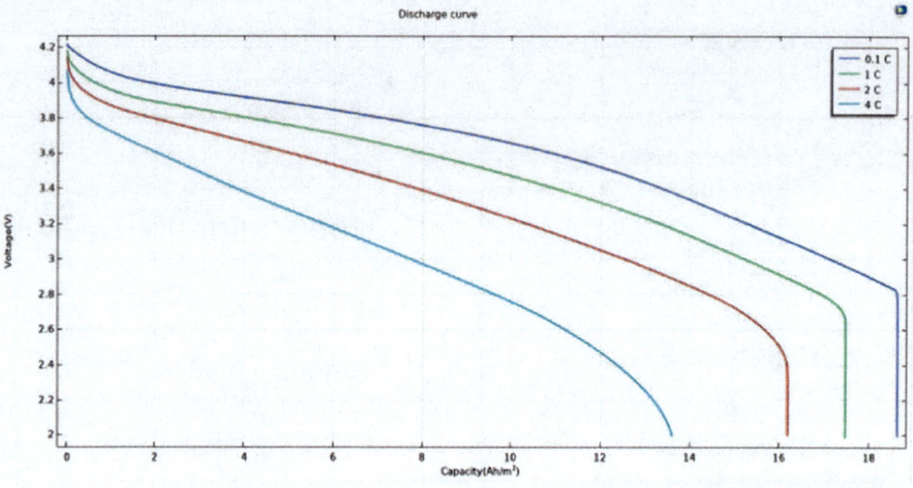

Fig. 7. The battery discharged at the rate of 0.1C, 1C, 2C, 4C.

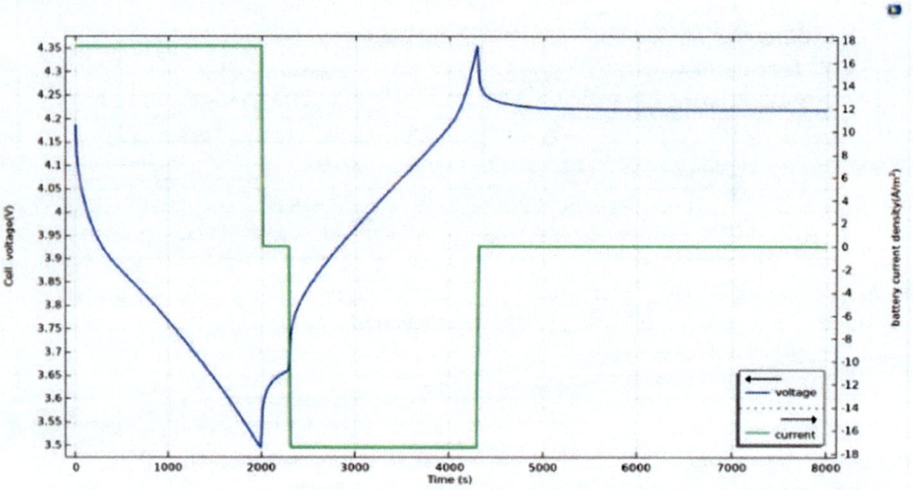

Fig. 8. The change curve of voltage and current in the process of battery charging and discharging.

the battery from overcharging, protects the battery well and improves the service life of the battery.

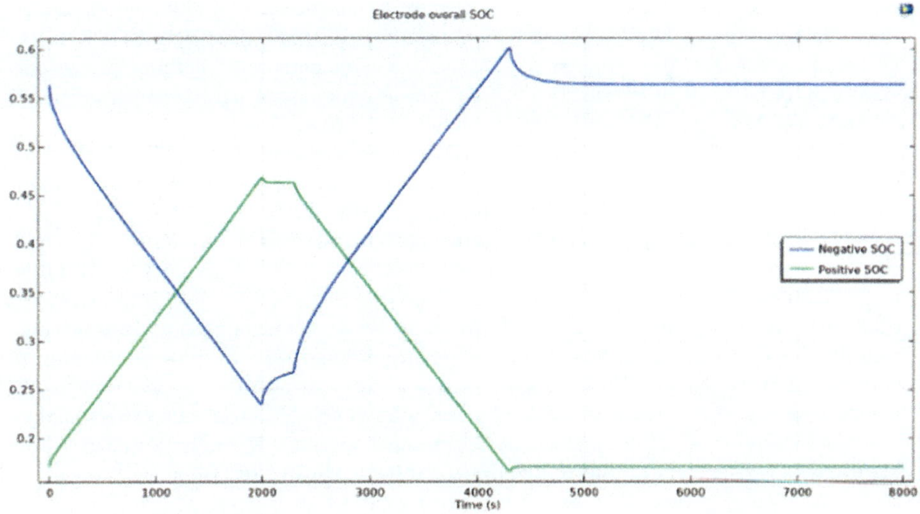

Fig. 9. SOC change curve after battery optimization algorithm.

6 Conclusion

The optimization of battery charging and discharging model based on genetic algorithm was discussed and analyzed in our work. The strategy could search data efficiently and reduce time complexity, that is, search important parameters of the charging and discharging model quickly and accurately. During the process of simulating calculation using genetic algorithm, it was found that the crossover operator and the mutation operator had a great influence on the final selector of the data.The optimized simulation results showed that the capacity degradation of the battery changed, and the SOC state had been optimized during the charging and discharging process.

References

1. Lee, D.-C., Lee, K.-J., Kim, C.-W.: Optimization of a lithium-ion battery for maximization of energy density with design of experiments and micro-genetic algorithm. Int. J. Precision Eng. Manuf.-Green Technol. **7**(4), 829–836 (2019). https://doi.org/10.1007/s40684-019-00106-4
2. Xu, M., Wang, R., Zhao, P., Wang, X.: Fast charging optimization for lithium-ion batteries based on dynamic programming algorithm and electrochemical-thermal-capacity fade coupled model. J. Power Sources, **438**, 227015 (2019). https://doi.org/10.1016/j.jpowsour.2019.227015
3. Deng, T., Ran, Y., Yin, Y., Liu, P.: Multi-objective optimization design of thermal management system for lithium-ion battery pack based on Non-dominated Sorting Genetic Algorithm II. Appl. Thermal Eng., 114394 (2019). https://doi.org/10.1016/j.applthermaleng.2019.114394
4. Cheng, L., Garg, A., Jishnu, A.K., Gao, L.: Surrogate based multi-objective design optimization of lithium-ion battery air-cooled system in electric vehicles. J. Energy Storage **31**, 101645 (2020). https://doi.org/10.1016/j.est.2020.101645

5. Hong, Y., Lee, C.W.: Pareto fronts for multiobjective optimal design of the lithium-ion battery cell. J. Energy Storage **17**, 507–514 (2018). https://doi.org/10.1016/j.est.2018.04.003
6. Cailian, L., Chun, Z.: Life prediction of battery based on random forest optimized by genetic algorithm. In: 2020 IEEE International Conference on Prognostics and Health Management (ICPHM) (2020). https://doi.org/10.1109/icphm49022.2020.9187060
7. Liu, H., Shi, H., Shen, H., Xie, G.: The performance management of a Li-ion battery by using tree-like mini-channel heat sinks: experimental and numerical optimization. Energy **189**, 116150 (2019). https://doi.org/10.1016/j.energy.2019.116150
8. Jiang, L., et al.: Optimization of variable-current charging strategy based on SOC segmentation for li-ion battery. IEEE Trans. Intell. Transp. Syst., 1–8 (2020). https://doi.org/10.1109/tits.2020.3006092
9. Choi, Y.Y., Kim, S., Kim, S., Choi, J.-I.: Multiple parameter identification using genetic algorithm in vanadium redox flow batteries. J. Power Sources **450**, 227684 (2020). https://doi.org/10.1016/j.jpowsour.2019.227684
10. Fatima Brondani, M., Sausen, A.T.Z.R., Sausen, P.S., Binelo, M.O.: Parameter estimation of lithium ion polymer battery mathematical model using genetic algorithm. Comput. Appl. Math. **37**(1), 296–313 (2017). https://doi.org/10.1007/s40314-017-0537-7
11. Zhang, X., Wang, Y., Liu, C., Chen, Z.: A novel approach of battery pack state of health estimation using artificial intelligence optimization algorithm. J. Power Sources **376**, 191–199 (2018). https://doi.org/10.1016/j.jpowsour.2017.11.068
12. Blaifi, S., Moulahoum, S., Colak, I., Merrouche, W.: Monitoring and enhanced dynamic modeling of battery by genetic algorithm using LabVIEW applied in photovoltaic system. Electr. Eng. **100**(2), 1021–1038 (2017). https://doi.org/10.1007/s00202-017-0567-6
13. Yeh, W.-C., et al.: New genetic algorithm for economic dispatch of stand-alone three-modular microgrid in DongAo Island. Appl. Energy **263**, 114508 (2020). https://doi.org/10.1016/j.apenergy.2020.114508
14. Lu, J., Chen, Z., Yang, Y., Lv, M.: Online estimation of state of power for lithium-ion batteries in electric vehicles using genetic algorithm. IEEE Access **6**, 20868–20880 (2018). https://doi.org/10.1109/access.2018.2824559
15. Suriya Narayanan, S.S., Thangavel, S.: Open Circuit Voltage Equation parameter identification of Lithium Ion Battery Model using Genetic Algorithm. In: 2020 International Conference on Inventive Computation Technologies (ICICT) (2020). https://doi.org/10.1109/icict48043.2020.9112455

An Enhanced Harmony Search Based on Quantum Mechanism

Maomao Liang[1,2], Ying Deng[1,2], Wen Xiao[1,2], Lijin Wang[1,2,3](✉), and Yiwen Zhong[1,2]

[1] College of Computer and Information Science, Fujian Agriculture and Forestry University, Fuzhou 350002, China
lijinwang@fafu.edu.cn

[2] Key Laboratory of Smart Agriculture and Forestry (Fujian Agriculture and Forestry University), Fujian Province University, Fuzhou 350002, China

[3] Digital Fujian Tourism Big Data Institute (Wuyi University), Wuyishan 354300, China

Abstract. Harmony search (HS), which is widely used in many fields, is a meta-heuristic algorithm. In each iteration, HS generates a new solution under the harmony search consideration rate (HMCR) from harmony memory (HM). And the new solution is fine-tuned through a pitch adjustment rate (PAR). The basic HS obtains a random value from HM for each variable of a solution, and adjusts it in a fixed range. For the sake of improving its performance, we present an enhanced harmony search based on quantum mechanism (QEHS), which incorporates quantum concepts and differential mutation operation into the harmony search algorithm. QEHS reinforces the exploration and exploitation capability in search space, with the use of wave function from Schrödinger formula to express the harmony in HM. For reflecting the effectiveness of QEHS, simulations are carried on 30 benchmark functions from CEC2014. To manifest its feasibility, we compared the results of QEHS with that of other HS variants and some popular algorithms. Facts turn out that QEHS is an efficient and competitive algorithm.

Keywords: Harmony search · Quantum mechanism · Differential mutation

1 Introduction

Stimulated by music improvisation process, Zong Woo Geem initially put forward harmony Search (HS) in 2001. It's a population-based meta-heuristic algorithm for obtaining approximate solutions to optimization problems with discrete variables at first, and then it was modified by Lee & Geem [2] for dealing with functions with continuous various and mixed various. Tuned by three key parameters and principle of probability, HS is convenient to implement and can solve various difficult problems in many fields, including multi-objective optimization [3], feature selection [4], medicine [5, 6], autonomous driving [7], and several others [8]. But HS still has some weaknesses: slow convergence; unsatisfactory accuracy; sensitive to controlled parameter, and frequently trap into regional optimal circumstances.

S.-C. Chu et al. (Eds.): ICGEC 2021, LNEE 833, pp. 47–59, 2022.
https://doi.org/10.1007/978-981-16-8430-2_5

Given these drawbacks, a lot of published survey articles [9, 10] and monographs [11, 12] have made many attempts. Structurally, lots of investigations are conducted to develop the initialization method for HM, or modify the improvisation procedures, or set new termination conditions. In specific, some researchers mix other heuristic in HS. Zhao et al. [13] hybridized the technique from particle swarm optimization(PSO) method [14] with the original HS to dynamically change parameter PAR and BW. Gheisarnejad [15] combined cuckoo search(CS) algorithm [16] into the basic HS to strengthen the ability of exploration and avert falling into the local optimal solution. Some scholars choose to change parameters in HS to achieve the purpose of improving the algorithm. Portilla-Flores et al. [17] utilized the maximum and minimum in HM to dynamically update BW for solving constrained numerical optimization problems. Amaya et al. [18] came up with a new HS variant which self-tune through the process of algorithm, and it requires fewer iterations than many others. Besides, there are many researchers working on introducing new operations or adaption operators in HS. Guo et al. [19] proposed an adaptive HS, called ABHS, which based on the best strategy to strengthen the search efficiency. Keshtegar et al. [20] presented an adaptive dynamic harmony search (ADHS). They improved the HS according to the data offered by HM for the improvisation process.

To overcome the mentioned problems, we consider introducing a new strategy to enhance the randomness of improvisation and the ability to transform breadth search into depth search. Thus we integrate the quantum concepts and differential mutation operation into HS, and an enhanced harmony search based on quantum mechanism (QEHS) is proposed in this paper.

Quantum computing (QC) is an active research field, including two directions: quantum algorithms and quantum computers. Many quantum algorithms have been presented over the years. At early stages, Shor [21] developed a quantum factoring algorithm in 1994. Since later 1990s, scholars commenced to mix quantum computing with meta-heuristic algorithm, and obtained many new algorithms, such as quantum dolphin swarm algorithm (QDSA) [22], non-homogeneous cuckoo search (NoCuSa) [23], differential evolution-crossover quantum particle swarm optimization (DE-CQPSO) [24], and quantum whale optimization algorithm (QWOA) [25], which are popular in terms of computer science, physics and mathematics.

Due to the strong randomization and uncertainty principle, the introduction of quantum mechanism into the meta-heuristic algorithm can increase the multiplicity of population, prevent convergence in advance, and even reduce its complexity and time.

The remainder of this paper is arranged as following. In Sect. 2, we briefly introduces the basic HS method. Then explains the proposed algorithm in detail in Sect. 3. Section 4 shows the comparison and analysis of experimental results. And a conclusion is contained in final Sect. 5.

2 Harmony Search Algorithm

Geem et al. [1] presented harmony search algorithm in 2001. For sake of expressing the structure of HS clearly, the calculation process is divided into 4 steps:

Step1: Initialize the problem and parameter values.

$$x_{m,n} = x_n^L + \left(x_n^U - x_n^L\right) \times rand, m = 1, 2, \cdots, HMS, n = 1, 2, \cdots, D \tag{1}$$

where $x^U{}_n$ and $x^L{}_n$ are the upper and lower limits of $x_{m,n}$ apart. And there are three principal parameters in HS: harmony memory considering rate (HMCR), harmony memory size (HMS), and pitch adjusting rate (PAR).

Step2: Generate a new harmony.

$$x_{new,n} = x_{r_1,n} \tag{2}$$

r_1 is a random number selected from 1 to *HMS*.

$$x_{current,n} = x_{current,n} \pm r_2 \times BW \tag{3}$$

where r_2 is a random number uniformly generated from [0,1].

Step3: Refresh the harmony memory.

In case of $\boldsymbol{X_{current}}$ is superior to $\boldsymbol{X_{worst}}$, replace it. The procedure is stated as follows.

Algorithm 1. Procedure of improvising a new harmony

for each variable, $x_{m,n}$, $m = 1,2, \ldots, HMS$, $n = 1,2, \ldots, D$
    if *rand* <= *HMCR* then
        Harmony memory consideration using Eq.2
        if *rand* <= *PAR* then
            Pitch adjustment according to Eq.3
        end
    else
        Random selection using Eq.1
    end
end

Step4: Check whether conditions for the termination of the algorithm are met.

3 Enhanced Harmony Search with Quantum Mechanism

In previous work, we have used the differential mutation strategy to enhance HS. Due to the advantages it has, the proposed QEHS also use a little part of differential mutation operation, that is current-to-best/2:

$$X_{current} = X_{current} + F(X_{best} - X_{current} + X_a - X_b + X_c - X_d), \tag{4}$$

We applied current-to-best/2 to random selection (Eq. 1 in Algorithm 1) to speed up convergence in the start stage, and utilized the idea of quantum mechanism to modify pitch selection (Eq. 2 in Algorithm 1) and pitch adjustment (Eq. 3 in Algorithm 1) to achieve the global optima near the final stage of algorithm.

It is difficult to set the parameter BW in HS, because its value is taken from $[-\infty, +\infty]$, and there is no empirical rule. If the value of BW is inappropriate, the performance of HS will be degraded. So we considered eliminating this parameter and introducing a new method to instead.

Inspired by [23], we assume that harmony memory is a system that satisfies the basic assumptions of quantum mechanism, and each harmony can be represented by a wave function. Wave function Ψ(γ, T) means that it is at position γ and time T in quantum system, and it conforms the Schrödinger formula [26] in a physical system and is used to describe how quantum state evolves over time:

$$i\hbar\frac{\partial}{\partial_t}\Psi(\gamma,T)=\left[\frac{-\hbar^2}{2M}\nabla^2+V(\gamma,T)\right]\Psi(\gamma,T) \tag{5}$$

where i is imaginary unit; $\hbar$ indicates the reduced Plank constant, the mass of particle is denoted as M; ∇^2 represents the Laplacian operator.

When lack of time, the harmony i is subjected to a 1-D delta potential, and its state can be described by the wave function, namely:

$$\psi(\gamma)=\frac{1}{l}\exp\left(-\frac{|\gamma|}{l}\right) \tag{6}$$

And l means the characteristic extent of the potential well.

In order to get the location of harmony, we used the Monte Carlo method. Combining the above formulas, we get some new equations that produce new harmony:

$$X_{current}=\overline{X}+\delta\times\exp(\eta)\times\left(X_{BEST}-\overline{X}\right) \tag{7}$$

$$X_{current}=\overline{X}+\delta\times\ln^{(1.0/\eta)}\times\left(\overline{X}-X_{r,j}\right) \tag{8}$$

$$\overline{X}=\frac{1}{n}\sum_{i=1}^{n}Xi \tag{9}$$

where δ is a parameter that controls the length of the difference between the current harmony and $\bar{X}$, X_{BEST}; η which obeys a uniform distribution on [0,1] is a random number.

$$X_{\text{current}}=X_{current}+r_1\times(X_{best}-X_{current})+r_2\times F\times(X_a-X_b+X_c-X_d) \tag{10}$$

where F is a mutation factor that obeys a normal distribution, set the mean to be 0.3 and the variance to be 0.5; r_1, r_2, r_a, r_b, r_c, r_d are all independent random variables belonging to [0,1].

Through Eq. 7, we will get a new individual that located between the average position and the optimal position, and it is biased towards the optimal. By Eq. 8, we can achieve a new position that is away from the center of the population, which helps prevent algorithm falling in a crowded position and enhance the multiplicity of population. Put another way, individuals in QEHS can learn information from the best one and averaged value in HM, so that the algorithm actually strengthens the global search and local development capabilities.

Besides, the structure of QEHS is shown in Algorithm 2.

Algorithm 2. Procedure of improvising a new harmony

```
for each variable, x_{m,n}, m = 1,2, ..., HMS, n = 1,2, ..., D
    if rand <= HMCR then
        Harmony memory consideration using Eq.7
        if rand <= PAR then
            Pitch adjustment according to Eq.8
        end
    else
        Random selection using Eq.10
    end
end
```

4 Experimental Section

In experiments, QEHS was run on 30 benchmark functions based on [27]. We introduced two evaluation criteria: *Error* and *Convergence graphs.* Average fitness and standard deviation of *Error* are recorded as "$AVG_{Er} \pm STD_{Er}$".

For the sake of minimizing *Error*, we set parameters in QEHS and show its comparison with variants of HS(i.e. HSDM [28], ESGHS [29], SGHS [30], NGHS [31], LHS [32], HD/Best [33], IDHSLDD [34]) in Table 1.

Table 1. Parameters of different HS variants

Algorithm	Parameter
HS	HMS = 5, HMCR = 0.98, PAR = 0.3, BW = 0.01
HSDM	HMS = 50, HMCR = 0.98
ESGHS	HMS = 100
SGHS	HMS = 5, HMCR = 0.98, PAR = 0.9, BW = 0.0005
NGHS	HMS = 5, Pm = 0.005
LHS	HMS = 5, HMCR = 0.99
HD/Best	HMS = 50, HMCR = 0.99, PAR = 0.9
IDHSLDD	HMS = 30, HMCRmin = 0.3, HMCRmax = 0.99, PARmin = 0.9, PARmax = 0.99
QEHS	HMS = 50, HMCR = 0.8, PAR = 0.1, delta = 1.6

4.1 Effectiveness of QEHS

We ran our algorithm at $D = 30$ where D is population dimension, $NP = 50$ where NP is the scale of population, $MaxFES = 50000$ where maximum times of iterations is $MaxFES$. Table 2 gives the results, those with better average fitness value are indicated in bold. The statistics of the comparison between HS and QEHS are put in the last line, it based on the Wilcoxon signed rank test as well as $\alpha = 0.05$. The signal " +" means QEHS is significantly superior to HS, "-" is opposite, while " =" means that there is no obvious gap between the two.

Table 2. Average error values obtained by QEHS for 30 benchmark functions at $D = 30$.

Fun	HS		*p*-value	QEHS
F_1	4.32e+07±2.90e+07	+	0.000002	**1.50e+06±8.80e+05**
F_2	1.63e+06±1.44e+06	+	0.000002	**8.20e+03±6.31e+03**
F_3	2.83e+04±2.74e+04	+	0.000115	**6.66e+03±5.27e+03**
F_4	1.35e+02±3.50e+01	+	0.000004	**7.89e+01±2.17e+01**
F_5	2.00e+01±5.04e-06	-	0.000002	2.10e+01±6.77e-02
F_6	1.80e+01±2.98e+00	+	0.000002	**3.48e+00±1.75e+00**
F_7	6.43e-01±2.59e-01	+	0.000002	**4.76e-03±8.29e-03**
F_8	1.58e-01±2.11e-01	-	0.000002	4.45e+01±4.72e+01
F_9	9.10e+01±2.46e+01	-	0.000028	1.52e+02±4.03e+01
F_{10}	6.94e-01±5.32e-01	-	0.000002	6.23e+03±4.14e+02
F_{11}	2.51e+03±4.13e+02	-	0.000002	7.23e+03±3.37e+02
F_{12}	2.34e-01±6.17e-02	-	0.000002	2.97e+00±3.63e-01
F_{13}	5.78e-01±1.02e-01	+	0.000002	**3.90e-01±7.07e-02**
F_{14}	4.82e-01±2.31e-01	=	0.909931	**4.25e-01±1.18e-01**
F_{15}	1.52e+01±5.40e+00	=	0.370935	1.59e+01±1.56e+00
F_{16}	1.07e+01±6.29e-01	-	0.000002	1.26e+01±3.79e-01
F_{17}	1.12e+07±4.77e+06	+	0.000002	**3.85e+05±2.35e+05**
F_{18}	4.71e+03±6.07e+03	+	0.000283	**7.07e+02±6.80e+02**
F_{19}	3.93e+01±3.33e+01	+	0.000002	**7.09e+00±9.07e-01**
F_{20}	4.65e+04±2.45e+04	+	0.000003	**9.43e+03±4.58e+03**
F_{21}	2.09e+06±1.48e+06	+	0.000002	**2.14e+05±1.01e+05**
F_{22}	6.99e+02±2.19e+02	+	0.000174	**3.30e+02±2.28e+02**
F_{23}	3.19e+02±2.40e+00	+	0.000002	**3.15e+02±7.51e-10**
F_{24}	2.36e+02±6.88e+00	+	0.000089	**2.27e+02±4.63e+00**
F_{25}	2.14e+02±3.65e+00	+	0.000002	**2.08e+02±2.02e+00**
F_{26}	1.76e+02±5.69e+01	+	0.000003	**1.04e+02±1.82e+01**
F_{27}	7.20e+02±2.13e+02	+	0.000008	**3.87e+02±4.51e+01**
F_{28}	1.53e+03±7.38e+02	+	0.000002	**8.67e+02±3.50e+01**
F_{29}	1.16e+06±3.00e+06	=	0.262299	**1.47e+03±4.34e+02**
F_{30}	1.11e+04±5.79e+03	+	0.000002	**2.47e+03±9.09e+02**
+/=/-			20/3/7	

By observing Table 2, we can say that QEHS performs better on functions F_1, F_2, F_3, F_4, F_6, F_7, F_{13}, F_{14}, F_{17}, F_{18}, F_{19}, F_{20}, F_{21}, F_{22}, F_{23}, F_{24}, F_{25}, F_{26}, F_{27}, F_{28}, F_{29}, F_{30} than that of HS, the performance of QEHS is slightly worse on functions F_5, F_8, F_9, F_{10}, F_{11}, F_{12}, F_{15}, F_{16}. According to the record of " $+/=/-$", QEHS is better than HS in 20 out of 30 functions, is worse in 7 functions, and the two are similar in 3 functions.

We give partial convergence graphs in Fig. 1. From it, we can see that QEHS converges earlier than HS, and the precision of solutions obtained by QEHS are also higher. Therefore, it can be say that QEHS is available.

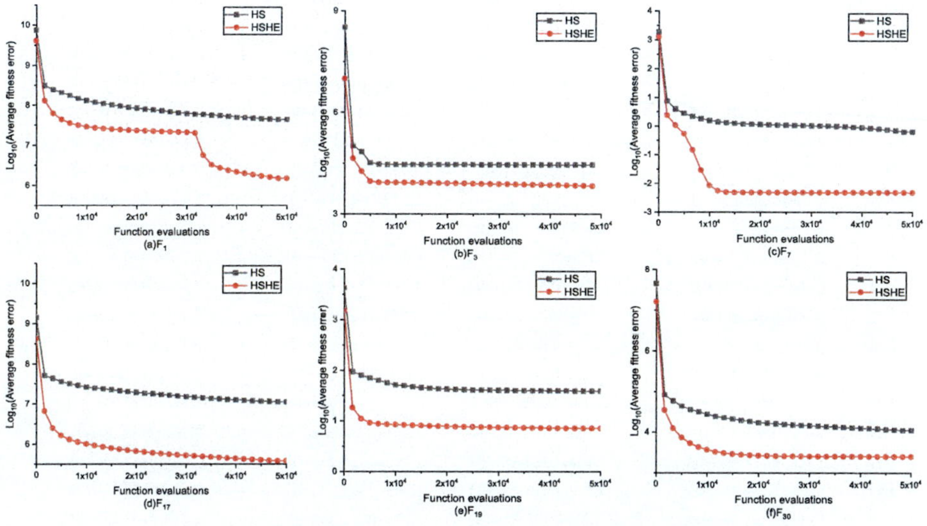

Fig. 1. The convergence curves of HS and QEHS

4.2 Scalability Study

We execute QEHS on 30 benchmark functions of CEC2014, set $D = 10$ and $D = 50$ respectively. In these dimensions, the scale of population NP is all set to 50.

As observed from Table 3, we come to a conclusion that the performance of QEHS is still steadily better than HS when the dimension of problem changes.

4.3 Comparison with Other HS Variants

In this section, experiments have performed on 30 benchmark functions with $D = 30$. Table 4 shows the average error, the outcome of test are given in Table 5 and Table 6.

Table 3. Average fitness error obtained by HS and QEHS for benchmark functions

Fun	$D = 10$			$D = 50$		
	HS		QEHS	HS		QEHS
F_1	1.01e+07±1.28e+07	+	**2.74e+04±2.95e+04**	3.70e+07±1.81e+07	+	**7.91e+06±2.48e+06**
F_2	4.69e+03±3.56e+03	+	**9.80e+02±1.50e+03**	1.17e+07±4.83e+06	+	**3.82e+03±3.64e+03**
F_3	9.09e+03±7.61e+03	+	**5.43e+02±4.57e+02**	**2.96e+04±1.40e+04**	+	6.98e+04±2.68e+04
F_4	**2.83e+01±2.45e+01**	=	3.25e+01±8.80e+00	1.82e+02±4.49e+01	+	**9.87e+01±8.95e+00**
F_5	**1.94e+01±3.18e+00**	-	2.04e+01±8.87e-02	**2.00e+01±1.55e-06**	-	2.12e+01±3.05e-02
F_6	4.64e+00±1.38e+00	+	**5.05e-01±7.28e-01**	3.42e+01±3.37e+00	+	**4.64e+00±2.31e+00**
F_7	1.87e-01±1.42e-01	+	**3.28e-02±2.11e-02**	1.15e+00±9.17e-02	+	**3.38e-03±5.68e-03**
F_8	**2.76e-10±2.67e-10**	-	2.45e+00±1.42e+00	**2.21e+00±7.06e-01**	-	2.66e+02±4.18e+01
F_9	1.35e+01±7.06e+00	+	**5.62e+00±7.00e+00**	**1.89e+02±3.82e+01**	-	3.74e+02±1.96e+01
F_{10}	**2.14e-01±7.64e-02**	-	8.07e+02±3.50e+02	**6.65e+00±2.51e+00**	-	1.29e+04±5.49e+02
F_{11}	**5.96e+02±1.73e+02**	-	9.06e+02±7.66e+02	**4.78e+03±7.85e+02**	-	1.38e+04±5.17e+02
F_{12}	**2.46e-01±1.26e-01**	-	1.31e+00±2.00e-01	**2.85e-01±6.46e-02**	-	4.12e+00±3.29e-01
F_{13}	4.41e-01±1.59e-01	+	**1.27e-01±2.92e-02**	6.75e-01±1.23e-01	+	**5.92e-01±6.68e-02**
F_{14}	3.90e-01±2.37e-01	=	3.09e-01±8.84e-02	**4.77e-01±2.63e-01**	=	5.29e-01±1.80e-01
F_{15}	2.47e+00±1.67e+00	+	**1.39e+00±5.14e-01**	3.52e+01±1.69e+01	=	**3.50e+01±2.02e+00**
F_{16}	2.71e+00±5.12e-01	+	**2.04e+00±5.37e-01**	**1.86e+01±6.86e-01**	-	2.26e+01±3.39e-01
F_{17}	2.70e+06±3.91e+06	+	**5.14e+03±4.12e+03**	1.28e+07±6.21e+06	+	**6.76e+05±3.12e+05**
F_{18}	1.03e+04±9.76e+03	=	**7.72e+03±3.40e+03**	1.49e+03±1.43e+03	+	**1.08e+03±9.27e+02**
F_{19}	1.96e+00±1.31e+00	+	**5.97e-01±8.22e-01**	6.22e+01±2.35e+01	+	**4.44e+01±1.68e+01**
F_{20}	8.92e+03±8.20e+03	+	**4.44e+03±2.66e+03**	7.64e+04±3.02e+04	+	**1.99e+04±7.01e+03**
F_{21}	1.71e+06±2.25e+06	+	**3.44e+03±2.98e+03**	1.02e+07±5.79e+06	+	**7.84e+05±4.37e+05**
F_{22}	6.72e+01±8.47e+01	=	**2.57e+01±1.26e+01**	**1.42e+03±4.01e+02**	+	1.47e+03±3.11e+02
F_{23}	**3.19e+02±6.03e+01**	+	3.29e+02±2.17e-13	3.71e+02±1.83e+01	+	**3.44e+02±1.48e-07**
F_{24}	1.35e+02±8.94e+00	+	**1.16e+02±1.80e+01**	**2.68e+02±4.89e+00**	+	2.69e+02±4.76e+00
F_{25}	**1.87e+02±2.43e+01**	=	1.98e+02±1.36e+01	2.22e+02±4.10e+00	+	**2.14e+02±1.70e+00**
F_{26}	1.07e+02±2.53e+01	+	**1.00e+02±2.44e-02**	2.06e+02±9.30e+01	+	**1.37e+02±5.61e+01**
F_{27}	3.93e+02±1.11e+02	+	**3.42e+02±9.86e+01**	1.31e+03±1.35e+02	+	**4.81e+02±5.99e+01**
F_{28}	5.19e+02±1.06e+02	+	**4.15e+02±6.02e+01**	2.37e+03±6.47e+02	+	**1.20e+03±5.35e+01**
F_{29}	6.02e+05±8.70e+05	+	**5.79e+04±3.15e+05**	1.26e+06±6.87e+06	=	**1.98e+03±5.52e+02**
F_{30}	1.22e+03±4.10e+02	+	**7.19e+02±1.76e+02**	2.34e+04±8.89e+03	+	**1.19e+04±1.93e+03**
+/=/-		20/5/5			17/5/8	

Table 4. Average error obtained by QEHS and HS variants at $D = 30$

Fun	HSDM	ESGHS	SGHS	NGHS	LHS	IDHSLDD	DH/Best	QEHS
F_1	1.04E+08	9.96E+07	5.96E+06	1.00E+08	4.19E+07	2.92E+06	1.67E+06	**1.50E+06**
F_2	**8.20E+03**	1.04E+04	1.35E+04	1.10E+07	7.50E+05	1.01E+04	1.01E+04	**8.20E+03**
F_3	2.18E+04	1.55E+04	1.65E+04	1.66E+04	2.03E+04	9.13E+03	2.25E+04	**6.66E+03**
F_4	1.18E+02	1.15E+02	8.27E+01	1.88E+02	1.40E+02	8.75E+01	**6.64E+01**	7.89E+01
F_5	2.10E+01	2.10E+01	**2.00E+01**	2.04E+01	2.06E+01	2.09E+01	2.10E+01	2.10E+01
F_6	4.15E+00	1.44E+01	1.74E+01	2.19E+01	1.97E+01	1.55E+01	1.59E+01	**3.48E+00**
F_7	2.55E-04	**1.71E-04**	5.16E-02	1.08E+00	5.71E-01	1.40E-02	2.19E-02	4.76E-03
F_8	7.33E-02	1.23E+00	1.01E+00	8.02E-01	2.38E+00	3.54E+00	**3.32E-02**	4.45E+01
F_9	1.83E+02	1.92E+02	1.37E+02	1.25E+02	1.13E+02	**8.79E+01**	9.98E+01	1.52E+02
F_{10}	2.25E+00	2.73E+01	5.01E+00	6.12E+00	1.07E+01	1.17E+01	**1.95E+00**	6.23E+03
F_{11}	7.34E+03	7.33E+03	2.84E+03	2.83E+03	**2.74E+03**	4.54E+03	3.20E+03	7.23E+03
F_{12}	2.85E+00	2.79E+00	**1.52E-01**	3.39E-01	4.48E-01	1.92E+00	2.77E+00	2.97E+00
F_{13}	4.15E-01	4.49E-01	4.97E-01	4.63E-01	4.52E-01	6.51E-01	6.62E-01	**3.90E-01**
F_{14}	5.00E-01	4.57E-01	3.96E-01	4.45E-01	**3.12E-01**	7.18E-01	7.13E-01	4.25E-01
F_{15}	1.66E+01	1.65E+01	1.02E+01	4.22E+01	3.23E+01	**8.80E+00**	1.34E+01	1.59E+01
F_{16}	1.31E+01	1.32E+01	**1.08E+01**	1.12E+01	1.10E+01	1.22E+01	1.19E+01	1.26E+01
F_{17}	4.83E+06	4.82E+06	1.41E+06	2.83E+07	8.63E+06	6.48E+05	**3.64E+05**	3.85E+05
F_{18}	1.80E+03	2.06E+03	5.07E+03	6.75E+05	5.87E+03	8.64E+03	5.52E+03	**7.07E+02**
F_{19}	7.62E+00	7.59E+00	2.41E+01	6.17E+01	4.04E+01	1.88E+01	1.52E+01	**7.09E+00**
F_{20}	2.19E+04	2.86E+04	3.43E+04	5.38E+04	4.23E+04	2.03E+04	3.92E+04	**9.43E+03**
F_{21}	8.60E+05	4.24E+05	8.50E+05	5.09E+06	1.25E+06	4.76E+05	**1.94E+05**	2.14E+05
F_{22}	4.04E+02	3.92E+02	7.39E+02	9.00E+02	7.20E+02	4.71E+02	7.27E+02	**3.30E+02**
F_{23}	**3.15E+02**	**3.15E+02**	**3.15E+02**	3.34E+02	3.19E+02	**3.15E+02**	**3.15E+02**	**3.15E+02**
F_{24}	2.28E+02	2.24E+02	2.34E+02	2.34E+02	2.33E+02	**2.19E+02**	2.37E+02	2.27E+02
F_{25}	2.38E+02	2.34E+02	**2.05E+02**	2.21E+02	2.12E+02	2.08E+02	2.07E+02	2.08E+02
F_{26}	**1.00E+02**	**1.00E+02**	1.76E+02	1.96E+02	1.96E+02	1.01E+02	1.82E+02	1.04E+02
F_{27}	1.15E+03	6.20E+02	6.98E+02	7.99E+02	7.55E+02	5.51E+02	7.37E+02	**3.87E+02**
F_{28}	8.63E+02	**8.41E+02**	1.72E+03	2.18E+03	2.04E+03	9.92E+02	1.11E+03	8.67E+02
F_{29}	1.12E+06	2.13E+03	**2.14E+02**	5.78E+05	2.90E+05	1.66E+03	5.70E+05	1.47E+03
F_{30}	3.59E+03	5.83E+03	**8.48E+02**	2.32E+04	1.47E+04	3.47E+03	3.48E+03	2.47E+03
+/=/-	12/16/2	15/10/5	16/2/12	22/1/7	21/1/8	16/5/9	15/7/8	

Table 5. Results of the multiple-problem Wilcoxon's test for QEHS and HS variants

Algorithm	R^+	R^-	*p*-value	$\alpha = 0.05$	$\alpha = 0.1$
QEHS *vs.* HSDM	**375**	90	0.003379	+	+
QEHS *vs.* ESGHS	**380**	85	0.002415	+	+
QEHS *vs.* SGHS	**307**	158	0.125438	=	=
QEHS *vs.* NGHS	**385**	80	0.001709	+	+
QEHS *vs.* LHS	**380**	85	0.002415	+	+
QEHS *vs.* IDHSLDD	**334**	131	0.036826	+	+
QEHS *vs.* DH/Best	**295**	170	0.198610	=	=

From Table 4, we can know that QEHS outperforms the other 7 HS variants. In all comparisons, the R^+ value of QEHS is greater than the R^- value in Table 5 . And the average ranking of QEHS in Table 6 is the first. Hence, we conclude that QEHS is a competitive algorithm, and superior to these compared harmony search algorithms.

Table 6. Average ranking of QEHS and HS variants by Friedman test for 30 functions

Algorithm	HSDM	ESGHS	SGHS	NGHS	LHS	IDHSLDD	DH/Best	QEHS
Ranking	4.7	4.37	3.77	6.27	5.4	3.97	4.3	**3.2**

Table 7. Average error obtained by QEHS and other algorithms at $D = 30$

Fun	CLPSO	GSO	CoDE	QEHS
F_1	1.20E+08	2.20E+08	7.81E+06	**1.50E+06**
F_2	3.15E+09	1.44E+10	3.28E+07	**8.20E+03**
F_3	4.04E+04	9.96E+04	**2.20E+02**	6.66E+03
F_4	4.96E+02	1.84E+03	1.71E+02	**7.89E+01**
F_5	2.10E+01	**2.04E+01**	2.08E+01	2.10E+01
F_6	2.85E+01	1.26E+02	3.01E+01	**3.48E+00**
F_7	2.92E+01	9.62E+01	**1.30E+00**	4.76E-03
F_8	1.62E+02	4.17E+02	1.05E+02	**4.45E+01**
F_9	2.51E+02	8.09E+02	2.00E+02	**1.52E+02**
F_{10}	4.26E+03	7.66E+03	**3.34E+03**	6.23E+03
F_{11}	6.30E+03	1.76E+04	**6.17E+03**	7.23E+03
F_{12}	2.47E+00	**1.14E+00**	1.72E+00	2.97E+00
F_{13}	6.79E-01	5.54E-01	5.86E-01	**3.90E-01**
F_{14}	5.10E+00	5.59E+00	**4.00E-01**	4.25E-01
F_{15}	8.27E+02	2.98E+03	2.07E+01	**1.59E+01**
F_{16}	1.29E+01	4.48E+01	**1.26E+01**	**1.26E+01**
F_{17}	4.31E+06	4.50E+07	**1.64E+04**	3.85E+05
F_{18}	1.12E+07	8.16E+06	5.43E+03	**7.07E+02**
F_{19}	4.87E+01	2.32E+02	1.30E+01	**7.09E+00**
F_{20}	2.06E+04	1.26E+05	**1.13E+02**	9.43E+03
F_{21}	7.15E+05	1.88E+07	**2.67E+03**	2.14E+05
F_{22}	4.98E+02	3.05E+03	4.89E+02	**3.30E+02**
F_{23}	3.37E+02	4.26E+02	3.16E+02	**3.15E+02**
F_{24}	2.70E+02	4.03E+02	2.41E+02	**2.27E+02**
F_{25}	2.22E+02	3.35E+02	**2.08E+02**	**2.08E+02**
F_{26}	**1.01E+02**	2.05E+02	**1.01E+02**	1.04E+02
F_{27}	5.00E+02	3.55E+03	5.07E+02	**3.87E+02**
F_{28}	1.61E+03	1.99E+04	1.14E+03	**8.67E+02**
F_{29}	9.25E+04	8.13E+05	7.30E+03	**1.47E+03**
F_{30}	3.02E+04	7.73E+05	5.57E+03	**2.47E+03**

4.4 Comparison with Other Algorithms

In this part, we make contrast between QEHS and several popular optimization methods, namely CLPSO, ABC, GSO, EPSDE and CoDE.

Table 8. Average ranking of QEHS and other algorithms by Friedman test at $D = 30$

Algorithm	CLPSO	GSO	CoDE	QEHS
Ranking	2.88	3.70	1.82	**1.60**

From Table 7 and Table 8, we can understand that QEHS is superior to the others.

5 Conclusion and Future Work

We proposed an enhanced harmony search based on quantum mechanism (QEHS) in this paper. Experiments confirm the good performance of it. When $D = 30$, QEHS is not only better than HS, but also superior to 7 HS variants and 3 meta-heuristics.

Later on, we intend to further improve QEHS to make it more competitive in comparison with other algorithms. In addition, we consider introducing quantum mechanism into other improved HS to prove the validity of this approach.

Acknowledgements. The writer is exceedingly grateful to all the reviewers and editors who spent their time and energy on this paper. This study is sustained by the Nature Science Foundation of Fujian Province of P. R. China (No. 2019J01401, No. 2021J01127), and the Special Fund for Scientific and Technological Innovation of Fujian Agriculture and Forestry University (No. CXZX2019117S, No. CXZX2020148C, No. CXZX2020150C), and Educational Research Project for Young and Middle-aged Teachers of Fujian Provincial Department of Education (JA170179), and in part by the Open Project Program of Digital Fujian Tourism Big Data Institute under Grant DFJTBDRI2020103.

References

1. Geem, Z.W., Kim, J.H., Loganathan, G.V.: A new heuristic optimization algorithm: harmony search. Simulation **76**(2), 60–68 (2001)
2. Lee, K.S., Geem, Z.W.: A new structural optimization method based on the harmony search algorithm. Comput. Struct. **82**(9–10), 781–798 (2004)
3. Lee, K.S., Geem, Z.W.: A new meta-heuristic algorithm for continuous engineering optimization: harmony search theory and practice. Comput. Methods Appl. Mech. Eng. **194**(36–38), 3902–3933 (2005)
4. Moayedikia, A., Ong, K.-L., Boo, Y.L., et al.: Feature selection for high dimensional imbalanced class data using harmony search. Engineering Applications of Artificial Intelligence 5738-49 (2017)
5. Gandhi, T.K., Chakraborty, P., Roy, G.G., et al.: Discrete harmony search based expert model for epileptic seizure detection in electroencephalography. Expert Syst. Appl. **39**(4), 4055–4062 (2012)

6. Landa-Torres, I., Manjarres, D., Salcedo-Sanz, S., et al.: A multi-objective grouping harmony search algorithm for the optimal distribution of 24-hour medical emergency units. Expert Syst. Appl. **40**(6), 2343–2349 (2013)
7. Mahaleh, M.B.B., Mirroshandel, S.A.: Harmony search path detection for vision based automated guided vehicle. Robotics and Autonomous Systems 107156-166 (2018)
8. Chatterjee, A., Ghoshal, S., Mukherjee, V.: Solution of combined economic and emission dispatch problems of power systems by an opposition-based harmony search algorithm. Int. J. Electr. Power Energy Syst. **39**(1), 9–20 (2012)
9. Assad, A., Deep, K.: Applications of harmony search algorithm in data mining: a survey. In: Proceedings of Fifth International Conference on Soft Computing for Problem Solving, pp. 863–874. Springer, Cham (2016). Doi: https://doi.org/10.1007/978-981-10-0451-3_77
10. Zhang, T., Geem, Z.W.: Review of harmony search with respect to algorithm structure. Swarm and Evolutionary Computation 4831-43 (2019)
11. Geem, Z.W.: Harmony search algorithms for structural design optimization. Vol. 239. Springer (2009)
12. Wang, X., Gao, X.-Z., Zenger, K.: An introduction to harmony search optimization method. Springer (2015)
13. Zhao, F., Liu, Y., Zhang, C., et al.: A self-adaptive harmony PSO search algorithm and its performance analysis. Expert Syst. Appl. **42**(21), 7436–7455 (2015)
14. Kennedy, J., Eberhart, R.: Particle swarm optimization. In: Proceedings of ICNN'95-International Conference on Neural Networks, pp. 1942–1948. IEEE (1995)
15. Gheisarnejad, M.: An effective hybrid harmony search and cuckoo optimization algorithm based fuzzy PID controller for load frequency control. Applied Soft Computing 65121-138 (2018)
16. Yang, X.-S., Deb, S.: Cuckoo search via Lévy flights. In: 2009 World congress on nature & biologically inspired computing (NaBIC), pp. 210–214. IEEE (2009)
17. Portilla-Flores, E.A., Sánchez-Márquez, Á., Flores-Pulido, L., et al.: Enhancing the harmony search algorithm performance on constrained numerical optimization. IEEE Access 525759-25780 (2017)
18. Amaya, I., Cruz, J., Correa, R.: Harmony search algorithm: a variant with self-regulated fretwidth. Applied Mathematics and Computation 2661127–2661152 (2015)
19. Guo, Z., Wang, S., Yue, X., Yang, H.: Global harmony search with generalized opposition-based learning. Soft. Comput. **21**(8), 2129–2137 (2015). https://doi.org/10.1007/s00500-015-1912-1
20. Keshtegar, B., Hao, P., Wang, Y., et al.: Optimum design of aircraft panels based on adaptive dynamic harmony search. Thin-Walled Structures 11837–11845 (2017)
21. Shor, P.W.: Algorithms for quantum computation: discrete logarithms and factoring. In: Proceedings 35th Annual Symposium on Foundations of Computer Science, pp. 124–134. IEEE (1994)
22. Qiao, W., Yang, Z.: Solving large-scale function optimization problem by using a new meta-heuristic algorithm based on quantum dolphin swarm algorithm. IEEE Access 7138972–7138989 (2019)
23. Cheung, N.J., Ding, X.-M., Shen, H.-B.: A nonhomogeneous cuckoo search algorithm based on quantum mechanism for real parameter optimization. IEEE Trans. Cybern. **47**(2), 391–402 (2016)
24. Xin-Gang, Z., Ji, L., Jin, M., et al.: An improved quantum particle swarm optimization algorithm for environmental economic dispatch. Expert Systems with Applications 152113370 (2020)
25. Agrawal, R., Kaur, B., Sharma, S.: Quantum based whale optimization algorithm for wrapper feature selection. Applied Soft Computing 89106092 (2020)

26. Schrödinger, E.: An undulatory theory of the mechanics of atoms and molecules. Phys. Rev. **28**(6), 1049 (1926)
27. Liang, J.J., Qu, B.Y., Suganthan, P.N.: Problem definitions and evaluation criteria for the CEC 2014 special session and competition on single objective real-parameter numerical optimization. Computational Intelligence Laboratory, Zhengzhou University, Zhengzhou China and Technical Report, Nanyang Technological University, Singapore 635490 (2013)
28. Qin, A.K., Forbes. F.: Proceedings of the 13th annual conference on Genetic and evolutionary computation, pp. 545–552. ACM (2011)
29. Luo, K., Ma, J., Zhao, Q.: Enhanced self-adaptive global-best harmony search without any extra statistic and external archive. Information Sciences 482228-247 (2019).
30. Pan, Q.-K., Suganthan, P.N., Tasgetiren, M.F., et al.: A self-adaptive global best harmony search algorithm for continuous optimization problems. Appl. Math. Comput. **216**(3), 830–848 (2010)
31. Zou, D., Gao, L., Wu, J., et al.: Novel global harmony search algorithm for unconstrained problems. Neurocomputing **73**(16–18), 3308–3318 (2010)
32. Ouyang, H.-B., Gao, L.-Q., Li, S., et al.: Improved harmony search algorithm: LHS. Applied Soft Computing 53133-53167 (2017)
33. Abedinpourshotorban, H., Hasan, S., Shamsuddin, S.M., et al.: A differential-based harmony search algorithm for the optimization of continuous problems. Expert Systems with Applications 62317-62332 (2016)
34. Zhu, Q., Tang, X., Li, Y., et al.: An improved differential-based harmony search algorithm with linear dynamic domain. Knowledge-Based Systems 187104809 (2020)

Performance Evaluation of Three Intelligent Optimization Algorithms for Obstacle Avoidance Path Planning

Shumeng Chen[1], Chia-Hung Wang[1,2(✉)], Zhenyu Dong[1], Qigen Zhao[1], Qipeng Yang[1], Yi Wei[1], and Guoshu Huang[1]

[1] College of Computer Science and Mathematics, Fujian University of Technology, Fuzhou City, China

[2] Fujian Provincial Key Laboratory of Big Data Mining and Applications, Fujian University of Technology, Fuzhou City, China

Abstract. In this paper, we study an intelligent path planning problem on avoiding traffic obstacles in urban transportation networks. We apply three heuristic methods to solve the optimization model, including genetic algorithm, wolf pack algorithm, and a mixture method of genetic algorithm and wolf pack algorithm. In the numerical results, we compare the solutions and performance of these three intelligent algorithms. It shows that the optimal path lengths obtained by these three algorithms are very closed, whereas the convergence time of the genetic algorithm is the shortest among them. We also find that the mixture method is more efficient in updating the population, and its convergence speed is the fastest. Besides, it is found that the convergence time of the wolf pack algorithm is smaller than the others. By means of the analysis in our experiments, we evaluate the pros and cons of these algorithms when solving the studied optimization problem, which would be of great significance for the intelligent optimization of obstacle path planning in smart cities.

Keywords: Urban transportation network · Shortest path · Intelligent optimization algorithm · Traffic obstacle · Path planning problem

1 Introduction

In the field of intelligent transportation, the path planning problem [1] has always been a hot issue and has great practical application value. The evaluation criteria of optimal path planning are mostly based on the consideration of time cost, distance cost and fuel cost, find an optimal path from the starting point to the target point to avoid obstacles in the transportation network. The problem to be solved in this paper is to find the shortest path that can avoid obstacles in the urban transportation network with a given starting point and ending point. It is worth noting that this paper uses a combined method of directed graph [2] and raster graph [3] to approximate the urban traffic environment. By solving such

S.-C. Chu et al. (Eds.): ICGEC 2021, LNEE 833, pp. 60–69, 2022.
https://doi.org/10.1007/978-981-16-8430-2_6

problems, we compare and evaluate the performance of the genetic algorithm [4,5], wolf-pack algorithm [6] and wolf-pack-genetic fusion algorithm.

Path planning is a classical optimization problem [7], which has been studied for decades and has great application value. In the early stage of the research, researchers mainly conducted path planning algorithm research on fixed geographical environments. In recent years, with the wider coverage of 5G network base stations [8], the path planning of the urban transportation network can be updated in real time using network communication [9,10]. How to study and implement an efficient and accurate path optimization algorithm has always been the subject of attention. In addition, with the development of path planning problems, some new algorithms also follow [11].

The planning task of the urban traffic road network is to find a collision-free shortest path from the source point to the destination point in an environment with obstacles. The design of path planning algorithm is determined by the difficulty of planning tasks, the complexity of the planning environment and the requirement of planning quality [12]. It is obvious that both the planning quality and the computing speed are important indicators of the algorithm design. The algorithms for path planning would be used in a wide range of applications, such as path planning for robotic manipulators, aircraft path planning, traveling salesman problem (TSP), and Line laying of power grid [13], etc.

In this paper, we study an intelligent route optimization problem for avoiding traffic obstacles. The purpose of our work is to compare the (near) optimal solutions determined by three meta-heuristic algorithms, including the genetic algorithm, the wolf-pack algorithm and the wolf-pack-genetic fusion algorithm. The main contribution is to conduct a sensitivity analysis on the performance of three algorithms, which could provide several suggestions to improve the solving strategy for the studied optimization problem.

The rest of this paper is organized as follows. In Sect. 2, we introduce the problem and formulate the mathematical model. In Sect. 3, we introduce the basic principles and steps of three meta-heuristic algorithms, including the genetic algorithm, the wolf-pack algorithm, and the wolf-pack-genetic fusion algorithm. Then, we will conduct the numerical experiments and illustrate the experimental results in Sect. 4. Finally, the conclusions are summarized in Sect. 5.

2 Path Planning for Obstacle Avoidance

2.1 Problem Formulation

According to whether the environment changes over time, path planning can be divided into dynamic planning and static planning [14]. According to the scope of path planning, it can be divided into global path planning and local path planning. The category of path planning in this paper is a static global planning, that is, finding the optimal path from a given starting point to an end point in an urban environment with obstacles. There are three more important factors: clearly giving the starting point and ending point, avoiding all obstacles, and optimizing the path as much as possible.

In general, we have two solutions to this problem. One is to use graph theory to establish an optimization model, find the objective function and constraints, and then use an optimization algorithm to solve it. The other is to establish a weighting matrix based on the graph theory model, and then use a meta-heuristic algorithm to solve it. In order to facilitate the realization of the meta algorithm, we combined the grid graph to represent the directed graph from the starting point to the end point, and establish a weighting matrix based on the directed graph. Then, we use the genetic algorithm, the wolf pack algorithm, and the wolf pack-genetic fusion algorithm to solve the problem, and evaluate the performance of the three algorithms.

In this paper, we use the grid method to simulate the situation that may exist in the actual environment. The grid method is to decompose the environment into a number of regular units (usually square), that is, a grid, which describes the actual terrain through the feature information of the grid. This method can approximately describe the city's traffic environment. The moving object has four moving directions in the grid, which are up, down, left, and right. In this paper, we divide the urban terrain environment into a grid map, in which gray squares represent obstacles on the way, and white areas are the paths that can be passed. In order to facilitate the establishment of the matrix corresponding to the raster image, we will number the raster here. As shown in Fig. 1, we can specify the start and end points on the path planning.

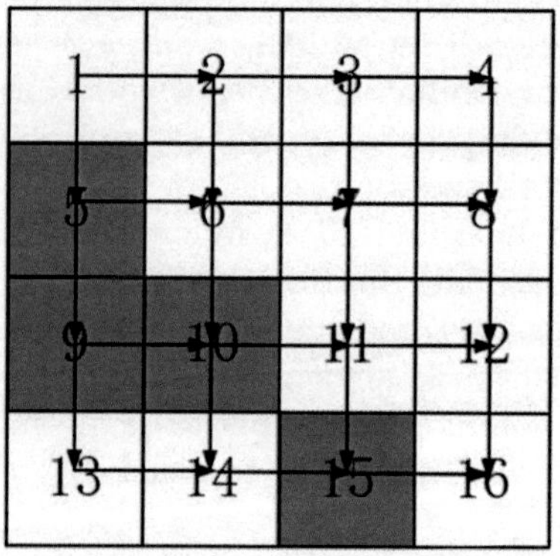

Fig. 1. Directed graph

Before formulating the optimization model, we have the following assumptions. Firstly, the movement area in the sports environment is a two-dimensional limited area. Secondly, there are a finite number of stationary obstacles in the environment, and their height does not affect the moving objects. Finally, the position and size of obstacles in the movement area never change.

2.2 An Optimization Model for Obstacle Avoidance Path Planning

As shown in Fig. 1, we use the center of each grid in the grid graph to represent the starting and ending points of the path. We denote the vertex set of the directed graph as V, where $V = \{V_1, V_2, ..., V_n\}$ and a total of n vertices. Let V_i and V_j represent the i-th and j-th vertices. The line between two adjacent vertices constitutes an edge, and we denote the set of all edges by E. Besides, we can assign a weighted value ω to a given edge according to transportation time, distance cost, etc. For example, the weight ω_{ij} represents the weight of the edge from the vertex V_i to vertex V_j. In this way, we can build a weighted directed graph $G = (V, E, \omega)$ in combination with the raster graph. In order to determine an end-to-end route, we can formulate an optimization model as follows:

$$\min f = \sum_{V_i, V_j \in E} \omega_{ij} X_{ij}, \tag{1}$$

subject to

$$\sum_{j=1}^{n} X_{ij} - \sum_{j=1}^{n} X_{ji} = \begin{cases} 1 & \text{for } i = 1 \\ -1 & \text{for } i = n \\ 0 & \text{for } i \neq 1, n \end{cases}, \tag{2}$$

where ω_{ij} represents the weight on the edge from vertices V_i to V_j. Our decision variables are X_{ij} for all i, j. When an edge is on the line from vertices V_i to V_n, it holds that $X_{ij} = 1$. Otherwise, when the edge is not on the line from vertices V_i to V_n, we have $X_{ij} = 0$. Objective function (1) represents the minimum value of the feasible path. Constraint (2) represents that the starting point of the next node must be the arrival point of the previous node. In this way, we can traverse all possible paths from V_i to V_n with the help of constraint (2).

3 Three Intelligent Optimization Algorithms

3.1 Genetic Algorithm

The main steps of Genetic Algorithm (GA) consist of selection, crossover, and mutation. In this paper, we use binary coding, roulette wheel selection for selection operation, and random number for crossover and mutation operation. The steps of our genetic algorithm are summarized as follows:

Step 1. Initialize the population size M, the maximum number of iterations Period, the crossover probability P_1 and the mutation probability P_2, etc.

Step 2. Determine the fitness of the individual and judge whether it meets the optimization criteria. If it is consistent, output the best individual and its optimal solution, and then end; otherwise, we proceed to the next step.

Step 3. Select individuals based on fitness.

Step 4. Perform crossover operation according to crossover probability P_1.

Step 5. Perform mutation operation according to mutation probability P_2, and then generate new individuals to join new groups.

Step 6. A new generation of the population is generated after crossover operation and mutation operation. Return to Step 2 until the number of iterations is reached or the optimal solution is found.

3.2 Wolf-Pack Algorithm

The wolf-pack algorithm(WPA) first initializes the position of a lead wolf. The algorithm continuously updates the position of the lead wolf by changing the position of the detective wolf. When the conditions are met, the fierce wolf launches a siege. The wolves continue to update the population through the survival of the fittest and maintain the optimal performance of the population. The algorithm steps are as follows:

Step 1. Initialize parameters such as the number of wolves M, Number of detective Wolves walks Try, the walking step length *Step*, the siege step length D, the maximum number of iterations *Period*, etc.

Step 2. Calculate the fitness to select the lead wolf, detective wolves and fierce wolves. Detecting the wolf performs wandering behavior.

Step 3. The fierce wolves obey the command of the lead wolf. After hearing the call of the lead wolf, it approaches the lead wolf and updates the leading wolf.

Step 4. If the distance between the fierce wolves and the leading wolf is less than the limit of d_{near}, then hunt prey.

Step 5. Renew the population after the hunting of prey ends.

Step 6. Judge whether the optimization goal is reached. If satisfied, output, otherwise continue to repeat the above steps.

3.3 A Mixture Method of Wolf-Pack and Genetic Algorithms

The Wolves-Genetic Fusion Algorithm (WPA-GA) is mainly realized by the adaptive evolution of individuals in the wolf pack. When the wolves update themselves, the selection, crossover, mutation and other operations in the genetic algorithm are integrated into the wolves algorithm. In this way, excellent individuals can be further selected, so that better genes are preserved. The wolf pack develops in a better direction, is more conducive to survival, and has more advantages in the process of solving. The main steps of the wolf-pack-genetic fusion algorithm are:

Step 1. Initialize the number of wolves M, detective wolves wandering times Try, Wandering step length *Step*, Siege step length D, Crossover probability: P_1, Mutation probability: P_2, Number of iterations: *Period*, etc.

Step 2. Select the leader wolf based on individual fitness.

Step 3. Detective Wolves start to walk, and updates the leader wolf.

Step 4. The call of the leader wolf, Fierce wolves began to attack and update the leader wolf.

Step 5. Siege the prey.

Step 6. Perform selection, crossover, mutation and other genetic operations within the wolf pack to regenerate individuals.

Step 7. Update the taboo table.

Step 8. Judge that the goal is reached, and output if it is met, otherwise continue to repeat the above steps.

4 Numerical Experiments

The parameters setting in those three algorithms are summarized as follows: the population size $M = 50$, the crossover probability $P_1 = 0.7$, the mutation probability $P_2 = 0.1$, and the number of iterations $Period = 50$. Besides, the number of wolves $M = 50$, the number of Detective Wolf wandering $Try = 4$, and the wandering step length $Step = 2$. For fierce wolves, we set the siege step length of fierce wolves $D = 0.5$. Our computational experiments are conducted by means of the software MATLAB R2020a on a computer with Intel(R) Core(TM) i5-6200U CPU@2.40 GHz with 4 GB RAM.

Several performance measures are to be evaluated in the numerical results, such as The running time, the shortest path, and the convergence time, etc. The running time is defined as the time required for the algorithm to complete the predetermined number of iterations. The shortest path is denoted as the length of the optimal path solved by the algorithm. The convergence time is defined as the number of iterations when the algorithm converges.

4.1 Effect of Varying the Problem Scale

Here, we set the population size as 50, and run 50 iterations. We compare the performance of three solving algorithms, as the grid sizes vary from 10 by 10, 50 by 50, 100 by 100, 150 by 150, 200 by 200, until 250 by 250.

In Fig. 2, the X coordinate represents the size of the problem, and the Y coordinate represents the calculated path length. In Fig. 2(a), we observe that the optimal path lengths determined by those three algorithms are very closed. Nevertheless, when increasing the problem scale, the advantages of the wolf pack-genetic fusion algorithm would gradually become prominent.

In Fig. 2(b), it can be found that the running time of the three algorithms is increasing as the problem size increases. But the wolf-pack algorithm and the wolf-pack-genetic fusion algorithm time increase faster, especially when the problem size exceeds 200. On the contrary, genetic calculations always take the least time and increase at a slower rate.

In Fig. 2(c), the Y coordinate represents the number of iterations for each algorithm, which can indicate the number of convergence. We find that the number of algorithm convergence starts to fluctuate as the problem size expands. It can be observed that the number of convergence of the wolf-pack-genetic fusion algorithm fluctuates less, and the number of iterations is also minimal. Besides, as time increases, the number of iterations required by the genetic algorithm becomes the largest.

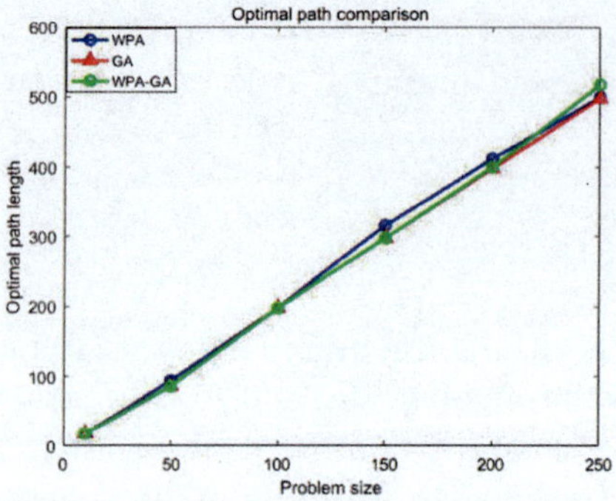

(a) Effect on the optimal path length

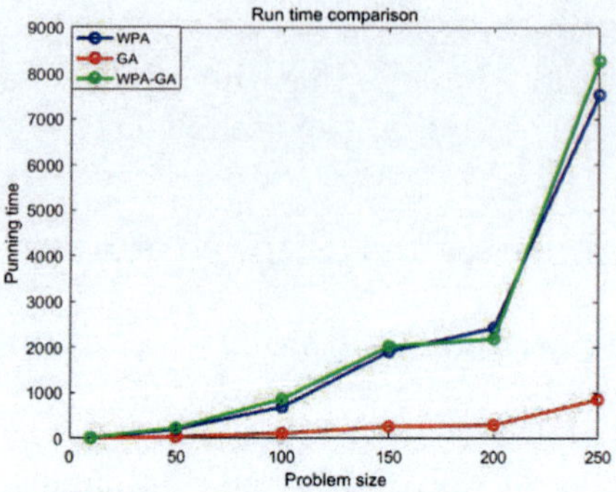

(b) Effect on the running time

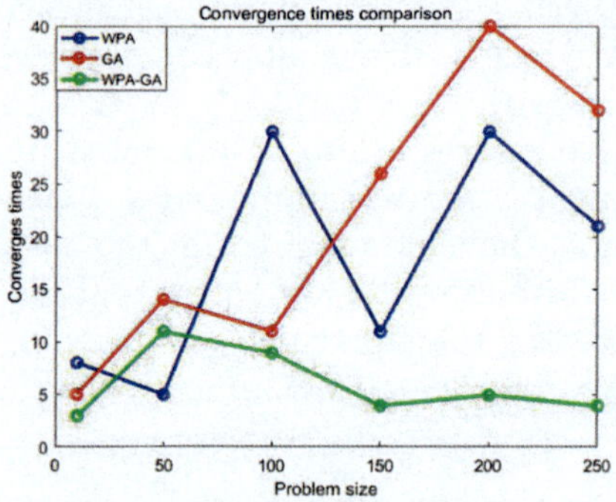

(c) Effect on the converges time

Fig. 2. Performance evaluation of three algorithms when varying the scale.

4.2 Effect of Varying the Population Size

In this experiment, we set the problem size to 100 * 100 and the number of iterations to 20. We set the population sizes as 10, 20, 30, 40, 50, 60, 70, 80, 90, and 100. Record the running time of the three algorithms, the optimal path, and the number of iterations. In Fig. 3, we illustrate the performance of those three solution algorithms.

In Fig. 3(a), we can intuitively understand that when the population size is less than 50, the solution result of the wolf-pack algorithm is relatively poor. However, when the population number exceeds 50, the three types have the same performance in calculating the path length.

In Fig. 3(b), we observe that as the population size increases, the running time of the three algorithms is increasing. Among them, the genetic algorithm increases the slowest. Besides, the wolf-pack-genetic fusion algorithm increases the fastest, and the time increase of the wolf pack algorithm is between them.

In Fig. 3(c), it can be seen that the convergence of the wolf-pack-genetic fusion algorithm is the smallest number of runs. With the increase of the population, the convergence times of the genetic algorithm and the wolf-pack algorithm show a

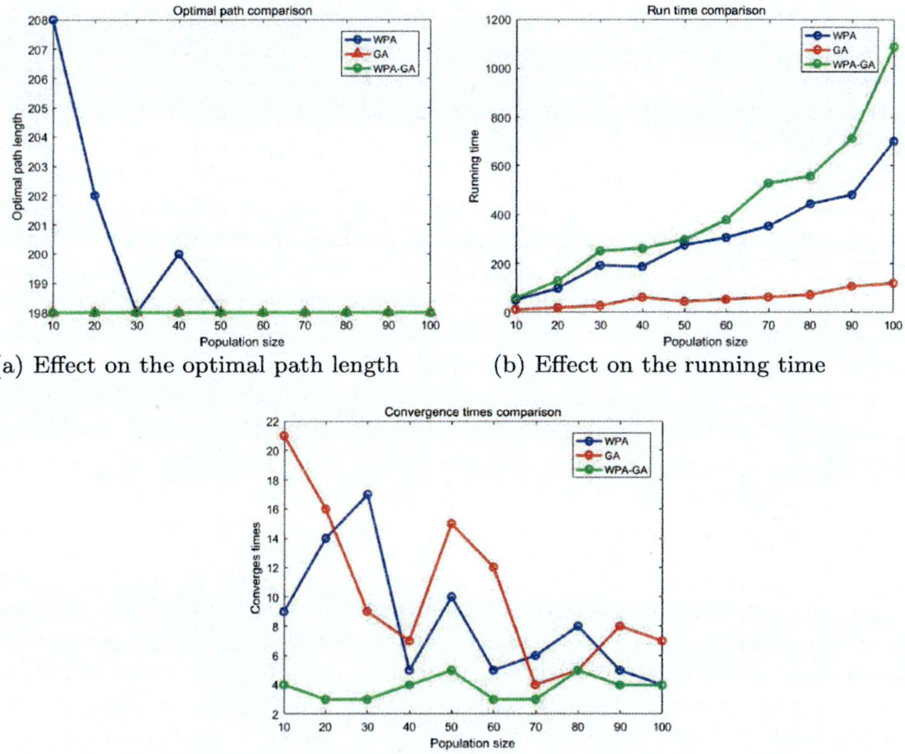

(a) Effect on the optimal path length (b) Effect on the running time

(c) Effect on the converges time

Fig. 3. Performance evaluation of three algorithms when varying the population size.

decreasing trend, but it is still larger than the wolf-pack-genetic fusion algorithm. This is also the main advantage of the wolf-pack-genetic fusion algorithm.

4.3 Discussion of Three Solution Algorithms

Through the above experiments, it can be seen that the biological intelligence algorithm is feasible in path planning and has good planning performance. On a grid topographic map with a given starting point and ending point and containing obstacles, the shortest path to avoid obstacles can be successfully planned. But each algorithm in this paper has its own advantages and disadvantages.

When only changing the size of the problem. With the expansion of the problem size, the running time required for the same number of iterations of the three algorithms becomes longer and longer. Among them, the genetic algorithm has a significantly shorter running time and has more advantages. Both the wolf-pack algorithm and the wolf-pack-genetic fusion algorithm run longer, and the gap is relatively small. Almost all three algorithms can find the optimal path. As for the number of iterations, the genetic algorithm needs more iterations to find the optimal path. The wolf-pack algorithm is better than the genetic algorithm, and the wolf-pack-genetic fusion algorithm requires the least number of iterations.

When only changing the population size. With the increase of the population, the running time required for the same number of iterations of the three algorithms is increasing. However, the genetic algorithm takes the shortest time, the wolf pack-genetic algorithm takes the longest time, and the wolf-pack algorithm takes between the two. When the population is small, the length of the optimal path determined by the wolf-pack algorithm is unstable, but when the population reaches 50 and above, all three algorithms can find the optimal path. As for the number of convergences, it is obvious that the wolf pack-genetic fusion algorithm has the least number of convergence times, and the genetic algorithm has the largest number of convergence times. The convergence times of the wolf-pack algorithm are between the genetic algorithm and the wolf-pack-genetic fusion algorithm.

5 Conclusions

In this paper, we solved the path planning problem of specifying the starting point and ending point in the urban road network with traffic obstacles. We applied three intelligent optimization algorithms to determine route planning for avoiding traffic obstacles. The contribution of this paper is to compare the pros and cons of those three algorithms, that is, genetic algorithm, wolf pack algorithm and a fusion algorithm combined with the wolf pack and genetic algorithms.

In the proposed scheme, the grid method can be used to divide the urban road network and the topographic map corresponding to the obstacles in the city. This method directly affects the amount of information stored in the traffic environment and the search length of route planning. On the one hand, if the size of the grid unit is larger, the size of the representative urban terrain is simpler, the storage capacity of road traffic information is smaller, and the calculation time required for the proposed intelligent optimization algorithm is shorter. On the other hand, if the size of the grid unit is smaller, the size of the representative urban terrain is more complicated, the amount of information storage of the road traffic environment is larger, and the calculation time required is longer.

For potential applications, the proposed scheme in this paper could be applied to establish an intelligent (vehicle) traffic route database to provide support for a road traffic management platform that provides route planning information exchange. The vehicle path planning with the specified starting point and ending point proposed in this paper can help vehicles drive on the shortest road in the city while avoiding obstacles on the road, such as buildings, parks, lakes, etc. It can save unnecessary end-to-end driving time and vehicle fuel consumption for traffic routes.

Acknowledgment. This work was supported in part by Fujian Provincial Department of Science and Technology, China, under the grant number 2021J011070, the Education Department of Fujian Province, China, under Grant No. JAT190404, and the Research Fund from Fujian University of Technology under Grant No. GY-Z18148.

References

1. Wang, F., Wang, Z.W., Li, M.J.: Robot path planning based on improved particle swarm optimization. IEEE, pp. 887-891 (2021)
2. Li, Y.P., Wei, k., Wang, D.: The algebraic algorithm for solving the path planning problem of a weighted directed graph. Appl. Mech. Mater. 3468, 1648-1653 (2014)
3. Bahar, M.R.B., Ghiasi, A.R., Bahar, H.B.: Grid roadmap based ANN corridor search for collision free path planning. Scientia Iranica **19**, 1850–1855 (2012)
4. Li, Y., Yuan, H.Y., Yu, J.Q., Zhang, G.W.: Overview of the application of genetic algorithms in optimization problems. Shandong University of Technology, pp. 242-244 (2019)
5. Hao, K., Zhao, J., Wang, B.B., Liu, Y.L., Wang, C.Q.: The application of an adaptive genetic algorithm based on collision detection in path planning of mobile robots. Computational Intelligence and Neuroscience, pp. 19-34 (2021)
6. Ma, X.R., Li, H.X., Zhou, Z., Chen, B.: Fleet maintenance strategy planning with time windows integrated with multi-agent and wolf pack reinforcement learning. In: IOP Conference Series, vol. 1043, pp. 1–9 (2021)
7. Huang, Q.: Research on path planning of mobile robot based on particle algorithm. Int. Core J. Eng. **7**, 423–431 (2021)
8. Wang, C.-H., Lee, C.-J., Xiaojing, W.: A Coverage-Based Location Approach and Performance Evaluation for the Deployment of 5G Base Stations. IEEE Access **8**, 123320–123333 (2020). https://doi.org/10.1109/ACCESS.2020.3006733
9. Hung, M.-H., Wang, C.-H., and He, Y.: A real-time routing algorithm for end-to-end communication networks with QoS requirements. In: Proceedings of the 3rd International Conference on Computing, Measurement, Control and Sensor Network (CMCSN2016), Matsue, Shimane, Japan, pp. 186-189 (2016)
10. Wang, C.-H., Lee, C.-J., Chung, W.-H., Wu, M.-E.: An atomic routing game for multi-class communication networks with quality of service requirements. In: Proceedings of the 24th Wireless and Optical Communication Conference (WOCC2015), Taipei, Taiwan, pp. 206-210 (2015)
11. Lin, H.Y., Huang, Y.C.: Collaborative complete coverage and path planning for multi-robot exploration. Sensors **21**, 4–15 (2021)
12. Dong, Y.D., Ser, K.G.: Path planning for mobile-anchor based wireless sensor networks localization: obstacle-presence schemes. Sensors **21**, 1–8 (2021)
13. Wang, C.-H., Nguyen, T.-T., Pan, J.-S., Dao, T.-k.: An optimization approach for potential power generator outputs based on parallelized firefly algorithm. Advances in Intelligent Information Hiding and Multimedia Signal Processing, vol. 64, pp. 297-306, Springer International Publishing AG (2016)
14. Ben, B.K.A., Ryad, C., Albert, Y.A.: An overview of nature-inspired, conventional, and hybrid methods of autonomous vehicle path planning. J. Adv. Transp. **2018**, 1–24 (2018)

Software Development and Reliability Systems

Experimental Data Acquisition and Processing System Based on the Internet of Things

Hang Liu(✉), Jun Wang, Mingtai Liu, Xiukai Jian, Yuhong Zhang, and Yuli Li

School of Electrical and Computer Engineering, Jilin Jianzhu University, Changchun 130118, China
liuhang@jlju.edu.cn

Abstract. With the development of the Internet of Things (IoT), the data acquisition system is more and more important for the follow-up of experimental data. In this paper, two data collecting platforms based on IoT technology are built for collecting and processing experimental data, so that the processing of experimental data can get rid of local restrictions more conveniently. One of the data platforms in this paper is based on AL Cloud IoT Platform, and the other one is a self-built cloud server architecture. The WeChat Mini program and the PC webpage are designed to monitor the state of the device and display the data visually. MySQL is used to store experimental data.

Keywords: Internet of Things · Data acquisition · WeChat Mini Program · MySQL

1 Introduction

Nowadays, the IoT is a hot topic. All the objects, machines, animals and people are connecting by IoT, which collect data from the terminals and process them to extract valuable information [1]. The IoT technology can acquire the characteristics and data of the display object through the perceptual effects of intelligent devices. And it can transfer data to other devices to analyze it, so as to facilitate the control and management of external things [2]. Benefiting from the continuous accumulation and rapid upgrading of technologies such as the IoT, the big data and artificial intelligence, IoT industry chain are gradually improving. Therefore, the information collection technology is gradually developing toward intelligence.

At present, the laboratory management is mostly in the state of decentralization. The computers and oscilloscopes which are used to monitor the instrument status and manage the working data, are placed locally. The decentralized control of equipment is not conducive to unified control of equipment, but also to the network collection of data, which poses great difficulties for the unified management of data and subsequent integration and analysis [3]. The data acquisition system of IoT becomes more and more important for subsequent data application processing of experiments [4]. The application of IoT technology in the subsequent data acquisition of semiconductor experiments can

S.-C. Chu et al. (Eds.): ICGEC 2021, LNEE 833, pp. 73–81, 2022.
https://doi.org/10.1007/978-981-16-8430-2_7

reduce the space needed for original acquisition equipment and make it more convenient for data transfer and processing by multi-mobile devices.

In this paper, two data acquisition platforms based on IoT are designed to collect the concentration and resistivity of semiconductor transducer, and a WeChat Mini Program, PC web terminal and database are designed to support them. The two systems can display data collected by hardware devices in real time from a distance and display them visually on mobile devices such as mobile phones and computers, enabling multi-device data monitoring.

2 MQTT Protocol

The development of an IoT device requires some server developments. Such as computers and mobile phones, which are used to connect the server and transmit information, as shown in Fig. 1. The server is composed by using the Message Queuing Telemetry Transport (MQTT) protocol. The MQTT protocol is a network protocol with using to transfer basic implementation information from equipment to equipment in the IoT. It is a lightweight publish/subscribe messaging protocol. It is widely used due to its simple specification and low traffic consumption [5]. These advantages make this protocol to become an ideal choice for designing a monitoring system in this paper.

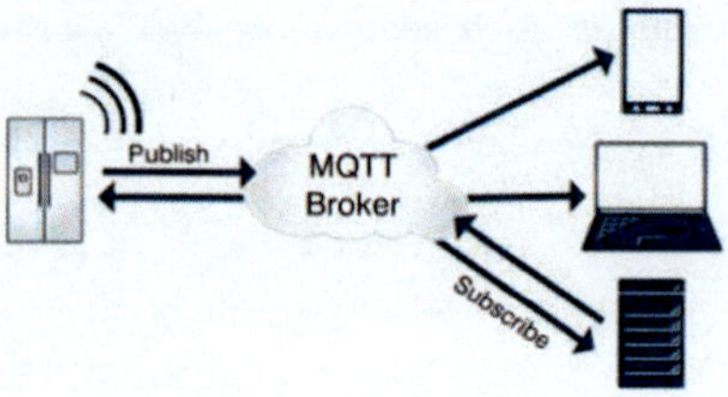

Fig. 1. Schematic diagram of MQTT protocol

3 Data Acquisition System Design

3.1 Cloud Platform Selection

The system is developed with using the classical C-S system, where the client and server carry out their respective tasks to complete the overall development of the monitoring system [6, 7]. There are a large number of cloud service platforms currently. The requirement of this system needs to transmit data for infinite distance. By comparing with several cloud platforms and actuating experiments. The current popular IoT platform, such as Ali Cloud, is simple to build and come with their own software, which can effectively save development costs and development time. But the software of the IoT platform cannot meet certain needs of the development system, so the server development needs to be self-built. The advantage of self-building server is that it can customize the system in depth by only subscribing to the cloud server on WeChat Mini Program or

PC web monitoring side for accessing the data. The data is not restricted by third party platforms and makes the system completely independent. Because of the advantages and disadvantages of these two methods, two data collection systems are designed in this paper. One system builds its own data collection system platform with the cloud server, the other sets up the data-collection platform for various needs with Ali Cloud platform.

3.2 Private Cloud Server to Build Data Collection Platform

The system overall design framework is shown in Fig. 2. The hardware devices to Wi-Fi, then uploads the detected data to the cloud server through the MQTT protocol. The Tencent cloud server is selecting to build a private data collection platform. The WeChat Mini Program or PC web terminal and MYSQL database are designed to obtain data by subscribing to the cloud server. And the WeChat Mini Program can remotely control the hardware devices, and complete the overall data collection platform.

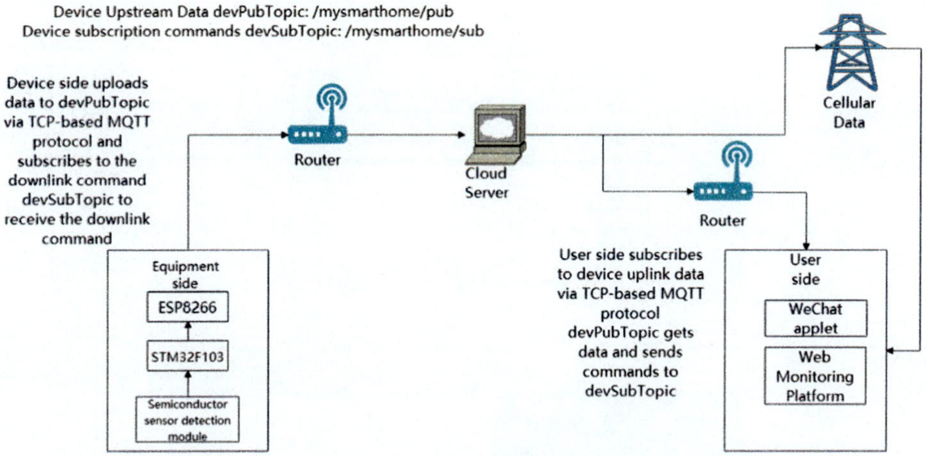

Fig. 2. System block diagram

WeChat Mini Program Design. In this paper, a WeChat Mini Program written by Uniapp is chosen as one of the data collection and acquisition systems. The WeChat Mini Program is an application which uses little memory and can be used quickly. It has some convenient functions for the monitoring of the system. The Uniapp can compile application code for iOS, Android, H5 and various small application platforms, and can be published on multiple platforms by talking to multiple end languages with a specific integration language, saving costs and development time.

In this system, the significance of designing WeChat Mini Program is able to remotely control the equipment. It can check the state of hardware equipment, which detects resistance value of semiconductor sensor. Thereby, the gas concentration is detected in real time. The design completes its workflow for WeChat applet development. The

WeChat Mini Program interface is shown in Fig. 3. After the hardware is powered up and connected to WIFI, the WeChat Mini Program start networking and subscribes to the cloud server theme. The WeChat Mini Program automatically captures the resistance and harmful gas concentration of semiconductor sensors monitored by hardware devices and controls the switching state of the beeper and light-emitting diode lights (LED). It can also graph the uploaded data on the image page. WeChat Mini Program is shown in Fig. 4.

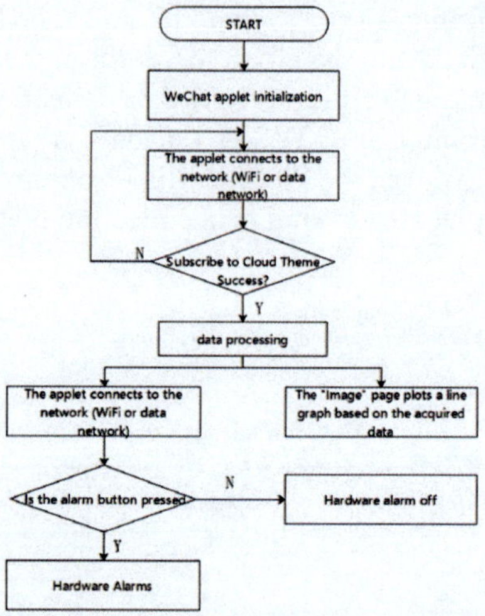

Fig. 3. WeChat Mini Program workflow

Fig. 4. WeChat Mini Program interface

Design of Computer Web Collection Terminal. For the monitoring system on the whole platform, only one WeChat Mini Program doesn't match the hardware because the small program can't receive data transfer from more than one device. So, the system is designed to transfer data from the PC web side to the cloud server for obtaining the environmental gas concentration measured by the hardware indirectly.

The Node-RED is a simple and powerful tool for building IoT applications. The point of Node-RED can be applied the "connection" of code blocks for perform system tasks. The developers execute tasks through connecting predefined chunks of code (called "nodes") with visual programming. The input nodes, processing nodes and output nodes are connected together, then form a "Flows" [8]. As shown in Fig. 5, it is used to design web nodes.

The system is programmed by Node-RED. The monitoring web page is created in Fig. 6. In this system the data are uploaded by hardware devices through TCP-based

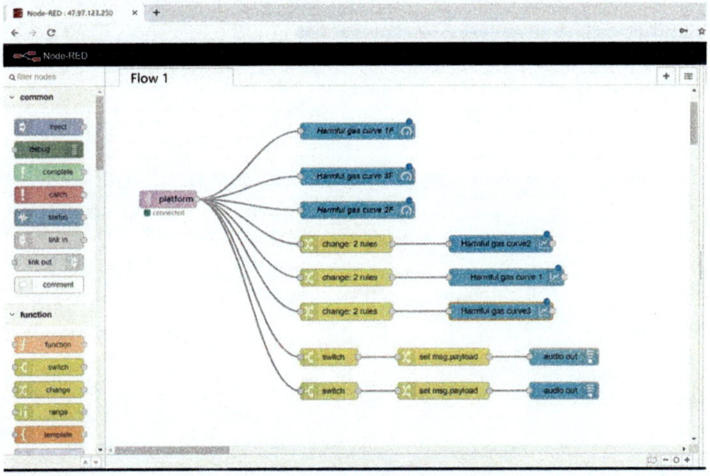

Fig. 5. Node flow design

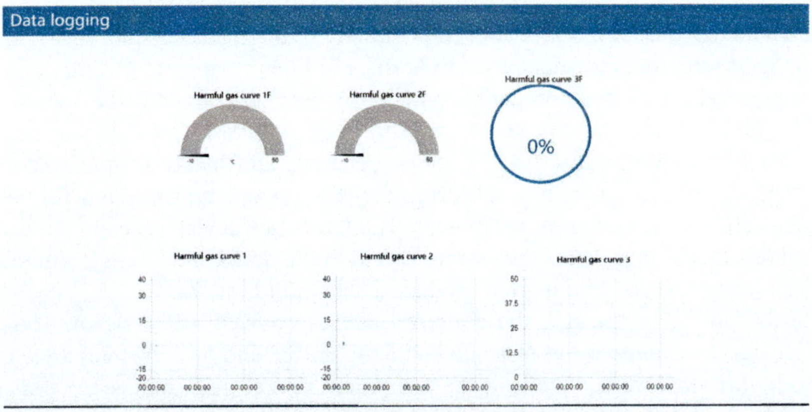

Fig. 6. Monitoring of the web side

MQTT protocol subscriber's uplink data, and the data image can be drawn. The system can store data for 48 h. Once the harmful gas concentrations on any floor exceed the warning value, a voicemail alert will be automatically generated.

MySQL Design. In order to realize the centralized storage of equipment and experimental data, the MySQL is selected as the database of collecting system. MySQL is an open-source relational database management system. It has the advantages of small size and fast speed. It also allows data to be stored in different tables instead of one place, so it can increase the flexibility of data access [9]. The database itself is also a TCP server. First the client and database connect to the server. Then connect to the database stand according to the database protocol. One database server only can establish limited connection values at the same time. The druid connection pool is an open-source database connection pooling project by Alibaba. In order to improve the response time of requests and server performance, the Druid connection pooling is used to manage objects and connect the database. The main data tables are designed as follows (Table 1):

Table 1. Data attributes

Name	Data type	Length	Decimal	Null
Id	int	11	0	N
Data time	timestamp	0	0	Y
Client	varchar	255	0	Y
Concentra	float	0	0	Y
Resistance	float	0	0	Y

3.3 Data Collection Platform Based on AL Cloud IoT Platform

The AL Cloud IoT platform can be logged in directly through an account. The users can create the required products, such as smart home, intelligent temperature and humidity, etc. AL Cloud IoT platform is compatible with WIFI, data network and other networking methods. The hardware devices upload data through A-link JSON data format. The data format is common with other AL cloud products and it has a large amount of data storage. This data collection platform adopts a custom transmission format for facilitating subsequent system development. The hardware device uploads the data to the IoT platform and then sends the resulting data to the data collection system via the server. The flow chart for creating the product is shown in Figs. 7a.

Once the project is created, the collected data is added in the system. The data mainly include resistance value of semiconductor sensor and gas concentration value detected. In this way, we can monitor the equipment in real time, reduce the difficulty and speed up the development progress. Once all the attribute data has been added, the corresponding public version of the APP can be automatically generated, as shown in

Figs. 7b. The IoT platform will then generate its own matching ternary group, which is equivalent to a verification code. The device must have the correct code to access the platform. The triad is written into the program and burned into the hardware device. Finally, the hardware device is connected with WIFI and transmit the data to the IoT platform through data transmission, then forward it to APP of the mobile phone. The design of data acquisition platform is completed.

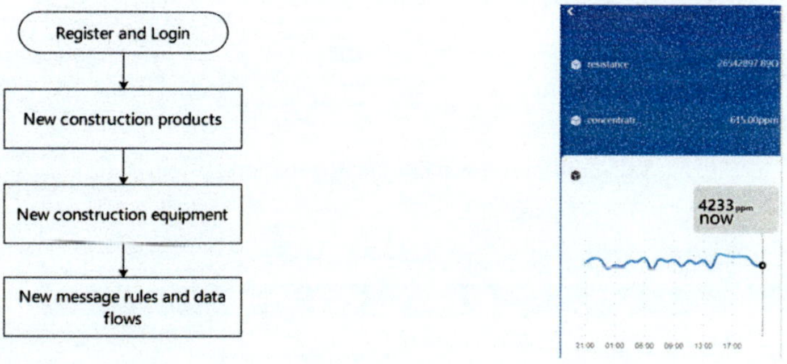

(a)Product Creation Flow Chart (b)Cloud Platform App Display Interface

Fig. 7. Cloud platform app design

4 Experiments

The resistance value of the semiconductor gas sensor is measured by MCU equipment. And the MCU connects the data collection system with the device. Devices are powered up and connected to WIFI. The MCU program automatically subscribes to cloud platform themes and uploads detected data. The WeChat Mini Program is opened. The device uploads the collected ambient gas data to the widget and webpage respectively via WIFI, as shown in Figs. 8a and 8b. The database data is shown in Fig. 9. It can meet remote monitoring data requirements.

After the data program in the MCU is changed into the triples of the Ari Cloud Networking Platform, the device restarts and connects with WIFI. The AL Cloud platform system's own APP is opened as shown in Fig. 10, Meeting remote monitoring data requirements.

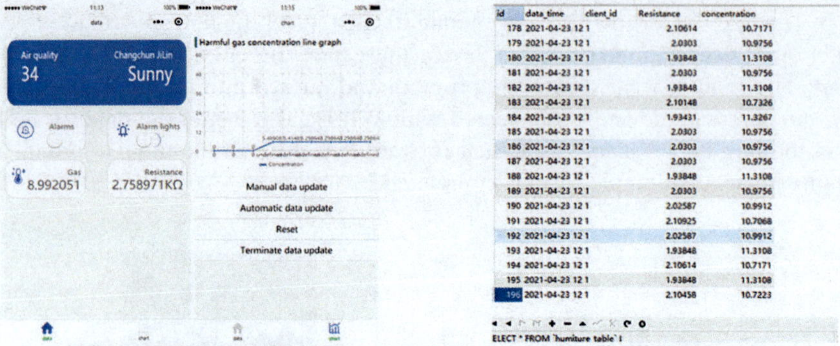

id	data_time	client_id	Resistance	concentration
178	2021-04-23 12	1	2.10614	10.7171
179	2021-04-23 12	1	2.0303	10.9756
180	2021-04-23 12	1	1.93848	11.3108
181	2021-04-23 12	1	2.0303	10.9756
182	2021-04-23 12	1	1.93848	11.3108
183	2021-04-23 12	1	2.10148	10.7326
184	2021-04-23 12	1	1.93431	11.3267
185	2021-04-23 12	1	2.0303	10.9756
186	2021-04-23 12	1	2.0303	10.9756
187	2021-04-23 12	1	2.0303	10.9756
188	2021-04-23 12	1	1.93848	11.3108
189	2021-04-23 12	1	2.0303	10.9756
190	2021-04-23 12	1	2.02587	10.9912
191	2021-04-23 12	1	2.10925	10.7068
192	2021-04-23 12	1	2.02587	10.9912
193	2021-04-23 12	1	1.93848	11.3108
194	2021-04-23 12	1	2.10614	10.7171
195	2021-04-23 12	1	1.93848	11.3108
196	2021-04-23 12	1	2.10458	10.7223

(a) WeChat Mini program collects data (b) Database interface

Fig. 8. Self-built cloud server data acquisition system

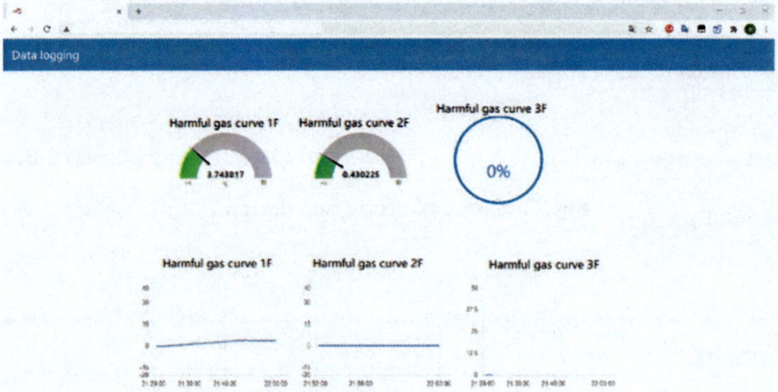

Fig. 9. Web-side monitoring interface

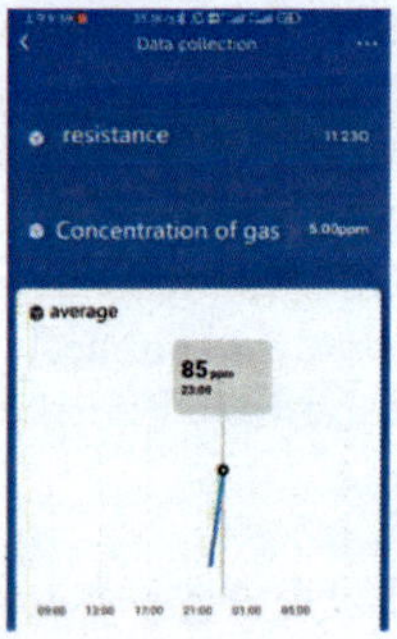

Fig. 10. AL cloud app interface

5 Conclusion

In this article, two data collection platforms are created using the IoT technology. One is built using the AL Cloud IoT platform, and the other uses the self-built cloud server architecture. The two data platforms have their advantages and disadvantages. The data collection system based on the AL cloud platform is simple, time saving and labor-saving, but lacks customization. The data collection system built by the self-built cloud platform has high flexibility, but it is troublesome to set up, so we need to design our own software to cooperate with it. For this purpose, we design a WeChat Mini Program and the computer webpage end. The MySQL database is also designed for better data storage and processing. Experiments have shown that both data collection systems are functional and capable of remotely monitoring laboratory data.

Acknowledgments. This work was supported by Science Foundation of Jilin Province Education Department (JJKH20190853KJ); Project of Jilin Provincial Science and Technology Department (No: 20190303064SF, 20200403072SF); Project of Jilin Province Development and Reform Commission (2019C048-4, 2020C021-5).

References

1. Sima, K., Kalas, D., Pretl, S., et al.: Resistance measurement method for IoT and low power application. In: 2020 43rd International Spring Seminar on Electronics Technology (2020)
2. Yang, J.: The Design and Implementation of Campus Energy Consumption Monitoring System Based on Internet of Things Technology. Shandong University (2020)
3. You, H., Bin, W., Ming-Ming, L.: Research on the analysis and design of automatic collection system for laboratory instrumental data. China Manage. Informatization. **23**(18), 186–189 (2020)
4. Ji-Xiang, X.: Software Design of Data Acquisition System Based on Internet of Things. University of Electronic Science and Technology of China (2015)
5. Gao-Yuan, Z., Shi-Gang, Y.: Design of high precision ultrasonic gas flow monitoring system based on internet of things. Instrum. Tech. Sens. **02**, 65–70 (2021)
6. Meng, G.: Research on Key Technology of Remote Monitorin System Development Based on Gizwits. XI'AN University of Technology, Xi an (2019)
7. Jian-Qing, Z.: Design of Multifunctional Air Quality Monitoring System Based on STM32. University of Electronic Science and Technology of China (2019)
8. Hong-Yang, W., Wei, X., Yun-Hui, S., Wang, X., Yang, Y.: The application of OPC UA and Node-red technology in IoT2040IoT-gateway. Manuf. Autom. **40**(07), 58–60 (2018)
9. Ohyver, M., Moniaga, J.V., Sungkawa, I., et al.: The comparison firebase realtime database and MySQL database performance using wilcoxon signed-rank test. Procedia Comput. Sci. **157**, 396–405 (2019)

A Mill Control System Based on GA-BP Network for Output Prediction

Hongwei Ren(✉), Sheng Zheng, and Xinyu Li

Taiyuan University of Technology, Taiyuan 030024, Shanxi, China

Abstract. Due to the fact that the mill is often not working with the best conditions, it leads to increase the energy consumption of the mill system, reduce the quality of the ink produced, and greatly reduce the production efficiency. The height of the material level is a kernel factor that affects the production efficiency and quality of the mill. However, it is difficult to measure the material level accurately. In the paper, the soft measurement method is used to construct a BP neural network prediction mode, and then the material level of the mill is obtained. In addition, the BP network is easy to fall into a local optimum. In order to address this task, the genetic algorithm is used to optimize the threshold and weight of the network. MATLAB simulation is adopted to demonstrate the feasibility and effectiveness of the proposed method. The experiment results show that the production efficiency and quality of the ink can be improved by GA-BP.

Keywords: Mill level · Soft measurement · BP neural network · Genetic algorithm · MATLAB

1 Introduction

As an important material for printing, ink is used to print text and pictures on paper by printing to form currency, magazines, newspapers, books, billboards, etc. These finished products are common necessities in our lives [1]. Economic development, cultural exchanges and dissemination are closely linked to our lives. The ink industry is an important industry that is indispensable in the process of economic development. Whether it is a developing or developed country, there is continuous demand for ink products, which makes the ink industry have a good development prospect. Our country's printing industry has been continuously improving by learning from foreign ink production technology and continuous innovation. The quantity, type, quality and production technology of ink have been greatly improved [2]. At present, our country can produce many types of inks to meet different needs. Our country should seize the opportunity of development, increase the upgrading and transformation of ink enterprises, enhance competitiveness, and promote the development of the ink industry. Under the premise of the development of the green economy, environmentally friendly inks have developed rapidly. Its application has reduced the environmental pollution of the printing industry and has achieved sustainable development [3]. However, there will be waste of raw materials and low utilization in the ink production process, so we should use advanced and high-performance

S.-C. Chu et al. (Eds.): ICGEC 2021, LNEE 833, pp. 82–91, 2022.
https://doi.org/10.1007/978-981-16-8430-2_8

sand mill equipment and optimized mill control in the ink production process to improve the production efficiency of ink and the production of ink quality [4].

The accuracy of mill level measurement is extremely important for automatic control, safety production, energy saving and consumption reduction. Traditional mill level measurement methods include differential pressure method, noise method, power method, etc. The characteristics of mill system include linearity, strong coupling, high delay. However, single parameter detection can not effectively obtain the expected results of the level of the mill. This paper uses soft measurement technology to collect data on the mill system, which includes the inlet and outlet temperature, motor current, and feed pressure of the mill system. These are used to build a BP neural network model to predict the material level under the best working conditions [5–7]. In order to improve the BP neural network function with slower convergence speed, weak generalization ability, and prone to local optimal problems, genetic algorithm is used to optimize its threshold and weight to make it have better predictive ability. Through a comprehensive analysis of the influencing factors of the mill level measurement, the main influencing factors are established. The BP neural network model optimized by genetic algorithm is constructed to predict the mill level, which is great significance for reducing the loss of raw materials in ink production [8–10].

At present, the method has not been applied to the ink production industry. The research in this paper hopes that the intelligent control methods is applied to ink production to improve ink production efficiency and production quality.

2 Introduction of Mill Control System

The mill system includes lubricating oil pumps, cooling water pumps, feed pumps, discharge pumps, and the mill itself. Factors that affect mill production efficiency and production quality include grinding bead density, cooling water temperature, and mill material level. In this paper, the mill material level and piston position are controlled to achieve the optimal working condition of the mill [11]. Figure 1 shows the principle diagram of mill level control.

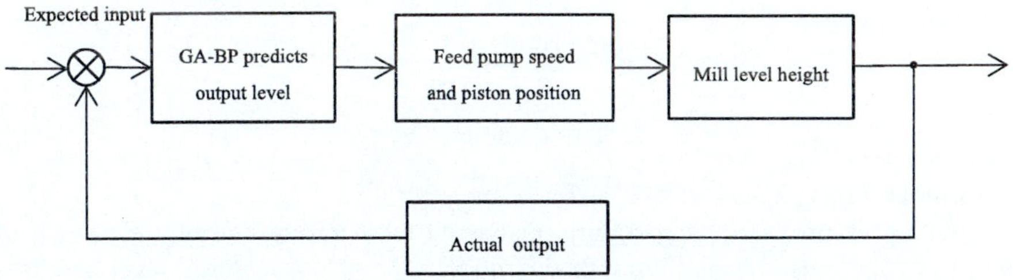

Fig. 1. Schematic diagram of mill material level control

Control idea: The relevant data of the mill is measured through soft-sensing technology, it includes the feed pressure, feed temperature, discharge temperature and motor current. They are used as the prediction network input. The GA-BP network is used to obtain the actual output level of the mill. Compared with expected level of the mill, the error is used as the input signal of the reverse propagation of the network to continuously adjust the weight and threshold of the network until the conditions are met, and the optimal material level is predicted and output. The material level is controlled by controlling the feed pump speed and piston position.

3 Model Construction of BP Neural Network Optimized Based on Genetic Algorithm

3.1 Introduction to BP Neural Network

The mill system has the characteristics of nonlinear and large lag. The BP neural network model can fit the nonlinear function well, and can effectively predict the mill level. The mill level prediction model is a 3-layer network [12].

By optimizing the weights and thresholds of the BP network model, the predicted material level value of the network is closer to the expected material level value. The following is the adjustment process of weights and thresholds [13].

The weight is updated.

Adjust and update the network connection weight ω_{ij} and ω_{jk} according to the network prediction error *e*, as shown in Eqs. 1 and 2.

$$\omega_{ij} = \omega_{ij} + \eta H_j \big(1 - H_j\big) x(i) \sum\nolimits_{k=1}^{m} \omega_{jk} e_k \tag{1}$$

$$\omega_{jk} = \omega_{jk} + \eta H_j e_k \tag{2}$$

In the formula: j = 1, 2, ⋯ , l; j = 1, 2, ⋯ , l; k = 1, 2, ⋯ , m.

The threshold is updated.

Update the network node threshold *a*,*b* according to the network prediction error *e*, as shown in Eqs. 3 and 4.

$$a_j = a_j + \eta H_j \big(1 - H_j\big) \sum\nolimits_{k=1}^{m} \omega_{jk} e_k \tag{3}$$

$$b_k = b_k + \eta e_k \tag{4}$$

In the formula: $j = 1, 2, \cdots, l; k = 1, 2, \cdots, m$

The feed pressure, feed temperature, discharge temperature and motor current of the mill system are used as the input of the model. The actual material level height is obtained through the hidden layer and the output layer. The calculated error is used to propagate the signal backwards. The weights and thresholds are continuously modified according to formula 1 to formula 4.

Calculate hidden layer nodes with empirical formula, as shown in Eq. 5.

$$l = \sqrt{m + n} + p \tag{5}$$

In the formula: m is the number of neural network input layers; l is the number of nodes in the hidden layer of neural network; n is the number of neural network output layers, p is a random number from 0–10. It is 8 in the text, so the number of hidden layer nodes l is 10. The BP neural network is a 4-10-1 network structure, as shown in Fig. 2.

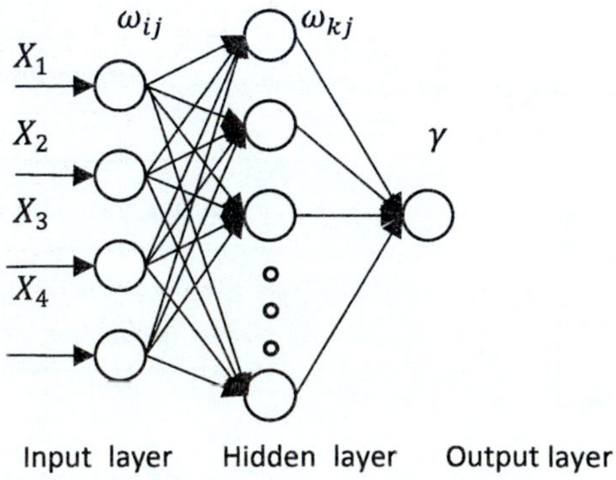

Fig. 2. BP neural network structure diagram

3.2 Optimized BP Network Based on Genetic Algorithm

The genetic algorithm can be described as:

$$GA = (P(0), N, l, s, g, P, f, T) \tag{6}$$

In the formula: $P(0)$ represents the initial population; N represents the population size; l represents the length of the code string; s represents the selection strategy; g represents the genetic operator; P represents the operation probability of the genetic operator; f represents the fitness function; T is the termination criterion.

Using BP neural network alone is very easy to fall into local extremes. It requires long training time, many iterations of the algorithm, and slow convergence speed. The genetic algorithm is used to improve the BP neural network to obtain the optimal threshold and weight value. The improved weights and thresholds are used to give the BP network to calculate the output error. Then the neural networks judges whether the error meets the conditions, if not, GA continuously optimizes the weights and thresholds. Then it recalculates the error until the conditions are met. The optimal weights and thresholds can be obtained. It effectively avoids the local optimal situation of the BP network, improves the prediction accuracy, and achieves the goal of optimization [14–16].

The genetic algorithm optimization BP neural network process is shown in Fig. 3:

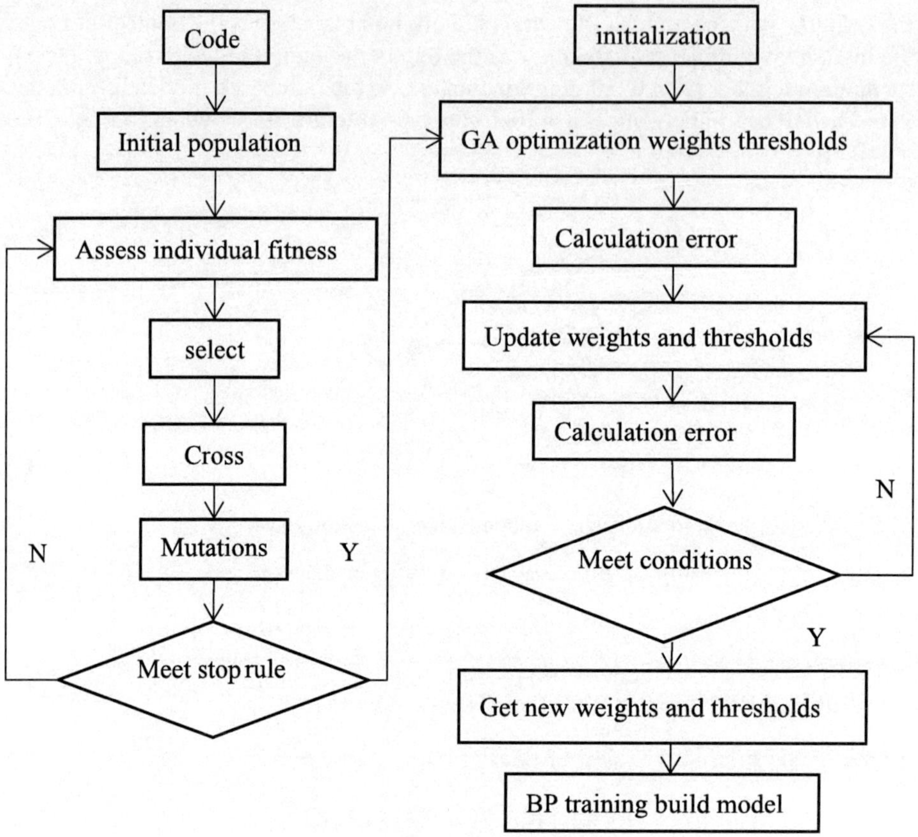

Fig. 3. Genetic algorithm to optimize BP neural network flow chart

The design steps of the genetic algorithm to optimize the BP neural network are as follows:

Normalization of data

The experimental data is collected through the relevant sensors of the mill system. The data obtained by the experiment due to accidental factors will affect the results of the mathematical model predictive analysis. The original data needs to be normalized in order to eliminate the influence of abnormal data. The processed data can be used to build and test the model. The normalization function selects the mapminmax function in MATLAB, as shown in Eq. 7.

$$q = \frac{x_n - x_{min}}{x_{max} - x_{min}} \tag{7}$$

In the formula: x_{max} is the maximum value of experimental data; x_{min} is the minimum value of experimental data. After the original data x_n is linearly transformed, the original data with the judgment value q in between $[0 - 1]$ are retained, and the ones out of the range are eliminated.

Chromosome coding

The binary code must determine the length of the binary string, which is related to the domain of the variable and the calculation accuracy of the problem. The advantage of binary coding is that it is easy to implement. But there is a Hamming cliff. Gray coding requires that only one code point is different between the codes of two consecutive integers. It effectively solves the Hamming cliff problem in binary coding.

The coding method in this paper adopts real number coding. The chromosome of each individual in the population is represented by a real number in the range [−3 3]. The code length is the number of variables required to solve the problem. The input layer to the hidden layer weight ω_{ij}, the hidden layer to the output layer weight ω_{jk}, the hidden layer threshold a, and the output layer threshold b are encoded. The encoding length L is:

$$L = ml + l + ln + n \tag{8}$$

In the formula: m is the number of neurons in the input layer; l is the number of neurons in the hidden layer; n is the number of neurons in the output layer.

The encoding result is −3 plus 6 multiplied by a random number. Then the encoding result is stored in a matrix with 1 row and s columns.

Initialization

After the chromosomes are encoded, individuals are formed, and then multiple individuals are randomly generated to form a population $P(0)$. The evolutionary algebra, population size N, and genetic operators g are initialized.

Evaluation of fitness function

The original fitness function can directly reflect the initial solution goal of the problem to be solved, but the fitness value may be negative. The standard fitness function is obtained by transforming the original fitness function.

The fitness function is used to determine the size of individual fitness. Each individual has its own characteristics. The difference in individual characteristics is used to determine its adaptability. The size of fitness can determine the difficulty of inheritance. The greater the fitness, the easier it is to inherit. The smaller the fitness, the more difficult it is to inherit. The fitness F_i is the sum of the absolute value of the error between the network training output value and the expected output value. The calculation formula is:

$$F_i = \sum_{i=1}^{n} |y_i - o_i| \tag{9}$$

In the formula: n is the number of training samples; y_i is the actual output value of the material level; o_i is the expected output value of the material level.

Genetic manipulation

Genetic operations include selection, crossover and mutation. The selection operation is based on the fitness of individuals. The individuals with high fitness are selected and inherited to the next generation. The operation selection in the article uses the roulette method, through which 30 individuals are selected to be inherited to the next generation.

The relative fitness of each individual is:

$$P(\chi_i) = \frac{f(\chi_i)}{\sum_{j=1}^{n} f(\chi_j)} \quad (10)$$

In the formula: $P(x_i)$ is the relative fitness of individual; $f(x_i)$ is the initial fitness of individual x_i; $\Sigma f(x_j)$is the cumulative fitness of the population.

The chromosomes selected by the roulette method may be duplicated, that is the individuals in the population may be the same.

Both crossover and mutation act on the population. The core of the algorithm is crossover. New individuals are inherited to the next generation through crossover. Whether to crossover depends on the crossover probability. The real number crossover method is used in the article to randomly select the position for crossover. It is the same position where two chromosomes cross. If the random number is greater than the crossover probability, then it returns to the loop. If the random number is less than the crossover probability, then it starts the crossover operation.

Mutation is to change the value of a certain gene of an individual to produce individuals with different characteristics. The chromosome and the mutation position are random. The mutation operation is determined by the mutation probability. New populations are generated after selection, crossover and mutation operations.

Judgment of termination conditions

If the termination condition is met, the maximum fitness is outputted. Otherwise, the population individual fitness and genetic operations are re-evaluated until the termination condition is met [17–19].

4 Simulation Results and Analysis

The construction of the network model is to use the editor in MATLAB to code. Then it realizes the simulation of the optimized BP neural network based on genetic algorithm. The raw data of the simulation was collected in the black ink line production equipment of an ink factory. Part of the data is selected from the original data for simulation. 500 sets of data are used as the training set, 20 sets of data are used as the test set. The prediction model of the neural network is a topological structure of 4-10-1.

The traditional BP neural network is easy to fall into the local optimum. In order to avoid this defect, the genetic algorithm is used to optimize the weights and thresholds of the neural network. In the program, the population size is 30, the evolution generation is 150, the crossover probability is 0.4, and the mutation probability is 0.1. The material level prediction of the mill is realized by programming in MATLAB R2016b. The output prediction of the GA-BP network is shown in Fig. 4.

The experimental result in Fig. 4 is the comparison between the predicted value of the material level and the original data. From the comparison of the experimental data and predicted data of the mill material level in Table 1, it can be seen that the average value of the original data level is 36.16, and the average value of the predicted value is 35.98. The average relative error of the GA-BP network to predict the mill level is 0.18, which is within the allowable error range. It can be seen that the prediction model is

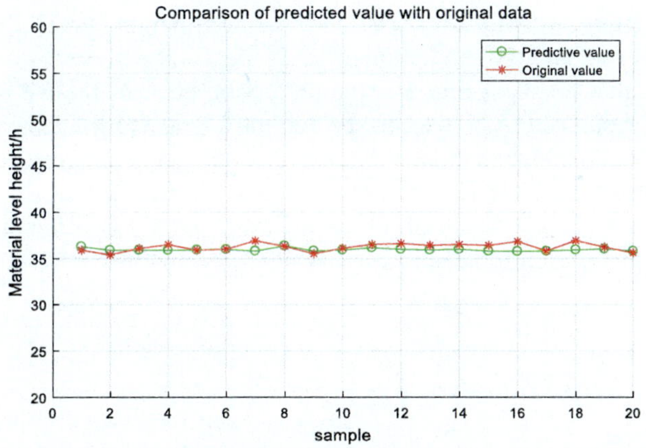

Fig. 4. Forecast effect

Table 1. Comparison of original data and predicted data of mill material level

Serial number	Experimental data	Forecast data
1	35.8	35.9
2	35.9	35.9
3	36.8	36.0
4	35.8	36.0
5	36.8	35.9
6	36.4	36.1
7	35.5	35.9
8	36.3	36.0
9	36.4	36.1
10	35.9	36.0

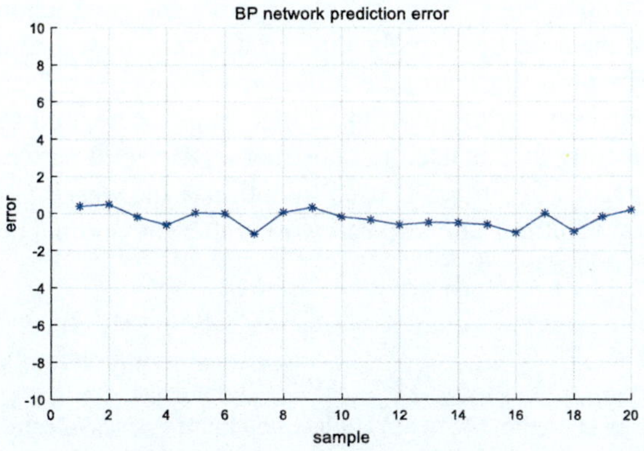

Fig. 5. Network prediction error

feasible for predicting the mill level. The material level height curve of this prediction model tends to be flat, and the prediction effect is ideal.

The experimental result of Fig. 5 is the prediction error of the BP network. From the experimental result of Fig. 5, we can see that the average absolute error is 0.37, and the maximum error is 1.04, which is within the allowable error range. The mill material level changes little. The material level should be controlled within this range. It can improve the efficiency and quality of ink production.And it meets the requirements of actual industrial production forecasts.

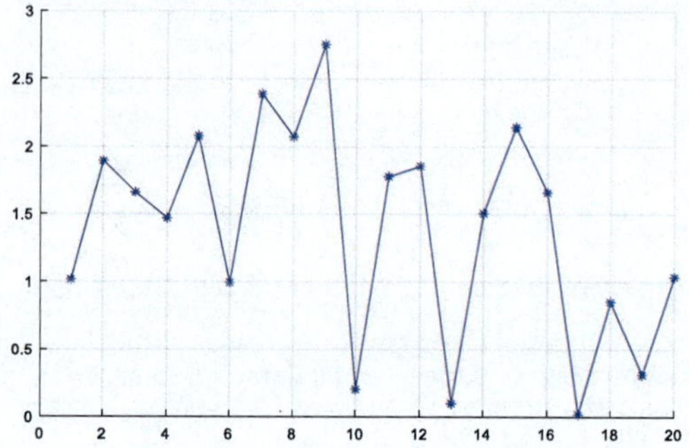

Fig. 6. Percentage of network error

The experimental result in Fig. 6 is the error percentage of the neural network. From the experimental results in Fig. 6, we can see that the average absolute error percentage is 1.39%, and the maximum error percentage is 2.7%, which is within the allowable range of error. Therefore, the optimized BP neural network based on genetic algorithm is effective for mills. The system material level forecast is accurate.

5 Conclusion

This paper controls the material level of the mill system to keep the material level under the optimal working conditions. Then it can improve the production efficiency and production quality of the mill. Because the direct measurement of the mill material level is inaccurate and difficult to measure, the mill material level is reflected here through the inlet and outlet pressure of the mill, the outlet temperature, and the current of the mill motor. The collecting data is used to construct a BP neural network to predict the material level. The genetic algorithm is used to optimize the weights and thresholds of BP to find the optimal solution. The mill can work under the optimal conditions, which can improve corporate efficiency.

References

1. Dazhi, W., Qiang, W., Qingbo, M., et al.: High temperature-assisted electrohydrodynamic jet printing of sintered type nano silver ink on a heated substrate. J. Micromechanics Microeng **29**(4), 045012 (2019)

2. Johnson Chelsea, E., Martin, P., Roberts Katherine, A., et al.: The Capability of Raman Microspectroscopy to Differentiate Printing Inks. J. Forensic Sci. **63**(1), 66–79 (2018)
3. Gu, W., Li, Y., Zhang, X.: Printing industry and the environment. Adv. Mater. Res. **663**, 759–762 (2013)
4. Gao, P., Zhou, W., Han, Y., et al.: Enhancing the capacity of large-scale ball mill through process and equipment optimization: an industrial test verification. Adv. Powder Technol. **31**(5), 2079–2091(2020)
5. Liu, Z., Zhen, C.: Modeling of complex equipment coal mill in power plant. Int. Core J. Eng. **5**(7), 10–16 (2019)
6. Xu, J., Jun, T., Zhao, T., et al.: Research on intelligent prediction and forecast model for construction period of transmission and transformation engineering based on bp neural network". In: IOP Conference Series: Earth and Environmental Science, vol. 687(1), p. 012154 (2021)
7. Zhu, W., Wang, H., Zhang, X.: Synergy evaluation model of container multimodal transport based on BP neural network. Neural Comput. Appl. **33**(9), 4087–4095 (2021)
8. Dou, K., Sun, X.: Long-term weather prediction based on GA-BP neural network. In: IOP Conference Series: Earth and Environmental Science, vol. 668(1), p. 012015 (2021)
9. Huang, D.J., Tian, C.C., Jiang, J.Y., et al.: Application of GA-BP neural network model for small watershed flood forecasting in Chun'an county, China. In: IOP Conference Series: Earth and Environmental Science, vol. 612(1), p. 012066 (2020)
10. Liang, H., Wei, Q., Lu, D., et al.: Application of GA-BP neural network algorithm in killing well control system. Neural Comput. Appl. **33**(3), 1–12 (2020)
11. Bortnowski, P., Gładysiewicz, L., Ozdoba, M., et al.: Energy efficiency analysis of copper ore ball mill drive systems. Energies **14**(6), 1786–1786 (2021).
12. Huang, L., Xie, G., Zhao, W., et al.: Regional logistics demand forecasting: a BP neural network approach. Complex Intell. Syst. (C31), 1–16 (2021)
13. Qiao, X., Guo, F., Zhang, R., et al.: Short-term tidal current prediction based on GA-BP neural network. In: IOP Conference Series. Earth and Environmental Science, vol. 513(1), p. 012061 (2020)
14. Han, Q.Y., Qian, L.J., Chu, X.Y.: Study on energy consumption prediction of liquor-making based on GA-BP neural net. Appl. Mech. Mater. **3485**, 1681–1687 (2014)
15. Zhang, S., Hu, Q.: Students' comprehensive quality evaluation based on BP neural network optimized by genetic algorithm. Xi'an institute of posts and telecommunications. In: Proceedings of the 2nd International Conference on Education, E-learning and Management Technology, pp. 6–7. Xi'an Institute of Posts and Telecommunications (2017)
16. Tan, T., Yang, Z., Chang, F., et al.: Prediction of the first weighting from the working face roof in a coal mine based on a GA-BP neural network. Appl. Sci. **9**(19), 4159 (2019)
17. Zheng, L., Xin, L., Kan, W., et al.: GA-BP neural network-based strain prediction in full-scale static testing of wind Turbine blades. Energies **12**(6), 1026 (2019)
18. J, Tang., W X, Li., B, Zhao.: The application of GA-BP algorithm in prediction of tool wear state. In: IOP Conference Series: Materials Science and Engineering, vol. 398(1) (2018)
19. Guo, B., Xu, J., Ling, C., et al.: Prediction of the heat load in central heating systems using GA-BP algorithm. Int. J. Adv. Network Monit. Controls **2**(4), 137–141 (2018)

State Evaluation and Attack Signal Reconstruction of Power Cyber-Physics System Based on Intermediate Variable Observer

Lin Lin[1], Zhipeng Cong[2], Qigui Nie[3], Xin Liu[3], Weiting Wang[3], Jian Li[4], and Defu Yang[4](✉)

[1] State Grid Shandong Electric Power Company, Jinan 250000, China
[2] State Grid Yantai Power Supply Company, Yantai 264000, China
[3] State Grid Shandong Electric Power Company Electric Power Research Institute, Jinan 250003, China
[4] School of Automation Engineering, Northeast Electric Power University, Jilin 132012, Jilin, China

Abstract. This paper considers the state assessment and attack signal reconstruction of the power cyber-physical systems (CPSs). First of all, we have established a power CPS model that includes false data injection (FDI) attacks. Secondly, an intermediate variable observer has been designed, the state of the power CPS system can be estimated by the intermediate variable observer. Furthermore, based on the intermediate variable observer, the reconstructed attack signal has been reconstructed. Moreover, a sufficient conditions to make the error system stable of the intermediate variable observer has been given. Finally, for convenience, a standard example of IEEE 9-node is used to verify the effectiveness of the intermediate variable observer.

Keywords: Cyber-physical systems · FDI attacks · Intermediate variable observer · Attack reconstruction

1 Introduction

Smart components and network technology have made great progress, the basic theory of combining physical production and network transmission has broad prospects for development. These advancements have promoted the development of cyber-physical systems (CPSs). Typical CPS systems include smart grids and drones. As an important infrastructure, the power CPS system is always facing threats from all aspects. Malicious attacks on the power system in recent years, such as the large-scale power outage in Ukraine in 2015 and the ransomware attack on American company Colonial Pipeline in 2021, indicate that the CPS system is inevitably subject to various attacks. Once an attacker successfully invades the CPS system, this not only affects industrial production, but may

S.-C. Chu et al. (Eds.): ICGEC 2021, LNEE 833, pp. 92–101, 2022.
https://doi.org/10.1007/978-981-16-8430-2_9

also cause the system to crash. Therefore, the stability research of CPS system is very important.

Smart grids work together by integrating advanced communication technology, control technology, and power infrastructure [1]. Because each part is highly coupled, ensuring the reliable operation of the power CPS system is critical and extremely urgent. Because all control commands and information interaction depend on the communication network, network communication are vital in the security of the CPS system. [2] proposes an integrity protection mechanism based on prekeying to protect critical smart grid communication channels that are usually not protected due to time constraints. A new methods of attack was been proposed in [3] to analyze the behavior of attackers in the energy internet, and proposed an anomaly detection method. [4] researched the possibilistic-probabilistic self-scheduling problem, and proposed a fuzzy Markov model to simulate the impact on smart grid performance and self-dispatch under different conditions.

Intruding into a communication network and injecting false measurements is called a FDI attack. For FDI attacks, the attacker can manipulate the information transmission of the communication network by injecting false data into the smart grid communication network. Specifically, the controller, actuator, and sensor are connected through a communication network, and the attacker can replace the original command with wrong information. By simulating multiple microgrids with a multi-agent system, [5] solves the problem of multiple microgrids facing FDI attacks. The problem of cooperative tracking has been discussed in [6] for multi-agent network systems, and solves the cascading problem of two network attacks. By coordinating network parameters, [7] proposes an FDI attack way, which provided an condition that can decrease the attacked measured values. [8] proposes a hybrid detection scheme, which take advantage of detection ideology which come from watermarking and moving, respectively. Compared with the FDI attack, the dynamic load altering attacks (DLAA) only needs to invade the load-side demand management program. [9] proposed a DLAA attack model and designed a defense strategy based on the pole configuration method. [10] proposed a dynamic distributed energy and load control algorithm under DLAA attack. [11] studied the problem of detecting DLAA attacks from the original data of smart meters when the dynamic characteristics of the power system are unknown. Based on the sliding mode observer, [12] proposes a scheme that can detect DLAA attacks and reconstruct DLAA attack signals.

At present, there have been many studies on attack signal estimation. For FDI attacks defense method, [13] studied the different detection methods for linear time-invariant systems, other sensor attacks, nonlinear systems, and noise systems, respectively. Under denial-of-service attacks, network attack reconstruction and detection methods in power CPS systems were proposed in [14]. [15] studied the reliable control problem, and used the observe to estimate the system state, the existing static output feedback controller has been replaced by reliable controller. [16] studied the event trigger control problem of the observer-based for CPS system under denial-of-service attacks. Based on distributed observer, [15] proposed a detection method which can detect whether there is an attack on each physical layer.

The stable operation of the CPS system is a hot issue, many scholars have studied excellent defense strategies. [17] studied the regional shock of the wide-area power system caused by FDI attacks, and a sliding mode controller has been proposed to defend against malicious attacks and regional shocks. Reliable control has been studied in [18] for CPS system based on dissipative properties under DoS attacks. For a class of attacks that obey Markov jumps, [19] proposed a robust sliding mode control strategy. For a Markov jump CPS system that requires transient performance, such as rapidity, the problem of finite-time sliding mode control have been studied in [20]. A sliding mode observer is designed in [21], the observer has good H_∞ performance for disturbances, but has high accuracy for the detection and identification of attack signals.

The contributions of this paper are three-folds:

First of all, we have established a power CPS model that includes false data injection attacks.

Secondly, when the system is observable, an intermediate variable observer has been designed, which can ensure that the error between the state estimation and the actual state converges to 0. Furthermore, based on the intermediate variable observer, we get the reconstructed attack signal. Moreover, we give sufficient conditions to stabilize the error system of the intermediate variable observer.

Finally, for convenience, a standard example of IEEE 9-node is used to verify the effectiveness of the intermediate variable observer.

Organizational structure of this paper are as follows. Section 2 gives the considered system model, and the designed attack estimation method is introduced in Sect. 3, then an example with simulation results is given in Sect. 4, Sect. 5 is conclusion.

Notation: For a real matrix S, S^T denotes its transpose, $\mathbb{R}^n$ is n-dimensional space. Meanwhile, positive definite and negative definite can be described $S > 0$ and $S < 0$, respectively. The symbol $*$ represents the symmetric item in the symmetric matrix.

2 System Description and Problem Statement

Suppose there is a power system, which have m generators and n load buses. Moreover, we can denote $R = \begin{bmatrix} R_{gg} & R_{gl} \\ R_{lg} & R_{ll} \end{bmatrix}$, which is the imaginary part of the

Y-Bus matrix of the power system. According to the smart grid model proposed by [9], the attacked power CPS system model can be expressed as

$$\begin{bmatrix} \dot{\delta} \\ \dot{\omega} \end{bmatrix} = \begin{bmatrix} 0 & I \\ \Sigma_1 & \Sigma_2 \end{bmatrix} \begin{bmatrix} \delta \\ \omega \end{bmatrix} + \begin{bmatrix} 0 \\ M_g^{-1} \end{bmatrix} u(t) + H\varpi(t) \tag{1}$$

where $\Sigma_1 = M_g^{-1}\left(R_{gl}R_{ll}^{-1}R_{lg} - R_{gg}\right)$, $\Sigma_2 = -M_g^{-1}D_g$, inertia coefficient is represented by $M_g \in R^{m\times m}$ and damping coefficient is represented by $D_g \in R^{m\times m}$. $\delta \in R^m$ is the generator rotor angles at generation buses, $\omega \in R^m$ is rotor angular frequency deviation at generation buses, $\varpi(t) \in R^r$ is FDI attack signal, H is attack distribution matrix.

Therefore, the attacked power CPS system dynamic mold can be represented by

$$\begin{aligned} \dot{x}(t) &= Ax(t) + Bu(t) + H\varpi(t) \\ y(t) &= Cx(t) \end{aligned} \tag{2}$$

where $x(t) = \begin{bmatrix} \delta \\ \omega \end{bmatrix}$ represents the states vector, the system matrices $A = \begin{bmatrix} 0 & I \\ \Sigma_1 & \Sigma_2 \end{bmatrix}$, $u(t) \in R^m$ is the control input, $B = \begin{bmatrix} 0 \\ M_g^{-1} \end{bmatrix}$ is input matrix, $y(t) \in R^p$ is the measured output vector; $\varpi(t) \in R^r$ is the attack vector; N is an indicator matrix, and C is output matrix.

Next, we give the following assumptions

Assumption 1. *The FDI attacks vector $\varpi(t)$ satisfies $\dot{\varpi}(t) \leqslant \varrho$ with $\varrho \geqslant 0$.*

Assumption 2. *H is a full-rank matrix with appropriate dimensions, $rank(H) = r$.*

Assumption 3. *For any λ without negative real part satisfies*

$$rank \begin{bmatrix} A - \lambda I & H \\ C & 0 \end{bmatrix} = n + rank(H)$$

3 Design of the Intermediate Variable Observer

In practical applications, most system states cannot be directly measured. Therefore, observers are used to assess system states in real time. In this section, an intermediate variable state observer will be designed. In addition, sufficient conditions to ensure the asymptotic stability of the error system are proved.

3.1 Introduction of Intermediate Variable Observer

The form of the observer is

$$\nu(t) = \varpi(t) - \zeta H^T x(t) \tag{3}$$

where $\nu(t)$ denotes intermediate variable, ζ is a variable to be designed, from (2) and (3), we have

$$\dot{\nu}(t) = \dot{\varpi}(t) - \varrho H^T(Ax(t) + Bu(t) + H\nu(t) + \zeta HH^T x(t)). \tag{4}$$

Accroading to (2) and (4), we have

$$\dot{\hat{x}}(t) = A\hat{x}(t) + Bu(t) + H\hat{\varpi}(t) + L(y(t) - \hat{y}(t)) \tag{5}$$

$$\dot{\hat{\nu}}(t) = -\zeta H^T N\hat{\nu}(t) - \zeta H^T(A\hat{x}(t) + Bu(t)) + \zeta HH^T\hat{x}(t)) \tag{6}$$

$$\hat{y}(t) = C\hat{x}(t) \tag{7}$$

$$\hat{\varpi}(t) = \hat{\nu}(t) + \zeta H^T\hat{x}(t) \tag{8}$$

where $\hat{x}(t)$ represents the estimates of $x(t)$, $\hat{\nu}(t)$ is the estimates of $\nu(t)$, $\hat{y}(t)$ is the output estimates of $y(t)$, and $\varpi(t)$ is the estimation of attacks.

If we denote $\bar{x}(t) = x(t) - \hat{x}(t)$, $\bar{\nu}(t) = \nu(t) - \hat{\nu}(t)$, and $\bar{\varpi}(t) = \varpi(t) - \hat{\varpi}(t)$.

The error between the observer state evaluation and the actual state can be expressed as

$$\dot{\bar{x}}(t) = (A - LC + \zeta HH^T)\bar{x}(t) + He_\xi(t) \tag{9}$$

$$\dot{\bar{\nu}}(t) = \dot{\varpi}(t) - \zeta H^T H\bar{\nu}(t) - \zeta H^T\left(A + \zeta HH^T\right)\bar{x}(t) \tag{10}$$

3.2 Stability Analysis

Theorem 1. *Suppose Assumption 1–3 are hold, if there exists matrices $P_1 > 0$ and $P_2 > 0$, a positive constant $\varepsilon > 0$, and a matrix F that satisfies*

$$\begin{bmatrix} He(P_1A - FC + \zeta P_1HH^T) & P_1H - \zeta(A + \zeta HH^T)^T HP_2 & 0 \\ * & -\zeta He(P_2H^T H) & P_2 \\ * & * & -\varepsilon \end{bmatrix} < 0 \tag{11}$$

then, the error system is uniformly asymptotically stable, and gain matrix L in (5) can be obtained from $L = P^{-1}F$.

Proof. The Lyapunov function is taken as

$$V(t) = \bar{x}^T(t)P_1\bar{x}(t) + \bar{\nu}^T(t)P_2\bar{\nu}(t) \tag{12}$$

The derivative of $V(t)$ are

$$\begin{aligned} \dot{V}(t) &= 2\bar{x}^T(t)P_1((A - LC + \zeta HH^T)\bar{x}(t) + N\bar{\nu}(t)) \\ &+2\bar{\nu}^T(t)P_2(\dot{\varpi}(t) - \zeta H^T N\bar{\nu}(t) - \zeta H^T\left(A + \zeta HH^T\right)e_x(t)) \end{aligned} \tag{13}$$

Furthermore, the following inequalities always hold:

$$2\bar{\nu}^T(t)P_2\dot{\varpi}(t) \leqslant \frac{1}{\varepsilon}\bar{\nu}^T(t)P_2P_2\bar{\nu}(t) + \varepsilon\varrho^2 \tag{14}$$

We can know from (13) and (14) that

$$\begin{aligned}\dot{V}(t) &\leqslant 2\bar{x}^T(t)P_1(A-LC+\zeta HH^T)\bar{x}(t)+2\bar{x}^T(t)P_1H\bar{\nu}(t)\\&-2\zeta\bar{\nu}^T(t)P_2H^T(A+\zeta HH^T)\bar{x}(t)-2\zeta\bar{\nu}^T(t)P_2H^TH\bar{\nu}(t)\\&+\frac{1}{\varepsilon}\bar{\nu}^T(t)P_2P_2\bar{\nu}(t)+\varepsilon\varrho^2\end{aligned} \tag{15}$$

Defining $\tilde{x}=[\bar{x}^T(t)\quad \xi^T(t)]^T$, we can get

$$\dot{V}(t)\leqslant\tilde{x}^T\Xi\tilde{x}+\varepsilon\varrho^2 \tag{16}$$

where

$$\Xi=\begin{bmatrix}He(P_1A-FC+\zeta P_1HH^T) & P_1H-\zeta(A+\zeta HH^T)^THP_2\\ * & -\zeta He(P_2H^TH)+\frac{1}{\varepsilon}P_2P_2\end{bmatrix} \tag{17}$$

Furthermore, from (12)

$$\begin{aligned}V(t)&\leq\mu_{\max}(P_1)\left\|\bar{x}(t)\right\|^2+\mu_{\max}(P_2)\left\|\bar{\nu}(t)\right\|^2\\&\leq\Gamma\left(\left\|\bar{x}(t)\right\|^2+\left\|\bar{\nu}(t)\right\|^2\right)\end{aligned} \tag{18}$$

where μ_{max} denotes the maximum eigenvalue, $\Gamma=\max\left[\mu_{\max}(P_1)\right],\mu_{\max}(P_2)]$.

Let $\bar{\Xi}=-\Xi$, we can know that if $\Xi<0$, then $\bar{\Xi}>0$, we have

$$\dot{V}\leq-\alpha V+\beta \tag{19}$$

where

$$\alpha=\frac{\mu_{\min}\left(\bar{\Xi}\right)}{\Gamma},\quad \beta=\varepsilon\varrho^2 \tag{20}$$

Denote the set of σ as follows

$$\sigma=\{\tilde{x}|\psi\leqslant\frac{\beta}{\alpha}\} \tag{21}$$

where μ_{min} denotes the minimum eigenvalue, $\psi = \mu_{\min}(P_1)\left\|\bar{x}(t)\right\|^2 + \mu_{\min}(P_2)\left\|\bar{\nu}(t)\right\|^2$.

We can denote $\tilde{\sigma}$ as the supplementary set of σ, we have

$$V(t)\geq\psi>\frac{\beta}{\alpha} \tag{22}$$

then, when $\bar{x}\in\tilde{\sigma}$, we can know from (21) and (22)

$$\dot{V}(t)\leq 0 \tag{23}$$

4 Simulation Results

In this section, a standard example of IEEE 9-node is used to verify the effectiveness of the intermediate variable observer. The parameters and initial conditions are as follows.

4.1 System Setup

The initial parameters are designed as follows $\hat{x}(0) = \begin{bmatrix} 0\,0\,0\,0\,0\,0 \end{bmatrix}$, $\hat{\varpi}_1(0) = \hat{\varpi}_2(0) = \hat{\varpi}_3(0) = 0$, $x(0) = \begin{bmatrix} -0.045\ -0.058\ -0.07\ 0\ 0.1\ 0.001 \end{bmatrix}$, $\zeta = 55$, $\varepsilon = 1$. The controllable input vector $u(t)$ is designed as $u(t) = K\hat{x}$. The inertia matrix, damping coefficients matrix, and the admittance matrix can be getted form [10].

In the simulation, the attack singal $\varpi(t)$ is created as Table 1.

Table 1. Illustration of the attack mode

t	[0,10)	[10,20)	[20,30)	[30,45)	[45,50]
$\varpi(t)$	0	$\begin{bmatrix} 0.2t \\ 0 \\ 0 \end{bmatrix}$	$\begin{bmatrix} 2 \\ sint \\ 0 \end{bmatrix}$	$\begin{bmatrix} -sin(t) \\ 1 \\ cost \end{bmatrix}$	0

The attack distribution matrix $H = \begin{bmatrix} 0\,0\,0\,1\,0\,0 \\ 0\,0\,0\,0\,1\,0 \\ 0\,0\,0\,0\,0\,1 \end{bmatrix}^T$

A feasible L of linear matrix inequality (17) is obtained as follows

$$L = \begin{bmatrix} 0.4330 & -1.0022 & -0.5318 \\ -2.1414 & 1.7992 & -0.2772 \\ -1.7678 & -0.1243 & 1.4066 \\ 51.8978 & -0.7537 & -2.7952 \\ -4.4681 & 65.8382 & -28.6998 \\ 85.9922 & -36.6504 & 135.8331 \end{bmatrix}$$

4.2 Simulation Results

In this subsection, the simulation results will be shown as follows. In [10, 20), only the first actuator channel is attacked; in [20, 30), the first channel and the second channel are attacked at the same time; in [30, 45), all actuator channels are attacked.

Figure 1 shows the state response curve and state evaluation curve of the power CPS system. The actual state of the power system has been represend by been represend by blue solid line, and the state evaluation curve has been represend by the red dashed line. Obviously, the observer we designed has good accuracy for tracking the state of the system.

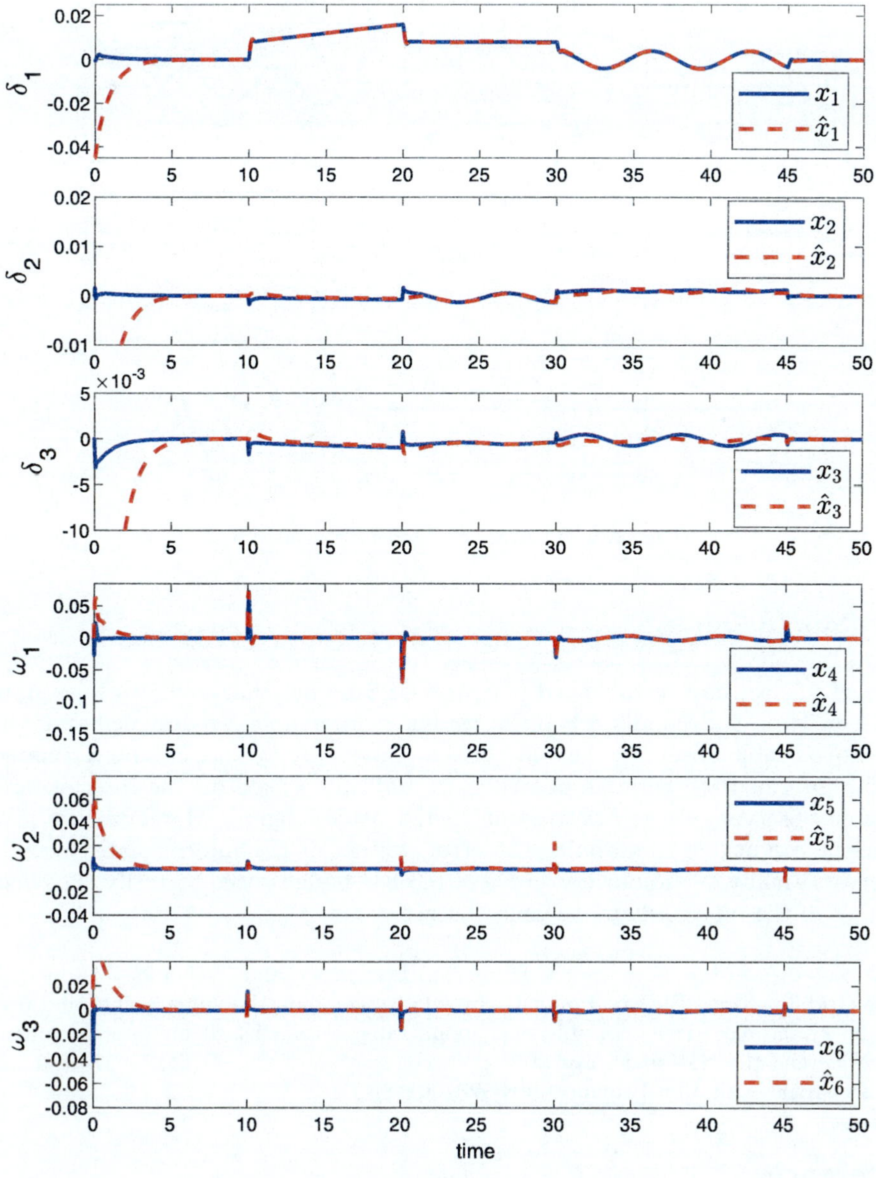

Fig. 1. The results of system state

Figure 2 shows the FDI attack signal and the reconstructed attack signal. The actual attack has been plotted by blue solid line, and the reconstructed attack signal has been represend by red dashed line. We can know from Fig. 2 that the intermediate variable observer can accurately reconstruct FDI attack signals.

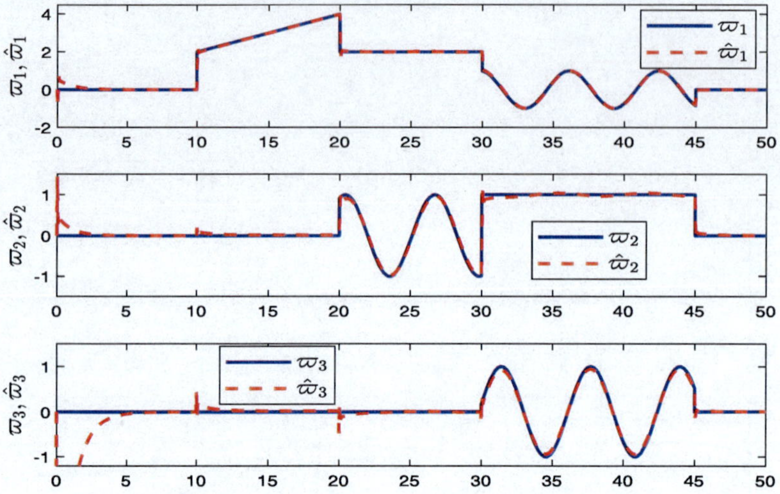

Fig. 2. The results of adaptive parameter

5 Conclusion

First of all, we have established a power CPS model that includes false data injection attacks. Secondly, when the system is observable, we designed an intermediate variable observer, the state of the power CPS system can be estimated by the intermediate variable observer. In addition, based on the intermediate variable observer, we get the reconstructed attack signal. Moreover, we give sufficient conditions to stabilize the error system of the intermediate variable observer. Finally, a standard example of IEEE 9-node is used to verify the effectiveness of the intermediate variable observer.

Acknowledgments. This work is supported by Science and Technology Project of State Grid Shandong Electric Power Company: Research on the vulnerability and prevention of the electrical cyber-physical monitoring system based on interdependent networks, and the National Natural Science Foundation of China (61873057), Education Department of Jilin Province(JJKH20200118KJ).

References

1. Nawaz, R., Akhtar, R., Shahid, M.A., Qureshi, I.M., Mahmood, M.H.: Machine learning based false data injection in smart grid. Int. J. Electr. Power Energy Syst. **130**, 106819 (2021)
2. Pal, A., Jolfaei, A., Kant, K.: A fast prekeying-based integrity protection for smart grid communications. IEEE Trans. Ind. Inform. **17**(8), 5751–5758 (2021)
3. Wang, H., Wen, X., Huang, S., Zhou, B., Wu, Q., Liu, N.: Generalized attack separation scheme in cyber physical smart grid based on robust interval state estimation. Int. J. Electr. Power Energy Syst. **129**, 106741 (2021)

4. Sanjari, I., Abdollahi, A., Rashidinejad, M., Afzali, P., Arasteh, H.: A new fuzzy model for investigating the effects of lightning on the risk-based self-scheduling strategy in a smart grid. Int. J. Electr. Power Energy Syst. **129**, 106771 (2021)
5. Yassaie, N., Hallajiyan, M., Sharifi, I., Talebi, H.: Resilient control of multi-microgrids against false data injection attack. ISA Trans. **110**, 238–246 (2021)
6. Tahoun, A., Arafa, M.: Cooperative control for cyber-physical multi-agent networked control systems with unknown false data-injection and replay cyber-attacks. ISA Trans. **110**, 1–14 (2021)
7. Liu, C., Liang, H., Chen, T.: Network parameter coordinated false data injection attacks against power system ac state estimation. IEEE Trans. Smart Grid **12**(2), 1626–1639 (2021)
8. Ghaderi, M., Gheitasi, K., Lucia, W.: A blended active detection strategy for false data injection attacks in cyber-physical systems. IEEE Trans. Control of Network Syst. **8**(1), 168–176 (2021)
9. Amini, S., Pasqualetti, F., Mohsenian-Rad, H.: Dynamic load altering attacks against power system stability: attack models and protection schemes. IEEE Trans. Smart Grid **9**(4), 2862–2872 (2018)
10. Amini, S., Pasqualetti, F., Mohsenian-Rad, H.: Detecting dynamic load altering attacks: a data-driven time-frequency analysis. In: 2015 IEEE International Conference on Smart Grid Communications (SmartGridComm), pp. 503–508 (2015)
11. Baron-Prada, E., Osorio, E., Mojica-Nava, E.: Resilient transactive control in microgrids under dynamic load altering attacks. In: 2017 IEEE 3rd Colombian Conference on Automatic Control (CCAC), pp. 1–5 (2017)
12. Su, Q., Li, S., Gao, Y., Huang, X., Li, J.: Observer-based detection and reconstruction of dynamic load altering attack in smart grid. J. Franklin Inst. **358**(7), 4013–4027 (2021)
13. Tan, S., Guerrero, J.M., Xie, P., Han, R., Vasquez, J.C.: Brief survey on attack detection methods for cyber-physical systems. IEEE Syst. J. **14**(4), 5329–5339 (2020)
14. Li, L., et al.: Cyber attack estimation and detection for cyber-physical power systems. Appl. Math. Comput. **400**, 126056 (2021)
15. Zhang, C.-L., Yang, G.-H., Lu, A.-Y.: Resilient observer-based control for cyber-physical systems under denial-of-service attacks. Inf. Sci. **545**, 102–117 (2021)
16. Lu, A.-Y., Yang, G.-H.: Observer-based control for cyber-physical systems under denial-of-service with a decentralized event-triggered scheme. IEEE Trans. Cybern. **50**(12), 4886–4895 (2020)
17. Li, M., Chen, Y.: Wide-area robust sliding mode controller for power systems with false data injection attacks. IEEE Trans. Smart Grid **11**(2), 922–930 (2020)
18. Ma, R., Shi, P., Wu, L.: Dissipativity-based sliding-mode control of cyber-physical systems under denial-of-service attacks. IEEE Trans. Cybern. **51**(5), 2306–2318 (2021)
19. Li, M., Chen, Y., Liu, Y.: Sliding-mode secure control for jump cyber- physical systems with malicious attacks. J. Franklin Inst. **358**(7), 3424–3440 (2021)
20. Cao, Z., Niu, Y., Song, J.: Finite-time sliding-mode control of markovian jump cyber-physical systems against randomly occurring injection attacks. IEEE Trans. Autom. Control **65**(3), 1264–1271 (2020)
21. Ye, L., Zhu, F., Zhang, J.: Sensor attack detection and isolation based on sliding mode observer for cyber-physical systems. Int. J. Adaptive Control Signal Process. **34**(4), 469–483 (2020)

Risk Assessment Method of Power Monitoring System Vulnerability Based on Attack Graph

Hongbo Ye[1], Kexiang Qian[2](✉), Minghui Jin[1], Xiao Han[2], and Fei Xiao[1]

[1] State Grid Shanghai Municipal Electric Power Company, Shanghai, China
[2] State Grid Key Laboratory of Information and Network Security, Institute of Information and Telecommunication, Global Energy Interconnection Research Institute, Beijing, China
qiankexiang@geiri.sgcc.com.cn

Abstract. In recent years, incidents threatening the network security of the power system industry have become more frequent, hackers' attack methods have become more and more complex, and the difficulty of power system network security protection has continued to increase. Aiming at the problems of complex attack strategies and irrational attackers in the actual power system environment, a vulnerability risk assessment method for power monitoring systems based on attack graphs is proposed. First, according to the dependency between the vulnerabilities, based on the Common Vulnerability Scoring System (CVSS) framework of industrial control vulnerabilities, the attack path and attack graph model are established, and then the possibility of the vulnerability being exploited and the overall impact of the exploit are calculated. The damage degree of the power system network, and on this basis, the risk value of the vulnerability is calculated. The vulnerability risk calculated by this method can accurately reflect the impact of the vulnerability on the entire power system network. The experimental results show that the vulnerability risk assessment method of power monitoring system based on attack graph proposed in this paper can effectively quantitatively evaluate industrial control vulnerabilities and improve the level of industrial control security protection.

Keywords: Attack graph · Common vulnerability scoring system · Power supervisory system · Vulnerability assessment

1 Introduction

As a pillar industry in our country, the power industry has constructed our country's key infrastructure. Links such as integrated dispatching, power utilization, power distribution, power transmission, transformation, and power generation support and cooperate with each other to form the power industry value chain, and each link requires the application of a large number of industrial control equipment. The industrial control equipment used in the electric power industry has a high degree of integration. Therefore, the manufacturers of these equipment are complex and diverse. Most of the industrial control systems developed are mainly aimed at realizing functional interfaces, and generally lack more in-depth safety performance testing. These problems have brought a large

S.-C. Chu et al. (Eds.): ICGEC 2021, LNEE 833, pp. 102–113, 2022.
https://doi.org/10.1007/978-981-16-8430-2_10

number of network information security vulnerabilities to the power industrial control system, including the hardware backdoors implanted in the manufacture of equipment, the system software vulnerabilities caused by irregular behaviors in the development of programs [1], providing for hackers to invade Conditions. Due to the lack of assessment methods for network information security vulnerabilities in industrial control systems, security flaws in industrial control systems cannot be discovered in time, so that there are still many hidden security risks after the system is online.

In response to the above problems, relevant personnel have carried out in-depth research. Industrial control system vulnerabilities are the key point of the overall security vulnerability of industrial control systems [2]. Most of the existing related studies focus on system-level quantitative risk assessment and integrate industrial control systems. The vulnerability risk assessment is included in the overall assessment scope [3]. [4] Found the difficulty of exploiting vulnerabilities and the hazards of vulnerabilities in industrial control systems. Starting from the two dimensions of nature, comprehensively considering the defense strength, attack strength, physical loss, information loss, etc., a series of vulnerability quantitative indicators are proposed, and the vulnerability analysis of possible attack paths is carried out in combination with the attack graph. However, this method Focusing on the actual characteristics of the system and ignoring the inherent attributes of vulnerabilities, the risk assessment results of industrial control vulnerabilities do not have good portability [5].

In summary, the method in this paper is based on the dependency between vulnerabilities and the general vulnerability scoring system framework of industrial control vulnerabilities, establishes attack paths [6] and attack graph models [7], and then calculates the possibility of vulnerability being exploited and the impact on the entire power system network after being exploited. The degree of harm, and on this basis, the risk value of the vulnerability is calculated. This article has the following two characteristics:

(1) This paper proposes a general vulnerability scoring system framework to measure power system network security. It summarizes vulnerability scoring scores from three aspects, namely: basic score, life cycle score and environmental score.
(2) Based on the attack graph, this paper designs an algorithm to calculate the basic metric value of the improved general vulnerability scoring system. This algorithm enables this article to obtain more accurate evaluation scores based on the actual environment.

2 Attack Graph

2.1 The Concept of Attack Graph

Attack graph [8] is a directed graph. According to the specific meaning of points and edges in the directed graph, attack graphs can be divided into state enumeration graphs and penetration dependency graphs.

This paper uses a penetration dependency graph to represent the dependency between each security attribute, as shown in Fig. 1. Different from the state enumeration notation, each node in the attack graph represents a security attribute of the system. The directed arc represents a penetration, which points from the preconditions of penetration to the results of penetration. The main difference between the two attack graphs lies in the meaning of the nodes. Each node in the state enumeration graph represents a security state of the entire network, and each node in the penetration dependency graph only represents the security attributes of the network. The state enumeration diagram is similar to a state transition diagram. It describes all possible penetration sequences of an attacker, but does not describe the dependencies between security attributes. In the penetration dependency graph, the dependency between security attributes is clearly expressed by directed edges.

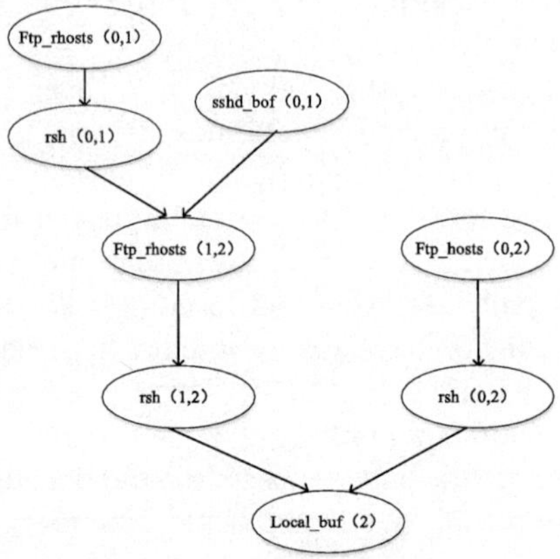

Fig. 1. Penetration dependency graph.

It essentially converts all the attributes in the attribute attack graph into edges, so that the graph only contains atomic attack nodes. It can clearly show the dependency between atomic attacks (penetration). The penetration dependency graph only expresses the dependency relationship between atomic attacks in the "or" relationship, for example, the atomic attack $rsh(0, 1)$ and $sshd\_bof\,(0, 1)$ in the graph indicate the "or" relationship.

3 Common Vulnerability Scoring System Framework

This paper proposes a general vulnerability scoring system framework, which can get three types of scores, namely basic score, life cycle score and environmental score. The basic score of industrial control vulnerabilities reflects the inherent attributes of industrial control vulnerabilities, which run through the entire life cycle of the vulnerability and exist in the information system. The life cycle score of industrial control vulnerabilities reflects the characteristics of vulnerability attributes that change over time [9]. The industrial control vulnerability environment score reflects the vulnerability attributes deeply associated with the industrial control environment, and the vulnerability risk assessor can specifically evaluate the potential impact of the industrial control vulnerability according to the actual environment [10].

3.1 Base Score

In order to obtain the basic score of industrial control vulnerabilities, the basic indicators of industrial control vulnerability risk assessment are taken as consideration factors [11], and the basic attributes of various aspects of industrial control vulnerabilities are comprehensively and in-depth analyzed. The basic indicators of industrial control vulnerability risk assessment include vulnerability exploitation indicators (attack vector, attack complexity, privilege requirements, user interaction), vulnerability impact range, vulnerability impact indicators (confidentiality, integrity, availability, visibility, controllability).

The vulnerability basic score is *BS*, the vulnerability impact score is *ISC*, the vulnerability exploit score is *ESC*, and the vulnerability impact basic score is *ISCBase*.

The basic scoring formula for vulnerability impact is:

$$ISCBase = 1 - ((1 - Conf) \times (1 - I) \times (1 - A) \times (1 - V) \times (1 - Cont)) \quad (1)$$

Among them, *Conf* represents confidentiality, *I* represents integrity, *A* represents availability, *V* represents visibility, and *Cont* represents controllability.

If the scope index is unchanged, the vulnerability impact score formula is:

$$ISC = 6.42 \times ISCBase \quad (2)$$

If the scope index is changed, the vulnerability impact score formula is:

$$ISC = 7.52 \times (ISCBase - 0.029) - 3.25 \times (ISCBase - 0.02)^{15} \quad (3)$$

he vulnerability exploitation score formula is:

$$ESC = 8.22 \times AV \times AC \times PR \times UI \quad (4)$$

Among them, *AV* represents the attack vector, *AC* represents the attack complexity, *PR* represents the privilege requirement, and *UI* represents the user interaction.

If $ISC < 0$, then

$$BS = 0 \quad (5)$$

If the scope index is unchanged, the basic scoring formula for the vulnerability is:

$$BS = RoundUp(Min((ISC + ESC), 10)) \quad (6)$$

Among them, *RoundUp* function: retain one decimal place of the value, round up, for example *RoundUp*(8, 13) = 8.2; *Min* function: Take the minimum of the two.

If the scope index is changed, the basic scoring formula for the vulnerability is:

$$BS = RoundUp(Min(1.08 \times (ISC + ESC), 10)) \quad (7)$$

At this point, the basic score of industrial control vulnerabilities is obtained.

3.2 Life Cycle Score

To calculate the life cycle score of industrial control vulnerabilities, it is necessary to analyze industrial control vulnerabilities from four indicators: vulnerability exploit code maturity, repair level, vulnerability report confidence, and exploit target service status.Referring to the life cycle indicator description table (Table 1), each industrial control vulnerability life cycle indicator corresponds to different levels of indicator values (for example, undefined, high, medium, and low) and corresponding indicator descriptions. If the vulnerability feature meets a certain indicator description, the corresponding indicator value is recorded.

According to the index values obtained in Table 1, since the life cycle score is a further modification on the basis of the vulnerability basic score, the score cannot be calculated independently. Put each score into formula (8), where the vulnerability life cycle score is *TS*.

$$TS = RoundUp(BS \times E \times RL \times RC \times SS) \tag{8}$$

Among them, *E* represents the maturity of the exploit code, *RL* represents the repair level, *RC* represents the confidence of the report, and *SS* represents the service status of the exploit target.

So far, the life cycle score of industrial control vulnerabilities is obtained.

Table 1. Life cycle indicator description table.

Indicator name	Index value	Indicator description
Exploit code maturity(*E*)	High	Vulnerability exists
	Feature	Functional vulnerabilities exist
	Poc	Theoretically exist
	No code	No loopholes exist
Repair level (*RL*)	No	No plan
	Temporary plan	Unofficial
	Temporarily resolved	Temporary plan
	Official fix	Official plan
Report confidence (*RC*)	No	Untrustworthy
	Reasonable	Partial fault
	Proven	Official verification
Service status of exploit targets (*SS*)	Very common	Completely affected
	Common	Partially affected
	Few	Rarely affected

3.3 Environmental Score

According to the index values obtained in Table 2, bring each score into the formula (9). Each variable in the formula is the result of the correction of the corresponding original variable, which is indicated by adding $M_$prefix in front of the original variable name. Among them, the vulnerability environment score is ES, the confidentiality requirement is $ConfR$, the integrity requirement is IR, the availability requirement is AR, the visibility requirement is VR, and the controllability requirement is $ContR$.

$$M\_ISCBase = Min(1 - ((M\_Conf \times ConfR)$$

$$\times(1 - M\_I \times IR) \times (1 - M\_A \times AR)$$

$$\times(1 - M\_V \times VR) \times (1 - M\_Cont \times ContR)$$

$$(1 - M\_V \times VR) \times (1 - M\_Cont \times ContR), 0.915) \tag{9}$$

Among them, M_Conf represents the confidentiality after correction, M_I represents the integrity after correction, M_V represents the visibility after correction, M_A represents the usability after correction, and M_Cont represents the controllability after correction.

Table 2. Description of environmental indicators.

Indicator name	Index value	Indicator description
Safety requirements	High	Serious impact
	Medium	Moderate impact
	Low	A certain degree of influence

At this point, the industrial control vulnerability environment score is obtained.

For the basic scores, life cycle scores, and environmental scores of industrial control vulnerabilities obtained above, according to the industrial control system vulnerability risk classification table (Table 3), the vulnerabilities can be classified into none, low risk, medium risk, high risk, very high five levels. According to the CVSS v3.0 standard and other vulnerability risk assessment methods widely used internationally, dividing the vulnerability risk level into four or five is a more effective means of grading. With the improvement of the quantification score of the vulnerability, the length of the interval of different levels decreases. This method can provide a more fine-grained distinction for high-risk vulnerabilities.

Table 3. Vulnerability risk classification table of industrial control system.

Grade	Indicator description
No risk	0.0
Low risk	0.1–4.0
Moderate risk	4.1–6.0
High risk	6.1–9.0
Extremely high risk	9.1–10.0

4 Attack Graph Model

4.1 Establish Attack Path and Attack Graph Model

Attack graphs can show vulnerabilities, dependencies between nodes and prior knowledge of power system network information [12]. The concept of the proposed attack graph is formalized to simulate the impact of different security conditions and exploit vulnerabilities when the power system network is attacked.

Definition 1 Attack graph G is a directed graph.

$$G{=}(E \cup C, R_r \cup R_i) \tag{10}$$

Among them: $E \cup C$ is the set of vertices, $R_r \cup R_i$ is the set of edges, E is the set of vulnerabilities, and C is the set of conditions.$R_r = \{C \times E\}$, And $R_i = \{E \times C\}$.

To model the vulnerabilities and explicitly model all the vulnerabilities in the attack graph, the unique characteristics of the vulnerabilities must be deleted.

Definition 2 Vulnerability exploit $e \in E$ is a triplet.

$$e = \langle v, h_s, h_d \rangle \tag{11}$$

Among them: v represents the vulnerability; h_s and h_d represent the host in the power system network. The triple means that the attacker exploits the vulnerability h_s on the target host h_d from the source host v.

In this article, preconditions are used to describe the conditions required to exploit the vulnerability, and postconditions are used to describe the consequences of exploiting the vulnerability.

Definition 3 Condition C.

$$C = (pre(E) \cup post(E)) \tag{12}$$

$c \times e \in R_r, c \in pre(e); e \times c \in R_i, c \in post(e); e \in E, c \in C$, Among them, $pre()$ and $post()$ respectively represent the function of mapping the exploit to its precondition set and the function of mapping the exploit to its post condition set.

4.2 Definition of Dependency

In fact, attackers usually use a series of vulnerabilities to destroy the power system network to achieve their ultimate goal. These attacks can be arranged in chronological order in which the attacker completed it. Therefore, the vulnerabilities and the sequence of exploits form a sequence that can indicate the attacker, and is called the attack path. These attack paths make up the attack graph.

Definition 4 Attack path:

$$AP = \{ap_1, ap_2, \cdots, ap_n\} \tag{13}$$

$ap_i = \left(E' \cup R_c\right), E' \subseteq E$, Among them, ap_i refers to the i-th attack path; E' is the vulnerability exploit subset; R_c is the directed edge set. This means that the attack path is a sequence combination of several attacks: $e_1, e_2, \cdots, e_n (n \geq 2)$, Satisfy $\forall i \in [2, n], \exists c, c \in pre(e_i) \wedge \left(c \in post\left(e_j\right)\right), j \in [1, i-1], c \in C$.

The different ways of attackers in the attack path lead to different dependencies between attack nodes. First define a dependency template, and identify these network attributes based on this template for further analysis. From the NVD or CVSS database, the basic measurement value of each vulnerability in the exploit can be obtained.

Definition 5 Dependency.

Given attack graph G and attack path set AP. For each attack path $ap \in AP$, $ap = \left(E', \mathrm{Re}\right)$ and $e_i, e_j \in E'$, if $\exists m \in \{AV, AC, PR, UI\}$,

$$em(e_i)[m] \in post\left(e_j\right) \tag{14}$$

It can be said that e_i and e_j have a dependency relationship, which means $r_d\left\langle e_j, e_i \right\rangle$.

The dependency between the two vulnerabilities is simply expressed as a directed edge in the attack graph [13], and the dependency is identified from the attack graph, as shown in Fig. 2. In this attack diagram, polygons are used to represent exploits, and ellipses are used to represent conditions. The edge from the ellipse to the polygon indicates that the condition is *a* precondition for exploiting the vulnerability. On the contrary, the edges from the polygon to the ellipse show the post-conditions of the exploit. Consider the given condition a as an asset, and the symbol $ap(a)$ represents any attack path of asset a. Choose the first path of the attack graph as the starting point. First, find the pre-conditions and post-conditions for exploiting these vulnerabilities in the chosen attack path. Next, define the basic metric as a vector bm of eight values for each exploit. Search all the relationships between the availability metric and the post-condition for each node metric. Finally, based on the previous steps and definition 5, the dependency between attack nodes in the attack path can be found and appear as directed edges in the attack graph. By performing this work on all attack paths, the dependency between the vulnerabilities in the attack graph in Fig. 2 can be revealed.

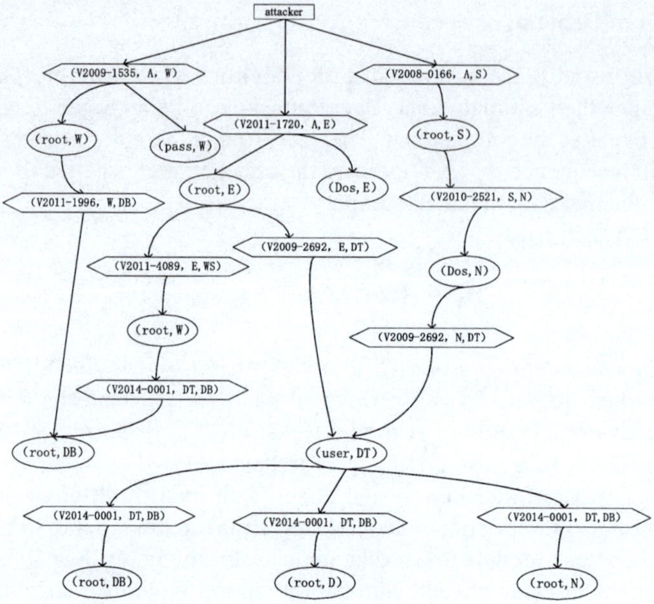

Fig. 2. Attack diagram display.

5 Experiment Analysis

The simulation network topology test chart used in this article is shown in Fig. 3. The test network consists of 8 hosts located in 2 subnets. A subnet is the DMZ (DeMilitarized Zone) network, which is separated from the local network by a firewall and includes a Web server, a mail server, and a DNS (Domain Name System) server. The firewall located in front of the DMZ has a set of policies to control remote access to internal hosts. Generally, all hosts in the DMZ network (except the Web server) do not actively connect to other external hosts, they passively receive service requests and respond to requests as needed. The Web server needs to obtain information from the SQL (Structured Query Language) server in the trusted network, so it is allowed to connect to the SQL server and actively send SQL queries to it. In addition to the SQL server, NAT (Network Address Translation) gateway server, administrator server and some local machines are also located in the trusted network. The administrator server and local computer are located behind the NAT gateway server, and all communications with the external network need to pass through the gateway server.

Calculate the attack probability of each attack path through different methods, and evaluate which method is closer to the baseline. In Fig. 4, the horizontal axis is the number of attacks in the attack path (attack path length), and the vertical axis is the percentage of successful attackers or the probability of attack. Each percentage or probability is the average value of simulated attack paths of different lengths. In Fig. 4, the curve simulation shows the percentage of successful simulated attackers and is used as a baseline for comparison. The CVSS basic score shows the aggregate score of the basic

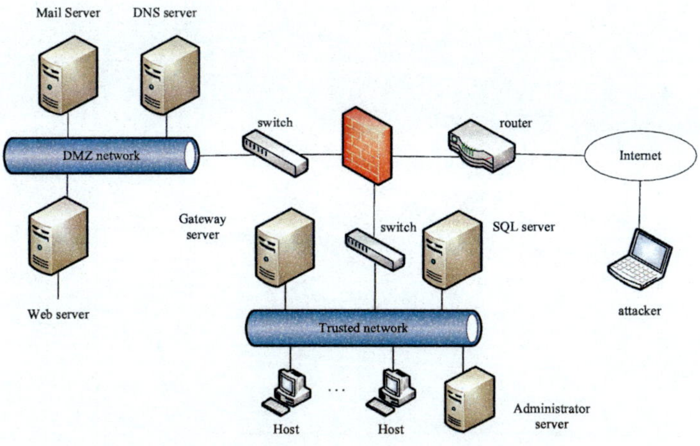

Fig. 3. Simulated network topology.

indicators of all vulnerabilities in the attack path. It can be seen from Fig. 4 that the results of this method are closer to the baseline than the CVSS method.

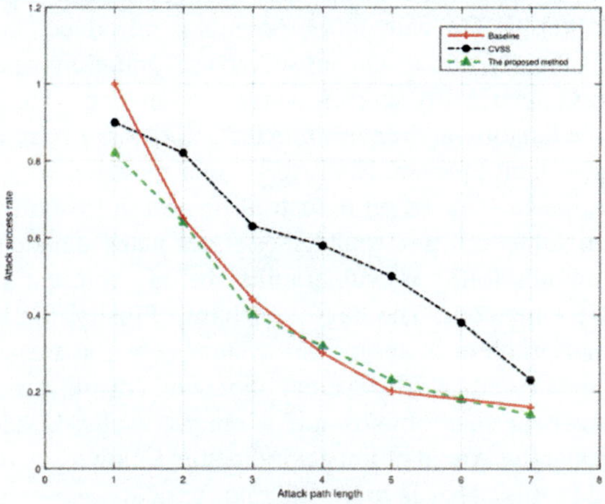

Fig. 4. The relationship between attack success rate and attack path length.

Figure 5 shows a similar simulation that calculates the sum of the basic metric scores of the vulnerabilities used in the attack path divided by 10. In Fig. 5, the horizontal axis is the total score divided by 10, and the vertical axis is the total score of the attack probability or the percentage of successful attackers. It can be seen from Fig. 5 that the method in this paper is still very close to the baseline.

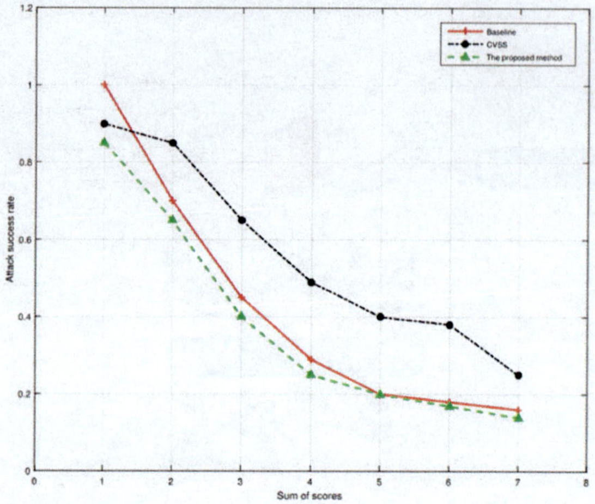

Fig. 5. The relationship between attack success rate and total score.

6 Conclusions

This paper proposes a vulnerability risk assessment method for power monitoring systems based on attack graphs. The calculation process is introduced and the simulation results are shown. Specifically, this article has solved the limitations of existing methods, and proposed a general vulnerability scoring system framework, which defines the basic indicators, life cycle indicators, and environmental indicators of the industrial control vulnerability risk assessment process. Next, this paper proposes a general vulnerability scoring system framework based on industrial control vulnerabilities based on the dependencies between vulnerabilities, establishes attack paths and attack graph models, and then calculates the possibility of vulnerability being exploited and its impact on the entire power system network. The degree of harm. Finally, the proposed method is applied to the actual problem of the power system network, the probability of the simulated attacker is calculated and the evaluation result is displayed. The result shows that the method in this paper can obtain more accurate results. In addition, the paper also discusses the comparative trend of impact indicators following actual conditions. In summary, the method in this paper is more suitable for actual power system networks, and has the characteristics of accuracy, flexibility, and high applicability.

Acknowledgements. This work is supported by the science and technology project of State Grid Corporation of China "Research on Vulnerability Analysis and Threat Detection Key Technology of Power Monitoring System in Cyberspace" (Grand No.5108-202117055A-0-0-00).

References

1. Hu, H.: Method for detecting network vulnerability of power industrial control system based on feature library. Autom. Instrum. **11**, 181–184 (2019)

2. Liang, N.: Research on the technology of vulnerability investigation and risk management of electric power monitoring system. Autom. Expo. **36**(S2), 41–45 (2019)
3. Ying, H., Liu, S., Han, L.: Overview of power industry control system security technology. Electr. Power Inf. Commun. Technol. **16**(3), 56–63 (2018)
4. Dong, J., Han, P.: Research on application of protocol analysis method for industrial control. Inf. Technol. Network Secur. **40**(02), 1–6 (2021)
5. Tao, Y.: Industry control system vulnerability risk assessment method. Small Microcomput. Syst. **41**(03), 603–609 (2020)
6. Holm, H., Karresand, M.,Vidstrom, A.: A survey of industrial control system testbeds. In: Secure IT Systems (LNCS), vol. 9417, pp.11–26. Springer, Cham (2015). https://doi.org/10.1007/978-3-319-26502-5_2
7. Cook, A., Smith, R., Maglaras, L.: Measuring the risk of cyber attack in industrial control systems. In: International Symposiumfor Ics & Scada Cyber Security Research, pp. 156–162 (2016)
8. Henry, M., Layer, R., Snow, K.: Evaluating the risk of cyber attacks on SCADA systems via petri net analysis with application to hazardous liquid loading operations. In: IEEE Conference on Technologies for Homeland Security, pp. 1022–1029 (2009)
9. Knowles, W., Prince, D., Hutchison, D.: Towards real-time assessment of industrial control systems (ICSs): a framework for future research. In: International Symposium on Ics & Scada Cyber Security Research, pp. 2120–2125 (2013)
10. Yasakethu, S., Jiang, J.: Intrusion detection via machine learning for SCADA system protection. In: International Symposium on Ics & Scada Cyber Security Research, pp. 523–529 (2013)
11. Deng, Z., An, J.: Research on grid event monitoring based on random matrix theory. J. Northeast Electr. Power Univ. **40**(02), 7–12 (2020)
12. Li, S., Xia, Q., Zhao, J., et al.: Analysis of inrush current characteristics of transformer in photovoltaic power generation system. J. Northeast Electr. Power Univ. **40**(02), 28–36 (2020)
13. Chen, Z., Yang, D., Zhao, F.: Two areas system inertia estimation based on ambient data. J. Northeast Electr. Power Univ. **40**(04), 48–54 (2020)

Dynamic Multiple Indicators Matching Processing for Power Load Forecasting System

Tiehua Zhou[1], Jiayu Hao[1], Futao Ma[2], and Ling Wang[1](✉)

[1] Department of Computer Science Technology, School of Computer Science, Northeast Electric Power University, Jilin, China
{thzhou,smile2867ling}@neepu.edu.cn

[2] China Mobile Group Hebei Co., Ltd. Langfang Branch, Langfang, China

Abstract. Electrical load forecasting is an important part of power system research, and improving the accuracy of load forecasting is the crucial. In order to improve the accuracy of load forecasting, this paper proposed a dynamic multiple indicators matching method for real-time power load forecasting system, which establishes an indicator structure by selecting key indicators. This method fully considers the direct and indirect effects of the attributes on the load, and the weights of each factor will change with time, making the final weight assignment more scientific and reasonable. Thus, it could effectively reduce the workload of load forecasting as well as improving the accuracy and reliability of load forecasting. We processed A-GPR real-time power load forecasting model to guarantee the timeliness of power supply and effective utilization of new energy over smart grid. Dynamic multiple indicators matching method is used to analyze the influencing factors, and establish a comprehensive index structure. Experiments show good prediction accuracy and prediction results.

Keywords: Electrical load forecasting · Multi-indicator dynamic matching · A-GPR model · Prediction analysis

1 Introduction

Electrical load forecasting [1] refers to the prediction of future loads using historical data, taking into account the operating characteristics of the power system, capacity increases, and natural conditions. Load forecasting is an important function of the grid energy management system and is the basis for the economic, safe and reliable operation of the smart grid system. The load of the power system has its own inherent cyclical pattern, which is also influenced by many factors, such as climatic conditions and economic factors. Due to the difference of load characteristics in different regions, the load forecasting work for different regions should be combined with the actual situation of the region, considering the influencing factors of the load on the basis of the load characteristics, and then choosing a suitable method for forecasting, in order to improve the accuracy of forecasting. The accuracy of load forecasting [2] directly affects the safety, economy and power supply quality of the power system, and has an important impact on

S.-C. Chu et al. (Eds.): ICGEC 2021, LNEE 833, pp. 114–124, 2022.
https://doi.org/10.1007/978-981-16-8430-2_11

the investment, scheduling, layout and the scientific nature of the operation of the power grid, and is an important basis for development planning as well as real-time control, and with the development of informationization of the power system, higher requirements are put forward for the accuracy of load forecasting [3]. Therefore, how to improve the prediction accuracy is the focus of current research on load forecasting technology.

The commonly used load forecasting methods include time series methods [4], regression analysis methods [5], gray forecasting methods [6], expert system methods [7], support vector machines [8], neural network algorithms [9], and so on. Among them, the time series method is the most widely used, which is a one-dimensional time series data to infer the load data for a future period based on the historical load data. Meanwhile, neural network algorithms are also important methods for load forecasting.

In this paper, we propose a real-time power load forecasting system with dynamic matching of multiple indicators, which has the functions of data collection, data analysis, dynamic matching of multiple indicators and real-time power load forecasting, etc. It can realize real-time extraction of key influencing factor sets for short-term and ultra-short-term power load forecasting; fully consider multiple indicators affecting the load and screen out the key indicators affecting the load, so that the changing characteristics of the load can be more easily grasped. The A-GPR method in the power load forecasting module is also proposed to improve the adaptability and robustness of the model.

2 Related Work

With the rapid development of science and technology, the enthusiasm for research on load forecasting in power systems is gradually rising around the world [10]. The gray forecasting method requires only a small amount of historical data to discover the laws affecting the load, and the calculation is small. Wang et al. [11] proposed an improved gray model that can adapt well to the changing patterns of electric loads, and proved that this model could achieve high forecasting accuracy by forecasting electric loads. In 2016, Zhao et al. [12] used a hybrid optimized gray model to forecast the electric loads of two cities with good results.

Neural network algorithms are commonly used for electric load forecasting due to their self-learning capability and ability to handle complex nonlinearities. R. Ballinip [13] in the early part of this century proposed a time series forecasting method based on ordered weighted average operator (OWA) [14], creating the best forecasting model ARIMA [15] and neural network model, which is highly adaptable in short-term forecasting of electric loads. In 2020, Park et al. [16] proposed a two-stage short-term power load forecasting scheme, which performs deep neural networks in the second stage with more outstanding results.

New integrated forecasting methods have emerged in recent years, providing a backup for research on the power load forecasting problem [17]. In 2018, Zheng et al. used Hidden Markov model [18] as well as Baum-Welch algorithm [19] and Viterbi algorithm [20] for decoding prediction process, which finally obtained the optimal model for load characteristic prediction.In 2020, Zhang [21] proposed an integrated prediction method based on empirical wave transformation (EWT) and improved density-based spatial clustering of noise application (IDBSCAN), where low and medium frequency components are predicted by long and short-term memory (LSTM) [22], and the final prediction results with high accuracy are obtained by superposition.

Although the load forecasting models established by previous forecasting methods have their own advantages, there are difficult to cope with the fluctuating electric load. Similarly, many load forecasting models use neural network algorithms, but the structure of neural networks and network parameters mostly need to be determined by subjective experience, which makes it difficult to guarantee the accuracy of forecasting results, and some methods do not consider the influence of local meteorological and economic factors on load forecasting, resulting in the absence of important information. To address the shortcomings of existing load forecasting methods, this paper proposes a real-time power load forecasting system based on dynamic multiple indicators matching. The model of this system integrates the influence of multiple indicators on load, the weights of each factor will change with time as the influencing factors of load are different and time changes in the process of processing load data.

3 Power Load Forecasting System Design

This paper proposes a real-time power load forecasting system based on dynamic multiple indicators matching, which has a reasonable structure with high accuracy and reliability. An Adjustable-Gaussian Process Regression (AGPR) model is also established to achieve timely power supply and effective utilization of new energy sources under the smart grid.

One of the solutions used to achieve the purpose of this system is to develop a real-time power load forecasting system based on multi-indicator dynamic matching, whose features include the following: a data collection module to collect load, meteorological and economic data and manage them in a categorized manner to build a system database; a data collection module to calculate the correlation between influencing factors and load based on the MIDM algorithm, while ranking them and filtering out key attributes to construct a comprehensive index structure; a data analysis module capable of transforming fixed inertia factors in the model into dynamically changing inertia factors according to random changes in different regions and times, and constructing a multi-indicator dynamic matching module of the A-GPR model; a real-time power load forecasting module capable of making real-time power load forecasts according to the constructed comprehensive index system and the A-GPR model; said data acquisition module is signal-connected to the data analysis module, the data analysis module is signal-connected to the multi-indicator dynamic matching module, and the multi-indicator dynamic matching module is signal-connected to the real-time power load forecasting module.

Another technical solution adopted to achieve the purpose of this system is to implement a real-time power load forecasting method based on dynamic matching of multiple indicators, the features of which include the following.

First, using the respective functions of the meteorological data unit, load data unit and economic data unit of the data acquisition module for data acquisition, the meteorological data unit is responsible for real-time collection of temperature, humidity, precipitation, visibility, wind direction, wind speed and weather data of the required area for data integration and transmission to the system database to generate meteorological data tables; the load data unit is responsible for real-time collection of industrial electrical load, agricultural electrical load, municipal electrical load, domestic electrical load and commercial electrical load for data integration and transmission to the system database to generate load data tables; the economic data unit is responsible for real-time collection of industrial electricity The load data unit is responsible for real-time collection of industrial electrical load, agricultural electrical load, municipal electrical load, postal electrical load, transportation electrical load, domestic electrical load and commercial electrical load for data integration and transmission to the system database to generate load data tables; the economic data unit is responsible for real-time collection of economic data from the government's annual economic report for data integration and transmission to the system database to generate economic data tables.

Secondly, based on the functions of the data analysis module, the Multi-Index Dynamic Matching (MIDM) algorithm is used to analyze the correlation of each indicator, analyze the correlation degree between the load data and the corresponding meteorological data and economic data, so as to determine the key indicators affecting the desired area.

Thirdly, using the function of multi-indicator dynamic matching module, a comprehensive dynamic indicator system is established based on the real-time correlation of load data and key indicators in different regions; the A-GPR model is established using the selected load data and the comprehensive indicator system.

Finally, using the function of real-time power load forecasting module, obtain meteorological data and economic data at the time to be forecasted, establish a comprehensive index system, and input the comprehensive index system and the corresponding load data into the A-GPR model as input vectors, whose output is the load value at the time to be forecasted.

4 Adjustable-Gaussian Process Regression (A-GPR) Model

4.1 Multi-indicator Dynamic Matching Algorithm

Multi-indicator dynamic matching algorithm can quickly analyze the correlation of each indicator, analyze the degree of correlation, and finally determine the key indicators, and its algorithm steps are as follows.

(1). Calculate the projected distance of each indicator curve with respect to the load,

$$x_{qi} = \frac{\left(k_{qi}\left(b_1(2) - b_{qi}\right) + b_1(1)\right)}{\left(k_{qi}^2 + 1\right)} \tag{1}$$

$$y_{qi} = k_{qi}x_{qi} + b_{qi},\ (1 \leq q \leq 24) \tag{2}$$

$$r_{qi} = \sqrt{(x_{qi_2} - x_{qi_1})^2 + (y_{qi_2} - y_{qi_1})^2} \tag{3}$$

$$R_{\mathrm{t}} = \sqrt{(x_{t_2} - x_{t_1})^2 + (y_{t_2} - y_{t_1})^2} \tag{4}$$

where x_{qi} represents the projected horizontal coordinate of the point on the curve of the influence indicator on the load curve, k_{qi} represents the slope of this point on the load curve, $b_1(2)$ represents the horizontal coordinate of the point on the curve of the influence indicator, $b_1(1)$ represents the vertical coordinate of the point on the curve of the influence indicator, b_{qi} represents the intercept of the load curve at this point, y_{qi} represents the projected vertical coordinate of the point on the curve of the influence indicator on the load curve, q represents the 24 points of the influence indicator on the curve for 24 h a day, *i* represents a certain influence indicator, r_{qi} represents the distance of the influence indicator to the load, x_{qi2} represents the horizontal coordinate of the point after the projection point of the influence indicator curve on the load curve, x_{qi1} represents the horizontal coordinate of the point before the projection point of the influence indicator curve on the load curve, y_{qi2} represents the horizontal coordinate of the point after the projection point of the influence indicator curve on the load curve, y_{qi2} represents the vertical coordinate of the point after the projection point of the influence indicator curve on the load curve. y_{qi1} represents the vertical coordinate of the previous point of the influence indicator curve at the projection point of the load curve, R_t represents the fluctuation distance of the load between two points, *t* represents 24 points of the load on the curve 24 h a day, y_{t2} represents the vertical coordinate of the latter point on the load curve, y_{t1} represents the vertical coordinate of the previous point on the load curve, x_{t2} represents the horizontal coordinate of the latter point on the load curve horizontal coordinate, and x_{t1} represents the vertical coordinate of the previous point on the load curve.

(2). Find the weights of each indicator,

$$\mathrm{w}_i = \frac{r_{qi}}{R_t} \tag{5}$$

where w_i represents the influence weight of each influence factor on the load, *i* represents a certain influence index, r_{qi} represents the projection distance of the influence factor on the load, and R_t represents the fluctuation distance of the load between two points.

(3). Calculation of the weights of the pooled indicators,

$$W_{i,j} = \frac{\sum_{k=i}^{n} W_k + \sum_{l=j}^{n} W_l}{n}, \quad i,j = 1,2,\ldots,n \text{且} i \neq j \tag{6}$$

$$w_i' = \frac{\sum_{j=1, j\neq i}^{n} w_{i,j} - w_i}{n-1}, \quad i = 1, 2, \ldots, n \tag{7}$$

where *n* represents the number of indicators, $w_{i,j}$ represents the indirect weights of the combined indicators, *i*, *j* represents an impact indicator, and $w_i{'}$ represents the weight of the aggregated indicators, which fully expresses the direct and indirect impact of each indicator on the load.

4.2 Real-Time Adaptive A-GPR Model

The robustness of the model is further improved by establishing an A-GPR model so that the originally fixed model parameters can be changed adaptively in real time according to the inputs of different indicators in the following steps,

(1). Initialized population size N, maximum number of iterations T_{max};
(2). $\vec{x}$, the vector of the current position of the particle;
(3). *Fitness*, the fitness value of the vector x;
(4). $\vec{v}$, the velocity of the particle, which has the same dimension as that of the vector $\vec{x}$;
(5). *pbest*, the fitness value corresponding to the best position encountered during the flight of each particle;
(6). $\vec{p}$, which is used to record the best position encountered during the flight of the particle, and the dimension of *pbest* is the same as the dimension of the vector $\vec{x}$;
(7). The update formula for the particle position,

$$v_i = K\left[wv_i + \phi_1 rand()\left(P_{best_i} - x_i\right) + \phi_2 rand()\left(g_{best_i} - x_i\right)\right] \tag{8}$$

$$\mathrm{x}_i = \mathrm{x}_i + \mathrm{v}_i \tag{9}$$

Where v_i represents the velocity of the particle, x_i represents the current position of the particle, i represents the current number of iterations per particle update, g_{besti} represents the optimal position of the population as a whole, p_{besti} represents the optimal position of the individuals in the population, w represents the inertia weight factor, $\varphi 1$ represents the individual cognitive learning rate, $\varphi 2$ represents the social learning rate, *rand()* is a random number between the interval [0,1], K represents the convergence factor, and K is described as follows,

$$K = \frac{2}{|2 - \phi - \sqrt{\phi^2 - 4\phi}|}, \phi = \phi_1 + \phi_2, \phi > 4. \tag{10}$$

The inertia weight factor is generally set in the standard particle swarm algorithm: $w_t + 1 = w_t = 1$. Since the inertia weight factor is a variable that affects the current particle velocity, a larger value is beneficial for global search and a smaller value is beneficial for local search, an A-GPR model is proposed to better balance the search capability,

$$w_t = w_{\max} - (w_{\max} - w_{\min}) \exp\left(-\frac{t}{T}\right) \tag{11}$$

Where *wt* represents the inertia weight factor for the current number of iterations, w_{max} is the inertia maximum, w_{min} is the inertia minimum, t is the current number of iterations, and T is the maximum number of iterations, so that the particle position is updated as follows,

$$v_i = K\left[w_t v_i + \phi_1 rand()\left(P_{best_i} - x_i\right) + \phi_2 rand()\left(g_{best_i} - x_i\right)\right] \tag{12}$$

$$\mathrm{x}_i = \mathrm{x}_i + \mathrm{v}_i \tag{13}$$

The ability of the algorithm to balance the global and local search due to the fact that the inertia weight factor is always changing, resulting in the velocity and position of the particles being constantly updated at both large and small ranges of the population all the time;

(8). The fitness function E of the particle is recalculated based on the position of the updated particle, and the fitness function is as follows,

$$E = \frac{1}{\mathrm{m}} \sum_{i=1}^{m} (y_i - y_{i-1}) \tag{14}$$

Where m is the number of input samples, y_i denotes the process output value of the current A-GPR model training, y_{i-1} is the output value of the previous generation model, and i represents the current number of iterations;

(9). Determine whether the progress requirements are met or whether the number of iterations is reached, and if so, terminate the search; otherwise, proceed to the next search;

(10). Output the information of the final particle, which is the parameter of the A-GPR model.

5 Discussion and System Implementation

In this paper, the actual power grid of a region is used as an example, and the historical load data of the region from 2014–2017, as well as corresponds to the historical meteorological data, where the influencing factor data include temperature, humidity, atmospheric pressure, rainfall, wind direction, wind speed, visibility, and GDP, are collected and analyzed using the multi-indicator dynamic matching method (Table 1).

Table 1. Correlation coefficients of load and each influencing factor

Influencing factors	Spring	Summer	Autumn	Winter
Temperature/u_1	0.3526	0.2148	0.6539	0.8521
Humidity/u_2	0.2548	0.7548	0.4125	0.4178
Rainfall/u_3	0.2587	0.8542	0.2365	0.5261
Visibility/u_4	0.1495	0.3659	0.4187	0.4236
Wind direction/u_5	0.4785	0.5214	0.3298	0.2148
Wind speed/u_6	0.2631	0.4582	0.5248	0.4163
Weather conditions/u_7	0.7451	0.6321	0.3658	0.5214
GDP/u_8	0.5269	0.2154	0.3241	0.6395

Correlation analysis was performed with data from 2016, where the correlation of each meteorological factor was varied throughout the year, with the greatest correlation between weather conditions and GDP in spring, rainfall and humidity in summer, wind speed and temperature in autumn, and temperature and GDP in winter. The selection of influencing factors is made according to the date of the forecast day, and a comprehensive index system is established.

The GPR prediction model is as follows,

$$y' = K(x^*, X)(K(X, X) + \delta_n^2 I)^{-1} y \tag{15}$$

where y' represents the output of the test sample, y represents the output of the training sample, $K(x^*,X)$ is the n × 1 order covariance function matrix of the training input variable x and the test input variable X^*,$K(X,X)$ is the n × n order covariance function matrix of the training input variable X, I is the unit order matrix, and $\delta_n{}^2$ is the hyperparameter of the model.

The difficulty of building the model lies in solving the hyperparameters of the model, which mainly exist in the covariance function and white noise; therefore, in order to solve the hyperparameters of the model, the specific form of the covariance function should be determined first. The specific form of the covariance function of the Gaussian process is as follows,

$$K(x_i, x_j) = \sigma_p^2 \exp(-\frac{|x_i - x_j|^2}{2l^2}) + \sigma_n^2 \sigma_{i,j} \tag{16}$$

$\sigma_{i,j}$ is Kronecher's constant, σ_p^2, l, σ_n^2 are the hyperparameters of the model, σ_n^2 is the variance of the kernel function of the model, l is the characteristic width, and i,j represent a column of the matrix.

To optimize the parameters of the model, the A-GPR model is constructed with the following steps,

Step 1: Initialize particle swarm information, such as flight speed, current position, etc., based on hyperparameter constraints with variance as the adaptation degree;

Step 2: The hyperparameter velocity and position information, combined with training samples and test samples of influencing factors are input to the GPR regression model, and the test samples of load values are rolled up and predicted based on Eqs. (15) and (16), and the fitness of the particles is calculated;

Step 3: Update the particle individual position optimal value and the population position optimal value, and determine whether the population optimal fitness satisfies the requirement. If it is satisfied, the iteration ends, otherwise go to step 4;

Step 4: Update the velocity and position information of the particle according to Eqs. (12) and (13), and go to Step 2.

Set the population size $N = 200$, the maximum number of iterations $T = 150$, $w_{max} = 0.9$, $w_{min} = 0.4$. The end of the algorithm is the optimal information of the model parameters, and then, the trained GPR model is tested to build the A-GPR load prediction model.

By building the A-GPR real-time load forecasting model, the historical load data and the combined meteorological factors at the forecast moment were used as the input quantity of the improved A-GPR model, and the predicted load values at the forecast moment were used as the output quantity of the model, and April 20, 2016 in the region was selected as the forecast date. The data from January 1 to April 19 were used as the training sample set to train and test the A-GPR model, and Table 2 shows the error analysis table of the load values predicted by the A-GPR model and the predicted values of the GPR model with the true values.

Table 2. Error analysis table for each time point

Time point	GPR		A-GPR		Actual value
	Forecast value/MW	Error/%	Forecast value/MW	Error/%	
01:00	33654.01	5.6%	35549.28	0.31%	35661
02:00	33291.22	5.9%	35357.37	0.36%	35488
03:00	31935.19	3.3%	33163.69	0.45%	33013
04:00	29236.32	7.8%	31746.77	0.05%	31731
05:00	29488.21	5.3%	31255.29	0.37%	31139
06:00	29094.12	6.0%	31275.46	1%	30956
07:00	29702.02	5.7%	31585.3	0.33%	31481
08:00	30717.94	4.9%	32448.87	0.48%	32291
09:00	31539.83	4.0%	33043.58	0.58%	32852
10:00	31021.45	6.6%	33387.89	0.51%	33216
11:00	31577.56	5.9%	33984.86	1.2%	33566
12:00	35842.43	6.8%	33922.88	1.1%	33552
13:00	30303.33	6.1%	32474.22	0.65%	32262
14:00	31123.39	6.6%	29617.27	1.4%	29205
15:00	28162.24	4.4%	26972.38	0.5%	26987
16:00	27343.31	5.1%	25976.4	0.25%	26025
17:00	26992.33	4.4%	25909.94	0.17%	25862
18:00	24523.75	6.9%	26460.56	0.4%	26354
19:00	25992.05	4.6%	27029.58	0.89%	27271
20:00	26843.75	6.8%	27859.21	3.2%	28809
21:00	29364.04	4.9%	29416.55	4.7%	30869
22:00	31621.14	5.0%	31980.06	3.9%	33296
23:00	33846.06	3.6%	34935.55	0.48%	35107
24:00	38411	6.4%	36825.39	2.0%	36086

From Table 2 and Table 3, we can see that the real-time load forecasting model based on dynamic matching of multiple indicators has good forecasting accuracy and can achieve good forecasting results.

Table 3. Average error analysis table

Category	Average error
GPR	5.7%
A-GPR	1.4%

6 Conclusion

The A-GPR method in the real-time power load forecasting module of this article enables the originally fixed model parameters to change adaptively in real time according to the input of different indicators, which improves the adaptability and robustness of the model and is a new development direction for future power system electrical load forecasting, which will be more beneficial to the development of smart grid and energy Internet.

Acknowledgements. This work was supported by the Science and Technology Development Plan of Jilin Province, China (No.20190201194JC).

References

1. Hu, B., Pang, C., Wang, L., Chu, H., Mao, C.: Big data management and application research in power load forecasting and power transmission and transformation equipment evaluation. J. Phys. Conf. Ser. **1069**(1), 012084 (2018)
2. Carvallo, J.P., Larsen, P.H., Sanstad, A.H., Goldman, C.A.: Long term load forecasting accuracy in electric utility integrated resource planning. Energy Policy **119**, 410–422 (2018)
3. Zhang, W., Qin, J., Mei, F., Fu, J., Dai, B., Yu, W.: Short-term power load forecasting using integrated methods based on long short-term memory. Sci. China Technol. Sci. **63**(4), 614–624 (2020). https://doi.org/10.1007/s11431-019-9547-4
4. Ding, N., Bésanger, Y., Wurtz, F.: Next-day MV/LV substation load forecaster using time series method. Electric Power Syst. Res. **119**, 345–354 (2015)
5. Chen, C., Zhou, J.N.: Application of regression analysis in power system load forecasting. Adv. Mater. Res. **960**, 1516–1522 (2014)
6. Cai, W.Z., Wang, D.T., Wang, Y.S., Yang, Y., Gao, Z.L.: Study of short-term wind power forecasting based on adaptive grey prediction method. Appl. Mech. Mater. **734**, 697–700 (2015)
7. Hu, L.Q., He, C.F., Cai, Z.Q., Wen, L., Ren, T.: Track circuit fault prediction method based on grey theory and expert system. J. Vis. Commun. Image Represent. **58**, 37–45 (2019)
8. Guo, L., Chen, J., Wu, F., Wang, M.: An electric power generation forecasting method using support vector machine. Syst. Sci. Control Eng. **6**(3), 191–199 (2018)
9. Lan, Y., Xue, L., Liao, X.: Short-term power load forecasting based on RBF neural network. Int. Core J. Eng. **7**(5), 51–59 (2021)
10. Pan, J., Qi, M.: Study on short-term load forecasting of distributed power system based on wavelet theory. In: 2018 10th International Conference on Measuring Technology and Mechatronics Automation (ICMTMA), pp. 170–173 (2018)
11. Wang, Y.L., Li, Q.Y.: Application of the improved grey model in power load forecasting. Appl. Mech. Mater. **575**, 658–661 (2014)

12. Zhao, H., Guo, S.: An optimized grey model for annual power load forecasting. Energy **107**, 272–286 (2016)
13. Ballini, R., Yager, R.R.: OWA filters and forecasting models applied to electric power load time series. Evol. Syst. **5**(3), 159–173 (2014). https://doi.org/10.1007/s12530-014-9112-2
14. Linares-Mustarós, S., Ferrer-Comalat, J.C., Corominas-Coll, D., Merigó, J.M.: The ordered weighted average in the theory of expertons. Int. J. Intell. Syst. **34**(3), 345–365 (2019)
15. Ho, C.H.: Forecasting a point process with an ARIMA model. Commun. Stat. Theory Methods **45**(17), 5066–5076 (2016)
16. Park, S., Moon, J., Jung, S., Rho, S., Baik, S.W., Hwang, E.: A two-stage industrial load forecasting scheme for day-ahead combined cooling, heating and power scheduling. Energies **13**(2), 443 (2020)
17. Veeramsetty, V., Chandra, D.R., Salkuti, S.R.: Short-term electric power load forecasting using factor analysis and long short-term memory for smart cities. Int. J. Circuit Theory Appl. **49**(6), 1678–1703 (2020)
18. Vobbilisetty, R., Di Troia, F., Low, R.M., Visaggio, C.A., Stamp, M.: Classic cryptanalysis using hidden Markov models. Cryptologia **41**(1), 1–28 (2017)
19. Steeneck, D., Eng-Larsson, F.: The Baum-Welch algorithm with limiting distribution constraints. Oper. Res. Lett. **46**(6), 563–567 (2018)
20. Hanif, M.K., Zimmermann, K.-H.: Accelerating Viterbi algorithm on graphics processing units. Computing **99**(11), 1105–1123 (2017). https://doi.org/10.1007/s00607-017-0557-6
21. Zhang, Q., Zhang, J.: Short-term load forecasting method based on EWT and IDBSCAN. J. Electric. Eng. Technol. **15**(2), 635–644 (2020)
22. Smagulova, K., Krestinskaya, O., James, A.P.: A memristor-based long short term memory circuit. Analog Integr. Circ. Sig. Process. **95**(3), 467–472 (2018). https://doi.org/10.1007/s10470-018-1180-y

Design of Cyber Security Monitoring System for Smart Substation

Qiang Ma[1], Zhipeng Cong[2], Zhenyu Shi[2], Fen Chen[2], and Jianpo Li[3](✉)

[1] State Grid Shandong Electric Power Company, Jinan 250000, China
[2] State Grid Yantai Power Supply Company, Yantai 264000, China
[3] Northeast Electric Power University, Jilin 132012, China
lijianpo@neepu.edu.cn

Abstract. With the development of automation in smart substation, it is facing huge cyber security risks under the severe background. In order to improve the capability of data association between cross-layer operations, a cross-layer data association method for operation instructions in smart substation is proposed. The method can find out the correlated data messages corresponding to the operation behavior. It can be used to help user effectively recover the communication link of related operation instructions, and improve the early warning capability for abnormal behaviors and network attacks. Considering the large number of smart substations and their scattered locations, a cyber security monitoring system based on the data association method is designed. Through the distributed data acquisition and centralized monitoring framework, it can help users effectively manage multiple smart substations, improve management efficiency and ensure the network security of smart substations.

Keywords: Smart substation · Cyber security · Attack path analysis

1 Introduction

The industrial control system is the core of the production process control. It can obtain the status of the industrial devices in real time, and generate control signals according to the existing strategy, so as to maintain the closed-loop control of the production process [1]. Aiming at the application scenarios of smart substations, its industrial control system is mainly designed in accordance with the International Electro technical Commission (IEC) 61850 standard and adjusted based on the hierarchical structure model [2]. From top to bottom, it can be divided into station layer, bay layer and process layer, which is similar to the field layer in the hierarchical structure model. Smart substations mainly use the three communication protocols including Generic Object Oriented Substation Event (GOOSE), Sampled Value (SV) and Manufacturing Message Specification (MMS) specified in the IEC 61850 standard for data communication.

In order to solve the problem of cross-layer attack behavior analysis and attack path recovery in smart substations, it proposes the cross-layer data association method for operation instructions based on the deep packet inspection technology and the communication protocol in IEC 61850 standard [3]. The cross-layer data association method,

S.-C. Chu et al. (Eds.): ICGEC 2021, LNEE 833, pp. 125–136, 2022.
https://doi.org/10.1007/978-981-16-8430-2_12

which correlates the data messages of the same attack behavior at different layers, identifies the attack origin and target, the device nodes accessed during the process [4]. This method can enhance the awareness capability of user, and find out weak points of the protection system and high-value targets in the substations, which can guide user to protect intelligence substation better. It also carries out a systematic implement based on the proposed method so that the system can automatically complete the whole process functions such as data collection, data processing, security analysis and visual presentation.

The remainder of this article is organized as follows. The IEC 61850 standard including its communication protocol and the Substation Configuration Description (SCD) file structure analysis is introduced in Sect. 2. In Sect. 3, it will introduce the technical analysis and implementation of the cross-layer data association method for operation instructions. In Sect. 4, the system framework and its deployment will be presented. Finally, the conclusion is given.

2 Related Work

2.1 IEC 61850 Standard

Because the equipment of the traditional Supervisory Control And Data Acquisition (SCADA) system running on the Modbus or Distributed Network Protocol 3 (DNP3) protocol is not enough to support the new generation of intelligent electronic equipment in substations, even with the Internet Protocol (IP) protocol conversion technology. It lacks deployment flexibility and ultimately relies on the serial communication on the Remote Terminal Unit (RTU). In order to unify the communication of substations, the Technical Committee 57 (TC57) of the International Electrotechnical Commission formulate the IEC 61850 standard [5].

IEC 61850 standard defines the communication protocol specification for the Intelligent Electronic Device (IED) in the substation [6]. The protocol is based on Transmission Control Protocol/Internet Protocol (TCP/IP) and can run over the substation LAN at high speed. The abstract data model defined in the IEC 61850 standard can be mapped to MMS, GOOSE and SV protocols.

2.2 Communication Protocol Analysis

Industrial control systems use specific industrial communication protocols for data exchange [7]. The deep packet inspection technology analyzes the content of captured data message according to the corresponding format specification of the protocol, and obtains information related to the communication process and the control process [8].

MMS Protocol. MMS protocol is a messaging mechanism used to model real devices and functions, exchange information, process data and control information between network devices or applications in real-time conditions [9].

As shown in Table 1, MMS protocol is designed based on the Open System Interconnection (OSI) model and uses International Organization for Standardization (ISO) basic protocol for layer information encapsulation [10].

On the network layer, MMS protocol uses TCP 102 port. On the transport layer, MMS protocol uses Transport Service ontop of the TCP (TPKT) and Connection-Oriented Transport Protocol (COTP) protocols for encapsulation. On the session layer, MMS protocol initiates an Association Control Service Element (ASCE) service request and response after establishing a COTP session.

Table 1. OSI model of MMS protocol based on TCP/IP

Layer	PDU	Protocols
Application (L7)	APDU	MMS: ISO 9506
		ACSE: ISO/IEC 8650/X.227
Presentation (L6)	PPDU	OSI Connection Oriented Presentation ISO 8823/X.226
Session (L5)	SPDU	OSI Connection Oriented Session: ISO 8327/X.225
Transport (L4)	TPDU	COTP: ISO/IEC 8073/X.224
		TPKT: RFC 1006
		TCP: RFC 793
Network (L3)	NPDU	IP: RFC 791
Data Link (L2)	Date Frame	Ethernet: ISO/IEC 8802–3
Physical (L1)	Bits	

On the presentation layer, MMS protocol uses the presentation service defined by the ISO 8822/X.216 standard.

On the application layer, MMS protocol implements two types of communication according to the ISO 9506 standard.

Confirmed MMS service defines 87 kinds of services, such as getNameList, read, write, getVariableAccessAttributes. The service is requested through the Confirmed-Request Protocol Data Unit (PDU), and confirmed by the Confirmed-Response PDU or Confirmed-Error PDU.

Unconfirmed MMS service defines three different unconfirmed services including informationReport, unsolicitedStatus and eventNotification. It will not receive Confirmed-Response PDU or Confirmed-Error PDU for confirmation.

GOOSE Protocol. GOOSE protocol is used for the transmission of time-sensitive events. The IEC 61850 standard defines two sets of communication services: a client/server model and a peer-to-peer model. Compared with MMS protocol, GOOSE protocol does not fully implement all the functions of the OSI model. It only implements the functions of the physical layer, the data link layer and the application layer, so it is simpler than MMS protocol.

On the data link layer, GOOSE messages are encapsulated by 802.3 Ethernet data frames. GOOSE protocol defines four fixed-length fields on this layer, which are the Application Identifier (APPID) (2 bytes), Length (2 bytes), and two 2-byte reserved fields. GOOSE protocol content is encapsulated in ApplicationProtocolDataUnit (APDU). The maximum length of APDU is 1496 bytes.

On the application layer, APDU field is the payload of GOOSE protocol, which is composed of GoCBRef, TimeAllowedtoLive, DatSet, GoID, T, StNum, SqNum, Simulation, ConfRev, NdsCom, NumDatSetEntries, and allData field.The T field is a timestamp, indicating the event time when the StNum field has changed. The StNum field is a counter, which will increase every time a GOOSE message is sent and the value change is detected in the DataSet specified by the DatSet field. The initial value of StNum is 1. The SqNum field represents the sequence number of the current GOOSE protocol message. Whenever a GOOSE message is sent, its value is increased by 1. After the StNum changes, SqNum will be set to 0.

2.3 SCD File Analysis

The description language of the SCD file is the Structured Control Language (SCL), and it is based on the Extensible Markup Language (XML) format. The nodes of the SCD file are organized in a tree-like hierarchy. As shown in Fig. 1, the complete SCD file consists of five parts: Header, Substation, Communication, IED, and DataTypeTemplates.

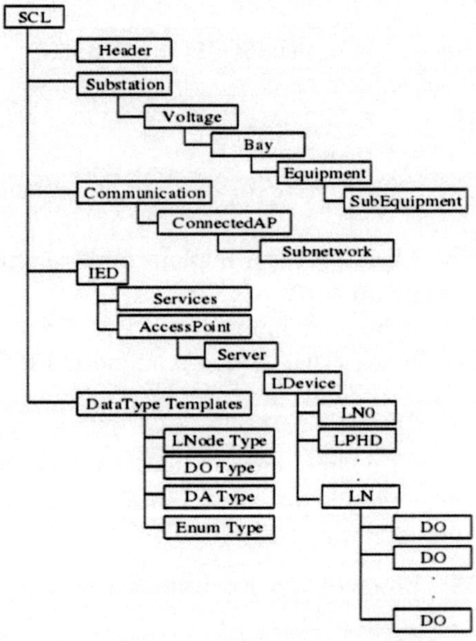

Fig. 1. Structure of SCD file

Based on the structural characteristics of SCD files, the problem of parsing SCD files can be transformed into a problem of parsing XML-based files. XML document is a kind of semi-structured data. For the parsing of XML files, there are currently four mature and mainstream methods. Among them, the Document Object Model (DOM) method has been widely adopted because of its characteristics of supporting two-way

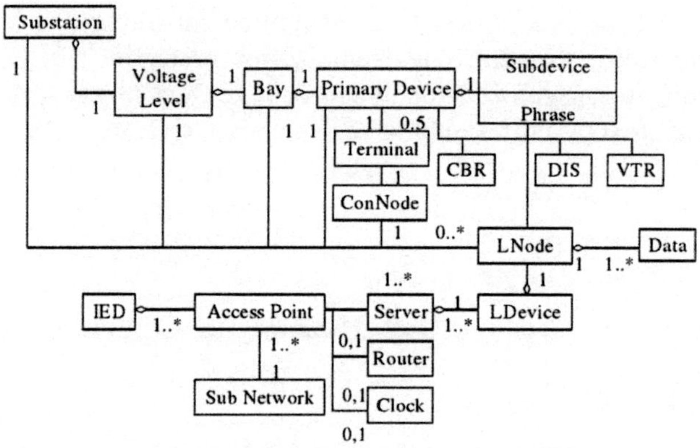

Fig. 2. DOM example of smart substation

access and navigating in the structure tree at any time. It implements the analysis of SCD files based on the DOM method, which is shown in Fig. 2.

The SCD file can be read in the form of a DOM tree by searching for nodes in a traversal method of the document. Any operations on nodes include adding, deleting and replacing. All of the above operations need to rebuild the DOM tree. Based on the information obtained in the above process, building an SCD document XML content index or visualizing the associated structure can show the configuration structure of the substation more clearly. Take the IED field in the SCD file as an example. According to its function, the information that can be obtained by analyzing the SCD file is described in detail. The content structure of the IED field is shown in Fig. 3. The specific content of each smart electronic device, SV, the publication and subscription of GOOSE are described in this tag.

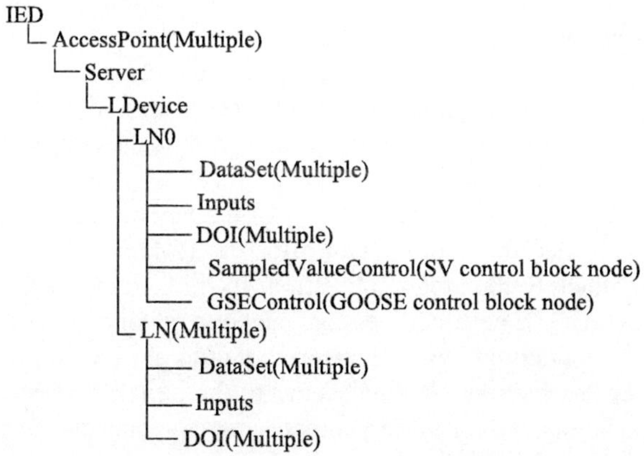

Fig. 3. Context structure of IED field

The subscription of SV and GOOSE is defined under the input node. The input node is composed of multiple external reference entries ExtRef. Each external reference entry specifies the address of the internal input virtual terminal intAddr and the external output virtual terminal address (consisting of iedName, ldInst, prefix, lnClass, lnInst, doName, daName). According to this information, it can find the Data Object (DO) and even the specific Data Attribute (DA) corresponding to the internal and external virtual terminals.

Similarly, it can parse the content of Header, Substation, Communication, DataType-Templates field in the SCD file to achieve a structured representation of the internal configuration of the smart substation.

3 Cross-Layer Data Association Method for Operation Instruction

3.1 Technical Analysis

In the smart substation scenario, the system is divided into three layers and two internal networks. The operation instructions initiated from the station layer to the process layer will form multiple data messages. The existing security audit methods mainly aim at analyzing the content of a single message. Therefore, a cross-layer data association method for operating instruction is proposed. Based on the internal configuration information of the smart substation obtained from the analysis of the SCD file, combined with the design characteristics of the communication protocol between each layer, multiple data messages corresponding to the same operating instruction are carried out. It achieves command-level operation behavior audit, which can effectively improve the capability of detecting abnormal operation behavior in smart substations.

Since the station layer is the outermost layer of the smart substation network, it is usually the most vulnerable to hacker attacks. In a typical attack scenario, the attacker first obtains the operating authority of the station layer device, such as monitoring host or Human Machine Interface (HMI) device. Through these devices, the attacker can send malicious attack instructions to the process layer devices, such as forcibly turning off a certain switch, causing physical damage to the substation. Since the operation of the smart substation mainly involves three different communication protocols, the attack forms different data messages between different internal networks of the smart substation. The entire attack path poses a huge challenge to the follow-up security audit. In this data tampering scenario, the attack behavior first occurs in MMS protocol message packet, and then based on the substation operation settings, the corresponding GOOSE or SV protocol message packet is generated.

If the attacker obtains the operating authority of the bay or process layer equipment through penetration means, the attacker can also use a Man-In-The-Middle (MITM) attack to launch an attack to the smart substation. In this attack scenario, the attacker may tamper with the GOOSE protocol message packet content to generate wrong status data. After the data is transmitted back through the internal network of the substation, it may lead to wrong instructions. In this scenario, the attack first occurs in GOOSE protocol message packet, and then MMS protocol message packets generated.

In the above two typical scenarios, due to the broadcast communication mechanism of the IEC 61850 standard communication process, multiple different devices may be involved. The key to data association is how to determine which data messages between different protocols are related. In MMS protocol message packet, there is a dataset field, which has a one-to-one correspondence with the APPID field in the GOOSE and SV protocols. For protocols such as GOOSE and SV that communicate through the broadcast mechanism, the information of publishing and subscription is recorded in the SCD file, which can also be used for data association.

Based on the above analysis, through the methods of information comparison and information mining, the goal of data association between MMS and GOOSE or SV protocols can be realized.

3.2 Algorithm Implement

Aiming at the above two types of typical attack scenarios, two different data association algorithms are implemented to meet the characteristics of different application scenarios.

Algorithm for Data Tampering Scenario. Obtain MMS protocol message packet sent by device A at the station layer to device B at the bay layer. Parse the dataset field content corresponding to device B from MMS protocol message packet through the deep packet inspection method. At the same time, obtain the GOOSE or SV protocol data message broadcast by device B, and use the deep packet inspection method to parse the APPID field from the message packet. According to the analysis, there is a one-to-one correspondence between APPID and dataset name. Through information comparison method, GOOSE or SV protocol message packets which are associated with device A can be filtered out. Since the GOOSE and SV protocols are broadcast communications, in order to further obtain the process-level related device information, by parsing the SCD file and analyzing the content of the Input tag, we can obtain the device C at the process layer which have subscribed device B. The number of device C may be larger than 1. The above-mentioned device A of station layer, device B of bay layer, and device C of process layer can form a cross-layer data association of operation instruction, and the corresponding MMS, GOOSE and SV protocol message packets are associated in the form of tags.

Algorithm for MITM Attack. Obtain GOOSE protocol message packet sent by device B at the bay layer to device C on the process layer. Each GOOSE protocol message packet has a corresponding APPID and StNum field, where the APPID field identifies the subscription to the traffic data packet. The StNum field is used to indicate whether the broadcast data has changed. If the data does not change, the StNum field remains unchanged. If the data changes, the StNum field is increased by 1.

4 Substation Cyber Security Audit System

In order to achieve the automation process, with the cross-layer data association method of operation instruction as the core, it designs a cyber security monitoring system for smart substation. At the same time, in order to achieve centralized monitoring of multiple smart substations, it split the system into two parts, perception platform and an integrated management system, based on a distributed technical concept.

4.1 System Framework

As shown in the Fig. 4, the system functions can be divided into four parts: data acquisition, data processing, security analysis, and visual presentation from the bottom to the top. Among them, the perception platform is distributed nodes responsible for data collection and data processing. The integrated management system is mainly responsible for summarizing the data uploaded by each perception platforms, executing security analysis and visualizing it.

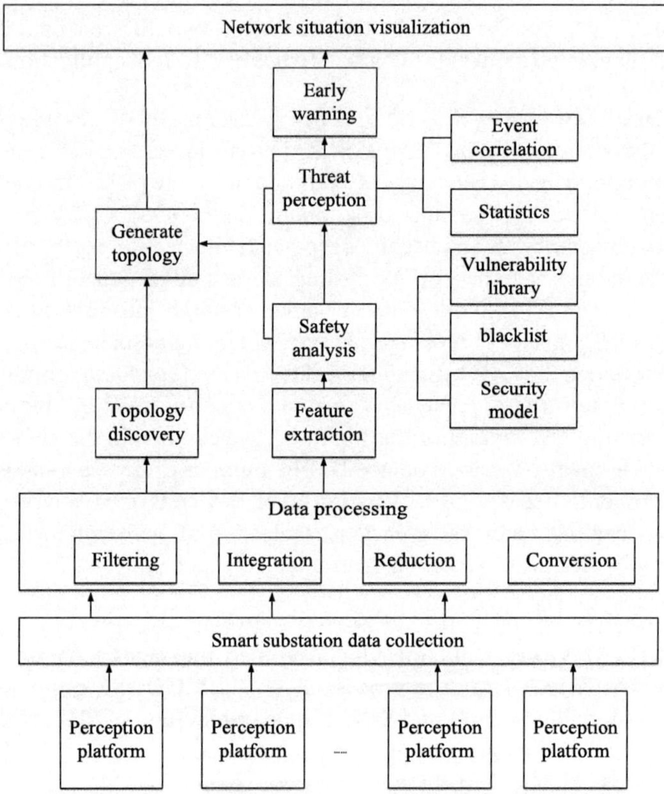

Fig. 4. System framework overlook

Data Acquisition. The perception platform is used as data collection probe, and it is connected to the smart substation network environment by bypass deployment, collects the traffic packets of the system network. There is no block, no delay, no modification to the information flow and data, and no affect to operation of substation. The perception platform is connected to the mirror port of the switch, and all the communication packet flow of the system is collected.

Data Processing. The packet flow of the substation system is obtained through the data acquisition module. Some of these data packets are generated by the communication protocol or the basic control functions of the devices. These data packets and do not contain useful control instructions payload. Therefore, these packets need to be filter out. The content quality of the data can be improved through data filtering, data integration, data reduction and data conversion.

Security Analysis. Based on the in-depth analysis of SCD files and protocol packet parse, a dedicated program module is designed to perform topology discovery and feature extraction on the processed data. Through topology discovery, the topology of the smart substation network can be obtained, information and location of network equipment and smart equipment can also be determined. Based on the obtained feature vectors, the analysis module uses streaming big data technology to compare and evaluate dimensions such as security baselines, vulnerability databases, whitelists, security behavior databases, and relevance, so as to determine whether there are security risks in the control network along with its degree of danger. The alarm management module gives corresponding alarms according to the judgment result produced by the analysis module.

Visual Presentation. This module is mainly to analyze and detect abnormal network behaviors detected and display the security situation implied to users through visual graphics. Statistical data such as network abnormal events, alarm events, protocols and data flows, and integrated information are gathered from the substation through the dispatching network to the integrated management system in a certain format. Real-time analysis and detection of abnormal network behaviors are realized through real-time updated visualized pages. The purpose of the security situation visualization system is to generate a comprehensive network security situation map that presents the network security status in a multi-view, multi-angle, and multi-scale manner. This method combines the advantages of computer and security personnel's processing capabilities in image processing, which can effectively reduce response time, improve system detection efficiency, and have the ability to predict security status.

4.2 System Deployment

The system deploys perception platforms in multiple smart substations, mainly through Virtual Local Area Network (VLAN) technology, relying on the existing physical network, to network the perception platforms located in different smart substations. The perception platform separately organizes the data collected in each smart substation control network, and summarizes the data in the integrated management system of the Provincial Information Center for data classification, statistics and information integration, and finally passes the data. The server performs visual presentation and releases in real time.

According to the network structure of the smart substation control system and the network structure between the plant station and the dispatching center, combined with the deployment conditions of the smart substation network security monitoring system, the network topology structure is designed and shown as Fig. 5.

The perception platform bypass is deployed in the control system network of the smart substation. It is mainly used to collect and process the flow data of the substation

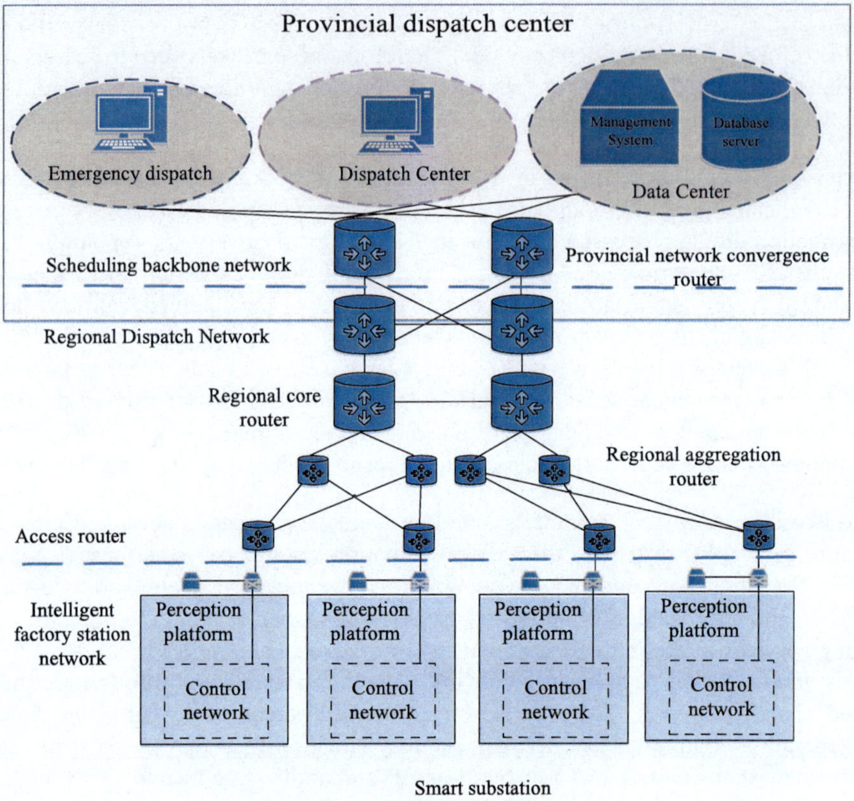

Fig. 5. System topology diagram

control network where it is located, and perform analysis and processing on the data such as topology analysis, feature extraction, statistical analysis, event correlation, threat judgment and assessment. For abnormal phenomena and potential network threats, the alarm module will trigger an alarm signal in time to remind dispatchers or security managers to respond or predict the development trend of the network situation. The processed data of each substation is collected through the power dispatch network to the integrated management system deployed by the information center, and presented in the form of visualized statistical graphs, topological graphs, data flow lines, and location graphs.

The network communication between the perception platform (plant site side) and the integrated management platform (information center side) is mainly planned to be achieved through the following two technologies.

Communication Based on API Interface. Between devices of the same manufacturer or devices with an API interface open, the functions in the device can be compiled into an API interface to provide an external call interface, such as obtaining and managing the operating status of the device, remotely collecting and issuing security policies, and obtaining security incidents of stations in a timely manner. The communication method based on the API interface has good information integrity and stability, and can quickly, effectively, safely, reliably and completely communicate with each other at the site.

Communication Based on Syslog Protocol. The integrated management system can receive log information sent from different manufacturers or devices that do not open API interfaces through the Syslog protocol. The server receiving the Syslog can store the messages of multiple devices in a unified manner, and parse the contents for corresponding processing. The Syslog log contains limited information: the program module that generated the log, severity level, time, host name or IP, process name, process Identity document (ID), and text.

However, due to the lack of a unified standard format in the Syslog protocol, the program may not be able to parse Syslog messages. At this time, it is difficult to achieve truly effective communication between devices of different manufacturers.

5 Conclusion

The designed smart substation cyber security monitoring system aims at solving the problem that the existing security audit method can only analyze single message packet. Through the packet parser of MMS, GOOSE, SV protocol and the in-depth analysis of the substation SCD file, according to the communication features, a cross-layer data association method for operation instruction is designed to achieve the goal of multiple data messages association corresponding to the same operation behavior. By this way, the capability of security audit is improved, and the users can more clearly understand the security situation of the substation. At the same time, the system is designed with a distributed data acquisition and centralized monitoring deployment method to help users grasp the network security status of multiple smart substations in real time, and improve the efficiency of network security operation, maintenance and management.

References

1. Badal, F.R., Das, P., Sarker, S.K., Das, S.K.: A survey on control issues in renewable energy integration and microgrid. Prot. Control Mod. Power Syst. **4**(1), 1–27 (2019). https://doi.org/10.1186/s41601-019-0122-8
2. Anand, S., Fernandes, B.G.: Reduced-order model and stability analysis of low-voltage dc microgrid. IEEE Trans. Industr. Electron. **60**(11), 5040–5049 (2012)
3. Che, L., Shahidehpour, M., Alabdulwahab, A.S., Al-Turki, Y.: Hierarchical coordination of a community microgrid with AC and DC microgrids. IEEE Trans. Smart Grid **6**(6), 3042–3051 (2015)
4. Lisovich, M., Wicker, S.: Privacy concerns in upcoming residential and commercial demand-response systems. IEEE Proc. Power Syst. **1**(1), 1–10 (2008)
5. Li, S., Xue, K., Yang, Q., Hong, P.: PPMA: Privacy-preserving multisubset data aggregation in smart grid. IEEE Trans. Industr. Inf. **14**(2), 462–471 (2018)
6. Xue, K.B., Zhu, Q., Yang, D., Wei, S., Guizani, M.: An efficient and robust data aggregation scheme without a trusted authority for smart grid. IEEE Internet Things J. **7**(3), 1949–1959 (2019)

7. Zheng, B., et al.: Scalable and privacy-preserving data sharing based on blockchain. J. Comput. Sci. Technol. **33**(3), 557–567 (2018)
8. Chen, M.F., Wang, J.H., Yang, Q.S., Xia, L.: An Improved topological hybrid passivity control of UPQC. J. Northeast. Electr. Power Univ. **39**(6), 78–84 (2019)
9. Wang, S.P.: Prediction and analysis of systemic network security model based on data mining. J. Northeast. Electr. Power Univ. **39**(6), 91–93 (2019)
10. Lin, C.Q., Tong, H., Yu, W.F., Xin, Y.C., Zhang, R.F.: Distribution grid intelligent state monitoring and fault handling platform based on ubiquitous power internet of things. J. Northeast. Electr. Power Univ. **39**(4), 1–4 (2019)

Bioinformatics and Biomedical Analysis

Multi-attention and Multi-layer Hashing for Cross-Modal Retrieval

Zhiyou Wang, Meijing Li(✉), and Tianjie Chen

College of Information Engineering, Shanghai Maritime University, Shanghai 201306, China
mjli@shmtu.edu.cn

Abstract. The research on cross-modal retrieval has broadened the access to various forms of data resources, among which cross-modal hashing methods have gained widespread attention owing to their excellent effects. However, existing hashing-based methods cannot to establish a deep inter-model correlation and fully employ the semantic information at the same time, and single-layer hashing may cause the hash representations not robust enough. To deal with above issues, we come up with a new method with multi-attention and multi-layer hashing to achieve the goal of mutual retrieval between different forms of data. Firstly, we apply the modal attention of multi-attention to capture the bit-level dependencies between cross-modal features to build a deeper inter-modal correlation. Meanwhile, the part of semantic attention is applied to retain common semantic information, which will improve the accuracy of retrieval tasks. Secondly, to learn a more robust hash representations, multi-layer hashing is used to jointly complete the learning task of hash representations and avoid the shortcomings of single-layer hashing. Through a series of experiments on two cross-modal datasets, it is indicated that the model we presented is better than the general approach in most evaluation metrics.

Keywords: Multi-modal · Cross-modal retrieval · Hashing method · Attention

1 Introduction

The advancement of internet-related technologies has brought about a huge amount of data resources in various forms [1–3], and the value of data has been given unprecedented attention. Information retrieval is an important way for people to obtain data, but traditional retrieval methods can only achieve mutual retrieval of data in specific forms. Based on this situation, people are becoming to turn their perspectives to the study of cross-modal data retrieval [4–6].

What is worthy of our attention, the hashing methods are extensively utilized in cross-modal retrieval tasks on account of the advantages of small storage space and fast retrieval speed. Cross-modal retrieval methods based on hashing techniques can be simply subdivided into two types. The shallow hashing methods rely on hand-crafted features, then, the final hash codes are obtained by projecting the features into the common Hamming space. However, such methods are difficult to retain the rich semantics in the hash codes. Methods such as CCA [7], CCQ [8], and SePH [9] separate feature

S.-C. Chu et al. (Eds.): ICGEC 2021, LNEE 833, pp. 139–149, 2022.
https://doi.org/10.1007/978-981-16-8430-2_13

extraction and hash function learning, failing to achieve well results in retrieval tasks. The deep hashing methods learn the representation of features through a deep neural network which could alleviate errors caused by hand-crafted, so it behaves better in cross-modal retrieval. For example, DCMH [10] proposed an end-to-end model that integrates the learning task of feature representations and hash representations into a framework which could strengthen the connection between features and hash codes. Towards SSAH [11] model, the idea of GANs [12] is introduced into the network structure to discriminate the features of the two modalities and explore the common semantic space better. AGAH [13] uses a multi-label semantic attention mechanism to strengthen the connection between features and semantic information. SPHMF [14] proposed a multi-scale pooling method to extract multi-level semantics, because enriching semantic information helps to increase the accuracy of the retrieval [15, 16]. Different from traditional hashing methods, the deep hashing methods behave better in terms of retrieval accuracy, but it still cannot fully explore the common representation space of cross-modal data.

The methods mentioned above, only start from start from one aspect of semantic retention or heterogeneity elimination, which cannot build deep-level inter-modal relevance and fully employ the semantic information of cross-modal data. In addition, only one hash layer is utilized for the learning task of hash representations in these models, which increases the learning difficulty of the hash layer. To overcome these shortcomings, we present a novel cross-modal retrieval method with multi-attention and multi-layer hashing (MAMH). First, to settle the questions of heterogeneity of cross-modal data and insufficient utilization of semantic information, a multi-attention strategy is applied to it. The modal attention is responsible for capturing capture the interdependence between modal features, and building deep-level inter-model relevance. Semantic attention is partly used to deepen the relationship between features and semantic information, and explore the common representation space better. Secondly, this paper uses multiple hash layers to learn hash functions together for obtaining more robust hash representations [28].

2 Proposed Method

Figure 1 illustrates the entire flow of MAMH model. The feature extraction section of MAMH is responsible for extracting 512-dimensional modal features, and hash learning part is responsible for mapping the features into hash codes. In the process of feature learning, through adversarial learning, we learn a more consistent distribution of feature representations, and the multi-attention mechanism is responsible for mining inter-model dependencies and semantic information in features to establish a more effective common representation space. Finally, we complete the cross-modal retrieval task by calculating and sorting the hamming distance of the hash codes output by the multi-layer hashing.

2.1 Problem Definition

The following describes the related symbols and formulas used in this article. Assuming, there are n image-text pairs now, $X = \{x_i\}_{i=1}^{n}$ is the input of the image modal, x_i is the i-th image sample. $Y = \{y_i\}_{i=1}^{n}$ is the input of the text modal, y_i is the i-th text sample.

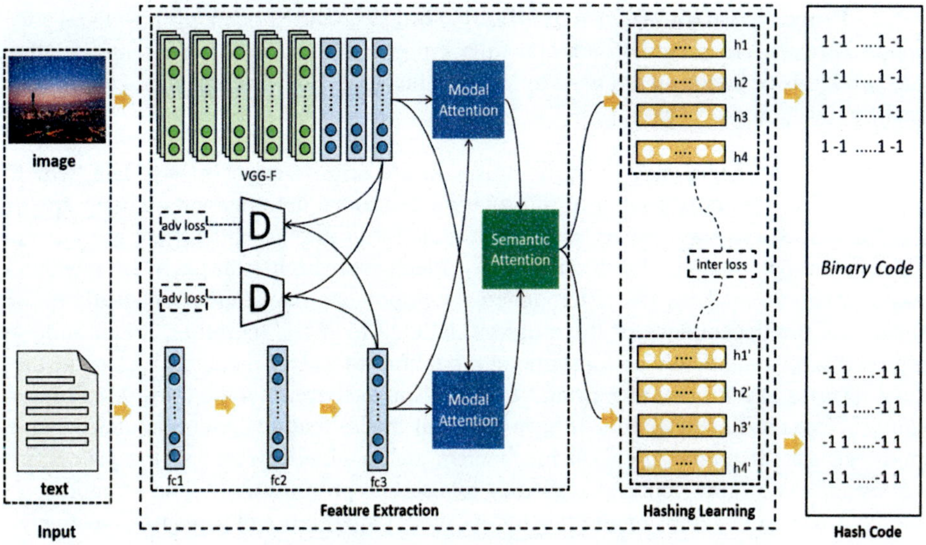

Fig. 1. The framework of the MAMH model

$\mathrm{L} = \{l_i\}_{i=1}^n$ represents a set of labels. Specifically, l_i is used to represent the label shared by the image-text pair of x_i and y_i, and l_i is a C-dimensional vector composed of 0 or 1. S_{ij} represents the similarity of an image-text pair. If $S_{ij} = 1$, the image-text pair has at least one common category, otherwise $S_{ij} = 0$.

2.2 Feature Extraction

Adversarial Learning. In the adversarial learning part, 4096-dimensional image features are obtained through the first seven layers in VGG-F [17, 18], then they are input to the eighth fully connected layer to obtain 512-dimensional features. For text modal, text features are extracted through three fully connected layers. We denote the 512-dimensional features as f^x and f^y respectively. To enhance the similarity of the image and text modal features, we use D_x and D_y two discriminators to distinguish the features. Towards D_x, the image features are judged as true, the text features are judged as false, and D_y is the opposite of D_x. The discriminant results of discriminant networks will be applied to guide the learning of the corresponding generative networks to improve the consistency between the two kinds of data in adversarial learning. The loss function of the adversarial learning component can be expressed by the following equation.

$$\begin{aligned} L_{adv} &= L_{adv}^x + L_{adv}^y \\ &= -\frac{1}{n}\Bigg[\sum_{i=1}^{n}\left(\log D_x\left(f_{\mathrm{i}}^x;\theta_{D_x}\right)\right) + \left(\log\left(1 - D_x\left(f_i^y;\theta_{D_x}\right)\right)\right) \\ &\quad + \sum_{i=1}^{n}\left(\log D_y\left(f_{\mathrm{i}}^y;\theta_{D_y}\right)\right) + \left(\log\left(1 - D_y\left(f_i^x;\theta_{D_y}\right)\right)\right)\Bigg] \end{aligned} \tag{1}$$

Feature Processing with Multi-Attention. In order to build a deeper inter-modal correlation between cross-modal data and fully employ the semantic information of the data, two kinds of attention are used to achieve this goal. First, we introduce the modal attention mechanism.

Modal Attention: There may be some noises in the original cross-modal data, which will both affect the transmission of information and cause heterogeneity among cross-modal data in some ways. In order to improve these problems, this article proposes a modal attention approach. The modal attention mechanism highlights the interdependent components and weakens the heterogeneous components between modal features by considering the dependence of the cross-modal data features, therefore, it can build a deep-level inter-model connections among modalities to a certain extent. We denote the modal features attention matrix as $M^o \in R^{D\times D}$ and D represents the dimension of the features. Note that images and text have different modal features attention matrix. The output features of modal attention mechanism are obtained by adding the processed features and the original features according to a certain proportion. $f^{x'}$ and $f^{y'}$ represent the output of two different modalities by the modal attention mechanism respectively. The expression is as follows:

$$\mathrm{f}^{\mathrm{k}}_{\mathrm{m-mask}} = \frac{1}{\mathrm{p}_1}\sum_{d=1}^{D} sigmoid\left(f^k M_d^{oT}\right) \times M_d^o, \mathrm{k} \in \{\mathrm{x,y}\} \tag{2}$$

$$f^{k'} = \lambda_1 f^k + (1-\lambda_1)\mathrm{f}^{\mathrm{k}}_{\mathrm{m-mask}}, \mathrm{k} \in \{\mathrm{x,y}\} \tag{3}$$

where λ_1 is a hyperparameter, p_1 can be obtained by this formula:

$$\mathrm{p}_1 = \sum_{d=1}^{D} sigmoid\left(f^k M_d^{oT}\right) \tag{4}$$

Semantic Attention: The semantic attention mechanism is applied to retain the rich semantic information in the feature and improve the retrieval accuracy. For the semantic attention mechanism, image and text data share a semantic attention matrix $M^s \in R^{C\times D}$. Each row of the matrix represents the weight of the features for a certain semantic in the label, and the attention weight is assigned to the features for highlighting the semantic information of the multi-label. After that, we can obtain features that not only consider inter-model dependencies, but also fully preserve semantic information. The expression is as follows:

$$\mathrm{f}^{\mathrm{k}}_{\mathrm{s-mask}} = \frac{1}{\mathrm{p}_2}\sum_{\mathrm{c}=1}^{\mathrm{C}} sigmoid\left(f^{k'} M_c^{sT}\right) \times M_c^s, \mathrm{k} \in \{\mathrm{x,y}\} \tag{5}$$

where p_2 can be obtained by this formula:

$$\mathrm{p}_2 = \sum_{\mathrm{c}=1}^{C} sigmoid\left(f^{k'} M_{\mathrm{c}}^{sT}\right) \tag{6}$$

The two attention-processed features are added according to a certain ratio to obtain the final feature result, in which λ_2 is a hyperparameter.

$$f^{k''} = \lambda_2 f^{k'} + (1-\lambda_2)\mathrm{f}^{\mathrm{k}}_{\mathrm{s-mask}},\ \mathrm{k} \in \{\mathrm{x,y}\} \tag{7}$$

where $f^{x''}$ and $f^{y''}$ represent the image and text features output by the semantic attention mechanism respectively.

We use the following formula to update M^o and M^s:

$$\begin{aligned} M^s &= \frac{1}{2l_s}\left(L^T F^x + L^T F^y\right) \\ M^{o_x} &= \frac{1}{f_{ys}}\left(F^{yT} F^x\right),\quad M^{o_y} = \frac{1}{f_{xs}}\left(F^{xT} F^y\right) \end{aligned} \tag{8}$$

where F^x means the set of image features, correspondingly, F^y denotes the set of text features. M^{o_x} and M^{o_y} represent the modal features attention matrix of two different modalities respectively. l_s represents the summation vector of labels vectors, f_{*_s} represents the summation vector of features vectors.

2.3 Hash Learning

Multi-layer Hashing. To ameliorate the issue of insufficient learning ability of the single hash layer and learn a more comprehensive and robust hash representations, this paper uses four hash layers to jointly complete the learning task of hash representations. The hash codes are obtained by weighting and summing the outputs of the four layers. It can be expressed by the following formula, where h_i^k is the output of one hash layer and a_i is the corresponding weight parameter.

$$h^k = a_1 h_1^k + a_2 h_2^k + a_3 h_3^k + a_4 h_4^k,\quad \mathrm{k} \in \{\mathrm{x,\ y}\} \tag{9}$$

Loss Function. To alleviate the problem of heterogeneity between hash codes of different modalities, while we want to retain as much useful information in the hash codes as possible, some of following loss functions are applied to our method.

Cosine Triplet Loss. In general, we can adopt the Hamming distance as a metric to optimize the common representation space for cross-modal data, however, the AGAH [13] pointed out that the measurement effect of the hamming distance is affected by the length of the hash codes. To overcome this drawback, the cosine distance can be used as an alternative. The cosine triplet loss can be expressed by the following formula:

$$L_{tri} = \sum_{i,j,k} \max\left(\cos\left(v_i, t_k^-\right) - \cos\left(v_i, t_j^+\right) + m,\ 0\right) + \sum_{i,j,k} \max\left(\cos\left(t_i, v_k^-\right) - \cos\left(t_i, v_j^+\right) + m,\ 0\right) \tag{10}$$

Specifically, v_i represents an image hash code, t_i represents a text hash code, and m represents a hyperparameter. The " +" and "-" indicate similar and not similar.

Cosine Quantization Loss. Under the constraints of cosine triplet loss, the angle between the hash codes and the positive or negative examples can be optimized, but the actual positional relationships of the vectors have not been considered. Taking (v, t^+, t^-) triples as an example, in fact, we hope that the absolute value of the vectors v, t^+ are closer to 1, and the vector v^- is best located in another quadrant. In response to this problem, the following loss function is used:

$$L_{\cos} = \sum_i \frac{1}{1 + \exp(\cos(|v_i|), 1)} + \sum_i \frac{1}{1 + \exp(\cos(|t_i|), 1)} \tag{11}$$

Quantization Loss. The hash codes output through the hash layer are continuous values in the range of (−1,1) [19]. However, the binary hash codes which are ultimately used to retrieve only contain two discrete values, so the following formula is applied by us to bridge the difference between the discrete and continuous values, where B_i^* means the hash code with just two discrete values of −1 and 1.

$$L_{qua} = \sum_i \|v_i - B_i^x\|_F^2 + \|t_i - B_i^y\|_F^2 \tag{12}$$

Asymmetric Discrete Loss. For the propose of building a deeper association between hash code and semantics [13], learning from the single-modal asymmetric hash methods [27], a code map $Q = \{q_i\}_{i=1}^C, q_i \in \{-1, 1\}^K$ is used to process the binary hash codes to the same dimension as the labels. Plentiful semantic information will be retained by reducing the gap between the mapping results and the labels.

$$L_{asy} = \sum_{i=1}^n \left(B_i^x Q^T - K\hat{l}_i\right)^2 + \sum_{i=1}^n \left(B_i^y Q^T - K\hat{l}_i\right)^2 + \sum_{i=1}^C \sum_{j=1}^C \left(q_i q_j^T - KI_{ij}\right) \tag{13}$$

where $\hat{l}_i = (l_i \times 2 - 1) \in \{-1, 1\}^C$, K denotes the dimension of our predetermined hash codes. Towards I, if $i = j$, then $I_{ij} = 1$, otherwise it is −1.

By assembling the mentioned formulas (1), (10), (11), (12), (13), we could obtain the complete loss function, it can be expressed by the following formula:

$$\min_{\theta_{G_x}, \theta_{G_y}} L_1 = L_{tri} + \alpha L_{\cos} + \beta L_{qua} + \gamma L_{asy} - L_{adv} \tag{14}$$

$$\min_{\theta_{D_x}, \theta_{D_y}} L_2 = L_{adv} \tag{15}$$

where α, β, γ are hyperparameters.θ_{G_x} and θ_{G_y} represent the parameters of the two generation networks. Correspondingly, the parameters of the two discriminative networks are denoted by θ_{D_x} and θ_{D_y} respectively.

3 Experiment

3.1 Datasets

We conduct experiments on two frequently employed cross-modal datasets, including MIRFlickr-25K [20] and NUS-WIDE [21]. The MIRFlickr-25K dataset includes a total of 25,000 sample pairs, from which we use 20,015 pairs in the experiment. The train set is 10,000 and the query set is 2000.The NUS-WIDE dataset contains a total of 269,648 sample pairs, from which 195,834 sample pairs were used in experiment. We randomly selected 10,000 pairs as the training set and 2100 sample pairs as the test set.

3.2 Experiment Details

The MAMH model contains a total of six hyperparameters: $\alpha, \beta, \gamma, \lambda_1, \lambda_2, m$. When their values are 1, 1, 0.001, 0.6, 0.8, 0.4 respectively, the model achieves the highest MAP. We obtain the related experiment data of the AGAH model by running the public code of the AGAH and others are all quoted from the paper [13]. Cross-modal retrieval tasks often use the MAP and PR curve to evaluate the performance [22], therefore, this article uses the same indicators for evaluation.

3.3 Experiment Result

Comparison with Existing Methods. In the comparative experiment stage, we selected eight sets of methods as references to verify and analyze the performance of different methods. In Tables 1 and 2, the first four methods in the control group are retrieval methods based on deep learning techniques, while the last four methods are shallow cross-modal hashing retrieval methods.

By analyzing the experimental results in Table 1 and Table 2, we can discover that the MAP values of the shallow methods are lower than that of the deep hashing methods. Unlike traditional methods, deep hash methods have strong learning ability, which could use deep neural network to learn better feature representation and fully fit the data distribution of different modalities, so deep hashing methods perform better. In most retrieval tasks, MAMH has the highest MAP value. This is duo to MAMH establishes a deeper inter-modal correlation, which alleviates the heterogeneous problem of cross-modal data to a certain extent and reduces the difficulty of retrieval tasks. In addition, MAMH fully employ semantic information to assist retrieval, and learns better hash representations through multiple hash layers, therefore, it improves the discrimination and robustness of hash codes, and finally show a better performance in the cross-modal retrieval tasks.

Table 1. Comparison experiment results (MAP) of MIRFlickr-25K.

Method	I - > T			T- > I		
	16-bit	32-bit	64-bit	16-bit	32-bit	64-bit
MAMH	**0.7821**	**0.7977**	**0.8085**	**0.7861**	**0.7948**	0.8010
AGAH [13]	0.7713	0.7865	0.8069	0.7789	0.7921	**0.8049**
SSAH [11]	0.7709	0.7840	0.7924	0.7681	0.7737	0.7838
PRDH [27]	0.7221	0.7395	0.7550	0.7548	0.7638	0.7772
DCMH [10]	0.7201	0.7287	0.7418	0.7533	0.7593	0.7730
SCM [26]	0.6558	0.6698	0.6732	0.6653	0.6799	0.6862
CMSSH [25]	0.5847	0.5842	0.5724	0.5671	0.5688	0.5612
STMH [24]	0.5657	0.5800	0.5984	0.5600	0.5773	0.5965
CVH [23]	0.5656	0.5623	0.5611	0.5653	0.5623	0.5613

Table 2. Comparison experiment results (MAP) of NUS-WIDE

Method	I- > T			T- > I		
	16-bit	32-bit	64-bit	16-bit	32-bit	64-bit
MAMH	**0.6230**	**0.6599**	**0.6593**	**0.6337**	**0.6477**	**0.6461**
AGAH [13]	0.6189	0.6590	0.6510	0.6226	0.6430	0.6400
SSAH [11]	0.6020	0.6219	0.6463	0.6123	0.6369	0.6398
PRDH [27]	0.5929	0.6333	0.6243	0.5939	0.6098	0.6006
DCMH [10]	0.5657	0.6007	0.6001	0.5336	0.5871	0.5916
SCM [26]	0.5393	0.5528	0.5577	0.4747	0.4813	0.4895
CMSSH [25]	0.5114	0.5060	0.4932	0.4489	0.3890	0.3804
STMH [24]	0.5223	0.5288	0.5371	0.4963	0.5292	0.5321
CVH [23]	0.3831	0.3746	0.3691	0.3662	0.3598	0.3564

By analyzing the relative positions of the curves of each model in Fig. 2 and Fig. 3, we can find a regularity that the PR curve of the MAMH model is higher than other deep hashing methods. Such experimental results once again indicate that the MAMH model has learned a more effective common representation, so it improves the accuracy of retrieval and is better than general deep cross-modal hashing methods.

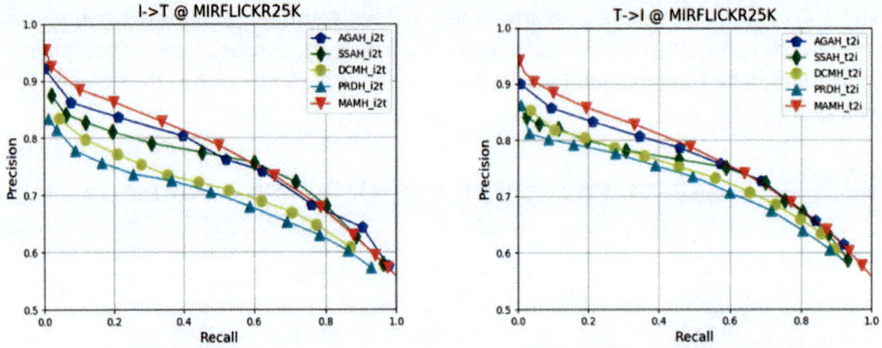

(a) image query text (b) text query image

Fig. 2. Precision-Recall curves on MIRFlickr-25K. Binary codes are 16bits.

Ablation Experiment. To confirm the authenticity of the two improvement strategies of MAMH model, we set up a set of ablation experiments to verify that the multi-attention mechanism and multi-layer hashing method can help to improve the performance of the model. Among them, MAMH-1 means removing multi-layer hashing from the model, MAMH-2 means that removes modal attention from the model, and MAMH means the complete model.

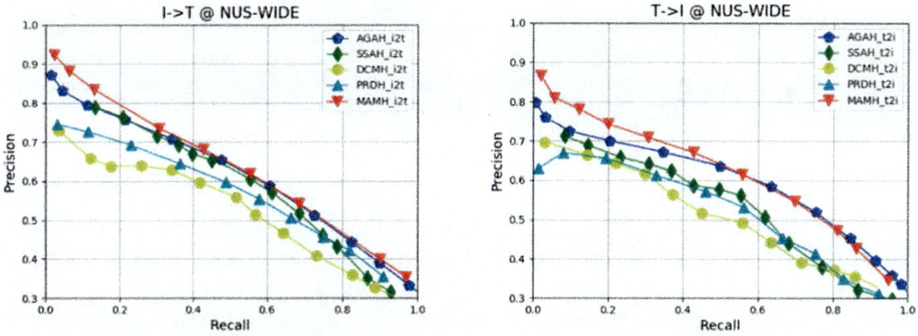

(a) image query text (b) text query image

Fig. 3. Precision- Recall curves on NUS-WIDE. Binary codes are 16bits.

As shown in Table 3, we can clearly see that when the modal attention or multi -layer hashing is missing, the MAP value is lower than the complete model. This phenomenon illustrates that the multi-attention mechanism used in this article improves the problem of insufficient connection between the modules, and the multi hash layer has learned a better hash representation, therefore, the two improvement strategies are real and effective.

Table 3. Mean average precision (MAP) of Ablation Experiment.

MAP of MIRFlickr25k with 16-bit hash codes		
Method	I->T	T->I
MAMH	0.7821	0.7861
MAMH-1	0.7800	0.7817
MAMH-2	0.7793	0.7805

4 Conclusion

In this work, a method with multi-attention and multi-layer hashing is presented for cross-modal retrieval. Starting from the trait of cross-modal data "underlying heterogeneity and semantic association", it performs better than general models in cross-modal retrieval tasks. Such a manifestation indicates that it is necessary to build deep-level inter-modal connections and fully semantic information in cross-modal retrieval. Therefore, in future research about cross-modal retrieval, the mining of deep correlations and dependencies among cross-modal data, and extracting valid information implied in the data to are still worthy of attention.

Acknowledgement. This study was supported by the National Natural Science Foundation of China (61911540482 and 61702324).

References

1. Hu, Y., Zheng, L., Yang, Y., Huang, Y.: Twitter100k: a real-world dataset for weakly supervised cross-media retrieval. IEEE Trans. Multimedia **20**(4), 927–938 (2017)
2. Huang, X., Peng, Y.: TPCKT: two-level progressive cross-media knowledge transfer. IEEE Trans. Multimedia **21**(11), 2850–2862 (2019)
3. Peng, Y., Qi, J., Huang, X., Yuan, Y.: CCL: cross-modal correlation learning with multi-grained fusion by hierarchical network. IEEE Trans. Multimedia **20**(2), 405–420 (2017)
4. Li, C., Liu, Z., Li, S., Lin, Z., Tian, L.: Variable length deep cross-modal hashing based on Cauchy probability function. Wireless Netw. **2**, 1–11 (2020)
5. Liu, X., Cheung, Y.M., Hu, Z., He, Y., Zhong, B.: Adversarial tri-fusion hashing network for imbalanced cross-modal retrieval. IEEE Trans. Emerg. Top. Comput. Intell. **5**(4), 607–619 (2021)
6. Zhu, L., Tian, G., Wang, B., Wang, W., Zhang, D., Li, C.: Multi-attention based semantic deep hashing for cross-modal retrieval. Appl. Intell. **51**(8), 5927–5939 (2021)
7. Hotelling, H.: Relations between two sets of variates. In: Kotz, S., Johnson, N.L. (eds.) Breakthroughs in Statistics. Springer Series in Statistics (Perspectives in Statistics). Springer, New York (1992). https://doi.org/10.1007/978-1-4612-4380-9_14
8. Lin, Y., Zheng, Z., Zhang, H., Gao, C., Yang, Y.: Bayesian query expansion for multi-camera person re-identification. Pattern Recogn. Lett. **130**, 284–292 (2020)
9. Han, Y., Fei, W., Jian, S., Qi, T., Zhuang, Y.: Graph-guided sparse reconstruction for region tagging. In: Computer Vision & Pattern Recognition, pp. 2981–2988. IEEE, Providence, RI, USA (2012)
10. Jiang, Q.Y., Li, W.J.: Deep cross-modal hashing. In: 2017 IEEE Conference on Computer Vision & Pattern Recognition, pp. 3270–3278. IEEE, Honolulu, HI, USA (2017)
11. Li, C., Deng, C., Li, N., Liu, W., Gao, X., Tao, D.: Self-supervised adversarial hashing networks for cross-modal retrieval. In: 2018 IEEE/CVF Conference on Computer Vision and Pattern Recognition, pp. 4242–4251. IEEE, Salt Lake City, UT, USA (2018)
12. Goodfellow, I.J., et al.: Generative adversarial networks. Adv. Neural. Inf. Process. Syst. **3**, 2672–2680 (2014)
13. Gu, W., Gu, X., Gu, J., Li, B., Xiong, Z., Wang, W.: Adversary Guided Asymmetric Hashing for Cross-Modal Retrieval. In: Proceedings of the 2019 on International Conference on Multimedia Retrieval, pp. 159–167. Association for Computing Machinery, Ottawa ON, Canada (2019).
14. Zhang, H., Pan, M.: Semantics-preserving hashing based on multi-scale fusion for cross-modal retrieval. Multimedia Tools Appl. **80**(11), 17299–17314 (2020). https://doi.org/10.1007/s11042-020-09869-4
15. Zhang, X., Lai, H., Feng, J.: Attention-aware deep adversarial hashing for cross-modal retrieval. In: Ferrari, V., Hebert, M., Sminchisescu, C., Weiss, Y. (eds.) ECCV 2018. LNCS, vol. 11219, pp. 614–629. Springer, Cham (2018). https://doi.org/10.1007/978-3-030-01267-0_36
16. He, J., Chen, S., Wang, Y., Qiao, Y.: Dual-supervised attention network for deep cross-modal hashing. Pattern Recogn. Lett. **128**, 333–339 (2019)
17. Deng, J.: ImageNet: A Large-Scale Hierarchical Image Database. In: 2009 IEEE Conference on Computer Vision and Pattern Recognition, pp. 248–255. IEEE, Miami, FL, USA (2009).
18. Chatfield, K., Simonyan, K., Vedaldi, A., Zisserman, A.: Return of the devil in the details: delving deep into convolutional nets. Comput. Sci. (2014)
19. Qiang, H., Wan, Y., Xiang, L., Meng, X.: Deep semantic similarity adversarial hashing for cross-modal retrieval. Neurocomputing **400**, 24–33 (2020)

20. Huiskes, M.J, Lew, M.S.: The MIR flickr retrieval evaluation. In: Proceedings of the 1st ACM International Conference on Multimedia Information Retrieval, pp. 39–43. Association for Computing Machinery, Vancouver, British Columbia, Canada (2008)
21. Chua, T.S., Tang, J., Hong, R., Li, H., Luo, Z.: NUS-WIDE: a real-world web image database from national university of Singapore. In: Proceedings of the ACM International Conference on Image and Video Retrieval, pp. 1–9. Association for Computing Machinery, Santorini, Fira, Greece (2009)
22. Xzab, D., Xw, A., Emb, C., Song, W.A.: Multi-label semantics preserving based deep cross-modal hashing. Signal Process. Image Commun. **93**, 116–131 (2021)
23. Kumar, R.S.: Learning hash functions for cross-view similarity search. In: Proceedings of the 23th International Joint Conference on Artificial Intelligence, pp. 1360–1365. AAAI Press, Barcelona, Catalonia, Spain (2012)
24. Wang, D., Gao, X., Wang, X., He, L.: Semantic topic multimodal hashing for cross-media retrieval. In: Proceedings of the 24th International Conference on Artificial Intelligence, pp. 3890–3896. AAAI Press, Buenos Aires, Argentina (2015)
25. Bronstein, M.M., Bronstein, A.M., Michel, F., Paragios, N.: Data fusion through cross-modality metric learning using similarity-sensitive hashing. In: 2010 IEEE Computer Society Conference on Computer Vision and Pattern Recognition, pp. 3594–3601. IEEE, San Francisco, CA, USA (2010)
26. Zhang D., Li W J.: Large-scale supervised multimodal hashing with semantic correlation maximization. In: Proceedings of the 28th AAAI Conference on Artificial Intelligence, pp. 2177–2183. AAAI Press, Québec City, Québec, Canada (2014)
27. Yang, E., Deng, C., Liu, W., Liu, X., Tao, D., Gao, X.: Pairwise relationship guided deep hashing for cross-modal retrieval. In: Proceedings of the 31th AAAI Conference on Artificial Intelligence, pp. 1618–1625. AAAI Press, San Francisco, California, USA (2017)
28. Lv, Y., Ng, W., Zeng, Z., Yeung, D.S., Chan, P.: Asymmetric cyclical hashing for large scale image retrieval. IEEE Trans. Multimedia **17**(8), 1225–1235 (2015)
29. Wang, X., Zou, X., Bakker, E.M., Wu, S.: Self-constraining and attention-based hashing network for bit-scalable cross-modal retrieval. Neurocomputing **400**, 255–271 (2020)

Check for updates

Quantitative Analysis Indicators and Risk Factor Assessment of HR-HPV

Ling Gong[1], Hua Xie[2], Ling Wang[3], Lu Zhang[2], Site Li[1], and Beibei Miao[1](✉)

[1] Department of Nursing, Beihua University, Jilin City, China
gongling@beihua.edu.cn
[2] Jilin Central General Hospital, Jilin City, China
[3] Department of Computer Science and Technology, School of Computer Science, Northeast Electric Power University, Jilin City, China
smile2867ling@neepu.edu.cn

Abstract. The purpose of this study is to explore the risk factors of HR-HPV (High-Risk Human PapillomaVirus) cervical infection, and to construct an evaluation model through quantification of indicators, so as to provide a basis for the identification of high-risk population, propose a reference for reducing and clearing HR-HPV infection of the cervix and determine a reasonable follow-up period. Computer science and technologies are widely used in the study to calculate and analyze clinical data. The outcome of the study can help to carry forward the prevention of HR-HPV infection, so as to achieve the secondary prevention of cervical cancer. Quantitative analysis indicators model can be used to identify risky people who are vulnerable to HR-HPV and improve the compliance of targeted population to participate in screening. And at the same time, the study can provide reference and basis in the decision of the interval between follow-up visits for HPV testing. Therefor the number of missing HPV screening among risky women is expected to decreased and psychological panic and unnecessary economic burden of patients caused by excessive testing could be avoided.

Keywords: HR-HPV · Risk factor · Quantitative analysis · Indicator

1 Introduction

Cervical cancer is the third leading cause of malignant tumor death in women whose cause can be determined as sustaining HR-HPV infection. But from the data in 2018, there are still about 311,365 deaths of cervical cancer occur every year worldwide. It is confirmed that cervical screening programs or cytology testing can prevent the occurrence of cervical cancer. Furthermore, the influences of HPV vaccination could emerge in decades [1]. The coverage of HPV vaccination and screening is not sufficient enough especially in less developed regions that leading to the incidence of cervical cancer accounts for 83% of the world. By contrast cervical cancer screening program has been organized by the Swedish National Health Service since the 1970s, which is free for most women of appropriate age, has reduced the incidence and mortality of cervical cancer by about 70 percent at that time. The high cost of screening and the lack

S.-C. Chu et al. (Eds.): ICGEC 2021, LNEE 833, pp. 150–158, 2022.
https://doi.org/10.1007/978-981-16-8430-2_14

of knowledge of cervical cancer prevention and control have affected the coverage of screening in developing countries, and thus the incidence of cervical cancer is at a high level.

HPV infection is mainly sexually transmitted and current researches show high incidence among sexual women [2]. But very few cases progress to persistent infection and cervical carcinoma eventually [3]. Spontaneous clearance of HPV is conducted by host's autoimmune function in a general way. Otherwise, persistence of HR-HPV continuously in cervical epithelial cells will proceed cervical intraepithelial neoplasia (CIN) and even malignancy in years. The main way to promote high-risk cervical HPV infection is to destroy the DNA of cells, which reduces the immune function and surveillance of cells. In gene study, oncogenes E6 and E7 expression with early gene E2 are considered to be causative to the virus transcription [4, 5]. Unfortunately, no effective methods by far are found to solve persistent HPV infection even HPV vaccine [6]. HR-HPV will continue to be a large threatening issue to global women's health.

2 Related Work

2.1 HPV and Cervical Cancer Screening

At present, human papillomavirus (HPV) is recognized as a small double-stranded DNA-virus that has more than 200 known subtypes, most of which are not very relevant to cancers [7]. Nevertheless, confirmed evidence shows that HR-HPV subtypes, such as HPV16, 18, 31, 33, 35, 39, 45, 51, 52, 56, 58, and 59, are responsible for cervical cancer and cervical intraepithelial neoplasia. Cervical infection of HPV 16 and 18 accounts for 70% of carcinoma of cervix globally [8]. HPV type 6 and 11 are low-risk types which mainly cause genital warts [9, 10].

Current screening methods include cytology, HR-HPV testing, and combined screening (cytology + HR-HPV testing). The new NCCN clinical Guidelines for Cervical Cancer Screening states that high risk HPV testing is an important part of cervical cancer screening and abnormal cytological management. At present, HPV detection method mainly adopts the second-generation hybridization capture technique (HC2). HC2 is one of the most advanced techniques for detecting HPV-DNA in the world, and is currently the only cervical cancer screening method approved by the US Food and Drug Administration (FDA) [11, 12]. Domestic and foreign studies have shown that the second-generation hybridization capture technology has very high sensitivity and specificity for the detection of HR-HPV, the sensitivity is about 95%, the specificity is about 89%, and the negative predictive value is about 99%, which can detect 13 types of HR-HPV at one time [8]. However time and space requirement is needed for participating in the screening, therefore self-sampling of HPV among hard-to-reach women is encouraged by advantages of convenience, privacy, cost-effectiveness to break barrier against participation in cervical cancer screening [13].

2.2 HPV Vaccination

HPV vaccination was initially approved to access to the market in 2006 aimed as an efficient prevention of HR-HPV16, 18 and low-risk HPV6,11 infections. United States of America activated National HPV Vaccination Program in the same year. Canada, Australia, Belgium and another five European countries followed closely behind USA. However, HPV Vaccination Program has not been introduced to national medical system worldwide especially Middle and Southern Africa regions where cervical cancer incidence is rather high because of probably economic and medical reasons [1].

Although nine-valent cervical cancer vaccine works against HPV subtypes 31,33,45,52,58 additionally including the four subtypes mentioned, precautioning cervical cancer (about 90%) and vulvar, oropharyngeal and other HPV-related cancers (about 50%), [10] these unvaccinated women are still exposed to HR-HPV infection and cervical cancer. On the other hand, study and practice on therapeutic vaccine of HPV is ongoing all around the world due to the poor and uneven coverage of prophylactic vaccination cross certain area and people's resistance to HPV prophylactic vaccination [9].

2.3 Risk Factors of HR-HPV Infection

HPV-DNA is more likely to be detected in female with high progesterone levels and cell-mediated immune function damage. The experimental results of Velema et al. show that women who have sex before the age of 16 are more likely to develop cervical intraepithelial neoplasia than those over the age of 20. The former accounts for more than half of the cases than the latter. The number among smokers in cervical cancer patients is significantly higher after the analysis of 14 pathologies from all over the world. Multiple sexual partners are verified as risk factors of HR-HPV. In addition, CD4 counts make a significant difference among HIV-positive women in obtaining HR-HPV infection [14]. Circumcision and the condoms using protect people from transmission of HPV [1]. Co-infection with virus, genetic and immunological factors are other potential factors that need to be verified.

HR-HPV infection reoccurs at a higher rate than 30%(average) in postmenopausal women who has accepted treatment of high squamous intraepithelial lesions (HSIL) for six months [15].

3 Our Study

The study is an interdisciplinary program in which researchers from university and clinical hospital are involved, supported by a scientific research platform. The convenience of collecting clinical data ensures the quantity of study samples. Clinical application and effect evaluation of the results will be carried out.

3.1 Purpose and Significances of the Study

With the establishment of the relationship between HR-HPV infection and cervical cancer, early-stage prevention of HR-HPV infection and effective control of cervical lesion development to cervical cancer are the main research contents of women's health care in the world. The study aims to build a high-risk type HPV risk factors assessment indicator system as a simple and reliable assessment tool to identify high-risk population of HR-HPV infection. Meanwhile the study makes an effort to excavate prediction indicators of cervical lesions. Follow-up period is expected to raise accurately and individually to reduce patients' burden and to realize more reasonable allocation of medical resources.

Currently the high cost of screening and lack of knowledge about prevention and control of cervical cancer affect the coverage of cervical cancer screening in developing countries, resulting in high levels of cervical cancer incidence. Under the background that HPV testing cannot be popularized [16], the results of this study are expected to provide a basis for relevant departments of the National Health and Family Planning Commission and clinicians to identify high-risk groups and standardize health education content, so as to reduce HR-HPV infection and prevent and control the occurrence of cervical cancer. It is preliminarily estimated that this study will save medical costs with remarkable social and economic benefits.

3.2 Content of the Study

Cervical cancer is regarded as one of preventable diseases. It develops slowly even if from precancer to cancer [17]. Though most of HPV infection is temporary, a small portion of the infected will present a state of persistent infection. Avoidance of persistent HR-HPV infection is a key point to maintain healthy cervical environment. HPV test is applied in many countries and regions to identify risky population in screening and to supply an evidence to help doctors to determine the following disease management such as therapy and follow-up period. Our study focuses on the prevention and management of HR-HPV infection and can be divided into three relational parts considering the consequence of disease development.

HR-HPV Risk Factor Assessment Indicators

Women may obtain HPV infection time after time throughout their whole life. Multiple subtypes are possible to invade simultaneously. Additionally, no obvious symptoms are shown along with the infection. Therefore, the risk factors of HR-HPV infections are difficult to estimate or evaluate by previous researches.

In this study, ontology knowledge base will be built based on the comprehensive personal medical records and other data of the HR-HPV positive patients in HPV detection. Big data mining and other unstructured data processing methods are applied to quantitatively analyze and evaluate the risk factors of different types of HR-HPV infection comprehensively. And machine learning method is used to carry out indicators quantification and weight analysis to build a complete evaluation model eventually.

The result of this part will provide a basis for the identification of HR-HPV infection risk groups and the evaluation of cervical cancer causes. We are expecting the improvement of high-risk groups compliance to participate in screening. Delayed diagnoses of precancer and invasive cancer can be declined. It makes more effective screening to achieve the secondary prevention of cervical cancer especially in the region that screening is not easily available. In the meanwhile, people are expected to realize the influence of the risk factors, change their risky behaviors and choose a healthy life style.

Follow-Up Period After HR-HPV Infection

Compared with other cancers, cervical cancer provides follow-up options and disease management opportunity for the characteristics of long-term cervical lesion development. Consensus on prevention of cervical cancer lie in precancer management and therapeutic methods [18].

Recommendations of follow-up period after HR-HPV infection suggested by different associations are varying. Up-dated information and individual infection cases are calling for considering discriminatively. In other words, patients having HR-HPV positive but normal cytology visit hospital in whether 6 or 12 months are basically depending on clinical experience of physician. Inevitably, inaccurate follow-up period and colposcope examination are preformed between times. As a consequence, frequent visiting and sampling, excessive economic and phycological burden are imposed upon patients. Or unexpected development of the disease occurs leading to the delayed diagnosis.

In the second part of the study, study group concentrates on persistent single HR-HPV infected patients untreated samples to estimate clearance rate and time of different types of HR-HPV. Analyze complex infection data then calculate reasonable follow-up of period according to the result of cytology, HPV subtypes infection and quantitative indicator model. Reified and individual follow-up period is raised to ensure the rational use of medical resources.

Shunt and Prediction

Patients with HR-HPV positive and ASC (atypical squamous cells) including ASCUS (atypical squamous cells of undetermined significance) and ASC-H (atypical squamous cells-cannot exclude HIS) or LSIL (Low-grade squamous intraepithelial lesion) are

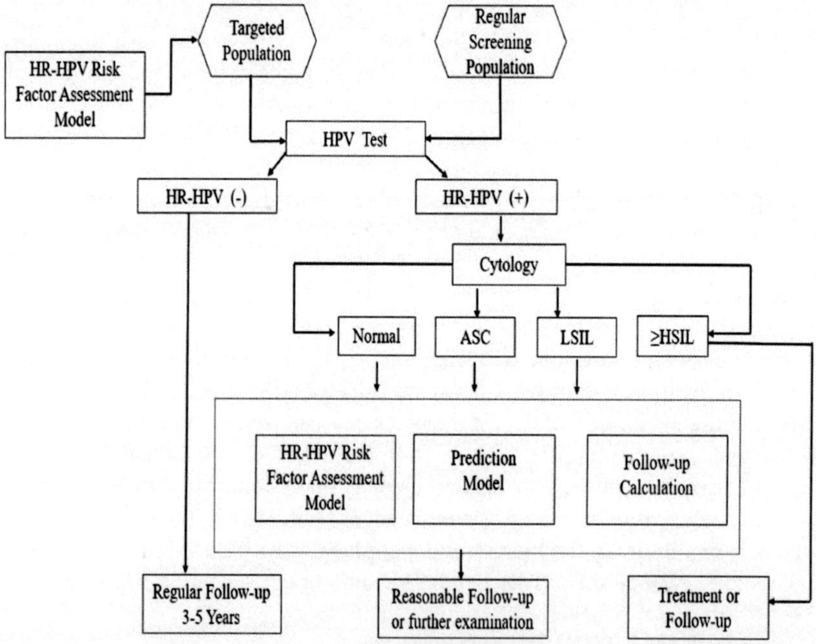

Fig. 1. Recommendation process for cervical cancer screening management from research achievement

potentially going to take further examination like colposcope and biopsy. These cytological results all have chance to turn to normal or higher grades lesion. But HSIL (High-grade squamous intraepithelial lesion) is definitely needed therapeutic methods such as conization cervix.

The correlation between risk factor evaluation indicator model, high risk HPV classification and pathological section of cervical intraepithelial neoplasia (CIN) was analyzed in the third part of the study. A prediction model will be built to forecast the track of the disease. Therefore, it can provide reference and prediction for the selection of further examination methods and reasonable follow-up period for HR-HPV positive and normal cytology, ASC or LSIL patients together with HR-HPV risk factor assessment model and follow-up calculation (see Fig. 1.).

3.3 Key Technologies and Difficulties

The construction of cervical cancer ontology knowledge base, the association analysis of the factors of text data such as electronic medical records, and the in-depth analysis of the causes of the disease induced by combining with the natural health status of patients constitute the main technologies of the study.

Big data analysis, machine learning and other computer technologies are relied to build a quantitative model of risk factor assessment indicators of HR-HPV, to achieve the accurate positioning of the disease and the rationality of follow-up control cycle judgment.

To obtain the results of cervical cancer causes and HR-HPV natural clearance may be difficult to overcome.

3.4 Methods

This project adopts theoretical research and experimental demonstration in parallel, and determines the research order according to the logical relationship between technical points.

The specific steps are as follows (see Fig. 2.):

① Prepare data in the early stage by searching and sorting out relevant data domestic and overseas. Electronic medical records, health histories, HPV test indicators and other relevant medical big data are also collected as a resource simultaneously.

② Construct the quantitative indicators model of risk factor assessment of HR-HPV and the ontology knowledge base. Determine the weight distribution of indicators among influencing factors.

③ Raise a reasonable follow-up time allocation table by calculating the natural clearance rate and clearance time of HPV high-risk types.

④ Causes of HR-HPV were evaluated and analyzed, and the shunt and prediction of ASCUS patients are conducted.

⑤Reliability and clinical validation are evaluated.

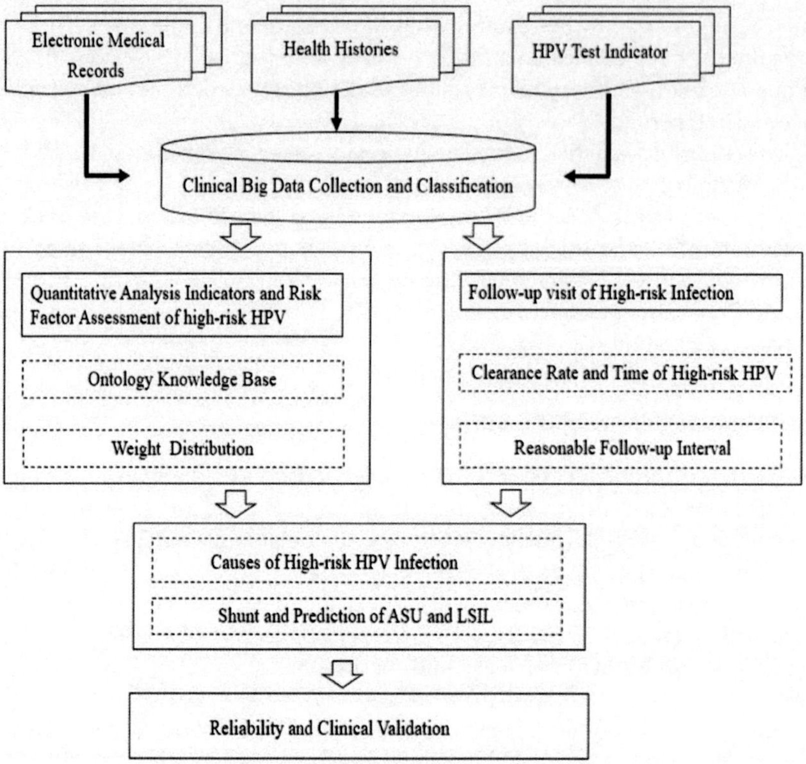

Fig. 2. Research roadmap

4 Discussion

Cervical cancers more possibly occur among women who do not take regular screening examination in most countries which carry out cervical cancer screening programs. Improvement of participation in screening is expected to play a significant role to reduce cervical cancer rates. For these non-participants, self-sampling turns into an alternative way to permit screening acquisition, which can increase the screening coverage to some extent [19].

Women's perception of screening and cervical cancer is another factor influencing screening rate. It also contributes to the missing follow-up and repeated HPV testing. A well-organized screening program should consider not only the equal opportunity of taking advantages of screening, but also the compliance of the patients. The reality is that women are involved in spontaneously in many countries and regions. Women should be sent invitations or reminder to push them into necessary review, provided that the accurate visiting interval is raised. Recently researches have suggested extending screening intervals. And screening related harm are existing in intensive screening.

Currently, despite there is limited data and research about optimal triage strategies for women with HR-HPV positive, triage strategies are receiving increasing and sustaining attention as an important element of the screening program. Triage strategies should be explored in order to bring benefits to physicians' decision making and patients.

5 Conclusion

A huge number of women are at a risk for persistent HR-HPV infection and cervical cancer. HPV vaccination is not a solution to existing continuous high-risk infection and cervical lesion but an effective way to protect the uninfected. Countries and regions are taking standard of treatment and screening program respectively that may result in overtreatment and excessive detection or targeted population missing in screening. The adverse effects of these excessive interventions include bleeding of the cervix, decreased sexual function [20, 21]. They contribute to a lower screening and follow-up compliance of the patient and to unnecessary costs and anxiety of the patients.

Aimed to improve women's cervical health issue, the study bases on a large amount of clinical data, utilizes science and computer technologies. The results of the study mainly include three modules: HR-HPV risk factor assessment model, prediction model and follow-up calculation module, which having realistic significance in identifying targeted population of HR-HPV infection, prevention strategy to HR-HPV infection, triage and prediction of HR-HPV infection and further examinations and individual accurate follow-up period.

Acknowledgments. This work was supported by the Education Department Foundation of Jilin Province (Grant No. JJKH20210066KJ).

References

1. Bruni, L., et al.: Human Papillomavirus and Related Diseases in the World (Summary Report 17 June 2019), ICO/IARC Information Centre on HPV and Cancer (2019)
2. Chesson, H.W., Dunne, E.F., Hariri, S., Markowitz, L.E.: The estimated lifetime probability of acquiring human papillomavirus in the United States. Sex. Transm. Dis. **41**(11), 660–664 (2014)
3. Winer, R.L., et al.: Early natural history of incident, type-specific human papillomavirus infections in newly sexually active young women. Cancer Epidemiol. Biomark. Prev. **20**(4), 699–707 (2011)
4. Tsakogiannis, D., et al.: Sites of disruption within E1 and E2 genes of HPV16 and association with cervical dysplasia. J. Med. Virol. **87**(11), 1973–1980 (2015)
5. van den Heuvel, C.N.A.M., et al.: RNA-based HR-HPV genotyping and identification of HR-HPV transcriptional activity in cervical tissues. Mod. Pathol. **33**(4), 748–757 (2020)
6. Hu, Z., et al.: TALEN-mediated targeting of HPV oncogenes ameliorates HPV-related cervical malignancy. J. Clin. Investig. **125**(1), 425–436 (2014)
7. Egawa, N., Egawa, K., Griffin, H., Doorbar, J.: Human papillomaviruses; epithelial tropisms, and the development of neoplasia. Viruses **7**(7), 3863–3890 (2015)
8. Koh, W.J., et al.: Cervical cancer, version 3.2019, NCCN clinical practice guidelines in oncology. J. Natl. Compr. Canc. Netw. **17**(1), 64–84 (2019)
9. Rumfield, C.S., Roller, N., Pellom, S.T., Schlom, J., Jochems, C.: Therapeutic vaccines for HPV-associated malignancies. Immuno Targets Therapy **9**, 167–200 (2020)

10. de Sanjosé, S., Brotons, M., Pavón, M.A.: The natural history of human papillomavirus infection. Best Pract. Res. Clin. Obstet. Gynaecol. **47**, 2–13 (2018)
11. Heideman, D.A.M., et al.: Clinical validation of the cobas 4800 HPV test for cervical screening purposes. J. Clin. Microbiol. **49**(11), 3983–3985 (2011)
12. Cui, M., et al.: Clinical performance of Roche Cobas 4800 HPV test. J. Clin. Microbiol. **52**(6), 2210–2211 (2014)
13. Madzima, T.R., Vahabi, M., Lofters, A.: Emerging role of HPV self-sampling in cervical cancer screening for hard-to-reach women: focused literature review. Can. Fam. Physician **63**(8), 597–601 (2017)
14. Mchome, B.L., et al.: HPV types, cervical high-grade lesions and risk factors for oncogenic human papillomavirus infection among 3416 Tanzanian women. Sex. Transm. Infect. **97**(1), 56–62 (2021)
15. So, K.A., Lee, I.H., Kim, T.J., Lee, K.H.: Risk factors of persistent HPV infection after treatment for high-grade squamous intraepithelial lesion. Arch. Gynecol. Obstet. **299**(1), 223–227 (2018)
16. Plummer, M., de Martel C., Vignat J., Ferlay J., Bray F., Franceschi S.: Global burden of cancers attributable to infections in 2012: a synthetic analysis. Lancet Global Health **4**(9), e609–e616 (2016)
17. McCredie, M.R.E., et al.: Natural history of cervical neoplasia and risk of invasive cancer in women with cervical intraepithelial neoplasia 3: a retrospective cohort study. Lancet Oncol. **9**(5), 425–434 (2008)
18. Wang, Z.Y., et al.: TruScreen detection of cervical tissues for high-risk human papillomavirus-infected women during the COVID-19 pandemic. Future Oncol. **17**(10), 1197–1207 (2021)
19. Verdoodt, F., Jentschke, M., Hillemanns, P., Racey, C.S., Snijders, P.J., Arbyn, M.: Reaching women who do not participate in the regular cervical cancer screening programme by offering self-sampling kits: a systematic review and meta-analysis of randomised trials. Eur. J. Cancer **51**(16), 2375–2385 (2015)
20. Flanagan, S.M., Wilson, S., Luesley, D., Damery, S.L., Greenfield, S.M.: Adverse outcomes after colposcopy. BMC Womens Health **11**, 2 (2011)
21. Graziottin, A., Serafini, A.: HPV infection in women: psychosexual impact of genital warts and intraepithelial lesions. J. Sex. Med. **6**(3), 633–645 (2009)

Prediction of Drug-Gene Interaction by Using Biomedical Subgraph Patterns

Guangjin Zhao, Meijing Li(✉), and Yingying Jiang

College of Information Engineering, Shanghai Maritime University, Shanghai 201306, China
mjli@shmtu.edu.cn

Abstract. The Drug-Gene Interaction (DGI) is a complex phenomenon which plays an important role in clinical treatment. Thus, prediction of DGI is fundamental. However, previous studies on DGI prediction are defective. Conventional machine learning methods unable to gain enough features, neural network methods commonly pay little attention on node information in the heterogeneous DGI graphs. To solve these problems, we use subgraph mining algorithms to extract features from graphs and transform the DGI prediction into a classification task. Firstly, we use three kinds of interaction data to construct heterogeneous graphs of DGIs. Secondly, to raise the amount of graph data, we use two strategies: group dividing and mode generation. Thirdly, to gain the features of DGIs, we use subgraph mining algorithms to find frequent patterns. Finally, different kinds of binary classifiers are constructed to predict DGIs. The experiment results show that our method performed well.

Keywords: Drug-Gene interaction · Subgraph mining · Classification

1 Introduction

Recently, the research on biological macromolecules is flourishing. The sequencing of specific biological macromolecules with big data technologies can effectively guide the research of biomedicine and clinical pharmacology [1]. In clinical practice, Adverse Drug Reaction (ADR) occurs commonly, which is closely related to allergy and drug rejection. Furthermore, these phenomena are caused by complex interaction between drug and gene, which is called the Drug-Gene Interaction. The side effect caused by the DGI is ubiquitous. Specially, the effect of using certain drug on different patients is various, from efficacy-delaying to life-threatening [2].

From the perspective of pharmacokinetics and pharmacogenomics, DGI is a comprehensives mutual influence between drug metabolic process and gene expression. Pharmacogenomics emerged from large-scale serialization of genes and proteins and other biological macromolecules, using high-throughput analysis pathway mapping and other methods to express physiological differences [3, 4]. Pharmacogenomics and pharmacokinetics show that there is a potential connection between the usage and dosage of drugs and gene expression, which can affect the effect of clinical treatment [5]. ADR are jointly affected by drugs and gene expression, so they have different performances

S.-C. Chu et al. (Eds.): ICGEC 2021, LNEE 833, pp. 159–168, 2022.
https://doi.org/10.1007/978-981-16-8430-2_15

in different individuals and different dosages, but they have certain characteristics in macroscopic data [6].

In summary, prediction of DGI is critical for precision medicine and biopharmaceuticals. DGI prediction methods mainly divided into statistical based methods and machine learning methods [7]. Statistical based methods rely on a large amount of biological experimental data. Paul Westervelt used Kaplan-Meier statistical method to analyze the relationship between CYP2C9, CYP450 and other gene clusters and different doses of drugs [8]. Machine learning prediction methods can divide into two categories. The first category is traditional machine learning methods. Ping Xuan et al. used gradient boosting decision trees to construct candidate target gene prediction models, which reduced the impact of imbalanced bands of data categories [9]. Kalpana Raja et al. used DGI data and DDI (Drug-Drug Interaction) data to construct a gene-drug relationship classification framework that uses traditional machine learning classifiers such as Random Forest and Naive Bayes, which is better than only using DGI data [10]. The second category is neural network methods. Zengrui Wu et al. used the NBI (Network Based Inference) method based on multiple similarities to predict the Drug-Target gene Interaction and gave an affinity index of drugs and genes. Furthermore, two improved weighted NBI methods (EWNBI and NWNBI) were derived [11]. Siyi Zhu et al. proposed a link prediction model Metapath2vec and combined the multi-similarity to predict the DGI [7]. Kexin Huang et al. presented a heuristic deep learning framework MolTrans based on molecular spatial structure and action sites to predict the relationship between drugs and target genes [12].

Statistical based methods require large amount of professional knowledge of pharmacokinetics and genetic molecular, while the number analyzed drugs and genes is very limited. Machine learning methods seldom pay attention to graph structure information and cannot extract node features effectively. Frequent subgraph mining algorithm is an important branch of data mining, and it is an effective means to obtain the features of graphs. Qiao et al. applied frequent subgraph mining algorithm to event extraction and build a model based on time series to predict social unrest events [13, 14]. Lee et al. used frequent subgraph mining algorithm in discovering ontology knowledge. In order to get information from the heterogeneous graph more comprehensively, we use a subgraph mining based method to extract features and use machine learning classifiers to predict DGI [15].

2 Methods

The heterogeneous graphs were constructed from three data source. To obtain more features, we first use data processing methods to expand the graphs. Group dividing and generating are used to preprocess the data extract from three databases. Then, the processed data will be divided into two categories and each of them can be identified by machine learning methods on the features extracted from heterogeneous graphs. AGM and gSpan are two kinds of subgraph mining algorithms. Finally, the DGI is classified through machine learning methods by viewing frequent patterns as features. The workflow of this paper is shown in Fig. 1.

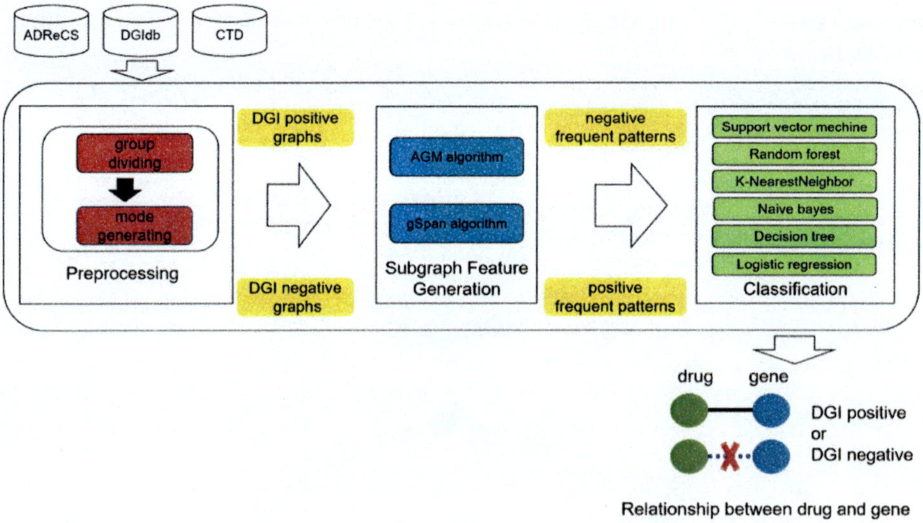

Fig. 1. The workflow of subgraph mining based DGI prediction method.

2.1 Preprocessing

The heterogeneous graphs constructed from the original data were in different sizes. To expand the graphs dataset, we normalize the heterogenous graphs by cutting and filling graphs according to the fixed standard size. However, the scale of the processed graphs is still not enough. In order to obtain as many frequent patterns of a DGI as possible, group dividing and mode generation methods were used to expand the graph data.

Group dividing is to split the graph structure into several subgraphs in the same size. To ensure that the information gap between the divided groups is as large as possible, the Levenshtein distance was used to measure the similarity of the node sequence. Levenshtein distance is a kind of edit distance, which represent the similarity between two sequences [16, 17]. Mode generation is a step to further generate subgraphs according to the neighbor selection rules, which is a kind of combination generation method by cartesian product.

$$Leva,b(i,j) = \begin{cases} max(i,j) & if\ min(i,j) = 0; \\ min\begin{cases} Leva,b(i-1,j)+1 \\ Leva,b(i,j-1)+1 \\ Leva,b(i-1,j-1)+1(ai \neq bj) \end{cases} & otherwise; \end{cases} \quad (1)$$

Take standard size = 50 as an example, which contains 24 ADR nodes and 24 disease nodes. Set unit size = 6, ADR and disease can be divided into 4 groups, 16 groups of data from group 1–1 to group 4–4 can be combined. In mode generation, if the neighbor selection rule set the maximum selection unit = 4, The ADR patterns set and disease patterns set generated. And the expanded subgraph is generated according to the Cartesian product combination method on these two patterns set. For example, ADR set use 3 adjacent nodes, disease set use 4 adjacent nodes, then mode 3–4 will generate 12 combinations to form 12 subgraphs. The process of group dividing and mode generation is shown in Fig. 2.

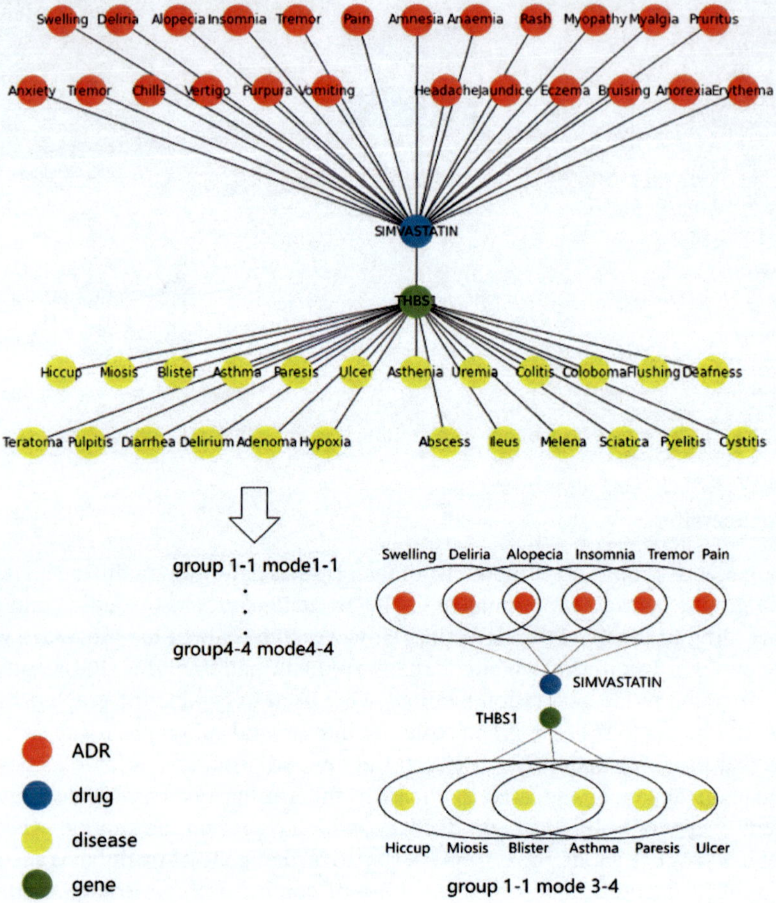

Fig. 2. Data preprocessing on heterogeneous graph with gene THBS1 and drug SIMVASTATIN. The red nodes represent ADRs, the blue nodes represent drugs, the yellow nodes represent diseases and the green nodes represent genes.

2.2 Subgraph Feature Generation

In the DGI heterogeneous graphs, a specific DGI is always associated with a specific frequent subgraph. For example, the DGI heterogeneous graph of THBS1 and SIMVASTATIN contains some unique combinations of ADRs and diseases. In other words, frequent subgraphs can be used to infer whether there is a relationship between a certain drug and a certain gene. In order to obtain data features effectively, this paper uses frequent subgraph mining algorithm to find frequent patterns in DGI graphs.

If the ADR-drug-gene-disease heterogeneous graph set is expressed as $Graphs = \{G_1, G_2, \ldots G_i \ldots G_n\}$, where G_{subi} is a subgraph of G_i. Then, the support of subgraphs can be defined as follows:

$$support = \frac{amount\ of\ G_{subi} \quad (G_{subi} \subset Gi \subset Graphs)}{n} \tag{2}$$

The absolute minimum support threshold only considers the frequency of the subgraphs. If the minimum support threshold can be marked as Sup_{min}, the absolute minimum support threshold can be defined as follows:

$$abs_{minsup} = Sup_{min} \times n \tag{3}$$

When the support value of a graph is larger than support threshold, than the subgraph is a frequent subgraph, and the pattern it represents is a frequent pattern of DGI. There is a connection between the frequent subgraphs generated from the heterogeneous graph structure and its core DGI.

AGM and gSpan, two subgraph mining algorithms were used to acquire frequent patterns of DGIs. The efficiency of subgraph mining algorithm mainly depends on two parts: frequent subgraph expanding approach and the number of isomorphic graphs. In graph mining field, more isomorphic graphs lead to longer running time [18]. The AGM algorithm is an early subgraph mining algorithm. The process of subgraph expansion and pruning was defective, which will generate a large amount of isomorphic graphs [19]. In order to obtain features more effectively, we use gSpan algorithm which is based on FP-growth strategy. Since the gSpan algorithm adopts Depth-First Search (DFS) and the rightmost path expansion rule, the process of searching and expanding are faster [20]. The number of isomorphic graphs generated during the subgraph expansion is much smaller than that of AGM [21, 22].

By using frequent subgraph mining, we construct a DGI positive frequent pattern database and a DGI negative frequent pattern database. For each pair of drug gene relationship, both positive and negative frequent patterns will be obtained. If we use the drug gene relationship graph to match the frequent patterns, a drug gene relationship can be marked [23].

2.3 Classification

The prediction result is basically influenced by the absolute minimum support threshold which is used to filter the frequent subgraphs of DGI patterns. Since DGI prediction can be viewed as a binary classification problem, we use machine learning method to predict DGI. In order to compare the performance of different types of classifiers, we select six classifiers to deal with the prediction task.

3 Experiment and Results

3.1 Data

There are three types of data used in this experiment: the DGI data, ADR-Drug relationship data, and Gene-Disease relationship data. DGI information is from the Drug-Gene

Interaction Database (DGIdb), which is derived from the Genome-wide association study (GWAS) project [24–27]. The ADR-Drug relationship information is from Adverse Drug Reaction Classification System (ADReCS), which is a system for adverse drug reaction reports. ADReCS is a comprehensive ADR ontology database that provides hierarchical ADR standardized terminology [28]. Gene-Disease relationship information is from the comparative toxicology database (CTD). CTD is a publicly available research resource which uses scientific data describing relationships between chemicals, genes, and human diseases. A primary goal of CTD is to advance the understanding of the effects of environmental chemicals on human health [29].

3.2 Evaluation Metrics

The Receiver Operating Characteristic Curve (ROC) takes the false positive rate (FPR) and the true positive rate (TPR) as the horizontal axis and vertical axis coordinates respectively, and can be used to adjust the classification results by changing the classification threshold. Normally, the number of negative examples of DGI is often greater than the number of positive examples [30]. ROC can not only effectively reflect the sensitivity of the classifier to the classification boundary, but also avoid the impact of data imbalance. Hence, this experiment uses the Area Under the ROC curve (AUC) to measure the performance of each classifier.

3.3 Results

Comparison of Classification Result on 6 Classifiers. The experiment used two frequent subgraph mining algorithms in feature extraction. The results show that the frequent subgraphs obtained by these two algorithms are totally the same in content and quantity when the absolute minimum support is 10. However, the time consumption of gSpan is 18.47% of the AGM spending. If we treat the subgraphs as features, binary classifiers can be constructed for each drug gene relationship which including the DGI positive data and DGI negative data.

In order to verify the influence of different absolute minimum support thresholds on the number of frequent features, this experiment uses different minimum support thresholds to filter frequent subgraph features, and uses several different types of classifiers for classification. As the support threshold increases, the amount frequent features decrease, and different classifiers respond variously. We select six classifiers: Random Forest, Naive Bayes, K-Nearest Neighbors, Decision Tree, Support Vector Machine, Logistic Regression for classification.

In the experiment, 70% of the data is selected as the training part, and the remaining 30% is used for testing. An important parameter of the experiment is the absolute minimum support threshold abs_{minsup}. By adjusting this parameter, frequent patterns can be screened, and the number of frequent patterns of positive and negative examples can be changed, which in turn affects the final classification result. The AUC metric of Support Vector Machine, Logistic Regression and K-Nearest Neighbors decrease significantly as the number of frequent patterns decreases.

The AUC metric of Support Vector Machine, Logistic Regression and K-Nearest Neighbors decrease significantly as the number of frequent patterns decreases. The classification results of Random Forest and Naive Bayes do not change significantly with the number of frequent patterns. Until abs_{minsup} up to 35, a sudden change of abs_{minsup} occurs. Decision trees perform the worst on this task. Obviously, different classifiers have different responses when the number of frequent patterns declines. But the AUC indicators of these six classifiers all drop sharply when the absolute minimum support threshold is greater than 35. The reason is that the number of frequent patterns after screening is too small. Consequently, the boundary between the two categories is not distinct enough. Figure 3 shows the changes in the frequent patterns of DGI positive and DGI negative with the increasing of the absolute minimum support threshold. Figure 4 shows the changes of classification result described by AUC of the six classifiers under the influence of different absolute minimum support thresholds. Table 1 shows the absolute minimum support threshold when the classifiers acquire the best AUC values.

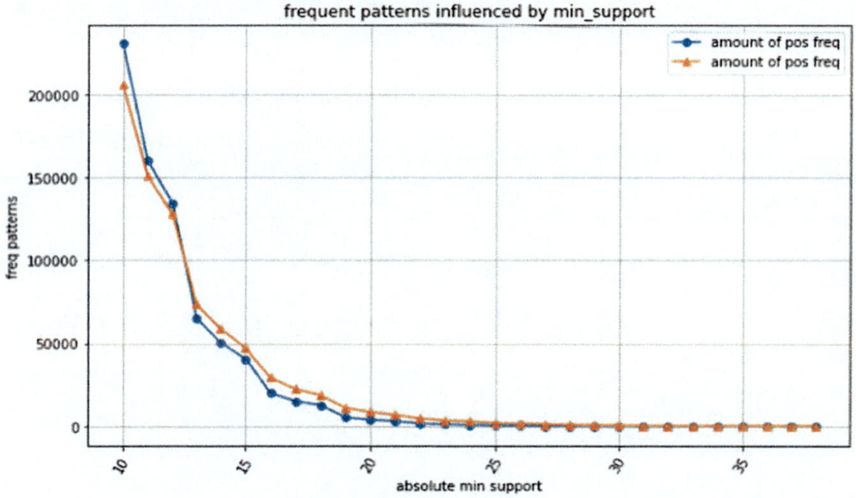

Fig. 3. The amounts of frequent patterns influenced by absolute minimum support

Comparison with Existing Methods. Experiments show that when the absolute minimum support is 10, using SVM for classification can obtain the best DGI prediction result. Previously, Siyi Zhu et al. used the same DGI data and proposed two link prediction methods based on network representation learning, metapath2vec and metapath2vec + +, to deal with the problem of Drug-Gene relationship prediction. The metapath2Vec + + mode is an improved version of metapath2vec which take the restart rank walk strategy into account on link prediction. Simultaneously, the best AUC of the experiment in this paper has reached 0.8944, which is better than 0.8093 of metapath2vec and 0.8367 of metapath2vec + +. This shows that frequent subgraph mining can better extract the characteristics of drug gene relationships. Figure 5 shows the best ROC curve of the method used in this article on the six classifiers and the ROC curve for link prediction using metapath2vec and metapath2vec + +.

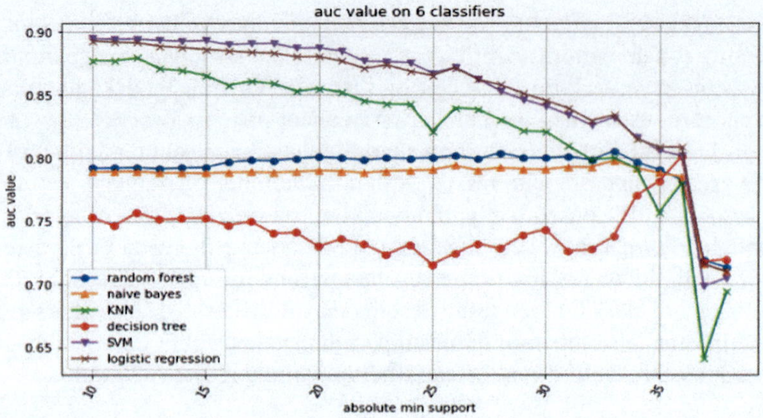

Fig. 4. The AUC values of 6 kinds of classifiers influenced by absolute minimum support

Table 1. The best AUC values and absolute minimum support parameter

Classifier	The best AUC value	abs_{minsup}
Random forest	0.8055	33
Naive bayes	0.7977	33
K-nearest neighbors	0.8791	12
Decision tree	0.8012	36
Support vector machine	0.8944	10
Logistic regression	0.8926	10

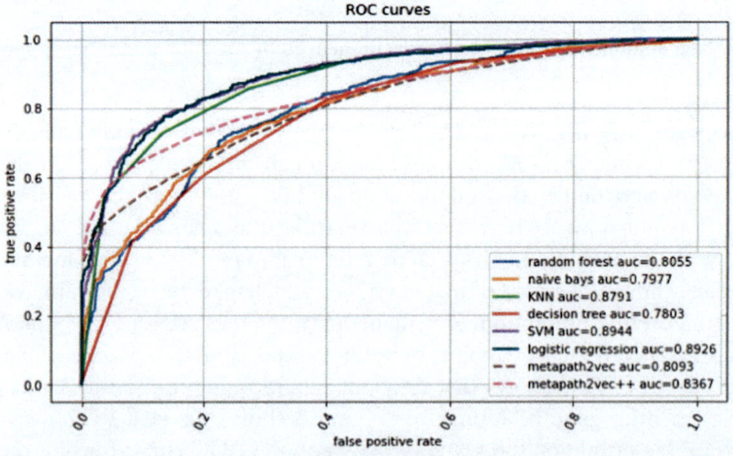

Fig. 5. Comparison of the result between subgraph mining based prediction method and metapath2vec based prediction method

4 Conclusion

In this paper, we predict DGI by using subgraph mining. We construct a heterogenous graph dataset and expand graphs by two strategies, group dividing method and mode generation method. We use subgraph mining algorithms to extract features and finally classify DGIs into two categories: DGI positive and DGI negative. The experimental results show that subgraph mining when the absolute minimum support threshold is 10 on this drug-gene relationship data set and use support vector machine in classification can achieve the best prediction performance. In the future, we hope to use more comprehensive data such as relationship on protein-gene, miRNA-disease and other associations to build a broader biological heterogenous information relationship network to improve the prediction of DGI.

Acknowledgement. This study was supported by the National Natural Science Foundation of China (61911540482 and 61702324).

References

1. Greene, C.S., Costello, J.C.: Biologically informed neural networks predict drug responses. Cancer Cell **38**, 613–615 (2020)
2. Hwang, Y., et al.: Identifying the common genetic networks of ADR (adverse drug reaction) clusters and developing an ADR classification model. Mol. Biosyst. **13**, 1788–1796 (2017)
3. Ram, P.T., Mendelsohn, J., Mills, G.B.: Bioinformatics and systems biology. Mol. Oncol. **6**, 147–154 (2012)
4. Benet, L.Z., Bowman, C.M., Koleske, M.L., Rinaldi, C.L., Sodhi, J.K.: Understanding drug–drug interaction and pharmacogenomic changes in pharmacokinetics for metabolized drugs. J. Pharmacokinet. Pharmacodyn. **46**(2), 155–163 (2019). https://doi.org/10.1007/s10928-019-09626-7
5. Lavecchia, A.: Machine-learning approaches in drug discovery: methods and applications. Drug Discov. Today **20**, 318–331 (2015)
6. Malki, M., Pearson, E.: Drug–drug–gene interactions and adverse drug reactions. Pharmacogenomics J. **20**, 1–12 (2019)
7. Siyi, Z., Jiaxin, B., Xiaoping, M., Chen, L., Xiangxiang, Z.: Prediction of drug-gene interaction by using Metapath2vec. Front. Genet. **9**, 248 (2018)
8. Westervelt, P., Cho, K., Bright, D., Kisor, D.: Drug–gene interactions: inherent variability in drug maintenance dose requirements. Pharm. Ther. **39**, 630–637 (2014)
9. Xuan, P., Sun, C., Zhang, T., Ye, Y., Shen, T., Dong, Y.: Gradient boosting decision tree-based method for predicting interactions between target genes and drugs. Front. Genet. **10**, 459 (2013)
10. Raja, K., Patrick, M., Elder, J.T., Tsoi, L.C.: Machine learning workflow to enhance predictions of adverse drug reactions (ADRs) through drug-gene interactions: application to drugs for cutaneous diseases. Sci. Rep. **7**, 3690 (2017)
11. Wu, Z., Li, W., Liu, G., Tang, Y.: Network-based methods for prediction of drug-target interactions. Front. Pharmacol. **9**, 1134 (2018)
12. Huang, K., Xiao, C., Glass, L.M., Sun, J.: MolTrans: molecular interaction transformer for drug–target interaction prediction. Bioinformatics **37**(6), 830–836 (2020)
13. Qiao, F., Pei, L., Xin, Z., Ding, Z., Cheng, J., Wang, H.: Predicting social unrest events with hidden markov models using GDELT. Discrete Dyn. Nat. Soc. **2017**, 1–13 (2017)

14. Qiao, F., Zhang, X., Deng, J.: Learning evolutionary stages with hidden semi-markov model for predicting social unrest events. Discrete Dyn. Nat. Soc. **2020**, 1–16 (2020)
15. Lee, K., Jung, H., Hong, J.S., Kim, W.: Learning knowledge using frequent subgraph mining from ontology graph data. Appl. Sci. **11**, 932 (2021)
16. Berger, B., Waterman, M.S., Yu, Y.W.: Levenshtein distance, sequence comparison and biological database search. IEEE Trans. Inf. Theory **67**, 3287–3294 (2021)
17. Rozinek, O., Mare, J.: The duality of similarity and metric spaces. Appl. Sci. **11**, 1910 (2021)
18. Ramraj, T., Prabhakar, R.: Frequent subgraph mining algorithms – a survey. Procedia Comput. Sci. **47**, 197–204 (2015)
19. Xian-Tong, L.I., Jian-Zhong, L.I., Gao, H.: An efficient frequent subgraph mining algorithm. **18**, 2469–2480 (2007)
20. Jiang, C., Coenen, F., Sanderson, R., Zito, M.: Text classification using graph mining-based feature extraction. In: Bramer, M., Ellis, R., Petridis, M. (eds.) Research and Development in Intelligent Systems XXVI, pp. 21–34. Springer, London (2010). https://doi.org/10.1007/978-1-84882-983-1_2
21. Jena, B.S., Khan, C., Sunderraman, R.: High performance frequent subgraph mining on transaction datasets: a survey and performance comparison. Big Data Min. Anal. **2**(3), 159–180 (2019)
22. Kashyap, N.K., Pandey, B.K., Mandoria, H.L., Kumar, A.: Graph mining using gSpan: graph-based substructure pattern mining. Int. J. Appl. Res. Info. Tech. Comput. **7**, 132–139 (2016)
23. You, Y., Sun, J., Chen, Y.W., Niu, C., Jiang, J.: Ensemble belief rule-based model for complex system classification and prediction. Expert Syst. Appl. **164**, 113952 (2020)
24. Cotto, K.C., et al.: DGIdb 3.0: a redesign and expansion of the drug-gene interaction database. Nucleic Acids Res. **46**(D1), D1068–D1073 (2017)
25. Freshour, S.L., Kiwala, S., Cotto, K.C., Coffman, A.C., Wagner, A.H.: Integration of the drug–gene interaction database (DGIdb 4.0) with open crowdsource efforts. Nucleic Acids Res. **49**, D1144–D1151 (2020)
26. Wagner, A.H., et al.: DGIdb 2.0: mining clinically relevant drug-gene interactions. Nucleic Acids Res. **44**, D1036–1044 (2016)
27. Yang, J., et al.: Conditional and joint multiple-SNP analysis of GWAS summary statistics identifies additional variants influencing complex traits. Nat. Genet. **44**, 1–3 (2012)
28. Cai, M.C., et al.: ADReCS: an ontology database for aiding standardization and hierarchical classification of adverse drug reaction terms. Nucleic Acids Res. **43**, D907–D913 (2015)
29. Mattingly, C.J., Rosenstein, M.C., Colby, G.T., Forrest, J.N., Jr., Boyer, J.L.: The Comparative Toxicogenomics database (CTD): a resource for comparative toxicological studies. J. Exp. Zool. Part A Comp. Exp. Biol. **305A**(9), 689–692 (2006)
30. Moses, L.E., Shapiro, D., Littenberg, B.: Combining independent studies of a diagnostic test into a summary ROC curve: data-analytic approaches and some additional considerations. Stat. Med. **12**, 1293–1316 (2010)

Medical Phrase Mining Method for Biomedical Text Processing

Ling Wang[1], Minglei Shan[1], Tong Li[2], Yingxuan Tang[1], and Tiehua Zhou[1](✉)

[1] Department of Computer Science and Technology, School of Computer Science, Northeast Electric Power University, Jilin, China
{smile2867ling,2202000690,thzhou}@neepu.edu.cn
[2] Jihua General Hospital, Jilin 132000, China

Abstract. With the rapid increase of biomedical data, mining high-quality phrases from these unstructured text data has become a hot research issue. Converting unstructured text into structured units will greatly reduce the semantic ambiguity in the text, which is very important for mining high-quality phrases. Phrase mining is one of the basic tasks of text processing. It aims to extract high-quality phrases from a corpus and has many downstream applications, such as information retrieval classification, document classification, entity recognition, topic modeling, etc. In this paper, we proposed a new model, which selects candidate sets from the corpus through phrase frequency and calculates the characteristics of candidate phrases to evaluate the quality. We collected text data of questions and answers in the medical field as a corpus for experiments. Experiments show that the model can mine medical professional phrases, and solve the problem of long phrases missing, with good accuracy.

Keywords: NLP · Biomedical text processing · Medical phrase mining · Data mining

1 Introduction

In recent years, the primary key of the Internet becomes a medium for exchanging medical knowledge and health advice. There are more and more medical-related websites and forums. In the forum, users could exchange information with experts in real-time and share medical information. As a result, the medical forum contains a large amount of information in the form of free documents, which contains valuable phrases such as disease-related traits, treatment drugs, and disease names. Mining valuable information from unstructured textual data and transforming unstructured text data into structural units is the important step. Extracting high-quality and professionally meaningful phrases is one of the basic steps of named entity recognition [1], topic modeling [2], and relationship extraction [3, 4]. Therefore, this paper studies the phrase mining of medical forum text data.

This paper proposes a model for extracting high-quality phrases from unstructured text data of medical forums. The model has the following advantages: (1) Based on

S.-C. Chu et al. (Eds.): ICGEC 2021, LNEE 833, pp. 169–178, 2022.
https://doi.org/10.1007/978-981-16-8430-2_16

phrase mining based on phrase frequency, phrase segmentation is added to correct the frequency of phrase occurrence. (2) By mining high-quality phrases in professional fields in the forum texts, we could better find the frequency and location of phrases related to diseases, to facilitate the statistics of patients with different diseases. (3) Experiments show that the model proposed in this paper has excellent accuracy in extracting quality phrases from the data set.

2 Related Work

Efficient and accurate mining of high-quality phrases in text data is a hot research topic at present. The existing phrase mining methods can be divided into two categories according to the extraction methods of the extracted candidate set, data-based phrase mining and grammar-based end and mining.

Data-based phrase mining uses N-gram to obtain phrase candidate sets from text and then evaluates the candidate phrases according to the set feature indexes. Shang et al. [5] proposed a phrase mining framework to evaluate the candidate phrases selected by N-gram to realize the automatic mining of high-quality phrases. Li et al. [6] based on data mining of high-quality phrases in the text, a data-driven approach can ensure that the frequency of mined phrases is relatively high. They also added some low-quality phrase sequences to expand the set of candidate phrases.

The grammatically based approach will obtain the candidate set according to the part of speech and context of the phrase in the text, to obtain the phrase that fits the relevant field of the text more. Gebhardt et al. [7] proposed to mix grammar and process complex phrases according to the mixed features. Jie et al. [8] proposed that text information should be structured by relying on the tree to improve the performance of entity recognition.

Many researchers have achieved high-quality phrase extraction in related fields. Wang et al. [9] realized the adaptive function of machine translation language time translation system by extracting high-quality phrases. Li et al. [6] solved the problem by combining phrase mining with topic extraction. Semantic analysis is also an important aspect of natural language processing. Researchers have made progress in semantic analysis, in different areas of expertise. Zhang et al. [10] carried out the semantic analysis of data in text-based on probability graph model and achieved excellent results in Both Chinese and English languages.

Many scholars apply word vector embedding to phrase mining and have done a lot of research. Mikolov et al. [11] proposed word vector embedding and proved its applicability. Pennington et al. [12] applied the logarithmic bilinear model to consider the combination of local information and global internal use. Stein et al. [13] discussed the applicability of word vectors in text processing tasks.

3 Motivation

As more and more medical knowledge is acquired online, more and more medical Q&A text data are accumulated in online forums. Mining high-quality phrases from such a large number of medical texts have become a hot research issue at present. The traditional method of phrase mining based on word frequency and context lacks efficiency and accuracy. On this basis, we propose CFB algorithm. The algorithm could add integrity and professional phrase features based on word frequency, improve the quality of phrases mined in the text by comprehensive evaluation of phrase features.

4 Context-Frequency-Based (CFB) Algorithm

The flow chart of this algorithm is shown in Fig. 1:

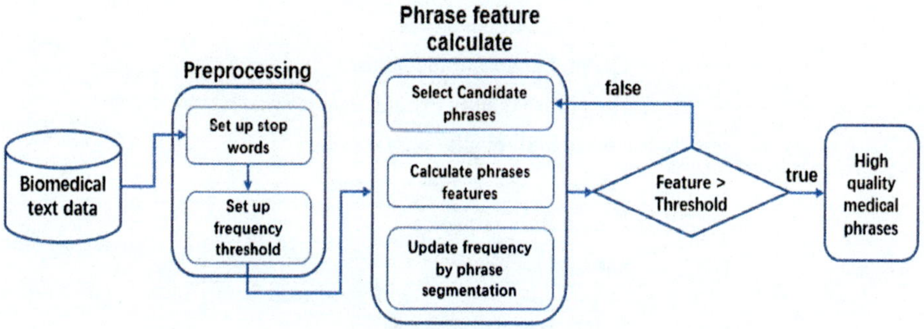

Fig. 1. The flow chart of CFB algorithm

The main step of this algorithm is as follows:

(1) Generating candidate set.
(2) Evaluate phrase quality based on consistency and informational requirements
(3) Frequency of phrase segmentation correction.
(4) Add the corrected frequency to the feature set.
(5) Screen for high-quality phrases.

By obtaining the initial frequency of each phrase in the text data set, the model selects the candidate set of the initial phrase through the threshold value. The judgment criterion of this part is the occurrence frequency of words and phrases in the text, without the participation of other features. The features of each phrase in the candidate set are calculated and compared. The features of each phrase in the candidate set are calculated by the context relation of the phrase, namely the probability of the common occurrence of context structure, and the IDF value of each candidate phrase. The phrases in the candidate set are filtered through the classifier to select the original high-quality phrases. Then, the frequency value of candidate phrases in the initial phrase candidates is corrected by the method of phrase sequence segmentation, and some high-quality phrases with low occurrence frequency are screened out, and those with low quality but high occurrence frequency are removed. We end up with the good set of phrases we need.

The notation table and details of CFB algorithm are as follows (Table 1):

Table 1. Notations table

Symbol	Definition
C	Candidate phrase set
M	Candidate multiword phrase set
c	Candidate phrase unit
m	Candidate multiword phrase
p_l	The left part of the divided phrase
p_r	The right part of the divided phrase
$f(c)$	Frequency of the phrase
$p(c)$	The probability of the phrase
$Q(c)$	The phrase quality of candidate phrase c
$\lceil c \rfloor$	c is for high-quality phrase events
S	The sequence of the keywords
P	The position sequence
L	The tags sequence

4.1 Candidate Phrases

We define that in a document, high-quality phrases should occur with sufficient frequency as a complete semantic unit. The detection task of candidate phrases can be defined as collecting all phrases in the corpus that meet the minimum threshold according to this requirement. In the actual filtering process, the minimum threshold τ and the maximum phrase program ω can be set. To more effectively mine the phrases in the document, we have two criteria: if a phrase does not meet the candidate requirement, then any phrase that contains the phrase does not meet the candidate requirement, so we can filter several long phrases. If a phrase meets the candidate requirement, then any subphases of the phrase also meet, so candidates can be obtained by extending the subphases.

According to the two criteria proposed above, our phrase candidate set screening can be carried out by setting ω wheel to screen the candidate phrase set from word phrase to ω word phrase. The words and phrases that are very frequent but not needed are deleted according to the stop word, and the phrases that the stop word expands are not considered in the subsequent extraction process. In each round of phrase extraction, the extended phrases of the candidate phrases extracted in the last round are taken into account, and the final set of candidate phrases is obtained.

The selection process is as follows Table 2:

4.2 Phrase Feature

Estimating phrase quality with only a small amont of training labels is not accurate because the generated candidate phrases can be confusing. We choose to calculate multiple ends and features for each phrase in the candidate set. Based on these features,

Table 2. Select frequency phrase

Algorithm 1: Select frequency phrase

Input: Corpus U, minimum threshold τ

Output: Raw frequency f of phrases

```
for i in 1 to |U| do
        index [U(i)] ← index[U(i)] ∪ i
end for
if index[U(i)] is not empty
    for v in index.key
            if |index[v]| ≥ τ
                    for j in index[v]
                            v' = u.join (U[j+1])
                            index'[v'] ← index'[v] ∪ j+1
                    end for
            else continue
    end for
else continue
end if
return f
```

the classifier was trained to estimate the quality of all phrases. For phrases not in the candidate set, the quality score is 0.

To determine whether a phrase belongs to a complete semantic unit, we set up a set of features to measure the consistency between the components of a phrase. Each candidate phrase in the candidate set is divided into two disjoint units to measure the consistency between the units. If the probability of a combination between the two parts is greater than expected, then the phrase is consistent.

The candidate set is C, and the candidate set of multiple phrases is M. For each candidate unit $c \in C$, the original frequency is set as $f(c)$, and the probability of the phrase is $p(c)$:

$$p(c) = f(c) / \sum\nolimits_{c' \in C} f(c') \tag{1}$$

For a multi-word phrase $m \in M$, we divide it into two parts that are most likely to be divided. $< p_l, p_r >$, p_l is the lift part of the phrase, and p_r is the right part of the phrase, the point-state mutual information, and divergence are used to measure the consistency characteristics: PMI(v), PKL(v)

$$PMI(m) = log\, p(m)/p(p_l)p(p_r) \tag{2}$$

$$PKL(m) = PMI(m) \cdot p(m) \tag{3}$$

To make the phrases mined can express corresponding meanings (such as the name of a drug, the symptoms of a disease, etc.), we use the characteristics of IDF and punctuation marks to measure whether the phrases provide useful information.

Average inverse document frequency (IDF) is often used in corpus statistics, and we use this feature to measure the informativeness of phrases. Calculate IDF values for words:

$$IDF(c) = log\, N/|\{i : c \in S_i\}| \tag{4}$$

The value of the IDF can be used to represent how much information the word can provide to retrieve a certain type of information. In general, the average value of IDF for high-quality phrases is relatively large.

In addition to IDF, we also judge whether a phrase is of high quality by the appearance of punctuation marks. If a phrase appears in parentheses, brackets, and curly braces, it may indicate that the phrase is more informative.

4.3 Classifier

The classifier used in our framework is the random forest, which can be well trained with a small amount of labeled data. When there is noise in the training sample, the random forest can reduce the influence of noise. In the generation stage of a decision tree, we randomly select phrase samples from high-quality phrases and inferior phrases respectively, and finally predict the phrase as the ratio of the number of decision trees of high-quality phrases to the total number of decision trees, namely, the original score of the phrase, which is used to represent the original quality of the phrase.

4.4 Update Frequency

It is not accurate to calculate the frequency of candidate phrases only based on the frequency of occurrence, because the frequency of high-quality phrases will decrease due to the long length of the phrase, while the frequency of some low-quality phrases will be high. We then examine the context of each word to determine whether it should be counted as a phrase. For example, there is a word sequence "$w_1 w_2 w_3 w_4 w_5$", where "$w_1 w_2 w_3$" is identified as a high-quality phrase to mine. If the frequency calculated by "$w_1 w_2$" does not meet the threshold condition, it will be excluded from the high-quality phrase set. The phrases "$w_1 w_2 w_3 w_4$" and "$w_1 w_2 w_3 w_4 w_5$" can't be mined.

Here, we propose a phrase segmentation method to correct the frequency of phrases. Combining with other phrase features in the previous part, some phrases with high frequency and low quality can be removed. We use the phrase segmentation method to divide the sentences in the document into multiple units that do not want to intersect and recalculate the frequency of the phrase as the corrected frequency. The corrected frequency can represent the number of times a phrase is divided into disjoint units.

The input of phrase segmentation is a sequence of key words $S = w_1, w_2, w_3 \ldots w_i$,this sequence is a sentence in this text, w_1 is the first keyword of the sequence, and a sequence of part of speech tags $L = p_1, p_2, p_3 \ldots p_i$, p_1 is the first speech tags of the sequence, we divide the S sequence into position sequence $P = d_1, d_2, d_3 \ldots d_j$, d_1 is the index of the first word, and we maximize the value of the following formula utilizing dynamic programming (Table 3).

$$P(S, L, P) = \prod\nolimits_{n=1}^{j} p(d_{n+1}|d_n, L)p(s_i|d_n, d_{n+1})p(\lceil s_i \rceil|s_i) \tag{5}$$

Table 3. Phrase segmentation

Algorithm 2: Phrase segmentation

Input: Word sequence S, phrase quality Q

Output: Optimal segmentation P

$x_0 \leftarrow 1$, $x_1 \leftarrow 0$

m←the max phrase length threshold

for i in 1 to n **do**

    for j in 1 to m

        if $h_i \cdot p(d_{t+1}=d_t+j,[w[i+1,i+j+1]|\, d_t]>h_{i+j})$

            $h_{i+j}= hi \cdot p(d_{t+1}=d_t+\mathrm{j},[w[i+1,i+j+1]|\, d_t]> h_{i+j})$

            $g_{i+j}= i$

        else continue

        end if

    end for

end for

$i \leftarrow n$

$k \leftarrow 0$

while $i>0$ **do**

    $k \leftarrow k+1$

    $p_m \leftarrow d_{g_{i+1}} d_{g_{i+2}} \ldots d_i$

    $i \leftarrow g_i$

return P

After the updated frequency is obtained by phrase segmentation above, we remove the part of phrase frequency below the threshold in the candidate set. The quality characteristics of the remaining phrases were recalculated, and a new random forest was trained based on this feature to obtain the quality score after the updated phrase, and the phrases above the scoring threshold were output as high-quality phrases as the final result.

5 Experiment

In this section, we compare our method with the other three phrase mining methods. To verify the accuracy of our method, we used textual records of questions and answers between patients and doctors in the medical community.

Our data set consists of transcripts of physician-patient questions and answers obtained from a healthcare community forum and contains 14,000 transcripts of questions and answers related to surgical conditions. The experimental environment is Based on Windows 10, GeForce GTX 1080Ti, and the processing language is Python. The comparison models are TF-IDF [14], SegPhrase [15] and TextRank [16] (Fig. 2).

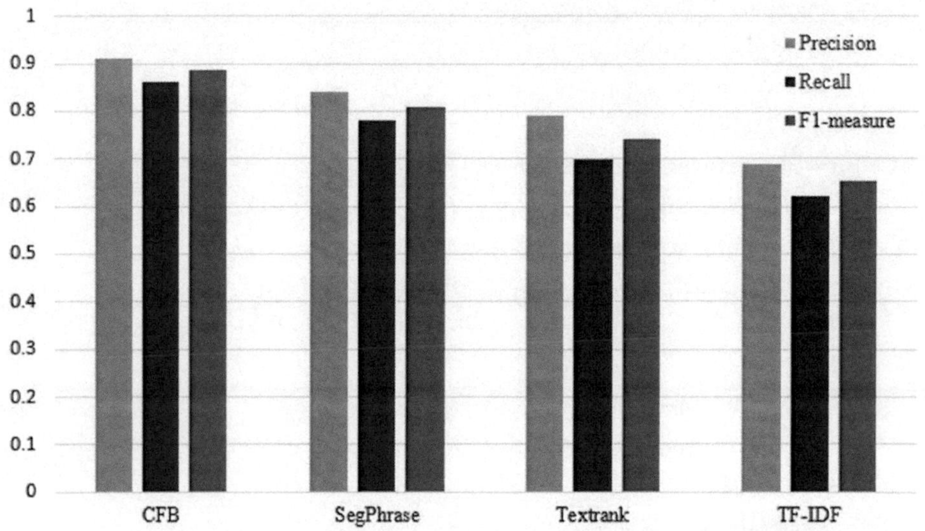

Fig. 2. The precision-recall curves among different algorithms on the dataset

By comparing with other phrase mining algorithms, the advantages of our proposed model are verified. What is accurate is the ratio of the number of high-quality phrases to the number of candidate phrases. The recall rate is the ratio of the number of high-quality phrases to the total number of high-quality phrases. F1-measure is a statistical indicator that accuracy rate and recall rate are calculated together.

$$F1Measure = (precision \cdot recall/precision + recall) \cdot 2 \tag{6}$$

In the experiment, data sets were drawn from medical community forums related to the medical field, and disease traits, treatments, drug names, and disease names were included in the range of high-quality phrases. For quality phrases that are not recognized in existing dictionaries, we do a Wikipedia search, and if we find a phrase mined under The Wikipedia Medical classification, it will be considered high quality and added to the phrase dictionary.

6 Conclusions

This paper introduces a model for extracting high-quality phrases from medical text data. The model evaluates phrase quality comprehensively from different endpoints and features. It also uses phrase segmentation and context analysis to correct the initial frequency of phrases to remove some low-quality frequently occurring phrases. Finally, phrases are filtered using the updated phrase frequency and other characteristics. The accuracy of our model was significantly improved when we tested the text data obtained in a medical forum. In the future, we will explore how to use good phrases extracted from the text for entity extraction and document categorization.

Acknowledgments. This work was supported by the Science and Technology Development Plan of Jilin Province, China (No. 20200403039SF).

References

1. Yu, J., Bohnet, B., Poesio, M.: Named entity recognition as dependency parsing. (2020) arXiv preprint arXiv:2005.07150
2. Wang, L., et al.: Mining infrequent high-quality phrases from domain-specific corpora. In: Proceedings of the 29th ACM International Conference on Information & Knowledge Management, pp. 1535–1544. Association for Computing Machinery, United States (2020)
3. Wang, Y., Yu, B., Zhang, Y., Liu, T., Zhu, H., Sun, L.: Tplinker: single-stage joint extraction of entities and relations through token pair linking (2020). arXiv preprint arXiv:2010.13415
4. El-Kishky, A., Song, Y., Wang, C., Voss, C., Han, J.: Scalable topical phrase mining from text corpora (2014). arXiv preprint arXiv:1406.6312
5. Jiang, M., Shang, J.: Scientific text mining and knowledge graphs. In: Proceedings of the 26th ACM SIGKDD International Conference on Knowledge Discovery & Data Mining, pp. 3537–3538. Association for Computing Machinery, United States (2020)
6. Li, B., Yang, X., Zhou, R., Wang, B., Liu, C., Zhang, Y.: An efficient method for high quality and cohesive topical phrase mining. IEEE Trans. Knowl. Data Eng. **31**, 120–137 (2019)
7. Gebhardt, K., Nederhof, M.-J., Vogler, H.: Hybrid grammars for parsing of discontinuous phrase structures and non-projective dependency structures. Comput. Linguist. **43**, 465–520 (2017)
8. Jie, Z., Muis, A.O., Lu, W.: Efficient dependency-guided named entity recognition. In: Proceedings of the Thirty-First AAAI Conference on Artificial Intelligence, pp. 3457–3465. AAAI, United States (2017)
9. Wang, R., Zhao, H., Lu, B.L., Utiyama, M., Sumita, E.: Connecting phrase based statistical machine translation adaptation, In: Proceedings of the 26th International Conference on Computational Linguistics, pp. 3135–3145. ACM, Japan (2016)
10. Zhang, Z., Zhao, H., Qin, L.: Probabilistic graph-based dependency parsing with convolutional neural network. In: Proceedings of the 54th Annual Meeting of the Association for Computational Linguistics, pp. 1382–1392. ACL, Germany (2016)
11. Mikolov, T., Sutskever, I., Chen, K., Corrado, G.S., Dean, J.: Distributed representations of words and phrases and their compositionality. In: Proceedings of the 27th Annual Conference on Neural Information Processing Systems, pp. 3111–3119. MIT Press, United States (2013)
12. Pennington, J., Socher, R., Manning, C.D.: Glove: global vectors for word representation. In: The Proceedings of the 2014 Conference on Empirical Methods in Natural Language Processing, pp. 1532–1543. ACL, Qatar (2014)

13. Stein, R.A., Jaques, P.A., Valiati, J.F.: An analysis of hierarchical text classification using word embeddings. Inf. Sci. **471**, 216–232 (2019)
14. Zhang, W., Yoshida, T., Tang, X.: A comparative study of TF* IDF, LSI and multi-words for text classification. Expert Syst. Appl. **38**(3), 2758–2765 (2011)
15. El-Kishky, A., Song, Y., Wang, C., Voss, C., Han, J.: Scalable topical phrase mining from text corpora (2014). arXiv preprint arXiv:1406.6312
16. Mihalcea, R., Tarau, P.: Text rank: bringing order into texts. In: Proceedings of 2004 Conference Empirical Methods Natural Language Process (2004)

BioBERT Based Efficient Clustering Framework for Biomedical Document Analysis

Khishigsuren Davagdorj[1], Kwang Ho Park[1], Tsatsral Amarbayasgalan[1], Lkhagvadorj Munkhdalai[1], Ling Wang[2], Meijing Li[3(✉)], and Keun Ho Ryu[4,5,6(✉)]

[1] College of Electrical and Computer Engineering, Chungbuk National University, Cheongju 28644, South Korea
khishigsurend@chungbuk.ac.kr, {khblack,tsatsral,lhagii}@dblab.chungbuk.ac.kr

[2] Department of Computer Science and Technology, School of Computer Science, Northeast Electric Power University, Jilin City, China
smile2867ling@neepu.edu.cn

[3] College of Information Engineering, Shanghai Maritime University, 1550 Haigang Avenue Pudong New Area, Shanghai 213, China
mjli@shmtu.edu.cn

[4] Faculty of Information Technology, Ton Duc Thang University, Ho Chi Minh City 700000, Vietnam
khryu@tdtu.edu.vn, khryu@ieee.org

[5] Biomedical Engineering Institute, Chiang Mai University, Chiang Mai 50200, Thailand

[6] Department of Computer Science, College of Electrical and Computer Engineering, Chungbuk National University, Cheongju 28644, Korea

Abstract. The large volumes of biomedical documents have been generating exponentially in modern applications. Document clustering methods play an important role in gathering textual content documents into a few meaningful coherent groups. However, clustering unstructured and unlabeled text is challenging to extract informative representations and find the relevant articles from the rapid growth biomedical literature. Therefore, traditional text document clustering methods often represent unsatisfactory results due to general non-contextualized vector space representations, which neglect the semantic relation between bio medical texts. Pre-trained language models have been gaining attention recently in variety of natural language processing tasks. In this paper, we propose a heavily pre-trained language representation BioBERT based clustering framework for biomedical document analysis in order to improve the clustering accuracy. In experimental architecture, we provide benchmarks of the pre-trained transformer model, statistical technique and word-embedding methods while incorporating with clustering algorithms. In order to distinguish the efficiency of the models, Fowlkes mallows score (FM), silhouette coefficient (SC), adjusted rand index (ARI), Davies-Bouldin score (DB) metrics are used. The comprehensive experimental results show that the BioBERT based K-means model achieves better clustering accuracies than other models.

Keywords: Biomedical literature · BioBERT · Language representation model · Document clustering

S.-C. Chu et al. (Eds.): ICGEC 2021, LNEE 833, pp. 179–188, 2022.
https://doi.org/10.1007/978-981-16-8430-2_17

1 Introduction

Active research in the biomedical field has generated an overwhelming amount of literature. PubMed, a search engine accessing primarily from the largest life sciences and biomedical literature database, has more than 32M articles as of June 2021. There are increasingly growing demands for the development of techniques to extract valuable knowledge from these documents [1]. Hence, document clustering, a field at the intersection between Natural Language Processing (NLP) and machine learning, is useful in many applications such as grouping search engine results, automatic categorization of documents, and others. The purpose of document clustering comprises main two phases such as document representation and clustering.

In traditional clustering methods, data enormous volumes of unstructured and unlabeled data points are divided by specific groups according to their similar characteristics [2]. However, text data contain uninformative features that affect the poor performance of document clustering while computing distance measures. Therefore, previous clustering studies focused on dimension reduction in order to improve clustering accuracy. In several existing studies, Term Frequency Inverse Document Frequency (TF-IDF) was used to examine the relative frequency of words in documents, and then transformation is performed using principal component analysis (PCA) and auto-encoders (AE) for reducing the high dimension of the data [3, 4]. TF-IDF vectors have relatively inadequate information compare to other embedding techniques because dependencies and relationships are not reflected in TF-IDF vectors. Word2Vec is one of the utmost commonly acknowledged pre-trained word-embedding models developed by Google [5]. Although many document clustering research works have been successfully done using efficient Word2Vec and Glove embedding models, biomedical domain specified document clustering is different from the clustering task in the general domain corpus. Lately, the Embeddings from Language Models (ELMo) and Bidirectional Encoder Representations from Transformers (BERT) have proved the value of contextualized representations produced by deeper structures like bidirectional language models for transfer learning [6, 7]. These advanced contextualized representations are reached considerable performances in various applications.

Researchers focused on document clustering in the machine-learning field [8–10]. In a study [8], Park et al., proposed a simple and effective clustering framework termed Advanced Document Clustering. They applied GloVe, fastText, BERT and ELMo embedding models are used in SQuAD 1.1, Yahoo Answers, REUTERS and FakeNewsAMT datasets. Moreover, they solved the high-dimensional data problem by applying techniques such as cosine similarity based clustering and the mini-batch centroids update rule. Moreover, unsupervised clustering accuracy and normalized mutual information metrics were used for cluster evaluation. In another study [9], Kong proposed a scientific paper recommendation system, entitled by VOPRec. In their experiment, Physical Review A was used, which contains 9151 citation relationships of 7547 papers collected between 2007 and 2009. In their proposed system, the text information was performed over word embedding techniques on papers of similar research interest. Doc2Vec, Struc2Vec and DeepWalk have represented the text as a vector. For finding papers of similar network topology, they converted structural identity to vectors. In a study [10], Luo and Shah introduced a biomedical text-clustering framework based on diseases concepts.

In order to do it, they extracted the phrases of diseases and then generated the concept embedding through neural networks. After that, they extract the representations based on a new weighting scheme. Then, document clustering has done by k-means + PCA + t-SNE. In another study, the authors used the Trecgen collection, which covers 4478 abstracts extracted from the TREC 2005 Genomics Track. Clustering evaluation results prove that both aggregated word embedding and intact concept embedding-based weighting schemes performed better results than the TF-IDF. In recent times, contextualized word embedding techniques such as ELMo and BERT have confirmed the efficiency of contextualized word representations. BERT reaches considerable performances on various NLP tasks. However, these techniques cannot reach higher performance in the biomedical domain because they are pre-trained on only general domain corpora. Thus, this study is motivated by domain-specific BioBERT that is based on the BERT model pre-trained on large-scale biomedical corpora [11].

In this paper, we present a heavily pre-trained language representation, BioBERT model-based efficient clustering framework for biomedical document analysis for improving the clustering accuracy. In experimental architecture, we provide benchmarks of BioBERT, Word2Vec, TF-IDF with PCA and TF-IDF with AE incorporating K-means [12] and expectation-maximization clustering (EM) [13] algorithms respectively. In order to determine the effectiveness of the models, FM, SC, ARI, DB metrics [14] are used. The comprehensive experimental results show that BioBERT based K-means model achieves better clustering results than other models. Overall, the major contributions of this paper are as follows:

1. Provide an empirical comparison study of advanced machine learning methods based on a deep clustering framework for biomedical document analysis.
2. Experimental results prove that pre-trained transformer model, BioBERT based clustering models enhance the evaluation performance in PubMed's dataset.
3. Empirically, practical models have solved biomedical domain specified document clustering problem, also our findings would be significantly important for exploring the similar articles according to their inherent characteristics.

The rest of this paper is organized as follows: Sect. 2 describes the procedure for the deep clustering framework for biomedical document analysis, including text data pre-processing, vector representation models, optimization of clustering algorithms and evaluation metrics. Section 3 presents the overall experimental analysis and results. Finally, Sect. 4 completes the findings of this paper and future work.

2 Experimental Framework

In this section, we introduce the architecture of the deep contextualized representations based clustering framework for biomedical document analysis. The overview of the experimental framework is presented in Fig. 1. In general, deep representations based clustering framework can be divided into three components: data pre-processing, vector representation models and optimization of clustering documents in an unsupervised manner.

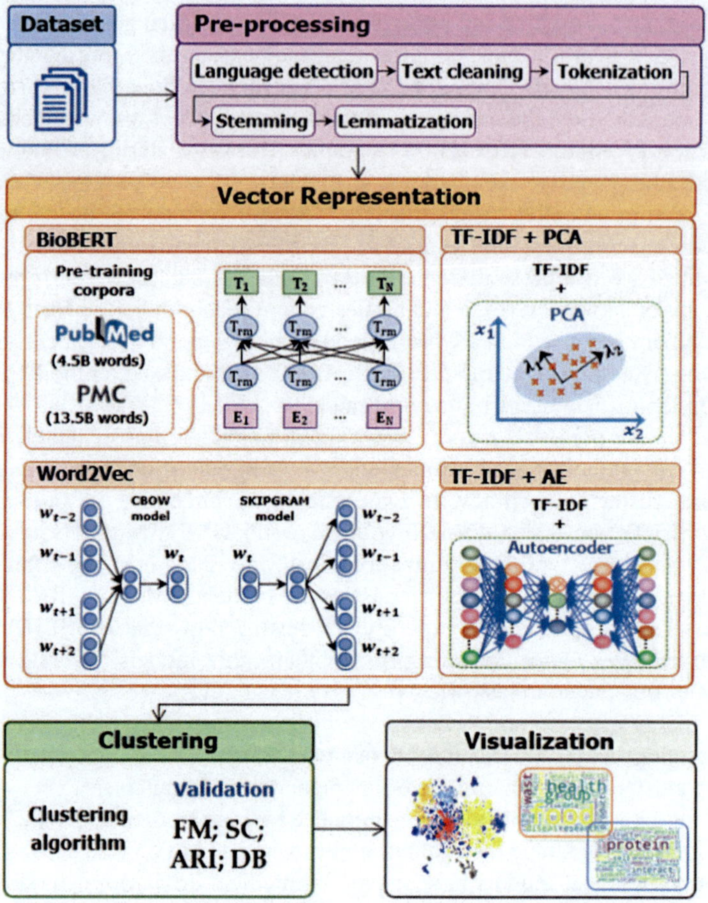

Fig. 1. Deep clustering framework for biomedical document analysis

2.1 Text Data Pre-processing

The real-world datasets cover valuable information; however, it is not in a form applicable for the required data analyzing process. Thus, it has to be cleaned in order to improve clustering effort [15]. In this study, we drop documents that detected any languages that are not English for handling multiple languages problems. The text-cleaning step considers removing stop words in English text, special characteristics, single numbers and noises. This cleaning step is done by using regular expressions to perform automatic searches of these specific patterns in the text to remove them. Common tasks are done based on tokenization, stemming and lemmatization. Tokenization technique splits the entire text document into singular words. Stemming and Lemmatization both generate the root form of the inflected words.

2.2 Vector Representations and Clustering Techniques

BioBERT: Bidirectional Encoder Representations from Transformers for Biomedical Text Mining (BioBERT) is a domain-specific language representation model pre-trained on large-scale biomedical corpora. BioBERT initialized with weights from BERT that was pre-trained on the general domain (English Wikipedia and BooksCorpus). Contrarily BioBERT is pre-trained on biomedical domain corpora of PubMed abstracts and PMC full-text articles. BioBERT outperforms various state-of-the-art models in various biomedical text-mining tasks.

Word2Vec: Developed by Mikolov [5], is one of the classical two-layer neural networks trained to generate a dense vector with a fixed dimension for each word. The skip-gram and continuous bag of words (CBOW) are utilized in the Word2Vec model. The skip-gram model predicts the context given a word. CBOW model is the exact opposite of the skip-gram model.

TF-IDF: Term Frequency-Inverse Document Frequency (TF-IDF) is a numerical statistic that estimates how relevant a word is to a document in a corpus. This technique is done by multiplying two metrics. Initially, it quantifies a word that appears in a document. Then compute a weight to each word in the document. In our study, we have used PCA and AE techniques for dimension reduction tasks in vectors produced from TF-IDF.

These models used in our study handle text in different ways, thus make different vector representations of a text document. The unit used by each model and the level at which they work is shown in Table 1.

Table 1. The vector representation models and their characteristics

Name	Unit	Level
BioBERT	Contextual string embedding	Sentences
Word2Vec	Words	Local context
TF-IDF	Words	Corpus

K-means: K-means is widely applied clustering for partitioning a given dataset into k clusters. K cluster centroids are initialized randomly, and then a collection of points is assigned to the cluster with the nearest mean, serving as a prototype of the cluster.

EM: Expectation-Maximization (EM) is an effective clustering algorithm for performing maximum likelihood estimation for parameters in probabilistic models. EM alternates between two steps of expectation (E) and maximization (M). EM computes probabilities according to each data points based on the probability distributions. Then, it maximizes the expected likelihood based on the new data assignments.

2.3 Evaluation Metrics

FM: Fowlkes-Mallows Score (FM) is an external evaluation metric that is defined as a mean of precision and recall values of a clustering:

$$FM = \sqrt{prec \cdot recall} = \frac{TP}{\sqrt{(TP + FN)(TP + FP)}}. \quad (1)$$

where: precision measures the fraction of a cluster that contains points of a specified class. While recall defines the extent to which a cluster comprises all points of a specified class.

SC: Silhouette Coefficient (SC) is the popular method that combines both cohesion and separation of clusters. It calculates the average distance to all other points in the cluster. Furthermore, when points are not contained in the cluster, calculate the point's average distance to all points in the closest cluster. Finding such value with respect to all clusters is calculated as follows:

$$S_i = \frac{\mu_{out}^{min}(x_i) - \mu_{in}(x_i)}{max\{\mu_{out}^{min}(x_i), \mu_{in}(x_i)\}} \quad (2)$$

where: $\mu_{in}(x_i)$ is the mean distance from x_i to points in its particular cluster, $\mu_{out}^{min}(x_i)$ - is the mean of the distances from x_i to points in the closest cluster. The S_i value of a point remains between $[-1, +1]$ interval. A value near to $+1$ specifies a well-defined clustering.

ARI: Adjusted Rand Index (ARI) calculates a similarity metric among two clusterings by estimating all points and counting points that are belonged in the same or different clusters in the predicted and correct clustering as following:

$$ARI = \frac{Index - Expected\ Index}{Max\ Index - Excepted\ Index} \quad (3)$$

Similarity score occurs between -1 and 1.0. The ARI closes 0.0 to 1.0 stands for good clustering.

DB: Davies Bouldin (DB) index defines how compact the clusters are compared to the distance between means of the clusters. The index is then calculated as follows:

Let μ_i indicate the cluster mean, known as:

$$\mu_i = \frac{1}{n_i} = \sum\nolimits_{x_j \in C_i} x_j \quad (4)$$

$$\sigma\mu_i = \sqrt{\frac{\sum_{x_j \in C_j} \delta(x_j, \mu_i)^2}{n_i}} = \sqrt{var(C_i)} \quad (5)$$

$$DB = \frac{1}{k} \sum\nolimits_{i=1}^{k} \max_{j \neq 1} \{DB_{ij}\} \quad (6)$$

where: $\sigma_{\mu i}$ indicates the dispersion of the points around the cluster means. $var(C_i)$ – total of variance. The minimum score is zero, with lower values indicating better clustering.

3 Experimental Results

In this paper, all experiments were completed on a computer with Intel Core i5-6600K, CPU 3.50 GHz and 32.0 GB RAM using a Microsoft Windows 10. The Python 3.5 programming language was employed. General machine learning and NLP tools were used from Gensim, NTLK, Tensor-flow, Scikit-learn and other libraries, to build the comparative experiments.

For mining biomedical document analysis, we have evaluated our deep contextualized representations based clustering framework on 5,750 abstract text data collected from the real-world PubMed. In an empirical comparison analysis, we have used BioBERT, Word2Vec, TF-IDF with PCA and AE for vector representations while incorporating K-means and EM clustering algorithms, respectively. It is well known that TF-IDF is a scoring method that assigns a score to each word in the sentence. Thus, we characterize documents by TF-IDF vectors, and then attain to reduce the dimension using PCA and AE techniques. Table 2 and Table 3 present the evaluation performances of K-means and EM-based clustering for biomedical documents and the best scores of each evaluation metric are marked in bold.

Table 2. Evaluation results of K-means based clustering for biomedical document.

Representations	Fowlkes mallows score	Silhouette coefficient	Adjusted rand index	Davies bouldin score
BioBERT	**0.7712**	**0.3041**	**0.4369**	**1.8507**
Word2Vec	0.5794	0.2395	0.1025	2.7911
TF-IDF with PCA	0.4672	0.0623	0.0719	3.8127
TF-IDF with AE	0.5531	0.0867	0.2758	3.4395

In Table 2, the elbow method is done by varying the number of clusters passed as input to K-Means between 2 and 15. K-means clustering technique with contextualized representations from BioBERT achieves the best FM of 0.7712, SC of 0.3041, ARI of 0.4369 and DB of 1.8507, significantly. Following it, the Word2Vec embedding based K-means model reaches FM of 0.5794, SC of 0.2395, ARI of 0.1025, and DB of 2.7911.

Table 3. Evaluation results of EM based clustering for biomedical document.

Representations	Fowlkes mallows score	Silhouette coefficient	Adjusted rand index	Davies bouldin score
BioBERT	**0.6798**	**0.2762**	**0.3258**	**2.2121**
Word2Vec	0.5339	0.2112	0.0875	4.8135
TF-IDF with PCA	0.4545	0.0752	0.0626	5.7447
TF-IDF with AE	0.5102	0.1118	0.0684	4.8371

As shown in Table 3, considering these evaluation performances of high-performing models based on EM clustering technique, BioBERT indicated the best vector representation than others. The best FM of 0.6798, SC of 0.2762, ARI of 0.3258 and DB of 2.2121 results are reached by contextualized representations from BioBERT with EM model. Meanwhile, TF-IDF with PCA representations based EM model presents the worst results for FM of 0.4545, SC of 0.0752, ARI of 0.0626 and DB of 5.7447.

Figure 2 illustrates the comparison results of biomedical document clustering in the PubMed dataset. Contextualized document representations from BioBERT outperform feature-based Word2Vec, TF-IDF with PCA, TF-IDF with AE representations among the overall evaluation performances. Therefore, we have applied K-means and EM clustering techniques to these representations. According to the evaluation metrics, it is clearly shown that K-means based results were better than the EM clustering technique.

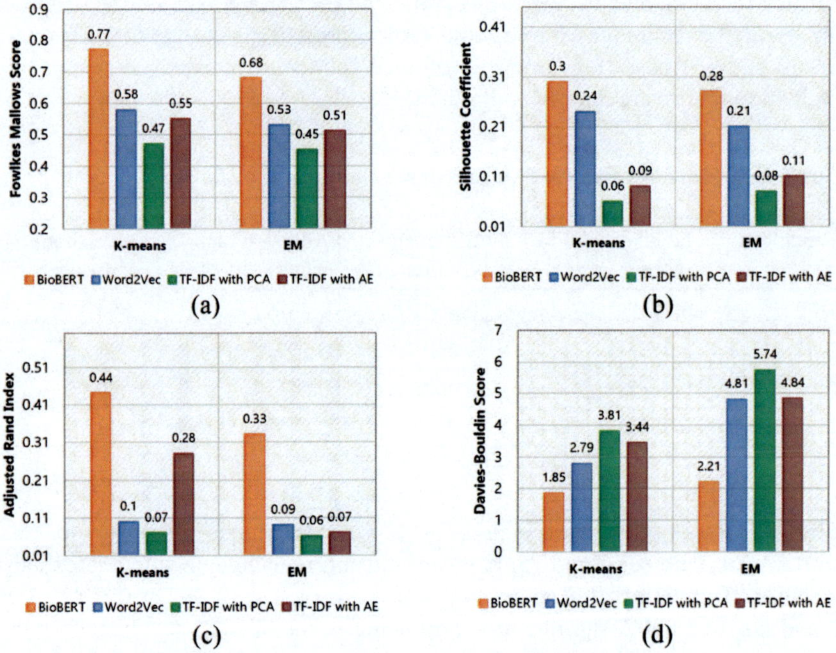

Fig. 2. Comparison results of biomedical document clustering: (a) Fowlkes Mallows score; (b) Silhouette coefficient; (c) Adjusted rand index; (d) Davies-Bouldin score

The best FM score is achieved by the BioBERT based K-means with the score of 0.7712, and following it, BioBERT based EM clustering with the score of 0.6798. Essentially, AE based TF-IDF vectors embedded slightly higher result than the PCA based TF-IDF models among FM, SC, ARI and DB evaluation metrics. Moreover, it should be noted that PCA based TF-IDF performs the lowest scores in FM of 0.4545, SC of 0.0623, ARI of 0.0626 and DB of 5.7447 when cooperating with the EM clustering technique. The reason of it, in TF-IDF, words are not obtained by sequentially, furthermore, dependencies and relationships between words cannot be reflected. Thus, TF-IDF offers less informative and ineffective representations.

Fig. 3. Word cloud representations for the clusters of biomedical documents

In Fig. 3, we provide visual representations for the clusters of biomedical documents using the "WordCloud" python library. Word cloud is a powerful visual representation that used to display the more frequently appeared words in text data. Otherwise, the most frequent words in a cluster appear to be bigger. Contextualized representations from BioBERT based K-means clustering model defined the best clustering model. Thus, we visualize the word clouds of clusters produced by that best clustering model.

4 Conclusion and Future Work

A huge amount of biomedical literature has been generated exponentially. Relevant articles can be found by simple keyword-based query, but which is a non-efficient and complex task on the large-scale database. Thus, clustering unstructured and unlabeled text is becoming increasingly important to generate documents into a few meaningful coherent groups. In this paper, we present pre-trained language representation from BioBERT model-based efficient clustering framework for biomedical document analysis in order to improve the clustering accuracy. In experimental architecture, we provide benchmarks for the combination of different variants of BioBERT, Word2Vec, TF-IDF with PCA and TF-IDF with AE methods while incorporating with K-Means and EM clustering algorithms. Experimental results prove that pre-trained transformer model, BioBERT based clustering models enhance the evaluation performance in Pub-Med's dataset among other models. In the future study, we will study the most suitable domain-specified document clustering techniques in order to enhance the clustering accuracy.

Acknowledgement. This research was supported by Basic Science Research Program through the National Research Foundation of Korea (NRF) funded by the Ministry of Science, ICT & Future Planning (No.2019K2A9A2A06020672) and, (No.2020R1A2B5B02001717) also by the

National Natural Science Foundation of China (Grant No. 61702324 and Grant No. 61911540482) in People's Republic of China.

References

1. Li, M., Hu, J., Ryu, K.H.: An efficient tool for semantic biomedical document analysis. In: Pan, J.-S., Li, J., Ryu, K.H., Meng, Z., Klasnja-Milicevic, A. (eds.) Advances in Intelligent Information Hiding and Multimedia Signal Processing. SIST, vol. 212, pp. 514–522. Springer, Singapore (2021). https://doi.org/10.1007/978-981-33-6757-9_63
2. Batbaatar, E., Pham, V.H., Ryu, K.H.: Multi-task topic analysis framework for hallmarks of cancer with weak supervision. Appl. Sci. **10**(3), 834 (2020)
3. Park, K.H., Batbaatar, E., Piao, Y., Theera-Umpon, N., Ryu, K.H.: Deep learning feature extraction approach for hematopoietic cancer subtype classification. Int. J. Environ. Res. Public Health **18**(4), 2197 (2021)
4. Amarbayasgalan, T., Jargalsaikhan, B., Ryu, K.H.: Unsupervised novelty detection using deep autoencoders with density based clustering. Appl. Sci. **8**(9), 1468 (2018)
5. Mikolov, T., Chen, K., Corrado, G., Dean, J.: Efficient estimation of word representations in vector space(2013). arXiv preprint arXiv:1301.3781
6. Peters, M.E., et al.: Deep contextualized word representations (2018). arXiv preprint arXiv: 1802.05365
7. Devlin, J., Chang, M.W., Lee, K., Toutanova, K.: Bert: pre-training of deep bidirectional transformers for language understanding (2018). arXiv preprint arXiv:1810.04805
8. Park, J., Park, C., Kim, J., Cho, M., Park, S.: ADC: advanced document clustering using contextualized representations. Expert Syst. Appl. **137**, 157–166 (2019)
9. Kong, X., Mao, M., Wang, W., Liu, J., Xu, B.: VOPRec: Vector representation learning of papers with text information and structural identity for recommendation. IEEE Trans. Emerg. Topics Comput. **9**, 226–237 (2018)
10. Luo, X., Shah, S.: Concept embedding-based weighting scheme for biomedical text clustering and visualization. Appl. Inf. **5**(1), 1–19 (2018)
11. Lee, J., et al.: BioBERT: a pre-trained biomedical language representation model for biomedical text mining. Bioinformatics **36**(4), 1234–1240 (2020)
12. Likas, A., Vlassis, N., Verbeek, J.J.: The global k-means clustering algorithm. Pattern Recogn. **36**(2), 451–461 (2003)
13. Langford, J., et al.: Expectation maximization clustering. In: Sammut, Claude, Webb, Geoffrey I. (eds.) Encyclopedia of Machine Learning, pp. 382–383. Springer US, Boston, MA (2010). https://doi.org/10.1007/978-0-387-30164-8_289
14. Aljarah, I., Habib, M., Nujoom, R., Faris, H., Mirjalili, S.: A comprehensive review of evaluation and fitness measures for evolutionary data clustering. In: Aljarah, I., Faris, H., Mirjalili, S. (eds.) Evolutionary Data Clustering: Algorithms and Applications. AIS, pp. 23–71. Springer, Singapore (2021). https://doi.org/10.1007/978-981-33-4191-3_2
15. Davagdorj, K., Lee, J.S., Pham, V.H., Ryu, K.H.: A comparative analysis of machine learning methods for class imbalance in a smoking cessation intervention. Appl. Sci. **10**(9), 3307 (2020)

Distantly Supervised Relation Extraction Based on Multi-head Self-attention and Gate Mechanism

Fengzhu Xu[1], Meijing Li[1](✉), and Keun Ho Ryu[2,3,4]

[1] College of Information Engineering, Shanghai Maritime University, Shanghai 201306, China
mjli@shmtu.edu.cn

[2] Data Science Laboratory, Faculty of Information Technology, Ton Duc Thang University, Ho Chi Minh 700000, Vietnam

[3] Biomedical Engineering Institute, Chiang Mai University, Chiang Mai 50200, Thailand

[4] Department of Computer Science, College of Electrical and Computer Engineering, Chungbuk National University, Cheongju 28644, Korea

Abstract. In the field of distant supervision relation extraction, PCNN (piecewise convolution neural network) is normally involved to trap local features of sentences, and has achieved good results. However, the existing PCNN-based methods are unable to capture the features of long-distance interdependence in sentences and cannot distinguish the influence of the three segments of PCNN on relation classification, which probably results in missing some vital information. To address the above two issues, we come up with a new model with multi-head self-attention and gate mechanism for distant supervised relation extraction. Firstly, to capture the features of long-distance interdependence in sentences, we employ an internal multi-head self-attention in PCNN model, which can grab information in different representation subspaces. Secondly, to distinguish the influence of the three segments of the piecewise max-pooling output on the relation classification, the gate mechanism is introduced to assign different weights to the three segments and highlight important segments. After a series of experiments, it is proven that the model we presented is better than the previous approaches on each evaluation criterion.

Keywords: Multi-head self-attention · Gate mechanism · Relation extraction · Distantly supervised

1 Introduction

With the explosion of information, extracting useful information from massive unstructured text becomes particularly important. For structural or non-structural data, information extraction extracts useful semantic information from it and is the core of constructing knowledge base. At the same time, relation extraction, as an important part of information extraction, has gained the attention of several researchers. Its goal is to extract the relationship that exists between the entity pairs of the sentence. Among the relation

S.-C. Chu et al. (Eds.): ICGEC 2021, LNEE 833, pp. 189–199, 2022.
https://doi.org/10.1007/978-981-16-8430-2_18

extraction methods, supervised learning is primarily targeted at labeled data and requires a high level of quality, but the cost of obtaining labeled data is large and inefficient.

To cope with these issues, Mintz proposed to align text with the knowledge base to quickly construct labeled data [1]. They proposed a hypothesis: If the relation between the head entity and the tail entity is determined in the knowledge base, so any sentences that have both entities express this relation. However, in fact, many sentences that have the same entities may not be consistent in their relations, which may cause some noise data.

To alleviate noise data problem, the current methods are mainly divided into rule statistics, multi-instance learning and reinforcement learning. The denoising methods based on rule statistical mine explicit structural information between entities [2–5]. However, it relies too much on many features engineering, which makes the relation classification effect poor. Therefore, Hoffmann proposed a multi-instance method [6], which means that the entity pairs in a bag are all same, and the prediction is made by the bag as unit. Based on multi-instance learning, to mitigate the effect of noisy labels on relation extraction, Surdeanu proposed a multi-instance multi-label learning approach (MIML) [7]. To be able to automatically capture contextual features of sentences, piecewise convolution neural network was introduced [9, 10]. Recent works based on attention and multi-instance learning framework have achieved good results [8, 11, 13–18, 27–30]. The denoising methods based on reinforcement learning improve the quality of corpus [19–21]. In addition, some researchers designed the entity-aware embedding representation layer to seize important features in sentences more effectively [22, 23]. However, existing methods still have two issues: (1) PCNN-based methods are poor at processing long sentences, since these methods are unable to learn the features of long-distance interdependence of sentence wholeness. (2) Existing methods ignore the varied effects of three segments in piece-max-pooling on relation classification.

To address the shortcomings of existing approaches, we come up with a new approach (MaG) to relation extraction. Our main contributions are summarized as: (1) to capture the long-distance interdependence features and internal structure in sentences, we employ an internal multi-head self-attention mechanism in PCNN model. (2) to distinguish the influence of three segments of piecewise max-pooling output on the relation extraction, gate mechanism is employed to assign more weight to the important parts and less weight to the noisier parts.

2 Proposed Method

We put forward a new approach to distant supervision relation extraction, including Entity-Aware Embedding module, PCNN with Multi-head Self-attention module, Gate Mechanism and Bag Representation. Figure 1 provides the overall architecture of the MaG.

2.1 Entity-Aware Embedding

To capture syntactic information between two entities, we use word2vec [26] to map the words in the sentences $s = [w_1, w_2, \ldots, w_n]$ to a low-dimensional dense vector to obtain word embedding representation and a sequence $v = [v_1, v_2, \ldots, v_n] \in R^{n \times d_w}$, where n and d_w denote sentence length and dimensionality of the word embedding. What's more, to express the relative position information, our study introduces the position vector. The two relative positions of each word in the sentence to the head and tail entities are converted into two low-dimensional vectors also known as position vectors. $r^{ph}, r^{pt} \in R^{d_p}$. Finally, we obtain the position embedding by splicing the two position vectors with the word embedding vector. $X^p = [x_1^{(p)}, x_2^{(p)}, \ldots, x_n^{(p)}] \in R^{n \times (d_w + 2d_p)}$, where $x_i^p = [v_i; r_i^{ph}; r_i^{pt}]$, $1 \le i \le n$, d_p is the dimension of position embedding.

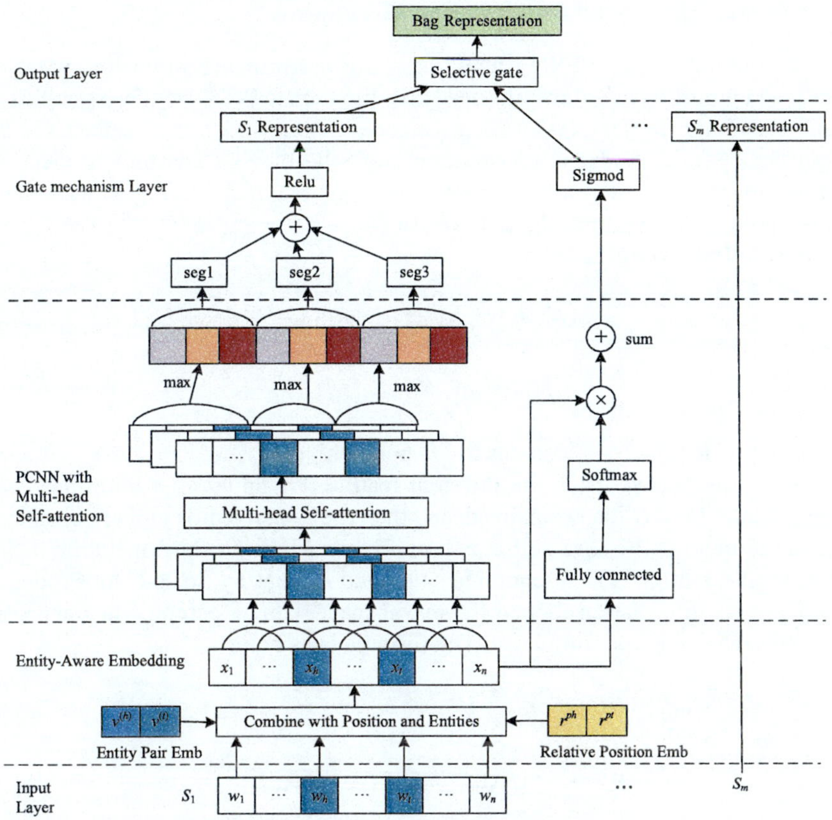

Fig. 1. MaG model architecture schematic.

For better enrichment of the relevant information in word embeddings, this work uses entity pair embedding. Entity pair embedding is defined as $X^{(e)} = [x_1^{(e)}, x_2^{(e)}, \ldots, x_n^{(e)}] \in R^{n \times 3d_w}$, where $x_i^{(e)} = [v_i; v^{(h)}; v^{(t)}]$,$1 \le i \le n$. Finally, entity-aware embedding representation is $X = [x_1, x_2, \ldots, x_n] \in R^{n \times d_x}$, d_x denotes entity-aware embedding dimension. These steps are written as

$$A = w^{(e)} X^{(e)} + b^{(e)} \tag{1}$$

$$M^{(e)} = sigmoid(\lambda \cdot A) \tag{2}$$

$$\widetilde{X}^{(p)} = \tanh(w^{(p)}X^{(p)} + b^{(p)}) \tag{3}$$

$$X = M^{(e)} \cdot X^{(e)} + (1 - M^{(e)}) \cdot \widetilde{X}^{(p)} \tag{4}$$

where $w^{(e)} \in R^{d_x \times 3d_w}$, $w^{(p)} \in R^{d_x \times d_p}$, λ is control the smoothness of the parameters.

2.2 PCNN with Multi-head Self-attention Mechanism

To solve this problem that PCNN usually has poor performance in dealing with long sentences, we employ an internal multi-head self-attention to PCNN model, which allows the MaG to capture the features of long-distance interdependence of sentences. This layer consists of three modules: convolution, multi-head self-attention and piecewise maximum pooling. Convolution is capable of learning features. Sliding window size is k and the output of convolution layer is $H^c = \{h_1^c, h_2^c, \ldots, h_n^c\} \in R^{n \times d_c}$. The formula for the convolution layer is as follows

$$h_i^c = w^{(c)} x_{i;i+k-1} + b^{(c)} \tag{5}$$

$$H^c = \{h_1^c, h_2^c, \ldots, h_n^c\} \tag{6}$$

where $w^{(c)} \in R^{d_c \times k \times d_x}$, d_c denotes filters number.

After convolution, we catch the different features by applying a multi-head self-attention module. First H^c is normalized, and then the model learns global dependency information through multi-head self-attention. Figure 2 indicates the operation flow of multi-head self-attention. The attention is computed h times by Scaled Dot-Product so that more relevant information about different subspaces can be learned. The final output is sent to piecewise max-pooling layer. The formula is as below

$$head_j = Attention(w_j^Q H^c, w_j^K H^c, w_j^V H^c), 1 \le j \le h \tag{7}$$

$$H^a = w^a[head_1; head_2; \ldots; head_h] \tag{8}$$

where h and d_h is the number and dimension of multi-head respectively. $w_j^Q, w_j^K, w_j^V \in R^{d_h \times d_c}$. To capture the structured features among entity pairs, we conduct a pooling approach for every segment through piecewise max-pooling layer. The equation is as below

$$q = [Pool(H_1^a); Pool(H_2^a); Pool(H_3^a)] \in R^{3 \times d_c} \tag{9}$$

where $Pool$ is piecewise max-pooling operation.

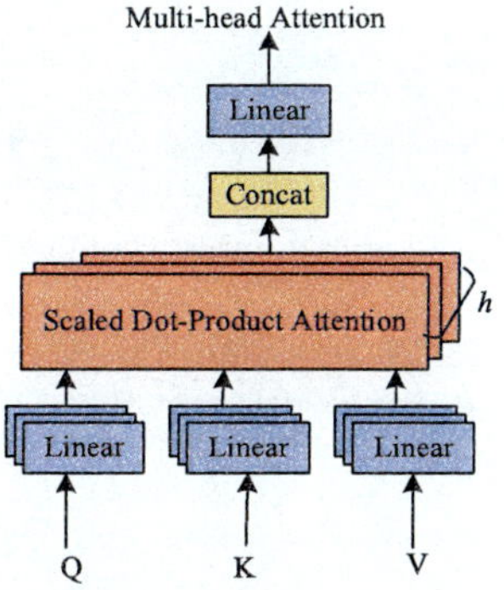

Fig. 2. This is the workflow diagram of multi-head self-attention.

2.3 Gate Mechanism

For purpose of distinguishing between the three parts of the piecewise max-pooling layer on relation classification, gate mechanism is used to assign a large weight to the important segments, and obtain the sentence representation. The formula of sentence representation is as below

$$g^i = \sigma(w^s q^i + b^s) \tag{10}$$

$$p^i = g^i \cdot q^i \tag{11}$$

$$s = relu([p^1; p^2; p^3]) \in R^{n \times 3d_c} \tag{12}$$

where σ denotes the sigmod activation function, g^i denotes a gate vector, and $w^s \in R^{d_c \times d_c}$ is a learning parameter.

2.4 Bag Representation

To help PCNN capture rich dependency information, according to previous studies [22], we use a lightweight self-attention to obtain a sentence-level vector representation m. These steps are described as

$$D = w^{a2}\sigma(w^{a1}x + b^{a1}) + b^{a2} \tag{13}$$

$$p^{(D)} = softmax(D) \tag{14}$$

$$m = \sum p^{(D)} \cdot X \tag{15}$$

where w^{a1} and $w^{a2} \in R^{d_x \times d_x}$, $p^{(D)}$ is the attention probability matrix. We use selective gate to combine s and m to get bag representation. The formula is as follows

$$g_i = \sigma(w^{g1}\sigma(w^{g2}m_j + b^{g2}) + b^{g1}) \in (0, 1), 1 \le j \le n \tag{16}$$

$$b = \frac{1}{m}\sum_{j=1}^{n} g_i \cdot s_i \tag{17}$$

where $w^{g1} \in R^{3d_c \times d_x}$, $w^{g2} \in R^{d_x \times d_x}$. Then the relation between the head and tail entity is determined by softmax classifier.

$$p = softmax(MLP(b)) \tag{18}$$

2.5 Loss Function

When training the model, we use negative log likelihood loss functions and L2 regularization. Training formula is as below

$$L = -\frac{1}{|D|}\sum\nolimits_{k=1}^{|D|} logp^{k}_{(i=y^k)} + \beta||\theta||_2^2 \tag{19}$$

where D is the dataset, p^k represents the prediction distribution, y^k means the predicted label, and θ is a generalized representation of the parameters of the model.

3 Experiments

3.1 Datasets, Evaluation Criteria and Parameters Settings

In evaluating the model, we use the New York Times (NYT) dataset [3], which is broadly employed in the field of distant supervision. It contains 53 relations, with 570k sentences in the training set and 172k in the test set. As with previous studies [9, 22], held-out evaluation method is conducted in our study to demonstrate the validity of the model. Specific evaluation indicators, such as PR (Precision-Recall) curves, AUC values and P@N. Based on Lin's parameter settings [18], we apply SGD and prevent overfitting by using a dropout strategy [24, 25]. d_w, d_p, d_x are set to 50, 5 and 150, respectively; The number of head, dropout rate, number of convolutional kernels, learning rate and d_c are set to 5, 0.5, 3, 0.1 and 230, respectively.

3.2 Baseline Models

This experiment compares MaG with previous models in attempt to demonstrate the effectiveness of our model. Including, (1) Mintz [1]: the original distant supervision method. (2) MultiR [6]: a graph structure that belongs to multi-instance learning. (3) MIMLRE [7]: according to MultiR, this method adds a multi-label approach to reduce noise. (4) PCNN [12]: unlike CNNs, this approach segments entities based on their locations and then pools them separately. (5) PCNN + ATT [11]: using a selective attention after the PCNN. (6) SeG [22]: self-attention-enhanced PCNN and entity-aware embedding are used to represent sentences, and the selective gate is used to denoise.

3.3 Comparison with Traditional Methods

Figure 3 and Table 2 indicate the PR curves and AUC values of the MaG model and the baseline models, respectively.

From the PR curves, we can discover a phenomenon that MaG model performs the best. (1) When the recall value starts to exceed 0.1, the precision values of the first three models start to decrease rapidly, which indicates that the feature-based approach will not capture the features accurately. (2) Compared with the latter three models, our MaG model has a clear advantage, which shows that multi-head self-attention enables the MaG to capture global features in sentences more effectively. At the same time, it also proves effectiveness of the gate mechanism for relation extraction. In the AUC values, the MaG model reaches 0.515, which is higher than SeG model by 0.058.

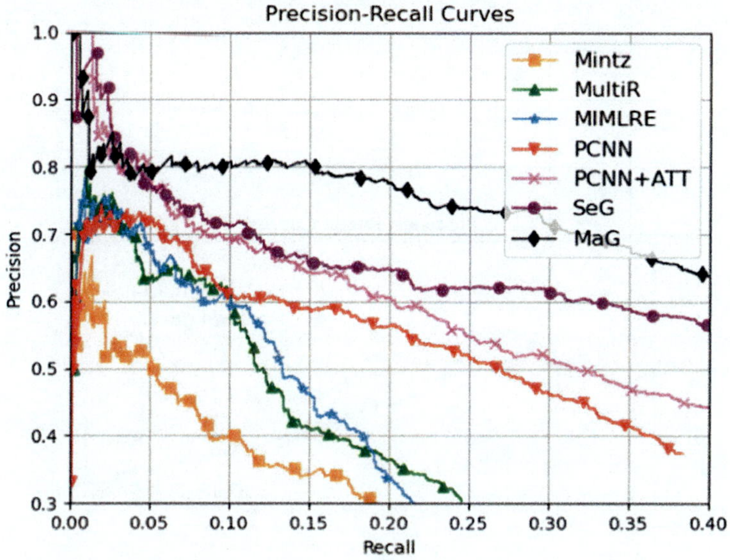

Fig. 3. Precision-recall curves of MaG and comparative models.

To avoid the influence of False Negative on relation prediction evaluation, we use Precision@N test. One or two or all sentences from the bag are randomly selected for testing. Table 1 represents the values of P@N. From the table we can notice that MaG has the highest value compared to the previous models. As opposed to SeG, the P@N value increases by 4.4%.

3.4 Ablation Experiment

Since it was necessary to confirm the validity of each part of the MaG model, we carry out ablation experiment (Fig. 4).

Compared with all variant models, the MaG results are significantly better. When MaG w/o multi-head self-attention, AUC value decreases to 0.461, while the P@mean

Table 1. P@N results of comparative approaches.

P@N (%)	One				Two				All			
	100	200	300	Mean	100	200	300	Mean	100	200	300	Mean
PCNN	73.3	64.8	56.8	65.0	70.3	67.2	63.1	66.9	72.3	69.7	64.1	68.7
PCNN + ATT	73.3	69.2	60.8	67.8	77.2	71.6	66.1	71.6	76.2	73.1	67.4	72.2
SeG	89.0	86.0	82.0	85.7	90.0	87.5	84.7	87.6	90.0	88.0	84.7	87.6
MaG	**95.0**	**88.5**	**83.7**	**89.1**	**95.0**	**90.5**	**86.7**	**90.7**	**96.0**	**92.0**	**88.0**	**92.0**

Table 2. AUC values of comparative approaches.

Approach	AUC
PCNN + ATT	0.390
SeG	0.457
MaG	**0.515**

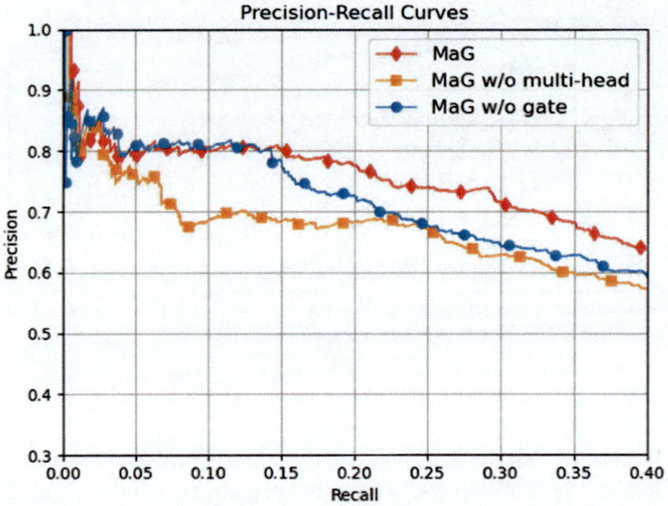

Fig. 4. Precision-recall results of MaG, w/o multi-head and w/o gate.

decreases by 6.1%, which proves the role of multi-head self-attention in capturing global dependency in sentences. When we remove the gate mechanism, AUC value declines to 0.488 and P@mean is reduced by 2.0%, which proves that gate mechanism is effective for relation classification (Tables 3 and 4).

Table 3. Precision@N results of MaG, w/o multi-head and w/o gate.

P@N (%)	One				Two				All			
	100	200	300	Mean	100	200	300	Mean	100	200	300	Mean
MaG	95.0	88.5	83.7	89.1	95.0	90.5	86.7	90.7	96.0	92.0	88.0	92.0
w/o multi-head	89.0	87.0	82.0	86.0	89.0	87.5	83.7	86.7	87.0	87.5	83.3	85.9
w/o gate	93.0	88.0	83.7	88.2	94.0	89.5	86.0	89.8	95.0	89.0	86.0	90.0

Table 4. AUC values of MaG and all variant models.

Approach	AUC
MaG	**0.515**
MaG w/o multi-head	0.461
MaG w/o gate	0.488

4 Conclusion

For this study, a distant supervision approach using multi-head self-attention and gate mechanism is presented to extract relations. In particularly, multi-head self-attention has been able to capture the features of long-distance interdependent and internal structures in sentences. The gate mechanism is applied to assign more weight to the important parts to distinguish the different contributions to relation extraction. Several experiment results confirm that our MaG approach is beyond the previous models in all evaluation criteria. In the future, we will introduce external knowledge base to eliminate noise.

Acknowledgement. This study was supported by the National Natural Science Foundation of China (61911540482 and 61702324).

References

1. Mintz, M., Bills, S., Snow, R., Jurafsky, D.: Distant supervision for relation extraction without labeled data. In: Proceedings of the Joint Conference of the 47th Annual Meeting of the ACL and the 4th International Joint Conference on Natural Language Processing of the AFNLP, pp. 1003–1011. ACL, Suntec (2009)
2. Zelenko, D., Aone, C., Richardella, A.: Kernel methods for relation extraction. J. Mach. Learn. Res. **3**(Feb), 1083–1106 (2003)
3. Riedel, S., Yao, L., McCallum, A.: Modeling relations and their mentions without labeled text. In: Balcázar, J.L., Bonchi, F., Gionis, A., Sebag, M. (eds.) ECML PKDD 2010, LNCS, vol. 6323, pp. 148–163. Springer, Berlin, Heidelberg (2010)
4. Takamatsu, S., Sato, I., Nakagawa, H.: Reducing wrong labels in distant supervision for relation extraction. In: Proceedings of the 50th Annual Meeting of the Association for Computational Linguistics, vol. 1: Long Papers, pp. 721–729. ACL, Jeju Island (2012)

5. Zhang, Q., Wang, H.: Noise-clustered distant supervision for relation extraction: a nonparametric bayesian perspective. In: Proceedings of the 2017 Conference on Empirical Methods in Natural Language Processing, pp. 1808–1813. ACL, Copenhagen (2017)
6. Hoffmann, R., Zhang, C., Ling, X., Zettlemoyer, L., Weld, D.S.: Knowledge-based weak supervision for information extraction of overlapping relations. In: Proceedings of the 49th Annual Meeting of the Association for Computational Linguistics: Human Language Technologies, pp. 541–550. ACL, Portland (2011)
7. Surdeanu, M., Tibshirani, J., Nallapati, R., Manning, C.D.: Multi-instance multi-label learning for relation extraction. In: Proceedings of the 2012 joint conference on Empirical Methods in Natural Language Processing and Computational Natural Language Learning, pp. 455–465. ACL, Jeju Island (2012)
8. Xiao, Y., Jin, Y., Cheng, R., Hao, K.: Hybrid Attention-Based Transformer Block Model for Distant Supervision Relation Extraction (2020). arXiv e-prints, arXiv:2003.11518
9. Zeng, D., Kang, L., Chen, Y., Zhao, J.: Distant supervision for relation extraction via piecewise convolutional neural networks. In: Proceedings of EMNLP, pp. 1753–1762. ACL, Lisbon (2015)
10. Zeng, D., Liu, K., Lai, S., Zhou, G., Zhao, J.: Relation classification via convolutional deep neural network. In: Proceedings of COLING 2014, The 25th International Conference on Computational Linguistics: Technical Papers, pp. 2335–2344. DCU and ACL, Dublin (2014)
11. Lin, Y., Shen, S., Liu, Z., Luan, H., Sun, M.: Neural relation extraction with selective attention over instances. In: Proceedings of the 54th Annual Meeting of the Association for Computational Linguistics, vol. 1: Long Papers, pp. 2124–2133. ACL, Berlin (2016)
12. Vaswani, A., et al.: Attention is all you need. In: Proceedings of the 31st International Conference on Neural Information Processing Systems, pp. 6000–6010. Curran Associates Inc., Long Beach (2017)
13. Du, J., Han, J., Way, A., Wan, D.: Multi-level structured self-attentions for distantly supervised relation extraction. In: Proceedings of the 2018 Conference on Empirical Methods in Natural Language Processing, pp. 2216–2225. ACL, Brussels (2018)
14. Li, X., Taheri, A., Tu, L., Gimpel, K.: Commonsense knowledge base completion. In: Proceedings of the 54th Annual Meeting of the Association for Computational Linguistics, vol. 1: Long Papers, pp. 1445–1455. ACL, Berlin (2016)
15. Ye, Z.-X., Ling, Z.-H.: Distant supervision relation extraction with intra-bag and inter-bag attentions. In: Proceedings of the 2019 Conference of the North American Chapter of the Association for Computational Linguistics: Human Language Technologies, vol. 1 Long and Short Papers, pp. 2810–2819. ACL, Minneapolis (2019)
16. Ji, G., Liu, K., He, S., Zhao, J.: Distant supervision for relation extraction with sentence-level attention and entity descriptions. In: Proceedings of the Thirty-First AAAI Conference on Artificial Intelligence, pp. 3060–3066. AAAI Press, San Francisco (2017)
17. Han, X., Yu, P., Liu, Z., Sun, M., Li, P.: Hierarchical relation extraction with coarse-to-fine grained attention. In: Proceedings of the 2018 Conference on Empirical Methods in Natural Language Processing, pp. 2236–2245. ACL, Brussels (2018)
18. Vashishth, S., Joshi, R., Prayaga, S.S., Bhattacharyya, C., Talukdar, P.: Reside: improving distantly-supervised neural relation extraction using side information. In: Proceedings of the 2018 Conference on Empirical Methods in Natural Language Processing, pp. 1257–1266. ACL, Brussels (2018)
19. Xiao, Y., Tan, C., Fan, Z., Xu, Q., Zhu, W.: Joint entity and relation extraction with a hybrid transformer and reinforcement learning based model. Proc. AAAI Conf. Artif. Intell. **34**, 9314–9321 (2020)
20. Wang, X., Yang, J., Wang, Q., Su, C.: Threat intelligence relationship extraction based on distant supervision and reinforcement learning. In: SEKE, pp. 572–576. Knowledge Systems Institute Graduate School, Skokie (2020)

21. Chen, T., Wang, N., He, M., Sun, L.: Reducing wrong labels for distantly supervised relation extraction with reinforcement learning. IEEE Access **8**, 81320–81330 (2020)
22. Li, Y., et al.: Self-attention enhanced selective gate with entity-aware embedding for distantly supervised relation extraction. Proc. AAAI Conf. Artif. Intell. **34**(05), 8269–8276 (2020)
23. Wen, H., Zhu, X., Zhang, L., Li, F.: A gated piecewise CNN with entity-aware enhancement for distantly supervised relation extraction. Inf. Process. Manag. **57**(6), 102373 (2020)
24. Cotter, A., Shamir, O., Srebro, N., Sridharan, K.: Better mini-batch algorithms via accelerated gradient methods. In: Proceedings of the 24th International Conference on Neural Information Processing Systems, pp. 1647–1655. Curran Associates Inc., Red Hook (2011)
25. Srivastava, N., Hinton, G., Krizhevsky, A., Sutskever, I., Salakhutdinov, R.: Dropout: a simple way to prevent neural networks from overfitting. J. Mach. Learn. Res. **15**(1), 1929–1958 (2014)
26. Mikolov, T., Sutskever, I., Chen, K., Corrado, G.S., Dean, J.: Distributed representations of words and phrases and their compositionality. In: Proceedings of the 26th International Conference on Neural Information Processing Systems, vol. 2, pp. 3111–3119. Curran Associates Inc., Red Hook (2013)
27. He, Z., Chen, W., Wang, Y., Zhang, W., Wang, G., Zhang, M.: Improving neural relation extraction with positive and unlabeled learning. Proc. AAAI Conf. Artif. Intell. **34**, 7927–7934 (2020)
28. Zhou, Y., Pan, L., Bai, C., Luo, S., Wu, Z.: Self-selective attention using correlation between instances for distant supervision relation extraction. Neural Netw. **142**, 213–220 (2021)
29. Chen, T., Shi, H., Tang, S., Chen, Z., Wu, F., Zhuang, Y.: CIL: Contrastive Instance Learning Framework for Distantly Supervised Relation Extraction (2021). arXiv e-prints, arXiv:2106.10855
30. Deng, L., Yang, B., Kang, Z., Yang, S., Wu, S.: A noisy label and negative sample robust loss function for DNN-based distant supervised relation extraction. Neural Netw. **139**, 358–370 (2021)

Image and Video Processing

Research on Small Target Pedestrian Detection Algorithm Based on Improved YOLOv3

Shuaishi Liu(✉), Yining Geng, Yihu Song, Wenyang Yan, and Yufeng Lian

College of Electrical and Electronic Engineering, Changchun University of Technology, Changchun 130000, China
liushuaishi@ccut.edu.cn

Abstract. Aiming at the problem that the target pedestrians' environment in the intelligent driving system is complex and the detection accuracy is not high due to large size changes, the YOLOv3 algorithm is improved in this paper and the K-means clustering method to directly calculate the anchor frame size of the data set is used. Perform cluster analysis to reduce network training time.The multi-scale fusion of shallow features and deep features improves the feature extraction effect of the skeleton network; the SE module is introduced to enhance the spatial channel after the multi-level feature fusion, which promotes the ability of pedestrian multi-scale information mining, and integrates the DIoU loss function Introducing the training process speeds up the model convergence and running speed.The model was trained with VOC2007, VOC2012 and Caltech datasets, and the Caltech pedestrian dataset was used to verify the effectiveness of the improved algorithm. The results show that the improved YOLOv3 algorithm for small-scale pedestrian targets in the intelligent driving system has improved detection accuracy.

Keywords: Pedestrian detection · YOLOv3 network · Multi-scale fusion · SEnet

1 Introduction

Pedestrians, as a type of target that appears frequently in vehicle-mounted cameras, have flexible and changeable actions and can instantly change their own motion state. Therefore, accurate perception of pedestrians can effectively reduce the occurrence of traffic accidents and improve driving safety, which is a key technology in the intelligent driving system. In the urban road environment, the scale of pedestrian targets is changeable, and the edges of small targets are easily blurred. There is less relevant information for algorithm extraction of features. Therefore, pedestrian detection technology is required to have high robustness.

In recent years, thanks to the development of deep learning, neural network-based target detection algorithms can learn the characteristics of corresponding data differences through a large amount of data, so they have been widely used. Among them, the typical

This paper is supported by the Project of Jilin Province Science and Technology Department Plan (No. 20200703016ZP), the Projects in Science and Technology of the 13th Five-year Plan of Education Department of Jilin Province (No. JJKH20200671KJ)

S.-C. Chu et al. (Eds.): ICGEC 2021, LNEE 833, pp. 203–214, 2022.
https://doi.org/10.1007/978-981-16-8430-2_19

target detection methods based on candidate frames are R-CNN [1], DRI-RCNN [2], Fast R-CNN [3] and Faster R-CNN [4].

Different from the two-step operation of the R-CNN series, the SSD algorithm [5] is based on the idea of combining regions in the detection process of regression, and uses default boxes of different scales and aspect ratios to improve the detection results. Researchers have proposed many improved algorithms based on the SSD framework, such as the DSSD model proposed by Fu et al. [6], the FSSD model proposed by Li et al. [7]; the RSSD model proposed by Jeong et al. [8], and so on. They mainly improve the feature extraction ability of convolutional neural networks by improving the architecture of the basic network, and sacrifice the detection speed of the network when improving the accuracy of target detection.

YOLO (You Only Look Once) [9] is one of the typical single-step detection algorithms. The candidate box extraction branch is discarded, the network structure is simplified, and the category probability and position coordinates of the object are directly returned to calculate the detection rate to a certain extent., Can detect pedestrians in real time in a short time. Because YOLOv1 directly predicts the confidence of the appearance of the target and the position of the pedestrian in the network, the amount of calculation is too large. YOLOv2 [10] reduces the number of prediction frames, but its target position prediction layer convolution kernel is smaller, and the detection effect of small targets is weak. YOLOv3 [11] rebuilt the skeleton network, calculated the specific values of the length and width of the anchor frame, and accurately positioned the target. Based on YOLOv3, literature [12] proposes Tiny-YOLOv3, which has a much smaller number of network convolutional layers than YOLOv3, without occupying a large amount of existing. Reduce the dependence on hardware. Although the relatively simple network structure speeds up the detection speed, it loses part of the detection accuracy.

Aiming at the problems of multi-scale feature extraction in pedestrian detection and low accuracy of small target pedestrian recognition, a SE-YOLO pedestrian detection model is designed. As shown in Fig. 1, a multi-layer feature fusion module is added at the end of the skeleton network to make The neural network obtains different receptive fields. Afterwards, the SE module was introduced to enhance the features of important channels, improve the ability to mine small-scale pedestrian feature information, build an effective neural network structure, and balance the efficiency of pedestrian detection. The DIoU loss is introduced into the training process to optimize the confidence level and speed up the model convergence and operation speed.

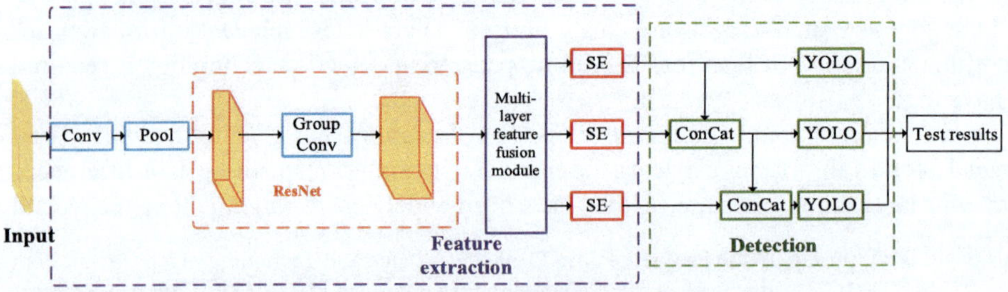

Fig. 1. Improved YOLOv3 pedestrian detection model

2 YOLOv3 Target Detection Algorithm

The YOLO network was proposed by the Joseph Redmon team in 2015. It is a single-stage pedestrian detection method that uses a single skeleton network to complete feature extraction and position prediction.The confidence level is used to evaluate whether each predicted target is accurate. It is comprehensively judged by two indicators, namely the probability of the existence of the target object and the accuracy of the bounding box. The accuracy of the bounding box is expressed by the intersection ratio IoU: $IoU = \frac{A \cap B}{A \cup B}$, Among them, $A \cap B$ represents the effective area and $A \cup B$ represents the total area occupied by the two bounding boxes.

Compared with Faster R-CNN and SSD, the YOLO method deletes the last fully connected layer of the network, speeds up detection by reducing the amount of parameter calculations and model size, and meets the real-time requirements of the intelligent driving system. Although YOLOv2 introduces the Anchor Box idea, it uses a priori anchor box to estimate the target position, reducing the number of prediction boxes. But its prediction layer convolution kernel is small, and its effect on small-scale target detection is weak. YOLOv3 introduces the residual module on the basis of Darknet-19 proposed by YOLOv2 to further deepen the network level. The improved network has 53 convolutional layers, named Darknet-53. Compared with YOLOv2 calculates the ratio of length to width of the anchor box, YOLOv3 calculates its specific value, which increases the accuracy of target positioning.

2.1 YOLOv3 Network Structure Configuration

YOLOv3 uses Darknet53 as the basic framework of the network. The structure is shown in Fig. 3. The basic unit of the network consists of a convolutional layer, a batch normalization layer (Batch normalization) and a Leaky ReLU activation function. Deepen the number of network layers and enhance the feature extraction capabilities. It also draws on the residual module in ResNet [13] and the network structure of Bottleneck [14], as shown in Fig. 2. Add layer jump operation to avoid the model degradation problem caused by too deep network layers. There are 5 residual modules in the network structure, and each residual module is denoted as Resn, which contains n residual units.

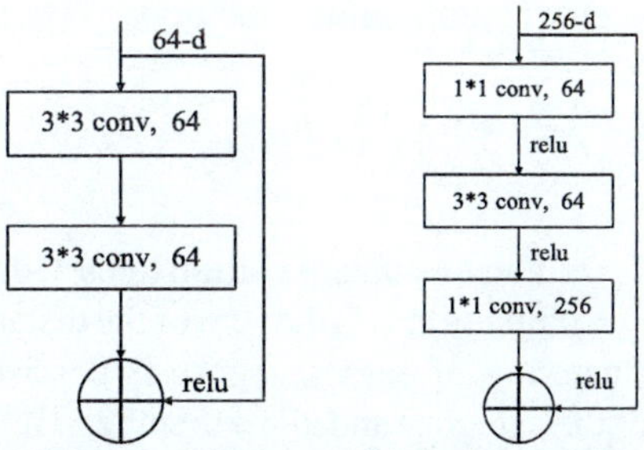

(a) Resnet Block (b) Bottleneck Block

Fig. 2. Module structure diagram

	Type	Filters	Size	Output
	Convolutional	32	3*3/1	416*416*32
	Convolutional	64	3*3/2	208*208*64
Res1: 1*	Convolutional	32	1*1/1	208*208*32
	Convolutional	64	3*3/1	208*208*64
	Residual			208*208*64
	Convolutional	128	3*3/2	104*104*128
Res2: 2*	Convolutional	64	1*1/1	104*104*64
	Convolutional	128	3*3/1	104*104*128
	Residual			104*104*128
	Convolutional	256	3*3/2	52*52*256
Res8: 8*	Convolutional	128	1*1/1	52*52*128
	Convolutional	256	3*3/1	52*52*256
	Residual			52*52*256
	Convolutional	256	3*3/2	52*52*256
Res8: 8*	Convolutional	256	1*1/1	26*26*256
	Convolutional	512	3*3/1	26*26*512
	Residual			26*26*512
	Convolutional	1024	3*3/2	13*13*1024
Res4: 4*	Convolutional	512	1*1/1	13*13*512
	Convolutional	1024	3*3/1	13*13*1024
	Residual			13*13*1024

Fig. 3. Darknet53 network structure

2.2 Anchor Frame Calculation

Inspired by FasterR-CNN, the YOLO detection layer introduces the anchor box of Anchorbox to estimate the target scale a priori, reduce the number of generated a priori boxes, and accelerate the speed of network inference. FasterR-CNN uses the RPN network to calculate the anchor frame, while YOLO uses the K-means clustering algorithm to directly calculate the size of the anchor frame in the data set, which reduces the network training time and improves the detection speed. This paper also uses the K-means clustering algorithm to calculate the a priori anchor frame information of the pedestrian benchmark data set, which is grouped into nine categories.

$$T = \sum_{i=1}^{k} \sum_{j \in c_k} \left(x^{(j)} - \mu_i \right)^2 \tag{1}$$

As shown in formula (1), μ_i is the cluster center position of the k-th anchor box category, $x^{(j)}$ is the j-th sample, and the cost function T is the sum of the distances between different categories. The degree of clustering of each category is described by the Euclidean distance between the center of the category and all its members. If the intra-class distance of the cluster is smaller, the cost function is smaller, and vice versa.

$$\alpha_j = \frac{1}{N\left(c_j\right)} \sum_{i \in c_j} x_i \tag{2}$$

$$label_i = \arg\min_{1 \le j \le k} \left\{ \sqrt{\sum_{i=1}^{n} (x_i - a_j)^2} \right\} \tag{3}$$

As shown in formulas (2) and (3), assuming that the initial random centers of the nine categories are $\alpha_1, \alpha_2, ..., \alpha_9$, for each sample size $x^{(j)}$ of the data set labeling frame, the category closest to the cluster center is used as its label $label_i$, Calculate the average distance between all points and the cluster center α_j to minimize the global label distance, and repeat until the maximum number of iterations k is reached.

In the experiments on the anchor frame clustering of the actual pedestrian benchmark data set, the results show that, as shown in Fig. 4, the width and height scale ratios of most anchor frames are (1:1, 1:2, 2:1) in line with the general goal Check the calculation result of the anchor frame. Assuming that the target has a small scale of 1–60 pixels, a medium scale of 60–120 pixels, and a large scale target of more than 120 pixels, then nine kinds of conventional anchor frames are obtained, and the YOLO layer uses the anchor frame as a scale for prediction.

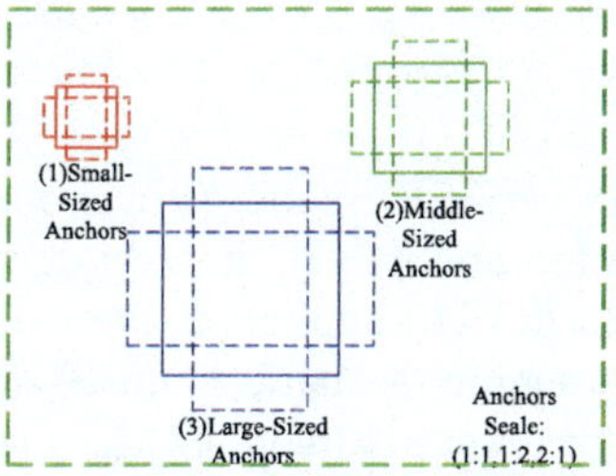

Fig. 4. Pedestrian dataset anchor box

3 Improve the Construction of YOLOv3 Network Model

3.1 Multi-layer Feature Fusion Module

In order to improve the comprehensive understanding of image semantics and underlying features of the feature map of the detection layer, a multi-layer feature fusion module of Feature Pyramid Network (FPN) [15] is used. The shallow feature map is suitable for detecting small targets because of its small downsampling multiple and small receptive field. The deep feature map (small-scale map) has a small resolution and cannot provide the required resolution information, so it is not suitable for the detection of small targets. Because of its large down-sampling multiple and the characteristics of wild feeling, it is suitable for detecting objects with large targets. In order to enhance the semantic features of different scales, a way of horizontally connecting the path from bottom to top and top to bottom is adopted for fusion.

Through the combined use of multiple up-sampling and down-sampling operations, multiple layers of feature information are combined to construct a deeper feature pyramid, and output in different feature layers to reduce the information error of the deep network. As shown in Fig. 5, first perform deep convolution on the input image; then construct feature pyramids through deeper layers to increase the robustness of features; finally, accumulate the processed low-level features and high-level features to make the low-level features can provide more accurate location information.

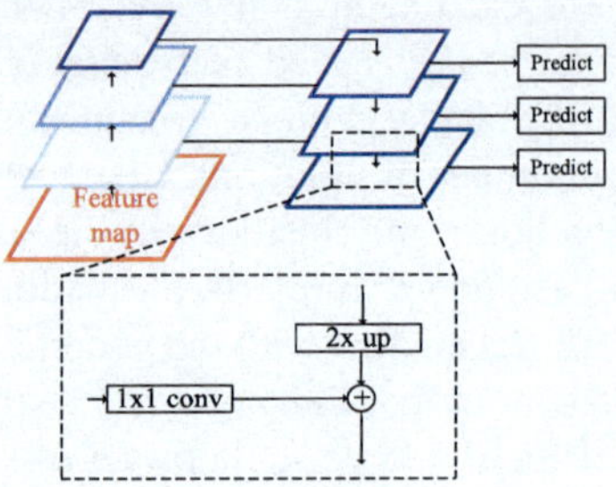

Fig. 5. Multi-layer feature fusion network structure

3.2 SENet Structure

SENet [16], as one of the main network architectures in the post-ResNet era, consists of three parts: Squeeze, Excitaion and Reweight. Its unique lightweight SE (Squeeze and Excitation) module can be flexibly applied to the existing network architecture. As an application of the Soft Attention mechanism, the SE module is different from the traditional method that refines the feature extraction task by adding feature convolutional layers. As shown in Fig. 6, SEnet explicitly constructs the interdependence between feature channels. Pay attention to the relationship between the channels, which improves the model's sensitivity to the channel dimensional characteristics. The operation of the SE module is also the same as its naming, which is divided into "Squeeze" operation and "Excitation" operation. The SE module first performs the Squeeze operation on the input feature map, as shown in formula (4):

$$y_c = F_{sq}(x_c) = \frac{1}{H \times W}\sum_{i=1}^{H}\sum_{j=1}^{W} x_c(i, j) \tag{4}$$

x_c is the c-th subset of the feature map x output after arbitrary processing by the convolution kernel, and the dimension of the feature map is $H \times W$. Squeeze operation usually adopts two algorithms of global maximum pooling or global average pooling in actual design, and the global spatial information is finally compressed into a channel dimension descriptor y. This operation will make the output feature no longer confined to the receptive field, and can be used outside the local range. SENet module structure The following information, and then obtain enough information to analyze the relationship between the channels.

The channel is activated. In order to reduce the complexity of the model and improve the generalization ability, the Excitation operation usually adopts the threshold mechanism of the Sigmoid function, as shown in formula (5):

$$s = F_{ex}(y, A) = \delta(g(y, A)) = \delta(A_2\delta(aA_1y)) \tag{5}$$

In the formula, δ represents the activation function such as Sigmoid, ReLU and SeLU, etc., and A is the parameter used to control the dimension of the feature map, where $A_1 \in R^{C\times\frac{C}{r}}$, $A_2 \in R^{\frac{C}{r}\times C}$. A_1 reduces the dimension of the input feature map through the attenuation factor r, and restores the feature map to the original dimension through A_2 after the input is activated, Excitation operation can flexibly calculate the nonlinear relationship between channels.

After multi-scale feature fusion, the SEnet branch network is introduced to first global pool the fused features to obtain the global feature information in the feature map, and then perform two fully connected operations on the global features, the first time the number of channels in the fully connected layer No change, used for the global weight of abstract features. For the second time, the number of channels in the fully connected layer is reduced to k times the previous value, which is used to extract important weights in the channels. Here, k is set to 8. Finally, after the sigmoid operation, the weight and the original fusion feature are multiplied layer by layer, thereby enhancing the features of important channels, weakening the less important channels, adaptively recalibrating the feature response of the channel, and automatically learning The relative importance of different channel features can be given different weights to enhance or suppress the corresponding features, thereby improving the recognition accuracy of the network model.

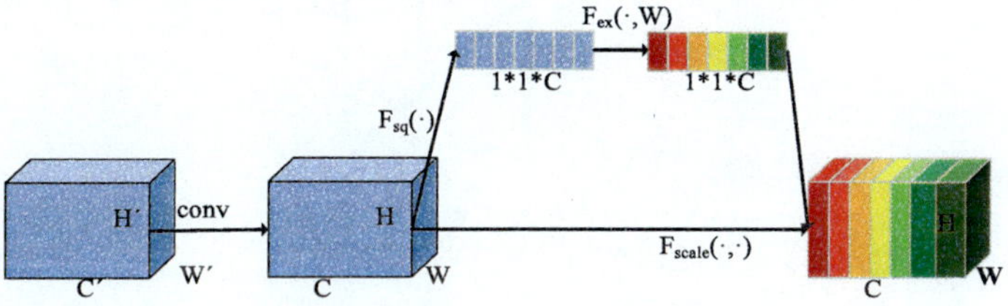

Fig. 6. SENet module structure

3.3 DIoU Loss Loss Function

In the actual detection, when the YOLOv3 algorithm is used for training and the same loss value is obtained, different results will occur, and the position between the predicted frame and the correct frame cannot be accurately reflected. Because IoU cannot well measure the distance and relative position of the two bounding boxes, directly using IoU as a loss function network will fail to converge. In order to eliminate this unstable influence and optimize the detection accuracy, this paper uses DIoU as the loss function.The IoU formula is shown in formulas (6) and (7):

$$IoU = \frac{(A \cap B)}{(A \cup B)} \tag{6}$$

$$IoU_{Loss} = 1 - IoU \tag{7}$$

In the formula, A represents the area of the correct calibration frame; B represents the area of the prediction frame. It can be seen that there are two situations (coincidence and non-coincidence) that can make IoU zero, so errors will occur when calculating the gradient.

Compared with IoU, $DIoU$ is more in line with the target frame regression mechanism. It takes into account the distance between the target and the anchor, the repetition rate and the scale, so that the training process can complete the target frame regression more stably, unlike c The same cannot converge during training. The $DIoU$ formula is shown in formula (8):

$$DIoU = IoU - \frac{\rho^2(p_1, p_2)}{l^2} \tag{8}$$

Among them, p_1 and p_2 represent the center points of the predicted frame and the real frame, respectively, and ρ refers to calculating the Euclidean distance between the two center points. l is the diagonal length of the minimum closure area of the two bounding boxes, as shown in Fig. 7.

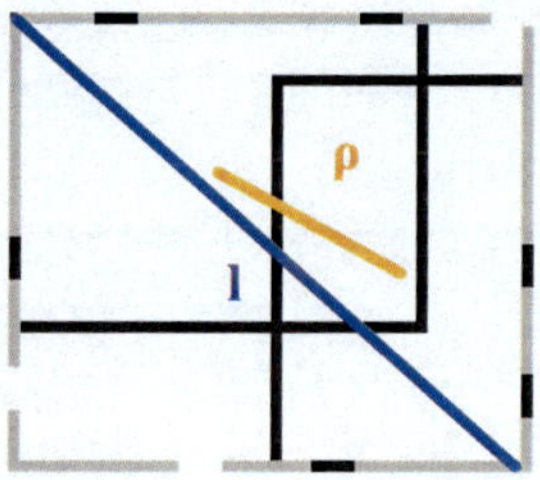

Fig. 7. DIoU schematic

4 Experiment and Result Analysis

For small target pedestrians in front of smart vehicles, the samples in this training are all internationally recognized pedestrian detection data sets: VOC2007, VOC2012, and Caltech data sets. There are 20 categories of label information in the VOC data set. Among them, the VOC2007 training set contains 5011 images, including 2007 pedestrians, and VOC2012 has 11530 pictures and 4372 pedestrians. The Caltech data set was compiled by a vision team composed of universities such as California Institute of Technology. They selected 137 min (approximately 250,000 images) and 2,300 pedestrians from a 10-h vehicle-mounted camera. The data set is divided into set00-set10, among which set00-set05 is the training set, and set06-set10 is the test set. The experiment was trained on the VOC2007, VOC2012 and Caltech mixed data sets. The original image is transformed into a size of 416 × 416 after image preprocessing operations such as size scaling and gray-scale filling, which are used as network input. Use 13 × 13, 26 × 26, and 52 × 52 feature maps at three designated locations as the feature output to get the prediction result.

In the training process, this paper uses the SGD stochastic gradient descent method for model training. The initial training rate is set to 0.0005, and the learning rate is reduced by 10 times at 15000 and 25000 iterations. Set the target at 1–60 pixels as a small scale, 60–120 pixels as a medium-scale target, and 120 pixels or more as a large-scale target.

Precision and recall are commonly used indicators for evaluating network quality. Accuracy refers to the proportion of the number of positive samples predicted to be "positive" to the total number of samples predicted to be "positive", and the recall rate refers to the proportion of the number of positive samples predicted to be "positive" to the actual number of positive samples in the test set. In multi-classification, when calculating the accuracy and recall rate of each category, it is necessary to regard each category as "positive" one by one, and all other categories as "negative". The calculation methods of precision and recall are shown in Eqs. (9) and (10) respectively. Among them, is the number of target types K that are correctly detected, and is the number of other types that are incorrectly detected as target type K The number is the number that has not been detected in the target category.

$$Precision = \frac{TP_K}{TP_K + FP_K} \tag{9}$$

$$Recall_k = \frac{TP_K}{TP_K + FN_K} \tag{10}$$

The PR (Precision-Recall) curve can reflect the accuracy and coverage of the target type under the specified IoU threshold. As shown in Fig. 8, it is the improved PR curve of YOLOv3 in this article.

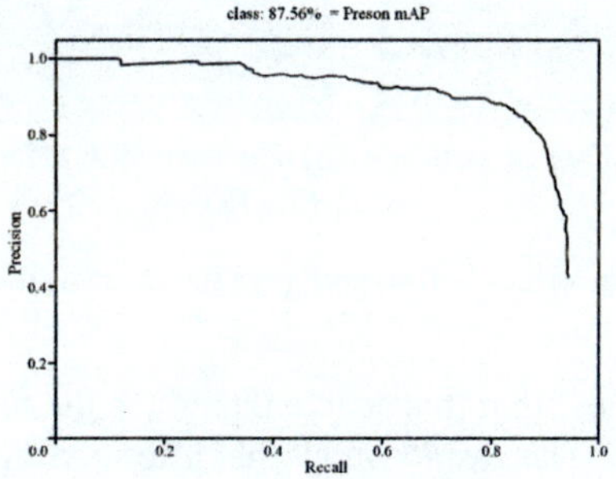

Fig. 8. PR curve icon

AP is the graph area between the curve and the horizontal axis. The formula is as shown in formula (11), where P is the accuracy and R is the recall rate.

$$AP = \int_0^1 P(R)dR \tag{11}$$

mAP refers to the average value of AP values of all categories, and the calculation formula is shown in formula (12):

$$mAP = \frac{1}{n}\sum_{K=1}^{n} AP_k \tag{12}$$

As shown in Table 1, compared with the original YOLOv3 algorithm, after the improvement of this article, the accuracy rate of large-size pedestrians has increased by 1.54%, medium-sized pedestrians has increased by 2.74%, and small-sized pedestrians has increased by 4.98%.

Table 1. Comparison of pedestrian detection accuracy rate (AP%) of different sizes

	Large	Medium	Small
YOLOv3	90.72	85.55	77.16
Ours	92.26	88.29	82.14

Figure 9 shows the comparison of pedestrian detection before and after the algorithm is improved. Compared with the original YOLOv3 algorithm, the algorithm in this paper improves the accuracy of small target pedestrian detection while improving the overall detection accuracy.

(a) The original YOLOv3 detection effect (b) The detection effect after improving YOLOv3

Fig. 9. Comparison of the effect before and after the improved YOLO V3 algorithm

It can be seen from Table 2 that this article integrates the multi-layer feature fusion module into the SE module. The results show that after adding the multi-layer feature fusion module, the missed detection rate is reduced by 1.6%, and the detection accuracy is increased by 0.5%. After the introduction of the SE module, the missed detection rate is reduced by 1.7% and the detection accuracy is increased by 1.13%. The results show that after adding SE module, multi-layer feature fusion module and DIoU loss, the missed detection rate of the model is further reduced, and the detection accuracy is also significantly improved.

In order to test the effectiveness of the algorithm in this paper, it is compared with other algorithms on the Caltech data set. As shown in Table 3, after adding a multi-layer

Table 2. Influence of different modules

	YOLOv3	YOLOv3-FPN	YOLOv3-SE	Ours
MR	13.9	12.3	12.2	9.9
mAP%	84.47	84.97	85.6	87.56

feature fusion module and introducing the SEnet structure, this paper reduces the missed detection rate by 8.47, 7.31 and 4 percentage points respectively compared with Faster R-CNN, SSD and the original YOLOv3 algorithm. The algorithm in this paper has a significant improvement in detection accuracy, which can better improve the missed detection of small target pedestrians.

Table 3. Performance comparison of different algorithms

	Faster R-CNN	SSD	YOLOv3	Ours
MR	18.37	17.21	13.9	9.9
mAP%	80.16	81.26	84.47	87.56

5 Concluding

This paper proposes an improved YOLOv3 method to detect pedestrian targets in urban road areas. Introduce the multi-layer feature fusion module, and add the SENet feature enhancement module on the basis of YOLOv3 to strengthen the network's feature extraction ability for small-scale targets. The priori anchor boxes are clustered on the pedestrian benchmark data set, and the location of special anchor boxes is predicted multiple times at different scales, which strengthens the network's ability to detect pedestrians in extreme situations. Introduce DIoU into the network model to strengthen the network's ability to detect pedestrians. This article improves the detection accuracy by improving YOLOv3.

References

1. Krizhevsky, A., Sutskever, I., Hinton, G.E.: ImageNet classification with deep convolutional neural networks. In: Proceedings of the 25th International Conference on Neural Information ProcessingSystems.Lake Tahoe, Nevada, USA, pp. 1097–1105 (2012)
2. Wen, Z., Du, Y.H., Yoshida, T., et al.: DRI-RCNN: an approach to deceptive review identification using recurrent convolutional neural network. Inf. Process. Manag. **54**(4), 576–592 (2018)
3. Girshick, R.: Fast R-CNN. In: 2015 IEEE International Conference on Computer Vision (ICCV), Santiago, pp. 1440–1448 (2015)
4. Ren, S., He, K., Girshick, R., et al.: Faster R-CNN: towards real-time object detection with region proposal networks. IEEE Trans. Pattern Anal. Mach. Intell. **39**(6), 1137–1149 (2017)

5. Liu, W., et al.: Ssd: single shot multibox detector. In: Leibe, B., Matas, J., Sebe, N., Welling, M. (eds.) Computer Vision – ECCV 2016: 14th European Conference, Amsterdam, The Netherlands, October 11–14, 2016, Proceedings, Part I, pp. 21–37. Springer International Publishing, Cham (2016). https://doi.org/10.1007/978-3-319-46448-0_2
6. Fu, C.Y., Liu, W., Ranga, A., et al.: DSSD: deconvolutional single shot detector. In: European Conference on Computer Vision and Pattern Recognition, pp. 910–925 (2017)
7. Li, Z., Zhou, F.: FSSD: feature fusion single shot multi-box detector. arXiv: 1512.02325 (2017)
8. Jisoo, J., Hyojin, P., Nojun, K.: Enhancement of SSD by concatenating feature maps for object detection. arXiv: 1705.09587 (2017)
9. Redmon, J., Divvala, S., Girshick, R., et al.: You only look once: unified, real time object detection. In: Proceedings of Computer Vision and Pattern Recognition, pp. 779–788. IEEE Press, Washington D.C. (2015)
10. Redmon, J., Farhadi, A.: YOLO9000: better, faster, stronger. In: Proceedings of the IEEE Conference on Computer Vision and Pattern Recognition, pp. 7263–7271 (2017)
11. Redmon, J.: YOLOv3: an incremental improvement (2018)
12. Zhang, Y., Shen, Y.L., Zhang, J.: An improved Tiny-YOLOv3 pedestrian detection algorithm. Optik **183**, 17–23 (2019)
13. Wang, G.W.: Multi-Scale Feature Fusion Improves Pedestrian and Vehicle Detection in YOLOv3 network. Nanjing University of Posts and Telecommunications, Nanjing (2019)
14. Shi, Z., Mao, L., Sun, J.: Multi-modal weighted fusion pedestrian detection algorithm based on YOLO. Comput. Eng., 1 (2021)
15. He, K.M., Zhang, X.Y., Ren, S.Q., et al.: Deep residual learning for image recognition. In: 2016 IEEE Conference on Computer Vision and Pattern Recognition, pp. 770–778 (2016)
16. Hu, J., Li, S., Albanie, S., et al.: Squeeze and excitation network (2018)

Zhuang Characteristic Culture Detection Based on Improved YOLOV3-SPP Network

Shu Guo, Baohua Qiang(✉), Wu Xie, Yijie Zhai, Ruidong Chen, and Hong Zheng

Guangxi Key Laboratory of Image and Graphic Intelligent Processing, Guilin University of Electronic Technology, Guilin 541004, Guangxi, China
qiangbh@guet.edu.cn

Abstract. In order to enhance the dissemination power of Zhuang culture and enhance the influence of traditional Chinese culture, this article proposes a data set containing the characteristic cultural products of the Zhuang nationality for the unique cultural development of the Zhuang nationality. The data set has a total of 1890 images and contains three categories, namely Zhuang Five-color glutinous rice, Zhuang Bronze drum, and Zhuang Hydrangea. At the same time, the YOLOV3-SPP network based on the darknet53 model is improved, and the ResNet101 structure is introduced to improve the robustness of the network. The verification results on the Zhuang characteristic data set show that the improved YOLOV3-SPP network has an accuracy rate of 92.78% on the Zhuang characteristic data set, which is 1.28% higher than the original YOLOV3-SPP network. The experimental results show that the improved YOLOV3-SPP structure can better realize the detection of the characteristic culture of the Zhuang nationality.

Keywords: Zhuang culture · Zhuang characteristic data set · YOLOV3-SPP model

1 Introduction

The Zhuang nationality is the most populous ethnic minority in China. In the long-term historical development, the Zhuang people have created their unique material and spiritual culture, which has increased the influence of Chinese traditional culture and played a very important role in the development of Chinese culture effect. Therefore, it is particularly important to lead Zhuang culture [1] to the world, and to classify, detect and recognize the products of Zhuang culture. In order to solve these problems in real life, some target detection algorithms have also been proposed.

In recent years, with the rapid development of deep convolutional neural networks, computer vision has become a research hotspot in many fields, and target detection [2] is one of the most basic and challenging tasks in computer vision. With the continuous development and optimization of target detection technology, it is widely used in practical life applications, such as robot navigation, face recognition, video surveillance, industrial detection, etc.

S.-C. Chu et al. (Eds.): ICGEC 2021, LNEE 833, pp. 215–226, 2022.
https://doi.org/10.1007/978-981-16-8430-2_20

Target detection algorithm based on deep learning [3], through a series of data enhancement technology, basic backbone network, feature extraction module, automatically learn hidden information [4] in network data, and extract abstract features in image data. Compared with traditional target detection algorithms, it can achieve better performance indicators.

In 2012, in the ImageNet image recognition competition, Hinton et al. won the championship for the first time through the convolutional neural network AlexNet [5] constructed by him, which greatly promoted the development of convolutional neural networks in the field of computer vision. As a result, algorithms to achieve the task of target detection have been proposed one after another. Target detection algorithms based on deep learning can be summarized into two categories: two-stage target detection algorithms and one-stage target detection algorithms. The two-stage algorithm uses the selective search method [6] to extract region suggestions, and uses two branches to implement classification and regression tasks respectively. In 2014, Girshick proposed the R-CNN [7] network based on the CNN structure. The detection performance greatly surpassed the current best traditional target detection algorithm, and the accuracy reached 58.5%. After that, the upgraded version of R-CNN Fast R-CNN [8] and Faster R-CNN [9] also came out one after another. Although their accuracy on the data set has been improved, there are many training steps and the training time is long.

In response to the common detection problems in the above-mentioned two-stage network, the first-stage network began to develop. The one-stage network does not need to provide an additional branch for classification, but performs classification and regression simultaneously, which further reduces training time. In 2016, Redmon et al. first proposed a one-stage detection network YOLOV1 [10]. The emergence of YOLOV1 pushed target detection to a climax. Since YOLOV1 does not have the complex detection steps of a two-stage network, the detection speed is much faster than that of a two-stage network, but the accuracy of YOLOV1 is relatively low due to more positioning errors. In 2016, Liu et al. proposed a network SSD on the basis of YOLOV1 [11]. SSD uses VGG16 [12] as the basic model. The difference from YOLOV1 is that SSD uses multi-scale feature mapping and performs detection on feature maps of different scales. The deep network has a larger receptive field and is suitable for detecting large targets. The shallow network pays attention to feature details and is suitable for detecting small targets. The multi-scale detection feature of SSD greatly enhances the robustness of detection. Compared with YOLOV1, the detection effect is better, but the accuracy rate can be further improved. In 2018, YOLOV3 [13] appeared, which used darknet53 [14] as the backbone network, using feature fusion technology to better extract image features, and the detection accuracy was further improved. Then the improved version YOLOV3-SPP [15] based on YOLOV3 was released. The SPP [16] structure of YOLOV3-SPP consists of three pooling operations, which are 5*5, 9*9, 13*13 maximum pooling, SPP pooling operation realizes the fusion of global features and local features [17], greatly enriching the learning ability of feature maps, and further improving the detection accuracy of the model.

The target detection model mentioned above still has the following shortcomings in Zhuang cultural detection:

- There is no data set on the cultural products of the Zhuang ethnic group for model training.
- Most of the existing target detection models have low detection accuracy.

In order to solve the above problems, the main contributions of this paper are as follows:

- A data set containing characteristic cultural products of the Zhuang nationality is proposed.
- Improved the detection model to better realize the detection of the characteristic culture of the Zhuang nationality.

2 Network Model

2.1 Framework

As shown in Fig. 1, the overall framework of the network proposed in this paper consists of three parts. First, the deeper ResNet101 model is used to extract the initial features of the picture with the input size of 416*416. Secondly, combined with the SPP maximum pooling operation, the expressive ability of the feature map is enriched, and the accuracy of the model is improved. Finally, using parallel multi-scale fusion technology, the feature maps of different sizes are respectively predicted, and the location and category information of the target are output.

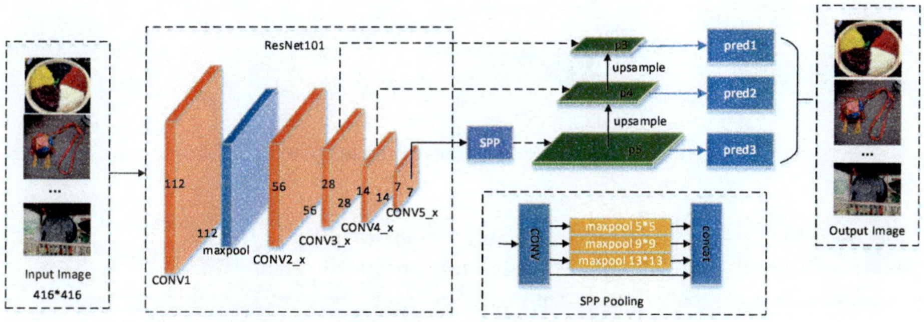

Fig. 1. The flow chart of the proposed network.

2.2 Residual Structure

The Resnet101 model is composed of the residual network structure of ResNet [18]. The network uses a large number of 1*1 and 3*3 convolution blocks to realize the jump connection of channels on the convolution structure, which avoids the difficulty of training and easy training of complex network models.

ResNet has two basic residual blocks, one is called Identity Block, as shown in Fig. 2. This kind of block requires the same input dimension and output dimension to realize jump connection. The other is called Conv Block, as shown in Fig. 2. This kind of block realizes that the input dimension and output dimension are inconsistent. When performing the addition operation, since the connection operation can only be realized when the number of channels of the main branch is the same as the number of channels of the side branch, it is necessary to perform a convolution operation on one of the branches to make the number of channels in the connection the same.

The difference between the residual network and the ordinary network is the introduction of jump connections. The ResNet101 model uses a residual structure, which better avoids the problem of model gradient explosion overfitting, and accelerates the convergence of the network.

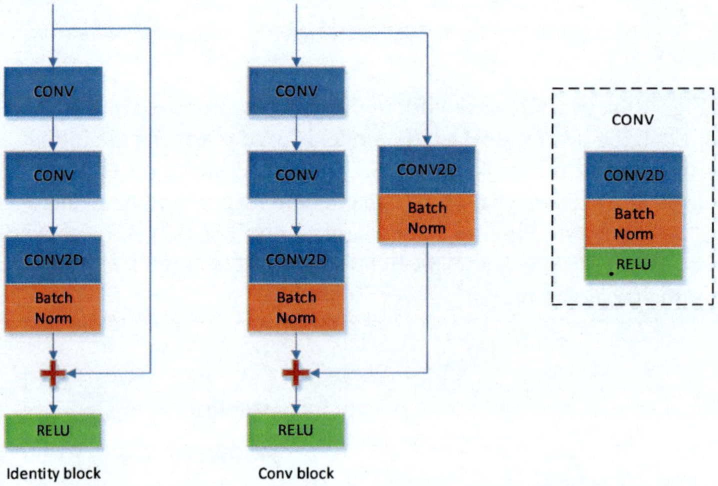

Fig. 2. Residual structure diagram of the proposed model.

Detailed ResNet-101 structure parameters are shown in Table 1. First, there is an input 7*7*64 convolution, followed by a 3*3 maximum pooling layer, and finally through 4 different channel convolution layers. Among them, conv2_x, conv3_x, conv4_x, conv5_x convolutional layers are all composed of 1*1 convolution, 3*3 convolution, and 1*1 convolution, which are repeated 3, 4, 23, and 3 times respectively. Remove the last fully connected layer, the actual resnet101 network used has 100 layers.

Table 1. Detailed parameters of resnet101 model.

Network layer	Kernel parameter	Repeat	Output size
Conv1	7*7 64	1	208*208
Max pool	3*3	1	104*104
Conv2_x	$\begin{bmatrix} 1*164 \\ 3*364 \\ 1*1256 \end{bmatrix}$	3	104*104
Conv3_x	$\begin{bmatrix} 1*1128 \\ 3*3128 \\ 1*1512 \end{bmatrix}$	4	52*52
Conv4_x	$\begin{bmatrix} 1*1256 \\ 3*3256 \\ 1*11024 \end{bmatrix}$	23	26*26
Conv5_x	$\begin{bmatrix} 1*1512 \\ 3*3512 \\ 1*12048 \end{bmatrix}$	3	13*13

2.3 Feature Prediction

The conv5_x, conv4_x, and conv3_x convolutions of the ResNet101 model have steps of 32, 16, and 8 respectively with respect to the input image. As shown in Fig. 3, the features output by conv5_x pass through the SPP pooling layer. The details of the SPP pooling layer are shown in Fig. 1. It includes three pooling operations, 5*5 pooling, 9*9 pooling, and 13*13 pooling. The features after the pooling operation are connected with the original features that have not been pooled to achieve the fusion of features.

The merged feature is convolved by CONV_F to obtain a new feature p5. On the one hand, p5 undergoes two convolution operations, and the obtained features are used to predict large objects. On the other hand, p5 performs an up-sampling operation after convolution and merges with the features of conv4_x. The merged feature is convolved by CONV_F to obtain a new feature p4. On the one hand, p4 undergoes two convolution operations, and the obtained features are used to predict the middle object. On the other hand, p4 performs an up-sampling operation after convolution and merges with the features of conv3_x. The merged features are then convolved with CONV_F to obtain new features p3, p3 is only operated by two convolutional layers, and the obtained features are used to predict small objects.

Among them, the CONV_F convolution operation is performed alternately by 1*1 convolution and 3*3 convolution, 1*1 convolution is used for dimensionality reduction, and 3*3 convolution is used for dimensionality increase. Repeated use of 1*1 convolution and 3*3 convolution can reduce the amount of model parameters and increase the interaction of different channel information.

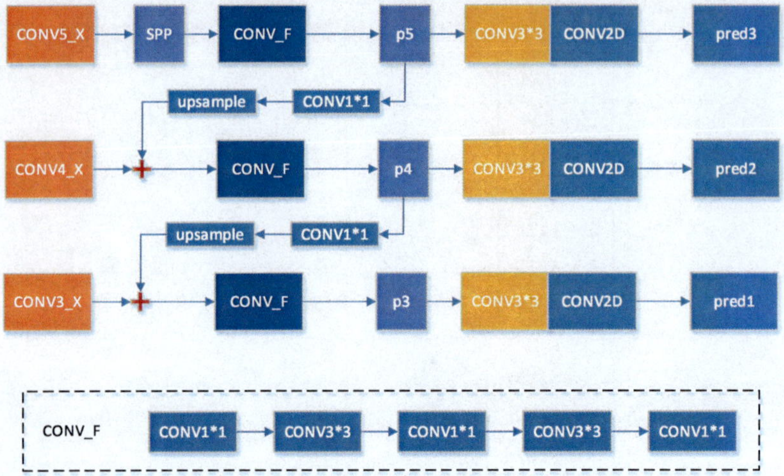

Fig. 3. Network feature prediction graph.

In summary, this article uses ResNet101 with a more complex structure as the basic network framework, replaces the final fully connected layer with a convolutional layer, introduces SPP pooling, parallel multi-scale fusion technology, and better realize the task of classifying and detecting Zhuang cultural products.

2.4 Model Loss Function

The loss of YOLOV3-SPP is composed of positioning error (Lbbox), confidence error (Lconf) and classification error (Lcls).

The positioning error is composed of two parts, the x and y parameters use the BCE loss, and the w and h parameters use the MSE loss. As shown in formula (1).

$$Lbbox = W_{box}BCE\left(tx, tx', ty, ty'\right) + W_{box}MSE\left(tw, tw', th, th'\right) \tag{1}$$

among them, *Wbox* represents the weight value of different boxes, *tx*, *tx'* represent the difference between the X coordinate of the real box and the predicted box, and the same is true for *ty*, *ty'*, *tw*, *tw'*, *th*, *th'*. Where *y* represents the Y coordinate, *w* represents the width, and *h* represents the height.

BCE stands for Binary Cross Entropy, and its calculation is shown in formula (2).

$$BCE = -ylog\left(y'\right) - (1 - y)\log\left(1 - y'\right) \tag{2}$$

where *y* is the true value and *y'* is the predicted value.

MSE stands for Mean Square Error, and its calculation is shown in formula (3).

$$MSE = \frac{1}{n}\sum_{i=1}^{n}(y'_i - y_i)^2 \quad (3)$$

where y' represents the predicted value, and y represents the target value.

The confidence error uses MSE loss.

Classification error uses cross entropy loss (Cross Entropy Loss Function).

3 Experiment and Result Analysis

3.1 Experimental Data Set

In order to promote the traditional culture of the Zhuang nationality and improve the influence of the traditional culture of the Zhuang nationality [19]. This experiment provides a brand-new Zhuang cultural data set, named Zhuang Yao. The data set contains three types of Zhuang cultural products, namely Five-color glutinous rice [20] (labeled as nuomi), Bronze drum [21] (labeled as tonggu), and Hydrangea [22] (labeled as xiuqiu).

The Five-color glutinous rice of the Zhuang nationality is named after the five colors of black, red, yellow, white and purple. It is a traditional delicacy of the Zhuang nationality. The Zhuang people regard it as auspicious and a symbol of abundance of grains. As shown in Fig. 4.

The surface of the bronze drum of the Zhuang nationality is often painted with sun patterns. In the hearts of the Zhuang people, the bronze drum is a collection of heaven, earth and gods, and a bridge between man and god. As shown in Fig. 5.

The Zhuang Hydrangea is a symbol of love in the Zhuang culture and a token of sentiment and determination between men and women of the Zhuang ethnic group. As shown in Fig. 6.

Fig. 4. Five-color glutinous rice.

The dataset Zhuangyao contains 1,890 pictures, including 630 pictures of Five-color glutinous rice, Bronze drums, and Hydrangea. The number of pictures in the training set and the number of pictures in the test set are divided according to a ratio of about 3:1. The total number of training pictures is 1,440 and the total number of test pictures is 450. The specific distribution is shown in Table 2.

Fig. 5. Bronze drum.

Fig. 6. Hydrangea.

Table 2. Data set allocation.

Data set label category	Total pictures	Train pictures	Test pictures
Five-color glutinous rice	630	480	150
Bronze drum	630	480	150
Hydrangea	630	480	150

3.2 Experimental Environment and Evaluation Indicators

(1) Experimental Environment

The experiments in this article are all done on Linux64-bit platform, Python version 3.7, using Pytorch to build the model. At the same time, image processing libraries such as opencv-python, numpy, matplotlib are used to process pictures. The hardware environment of the experiment is shown in Table 3.

The parameter settings of the model used in the experiment are shown in Table 4.

(2) Evaluation Index

In target detection tasks, precision and recall are usually used together as the standard to measure the quality of the model. The precision rate indicates the proportion of the true positive cases among the positive cases judged by the model. Recall rate indicates the proportion of positive cases correctly judged by the model in the total positive cases. Taking recall rate as the abscissa and precision rate as the ordinate, the P-R curve can be drawn on the coordinate axis, where AP is the area enclosed by the P-R curve on the X and Y axes, mAP is the average of AP values calculated from multiple target detection categories. Therefore, the experiment in this paper uses mAP as the metric of the model,

Table 3. Experimental hardware.

Hardware name	Hardware model
Operating system	Ubuntu 16.04
CPU	Intel Xeon E5–2603 v2
GPU	NIVDIA 1080Ti
RAM	32G
Video memory	11G

Table 4. Model parameters.

Parameters	Settings
Learning strategy	Step
Initial learning rate	0.0001
Batch size	16
Momentum	0.9
Data set format	XML
Weight decay	0.0005
Fixed number of iterations	150

Table 5. Prediction results of Zhuang cultural categories.

Category	AP	mAP/%
Five-color glutinous rice	0.971	92.78
Bronze drum	0.910	
Hydrangea	0.902	

and sets the IOU > 0.5 between the inspection frame and the real frame to indicate that the object is detected. The mAP calculation formula is shown in formula (4).

$$mAP = \frac{\sum_{i=1}^{n} AP_i}{N} \tag{4}$$

where N represents the number of categories of the data set.

3.3 Validation on the Data Set

We conduct experiments on the YOLOV3-SPP network based on ResNet101. The experimental results of each category are shown in Table 5.

We use the original YOLOV3-SPP network and the network proposed in this article for comparative experiments. In order to ensure the fairness of the experiment, the Zhuang Yao data set proposed in this paper is used, and the same model parameters are set. The results are shown in Table 6.

Table 6. Comparison results of different models.

Method	AP			mAP/%
	Five-color glutinous rice	Bronze drum	Hydrangea	
YOLOV3-SPP	0.932	0.909	0.904	91.50
Ours	0.971	0.910	0.902	92.78

It can be seen from Table 6 that compared to the YOLOV3-SPP method based on the Darknet53 model, the improved YOLOV3-SPP network based on ResNet101 has a higher accuracy rate. YOLOV3-SPP and the test results of this paper are shown in Fig. 7. It can be seen that there is deviation in YOLOV3-SPP positioning, and there is a situation of inspection error, as shown by the red arrow in Fig. 7. The method used in this article has a slight error in positioning, but the overall detection effect is good.

(a) YOLOV3-SPP detection effect

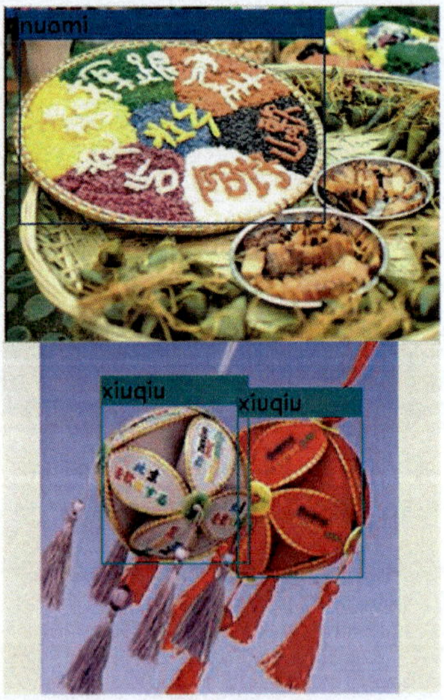

(b) Ours

Fig. 7. YOLOV3-SPP and the test results of this article.

4 Conclusion

The inheritance and development of the traditional culture of ethnic minorities are of great significance. To this end, this article focuses on the three types of Zhuang nationality characteristic cultural development products, Zhuang Five-color glutinous rice, Zhuang Bronze drum, and Zhuang Hydrangea. Research is based on deep learning, and the YOLOV3-SPP network based on the ResNet101 model is introduced. The experimental results show that the improved YOLOV3-SPP detection model has an accuracy increase from 91.5% to 92.78% compared to the original YOLOV3-SPP detection model. The experimental results prove that the model proposed in this paper can achieve better detection results under the self-defined Zhuang data set of Zhuang Yao.

Acknowledgments. This work was supported in part by the National Natural Science Foundation of China under Grant 61762025; in part by the Natural Science Foundation of Guangxi under Grants 2019GXNSFDA185007, and 2019GXNSFDA185006; in part by the Guangxi Key Research and Development Project under Grants AB18126053, AB18126063, AD18281002, and AD19110137; in part by the Guilin Science and Technology Development Program under Grants 20190211–17; in part by the Development Foundation of the 54th Research Institute of China Electronics Technology Group Corporation under Grant SXX18138X017; in part by the Guangxi Key Science and Technology Planning Project under Grants AA18118031 and AA18242028; in part by the Innovation Project of GUET Graduate Education under Grants 2020YCXS052.

References

1. Zhang, W., Li, Q., Fan, Z.: The application research of Guangxi bronze drum decoration in illustration design. Art Educ. Res. (09), 62–63 (2019). 张文柱,李秋旗,范紫薇.广西铜鼓纹饰在插画设计中的应用研究[J].美术教育研究 (09), 62–63 (2019)
2. Lu, F., Liu, H., Huang, C., Yang, Y., Xie, Y., Liu, C.: A review of target detection technology based on deep learning. Appl. Comput. Syst. **30**(03), 1–13 (2021). 陆峰,刘华海,黄长缨,杨艳,谢禹,刘财喜.基于深度学习的目标检测技术综述[J].计算机系统应用 **30**(03), 1–13 (2021)
3. Shi, C., Zhang, W., Chen, H., Ge, L.: A review of salient target detection based on deep learning. Comput. Sci. Explor. **15**(02), 219–232 (2021). 史彩娟,张卫明,陈厚儒,葛录录.基于深度学习的显著性目标检测综述[J].计算机科学与探索 **15**(02), 219–232 (2021)
4. Xue, X., Liu, J.: Collaborative ontology matching based on compact interactive evolutionary algorithm. Knowl.-Based Syst. **137**, 94–103 (2017)
5. Krizhevsky, A., Sutskever, I., Hinton, G.E.: Imagenet classification with deep convolutional neural networks. In: Advances in Neural Information Processing Systems, pp. 1097–1105 (2012)
6. Uijlings, J.R.R., Van De Sande, K.E.A., Gevers, T., et al.: Selective search for object recognition. Int. J. Comput. Vision **104**(2), 154–171 (2013)
7. Girshick, R., Donahue, J., Darrell, T., et al.: Rich feature hierarchies for accurate object detection and semantic segmentation (2014). arxiv preprint arXiv:1311.2524

8. Girshick, R.: Fast r-cnn (2015). arXiv preprint arXiv:1504.0808
9. Ren, S., He, K., Girshick, R., et al.: Faster R-CNN: towards real-time object detection with region proposal networks. IEEE Trans. Pattern Anal. Mach. Intell. **39**(6), 1137–1149 (2017)
10. Redmon, J., Divvala, S., Girshick, R., et al.: You only look once: unified, real-time object detection (2015). arXiv preprint arXiv:1506.02640
11. Liu, W., Anguelov, D., Erhan, D., et al.: SSD: Single shot multibox detector (2016). arXiv preprint arXiv:1512.02325
12. Xie, X., Xiao, H., Quan, L., et al.: Visualization and pruning of SSD with the base network VGG16. In: The 2017 International Conference. ACM (2017)
13. Redmon, J., Farhadi, A.: Yolov3: an incremental improvement (2018). arXiv preprint arXiv:1804.02767
14. Yi, X., Song, Y., Zhang, Y.: Enhanced Darknet53 combine MLFPN based real-time defect detection in steel surface (2020)
15. Liu, Z., Yuan, L., Zhu, M., Ma, S., Chen, L., Zhou, T.: YOLOv3 traffic sign detection combining SPP and improved FPN. Comput. Eng. Appl. **57**(07), 164–170 (2021). 刘紫燕,袁磊,朱明成,马珊珊,陈霖周廷.融合SPP和改进FPN的YOLOv3交通标志检测[J].计算机工程与应用 **57**(07), 164–170 (2021)
16. Dong, S., Ma, Y., Li, C.: Implementation of detection system of grassland degradation indicator grass species based on YOLOv3-SPP algorithm. J. Phys. Conf. Series **1738**, 012051 (2021)
17. Xue, X., Chen, J.: Using compact evolutionary tabu search algorithm for matching sensor ontologies. Swarm Evol. Comput. **48**, 25–30 (2019)
18. Phawinee, S., Cai, J.F., Guo, Z.Y., Zheng, H.Z., Chen, G.C., Hsieh, W.H.: Face recognition in an intelligent door lock with ResNet model based on deep learning. J. Intell. Fuzzy Syst. **40**(4), 8021–8031 (2021)
19. Shi, J.: Research on the protection and inheritance of intangible cultural heritage in lianshan zhuang yao autonomous county. Contemp. Tour. **19**(14), 10–11 (2021). 史锦珊.连山壮族瑶族自治县非物质文化遗产保护与传承探究[J].当代旅游 **19**(14), 10–11 (2021)
20. Luo, B.: A preliminary study on the color of five-color rice - a case study of the traditional five-color and wamy food of southern minorities. Popular Arts (02), 62–63 (2021). 罗冰.五色饭的色彩初探——以南方少数民族传统五色糯食为例[J].大众文艺 (02), 62–63 (2021)
21. Ding, J.: Innovative research on Guangxi bronze drum visual design art based on cultural inheritance. Tomorr. Fash. (06), 83–84 (2021). 丁静.基于文化传承的广西铜鼓视觉设计艺术创新研究[J].明日风尚(06), 83–84 (2021)
22. Jing, J., Luo, J.: The origin of the original ecological culture of the traditional sports of Zhuang nationality in Guangxi. Martial Arts Res. **5**(07), 96–99 (2020). 经建坤,罗建达.广西壮族传统体育抛绣球运动的原生态文化发展渊源[J].武术研究 **5**(07), 96–99 (2020)

A Method for Detecting Safety Helmet in Substation Operation Scene Based on Yolov3

Xiaoyu Chen(✉), Jianyou Yu, Dezhi Zhao, Jian Sun, Xianghong Chen, Mingli Yi, and Lin Xia

Baishan Power Supply Company of State Grid Jilin Electric Power Company Limited, Baishan, China

Abstract. Under the complex posture of the construction personnel, the existing detection methods for wearing helmets have the problems of high detection difficulty and low accuracy. This paper proposes a method for detecting wearing helmets based on head recognition. The head area is obtained through skin color feature recognition and head detection. And perform cross-validation to determine the head area. Use the Yolov3 target detection network to identify the helmet, and judge the wearing condition of the helmet based on the positional relationship between the helmet area and the head area. Finally, the performance of other helmet-wearing detection methods is compared and analyzed through experiments. Summarize the detection methods of helmet-wearing and put forward prospects.

Keywords: Target detection · Safety helmet detection · Head recognition · Yolov3

1 Introduction

The substation is an important link between the power plant and the user. It is the main electricity production site. In order to ensure the daily production process of the substation. It must strictly monitor the environment, equipment, and production process in the substation. The safety production regulations of the substation stipulate all personnel entering the substation. Appropriate safety helmets must be worn according to identity and work tasks. And these personnel often don't wear safety helmets as required. This will bring great hidden dangers to the safety of substation operations. We are in order to complete the safety production goal of the substation. It is particularly important to monitor whether the staff wear helmets or not [1].

After years of development, safety helmet-wearing detection. It has shifted from traditional machine learning methods to deep learning methods. Some scholars use popular target detection networks to detect the wearing of helmets. For example, Girshick R et al. used the improved Faster R-CNN to detect the wearing of helmets. This type of method has high detection accuracy but takes a long time and has poor robustness. The other is the One Stage method based on regression models, such as SSD, Yolo series, etc.Wang Qiuyu improved the YO-LO network structure. Proposing a helmet-wearing

S.-C. Chu et al. (Eds.): ICGEC 2021, LNEE 833, pp. 227–238, 2022.
https://doi.org/10.1007/978-981-16-8430-2_21

recognition algorithm based on semi-supervised learning. Fang Ming et al. added dense blocks on the basis of Yolov2 to quickly detect safety helmets and so on. This type of method uses an improved target detection network to detect the helmet. The problem of small targets and partial occlusion is solved, but the relative position relationship between the helmet and the human body is not solved.

Some scholars use the method of head positioning to detect the wearing of helmets. For example, Kang Li et al. proposed to quickly locate pedestrians based on the results of moving object segmentation. According to the pedestrian's head position, color space transformation and color feature discrimination, the detection of helmet-wearing is realized. In this method, the determination of the head position is based on one-fifth of the pedestrian's body ratio. Such a method is not universal.

The reason for the increased difficulty in recognizing helmets based on complex postures. This paper proposes a method for detecting wearing helmets based on head recognition. This method is cross-validated by facial feature recognition and head recognition. Precisely locate the head area of the operator. Solve the problem that the head position is difficult to determine in complex postures. Using Yolov3 target detection network for helmet-wearing detection. The relative position relationship between the helmet and the human body can be resolved, so as to determine the wearing condition of the helmet. This paper mainly solves the problem of low detection accuracy of helmet wearing under the complicated posture of construction workers.

The reason for the increased difficulty in recognizing helmets based on complex postures. This paper proposes a method for detecting wearing helmets based on head recognition. This method is cross-validated by facial feature recognition and head recognition. Precisely locate the head area of the operator. Solve the problem that the head position is difficult to determine in complex postures. Using Yolov3 target detection network for helmet wearing detection. The relative positional relationship between the helmet and the human body can be obtained. Then judge the wearing of the helmet. This article mainly solves the problem of low detection accuracy when the construction workers wear helmets in complex postures.

2 Safety Helmet Wearing Detection Method

The process of the safety helmet-wearing detection method designed in this paper is shown in Fig. 1. Position the head of the operator. Then, the head area is obtained by skin color feature recognition and head detection and recognition respectively. And perform cross-validation to determine the head area. Then, using the Yolov3 target detection network to identify the helmet. Finally, according to the characteristics of the data set, it is judged whether the helmet is worn correctly.

2.1 Yolov3 Target Detection Network

Enter the feature extraction network Darknet-53.Different levels of features are extracted through convolution operation. The Residual Block is used in this process. At the same time, the principle of Feature Pyramid Network (FPN) is used for reference. During detection, the input image is divided into $S \times S$ grid cells. If the center point of the target

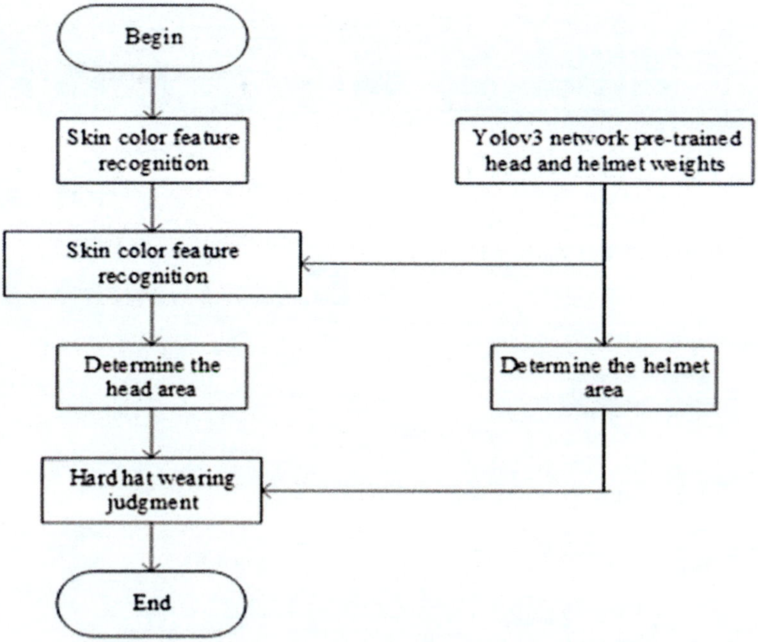

Fig. 1. Process of the detection method for wearing a helmet

is in a certain grid cell, then the grid cell is responsible for the prediction of the target. Each grid cell will predict 3 bounding boxes (bbox). The dimension of the network output is shown in formula (1):

$$\mathrm{S} \times \mathrm{S} \times \mathrm{B} \times (4 + 1 + \mathrm{C}) \tag{1}$$

In the formula, B represents the number of bbox predicted by each grid cell, and the default is 3; 4 represents the 4 parameters of the bbox, which are the center coordinates and width and height (x, y, w, h); 1 represents the confidence of the category; C represents the total number of categories. The predicted category confidence is shown in formula (2):

$$P_r(class_i \| P_r object) \times IOU_{pred}^{truth}) \tag{2}$$

In the formula, if there is a target in the grid cell. Then Pr(object) = 1, otherwise Pr(object) = 0; Pr(classilobject) represents the confidence of the i-th target; IoUtprruetdh represents the intersection ratio of the prediction box bbox and the ground truth box (Ground Truth, GT) (Intersection over U-nion, IoU), can be used to measure the accuracy of prediction; all results are filtered using NMS algorithm to get the final output (Fig. 2).

2.2 Head Positioning Method

The postures of substation operators are complex, and their head postures are diverse. A single detection network cannot accurately detect the head area. The head positioning

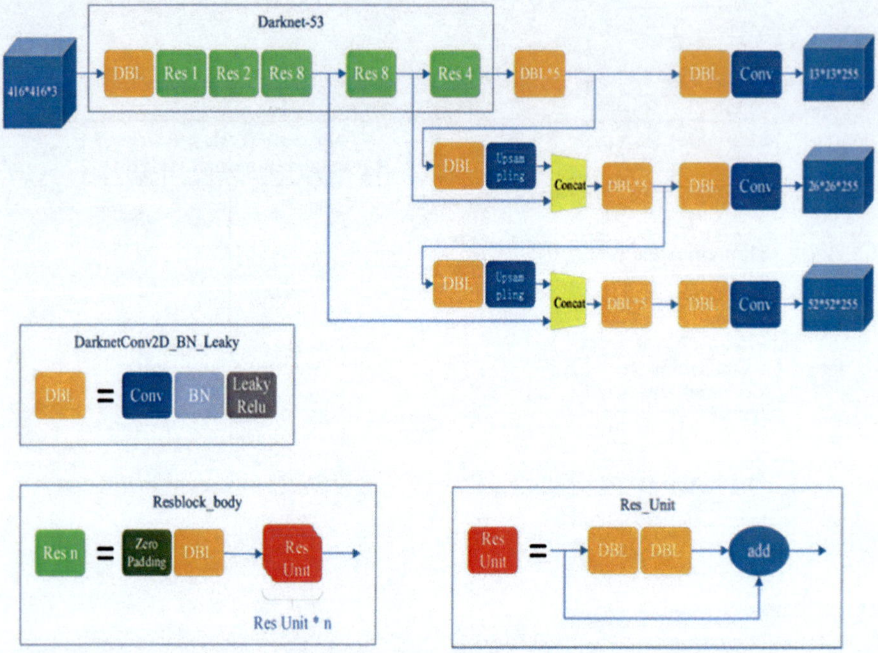

Fig. 2. Yolov3 network structure

method proposed in this article. It is used to accurately locate the head area of construction workers. Solve the problem that the head position is difficult to determine due to the complex posture of the construction personnel. The main steps are: 1) Recognizing the skin color feature and detecting the face area; 2) Pre-training weight detection head; 3) Cross screening of face area and head area; 4) Determine all head areas of the picture.

2.2.1 Skin Tone Detection

In the construction state of the operator, the areas exposed to the skin are basically only the face, neck, and hands. Therefore, this article first performs light compensation on the picture. Then cluster the workers' skin color on the YCbCr color space. Compared with RGB, YCbCr color space is more convenient for chromaticity analysis [2]. The conversion formula from RGB to YCbCr color space is shown in formula (3).

$$\begin{pmatrix} Y \\ C_b \\ C_r \end{pmatrix} = \begin{pmatrix} 16 \\ 128 \\ 128 \end{pmatrix} + \begin{pmatrix} 65.481 & 128.533 & 24.966 \\ -37.797 & -74.203 & 112 \\ 112 & -93.786 & -18.214 \end{pmatrix} \begin{pmatrix} R \\ G \\ B \end{pmatrix} \tag{3}$$

This paper uses the YCrCb color space range screening method to identify skin color. The experiment selects 100 pictures to perform k-means clustering on the RGB features of skin color. After conversion, the ranges of Y, Cr, and Cb are $85 < Y < 220$, $140 < Cr < 175$, and $100 < Cb < 120$, respectively. Secondly, the top hat transformation is mainly used to correct the influence of uneven lighting. The binary image after skin color detection contains a lot of noise. Therefore, this paper performs top hat transformation on the binary image after skin color detection. This can reduce the range of non-face areas. The top hat transformation formula is shown in Eq. (4). f is a grayscale image. $f^{\circ}b$ is the opening operation of the structure element b to the image f.

$$T_{hat}(f) = f - (f^{\circ}b) \tag{4}$$

Finally, the relevant morphological processing is performed on the image. Locate the larger part of the connected area. It is tentatively designated as the face area to facilitate subsequent cross-validation.

2.2.2 Head Detection

The posture of the construction personnel for substation operation is complicated. There are various head postures.Therefore, this article uses the HollywoodHeads dataset and the Yolov3 network to pre-train head detection weights [3]. These subsets do not overlap in terms of actions. Head images of people with poor visibility, such as strong occlusions and low-light conditions. These were marked with a "difficult" sign and were excluded from the evaluation. The HollywoodHeads data set is trained on the Yolov3 network, and the AP value reaches 95.4%. The pre-training weights have a good effect on head detection in normal postures (upright, walking, etc.).

2.2.3 Cross-Screening of the Head Area

This paper proposes a cross-screening method for the head region. This method is to solve the phenomenon of missing detection in the head detection network [5]. According to the characteristics of the data set in this article, it is difficult to detect some heads of construction workers in complex postures. However, facial skin tone characteristics can still be detected. Therefore, this paper compares and screens the detected skin color area and human head area. When performing comparative screening of skin color area and human head area. The skin color areas with larger connected areas are not all face areas. There are skin tone areas such as arms, hands, and neck [6]. When the skin color area overlaps with the human head area, the area is evaluated as the head area. The evaluation value is 10. If there is and only the human head area, the area is evaluated as the head area. The evaluation value is 8;

When there is and only a skin color area, the area is evaluated as a suspected head area. The evaluation value is 5. Set the skin color area to F and the human head area to H. The evaluation value is x. Then the cross-screening formula (5) of the head region is:

$$\begin{gathered} F \subset H, x = 10 \\ !\exists H, x = 8 \\ !\exists F, x = 5 \end{gathered} \tag{5}$$

By setting the evaluation value, the head area in the picture is further determined. This can improve the accuracy of the helmet wearing detection. This also provides support for the judgment method of wearing a helmet.

2.3 How to Judge the Helmet Wearing

This article uses the Yolov3 network to detect the helmet. According to the characteristics of the data set, if the helmet area is handed over to the head areas A and B with an evaluation value of 10 and 8, then it is determined that the constructor is wearing a helmet. It is determined that the construction personnel are wearing safety helmets. The head area includes areas of skin color, such as the face, neck, arms, and hands. The intersection of the helmet area A and B presents a dangerous situation of holding a helmet [7, 8]. According to the human body structure and the characteristics of the data set, it is found that only the head and hands exist in the helmet area. That is, the construction workers wear safety helmets and hand-held safety helmets on their heads [9–11]. Therefore, this paper calculates the area A and the center point of all skin color areas. The area calculation uses Seed Filling. The idea is to select a foreground pixel as the seed. Then experiment according to the two basic conditions of the connected region. Combine the foreground pixels adjacent to the seed into the same pixel set. The finally obtained pixel set is a connected region.

The connected region analysis method based on the seed filling method is given below:

(1) Scan image B until the current pixel point B(x,y) ==1:

   1) Use B(x,y) as the seed (pixel position) and give it a label. Then push all foreground pixels adjacent to the seed into the stack;
   2) Pop the top pixel of the stack and give it the same label. Then push all the foreground pixels adjacent to the top pixel of the stack into the stack;
   3) Repeat step b until the stack is empty; at this time, a connected region in image B is found. The pixel value in this area is marked as label;

(2) Repeat step 1) until the end of the scan; after the end of the scan, all the connected areas in image B can be obtained; the center point is calculated using Image Moments. Use the computational geometric distance algorithm for each connected region to get the center. The mathematical expression is:

$$M_{pq} = \int_{a1}^{a2} \int_{b1}^{b2} x^p y^p f(x, y) dx dy \tag{6}$$

$$i_c = \frac{M_{10}}{M_{00}}, j_c = \frac{M_{01}}{M_{00}} \tag{7}$$

3 Data Set

The hard hat data set in this article is collected by a web crawler. Selection of public data sets. A total of 8,000 photos were collected from the field. The data set contains pictures of various poses of a single person and multiple persons in various construction scenarios. Moreover, this article considers the balance between positive and negative samples. At the same time, this article has carried out data augmentation on the data set. The data set is expanded by operations such as flipping, blurring, and panning. An example of the data sample is shown in Fig. 3.

Fig. 3. Data sample

4 Experiment Analysis

Experimental environment configuration: GPU: NVIDIA TITAN Xp × 2, CUDA10.0, Ubuntu16.04, memory 12G.

This experiment uses precision, false positive rate (fp, false positive rate), miss rate (miss), intersection ratio (IOU) and average accuracy (mAP) to measure the effectiveness of the method. The calculation formula is shown in (8)–(11):

$$precision = \frac{TP}{TP + FP} \tag{8}$$

$$fp = \frac{FP}{FP + TN} \tag{9}$$

$$miss = \frac{FN}{TP + FN} \tag{10}$$

$$IOU = \frac{partacreage \cap overallacreage}{partacreage \cup overallacreage} \tag{11}$$

Among them: TP (true positive) represents the positive samples predicted by the model to be positive; FP (false positive) represents the positive samples predicted to be negative by the model; FN (false negative) represents the negative sample predicted by the model to be negative; TN (true negative) represents the negative samples predicted by the model to be positive. Partacreage is the area of the helmet that is framed after detection.overallacreage is the marked hard hat area.

4.1 Analysis of Training Performance of Helmet Detection Model Based on Yolov3

Because there is no official data set for the detection of helmet wearing. The data set of this research is to collect more than 8,000 pictures of "people wearing hard hats" and "people not wearing hard hats" from the Internet. Use the labeling tool LabelImg to label the target in the picture. Put "people wearing hard hats" in the "hat" category. "Persons who do not wear helmets" are classified as "person". The marking process is shown in Fig. 4.

Fig. 4. Annotation of the data set

In Fig. 5, increasing the brightness and decreasing the brightness can simulate the changes in the scene when the light is different. Changing the color tone can simulate helmets of different colors. In this way, the number of samples can be increased to improve the robustness of the model to different scenarios. And it uses multi-scale training.Randomly change the size of the sample before sending it to the network for training. This can improve the model's ability to detect targets of different scales [13].

Set the number of training batches (Batches) to 50200 times. Send the data set to Yolov3 and improved Yolov3 for training. In the training process, the smaller the loss value, the closer the trained model is to the real model of the data set.As shown in Fig. 6, the loss function curve (Loss Curve) of Yolov3 and improved Yolov3 during training.

In Fig. 6, batches represents the number of training batches.The ordinate avg_loss represents the average loss value.It can be seen from Fig. 6(a) and Fig. 6(b). The network loss during the two training sessions maintained a steady downward trend. It converges when iterates to about 50200 times.It shows that the training is basically smooth. At the same time, compared with the curve of Fig. 6(a), the curve of Fig. 6(b) converges faster.And the fluctuation range of the network becomes smaller and converges to a smaller value. It shows that the improved model is closer to the real model of the data set.

Fig. 5. Data enhancement example diagram

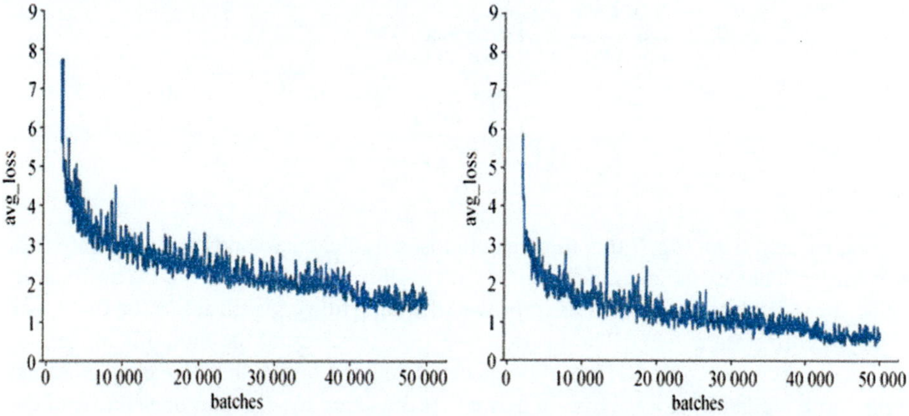

(a) Yolov3 loss curve

(b) Improve yolov3 loss curve

Fig. 6. Loss curve

4.2 Performance Analysis of Helmet Detection Model Test Based on Yolov3

In the field of target detection, mean Average Precision (mAP) is generally used to evaluate the effect of the model. AP represents the average accuracy of a single category.

mAP represents the mean value of AP of multiple categories [14, 15]. AP can be obtained from the confusion matrix of the model classification result.

The graph drawn with Recall as the horizontal axis and Precision as the vertical axis is called the PR curve. The PR curve integrates the information of precision rate and recall rate. The area enclosed by the curve and the two coordinate axes is the value of

AP. As shown in formula (12).

$$AP = \int_0^1 (R)d_R \tag{12}$$

The PR curves of the model before and after the improvement are measured on the test set, as shown in Fig. 9. Calculated from the figure YOLOv3. $AP_{hat} = 88.97\%$, $AP_{person} = 85.23\%$, mAP = 87.10%; Improve YOLOv3 $AP_{hat} = 90.59\%$, $AP_{person} = 93.30\%$, mAP = 91.94%.

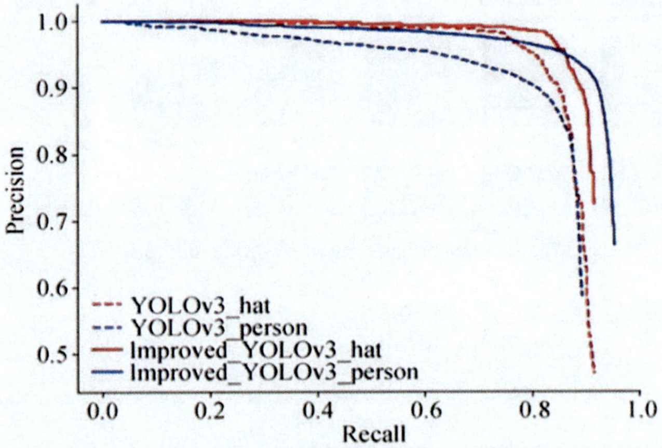

Fig. 7. PR curve

It can be seen from Fig. 7 that the area enclosed by the curve and the coordinate axis is also larger. That is, the average accuracy rate AP has been improved. It can be seen that the improved Yolov3 model has improved the detection capabilities of the two types of targets to different extents. The overall detection accuracy of the model is higher.

It can be seen from Fig. 8. Both models can detect the target "hat" wearing a helmet and the target "person" not wearing a helmet.At the same time, it can be seen from the first set of test results. The original Yolov3 missed some occluded targets and overlapping targets.It can be successfully detected in the improved Yolov3The original Yolov3 cannot determine whether the intermediate target is "hat" or "person". It is a difficult sample. The improved Yolov3 can distinguish correctly. The original Yolov3 cannot detect very small targets, while the improved Yolov3 can detect more small targets. It can be known by comparing the test results. The improved Yolov3 has improved the ability to detect occluded targets, difficult samples and small targets. The detection accuracy has been improved.

Fig. 8. Test comparison

5 Concluding Remarks

This article first uses the Yolov3 network to pre-train the weights of the head and helmet. Then use the YCbCr color space to detect the skin color of the constructors. Determine the head area through cross screening of the head area. It is determined by the position of the helmet area and the head area. You can judge the wearing of the helmet. Experimental results show. The method in this paper has good results for the detection of construction workers' helmet-wearing in different postures. Due to the delineation of the head area, the detection accuracy of helmet wearing is also better than previous methods.However, because the method uses skin color features and network detection complementary to detect the head region. Therefore, the effect is not ideal when the character's goal is extremely small. It will be studied in depth in future research. In order to improve the effect of helmet detection model based on Yolov3 substation operation scene.

Acknowledgment. This work was funded by the "Research and Application of Key Technologies for the Whole Process Management and Control of Substation Operations" (Project Number:SGJLBS00AJJS2000262).

References

1. Girshick, R., Donahue, J., Darrell, T., et al.: Region-based convolutional networks for accurate object de- tection and segmentation. IEEE Trans. Pattern Anal. Mach. Intell. **38**(1), 142–158 (2015)
2. Shick, G.R.: Fast R-CNN. In: IEEE International Conference on Computer Vision, pp. 1440–1448. IEEE, Piscataway (2015)
3. Ren, S.Q., He, K.M., Girshick, R., et al.: Faster R-CNN: towards real-time object detection with region proposal networks. In: Proceedings of the 28th International Conference on Neural Information Processing Systems, pp. 91–99. IEEE, Piscataway (2015)
4. Wang, C.Y., Liao, H.Y.M., Yeh, I.H., et al.: CSPNet: a new backbone that can enhance learning capability of CNN. In: IEEE/CVF Conference on Computer Vision and Pattern Recognition Workshops, pp. 390–391 (2020)
5. He, K., Zhang, X., Ren, S., et al.: Deep residual learning for image recognition. In: The IEEE Conference on Computer Vision and Pattern Recognition (CVPR), pp. 770–778. IEEE press (2016)

6. Lin, T.Y., Dollár, P., Girshick, R., et al.: Feature pyramid networks for object detection. In: 2017 IEEE Conference on Computer Vision and Pattern Recognition (CVPR), Honolulu, HI, pp. 936–944 (2017)
7. Liu, S., Qi, L., Qin, H., et al.: Path aggregation network for instance segmentation. In: 2018 IEEE/CVF Conference on Computer Vision and Pattern Recognition (CVPR). IEEE (2018)
8. Ren, S., He, K., Girshick, R., Sun, J.: Faster R-CNN: towards real-time object detection with region proposal networks. IEEE Trans. Pattern Anal. Mach. Intell. **39**(6), 1137–1149 (2017)
9. Lin, T.Y., Goyal, P., Girshick, R., et al.: Focal Loss for dense object detection. IEEE Trans. Pattern Anal. Mach. Intell. **1**, 2999–3007 (2017)
10. Redmon, J., Farhadi, A.: YOLOv3: an incremental improvement. In: IEEE Conference on Computer Vision and Pattern Recognition, pp. 89–95 (2018)
11. Jo, K., Im, J., Kim, J., Kim, D.: A real-time multi-class multi-object tracker using YOLOv2. In: IEEE International Conference on Signal and Image Processing Applications (ICSIPA), Kuching, pp. 507–511 (2017)
12. Zhang, K., Guo, Y., Wang, X., Yuan, J., Ding, Q.: Multiple feature reweight densenet for image classification. IEEE Access **7**, 9872–9880 (2019). https://doi.org/10.1109/ACCESS.2018.2890127
13. Vu, T.H., Osokin, A., Laptev, I.: Context-aware CNNs for person head detection. In: 2015 IEEE International Conference on Computer Vision (ICCV). IEEE (2016)
14. Tan, M., Chen, B., Pang, R., et al.: MnasNet: platform-aware neural architecture search for mobile. In: 2019 IEEEConference on Computer Vision and Pattern Recognition, pp. 2820–2828 (2019)
15. Redmon, J., Farhadi, A.: YOLO9000: better, faster, stronger. In: IEEE Conference on Computer Vision & Pattern Recognition, pp. 6517–6525. IEEE (2017)

Gray-Scale Image Color Migration Algorithm Based on Dual-Stream Convolutional Neural Network

Xiwen Zhang and Jin Duan(✉)

School of Electronic Information Engineering Changchun, University of Science and Technology, Changchun 130022, China
duanjin@vip.sina.com

Abstract. In order to solve the problems of insufficient image feature information extraction in image processing by traditional color migration algorithms, resulting images have obvious color mis-transmission and unnatural colors, a gray-scale image color migration algorithm based on dual-stream convolutional neural network is proposed. Firstly, the improved SE-ResNet-20 subnetwork with feature recalibration function and the improved VGG19 subnetwork are used to extract feature maps containing local information and global information respectively, and then the two types of feature maps are fused, and the fusion results are used as the input of the coloring network. Finally, the color transfer of gray image is realized by feature analysis combined with classification information. In the training process, two loss functions, pixel loss and classification loss, are used to train the model at the same time. The experimental results show that the algorithm in this paper effectively solves the problem of color mis-transmission. As a result, the color distribution of the image is more reasonable and natural, and the texture is clearer.

Keywords: Dual-stream convolutional neural network · Color migration · SE-ResNet · VGG19

1 Introduction

The image color migration algorithm refers to obtaining the color information of other images under the condition that the image style and texture characteristics remain unchanged, so that the image presents an unprecedented visual effect. With the wide application of deep learning technology, breakthroughs have been made in many fields. In order to free hands and improve the efficiency of coloring, people began to turn their attention to automatic coloring, and deep learning technology was also applied in the field of image coloring. In 2017, Zhu [1] proposed a color migration algorithm based on generating confrontation network, which mainly solves the migration without paired samples, and trains colored GAN networks through the cycle loss established between two opposite mapping functions and the confrontation loss of GAN networks. In 2018, Zhou [2] proposed a deep coloring model with edge thinning. The image is transformed

S.-C. Chu et al. (Eds.): ICGEC 2021, LNEE 833, pp. 239–248, 2022.
https://doi.org/10.1007/978-981-16-8430-2_22

into YUV color space, and then two neural networks are trained to obtain U value and V value respectively, and then combined into color image. In 2019, Zhang Na et al. [3] proposed to use VGG to obtain the classification information of the image and fuse the features extracted by dense neural network to calculate the coloring error and then train to obtain the coloring model. In the same year, Xiao et al. [4] introduced semantic segmentation in the CycleGAN generation network to effectively guide the consistency of coloring in the image area, and better realized the image coloring in unsupervised mode. In 2020, Patricia et al. [5] proposed ChromaGAN coloring algorithm, whose core idea is to integrate scene classification into generation confrontation network, and realize gray image coloring by perception of scene category and color information. In the same year, Su et al. [6] proposed to combine target detection with image coloring, intercept objects through target detection algorithm, then predict the color distribution of the whole image and objects respectively, and finally fuse their coloring effects.

At present, the proposed gray-scale image color migration method based on deep learning realizes the automation of image color migration, and avoids manual intervention to a certain extent. However, due to the simple network structure, the image information is lost in the process of extracting image features and the image features are not fully understood, which leads to the phenomenon of color misinformation.

2 Design of Network Architecture

In this paper, aiming at the problem of color misinformation caused by insufficient extraction of image feature information in the existing gray-scale image color migration algorithm based on deep learning, a gray-scale image color migration algorithm based on dual-stream convolutional neural network is proposed. The algorithm adopts dual stream structure, and the network extracts different features of the image through two branches: local feature extraction network and global feature extraction network. In the local feature extraction network, an improved se-resnet-20 (squeeze and excite residual network-20) is proposed to enhance the useful information of color offset. The global feature extraction network adopts the improved vgg19 [7] network. Firstly, the global and local features are fused, and then the fused features are further predicted and extracted by coloring network, and the classification information is combined by feature analysis. Before network training, the input image is mapped to lab color space, and the gray value of L channel is used as the input of the network. After training, combined with the AB channel chroma signal output by the network, the final output image is obtained. The network structure is shown in Fig. 1:

Among them, the global features of the image are extracted by the VGG19 network and then classified. The main structure of vgg19 is composed of 16 layer 3×3 convolution core replacing the larger convolution core in Alex Net [8] and other networks, and plus three layers of full connection layer.

This paper converts all standard convolutions except the first layer of convolution into depth separable convolution [9], which reduces the amount of calculation while ensuring the convolution effect. The depth separable convolution is realized by two convolutions: a depth convolution and a 1×1 point convolution. M convolutions of $D_k \times D_k$ are used to perform a convolution on the M channels of the input image to generate M feature

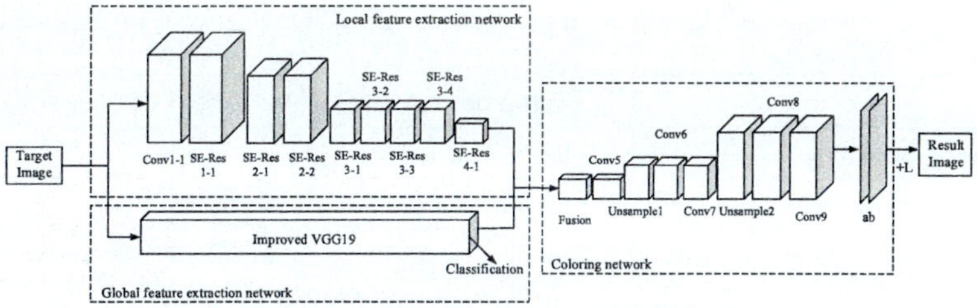

Fig. 1. Network structure diagram

maps. Among the M feature maps, through N convolution kernels of 1×1, respectively, $M \times N$ feature maps are generated. Finally, the feature maps corresponding to the M input channels are superimposed and merged into a feature map. The final parameter is the sum of the two, as shown in Eq. (1):

$$D = D_{\mathrm{k}} \times D_{\mathrm{k}} \times M \times C_{\mathrm{I}} \times C_{\mathrm{I}} + M \times N \times C_{\mathrm{I}} \times C_{\mathrm{I}} \tag{1}$$

When $D_k = 3$, the ratio of the parameters calculated by the depth separable convolution and the traditional convolution is:

$$T = D/C \approx 1/9 \tag{2}$$

Before entering the global feature network, the image will be scaled to 224×224 size. After 16 layers of depth separable convolution operation, the number of feature maps is 512, and then it will continue to pass through 3 layers of fully connected layers. The final output is the classification of grayscale images, finally uses the classification result of fc3 [10] and the real label of the image for loss. The global feature extraction network structure diagram in this paper is shown in Fig. 2.

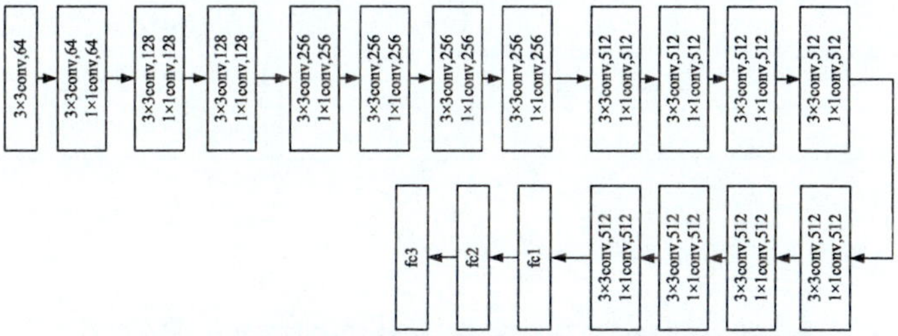

Fig. 2. Improved VGG19 network structure diagram

In order to reduce the performance gap between color images and grayscale images, this paper only uses the brightness channel of the image to retrain the VGG19 network. As shown in Table 1, the accuracy rate of gray image Top-5 classification is improved by the improved VGG19 network to be close to the accuracy rate of the original VGG19 network on color images.

Before coloring, the feature map output by the global feature extraction network fc1 layer is fused with the local features. After the fusion, the H/16 × W/16 × 512

Table 1. Comparison of classification performance of VGG19 network on ImageNet

	Top-5 correct rate (%)	Top-1 correct rate (%)
Color image test results	91.26	73.45
Grayscale image test results	81.43	60.24
Improved grayscale image test results	90.02	72.19

rectangular parallelepiped features are used as the input of the shading network, after processing the convolutional layer and the up-sampling layer, the output layer of the coloring network outputs ab value, which is finally merged with the initial gray image H × W × 1 and converted back to the RGB color space to complete the color migration of the gray image. The image coloring network configuration is shown in Table 2.

Table 2. Image coloring network configuration table

Name	Layer design	Output size
Fusion		H/16 × W/16
Conv5	3 × 3, 256	H/16 × W/16
Unsample1		H/8 × W/8
Conv6	3 × 3, 128	H/8 × W/8
Conv7	3 × 3, 128	H/8 × W/8
Unsample2		H/4 × W/4
Conv8	3 × 3, 64	H/4 × W/4
Conv9	3 × 3, 64	H/4 × W/4
Unsample3		H/2 × W/2
Conv10	3 × 3, stride1	H/2 × W/2
Output		H × W

3 Improved Local Feature Extraction Network Based on SE-ResNet-20

3.1 Principle of SENet

SENet [11] is composed of multiple SE Blocks. SE Block is mainly composed of three parts: Convolutional layer, Squeeze layer and Excitation layer. The first is the convolution operation, which performs the $F_{tr}: X \rightarrow U$ conversion operation, where $X \in R^{W' \times H' \times C'}$ represents the input of the SE Block and $U \in R^{W \times H \times C}$ represents the output of the convolution. Using the squeeze layer, the feature graphs of multiple spatial dimensions are aggregated to generate a channel descriptor. The channel descriptor is embedded into the global distribution of channel by channel features, which makes the bottom layer of the network have more information about the global receptive field. The above operation is to generate channel by channel statistics through the global average pooling layer. The output U of the convolution is compressed by the spatial dimension of $W \times H$ to obtain the statistical value $z \in R^C$, and the input of the dimension $W \times H \times C$ is converted into the output of $1 \times 1 \times C$. The calculation formula of the element c in z is shown in Eq. (3), and the specific network structure of SE Block is shown in Fig. 3:

$$z_c = F_{sq}(u_c) = \frac{1}{H \times W}\sum_{i=1}^{H}\sum_{j=1}^{W} u_c(i, j) \tag{3}$$

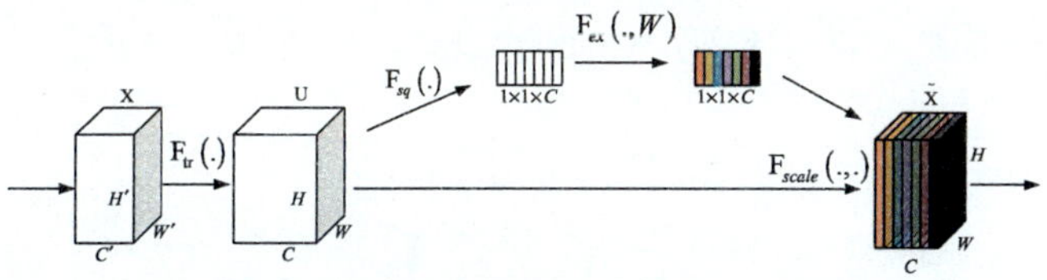

Fig. 3. SE Block network structure diagram

Capture the correlation between channels in the Excitation operation. The function that performs the Excitation operation should satisfy: the function is flexible and can learn a non-mutual exclusion to ensure the excitation state of the channel characteristics. The gate control mechanism using Sigmoid activation function is shown in Eq. (4):

$$s = F_{ex}(z, W) = \sigma(g(z, W)) = \sigma(W_2\delta(W_1 z)) \tag{4}$$

Where δ represents the ReLU activation function, σ represents the Sigmoid activation function, the dimensionality reduction layer parameter is $W_1 \in R^{\frac{C}{r} \times C}$, the parameter of the ascending layer is $W_2 \in R^{C \times \frac{C}{r}}$ and the ratio is r;

The output of SE Block is obtained by relocating the output U with activation by Eq. (5):

$$\underset{C}{\tilde{X}} = F_{scale}(u_C, s_C) = s_C \cdot u_C a \tag{5}$$

Where $\underset{C}{\tilde{X}}$ and F_{scale} represent the corresponding channel-by-channel product between the feature map u_C and the scalar s_C. u_C is the output of the convolutional layer, and s_C is used to adjust the weight of each channel. Through the operation of Excitation, SE Block obtains the weight parameters corresponding to each channel, so that useful information can be enhanced.

3.2 Residual Network

The performance of the convolutional neural network increases as the network depth increases. For this reason, He [12] et al. proposed a residual network (ResNet), by introducing residual mapping, the gradient information is still large in the backhaul, thereby overcoming the disappearance of the gradient in the network training. When the input of the network is x, the feature learned by the network is $H(x)$. Learn another kind of mapping-residual mapping: $F(x) = H(x) - x$ through stacked nonlinear layers. Therefore, the mapping $H(x)$ learned by the original network is transformed into $F(x)+x$, the formula $F(x) + x$ is implemented in the feedforward network through a shortcut connection. In the residual network, the shortcut connection is transformed into a simple identity mapping. The input is directly added by the identity map to the output of the stacked convolutional layer. No additional parameters were introduced, and the amount of calculation and complexity of the network did not increase. The residual module structure is shown in Fig. 4. The residual block in the figure contains two layers, and the residual mapping is $F = W\sigma_2(W_1x)$, where σ represents the ReLU activation function, which can be added element by element to get the network feature map.

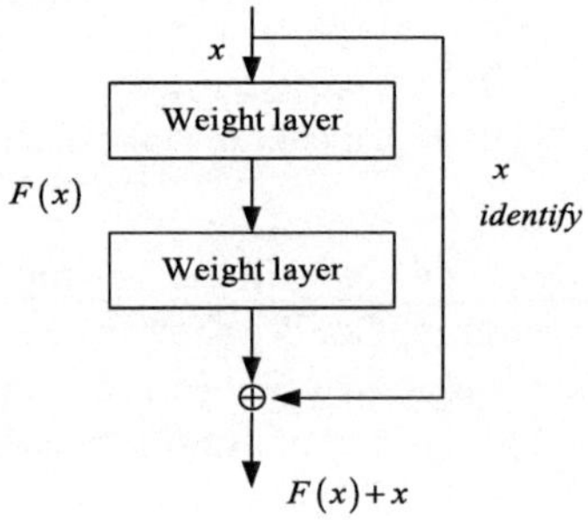

Fig. 4. Structure diagram of residual module

3.3 Improved Local Feature Extraction Network

This paper proposes a structure that adds SE Block on the basis of the ResNet-20 [13] model based on the previous algorithm research to build SE-ResNet-20. By introducing a global pooling layer to generate channel-by-channel feature distribution, the encoding of global information is realized; The correlation between each channel is captured by the ReLU activation function and the sigmoid gating mechanism to achieve feature recalibration. Fully connected layers are used at both ends of the ReLU activation function to reduce the complexity of model parameters. SE-ResNet significantly improves the model's ability to extract features by recalibrating the characteristics of different network channels. The structure of SE-ResBlock is shown in Fig. 5.

SE-ResNet-20 network configuration is shown in Table 3. Among them, $f_c[8, 64]$ represents the dimensionality reduction layer and the dimensionality increase layer of the two fully connected layers in SE Block, and the dimensions are 8, 64, respectively.

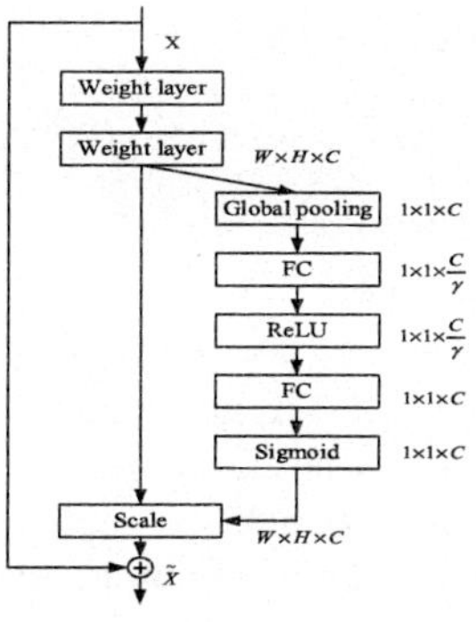

Fig. 5. SE-ResBlock structure diagram

Set the dimensionality reduction coefficient of the dimensionality reduction layer to 8. Therefore, the dimensions of the dimensionality reduction layer in SE Block are 8, 16, 32, and 64.

Table 3. SE-ResNet-20 network configuration table

Name	Layer design	Output size
Conv1	stride2	H/2 × W/2
Conv1–1	$\begin{pmatrix} 3 \times 3\text{conv},64 \\ 3 \times 3\text{conv},64 \\ fc[8, 64] \end{pmatrix} \times 1$	H/2 × W/2
Pool1		H/4 × W/4
Conv2-x	$\begin{pmatrix} 3 \times 3\text{conv},128 \\ 3 \times 3\text{conv},128 \\ fc[16, 128] \end{pmatrix} \times 2$	H/4 × W/4
Pool2		H/8 × W/8
Conv3-x	$\begin{pmatrix} 3 \times 3\text{conv},256 \\ 3 \times 3\text{conv},256 \\ fc[32, 256] \end{pmatrix} \times 4$	H/8 × W/8
Pool3		H/16 × W/16
Conv4–1	$\begin{pmatrix} 3 \times 3\text{conv},512 \\ 3 \times 3\text{conv},512 \\ fc[64, 512] \end{pmatrix} \times 1$	H/16 × W/16

4 Experiment and Result Analysis

4.1 Data Set Preprocessing

In order to enrich the learned color information and improve the generalization ability of the model, this paper uses the subset of ImageNet [14] data set as the training set. First, the input image is mapped to the Lab color space, and the L channel grayscale value is used as the input of the network. The ab channel chrominance signal is predicted through the learning of the network. After the training is over, the L channel grayscale value of the original image and the ab channel chromaticity value of the network output need to be merged, and finally the image is output. Finally output the image.

The loss during network training consists of two parts, one is pixel loss, and the other is classification loss. Pixel loss is used to measure the loss between the network output and the real image, as shown in Eq. (6):

$$L(f, f*) = \frac{1}{n}\sum_{k=1}^{n}\left[\frac{1}{WH}\sum_{i=1}^{W}\sum_{j=1}^{H}\left(f_k^*(i,j) - f_k(i,j)\right)^2\right] \tag{6}$$

Among them, n is the number of images in each training batch, W and H are the width and height of each sample respectively, f represents the color value of the a and b channels output, and f^* represents the color value of the original a and b channels.

The global feature network finally outputs the classification and the classification label corresponding to the image for cross-entropy loss, which constitutes the classification loss function, as shown in Eq. (7):

$$L = \frac{1}{n}\sum_{k=1}^{n}\left[-\sum_{k} L_k^{label}\log\left(L_k^{class}\right)\right] \tag{7}$$

In which n is the number of images in each training batch, L_k^{class} is the category loss function of the kth picture, and L_k^{label} is the classification label loss function of the kth picture.

4.2 Analysis of Image Color Migration Results

In order to verify the advantages of the algorithm in this paper, the algorithm in this paper is compared with several representative algorithms, and the results are shown in Fig. 6, Fig. 7. Table 4, Table 5 are the comparison of evaluation parameters.

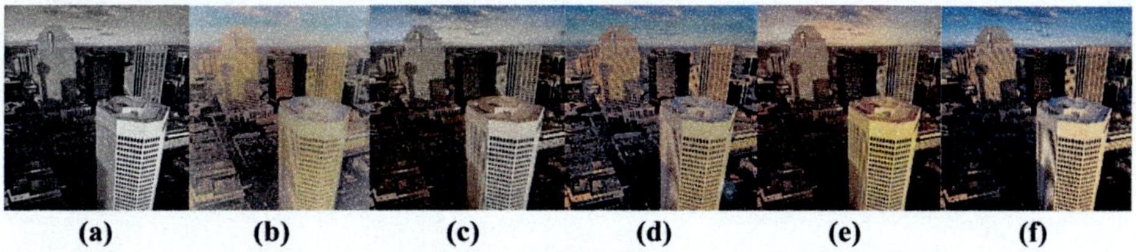

Fig. 6. Comparison of different algorithms (a) Grayscale image; (b) Original image; (c) Document [2] algorithm; (d) Document [7] algorithm; (e) Document [8] algorithm; (f) Results of this paper

Table 4. Comparison of evaluation parameters of various algorithms

	Standard deviation	PSNR	Average gradient	Color degree
Literature [2] results	59.586	24.501	25.452	6.3
Literature [7] results	62.134	28.758	28.145	7.5
Literature [8] results	62.287	28.845	27.128	8.2
Results of this paper	62.749	29.458	28.542	8.5

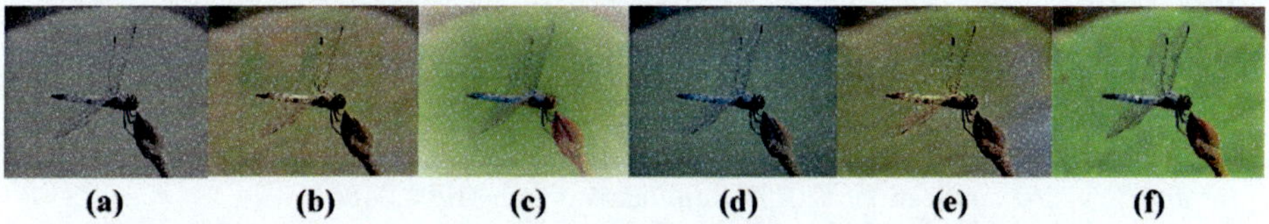

Fig. 7. Comparison of different algorithms (a) Grayscale image; (b) Original image; (c) Document [2] algorithm; (d) Document [7] algorithm; (e) Document [8] algorithm; (f) Results of this paper

Table 5. Comparison of evaluation parameters of various algorithms

	Standard deviation	PSNR	Average gradient	Color degree
Literature [2] results	54.124	26.457	12.548	6.4
Literature [7] results	54.587	28.753	13.785	8.5
Literature [8] results	54.507	25.789	10.532	7.1
Results of this paper	54.735	30.457	13.458	9.4

5 Conclusion

In this paper, dual-stream network is used to realize the color migration of the gray image. First, local features and global features of the image are extracted by local feature extraction network based on SE-ResNet-20 and global feature extraction based on improved VGG19. Then, the extracted local features and global features are fused by fusion layer, and the fused features are used as the input of the coloring network, and finally the color image is output. The gray-scale image color migration algorithm in this paper overcomes the problem of color mis-transmission by more effectively extracting image features and accurate classification effects, and obtains a better subjective visual effect.

References

1. Zhu, J., Park, T., Isola, P., et al.: Unpaired image-to-image translation using cycle-consistent adversarial networks. In: IEEE International Conference on Computer Vision (ICCV), pp. 2242–2251. IEEE, Venice (2017)
2. Zhou, S., Liang, X., Guo, J., et al.: An Edge-Refined vectorized deep colorization model for grayscale-to-color images. Neurocomputing **311**, 305–315 (2018)
3. Zhang, N., Qin, P.L., Zeng, J.C., et al.: Gray image coloring algorithm based on dense neural network. Comput. Appl. **39**(6), 1816–1823 (2019)
4. Xiao, Y.X., Jiang, A.W., Liu, C.H., et al.: Single image colorization via modified cyclegan. In: International Conference on Image Processing, pp. 3247–3251 (2019)
5. Patricia, V., Lara, R., Coloma, B.: ChromaGAN: adversarial picture colorization with semantic class distribution. In: Winter Conference on Applications of Computer Vision (WACV), pp. 2435–2443 (2020)
6. Su, J.W., Chu, H.K., Huang, J.B.: Instance-aware image colorization. In: CVPR, pp. 7968–7977 (2020)
7. Simonyan, K., Zisserman, A.: Very deep convolutional networks for large-scale image recognition. In: Proceedings of the 2015 International Conference Learning Representations, pp. 1–14. IEEE, San Diego (2015)
8. Howard, A.G., Zhu, M., Chen, B., et al.: Mobilenets: efficient convolutional neural networks for mobile vision applications (2017). http://arxiv.org/abs/1704.04861
9. Krizhevsky, A., Sutskever, I., Hinton, G.: ImageNet classification with deep convolutional neural networks. In: Proceedings of the 25th International Conference on Neural Information Processing System, pp. 1097–1105 (2012)
10. Zeiler, M.D., Fergus, R.: Visualizing and understanding convolutional networks. In: European Conference on Computer Vision, 818–833 (2014)
11. Hu, J., Shen, L., Sun, G.: Squeeze-and-excitation networks. In: Proceedings of the IEEE Conference on Computer Vision and Pattern Recognition, pp. 7132–7141 (2018)
12. He, K., Zhang, X., Ren, S., et al.: Deep residual learning for image recognition. In: Proceedings of the IEEE Conference on Computer Vision and Pattern Recognition, pp. 770–778. IEEE (2016)
13. Liu, W., Wen, Y., Yu, Z., et al.: SphereFace: deep hypersphere embedding for face recognition (2017). http://arxiv.org/abs/1704.08063
14. Deng, J., Dong, W., Socher, R., et al.: ImageNet: a large-scale hierarchical image database. In: 2009 IEEE Computer Society Conference on Computer Vision and Pattern Recognition, Miami, Florida, USA, 20–25 June 2009. IEEE (2009)

FM-Mnet Real-Time Object Detection Model for Video Feature Fusion Processing

Tiehua Zhou, Yifan Zhang, Yuan Li, and Ling Wang(✉)

Department of Computer Science and Technology, School of Computer Science, Northeast Electric Power University, Jilin, China
{Thzhou,2202000719,2202000701,smile2867ling}@neepu.edu.cn

Abstract. Current work on object detection mainly uses the method of one-stage or two-stage. The one-stage method regresses the probability of category and position coordinate value of the object, which results in lower accuracy. The two-stage method firstly generates a series of candidate boxes as samples by the algorithm, secondly, it classifies the samples through the convolutional neural network. Even though it has a high accuracy rate, but the complexity of training time of the model is relatively high. In this paper, we proposed a FM-Mnet model, which is based on the feature fusion method to addresses both limitations. Through this model, a relatively high accuracy rate can be obtained with feature fusion strategy. The experiment shows that this model could lead to a considerable performance in object detection when there are only about five videos in each category.

Keywords: Real-time object detection · Convolutional neural network · Feature fusion processing · Deep learning

1 Introduction

Among various fields, the deep learning method has been applied, especially Hinton proposed in 2015 to automatically learn advanced features in multimedia data through neural networks [1], which made deep learning begin to be applied in computer vision trends. As one of the research branches of computer vision, object detection has made great progress and could help a lot for the semantic understanding of images and videos. The essence of object detection is locating and classify the target, and the task of object detection is finding all the targets of interest in a given image accurately and efficiently, and the rectangular bounding box is applied to locate the position and size of the detected target. One of the most basic pieces of information: What objects are where? [2]. Object detection is a basic task in the field of computer vision, which enables scene analysis and understanding to proceed smoothly, while the target of this technology which is applied in the video is suitable to the task requirements of real scenes and is applied popularly in scientific development and actual industry, such as humans face recognition [3], video surveillance [4], autopilot [5], drone navigation [6], etc. In the object detection process, target distortion increases the difficulty of object detection, because of the differences in appearance, posture, shape, and quantity of different target objects, moreover, the

S.-C. Chu et al. (Eds.): ICGEC 2021, LNEE 833, pp. 249–258, 2022.
https://doi.org/10.1007/978-981-16-8430-2_23

illumination and occlusion. The object detection algorithm has been researched and explored.

The current mainstream object detection algorithms are mainly based on deep learning models, which could be roughly grouped into 2 categories: the one-stage object detection algorithm and the two-stage object detection algorithm. For the one-stage object detection algorithm, the region proposal stage is not needed. The category probability and coordinate of the position of the object could be generated through one stage, and YOLO [7] and SSD [8] are two of some typical algorithms. The two-stage object detection algorithm groups the detection task into 2 steps, in the first step, the candidate region is generated, which contains the approximate location coordinate of the target, in the second step, the position of the candidate region is classified and refined. R-CNN [9] and Fast R-CNN [10] are two of some typical algorithms. The values of detection accuracy and speed are usually chosen to metric the performance of the model. The accuracy mainly considers the positioning and classification accuracy of the object. In conclusion, these two algorithms have coressponding advantages, the two-stage algorithm outperforms in accuracy, while the one-stage algorithm outperforms in speed.

In this paper, we convert the videos data into corresponding frames and use classification categories to individually label the video frames. Then use the CNN network to perform feature extraction on the image collection of each video. Finally, the characteristics of each individual are merged to find the overall characteristics that can represent the action, which makes the video classification problem been transformed into a graphics classification problem.

2 Related Work

Traditional object detection methods are generally divided into three steps: firstly, selecting some candidate regions on a given image; secondly, extracting features from the selected regions; third, getting classification result from the trained classifier. Region selection is to locate the location of the target. The sliding window strategy is generally used to traverse the entire image [11]. The commonly used features for extracting features of the selected region are SIFT [12] and HOG [13]. Finally, some classification algorithms as SVM [14] or AdaBoost [15] are used to classify the features which are extracted in the second step.

In recent years, object detection algorithms have been shifted from traditional algorithms based on manual features to methods based on deep neural networks, there are two directions are focused on: two-stage algorithm and one-stage algorithm. About two-stage, Faster R-CNN [16] was proposed by Ren et al., who firstly complete the prediction of proposals by CNN, and lots of object detection algorithms rely on the idea of this network. The concept of position-sensitive occurred with the R-FCN series of networks [17, 18]. For the predicted feature map, the position-sensitive score map is introduced to enhance the position information of the sign, which could lead high detection accuracy. When it comes to the one-stage algorithm, OverFeat must be mentioned. Sermanet et al. proposed the OverFeat network to integrate classification, positioning, and detection functions into one network. The subsequent YOLO network series and SSD network are both classic one-stage detection algorithms. Redmon et al. improved the original YOLO network and then the YOLO9000 [19] was proposed. YOLO9000 can use small-scale detection and annotation data sets and large-scale classification and annotation data sets to train general object detection models. For the task of small object detection, Fu et al. introduced the top-down structure based on the improvement of SSD and proposed the DSSD model [20].

However, there are still many problems in object detection, such as the accuracy of small objects detection, detection of irregularly shaped objects, and occlusion of objects. For small object detection, Tsung Yi et al. proposed the top-down structure to improve the detection effect of small objects [21]. Jifeng et al. proposed deformable convolution and deformable ROI pooling to make up for CNN's shortcomings in geometric transformation, which solves the problem of detecting irregularly shaped objects [22].

3 Motivation

A high-accuracy object detection model requires abundant data as support. For the current mainstream CNN algorithm for object detection, the huge amount of data allows it to adjust parameters more accurately in training, which could improve the accuracy of the entire model. In Table 1, it shows the current data set commonly used in object detection, including the number of images and preselected boxes. The number of data sets unit *k* represents a thousand. When training the model, images and preselected boxes are used as the main input parameters. It is found that the minimum number of images is 80,000, and the maximum is 1.15 million. Regarding preselected boxes, its minimum size is 180,000 and the maximum is 14.81 million. The time cost of model training is increased, when images and preselected boxes are used for calculation at the same time.

Table 1. Current popular data sets for object detection.

Dataset	Images	Boxes
ImageNet all	477*k*	534*k*
ImageNet dense	80*k*	186*k*
COCO	123*k*	896*k*
OpenImages	1515*k*	14815*k*
Objects365	638*k*	10101*k*

Therefore, we propose the FM-Mnet model, which achieved a relatively good robust effect with a few datasets. The experiment showed that this model used about five videos as the training set for each classification, making its accuracy rate approximately 96%.

4 FM-Mnet Model

This section mainly discusses the main process of the FM-Mnet model. Firstly, video data is needed to turn into a sequential picture and those sequential pictures are labeled. Then train those pictures by the FM-Mnet model. Training a character's feature extraction model adopts a feature fusion strategy. The different features are merged. After the training, the model would be verified with all the video frames in the test set. Finally, obtain the top1 accuracy and top5 accuracy outputs on the full test set. In Fig. 1, it shows the whole process of model training which contains the core idea of the entire model.

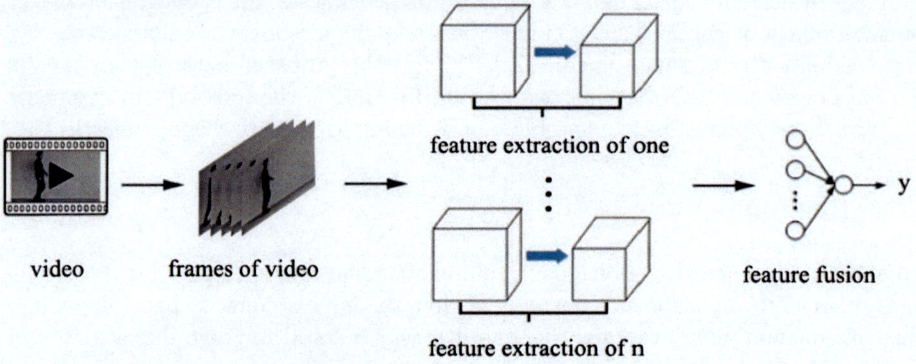

Fig. 1. The whole process of model training. y is the result of classification.

Table 2. Data preprocessing

```
Data Preprocessing
Input: Dataset (Video)
Output: Set (frames of Video)
for V_i in V do
        count=0
        for i in range(len(V_i)) do
                if i >count*k and i<count*(k+1) then
                    Vset← create([V_i[i])
                    count++
                else then
                    continue
                end if
        end for
FrameSet ← create(Vset, Lable)
end for
```

4.1 Data Preprocessing and Feature Extraction

Video data is not only a picture frame by frame, but also contains the connection between the previous frame and the next frame. This means that video is a time-series information dimension. Human action judgment tasks rely on the time series information of actions. Therefore, on the basis of reading the original video, continuous k frames are obtained and stored. Then the picture data of each frame is inverted and normalized, and its classification label is assigned. Finally, the data is processed in small batches. The detailed steps of data preprocessing are in Table 2.

Because the MobileNetV2 [23] network has the advantage of significantly reducing the required computing and memory resources while maintaining a high accuracy rate. Therefore, it can be considered to achieve the purpose of using fewer data to achieve higher accuracy. In the same category, multiple videos are mentioned separately. For a single video, take the pre-processed 160 * 160 * 3 tensor data in turn as input. Then pass a conv2d convolution. After 7 Bottleneck filters, pass a conv2d convolution. Finally, use a 7 * 7 pool. The layer pooling is used to obtain the desired 1 * 1 * 1280 feature, which can represent the action feature of the current video. In Table 3, it shows the entire feature extraction process. Finally, feature extraction is performed on all the videos of the same action in turn and saved as a set.

Table 3. The calculation process of feature extraction

Input	Output	Operator	t	c	n	s
$160^2 * 3$	$80^2 * 32$	Conv2d	–	32	1	2
$80^2 * 32$	$80^2 * 16$	bottleneck	1	16	1	1
$80^2 * 16$	$40^2 * 24$	bottleneck	6	24	2	2
$40^2 * 24$	$20^2 * 32$	bottleneck	6	32	3	2
$20^2 * 32$	$10^2 * 64$	bottleneck	6	64	4	2
$10^2 * 64$	$10^2 * 96$	bottleneck	6	96	3	1
$10^2 * 96$	$5^2 * 160$	bottleneck	6	160	3	2
$5^2 * 160$	$5^2 * 320$	bottleneck	6	320	1	1
$5^2 * 320$	$5^2 * 1280$	Conv2d 1 * 1	–	1280	1	1
$5^2 * 1280$	1 * 1 * 1280	Avgpool5 * 5	–	–	1	–

4.2 Feature Fusion

Each person is an independent individual, which leads to a certain difference in the same actions. To resolve this difference, a feature fusion strategy is adopted. The features of the same action from different individuals are taken as input, and the label of this action is taken as output. This means that the fused feature can cover all features, which will greatly improve the possibility of its recognition. In Fig. 2, it shows the fusion process of the entire model.

Different features of the same action can be obtained through feature extraction in 4.2. To reduce the difference of the data, all the features will be normalized, and the

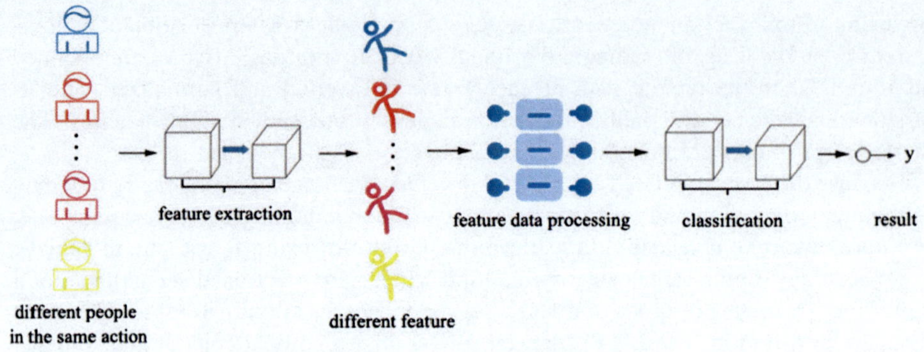

Fig. 2. Process of feature fusion strategy

processing formula is as follows:

$$F_{norm} = \frac{F - F_{min}}{F_{max} - F_{min}} \tag{1}$$

In the formula, F is the current eigenvalue. F_{min} is the minimum value of all eigenvalues. F_{max} is the maximum value of all eigenvalues. F_{norm} is the normalized eigenvalue. After normalization, All the features form a 1280 * 5 matrix as the input of CNN to train the final model. The detailed steps of data preprocessing are in Table 4.

Table 4. Algorithm flow of feature fusion

Algorithm flow of feature fusion
Input: *Dataset* (*FrameSet*)
Output: *Classification*
1: **for** F_i **in** *FrameSet* **do**
2: **if** F_i.*Lable* **is** *L* **then**
3: *featureSet* ← mobileNetV2(F_i.*Vset*)
4: **else then**
5: **continue**
6: **end if**
7: **end for**
8: **if len**(F_i **in** *FrameSet*) **not same then**
9: *processedFeatureSet* ← Normalize (*featureSet*)
10: **else then**
11: **continue**
12: **end if**
13: *Classification* ← mobileNetV2 (*processedFeatureSet*)

5 Experiment

The model evaluation has been performed on the Window10 system, which GPU is NVIDIA GeForce GTX 1080Ti. The program is implemented on the pycharm software in python language.

5.1 Dataset

The data set is the Human3.6M dataset [24]. This data set uses 4 RGB cameras, 1 time-of-flight sensor, and 10 motion cameras to capture in a 4 m × 3 m indoor space. It consists of motion capture data from seven actors performing 15 actions. In this article, five actions are used: boxing, handclapping, handwaving, running, and walking. The same train and test split subjects 1, 6, 7, 8, 9 are applied for training, and subjects 5, 11 are applied for testing.

5.2 Evaluation

This experiment compares the FM-Mnet model with the R-CNN network model [15]. The training and test set are the Human3.6M dataset. The accuracy of the five actions is shown in Table 5. It can be seen that the FM-Mnet model maintains an advantage in the detection of some actions.

Table 5. Comparison of accuracy between FM-Mnet model and R-CNN model

Method	Boxing	Handclapping	Handwaving	Running	Walking	Mean_accuracy
FM_Mnet	0.99	0.98	0.98	0.90	0.96	0.96
R-CNN	0.89	0.83	0.87	0.78	0.88	0.85

The accuracy and loss curves of the FM-Mnet model are shown in Fig. 3. The mapping of the tensorboard graph based on the TensorFlow framework shows that the FM-Mnet model has stabilized after repeated iterations. Figure 3(a) is the accuracy curve, in which the dark red curve represents the training trend of the model, and the light red curve represents the testing trend of the model. As the number of iterations represented by the abscissa increases, the accuracy of the model represented by the ordinate gradually stabilizes. Similarly, Fig. 3(b) shows the loss curve, in which the loss value decreases as the number of iterations increases, and eventually stabilizes.

To further evaluate the FM-Mnet model, we show the confusion matrix diagram. In Fig. 4, the ordinate is the label of predicted action, and the abscissa is the label of actual action. From Fig. 4, the FM-Mnet model performs well in object detection, basically reaching more than 90%. However, the running action performed poorly. Many running actions are recognized as walking. The entire model needs to be optimized in the feature extraction of similar actions.

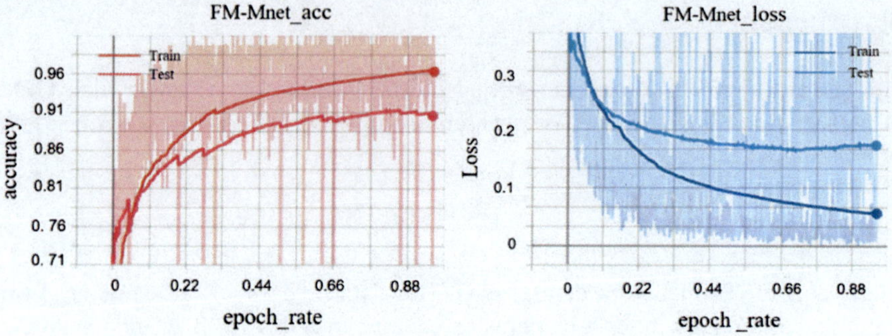

(a) the accuracy curve of FM-Mnet model **(b)** the loss curve of FM-Mnet model

Fig. 3. The accuracy and the loss curve of the FM-Mnet model

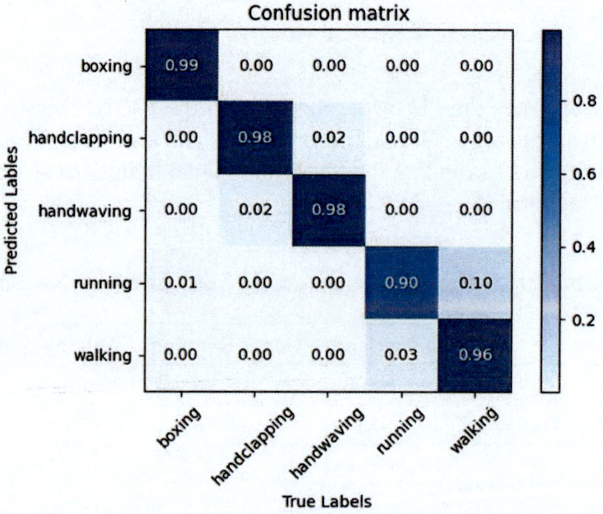

Fig. 4. The confusion matrix of the FM-Mnet model

6 Conclusion

In this paper, the FM-Mnet model was proposed to deal the task object detection. The core idea of this model is to extract video frames into pictures as a training set. During the extraction process, we kept all the frames of the video. Even if we keep all the frames, as a training dataset of the lightweight neural network, the training can still be completed in a short time. At the same time, experiments show that our model has a good performance in video object detection. However, the decimation method of preserving all frames is questionable, and preserving all may cause redundancy. In the next work, we may explore ways to extract frames and consider whether there is a better way to ensure accuracy while spending less training time. In addition, this model has some shortcomings in similar motion detection. Experiments show that 10% of Running is wrongly estimated as walking. The special diagnosis difference of similar actions is the focus of our next step.

Acknowledgement. This work was supported by the Science and Technology Development Plan of Jilin Province, China (Grant No. 20200403039SF) .

References

1. LeCun, Y., Bengio, Y., Hinton, G.: Deep learning. Nature **521**(7553), 436–444 (2015)
2. Zou, Z., Shi, Z., Guo, Y., Ye, J.: Object detection in 20 years: a survey (2020). arXiv Preprint arXiv:1905.05055
3. Byeon, Y.H., Pan, S.B., Moh, S.M., Kwak, K.C.: A surveillance system using CNN for face recognition with object, human and face detection. In: Information Science and Applications (ICISA), pp. 975–984. Springer, Singapore (2016)
4. Pérez-Hernández, F., Tabik, S., Olmos, R., Fujita, H., Herrera, F., Lamas, A.: Object detection binary classifiers methodology based on deep learning to identify small objects handled similarly: application in video surveillance. Knowl.-Based Syst. **194**, 105590 (2020)
5. Gao, X., Ge, D., Chen, Z.: The research on autopilot system based on lightweight Yolo-V3 target detection algorithm. In: Journal of Physics: Conference Series 2020, vol. 1486, pp. 3–032028. IOP (2020)
6. Zhai, X.H., Liu, K., Nash, W., Castineira, D.: Smart autopilot drone system for surface surveillance and anomaly detection via customizable deep neural network. In: International Petroleum Technology Conference, pp. 975–984. OnePetro, Dhahran (2020)
7. Redmon, J., Divvala, S., Girshick, R., Farhadi, A.: You only look once: unified, real-time object detection. In: Proceedings of the IEEE Conference on Computer Vision and Pattern Recognition, pp. 779–788. IEEE, Las Vegas (2016)
8. Liu, W., et al.: Ssd: single shot multibox detector. In: European Conference on Computer Vision, pp. 21–37. Springer, Cham (2016)
9. Girshick, R., Donahue, J., Darrell, T., Malik, J.: Rich feature hierarchies for accurate object detection and semantic segmentation. In: Proceedings of the IEEE Conference on Computer Vision and Pattern Recognition, pp. 580–587. IEEE, Columbus (2014)
10. Girshick, R.: Fast r-cnn. In: Proceedings of the IEEE Conference on Computer Vision and Pattern Recognition, pp.1440–1448. IEEE, Boston (2015)
11. Sudowe, P., Leibe, B.: Efficient use of geometric constraints for sliding-window object detection in video. In: Crowley, J.L., Draper, B.A., Thonnat, M. (eds.) ICVS 2011. LNCS, vol. 6962, pp. 11–20. Springer, Heidelberg (2011). https://doi.org/10.1007/978-3-642-23968-7_2
12. Lowe, D.: Distinctive image features from scale-invariant keypoints. Int. J. Comput. Vision **60**(2), 91–110 (2004)
13. Dalal, N., Triggs, B.: Histograms of oriented gradients for human detection. In: IEEE Computer Society Conference on Computer Vision and Pattern Recognition 2005, vol. 1, pp. 886–893. IEEE, San Diego (2005)
14. Malisiewicz, T., Gupta, A., Efros, A.A.: Ensemble of exemplar-svms for object detection and beyond. In: 2011 International Conference on Computer Vision. pp. 89–96. IEEE, Berlin (2011)
15. Ren, S., He, K., Girshick, R., Sun, J.: Faster R-CNN: towards real-time object detection with region proposal networks. Adv. Neural. Inf. Process. Syst. **28**, 91–99 (2015)
16. Kyrkou, C., Theocharides, T.: R-FCN-3000 at 30fps: decoupling detection and classification. IEEE Trans. Very Large Scale Integrat. (VLSI) Syst. **19**(6), 1034–1047 (2010)

17. Singh, B., Li, H., Sharma, A., Davis, L.S.: Ensemble of exemplar-svms for object detection and beyond. In: Proceedings of the IEEE Conference on Computer Vision and Pattern Recognition, pp. 1081–1090. IEEE, Colorado (2011)
18. Sermanet, P., Eigen, D., Zhang, X., Mathieu, M., Fergus, R., LeCun, Y.: Overfeat: integrated recognition, localization and detection using convolutional networks (2013). arXiv preprint arXiv:1312.6229
19. Redmon, J., Farhadi, A.: YOLO9000: better, faster, stronger. In: Proceedings of the IEEE Conference on Computer Vision and Pattern Recognition. pp. 7263–7271. IEEE, San Juan (2017)
20. Fu, C.Y., Liu, W., Ranga, A., Tyagi, A., Berg, A.C.: DSSD: deconvolutional single shot detector (2017). arXiv preprint arXiv:1701.06659
21. Lin, T.Y., Dollár, P., Girshick, R., He, K., Hariharan, B., Belongie, S.: Feature pyramid networks for object detection. In: Proceedings of the IEEE Conference on Computer Vision and Pattern Recognition. pp. 2117–2125. IEEE, San Juan (2017)
22. Dai, J., et al.: Deformable convolutional networks. In: Proceedings of the IEEE International Conference on Computer Vision, pp. 764–773. IEEE, Venice (2017)
23. Sandler, M., Howard, A., Zhu, M., Zhmoginov, A., Chen, L.C.: Mobilenetv2: inverted residuals and linear bottlenecks. In: Proceedings of the IEEE Conference on Computer Vision and Pattern Recognition, pp. 4510–4520. IEEE, Salt Lake City (2018)
24. Ionescu, C., Papava, D., Olaru, V., Sminchisescu, C.: Human3.6m: large scale datasets and predictive methods for 3D human sensing in natural environments. IEEE Trans. Pattern Anal. Mach. Intell. **36**(7), 1325–1339 (2013)

Modeling and Trajectory Tracking Control Simulation of 3-DOF Hoisting Manipulator

Mujun Xie, Shaoshi Qin(✉), Changhong Jiang, and Aoshuang Li

College of Electronic and Electronic Engineering, Changchun University of Technology, Changchun 130000, China

Abstract. To realize the precise control of the hoisting manipulator, carry out the modeling and control research of a 3-DOF hoisting manipulator with telescopic joints. Use the assumed model method to describe the elastic deformation of the horizontal telescopic link. According to the kinetic energy and potential energy of each link of the manipulator, combined with the Lagrange equation, gain the 3-DOF manipulator model. Based on the manipulator model set up by Adams, using Matlab and Adams joint simulation to collaborate the model correctness. Based on the model, trajectory tracking control carried out applying the full nonlinear integral sliding mode. The simulation results show the method is faster than the traditional sliding mode control tracking speed and has good performance.

Keywords: Hoisting manipulator modeling · Model confirmation · Full nonlinear integral sliding mode control · Trajectory tracking

1 Introduction

In recent years, as a large-scale manipulator, hoisting manipulators have widely used in the occasions of handling large loads. To set up the model of the hoisting manipulator and the research on the precise control of the manipulator tracking have certain practicable application value.

Experts and scholars at home and abroad have carried out research on the deformation modeling of rigid manipulators. Document [1] uses the assumed mode method to discretize the flexible manipulator, and gains the model equations based on the Lagrange equation and the principle of virtual work. Literature [2] proposes a general method for dynamic modeling of rigid-flexible coupling of N-DOF flexible arms, which reduces the complexity of the modeling process. Literature [3] set up the model of the flexible manipulator by combining the finite method and the Lagrange equation. In the document [4], the rotor torsion spring model with the concentrated mass used to describe the flexibility effect of the joint, and to equate the space double flexible manipulator set up the Hamilton principle. In reference [5] according to the ADAMS flexible body modeling theory. A model of multi-stage flexible telescopic boom based on virtual prototype set up. In the document [6], the axially moving beam adopts the assumed mode method to describe the deformation, and the Lagrange method used to set up to equate the axially moving beam.

S.-C. Chu et al. (Eds.): ICGEC 2021, LNEE 833, pp. 259–270, 2022.
https://doi.org/10.1007/978-981-16-8430-2_24

Various manipulator trajectory tracking control methods. Document [7] adopts the sliding mode control method and sets the adaptive compensation item for the neural network to the estimate error in the control law. Literature [8] designed a high-order sliding mode control based on a neural network. Document [9] application of nonlinear integral sliding mode controller RBF neural network, under the influence of the tracking error is small external disturbance, jitter amplitude reduced, but the greater volatility of the tracking error. Document [10] designed a smooth second-order sliding mode control method, which postpones the chattering in traditional sliding mode control, and has a faster convergence rate and strong robustness.

The hoisting manipulator studied in this paper contains two rotary joints and one horizontal telescopic joint. For lateral deformation of the horizontal telescopic arms during exercise, using assumed mode method to describe. Kinetic and potential energy-based link, combined Lagrange equation gained model of the manipulator, the boundaries of the model and ADAMS, the model for the accuracy of the confirmation simulation. Based on the set-up model, design a full nonlinear integral sliding mode controller, and research trajectory tracking control.

2 Modeling of Hoisting Manipulator

2.1 Hoisting Robot Arm Structure and Coordinate Transformation

According to the engineering background, simplify the hoisting manipulator studied in this paper to a 3-DOF manipulator with two rotary joints and one horizontal telescopic joint, as shown in Fig. 1. The link 2 is a horizontal telescopic link, which is prone to elastic deformation when the load is heavy. Set up the coordinate of each joint in the figure, specifying that axis X_i is the axis direction of link i, axis Y_i is in the direction at a right angle to axis X_i, and axis Z_i is the normal direction of the $X_i o Y_i$ plane and satisfies the right-hand rule. Define the base coordinate $X_0 o Y_0$ as the base coordinate, and the rest are the fixed coordinate after the joint in turn.

According to Fig. 1, combined with the D-H parameter method, the pose coordinate transformation of the hoisting manipulator is:

$$ {}^{0}T_1 = \begin{bmatrix} 0 & 1 & 0 & L_1 \\ c_1 & 0 & -s_1 & 0 \\ s_1 & 0 & c_1 & 0 \\ 0 & 0 & 0 & 1 \end{bmatrix} \quad {}^{1}T_2 = \begin{bmatrix} 1 & 0 & 0 & x(t) \\ 0 & 1 & 0 & \omega(x,t) \\ 0 & 0 & 1 & 0 \\ 0 & 0 & 0 & 1 \end{bmatrix} \quad {}^{2}T_3 = \begin{bmatrix} 0 & -c_1 & s_2 & 0 \\ 1 & 0 & 0 & L_3 \\ 0 & s_2 & c_2 & 0 \\ 0 & 0 & 0 & 1 \end{bmatrix} $$

Among them: $s_1 = \sin\theta_1$, $c_1 = \cos\theta_1$, $s_2 = \sin\theta_2$, $c_2 = \cos\theta_2$, θ_1 is the rotation angle of link 1, θ_2 is the rotation angle of link 4, $x(t) = L_2 + \Delta L$ is the link 2, L_2 is the link 2 base arm, ΔL is the change in link 2, L_3 is the link 3, $\omega(x, t)$ is the transverse elastic deformation of the link at $x(t)$.

2.2 Description of Elastic Deformation of Horizontal Telescopic Connecting Rod

In this study, simplify the horizontal telescopic link into a Euler-Bernoulli beam of constant cross section with one end fixed, the beam does not consider the moment of inertia

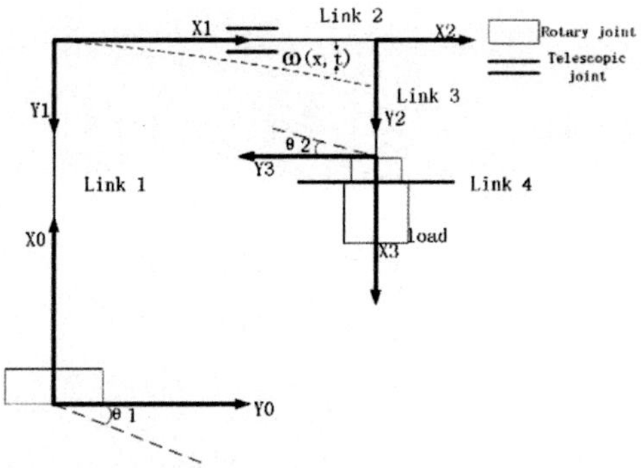

Fig. 1. Simplified model of hoisting manipulator

and shear deformation of the cross section, and the cross section of the beam is always at a right angle to the axis of the beam [7]. The elastic deformation of the horizontal telescopic link can get by multiplying each order modal and the modal coordinate. In practical engineering applications, the first-order mode plays a major role in elastic deformation [11]. So take the first-order to represent the transverse deformation of the horizontal telescopic link is:

$$\omega(x, t) = \varphi_1(x)q_1(t) \tag{1}$$

In the formula, $\varphi_1(x)$ is the first-order modal, $q_1(t)$ is the first-order modal coordinate.

2.3 Total Kinetic Energy of Hoisting Manipulator

The kinetic energy of link 1 is:

$$T_1 = \frac{1}{2}J_1\dot{\theta}_1^2 \tag{2}$$

The position point on the link 2 in the inertial coordinate is:

$${}^0\gamma_2 = {}^0T_1{}^2\gamma = \left| L_1 - \omega \; xc_1 \; xs_1 \; 1 \right|^{\mathrm{T}} \tag{3}$$

$${}^0\dot{\gamma}_2 = \left| -(\omega'\dot{x} + \dot{\omega}) \; \dot{x}c_1 - xs_1\dot{\theta}_1 \; \dot{x}s_1 + xc_1\dot{\theta}_1 \; 0 \right|^{\mathrm{T}} \tag{4}$$

The kinetic energy of link 2 is:

$$T_2 = \frac{1}{2}\rho_2 A_2 \int_0^x {}^0\dot{\gamma}_2^T \, {}^0\dot{\gamma}_2 d\gamma = \frac{1}{2}\rho_2 A_2 \int_0^x (\omega'\dot{x} + \dot{\omega})^2 + \dot{x}^2 + (\dot{\theta}_1 x)^2 dx \tag{5}$$

The position point on the link 3 in the inertial coordinate is:

$${}^0\gamma_3 = {}^0T_1 \, {}^1T_2{}^3\gamma = \left| L_1 - \omega - \gamma \; xc_1 \; xs_1 \; 1 \right|^{\mathrm{T}} \tag{6}$$

$$ {}^{0}\dot{\gamma}_3 = \left| -(\omega'\dot{x} + \dot{\omega})\ \dot{x}c_1 - xs_1\dot{\theta}_1\ \dot{x}s_1 + xc_1\dot{\theta}_1\ 0 \right|^{\mathrm{T}} \tag{7} $$

The kinetic energy of link 3 is:

$$ T_3 = \frac{1}{2}\rho_3 A_3 \int_0^{L_3} {}^{0}\dot{\gamma}_3^{T}\,{}^{0}\dot{\gamma}_3 d\gamma = \frac{1}{2}\rho_3 A_3 L_3((\omega'\dot{x} + \dot{\omega})^2 + \dot{x}^2 + (\dot{\theta}_1 x)^2) \tag{8} $$

The position point on the link 4 in the inertial coordinate is:

$$ {}^{0}\gamma_4 = {}^{0}T_1\,{}^{1}T_2\,{}^{2}T_3\,{}^{4}\gamma = \left| L_1 - \omega - L_3\ xc_1 - \gamma(s_1s_2 + c_1c_2)\ xs_1 + \gamma(c_1s_2 - s_1c_2)\ 1 \right|^{\mathrm{T}} \tag{9} $$

$$ \begin{aligned} {}^{0}\dot{\gamma}_4 = \big| & -(\omega'\dot{x} + \dot{\omega})\ \dot{x}c_1 - xs_1\dot{\theta}_1 + \gamma(\dot{\theta}_1 - \dot{\theta}_2)\sin(\theta_1 - \theta_2) \\ & \dot{x}s_1 + xc_1\dot{\theta}_1 + \gamma(\dot{\theta}_2 - \dot{\theta}_1)\cos(\theta_1 - \theta_2)\ 0 \big|^{\mathrm{T}} \end{aligned} \tag{10} $$

The kinetic energy of link 4 and the load is:

$$ \begin{aligned} T_4 &= \frac{1}{2}\rho_4 A_4 \int_{-\frac{L_4}{2}}^{\frac{L_4}{2}} {}^{0}\dot{\gamma}_4^{T}\,{}^{0}\dot{\gamma}_4 d\gamma + \frac{1}{2}M\,{}^{0}\dot{\gamma}_4^{\mathrm{T}}\,{}^{0}\dot{\gamma}_4 \\ &= \frac{1}{2}(\rho_4 A_4 L_4 + M)((\omega'\dot{x} + \dot{\omega})^2 + \dot{x}^2 + (\dot{\theta}_1 x)^2) + \frac{1}{2}(J_2 + J_3)(\dot{\theta}_1 - \dot{\theta}_2)^2 \end{aligned} \tag{11} $$

The total kinetic energy of the hoisting manipulator is:

$$ T = \sum_{i=1}^{4} T_i \tag{12} $$

Among them, J_1 is the moment of inertia of link 1, $J_2 = \frac{1}{6}\rho_4 A_4 L_4^3$ is the moment of inertia of link 4, J_3 is the moment of load moment of inertia, ρ_i, A_i is the density and cross-sectional area of the link, M is the load mass, and L_4 is the link 4, $\omega' = \frac{\partial\omega(x,t)}{\partial x}$, $\dot{\omega} = \frac{\partial\omega(x,t)}{\partial t}$.

2.4 Total Potential Energy of Hoisting Manipulator

The height point on the manipulator in the inertial coordinate is ${}^{0}\gamma_{ix}$.

The potential energy of link 1 is:

$$ V_1 = \rho_1 A_1 g \int_0^{L_1} \gamma d\gamma = \frac{1}{2}\rho_1 A_1 g L_1^2 \tag{13} $$

The potential energy of link 2 is:

$$V_2 = \rho_2 A_2 g \int_0^x {}^0\gamma_{2x} d\gamma + \frac{1}{2} EI \int_0^x \left(\frac{\partial^2 \omega(x,t)}{\partial x^2} \right)^2 dx = \rho_2 A_2 x g L_1 - \rho_2 A_2 g \int_0^x \omega dx + \frac{1}{2} EI \int_0^x \left(\frac{\partial^2 \omega(x,t)}{\partial x^2} \right)^2 dx \tag{14}$$

The potential energy of link 3 is:

$$V_3 = \rho_3 A_3 g \int_0^{L_3} {}^0\gamma_{3x} d\gamma = \rho_3 A_3 L_3 g (L_1 - \omega) - \frac{1}{2} \rho_3 A_3 g L_3^2 \tag{15}$$

The potential energy of link 4 and the load is:

$$V_4 = \rho_4 A_4 g \int_0^{L_4} {}^0\gamma_{4x} d\gamma + Mg {}^0\gamma_{4x} = (\rho_4 A_4 L_4 g + Mg)(L_1 - \omega - L_3) \tag{16}$$

The total potential energy of the hoisting manipulator is:

$$V = \sum_{i=1}^{4} V_i \tag{17}$$

2.5 Model of Hoisting Manipulator

Take the Lagrange function is:

$$L = T - V \tag{18}$$

According to the lagrange principle:

$$\frac{d}{dt}\left(\frac{\partial L}{\partial \dot{q}} \right) - \frac{\partial L}{\partial q} = \tau \tag{19}$$

Got total kinetic energy T, V the total potential energy into finishing Lagrange equation yields:

$$M(q)\ddot{q} + Kq + G(q, \dot{q}) = \tau \tag{20}$$

Among them: $q = \left(\theta_1 \; \theta_2 \; x \; q_1 \right)$ is a generalized variable, $\tau = \left(\tau_1 \; \tau_2 \; \tau_3 \; 0 \right)$ is a generalized moment, $M(q) \in R^{4\times4}$ is a positive definite symmetrical mass, $K \in R^{4\times4}$ is a stiffness, and $G(q, \dot{q})$ is a containing Coriolis force and centrifugal force.

$$M = \begin{vmatrix} M_{11} & M_{12} & 0 & 0 \\ M_{21} & M_{22} & 0 & 0 \\ 0 & 0 & M_{33} & M_{34} \\ 0 & 0 & M_{43} & M_{44} \end{vmatrix}, \; K = \begin{vmatrix} 0 & 0 & 0 & 0 \\ 0 & 0 & 0 & 0 \\ 0 & 0 & 0 & K_{34} \\ 0 & 0 & 0 & K_{44} \end{vmatrix}, \; G = \begin{vmatrix} g_1 & 0 & g_3 & g_4 \end{vmatrix}$$

$$M_{11} = J_1 + J_2 + J_3 + 1/3Dx^3 + Hx^2,\ M_{12} = M_{21} = -(J_2 + J_3),\ M_{22} = J_2 + J_3$$

$$M_{33} = D(q_1 I_5 + x) + H(\omega'^2 + 1),\ M_{34} = M_{43} = Dq_1 I_2 + H(\omega' \varphi_1(x))$$

$$M_{44} = DI_6 + H\varphi_1(x)^2,\ K_{34} = -(Dg\varphi_1(x) + Hg\dot{\omega}),\ K_{44} = EII_4,\ g_1 = (Dx^2 + 2Hx)\dot{x}\dot{\theta}_1$$

$$g_3 = D(2\dot{x}q_1\dot{q}_1 I_1 + 1/2(\omega'\dot{x})^2 + \dot{q}_1^2 I_2 + 1/2\dot{x}^2 - 1/2\dot{\omega}^2 - 1/2(\dot{\theta}_1 x)^2 + gL_1) + H(\omega'\omega''\dot{x}^2 + 2\omega'\dot{\omega}'\dot{x} - \theta_1^2 x) + 1/2EI\omega''^2$$

$$g_4 = D(\varphi_1(x)\omega'\dot{x}^2 + \varphi_1(x)\dot{\omega}\dot{x} - q_1\dot{x}^2 I_1 - gI_3) + H(\varphi_1(x)\omega''\dot{x}^2 + \dot{\varphi}(x)\omega'\dot{x} + \dot{\varphi}(x)\dot{\omega} + 2\varphi_1(x)\dot{\omega}'\dot{x} - \dot{\varphi}_1(x)\dot{\omega}\dot{x} - \dot{\varphi}_1(x)\omega'\dot{x}^2 - \varphi_1(x)g)$$

$$I_1 = \int_0^x \dot{\varphi}_1^2(x)dx,\ I_2 = \int_0^x \varphi_1(x)\dot{\varphi}_1(x)dx,\ I_3 = \int_0^x \varphi_1(x)dx,\ I_4 = \int_0^x \varphi_1''^2(x)dx$$

$$I_5 = \int_0^x {\varphi'}_1^2(x)dx,\ I_6 = \int_0^x \varphi_1^2(x)dx,\ \dot{\omega}' = \frac{\partial\dot{\omega}}{\partial x},\ \omega'' = \frac{\partial^2\omega(x,t)}{\partial x^2}$$

$$D = \rho_2 A_2,\ H = \rho_3 A_3 L_3 + \rho_4 A_4 L_4 + M$$

3 ADAMS Model Establishment and Manipulator Model Confirmation

3.1 Adams Model Set up

Based on the ADAMS2017 dynamic analysis software, the rigid model of the entire manipulator is first established. Add constraints and joint moments between each link. Then export the.x_t file to Ansys, mesh the horizontal link, define the rigid area, and select the first-order mode for flexibility. Exporting the.mnf file and importing it into ADAMS, the parametric model of the manipulator is shown in Fig. 2. The manipulator model exported from Adams controls to Matlab is shown in Fig. 3. The input variables to define the model are two rotation moments tol1, tol2 and horizontal drive tol3, and the output variables are defined as two rotation angles A_1, A_2, telescopic link displacement x and modal coordinates q1.

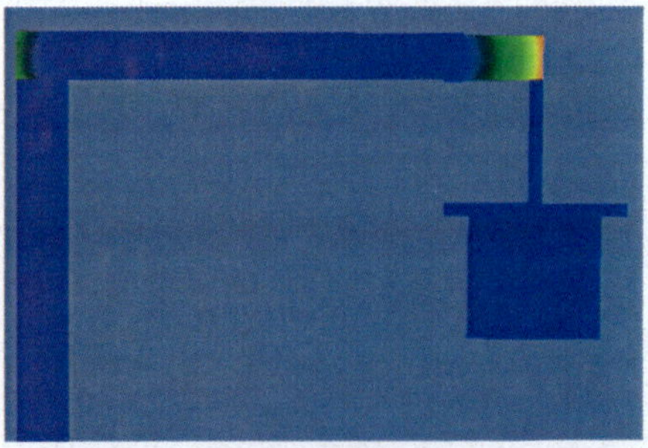

Fig. 2. ADAMS manipulator model

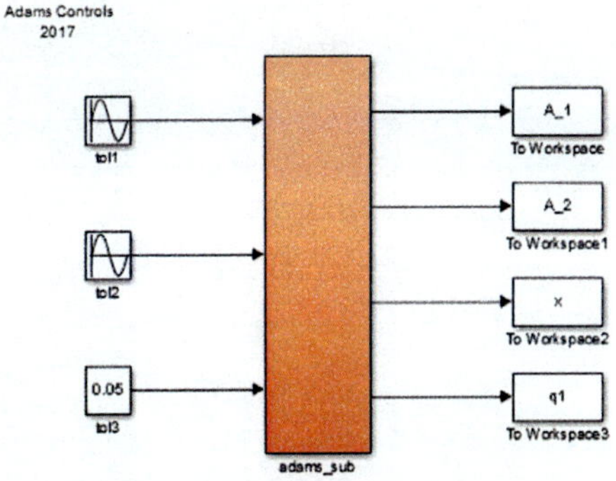

Fig. 3. Matlab Simulink diagram

3.2 Model Confirmation

Kinetic Model and hosting manipulator parametric model, and the parameters in Table 1 based on the manipulator, in a case where the driving torque applied to the same, whether the output is consistent observation, corroborated the correctness of the model.

Table 1. Manipulators parameters

Link 1 length L_1	3.0 m	Link 1 density ρ_1	$7.801 * 10^{-3}$ kg/m^3
Link 2 base arm length L_2	1.8 m	Link 2 density ρ_2	$7.801 * 10^{-3}$ kg/m^3
Link 2 telescopic range ΔL	0–2.0 m	Link 3 density ρ_3	$2.74 * 10^{-3}$ kg/m^3
Link 3 length L_3	0.5 m	Link 4 density ρ_4	$2.74 * 10^{-3}$kg/m^3
Link 3 length L_4	0.7 m	Load mass M	975.125 kg
Link 1 cross-sectional area A_1	0.04 m	Moment of inertia link 1 J_1	6.3 kg.m^2
Link 2 cross-sectional area A_2	0.04 m^2	Moment of inertia link 4 J_2	3.3 kg.m^2
Link 3 cross-sectional area A_3	0.002 m^2	Load Moment of inertia J_3	40.0 kg.m^2
Link 4 cross-sectional area A_4	0.035 m^2	Link 2 Elastic modulus E	$2.07 * 10^{11}$ N/m^2
		Link 2 moment of inertia section I	$2.2 * 10^{-6}$ m^4

The starting angle of arm joints 1 and 2 is 0, and the joint torque is 500 $\sin(\pi t)$ and 5 $\sin(\pi t)$. The first value of the telescopic manipulator is 1.8 m, the movement speed is 0.05 m/s, the simulation step is 0.01 s, and the simulation time is 2 s. This article only considers the variable of the three joints of the manipulator. The simulation results show in the figure below (Fig. 4).

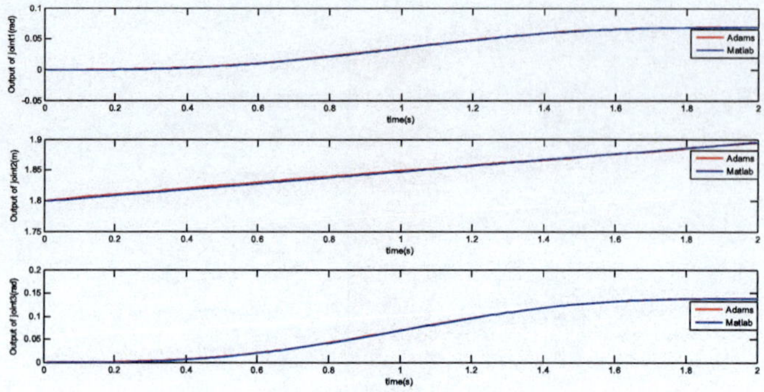

Fig. 4. Joint output simulation curve

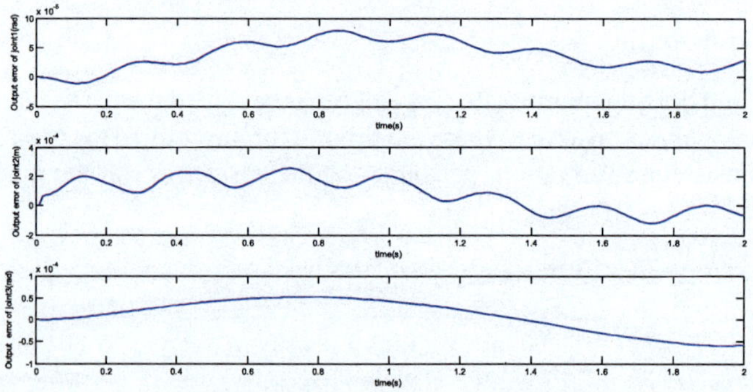

Fig. 5. Output error simulation curve

In Fig. 5 that under the same signal input, the maximum output error of each joint is 7.95×10^{-5} rad, 2.56×10^{-4} m, and 5.25×10^{-5} rad. The error is small. Adams and Matlab co-simulation corroborated the correctness of the model of the manipulator.

4 Controller

Based on the set up ideal model of the manipulator, combined with confuse the model in the project and external interference, the model used in this paper to study the trajectory tracking control is:

$$M(q)\ddot{q} + Kq + G(q, \dot{q}) = \tau + \rho(t) \tag{21}$$

In the formula:

$$\rho(t) = -\Delta M(q)\ddot{q} - \Delta Kq - \Delta G(q, \dot{q}) + d(t) \tag{22}$$

$\Delta M(q)$, ΔK, $\Delta G(q, \dot{q})$ is confuse the model, $d(t)$ is the external interference.

Assuming: the external disturbance and confusion term $\rho(t)$ in Eq. (21) has an upper bound and the upper bound is δ, then:

$$|\rho(t)| \le \delta \tag{23}$$

4.1 Design of the Full Nonlinear Integral Sliding Mode Controller

Define the expected trajectory of the manipulator tracking as, then:

$$e = q_d - q \to 0 \tag{24}$$

$$\dot{e} = \dot{q}_d - \dot{q} \to 0 \tag{25}$$

$e, \dot{e}$ is the position error and speed error of trajectory tracking.

The whole sliding surface can design as:

$$s = \dot{e} + \lambda_1 e - f^*(t) \tag{26}$$

Among them, $f^*(t)$ is the full sliding mode, which needs to meet three conditions.

(1) $f^*(0) = \dot{e}(0) + \lambda_1 e(0)$
(2) $t \to \infty, f^*(t) \to 0$.
(3) $f^*(t)$ first derivative exists and have bound.

The sliding surface of the whole design is:

$$s = \dot{e} + \lambda_1 e + \lambda_2 \int_0^t g(e) d\tau - (\dot{e}(0) + \lambda_1 e(0)) e^{-\theta t} \tag{27}$$

In the formula:$\lambda_1 = diag\{\lambda_{11}, \lambda_{12}, \lambda_{13}\}$, $\lambda_2 = diag\{\lambda_{21}, \lambda_{22}, \lambda_{23}\}$, $\lambda_{11}, \lambda_{12}, \lambda_{13}, \lambda_{21}, \lambda_{22}, \lambda_{23} > 0$, $e(0)$, $\dot{e}(0)$ is the position error and speed error at zero time, $\theta \in R^+$.

In formula (27), $g(e)$ is a nonlinear, its form is:

$$g(e) = \begin{cases} \beta \sin \frac{\pi e}{2\beta} & |e| < \beta \\ \beta & e \geq \beta \\ -\beta & e \leq -\beta \end{cases} \tag{28}$$

In the formula: $\beta \in R^+$.

The derivative of Eq. (27) is:

$$\begin{aligned} \dot{s} &= \ddot{e} + \lambda_1 \dot{e} + \lambda_2 g(e) + \theta(\dot{e}(0) + \lambda_1 e(0)) e^{-\theta t} \\ &= \ddot{q}_d - M^{-1}(\tau + \rho(t) - Kq - G(q, \dot{q})) + \lambda_1 \dot{e} + \lambda_2 g(e) + \theta(\dot{e}(0) + \lambda_1 e(0)) e^{-\theta t} \end{aligned} \tag{29}$$

Add a constant rate approaching law for sliding mode control:

$$\dot{s} = -D\text{sgn}(s) \tag{30}$$

In the formula: $D \in R^+$.

Combining Eq. (29) and Eq. (30) together can get the controller:

$$\tau = M(\ddot{q}_d + \lambda_1 \dot{e} + \lambda_2 g(e) + \theta(\dot{e}(0) + \lambda_1 e(0)) e^{-\theta t} + D\text{sgn}(s)) - \delta + Kq + G(q, \dot{q}) \tag{31}$$

4.2 Stability Analysis

Define the Lyapunov function as:

$$V = \frac{1}{2}s^{\mathrm{T}}s \tag{32}$$

Derivative of formula (32) is:

$$\begin{aligned}\dot{V} &= s\dot{s} \\ &= s(\ddot{q}_d - M^{-1}(\tau + \rho(t) - Kq - G(q, \dot{q})) + \lambda_1\dot{e} \\ &\quad + \lambda_2 g(e) + \theta(\dot{e}(0) + \lambda_1 e(0))e^{-\theta t})\end{aligned} \tag{33}$$

The formula (31) into equation to give:

$$\begin{aligned}\dot{V} &= s\dot{s} \\ &= s(\rho(t) - \delta) - D|s| \le 0\end{aligned} \tag{34}$$

Because of $\dot{V}$ is semi-negative definite, and $D \in R^+$, $\dot{V} \equiv 0$, $s \equiv 0$, the closed loop is asymptotically stable by LaSalle's theorem, and $t \to \infty$, $s \to 0$.

4.3 Trajectory Tracking Simulation Analysis

The wished trajectory is: $q_d = \left[0.5 + 0.1\sin(2\pi t)\ 0.1\cos(2\pi t)\ 1.8 + 0.1\sin(2\pi t)\right]^{\mathrm{T}}$.

The first value of the tracking position: $q_0 = \left[0.3\ 0.3\ 1.75\right]^{\mathrm{T}}$.

Selecting parameters for the sliding mode controller: $\lambda_1 = \mathrm{diag}(10,15,10)$, $\lambda_2 = \mathrm{diag}(10,20,10)$, $D = \mathrm{diag}(2.5, 2.5,2.5)$, $\beta = 0.05$, $\theta = 5$, $\delta = \left[10\ 10\ 10\right]$.

Select confusions and external disturbances is: $\rho(t) = \left[10\ \sin(0.5t)\ 10\ \cos(0.5t)\ 10\ \cos(0.5t)\right]^{\mathrm{T}}$.

The simulation results show in the figure below.

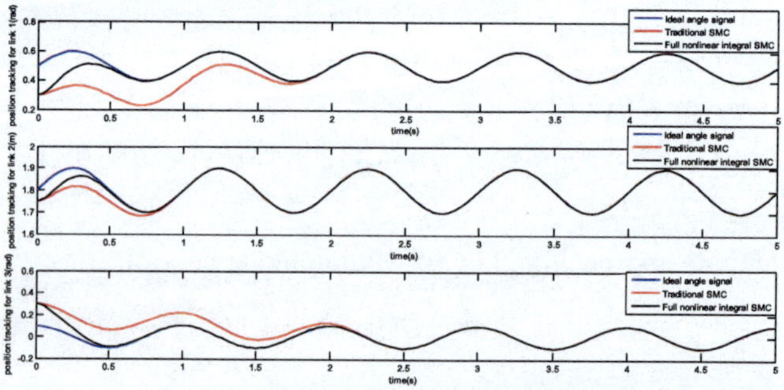

Fig. 6. Trajectory tracking simulation curve

Under the conditions of the same sliding mode surface parameter settings, from the trajectory tracking of joint 1, joint 2, and joint 3 in Fig. 6 and the error curve in Fig. 7,

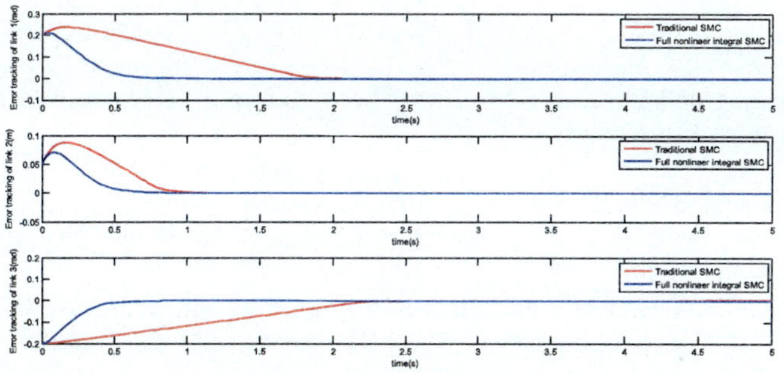

Fig. 7. Error simulation curve

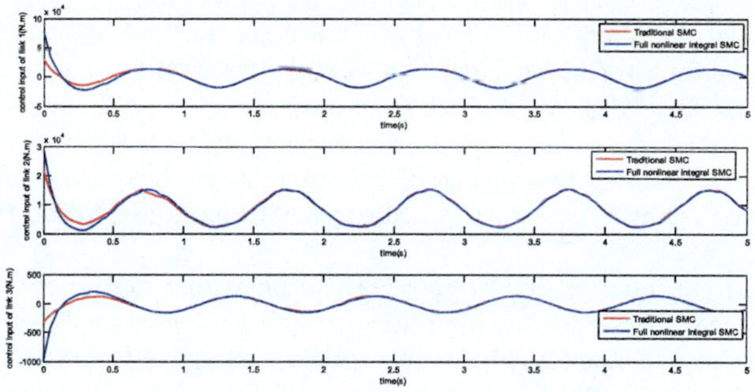

Fig. 8. Control torque simulation curve

it can see the entire nonlinear integral sliding mode control can track about 0.5 s. The expected trajectory is 1.3 s, 0.4 s and 1.8 s earlier than the traditional sliding mode, with fast response speed and good response. It can be seen from the torque output curve in Fig. 8 that the output torque of the entire nonlinear integral sliding mode control is greater than the output torque of the traditional sliding mode at the initial stage of tracking. In the process of trajectory tracking of the manipulator, the elastic vibration of the horizontal manipulator has little effect on the joint variables, and only affects the end posture, so its influence is not discussed in the trajectory tracking.

5 Conclusion

In this paper, study the 3-DOF hoisting manipulator in the following two: use the assumed model method to describe the elastic deformation of the telescopic joint, use the knot and Lagrange method to set up the model of the manipulator, set up the parametric model by Adams. The comparison corroborates the correctness of the model; based on the set up model, design a full nonlinear integral controller. The algorithm improves the early moment torque value by introducing nonlinear integrals on the sliding surface, which overcomes to interfere the sliding surface in the arrival stage, and at the same time reduces the steady state error. The simulation corroborates the effectiveness of the algorithm.

References

1. Pan, Y.: Dynamic modeling and simulation verification of space flexible manipulator. Electron. Test. **6**, 23–26 (2014)
2. Yu, F., Chen, X.: Research on the general dynamic modeling method of N-degree-of-freedom flexible manipulator. Vibr. Shock **39**(16), 103–111 (2020)
3. Wei, Y., Jiao, G.: Dynamic analysis of flexible manipulator based on MATLAB. Mech. Eng. Autom. **1**, 33–36 (2016)
4. Gu, Q., Li, A., Zhao, H.: Rigid-flexible coupling modeling and calibration error analysis of space dual-flexible manipulators. Chin. Space Sci. Technol. **37**(1), 33–40 (2017)
5. Du, W., Wu, P., Yao, X.: Dynamic modeling analysis and experiment of multi-stage flexible telescopic boom. Mach. Des. Manuf. **12**, 35–39 (2017)
6. Hua, H., Liao, Z., Zhang, X.: Efficient dynamic modeling and frequency response analysis of an axially moving cantilever beam. Acta Mechanica Sinica **49**(06), 1390–1398 (2017)
7. Qian, Q., Zhang, A.: Multi-joint manipulator trajectory tracking adaptive neural network sliding mode control. Autom. Instrument. **39**(12), 39–42+47 (2018)
8. Zhang, J., Li, Z., Yin, Y., Qi, C., Wu, K.: High-order sliding mode neural network adaptive control of multi-joint manipulators. Mech. Sci. Technol. **4**(27), 1–8 (2021)
9. Dong, J., Chen, L.: Research on trajectory error of nonlinear integral sliding mode control manipulator based on RBF neural network algorithm. Chin. J. Eng. Mach. **16**(02), 106–110 (2018)
10. Zhang, R., Chen, Q.: Trajectory tracking control of manipulator based on smooth second-order sliding mode. J. Syst. Simul. **4**(27), 1–8 (2021)
11. Liu, G., Li, J., Zhang, X., Liu, Y.: Modal truncation technology of flexible manipulator based on control system model reduction. J. Shandong Inst. Technol. **3**, 6–11 (1998)
12. Liu, J.: Sliding Mode Variable Structure Control Matlab Simulation. Tsinghua University Press, Beijing (2015)

Intelligent Multimedia Tools and Applications

Virtual Display and Interaction System Design of Bingxi Scroll Painting Based on Augmented Reality Technique

Yihang Du[1,2], Yan Pei[2], Lijun Wang[1], Chenzong Bai[2], and Ke Niu[3](✉)

[1] School of Design and Arts, Beijing Institute of Technology, Beijing, China
[2] New Media Arts Department, National Academy of Chinese, Theater Arts, Beijing, China
[3] Collaborative Innovation Center for HSR Driver Health and Safety, Zhengzhou Railway Vocational and Technical College, Zhengzhou, China

Abstract. The Bingxi scroll painting depicts the grand scene of royal ice activities in Qianlong period of Qing Dynasty. It adopts augment reality technique to synthesis the material background and three-dimensional dynamic figures, which could display the combined original painting replica and digital performance of ice activities in virtual space, so as to restore the simulation experience of royal scene in Qing Dynasty. The interactive method which abided by multi-sensory integration design philosophy could construct a realistic environment and historical culture charm for audience intuitively. Through the comprehensive evaluation of the system, it is found that the character modeling proportion is the most important factor which influenced virtual display. This work could invigorate the Chinese traditional ice sports and transform the multivariate culture in contemporary era.

Keywords: Bingxi scroll painting · Skating activities · Augmented reality

1 Introduction

Bingxi was an important royal ice collective activity in Qing Dynasty, which has a long history in China. The Bingxi scroll painting was painted during the Qianlong period of Qing Dynasty and depicted the grand scene of Emperor Qianlong participated in the Bingxi activities. It was painted by palace painters Zhang Weibang and Yao Wenhan and is now enshrined in the Palace Museum in Beijing. Bingxi is the general name of ice sports in Qing Dynasty and has certain military, competitive, performing and entertaining functions. Bingxi activities began with production and living needs of the northern minorities in the long-term snow environment. With the gradually merged mainstream culture of the Manchu and Chinese Central Plains culture, Bingxi was defined as a national folk activity under the advocacy of Qing Dynasty government. The creative transformation of the cultural heritage represented by the picture of ice activities will echo the theme of 2022 Beijing Winter Olympic Games, which is conducive to the inheritance and development of traditional sports and culture connotation.

The original works of Bingxi paintings are limited by material entities and two-dimensional plane, and could not exhibition all the original environment features and

S.-C. Chu et al. (Eds.): ICGEC 2021, LNEE 833, pp. 273–283, 2022.
https://doi.org/10.1007/978-981-16-8430-2_25

dynamic artistic expression in digital communication process. Augmented reality (AR) and interactive technology provide the possibility for digital creation and virtual display of traditional painting. The environment for AR supports users in mixing digital media and physical models in an immersive, multi projection set-up [1]. The grand scene of Chinese traditional ice and snow activities will be shaped in virtual space and reality integration.

2 Method

AR technology is the augmentation or superimposition of computer-generated virtual information into the real environment to enable users to observe a fusion of virtual and reality. Existing AR systems can be divided into wearable, handheld and spatial projection according to the display method [2–4]. The disadvantage of this type of display is that the device is heavy and needs to be worn on the operator's head, which will cause discomfort for a long time and is not suitable for the display of Bingxi scroll painting. Handheld AR systems usually use tablets or laptops, which do not provide an immersive experience for the user and tie up the operator's hands. Spatial projection AR systems use a projector to project virtual information onto a physical scene [5–7]. Compared to wearable and handheld AR technologies, projection systems do not require any handheld or wearable devices to provide augmented information to users. In terms of projective AR research, the RoomAlive system developed by Microsoft Research can project the room into an immersive experience environment [8] and allow the operator to interact with the computer, the projector images to all corners of the room, the Kinect sensor and its skeleton tracking technology can track and identify the operator's actions such as touching, stepping, dodging, and interact with the projected image [9, 10]. The projected image of RoomAlive is independent of the perspective of the experiencer and is mainly used in the field of games. Shi [11] used multiple depth cameras and projectors, and calibrated the camera and projector sets to accurately project graphic patterns on any surface and interact with human gestures in a flat or three-dimensional space. Michela [12] proposed a force sensor-based assembly guidance system for interacting with AR devices, which was used to provide correct assembly sequence guidance to operators. Funk [13] used Kinect sensors, projectors and button devices to build a projection-based assembly assistance system, which projects assembly guidance instructions to the work scene, monitors incorrect assembly operations and projects warning instructions, can significantly reduce the rate of assembly errors and improve assembly efficiency.

Currently, most of the research on projection-based AR technology focuses on the human-computer interaction between the operator and the projection screen, but neglects the research related to the projection screen and the viewpoint of the experiencer. AR technology uses graphics processing algorithm and multi-channel information display system to generate virtual simulation with immersion, interaction and imagination on the computer, providing visual, auditory, tactile and other sensory integration simulation in the virtual digital space for audience. An AR system were proposed for Bingxi scroll painting exhibition and comprised two parts: a conventional physical painting (such as an art print or original painting) and an AR-based digital component that is integrated with the physical painting in such a way that only experiencing the both parts at the same time with an AR-enabled device makes the entire painting complete and meaningful [14]. The interaction exhibition frame as shown in Fig. 1.

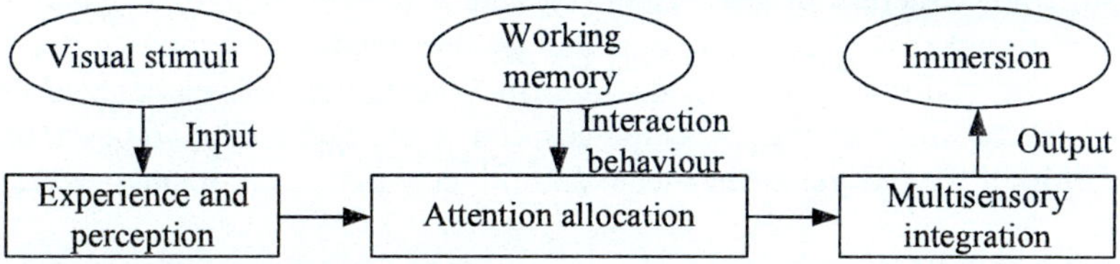

Fig. 1. Interaction exhibition method

2.1 Registration

The seamless integration of virtual image and real world in AR system can create an immersive sensory experience for users. The key technology is to splice and calibrate virtual and reality coordinates through 3D registration [15]. The camera calibration method of OpenCV development package was adopted to realize 3D registration, and the internal and external parameter matrix was established.

The parameter information captured by the camera can be expressed as follows:

$$\begin{bmatrix} x_c \\ y_c \\ z_c \end{bmatrix} = \begin{bmatrix} R & T \\ 0 & 1 \end{bmatrix} \begin{bmatrix} x_m \\ y_m \\ z_m \\ 1 \end{bmatrix} = \begin{bmatrix} r_1 & r_2 & r_3 & t_1 \\ r_4 & r_5 & r_6 & t_2 \\ r_7 & r_8 & r_9 & t_3 \\ 0 & 0 & 0 & 1 \end{bmatrix} \begin{bmatrix} x_m \\ y_m \\ z_m \\ 1 \end{bmatrix} \tag{1}$$

Where 0 means no value; R is circulant matrix; T is translation variable.

The transformation of virtual image data in real image can be expressed as matrix:

$$w \begin{bmatrix} u \\ v \\ 1 \end{bmatrix} \begin{bmatrix} f_x & s & u_0 \\ 0 & f_y & v_0 \\ 0 & 0 & 1 \end{bmatrix} \begin{bmatrix} x_c \\ y_c \\ z_c \end{bmatrix} = k \begin{bmatrix} x_c \\ y_c \\ z_c \end{bmatrix} \tag{2}$$

Where k is internal parameter of video acquisition device; f_x and f_y are the weight coefficients of video capture device on the two coordinate axes; (u_0, v_0) is the starting coordinate point; S is random variable factor.

2.2 Display and Output

The hardware of the projection-based AR assembly induction system mainly includes the Kinect sensor, projector and computer. The Kinect sensor is mounted above the display area and captures color and depth images of the user and the work area. The projector is mounted directly above the workbench and aligned vertically downward to the display area. The Kinect sensor and projector are connected to a computer, and includes a depth sensor and a color sensor. The depth sensor collects the depth data of the assembly scene, and the computer tracks the user's head position based on the depth data and recognizes the user's gestures to realize the freehand interaction. The projector is used to display the image of the painting to the exhibition area to guide the user operate.

The software structure of the system is mainly divided into two modules: gesture recognition and viewpoint tracking. The gesture recognition module determines the user's action and loads the guidance information through the recognized gesture. The viewpoint tracking module tracks the position of the user's fixation point, changes the orientation of the 3D guidance scene, and projects the image related to the user's viewpoint.

The real scene detection results and virtual images of skating will be visualized by AR display device simultaneously. The performance of the display device directly affects the immersion experience to audience. The design of virtual image of Bingxi painting at edge detection in different display environments, and edge detection was carried out repeatedly during display test. Video acquisition equipment was repeat the calibration of image output when the edge detection results were fuzzy or incomplete, until the automatic recognition, stitching and matching of contour were completed.

3 Original Feature Extraction and Visual Design

First, the two-dimensional image materials were extracted from the Bingxi original work, then the typical scenes and characters modeling were constructed in three-dimensional editing software, and finally import the two-dimensional and three-dimensional data into the AR development engine to realize the synthetic output. The two-dimensional data mainly rely on high-definition image acquisition system and Adobe Illustrator drawing software. The three-dimensional data was modeled by Maya and 3D Max software, and Unity3D kits were selected as the virtual synthesis platform.

The original painting of Bingxi is about 578.8 cm long and 36.5 cm wide (see Fig. 2), which combines traditional martial arts, theatre art and acrobatic performance. The characters are mainly composed of emperors, officials and soldiers. Soldiers are keeping with high consistency in modeling and involved in skating activities, but the differences reflected on costumes, props and movements. Seven representative groups of characters were selected to construct three-dimensional images, and the details of costumes, props and dynamic effects were restored as much as possible, including handstand, double skating, parent child skating, figure skating with sticks, flag dance, figure skating with long stick brandish and archery figure skating (see Fig. 3).

Autodesk Maya 2015 software were selected to model characters, scenes and props. It should be complied with connotation expression of original work and combined the physical properties of the material for secondary creation. Regular skating was the representative motion in Bingxi painting, the character modeling and dynamic effect were recreated on basis of referring to the records of ancient books and literature (a book entitled of "Life records of Beijing in Qing Dynasty"), and combined with the characteristics of modern speed skating (see Fig. 4).

The modeling of clothes was highlighted drooping effect of materials to the handstand posture (see Fig. 5). Children clothes modeling in parent child skating was decorated with the exclusive symbol for royal children to show the genetic relationship with court,

Fig. 2. Part of Bingxi painting original work

a. Handstand

b. Figure skating with sticks

c. Parent child skating

d. Archery figure skating

e. Figure skating with long stick

f. Double skating

g. Flag dance

Fig. 3. Typical character in Bingxi painting

Fig. 4. Regular skating character modeling

which was referred a book entitled "General study of Qing Dynasty documents" and optimized the original ambiguous pattern of dress.

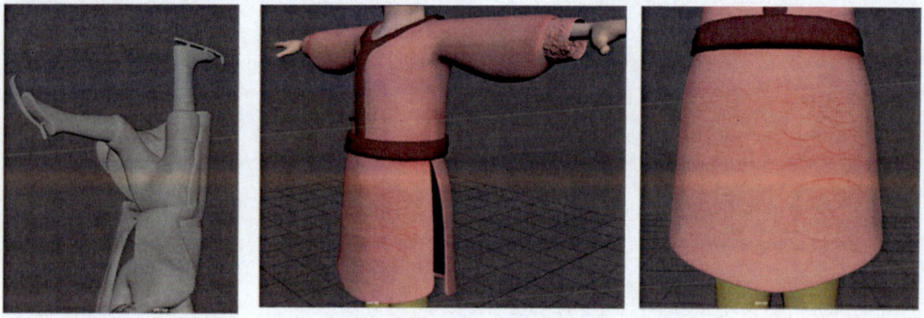

Fig. 5. Clothes modeling

There were two wide belts inside the hand-held shield in the Qing Dynasty, it was convenient to put the arm into belt when wearing, and hold the other edge by hand to realize the stability. In order to eliminate the negligence of original work, a short portable belt was remedied in the modeling and the historical physical form of the shield was restored (see Fig. 6).

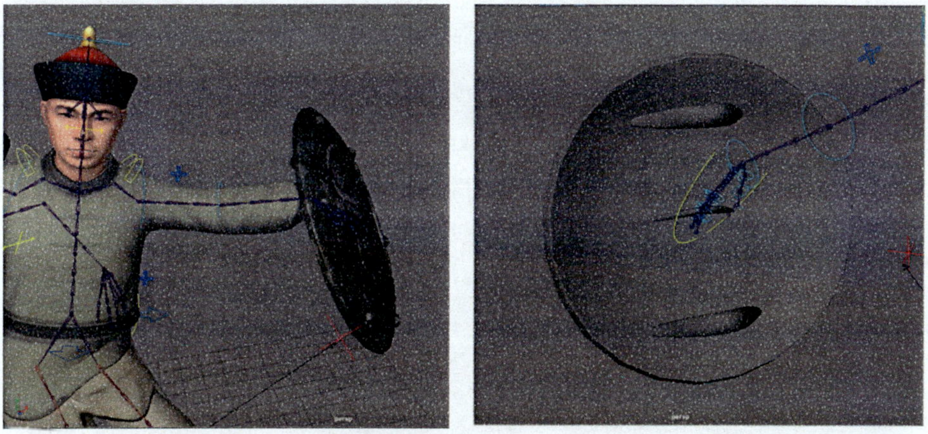

Fig. 6. Shield modeling

4 Design and Implementation of AR System

Unity3D was selected as the development platform of AR. Vuforia as a plug-in in Unity3D which supports the programming of various kinds of language programs. Virtual Agents can be placed in the real world using a uniquely identifying AR marker using the SDK. Firstly, the project was created by Vuforia and the SDK was generated. Then, the character materials in original work were edited as recognizable pictures and imported into Vuforia, and generated feature marks that can be detected. The 3D model of characters material in FBX format were transferred from Autodesk Maya 2015 import into Unity3D, and finally generated for AR system.

The system consists of six modules which contain scene acquisition, image recognition, 3D registration, virtual synthesis, human-computer interface and synthesis output. The target is obtained from image collection device in the scene acquisition module. The feature point set of the target image is obtained in the recognition module, which is matched with the feature points of the location recognition map, and then the matched real location is tracked and sent to 3D registration module. The spatial position of the virtual object in the real world is calculated in 3D registration module, and the virtual object fusion with physical world are realized in virtual synthesis module. The 3D modeling as shown in Fig. 7. In the on-site display, the user could trigger a series of hybrid image generation through the human-computer interaction interface and experience of virtual interactive in Bingxi activities in the synthesis output module.

A holographic projection module is set in the middle of semi-wrapped exhibition area to virtual display of Bingxi characters (see Fig. 8), and a 1:1 inkjet copy of the original work is set below. Audience can scan the physical image through the AR recognition software in mobile device to obtain the fusion dynamic Bingxi painting. The sound effects such as cheering and ice sliding were synthetized to build a simulation experience of Bingxi activities in Qing Dynasty.

Fig. 7. Typical character modeling

Fig. 8. Exhibition area and simulation

5 Comprehensive Evaluation of AR Exhibition

Comprehensive evaluation of the AR exhibition design before on-site exhibition is helpful to obtain the user's intuitive feelings and preferences for the works. To carry out design optimization according to the evaluation results will be advantageous to improve the user's interactive experience in the on-site exhibition. The comprehensive evaluation index construction of AR exhibition was carried out from usability and visual aesthetics.

The usability of AR system is related to the functional design of visual display terminal equipment and indirectly affects the aesthetic. A comprehensive evaluation system was established according [16–18] and learnability, controllability and interface comfort were selected as usability evaluation indexes. The visual aesthetic indexes included five indicators which contained integration of virtual and physical, proportion of 3D modeling, consistency of virtual and physical style, reality of texture and richness of layers.

Ten adults (seven females) were recruited from Chinese Theatre Art Academy and Beijing Jiaotong university, aged 18–27 years (mean age = 19.7, standard deviation [SD] = 2.4 years). Before the start of experiment, a questionnaire report on the basic situation of personal. Based on the established system for acoustic expression evaluation in previous, the sound scene and sound sequence of three works were comprehensively evaluated by all participants combined with field investigation. It contains two dimensions and nine items, seven-Likert score was used. The result as shown in Table 1 that character modeling proportion were the significant advantages of comprehensive evaluation system (see Fig. 9). The screen comfort was received the most negative evaluation, it might be due to the AR system were designed to mobile display device terminal.

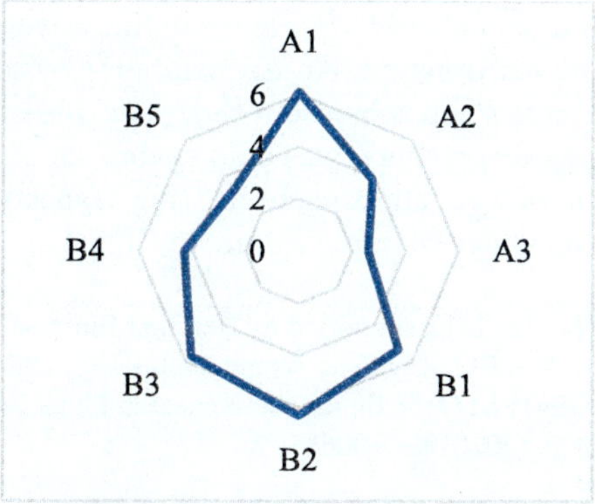

Fig. 9. Spider diagram of evaluation

Table 1. Evaluation results for AR system

Dimension	Indexes	Mark	Evaluation results
Usability	Learnable	A1	6.1
	Controllability	A2	3.9
	Interface comfort	A3	2.6
Aesthetics	Integration of virtual and physical	B1	5.3
	Proportion of 3D modeling	B2	6.3
	Consistency of virtual and physical style	B3	5.7
	Reality of texture	B4	4.3
	Richness of layers	B5	3.4

6 Conclusion

We presented digital exhibition method for ancient painting alive which can show information for dynamic skating performance. An AR exhibition system of Bingxi painting is built by above key techniques and achieved a certain degree of virtual interaction in digital space. By building total procedure, we could restore various augmented contents with Chinese traditional ice sports atmosphere in Qing Dynasty, and presentation of localization significance of Winter Olympic culture in China.

Acknowledgments. This work has been supported by Beijing Municipal Education Commission Science Program (KM202110049001), Key Science and Technology Program of Henan Province (202102210159, 212102210396) and Open Fund from Research Platform of Zhengzhou Railway Vocational & Technical College (20201KFJJ006).

References

1. Jacucci, G., Oulasvirta, A., Salovaara, A., Psik, T., Wagner, I.: Augmented reality painting and collage: evaluating tangible interaction in a field study. In: Costabile, M.F., Paternò, F. (eds.) INTERACT 2005. LNCS, vol. 3585, pp. 43–56. Springer, Heidelberg (2005). https://doi.org/10.1007/11555261_7
2. George, C., Karagiannis, P., Makris, S., et al.: Augmented reality applications for supporting human-robot interactive cooperation. Procedia CIRP **41**, 375–378 (2016)
3. Munoz, J.: An improved augmented reality system based on and AR. Comput. Rev. **57**(11), 694–694 (2016)
4. Henderson, S., Feiner, S.: Exploring the benefits of augmented reality documentation for maintenance and repair. IEEE Trans. Visual Comput. Graph. **17**(10), 1355–1368 (2011)
5. Qureshi, I., Parah, S.A., Lone, N.A., Hurrah, N., Qureshi, G.J.: Realization of a robust watermarking system in spatial domain. In: Singh, P.K., Singh, Y., Kolekar, M.H., Kar, A.K., Chhabra, J.K., Sen, A. (eds.) ICRIC 2020. LNEE, vol. 701, pp. 369–379. Springer, Singapore (2021). https://doi.org/10.1007/978-981-15-8297-4_30

6. Binetti, N., Cheng, T., Mareschal, I., et al.: Assumptions about the positioning of virtual stimuli affect gaze direction estimates during Augmented Reality based interactions. Sci. Rep. **9**(1), 22–25 (2019)
7. Joo-Haeng, L.: Spatial augmented reality as a new interaction channel to bridge real and virtual worlds. Commun. Korean Inst. Inf. Sci. Eng. **33**(11), 27–36 (2015)
8. Jones, B., Sodhi, R., Murdock, M., et al.: RoomAlive: magical experiences enabled by scalable, adaptive projector-camera units. In: 27th Symposium on User Interface Software and Technology. ACM, Hawaii (2014)
9. Difilippo, N., Jouaneh, M.: Characterization of different microsoft kinect sensor models. IEEE Sens. J. **15**(8), 1 (2015)
10. Madrigal, J., Negrete, J., Guerrero, R., et al.: 3D Motion tracking of the shoulder joint with respect to the thorax using marg sensors and data fusion algorithm. Biocybern. Biomed. Eng. **40**(3), 46–49 (2020)
11. Shi, H., Hwang, K., Zhang, R., et al.: Tracking and proximity detection for robotic operations by multiple depth cameras. Inf. Sci. **483**, 350–362 (2019)
12. Michela, D.: An integrated environment based on augmented reality and sensing device for manual assembly workstations. Procedia Cirp **18**(7), 148–156 (2016)
13. Horejsi, P., Novikov, K., Simon, M.: A smart factory in a smart city: virtual and augmented reality in a smart assembly line. IEEE Access **99**, 1–4 (2020)
14. Veen, M.: Crossroads of seeing: about layers in painting and superimposition in Augmented Reality. AI Soc. **23**(1), 1–8 (2020)
15. Hwang, J.-I., Sung, M.-H., Kim, I.-J., Ahn, S.C., Kim, H.-G., Ko, H.: Painting alive: handheld augmented reality system for large targets. In: Shumaker, R. (ed.) VAMR 2013. LNCS, vol. 8021, pp. 32–38. Springer, Heidelberg (2013). https://doi.org/10.1007/978-3-642-39405-8_4
16. Timothy, J.M., Dieck, C. (eds.): Augmented Reality and Virtual Reality. PI, Springer, Cham (2018). https://doi.org/10.1007/978-3-319-64027-3
17. Martins, V.F., Kirner, T.G., Kirner, C.: Subjective usability evaluation criteria of augmented reality applications. In: Shumaker, R., Lackey, S. (eds.) VAMR 2015. LNCS, vol. 9179, pp. 39–48. Springer, Cham (2015). https://doi.org/10.1007/978-3-319-21067-4_5
18. Zafar, S., Zachar, J.: Evaluation of Holo Human augmented reality application as a novel educational tool in dentistry. Eur. J. Dent. Educ. **24**(2), 35–38 (2020)

GPU Implementation of Staggered Period Estimation Algorithm Acceleration

Zhuojun Xu[1](✉), Hangwei Hu[2], Long Wang[3], Chengzhi Yang[3], and Yantao Tian[1]

[1] College of Communication Engineering, Jilin University, 5988 Renmin Street, Changchun, China
xuzhuojun@jlu.edu.cn

[2] Beijing Electro-Mechanical Engineering Institute, Beijing, China

[3] Aviation University Air Force, 2222 Nanhu Road, Changchun, China

Abstract. In order to realize the on-line processing of the staggered period estimation algorithm, the parallel processing architecture of GPU (graphics processing units) and redesigns the staggered period estimation algorithm is proposed in this paper. The experiment results show the parallel algorithm only takes 900 ms to deal with the mixed pulse with one hundred thousand pulses.

Keywords: GPU · Staggered period estimation · On-line processing

1 Introduction

There are five major characteristics of radar PDW (pulse description word): RF (radio frequency), PW (pulse width), PA (pulse amplitude), DOA (direction of arrival), and TOA (time of arrival) [1], RF, PW and DOA is usually used in pre-sorting of radar signal, the envelope of PA and the PRI (pulse repetition interval) of TOA is usually used in Radar signal sorting [2]. PRI is an important parameter of radar signal for period estimation. The period estimation algorithm of stagger interval designed in the previous work of this paper can directly estimate the hidden stagger interval sub-period in radar aliasing pulse [3]. However, due to the large amount of computation, it cannot meet the requirements of real-time computing. In terms of hardware acceleration algorithm, the performance of digital signal processing devices has been rapidly improved in recent years. This offers more choices for the engineering implementation of radar signal processing algorithm such as DSP (digital signal processor), FPGA (field-programmable gate array) or DPS + FPGA integrated hardware platform [4, 8]. Meanwhile, GPU (graphics processing units) is becoming more mature and shows more powerful parallel computing capabilities. In addition to image processing, GPU is also gradually emerging in the domain of general computing. GPU is a special chip for processing graphics tasks [5, 7]. The task of GPU is to synthesize millions of pixels and display them on the screen, which requires the ability to process millions of tasks at the same time. Therefore, GPU is designed to handle multi tasks in parallel instead of single task mode of CPU (central processing unit) [6].

S.-C. Chu et al. (Eds.): ICGEC 2021, LNEE 833, pp. 284–293, 2022.
https://doi.org/10.1007/978-981-16-8430-2_26

Considering that there are a large number of non-interference parallel computing modules in the staggered interval period estimation algorithm, GPU has huge parallel computing power at lower development cost. The purpose of this paper is to accelerate the staggered interval estimation algorithm based on GPU. Firstly, the basic knowledge of GPU and the programming architecture of CUDA (compute unified device architecture) are introduced. Secondly, the algorithm of stagger interval period estimation is redesigned based on the architecture characteristics of CUDA. Finally, experiments are designed to compare the recognition accuracy and running time of the algorithm before and after acceleration using the same TOA model.

2 Model of Stagger Radar

The PRI of a radar signal is a sequence of time intervals between adjacent pulses, which is reflected in the change of TOA [3]. Suppose the TOA model of n pulses denotes T (n is a positive integer).

$$T = [t_1, t_2, ..., t_n]^T \tag{1}$$

Where t_i denotes the i th TOA, integer $i \in [1, n]$.

The PRI of the stagger radar changes according to the law. For a stagger radar with m stagger numbers, the PRI is $[PRI_0, PRI_1, ..., PRI_{m-1}]$. Its repetition period is also called the skeleton period. The period within the skeleton period is called the stagger period, and the skeleton period is equal to the sum of all the stagger periods. The TOA model T of the stagger radar has a regularity:

$$t_{i+1} = t_i + PRI_{i\backslash m} \tag{2}$$

Where $PRI_{i\backslash m}$ represents the i th PRI, $i \in [1, n]$. and \ is denoted the remainder. The original stagger radar mainly includes the following three conditions.

(1) Radar pulses loss

If the miss rate of the TOA model is denoted η, the missing model T_{miss} of the radar pulses T can be represented as:

$$T_{miss} = rand(T, \lfloor n \times \eta \rfloor) \tag{3}$$

$\lfloor\rfloor$ represents taking an integer down. $rand(T, \lfloor n \times \eta \rfloor)$ represents any $\lfloor n \times \eta \rfloor$ non-repeating elements from the array T.

(2) Radar pulses fake

If the false rate denotes φ, the false model T_{false} between the time range $t_1 \sim t_2$ (unit μs) can be represented as:

$$T_{false} = rand([t_1, t_1 + 1, ..., t_2]^T, \lfloor \Delta t \times \varphi \rfloor) \tag{4}$$

Where Δt is denoted time range $\Delta t = t_2 - t_1 + 1$. $rand([t_1, t_1 + 1, ..., t_2]^T, \lfloor \Delta t \times \varphi \rfloor)$ represents taking $\lfloor \Delta t \times \varphi \rfloor$ non-repeating elements from the integer vector $[t_1, t_1 + 1, ..., t_2]^T$.

(3) Radar pulses aliasing

Radar pulses in complex electromagnetic environments are mixed by multiple radar pulses. Considering that multiple radars are mixed and the pulses with missing and false, the TOA model of a complex electromagnetic environment can be expressed as:

$$T = \left(\bigcup_{i=1}^{N} T_i \cup T_{false} \cup T_{miss} \right) - T_{miss} \tag{5}$$

Where N is denoted the number of mixed radars. $\cup$ is denoted a union and $\cap$ is denoted an intersection. $-$ represents removing elements from the set.

3 Staggered Period Estimation Algorithm

Phase is an important concept to describing a sine or cosine signal, which represents the position of a particular moment (crest or trough or special point) in loop. The initial phase represents the initial position at a particular moment in the 0th cycle. When the phase and the initial phase are applied to the TOA model in Sect. 2, each TOA of a radar represents the phase, and the 1th TOA represents the initial phase. The fixed interval period is equal to the phase difference between the initial phase and the 2th phase.

According to the staggered period estimation algorithm which proposed in paper [3]. A systemic frame model is designed in Fig. 1. The input of this algorithm-$\boldsymbol{T}(t)$ represents a mixed pulse data set at a given time t, Δt denotes the length of the mixed

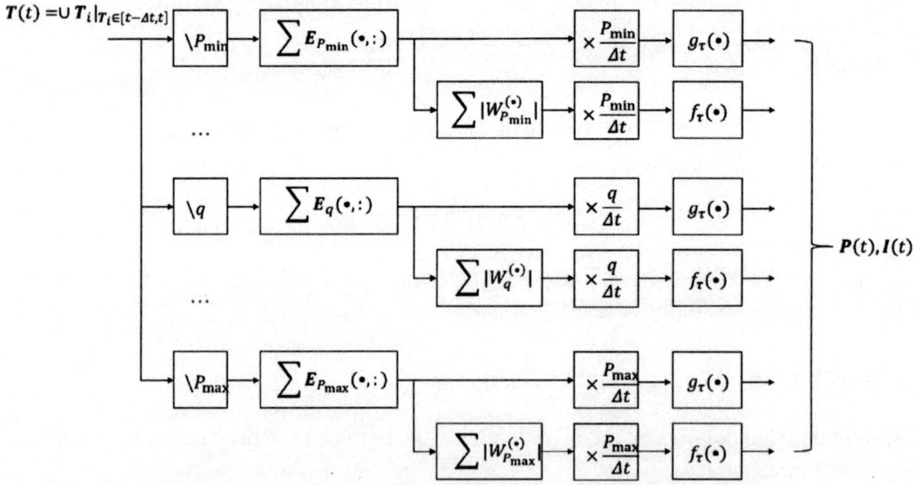

Fig. 1. Schematic of staggered period estimation algorithm

pulse. $\boldsymbol{P}(t)$ denotes the set of estimated interval sub-period at a given time t, $P_{\min}$ and $P_{\max}$ are respectively denoted the minimum and the maximum of the radar interval periods, interval period value $q \in [P_{\min}, P_{\max}]$, $q \in \boldsymbol{Z}$.

f_τ represents whether there are pulses at period q in the model T, g_τ is the phase threshold judgment function, $\boldsymbol{I}(t)$ denotes the initial phase set of estimated interval sub-period at a given time t. Staggered period at t can be obtained after harmonic processing of $\boldsymbol{P}(t)$ and $\boldsymbol{I}(t)$. If only executed once, let $t = 0$, $\Delta t = L$. If mixed pulse is stream data, divided the steam data into blocks as the length of Δt. Table 1 is the pseudo-code of staggered period estimation algorithm.

Table 1. Pseudo-code of staggered period estimation algorithm

Initialization
1. Set $P_{\min}, P_{\max}$
2. Set $\tau_p, \tau_r, \Delta t$
3. Set $t = t_{\text{begin}}$

repeat
4. Set $\boldsymbol{T}(t) = \cup \boldsymbol{T}_i |_{T_i \in [t, t+\Delta t]}$
5. For $q = [P_{\min}, P_{\max}]$
   $\boldsymbol{D}_q = [W_q^0, W_q^1, \ldots, W_q^{q-1}]$
   $\boldsymbol{N}_q = \sum_{i=1}^{n} \boldsymbol{E}_q(t_i \backslash q + 1, :)$
   $p_q = \left|\sum_{i=1}^{q} \boldsymbol{N}_q(i) \times \boldsymbol{D}_q(i)\right| \times q/\Delta t$
   $\boldsymbol{R}_q = \boldsymbol{N}_q \times q/\Delta t$
   If $p_q > \tau_p$ Then $q => \boldsymbol{P}(t)$
   If $\boldsymbol{R}_q > \tau_r$ Then $g_\tau(\boldsymbol{R}_q) => \boldsymbol{I}(t)$
   End For
6. Output: Clustering and Harmonic Processing on $\boldsymbol{P}(t)$ and $\boldsymbol{I}(t)$
7. $t = t + \Delta t$

Until $t = t_{\text{end}}$

4 Parallel Implementation of Algorithm Acceleration

The schematic diagram of the stagger interval period estimation algorithm is designed as shown in Fig. 2. By observing the schematic diagram, it can be found that the calculation of the standard energy value $\mathrm{p_q}$ and phase distribution $\mathrm{R_q}$ of each sub period q in the stagger interval period estimation algorithm has no relation with the calculation results of other sub periods. If the sum of each sub period is executed in parallel, the speed of the algorithm can be effectively improved.

Considering that the essence of GPU parallel calculation is to operate each element of the matrix in parallel, in order to realize the parallel processing of stagger interval

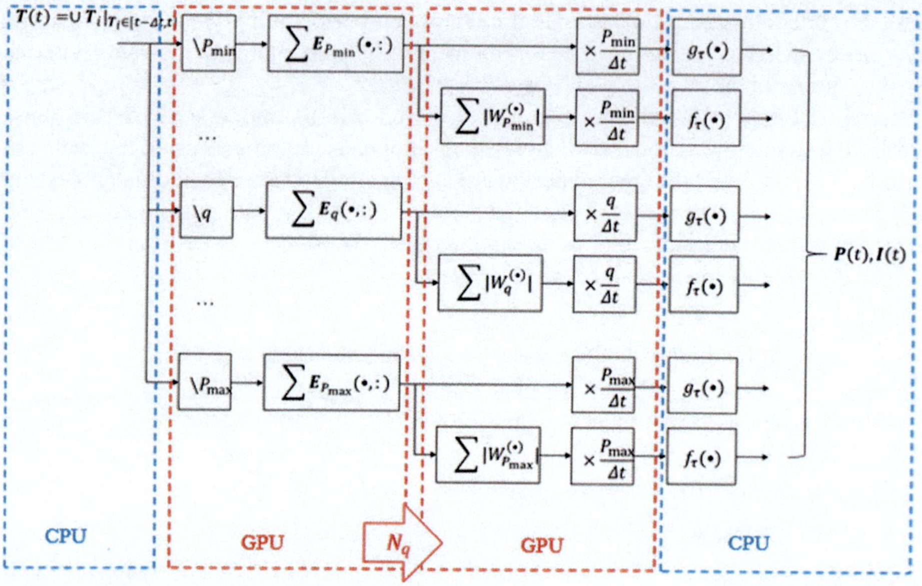

Fig. 2. Parallel execution of stagger interval period estimation algorithm

period estimation algorithm, first, the data of each sub period into the same matrix is established, and the indexes of rows or columns is taken as the characteristic parameter to identify each sub period. Next, according to the data dependency in the process of sub cycle calculation, it is necessary to divide the sub cycle calculation module into several sub modules, and realize the functions of each sub module with different kernel functions.

(1) Preprocess Module

In order to realize real-time calculation, the TOA model in formula (1) is actually a real-time input data stream. For the pulse flow of 1 s, the execution time of the algorithm is less than 1 s to meet the requirements of real-time calculation. According to the time period of one second, the function of the preprocessing module is to divide the pulse stream into several pulse streams, sending them to the stagger interval period estimation algorithm to estimate the interval period in this time period.

(2) Remainder calculation/remainder statistics module

After getting the pulse flow of the time, we need to divide the threads according to the maximum *Pmax* and minimum *Pmin* value of the interval period value and the number of input pulses n. A total of $\Delta \boldsymbol{q} \times \boldsymbol{n}$ thread needs to be processed in parallel, where $\Delta q = Pmax - Pmin + 1$. The remainder statistics results of the module output need to be calculated according to the capacity of the maximum cycle: $\Delta q \times Pmax$. Figure 3 shows the calculation process of remainder calculation/remainder statistics module.

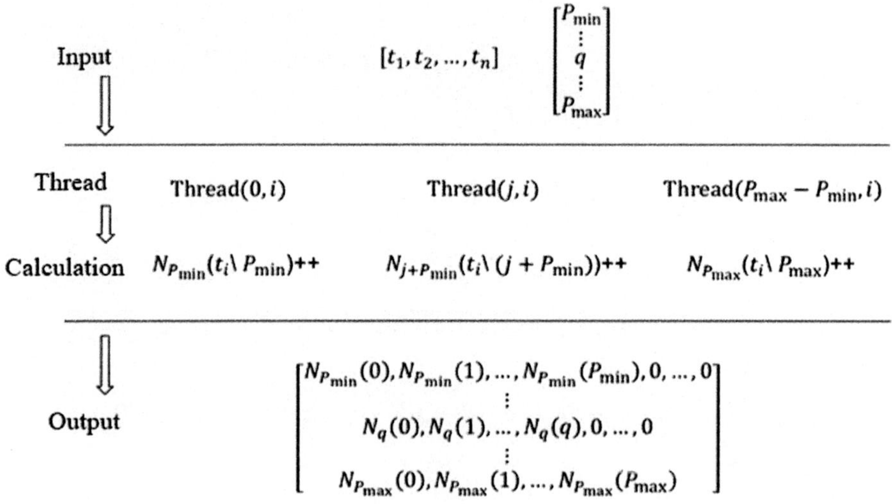

Fig. 3. Calculation process of remainder calculation/remainder statistics module

(3) Phase distribution/standard energy value calculation module

The statistical result N_q of remainder is the precondition for the calculation of phase distribution R_q and standard energy value P_q. The phase distribution is the result of equalization of the remainder statistical results, and the standard energy value is the result of energy mapping from the remainder statistical results to DFT sequence D_q and equalization. The purpose of equalization processing is to unify the energy value, ensure the unification of the standard energy value and phase distribution of different pulse flows in the interval period value, and facilitate using the same threshold processing results. Figure 4 is the calculation flow chart of the phase distribution/standard energy value calculation module.

(4) Threshold judgment module

After getting the phase distribution $\mathrm{R_q}$ and standard energy value $\mathrm{p_q}$, the parallel computing of GPU is finished, and then the CPU needs to judge the threshold value to get the set P of all the potential interval period values and the corresponding initial phase set I of the model $T(t)$.

(5) Interval period estimation module

The value and type of the hidden interval period of the mixed pulse can be obtained and reported as the algorithm processing result of the time t. Then we can estimate the period of the next time.

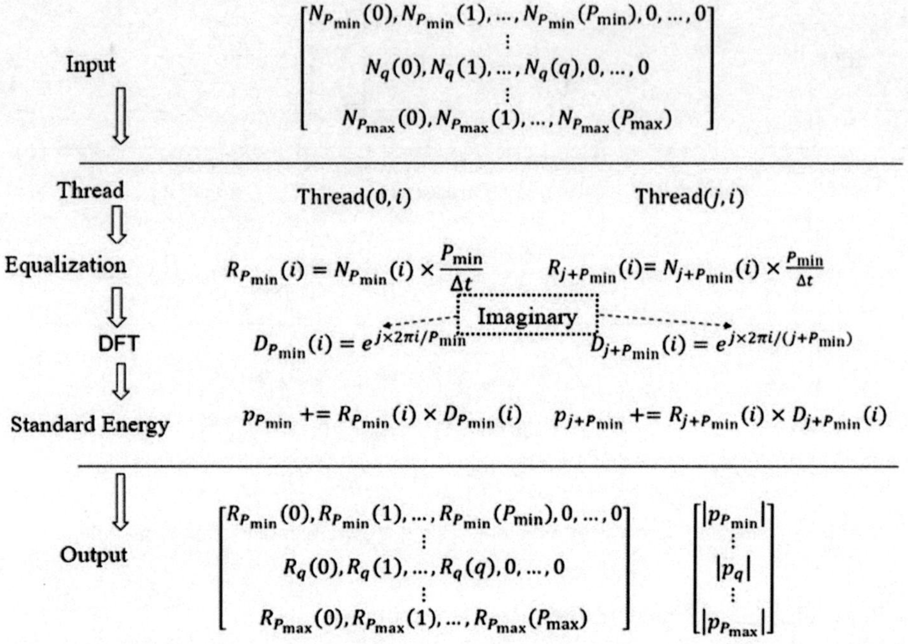

Fig. 4. Calculation process of phase distribution/standard energy value calculation module

5 GPU Parallel Implementation and Analysis

CPU + GPU heterogeneous platform is used as the experimental platform of stagger interval period estimation algorithm. Among them, the type of CPU is Intel i5-8300h, 4 cores, the main frequency is 2.3 GHz, the storage memory is 8 GB RAM, and the operating system is win10 64 bit operating system. GPU adopts GeForce GTX 1050 Ti of NVIDIA Company and Pascal architecture, and equips with 4G video memory and 768 stream processors to meet the calculation requirements of staggered interval period estimation algorithm. In addition, the stagger interval period estimation algorithm is tested by MATLAB 2014, and the parallel algorithm is tested by visual studio 2017 + CUDA 8.2.

5.1 Contrast Experiment

In this section, the purpose of the experiment is comparing the accuracy and running time of the stagger interval period estimation algorithm and the parallel algorithm in this chapter in processing the same mixed pulse stream. The TOA model settings for verification are shown in Table 2, where the jitter rate only is the parameters of jitter pulse, and the jitter rate of convention and stagger pulse is 0% by default.

Parameter setting in the experiment: maximum and minimum period are $Pmax = 5000$, $Pmin = 60$; the maximum stagger number is 5; the TOA range of mixed pulse is $\Delta t = 106$ (1 s); Threshold parameter setting $\tau_P = 0.75$, $\tau_P = 0.6$.

Table 2. TOA model setup

Experiment label	PRI type	False rate	Missing rate	Jitter rate
1	One group of stagger pulse	0%	0%	—
2	One group of stagger pulse	0%–90%	0%–40%	—
3	Two groups of mixed stagger pulse	0%	0%	—
4	Two groups of mixed stagger pulse	0%–90%	0%–40%	—
5	Three groups of mixed stagger pulse	0%	0%	—
6	Three groups of mixed stagger pulse	0%–90%	0%–40%	—
7	Four groups of mixed stagger pulse	0%	0%	—
8	Four groups of mixed stagger pulse	0%–90%	0%–40%	—

Table 3. Comparison of experimental results

Experiment label	Original algorithm		Parallel algorithm		Missing rate
	Accuracy	Time (ms)	Accuracy	Time (ms)	
1	100.0%	3700.813	100.0%	495.942	1000
2	98.8%	3672.627	98.4%	490.750	1000
3	99.9%	6768.805	99.9%	547.512	1000
4	97.8%	6592.018	98.0%	570.764	1000
5	99.9%	8235.647	99.9%	589.022	1000
6	95.2%	8313.157	95.6%	599.163	1000
7	99.9%	9974.195	99.9%	671.209	1000
8	95.5%	10386.238	94.9%	689.226	1000

By observing the running time of parallel algorithm in experiments 1 to 8 in Table 3, the running time of parallel algorithm increases with the rise of the number of radar groups in mixed pulse. By observing separately the running time of parallel algorithm in experiments 1/2, 3/4, 5/6, 7/8, the running time of parallel algorithm increases relatively in the environment of 0% to 90% false and 0 to 0.4 missing. While the number of radar groups, false and missing of mixed pulse affect the number of mixed pulse. Therefore, it is necessary to study the impact of the number of mixed pulse on the running time of parallel algorithm.

5.2 Influence of the Number of Mixed Pulse on Acceleration Performance

Except the influence of hardware device on acceleration performance of parallel algorithm, the number of mixed pulse also affects the acceleration performance of parallel algorithm: more mixed pulse means more threads. Considering the real complex electromagnetic environment, the maximum number of mixed pulses can reach one million pulses per second. Radar signal preprocessing initially sort mixed pulse according to the carrier frequency, pules width, angle of arrival and other parameters, the number of mixed pulse in the part of final arrival stagger period estimation algorithm. While the range of interval period estimated by interval period estimation algorithm is usually 60 μs to 5000 μs. And considering multiple groups pulse mix, it may be assumed the number of the maximum mixed pulse is one hundred thousand pulses per second, that is, the number of TOA model pulses in this section is set from 0 pulses per second to one hundred thousand pulses per second.

Parameter setting in the experiment: maximum and minimum period are *Pmax* = 5000, *Pmin* = 60; the maximum stagger number is 5; the TOA range of mixed pulse is $\Delta t = 106$ (1 s); Threshold parameter setting $\tau_P = 0.75$, $\tau_P = 0.6$.

The number of TOA model pulse is set 0–10^4 (the number of steps is 1000, each TOA model is tested 1000 times).

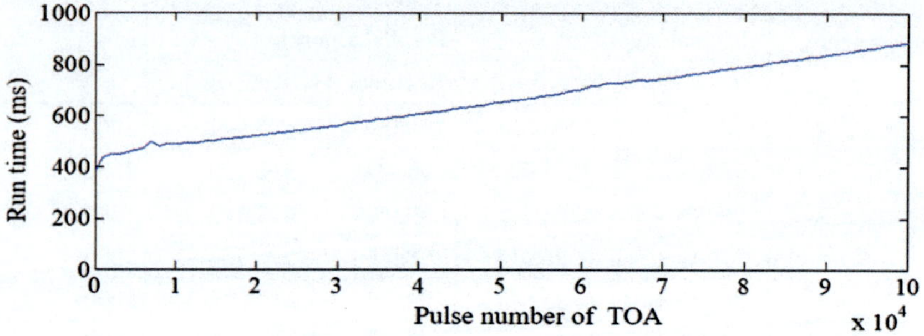

Fig. 5. Influence of the number of mixed pulse on the acceleration performance

TOA model setting: one group of stagger pulse, the false rate is set to 0%–90% missing rate is set to 0%–40%, jitter rate is set to 0%. One group of conventional pulse or one group of highly repetitive frequency pulse, the purpose of conventional pulse and highly repetitive frequency pulse is to increase the number of mixed pulse. In order to control the number of mixed pulse conveniently, their false rate, missing rate and jitter rate are set to 0%. The highly repetitive frequency pulse refers to the conventional pulse in the interval period of 3 μs–50 μs, the maximum number of pulses per second can reach three hundred thousand per second.

According to the experimental results in Fig. 5, the running time of parallel algorithm increases linearly with the number of mixed pulse. And even if the number of mixed pulse reaches one hundred thousand pulses per second, the running time of stagger interval period estimation algorithm based on GPU is only 871 ms, which meets the real-time requirement of one second.

6 Conclusion

Considering the lots of non-interference parallel computing models in stagger interval period estimation algorithm, in order to accelerate the stagger interval period algorithm, and meet the requirements of real-time computing of algorithm, the parallel processing instruction of GPU is introduced in this chapter. This chapter first introduces the basic knowledge of GPU and the programming instruction of CUDA; then, the stagger interval period estimation algorithm is redesigned by the instruction characteristics of CUDA. The experiment results show the parallel algorithm based on stagger interval period estimation algorithm of GPU, which not only ensures accuracy, but also meets the requirements of real-time computing of algorithm, it can estimate accurately the hidden interval period of multiple radar mixed pulses, and it only takes 900 ms to deal with the mixed pulse with one hundred thousand pulses. It shows that the parallel algorithm in this chapter not only ensures the accuracy, but also meets the requirements of real-time calculation.

References

1. Feng, X., Hu, X., Liu, Y.: Radar signal sorting algorithm of k-means clustering based on data field. In: 3th IEEE International Conference on Computer and Communications, ChengDu, China, pp. 2662–2666 (2017)
2. Li, C., Xu, Z., Yang, C., et al.: Virtual implementation of radar full pulse description word for battlefield electromagnetic environment. In: 30th Chinese Control And Decision Conference (CCDC) Proceedings on Proceedings, ShenYang, China, pp. 4040–4045 (2018)
3. Xu, Z., Hu, H., Xu, C., Yang, W., Li, C., Yang, C., et al.: Stagger period estimation algorithm for multiple sets of radar pulses. IET Radar Sonar Navig. **14**(6), 898–904 (2020)
4. Lu, Z.: Fast geocoding of spaceborne synthetic-aperture radar images using graphics processing units. J. Appl. Remote Sens. **6**(1), 063553 (2012)
5. Qin, H., et al.: Research on multi-GPU parallel technology in software radar signal processing. J. Xidian Univ. **40**, 145–151 (2013)
6. Jing, B.: Signal Processing of LFMCW Radar and Implementation Based on GPU. XiDian University, XiDian (2018)
7. Wu, J., Cheng, Y., Wei, Z., et al.: GPU acceleration of FSI simulations by the immersed boundary-lattice Boltzmann coupling scheme. Comput. Math. Appl. **78**(4), 1194–1205 (2019)
8. Xu, C., Jiang, S., Luo, G., et al.: The case for FPGA-based edge computing. IEEE Trans. Mob. Comput. (99), 1 (2020)

BioAR: Bio-Laboratory Automation Meets Augmented Reality

Zheng Zou[1,2,3] and Shi-Jian Liu[4(✉)]

[1] College of Computer and Cyber Security, Fujian Normal University, No. 8 Xuefu South Road, Shangjie, Minhou, Fuzhou, Fujian, China
[2] Fujian Provincial Engineering Technology Research Center for Public Service Big Data Mining and Application, Fujian Normal University, No. 8 Xuefu South Road, Shangjie, Minhou, Fuzhou, Fujian, China
[3] Digital Fujian Institute of Big Data Security Technology, Fujian Normal University, No. 8 Xuefu South Road, Shangjie, Minhou, Fuzhou, Fujian, China
[4] College of Computer Science and Mathematics, Fujian University of Technology, No. 3 Xueyuan Road, University Town, Minhou, Fuzhou, Fujian, China

Abstract. Modern laboratory is benefiting from science and engineering by adopting automated systems into labor-intensive lab routines. In this paper, a computer-aided system that involves augmented reality (AR) to promote the automatic levels of the bio-laboratory is introduced. The motivation of this work lies in the idea of offering visual guidance for users to pick a tube manually from the rack after the retrieval. Because it is difficult for people to tell the target tube from the others since they look almost the same. Design issues and implementation details are given after a careful analysis of the system requirements. The BioAR system is tested out and successfully applied to a bio-laboratory with tube related operations, whose performance is approved by the technicians. A video demo of the proposed system can be found at https://www.bilibili.com/video/BV1FX4y157WB/.

Keywords: Augmented reality · Laboratory automation · Sample management · Strain retrieval

1 Introduction

The idea of laboratory automation has been well accepted since clinical laboratories began using robotics in the early 1980s [1]. By applying automated systems into labor-intensive bio-laboratory tasks, biology is profiting from engineering [2].

A typical case of bio-laboratory automation is the sample storage and retrieval [1], where the tube-rack system has been popularly used nowadays. Figure 1 shows a tube-rack system named FluidX which is proposed by the Brooks Life Sciences[1]. With the purpose of sample storage, strains are saved into the tubes, which will be arranged into a rack and finally be stored together

[1] https://www.brookslifesciences.com/.

S.-C. Chu et al. (Eds.): ICGEC 2021, LNEE 833, pp. 294–305, 2022.
https://doi.org/10.1007/978-981-16-8430-2_27

into a $-89°$ refrigerator. Both the tubes and racks, as we can see in Fig. 1, are labeled with barcodes, which is useful for identifying and retrieving strains from a properly maintained digital database. And the last step of the sample retrieval is to pick the target tube from the rack in the real world by either a robot or a user.

Fig. 1. Demonstration of a commercial tube-rack system with the purpose of sample management.

Apart from automation tasks, where robotics is intensively used with little human intervention, there are also cases that user interactions are inevitable. In the case of human involved tasks, it is necessary to boost its efficiency by user-friendly automation methods as well. While this paper focuses on promoting automatic levels of bio-laboratory by augmented reality (AR) techniques.

Taking the above sample retrieval case as an example, during the last step, there has both a spatial switching and a temporary memory process for the user to first observe the retrieved results on the digital screen and then pick the target tube from the rack, which is prone to making mistakes.

To solve this problem, we would like to offer some intuitive guidance for users of bio-laboratory by a proposed AR system, which is the main contribution of this work. The rest of the paper is organized as follows. Section 2 gives some works related to ours. Section 3 presents the proposed system in detail. Experiments and results are given in Sect. 4. Section 5 concludes the paper.

2 Related Work

With the development of AR techniques, applications of AR in biology have been explored for many years. The most well-studied case is involving AR for the education of molecular structures since the structural information on biological macromolecules is an essential requirement for the understanding of the biological function [3] and AR can offer insight for complex spatial structures understanding.

AR can give users an illustration that they are seeing and interacting with the augmented objects in the real world. Traditionally, 3D molecular models may be generated [4,5] and viewed/interacted [6] through some specially developed software running on a computer, which is a simulated environment. By contrast, with the usage of AR, the molecular models could be manipulated as if they are in the palm of the user's hands through either a camera embedded laptop [7] (see Fig. 2(a)) or head-mounted devices (HMD) such as Microsoft HoloLens [8] (see Fig. 2(b)).

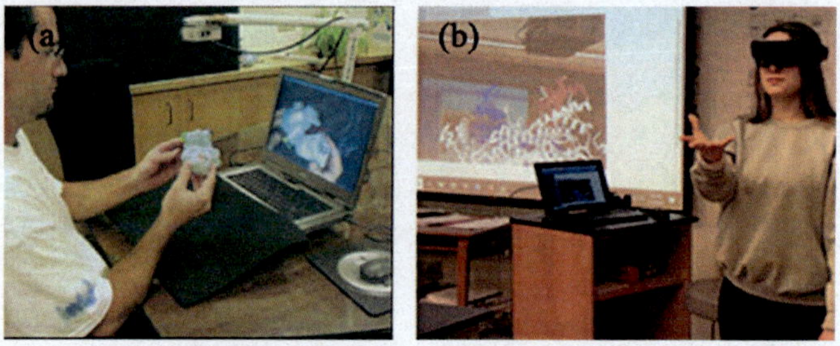

Fig. 2. Demonstrations of using a camera embedded laptop [7] (a) and a Microsoft HoloLens [8] (b) respectively for molecular structures education.

Comparing to the AR-based molecular structures studies, there have not so many reports on AR involved bio-laboratory. Sato et al. [9] proposed a nonoverlapped view management method for augmented reality by tabletop projection (see Fig. 3(a)), which can be installed above a workspace and pointed downward to present some supportive information. The projector-based AR solution is an alternation to the previously mentioned ones, which is interesting but the forms of the augmented contents are limited. To assist with the safe and independent chemistry experiments, a tangible experiment support system is proposed by Sokan et al. [10]. However, as can be seen in Fig. 3(b), lots of markers are used which may mess up the working bench.

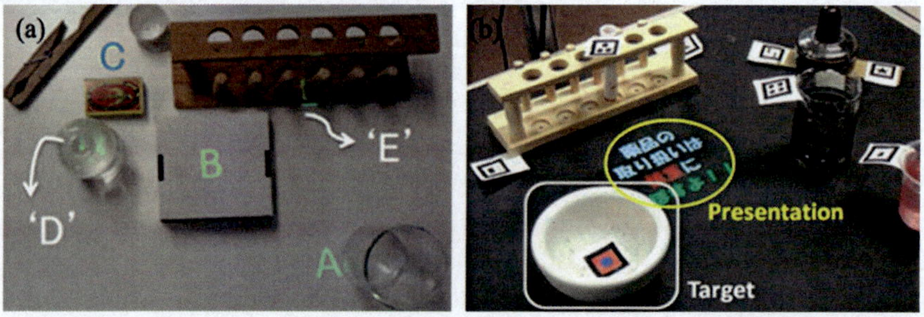

Fig. 3. The experiment support system proposed by Sato et al. [9] (a) and Sokan et al. [10] (b) respectively.

In this work, other than laptops and HMD, the mobile device is used considering its mobility and popularity. And to prevent involving extra objects into the experimenting environment, no marker is used.

3 The BioAR System

A prototype application referred to as the BioAR is developed to promote the automatic levels of the bio-laboratory by techniques including augmented reality.

3.1 Preliminary and System Requirements

This BioAR system is developed based on the commercial strain management solution named FluidX, which consists of not only tubes and racks but also an optical hardware named PerceptionHD for image data acquisition and a software named IntelliCode for bar-code decoding, as depicted in Fig. 4(a) and (b) respectively.

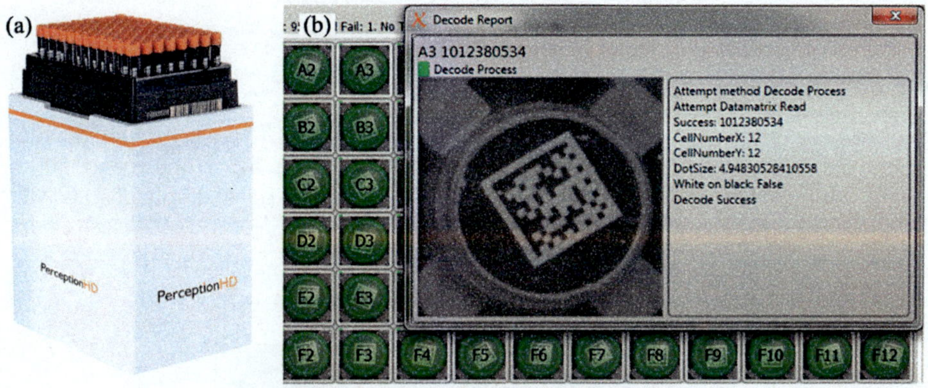

Fig. 4. The FluidX consists of the PerceptionHD SBS Whole Rack 2D Code Reader (a) and the barcode reading software named IntelliCode (b) respectively.

3.2 System Overview

The entire BioAR system can be divided into three major parts. As demonstrated in Fig. 5. The first part is the barcode scanning subsystem, which is enabled by the FluidX, as we mentioned earlier. The second part is a web-server subsystem, which connects to the FluidX and the mobile device through networks. One of the core functionalities of the server is maintaining a database where the latest digital data of the users and the strains are stored. It also obligates to send data to the mobile device which drives the data for AR visualization. The last part is the mobile subsystem where AR lives on. Since the mobile subsystem is the core of this paper, we will focus on this one in the following Sections. And the rest two parts will be mentioned only when necessary.

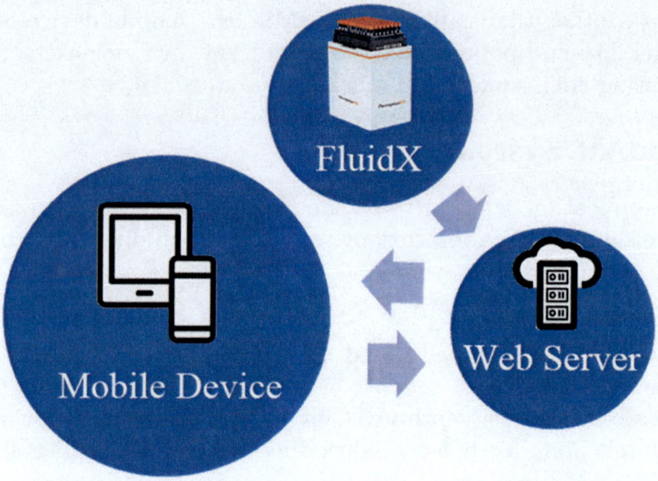

Fig. 5. Overview of the proposed BioAR system.

3.3 Design Issues and Decisions

According to the system requirements described in Sect. 3.1, we further divide the mobile subsystem into 3 levels and 5 modules (see Fig. 6), which responsible for communication, user management, barcode reading, data display, and AR guidance respectively.

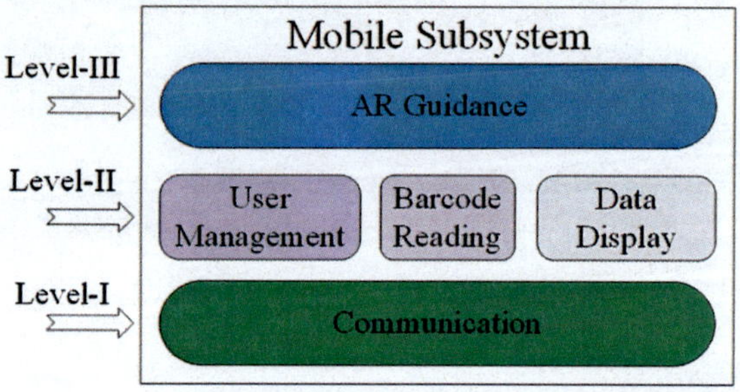

Fig. 6. Modules of the mobile subsystem.

Communication and User Management. Considering the limitations of energy consumption and computational ability, the mobile system is designed to focus on data visualization other than data management. Instead, all user and strain data are stored and managed by a web-server, and transmitted through the networks.

The main data transmitted between the web-server and the mobile device are those for user authentication and data display. Two kinds of networking techniques are involved, which are HTTP request/response and web-socket.

We assume that users are pre-registered in the user database which is managed by the web-server. To log into the mobile system for the first time, both username and password should be filled in by the graphical user interface (GUI) demonstrated in Fig. 7(a), and then sent to the server for user authentication, which will lead to the result shown in Fig. 7(b) when succeed. As shown in Fig. 7(b), each account could be identified by the thumbnails. The users could also add an account to or remove it from the system.

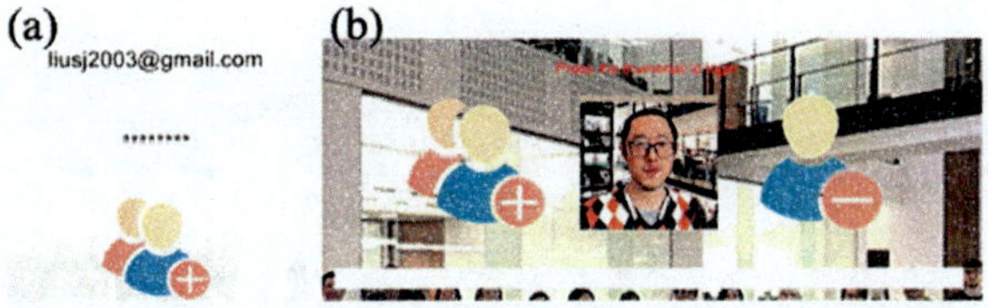

Fig. 7. Demonstration of user management. (a) The GUI for user login according to the supplied username and password. (b) The result of user login when succeed.

Barcode Reading and Data Display. One of the key elements of the tube-rack system is the usage of barcodes for item identification and trace[11,12]. There are 2 kinds of barcodes, including a 1D barcode and a 2D barcode, printed on the side (see Fig. 8(a)) and bottom of each tube and rack respectively. When fully placed, the bottom view of the rack appears like the image shown in Fig. 8(b), of which the IntelliCode takes as the input and then saves the decoded data into the database by barcode reading. The row and column index of each barcode, which stands for the position of the tubes, is saved as well.

Other than reading the whole 2D barcodes within the bottom view of a rack, our mobile system reads one barcode each time when the user scans a target barcode, as the one marked with the yellow rectangle in Fig. 8(a), using the mobile camera.

As shown in Fig. 9, the mobile system also is capable of displaying data of a rack and all tubes within it in real-time. The data appear on this GUI panel include statistical info of the rack (marked with the red rectangle), distributions of the tubes (marked with the green rectangle), details of a selected tube (marked with the blue rectangle), etc. All of them are saved in the database and can be acquired from the web-server through networking. If the database is modified during this period, the mobile system will update the display once it receives the notification from the server as introduced in Sect. 3.3.

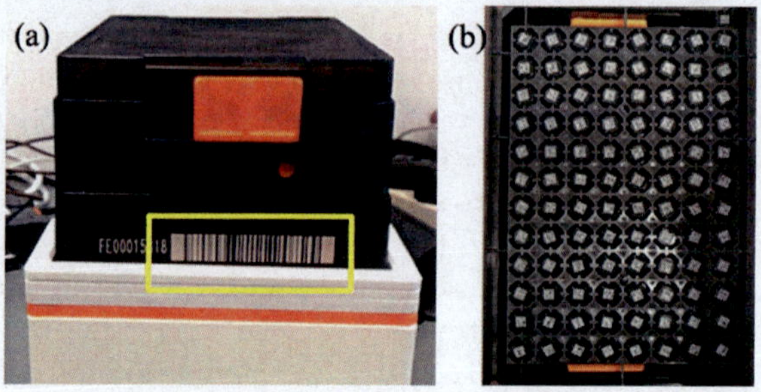

Fig. 8. Individual barcode printed on the side of a rack (a), and multiple ones viewing from the bottom of the rack (b).

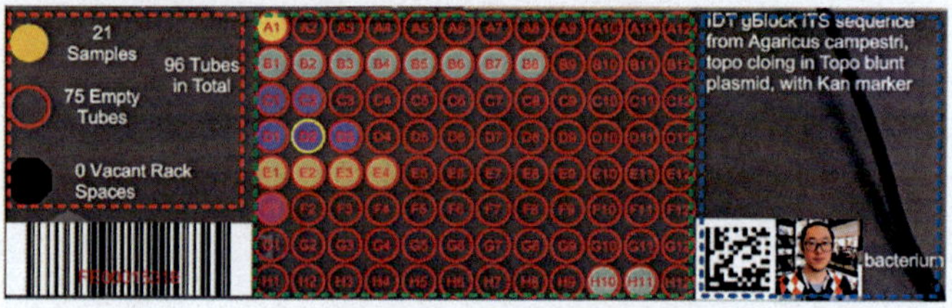

Fig. 9. Demonstration of data display of the mobile system.

AR Guidance. One of the AR guidance offers by our system is to help the users with some lab operations regarding the tubes. The reason is that when a rack is fully packed with tubes, which look almost the same except for the barcode distinguishable only by machine, it is difficult for people to tell where the target one is.

For example, the last step of strain retrieval from a tube-rack system is the target picking. If it is done by a human, the most straightforward way is finding the tube according to its row and column index which stands for the position within the rack. To achieve that, the user should first observe the retrieved results on the digital screen and then pick the target tube from the rack in the physical world, which is prone to making mistakes because there exits both a spatial switching of the sight and a temporary memory of the brain during the procedures.

With the help of AR technology, we could place virtual objects (e.g., the one marked with the yellow rectangle shown in Fig. 10) that always point to the targeted tubes. It is really helpful to mark the one that users are looking for, either of which is selected on the GUI of the mobile system (see the one marked with the dashed blue box shown in Fig. 10) or is retrieved from the database.

A marker-less AR method based on 3D shape recognition is adopted in our system. Though the tubes and racks are labeled with barcodes, which could be treated as the markers, the angle of view would be limited for the sake of barcode-based AR tracking. Since the selected rack is placed on the PerceptionHD reader to enable the run-time trace of the tubes in general, we decide to use its shape information for AR registration and tracking, which works pretty well in the experiments.

4 Experiment and Result

In this section, the implementation details of the BioAR system and experimental results are presented. An iPad Pro is used to run the mobile subsystem in our experiments.

4.1 Implementation Details

The mobile system is programmed using C# language on a common Apple computer with the Unity 3D engine, and then it is exported to the XCode for iPad deployment.

To ease the load of the web-server, a local copy of the data will be saved into the mobile device instead of requesting them from the server unless they have been modified. The JSON format is used to organize the data for its popularity in web applications nowadays.

The function of barcode reading is enabled by the famous ZXing[2] lib. To improve the reading efficiency, after a frame is captured by the camera, other than searching barcodes from the entire image, a specified region, as shown in Fig. 11, is used. And the user is responsible for making sure the target is located in that local region.

Regarding the programming for the AR guidance, Vofuria is adopted which is a state-of-the-art AR engine. To achieve that, 3D models of objects such as the tube, rack, barcode reader are constructed by 3ds Max in advance according to its physical shapes, which is mandatory for shape-based registration and tracking. The virtual objects for AR guidance are also modeled and placed properly in the virtual space, so it could be aligned with the physical objects accurately during the AR guidance.

4.2 Results

A couple of experiments are carried out to evaluate the performance of the proposed system.

Firstly, by changing the position of tubes placed in a rack and then re-scanning the rack using IntelliCode, the communication function is tested. For example, by pressing the "Go" button of the IntelliCode to read the rack, who has two tubes (i.e., the A1 and A2 shown in Fig. 12) swapped previously, Fig. 12 (a) and (b) shows the result before and after the data transmitted between the server and the mobile system respectively. The updating process finished shortly, and it works properly as shown in Fig. 12.

[2] https://github.com/zxing/zxing.

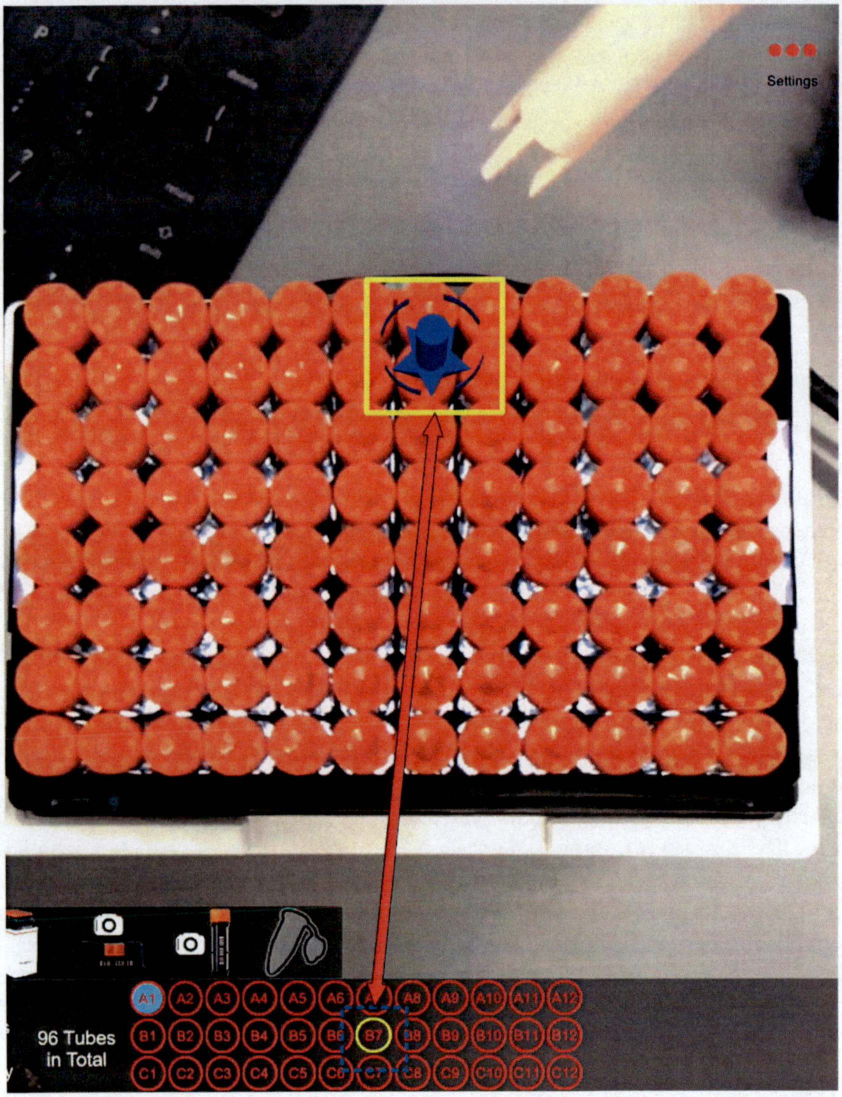

Fig. 10. Demonstration of using a virtual object (marked with the yellow rectangle) to identify the target tube selected in the GUI of the mobile system (marked with the dashed blue box).

The second experiment is about AR-based tube operations. It begins with testing the AR guidance for tubes selected on the mobile GUI. Along with the random press of GUI buttons which represent different tubes, the virtual object switches over different positions smoothly and accurately no matter the rack is with a cap on it or not.

Fig. 11. A local region of the frame is used to improve the efficiency of barcode reading.

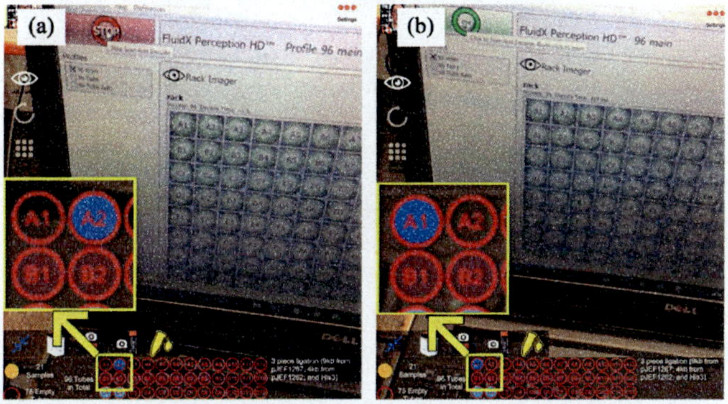

Fig. 12. Real-time data display before (a) and after (b) the data transmitted between the server and the mobile system resulting from database modification.

Then tubes are pulled out of the rack and scanned by the mobile system as illustrated in Fig. 13(a). As a result, the virtual object (marked by the dashed yellow box in Fig. 13(b)) appears immediately, which identifies the tube's position correctly.

5 Limitation and Conclusion

This paper presents a prototype system named BioAR, which involves augmented reality to promote the automatic levels of a bio-laboratory. The main functions of the proposed system include user management, barcode reading, data display, and AR guidance operated on a camera embedded mobile device. The data for user authentication and display is managed by a web-server and transmitted between subsystems through networks.

By seeing through the screen of the mobile device, the users can get help from virtual clues that overlay on the experimental instruments with the proposed system during the lab works. Specifically, the scenario of offering AR guidance for users to pick a target tube from the modern tube-rack system is explored. The system performs effective and efficient during the experiments, and it could easily be extended to fit other specific lab routines.

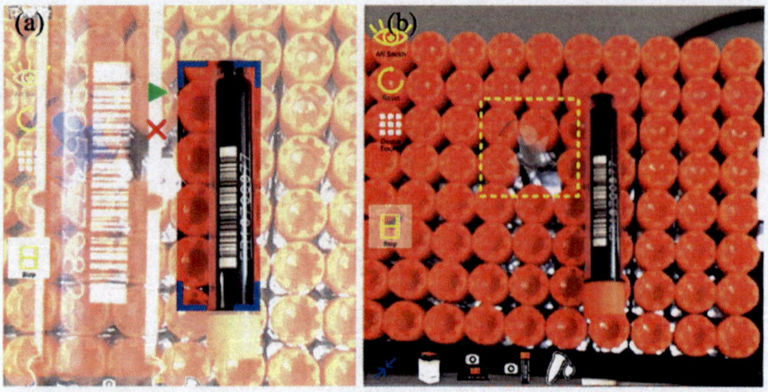

Fig. 13. Experiment of AR guidance for tubes retrieved by barcode scanning. (a) A tube is pulled out of the rack and scanned by the system. (b) A virtual object (marked by the dashed yellow box) appears immediately after the scanning.

A limitation of the BioAR system is that though the users are free to view the physical target from different perspectives using the mobile device, the system may make mistakes resulting from the geometric symmetry of the target (e.g., the rack) since the AR tracking is based only on the 3D shape recognition. One of the solutions is to recalibrate the coordinates by a special texture that appeared on the target. An HMD may be adopted to improve the experience of the BioAR system for future works.

Acknowledgment. This work is supported by the National Natural Science Foundations of China (No. 61872085), the Scientific Research Project of Fujian Education Department (JK2017029, JAT190069), and the Scientific Research and Development Foundation of Fujian University of Technology (GY-Z18181, GY-Z20068, XF-X19017).

Most of this work was done by Dr. Shi-Jian Liu during his visit at the University of Manchester, UK. The authors appreciate all the supports from Prof. Yizhi Cai and the members of the Cai Lab. The authors also gratefully acknowledge the helpful comments and suggestions of the reviewers, which have improved the presentation.

References

1. Chapman, T.: Lab automation and robotics: automation on the move. Nature **421**(6923), 661–663 (2003)
2. Kuncova-Kallio, J., Kallio, P.J.: Lab automation in cultivation of adherent cells. IEEE Trans. Autom. Sci. Eng. **3**(2), 177–186 (2006)
3. O'Donoghue, S.I., et al.: Visualization of macromolecular structures. Nat. Methods **7**(3 Suppl), S42 (2010)
4. Giard, J., Macq, B.: Molecular surface mesh generation by filtering electron density map. Int. J. Biomed. Imaging **2010**, 923780 (2010)
5. Giard, J., Rondao-Alface, P., Gala, J., Macq, B.: Fast surface-based travel depth estimation algorithm for macromolecule surface shape description. IEEE-ACM Trans. Comput. Biol. Bioinf. **8**(1), 59–68 (2011)

6. Goddard, T.D., Ferrin, T.E.: Visualization software for molecular assemblies. Curr. Opin. Struct. Biol. **17**(5), 587–595 (2007)
7. Gillet, A., Sanner, M., Stoffler, D., Olson, A.: Tangible interfaces for structural molecular biology. Structure **13**(3), 483–491 (2005)
8. Peterson, C.N., Tavana, S.Z., Akinleye, O.P., Johnson, W.H., Berkmen, M.B.: An idea to explore: use of augmented reality for teaching three-dimensional biomolecular structures. Biochem. Molec. Biol. Educ. **48**, 276–282 (2020)
9. Sato, M., Fujinami, K.: Nonoverlapped view management for augmented reality by tabletop projection. J. Vis. Lang. Comput. **25**(6), 891–902 (2014)
10. Sokan, A., Hou, M.W., Shinagawa, N., Egi, H., Fujinami, K.: A tangible experiment support system with presentation ambiguity for safe and independent chemistry experiments. J. Ambient Intell. Hum. Comput. **3**(2), 125–139 (2012)
11. Liu, S.J., Zhang, J., Pan, J.S., Weng, C.J.: A novel information imbedding and recovering method for QR code based on module subdivision. J. Inf. Hiding Multimedia Signal Process. **9**(2), 515–522 (2018)
12. Liu, S.J., Zhang, J., Pan, J.S., Weng, C.J.: SVQR?: a novel secure visual quick response code and its anti-counterfeiting solution. J. Inf. Hiding Multimedia Signal Process. **8**(5), 1132–1140 (2017)

3D Voxel Reconstruction Based on Shape Layer

Linlin Tang(✉), Shuaijie Shi, Shiyu Qin, Xin Huang, and Yijie Fan

Harbin Institute of Technology, Shenzhen, Shenzhen, China

Abstract. 3D reconstruction is a basic and important task in computer vision. The overall reconstruction quality evaluation of voxel based 3D reconstruc-tion method is not high currently. This paper mainly aims at the problem that the overall reconstruction quality evaluation of voxel based 3D reconstruction methods is not high. According to the good conversion relationship between shape layer and voxels, a new shape layer based 3D voxel recon-struction network is proposed to transform the prediction problem of 3D voxels into the prediction problem of 2D depth map. Experiments show that it is superior to other voxel based 3D reconstruction methods.

Keywords: 3D reconstruction · Shape layer · Neural network

1 Introduction

Goal of 3D reconstruction is to restore the missing dimensions of the corresponding 3D shape from the information contained in the 2D image, and infer the 3D geometric structure of objects and scenes. 2D images can be said to be a treasure house with a large number of 3D shape resources. The 3D reconstruction technology is the key to open this treasure house.

The traditional 3D reconstruction field has a relatively complete theoretical system, but it has problems such as the huge memory consumption and high price. With the emergence of deep learning methods, people have new ideas in the field of 3D reconstruction. Representation method of 3D reconstruction results affects the quality and efficiency of 3D reconstruction.

Research of 3D reconstruction based on voxels started early, the convolutional neural networks can be easily applied to the reconstruction of 3D voxels. Therefore, voxels as a representation of 3D shapes are used here. Although previous work can successfully reconstruct 3D voxels from 2D images, the overall reconstruction quality is not well evaluated. A network structure with a single image for 3D voxel reconstruction is introduced. By using the feature that the resolution of 32 × 32 × 32 3D voxels and shape layer almost completely converted, we transform the prediction results of the 3D reconstruction network from 3D voxels into a complex 2D depth map.

2 Related Works

In 3D computer graphics, voxels represent values on a regular grid in 3D space. Similar to a pixel in a 2D image, a voxel itself does not need to explicitly carry any position

S.-C. Chu et al. (Eds.): ICGEC 2021, LNEE 833, pp. 306–315, 2022.
https://doi.org/10.1007/978-981-16-8430-2_28

information, because it can implicitly infer the position of a voxel in 3D space through its position relative to other voxels.

In 2015, Wu et al. [2] modified output shape of the convolutional neural network so that it can finally reconstruct the voxel representation of the 3D shape through the input depth map. This achievement created a precedent for using deep learning for 3D reconstruction work.The 3D-R2N2 network proposed by Choy et al. [3] is an earlier autoencoder structure based on voxel representation. It uses a unified reconstruction network with the help of its 3D convolutional Long Short-Term Memory (LSTM) layer to handle the single-view and multi-view reconstruction tasks. The 2D image encoder of network is connected to the 3D decoder part through the 3D convolution LSTM layer, it uses the memory characteristics of LSTM unit to "carve" predictive model. A large number of experiments have shown that this method can overcome the problems of lack of texture and wide baseline in the traditional Structure from Motion (SFM) method [4–6].

Recent work tends to abandon the intermediate process and directly obtain the 3D voxel model. Tulsiani et al. [7] use a decoder composed of a 3D deconvolution layer to predict the probability of voxel occupancy. This method has achieved good results in performing 3D reconstruction tasks. But it takes very high computational complexity and memory requirements, which limits these methods to only run at low resolutions. In order to solve the problem of much calculation and memory, some researcher [8] propose a special voxel encoding method of octree. However, it requires the use of complex structures and mechanisms to propagate features through the decoder, which hinders the portability of the framework and the exploration of alternative network structures.

3 Proposed Approach

3.1 Shape Layers

Shape layer is a 3D shape representation method proposed in [9] that can be almost perfectly converted with voxels. As shown in Fig. 1, we can get six depth maps of 3D voxels in the 3D Cartesian coordinate system through projection. Assuming that the origin of the coordinate system is at the center of the cube and all views face the origin, each depth map *di* is an orthogonal projection obtained from the view position *vi*, which is located at the center of the *i* side of the unit cube. Then merge these six depth maps to obtain 3D shapes Sx, Sy, Sz, and then merge the shapes to obtain S, finally we restore the selfoccluded part through the nested shape layer.

3.2 Network Model Design

We use the feature that voxels and shape layers can be converted to each other almost without loss at low scores to convert the 3D voxel prediction problem [11] that consumes more memory and computing resources into a 2D forecasting problem that consumes less memory and computing resources. 3D reconstruction network structure based on the shape layer is roughly shown in Fig. 2. In this network, a network similar to the encoder and decoder structure as the backbone structure of reconstruction network is

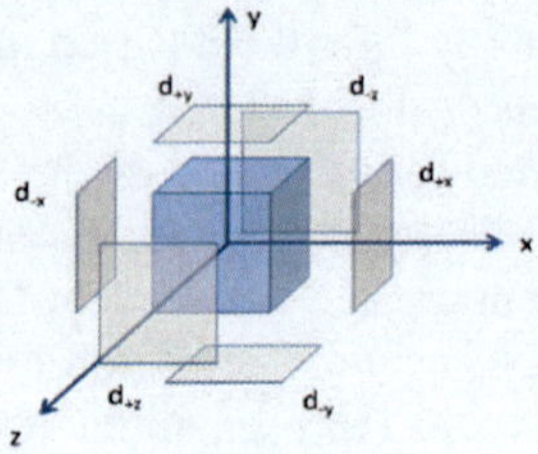

Fig. 1. Six depth maps corresponding to the cambium

used. What's more, a jump connection similar to the U-net network has been added into this network [10]. For each convolution module, we have added a residual design. In this hourglass-like network architecture, down-sampling promotes the continuous exploration and accumulation of 3D shape information contained in the 2D image. This information is then propagated to all parts of 3D reconstruction shape through overall upsampling.

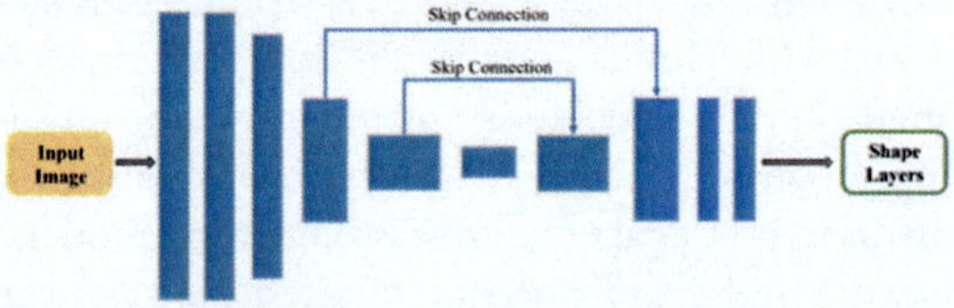

Fig. 2. Schematic diagram of 3D reconstruction network structure based on nested shape layer

2D image encoder used in this network increases the number of channels and reduces the resolution through a series of convolution operations. We finally get a series of feature maps of different scales, which are then input to the subsequent 3D decoder through a specific jump connection. Final output result of 3D decoder is not a voxel in the general sense, but a shape layer can be almost perfectly converted with the voxel. The shape layer is composed of complex depth map, so that the prediction problem of 3D voxel can be transformed into a prediction problem of complex 2D depth map.

Three types of convolution modules are mainly used in this network, that is, the equal-resolution convolution module that does not change the resolution of the feature map, the down-sampling convolution module that reduces the resolution, and the upsampling convolution module that improves the resolution. In order to avoid network degradation that the reconstruction quality decreases while increasing the depth of the network, corresponding residual convolution module for the above three modules is designed. It is shown in Fig. 3.

For the residual convolution module that does not change the resolution, the most classic residual module design is used. We call this module resSame Module. For the down-sampled residual convolution module, resolution of the output feature map is reduced, but number of channels is increased. In order to enable the jump connection to be added to output of the residual module, it is necessary to adjust the resolution and dimension of jump connection. Two different jump connection methods are used here.

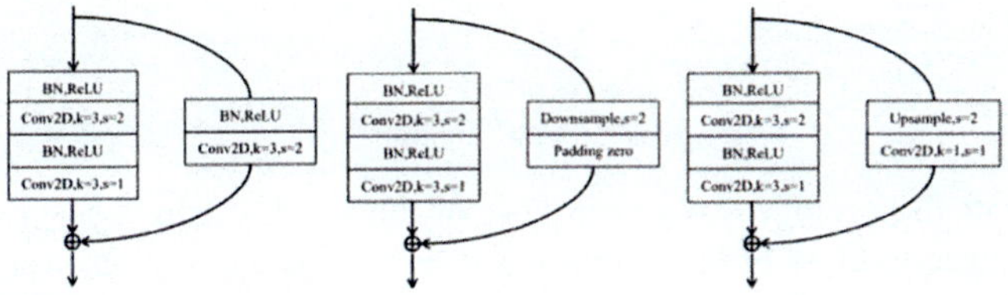

Fig. 3. Three different down-sampling residual convolution modules

As shown in Fig. 3 (left), it directly uses a convolution module in the jump connection can directly change the output resolution and channel number of the jump connection; as shown in Fig. 3 (middle), it uses in jump connection Downsampling changes the resolution of the feature map, and then uses the feature map with all zeros to fill in the missing feature map channel. For the upsampling residual convolution module, similar to the downsampling residual convolution module, the resolution and dimension of the jump connection need to be adjusted accordingly, as shown in Fig. 3 (right).

3.3 Loss Function

Since shape layer is used as the output prediction result in the reconstruction network, we hope to use features of voxels and shape layers to transform the 3D volume. The voxel prediction problem is transformed into a 2D prediction problem that consumes less memory and computing resources. So, this paper chooses a loss function that is different from traditional voxel prediction for training. For each pixel, the general L1 loss function formula is shown in Eq. (1):

$$L_1(x, y) = |x - y| \tag{1}$$

Where x and y are the prediction results and labels of each pixel in the depth map, where y is greater than or equal to 0. When y is greater than 0, it indicates that the pixel is the foreground, and when y is equal to 0, it indicates that the pixel is the background.

However, if the regression loss is simply applied to the entire depth map without distinguishing between foreground pixels and background pixels, the reconstruction network will be more biased towards background pixels. To avoid this problem, average operation of the loss of the foreground and background regions respectively is used, and then loss function by weighted addition is calculated. Loss for each pixel is shown in Eq. (2):

$$L(x, y) = \begin{cases} |x - y|, & y > 0 \\ \max(0, x + m) & otherwise \end{cases} \tag{2}$$

In order not to limit potential of reconstruction network, the idea of margin loss is used. For background pixels, in order to minimize the loss function, the reconstruction network will force the background pixels to obtain a value less than or equal to m during the training process. So, 3D reconstruction network will pay less attention to background pixels, thereby making better optimization of the foreground area. The loss function of the shape layer is shown in Eq. (3):

$$L = L_{fg} + \alpha L_{bg} \tag{3}$$

L_{fg} represents the average foreground loss of the whole shape layer, L_{bg} represents the average background loss of the whole shape layer, and α is custom weight, which is used to adjust the proportion of average foreground loss and average background loss in the shape layer loss function.

4 Experimental Results and Analysis

4.1 Data Sets

Data set used in this paper is a subset of the ShapeNet data set [2]. ShapeNet is a large, informationrich 3D model repository. It contains models spanning multiple semantic categories and provides a large set of annotations for each model. The ShapeNet data set is too informative. In field of 3D reconstruction, people generally use its processed subset for research. Therefore, we use a subset of ShapeNet processed by 3D-R2N2. ShapeNet-3DR2N2 is a subset of ShapeNe which contains13 main categories and 43783 3D voxels. For each 3D binary voxel, there are 24 2D images corresponding to it. The resolution of the dataset images provided by 3D-R2N2 is 137 × 137. In order to facilitate the construction of the reconstruction network, we changed its resolution to 128 × 128 in the preprocessing stage.

4.2 Ablation Study

4.2.1 Effect of Number of Shape Layers

Too many or too few shape layers are calculated will affect the memory consumption and reconstruction effect. In order to find the best number of shape layers, we first use the nested shape layer to transform the 3D shape voxels in the data set, then use the 3D voxel expression method to express the nested shape layer. By calculating the intersection over union (IoU) between the voxel representations of the 3D shape of the object before and after the transformation, the reconstruction quality of the nested shape layer is evaluated. The calculation formula of IoU is shown in formula (4).

$$IoU(A, B) = \frac{|A \cap B|}{|A \cup B|} \tag{4}$$

Here different layers of nested shape layers for conversion testing are used. If the calculated IoU between the voxels before and after the conversion is 100%, then the nested shape layer can completely reconstruct the voxels. Experimental results after conversion experiment on the 3D voxel with a resolution of 32 × 32 × 32 are shown in Table 1.

It can be seen from Table 1 that when we convert the nested shape layer and the 3D voxel with a resolution of 32 × 32 × 32 to each other, there is 5.2% the 3D voxels cannot be completely reconstructed in the case of using only 1 layer of shape layer. But the average IoU on the entire data set is approximately 100%, which means that 3D

Table 1. Reconstruction capability of nested shape layer at $32 \times 32 \times 32$ resolution

Number of cambium	1	2	3	4	5
Percentage of complete reconstruction (%)	94.8	100.0	100.0	100.0	100.0
Average IoU (%)	100.0	99.9	100.0	100.0	100.0

shape that cannot be completely reconstructed using only one shape layer. Although some details are lost in the process of transforming the 3D shape from the voxel expression method to the nested shape layer, the loss content is very small. Moreover, there are often data with very low modeling quality in the data set. As shown in Fig. 4, picture on the right is the car model obtained by Choy et al. [3] from the voxelization of the mesh model of the ShapeNet dataset, picture on the left is its corresponding 2D image. It can be seen that there is a big gap between the two pictures. The quality of the 3D voxel is low. Under the premise that there is a gap between the data set itself and the correct 3D shape, the small error generated by the conversion of the 3D voxel and the nested shape layer has little effect on the reconstruction network. Generally speaking, for a 3D reconstruction network that uses binary voxels with a resolution of $32 \times 32 \times 32$ as expression of the reconstruction result, it is completely feasible to use a nested shape layer with only 1 layer to replace 3D voxels as the output result of the reconstruction network for related research work.

Fig. 4. Poor quality label voxels

4.2.2 Effect of m and α

As shown in formula (2), a parameter m is used as the margin in margin loss function for the background pixel. To different margin values, related experiments are applied on these three categories of car, table and chair. Experimental results are shown in Fig. 5.

It can be seen from Fig. 5 that for the reconstruction of the two categories of car and table, when the margin is 1, the IoU of the reconstruction result is the highest. For the chair category, when the margin is 0.5, the IoU of the reconstruction result is the highest. For the average IoU of the reconstruction results on the three classes, when the margin is 1, the reconstruction result has the highest IoU.

Similarly, in the calculation process of the loss function,we use the weight α. For different values of α, related experiments are applied on the three categories of car, table and chair. The experimental results are shown in Fig. 6.

For the reconstruction of the two categories of car and table, when the weight α is 1, the IoU of the reconstruction result is the highest; for the chair category, when the weight α is 1.5, the IoU of the reconstruction result is the highest. For the average IoU of

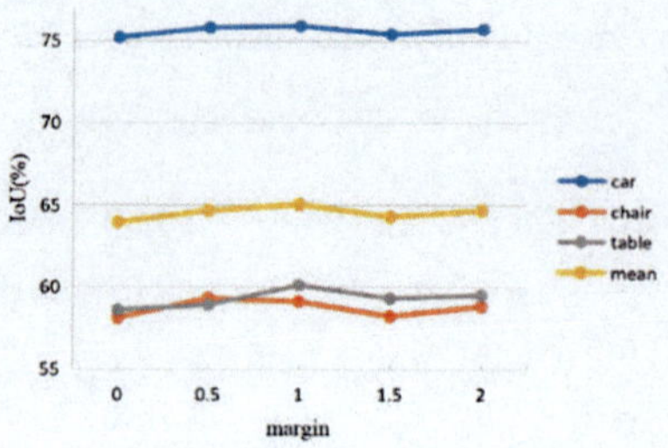

Fig. 5. The impact of the value of margin on the IoU (%) of the reconstruction results of the three categories of car, table and chair

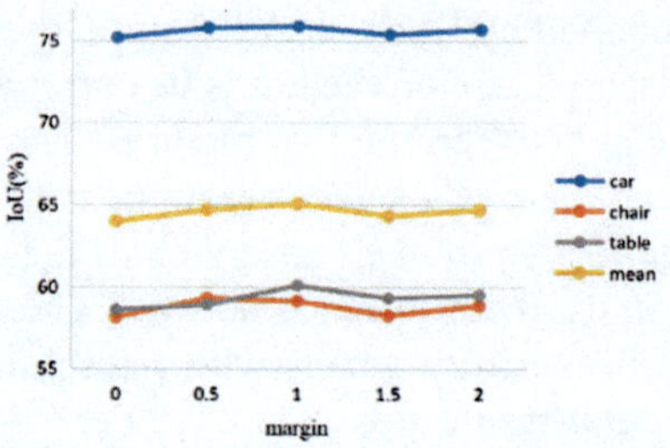

Fig. 6. The influence of the value of weight α on the IoU (%) of the reconstruction results of the three categories of car, table and chair

reconstruction results on three classes, when the weight α is 1, the reconstruction result has the highest IoU.

4.2.3 Effect of the Number of Residual Convolution Modules

This reconstruction network uses a large number of residual convolution modules that do not change resolution and channels number, namely the resSame module. Before multiple 2D feature maps of different scales are input to corresponding convolution module of the 3D decoder, we use the resSame module to convolve feature maps. Aiming at the number of resSame modules that the feature map passes through before inputting into the 3D decoder, reconstruction experiments are applied on the 13 main categories ShapeNet data set, and obtained the IoU evaluation of the reconstruction results. Experimental results are shown in Table 2:

It can be concluded from the table that the residual network is designed to alleviate the network degradation problem and the network performance decreases as the depth increases. However, this does not mean that the number of residual modules can be increased at will.

4.3 Evaluation Results

In order to evaluate performance of voxel reconstruction network in reconstructing 3D shape from a single color image, 3D-R2N2 [3], OGN [8] and Matryoshka [9] are chosen

Table 2. The influence of the number of resSame modules passed by the multi-scale feature map on the reconstruction quality of the 13 main categories of the ShapeNet dataset

Quantity			
Category	1	2	3
Airplane	**61.8**	61.0	60.1
Bench	**61.7**	60.7	61.6
Cabinet	**80.9**	79.2	78.2
Car	**83.9**	83.6	81.7
Chair	**49.7**	49.4	49.3
Monitor	**50.7**	48.4	49.5
Lamp	**55.5**	53.7	54.3
Speaker	**69.0**	68.6	67.5
Firearm	55.1	54.7	**57.3**
Couch	**67.3**	66.8	66.4
Table	**53.5**	52	52.1
Cellphone	**82.9**	70.5	78.9
Watercraft	**53.5**	51.8	51.8
Mean	**64.3**	61.6	62.2

as baseline. Average IoU of each category and average IoU of all categories are given here. Experimental results are shown in Table 3.

Table 3. The 3D reconstruction network based on shape layer and four baselines are on 13 main categories of shapenet dataset Quantitative comparison of IOU (%)

Category	3D-R2N2 [3]	OGN [8]	Matryoshka [9]	Ours
Airplane	51.3	58.7	**64.7**	61.8
Bench	42.1	48.1	57.7	**61.7**
Cabinet	71.6	72.9	77.6	**80.9**
Car	46.6	81.6	**85.0**	83.9
Chair	49.7	48.3	**54.7**	49.7
Monitor	46.8	50.2	**53.2**	50.7
Lamp	38.1	39.8	40.8	**55.5**
Speaker	66.2	63.7	**70.1**	69.0
Firearm	54.4	59.3	**61.6**	55.1
Couch	62.8	64.6	**68.1**	67.3

(*continued*)

Table 3. (*continued*)

Category	3D-R2N2 [3]	OGN [8]	Matryoshka [9]	Ours
Table	**60.6**	53.6	57.3	53.5
Cellphone	66.1	70.2	75.6	**82.9**
Watercraft	51.3	**63.2**	59.1	53.5
mean	56.0	59.6	63.5	**64.3**

Reconstruction network has achieved leading results in the average IoU evaluation indicators of all categories, and achieved the best results in four categories: bench, carbinet, lamp and cellpone. Figure 7 shows that the reconstruction result of a vehicle. Pictures from left to right are 2D images, label voxels and reconstructed voxels.

Fig. 7. Display of the reconstruction effect of the vehicle.

5 Conclusion

Aiming at problem that the overall reconstruction quality of the current voxel-based 3D reconstruction method is not high, according to the good conversion relationship between shape layer and voxel with a resolution of $32 \times 32 \times 32$, a new 3D voxel reconstruction network based on the shape layer is designed, which changes the prediction problem of 3D voxels into 2D depth Graph prediction problem. Experiments on the reconstruction network on the ShpaeNet dataset. Compared with other voxel-based 3D reconstruction methods, we have achieved much better results. Although the results of the 3D voxel reconstruction network based on the shape layer are overall improved compared with the previous methods, it still cannot solve the problem of poor performance of multiple categories in the evaluation index. This is the common defect of voxel-based methods.

Acknowledgement. This work was supported by the Shenzhen Foundational Research Funding under grant number JCYJ20180507183527919 and the Shenzhen Science and Technology Plan under grant number JCYJ20180306171938767.

References

1. Fu, K., Pemg, J., He, Q., et al.: Single image 3D object reconstruction based on deep learning: a review. Multimedia Tools Appl. **80**(1), 463–498 (2021)

2. Wu, Z., Song, S., Khosla, A., et al.: 3D Shapenets: a deep representation for volumetric shapes. In: The IEEE Conference on Computer Vision and Pattern Recognition, pp. 1912–1920 (2015)
3. Choy, C., Danfei, X., Gwak, J., Chen, K., Savarese, S.: 3D-R2N2: a unified approach for single and multi-view 3D object reconstruction. In: Leibe, B., Matas, J., Sebe, N., Welling, M. (eds.) ECCV 2016. LNCS, vol. 9912, pp. 628–644. Springer, Cham (2016). https://doi.org/10.1007/978-3-319-46484-8_38
4. Akhiter, I., Sheikh, Y., Khan, S., et al.: Nonrigid structure from motion in trajectory space. In: The International Conference on Neural Information Processing Systems, pp. 41–48 (2008)
5. Paladini, M., Del Bue, A., Xavier, J., et al.: Optimal metric projections for deformable and articulated structure-from-motion. Int. J. Comput. Vision **96**(2), 252–276 (2012)
6. Kumar, S., Cherian, A., Dai, Y., et al.: Scalable dense non-rigid structurefrom-motion: a grassmannian perspective. In: The IEEE Conference on Computer Vision and Pattern Recognition, pp. 254–263 (2018)
7. Tulsiani, S., Zhou, T., Efros, A.A., et al.: Multi-view supervision for single-view reconstruction via differentiable ray consistency. In: he IEEE Conference on Computer Vision and Pattern Recognition, pp. 2626–2634 (2017)
8. Tatarchenko, M., Dosovitskiy, A., Brox, T.: Octree generating networks: efficient convolutional architectures for high-resolution 3D outputs. In: The IEEE International Conference on Computer Vision, pp. 2088–2096 (2017)
9. Richter, S.R., Roth, S.: Matryoshka networks: predicting 3d geometry via nested shape layers. In: The IEEE Conference on Computer Vision and Pattern Recognition, pp. 1936–1944 (2018)
10. Ronneberger, O., Fischer, P., Brox, T.: U-net: convolutional networks for biomedical image segmentation. In: Navab, N., Hornegger, J., Wells, W.M., Frangi, A.F. (eds.) MICCAI 2015. LNCS, vol. 9351, pp. 234–241. Springer, Cham (2015). https://doi.org/10.1007/978-3-319-24574-4_28
11. Tolstikhin, I., Houlsby, N., Kolesnikov, A., et al.: Mlp-Mixer: an All-mlp architecture for vision (2021). arXiv preprint arXiv:2105.01601

Research on Weld Recognition and NURBS Curve Fitting Method Based on Improved OTSU Double Threshold Segmentation

Liu Shuaishi(✉), Zhou Siyu, Bai Rong, and Zhu Jiaqi

College of Electrical and Electronic Engineering, Changchun University of Technology, Changchun 130000, China
liushuaishi@ccut.edu.cn

Abstract. In the welding process based on machine vision, the precise identification of the weld seam and the fitting of the welding curve are the key factors of welding quality. In view of the uneven illumination of the collected images due to the material of the welding components, there are a large number of noise points, and the traditional algorithm takes a long time to calculate and has poor real-time performance. In this paper, we use the OTSU double threshold algorithm based on the improved genetic algorithm to perform foreground detection and image segmentation by graying and denoising the collected welding images. After the center line coordinates are extracted, the cubic NURBS (Non Uniform Rational B-spline) curve is used for accurate fitting. Finally, the accuracy of the image segmentation processing algorithm is verified by experiments. The experiment proves that the algorithm proposed in this paper is effective and the calculation time is significantly reduced.

Keywords: Weld recognition · Genetic algorithm · OTSU · NURBS

1 Introduction

In the process of welding automation, the precise identification and tracking of welds is the main purpose of weld image processing. In the process of image processing, how to segment the weld information from the background has become a key step in the process. The complete and correct segmentation and extraction of the weld information can be achieved in the following curve fitting steps to achieve high-precision fitting and achieve the purpose of reducing fitting errors.

In the weld image processing technology, image segmentation is the basis for image analysis, feature extraction and target recognition, and is the key technology to achieve accurate weld recognition. Japanese scholar Otsu once proposed the maximum between-class variance method, also known as the OTSU [1] algorithm, which is a global threshold

This paper is supported by the Project of Jilin Province Science and Technology Department Plan (No. 20200703016ZP), the Projects in Science and Technology of the 13th Five-year Plan of Education Department of Jilin Province (No. JJKH20200671KJ).

S.-C. Chu et al. (Eds.): ICGEC 2021, LNEE 833, pp. 316–327, 2022.
https://doi.org/10.1007/978-981-16-8430-2_29

algorithm for adaptive threshold determination. Once this algorithm was proposed, it has been widely used in image segmentation algorithms. Xiang et al. uses the OTSU algorithm to find the optimal threshold of the image instead of artificially setting the threshold [2, 3]. Tang et al. improved the OTSU algorithm to make the foreground and background quality of image segmentation better by introducing the height of the threshold neighborhood gray value and the average gray value of the area. In order to solve the problem of image binarization under uneven lighting conditions, a local threshold algorithm is generally used [4]. Chen et al. use the improved OTSU algorithm to calculate the local threshold according to the brightness distribution of the local area of the image to reduce the impact of uneven illumination [5] Song et al. obtains the upper and lower thresholds through the candidate background and candidate foreground pixel sets, and obtains the optimal threshold through calculation [6]. The optimal threshold is used as the basis for the selection of high and low thresholds, and the final double threshold connection is performed to achieve the purpose of accurate image segmentation. At present, the improved OTSU algorithm mostly addresses the problems of high time complexity and poor real-time performance of the OTSU algorithm. The running speed of the OTSU algorithm is improved. Qian et al. proposed and selected an improved OTSU double threshold segmentation algorithm based on particle swarm optimization algorithm to perform initial segmentation of the foreground and background of the image [7]. Zhang et al. proposed an improved wolf pack optimization two-dimensional OTSU threshold segmentation algorithm, which improves the operating efficiency of the OTSU algorithm [8].

In the above references, good results have been achieved in reducing the segmentation time, but there is still room for improvement in segmentation accuracy and algorithm stability. In view of the noise in the collected images and the time-consuming calculation of traditional algorithms and poor real-time performance, this paper proposes an OTSU double threshold image segmentation algorithm based on an improved genetic algorithm. Firstly, perform gray-scale processing and image denoising on the collected weld image, then segment the weld information from the background and extract the center line of the weld, and finally perform accurate fitting of the three NURBS curves. (see Fig. 1).

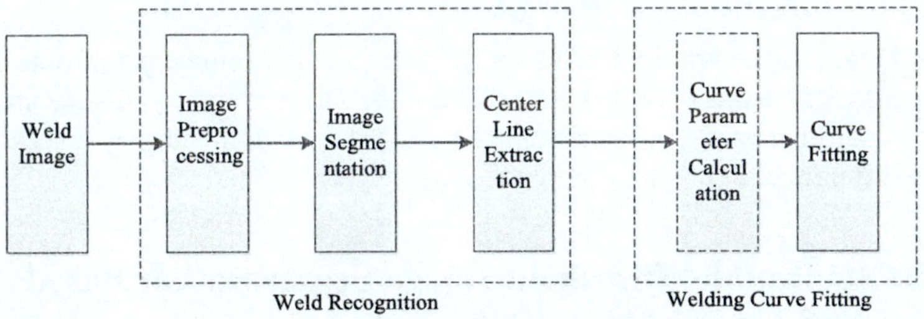

Fig. 1. The overall experiment process

2 Weld Image Preprocessing

2.1 Weld Image Gray-Scale

In this paper, the weighted average method with higher usage and better effect is used to grayscale the image. This method gives different weights A_R, A_G, A_B to each RGB channel when processing RGB color images. Calculated as follows:

$$I = A_R * R + A_G * G + A_B * B \tag{1}$$

Normally $A_R = 0.299, A_G = 0.587, A_B = 0.114$ convert the collected RGB image into grayscale image.

2.2 Weld Image Denoising

Due to various interference factors in the actual working environment, the collected images will contain a lot of noise, and because the surface of some workpieces to be welded is smooth and easy to reflect, it will seriously affect the image collection effect. Common methods [9] for image denoising include mean filtering, median filtering, Gaussian filtering, and bilateral filtering. This paper uses bilateral filtering improved on the basis of Gaussian filtering to remove image noise. Bilateral filtering [10] is a filter function that contains two Gaussians, and considers the Euclidean distance of the pixel, the spatial distance relationship, and the radiation difference in the pixel (range domain) R during sampling.

Assuming that the gray value of the original image at the point (m, n) is $I(m, n)$, and the gray value of the image at point (i, j) obtained after the bilateral filtering algorithm is $I'(i, j)$, the derivation formula of the bilateral filtering is as follows:

$$I'(i,j) = \frac{\sum_{(m,n)\in\Omega_{r,i,j}} \omega_d(m,n)\omega_r(m,n)I(m,n)}{\sum_{(m,n)\in\Omega_{i,j}} \omega_d(m,n)\omega_r(m,n)} \tag{2}$$

$$\omega_d(m,n) = \exp\left\{-\frac{(i-m)^2+(j-n)^2}{2\sigma_d^2}\right\} \quad \omega_r(m,n) = \exp\left\{-\frac{I'(i,j)-I(m,n)^2}{2\sigma_r^2}\right\} \tag{3}$$

Among them: r is the window radius of the filter, the coordinate set of pixels in the square area with side length $(2r + 1)$, ω_d, ω_r are the space weight and gray similarity weight at the coordinates; σ_d, σ_r are the space standard deviation and gray level standard respectively difference.

3 The OTSU Double Threshold Image Segmentation Based on Improved Genetic Algorithm

3.1 Improved OTSU Double Threshold Segmentation

In the traditional OTSU algorithm [6], the image is only divided into two parts: foreground target C0 category and background C1 category. After calculating the maximum value of variance, T =max (σ^2). However, in this paper, due to the interference of noise, the gray histogram has no obvious peaks and valleys, and the segmentation effect cannot achieve the optimal segmentation effect. In this paper, t_1 and t_2 are the two segmentation thresholds of the foreground target and the background. Not only the gray information of the image is considered, but also the relevant information of the neighborhood space is considered. It is assumed that the pixels in the image whose gray level is lower than t_1 constitute the C0 category. Pixels with a gray level of $[0, t_1]$ are classified as C0, and pixels with a gray level higher than t_1 and lower than t_2 constitute C1. Pixels with a gray level of $[t_1 + 1, t_2]$ are classified as C1, and the gray level is higher than t_2 pixels below $L - 1$ constitute category C2. Pixels with gray level $[t_2 + 1, L - 1]$ are classified as category C2, and the whole image is divided into three parts. The proportion of C0 category of the image is ω_0, the average gray value of C0 category is μ_0; the proportion of C1 category to the entire image is ω_1, the average gray value of C1 category is μ_1; the proportion of C2 category to the entire image is ω_2, C2 category The mean gray value is μ_2; the mean gray value of the entire image is μ_r.

$$\omega_0 = \sum_{i=0}^{t_1} p_i \quad \omega_1 = \sum_{i=t_1+1}^{t_2} p_i \quad \omega_2 = \sum_{i=t_2+1}^{L-1} p_i \tag{4}$$

$$\mu_0 = \sum_{i=0}^{t_0} \frac{p_i * i}{\omega_0} \quad \mu_1 = \sum_{i=t_1+1}^{t_1} \frac{p_i * i}{\omega_1} \quad \mu_1 = \sum_{i=t_2+1}^{L-1} \frac{p_i * i}{\omega_2} \tag{5}$$

$$\mu_r = \omega_0\mu_0 + \omega_1\mu_1 + \omega_2\mu_2 \tag{6}$$

$$\sigma^2 = \omega_0(\mu_0 - \mu_r)^2 + \omega_1(\mu_1 - \mu_r)^2 + \omega_2(\mu_2 - \mu_r)^2 \tag{7}$$

When the OTSU method is extended to double threshold segmentation, the optimal threshold value of T_1, T_2 should maximize the variance between the three categories, so that the problem of image segmentation is resolved into the problem of solving the optimal threshold value T_1, T_2.

In order to simplify the calculation process and reduce the calculation time, when the maximum value of the variance between the three categories is calculated, T_1 is equal to the maximum value.When T_2 = a*T_1, the experimental results show that when the illumination is uneven and the effect of noise cannot be completely eliminated, the ideal effect is obtained when a ∈(0.5,0.6) is set, and the segmented binary image can effectively reflect the shape of the weld.

3.2 Improved Genetic Algorithm

In the process of solving the threshold, the traditional genetic algorithm usually sets the crossover probability and mutation probability as fixed values. If the value is improper, the search result will appear premature, and it is easy to fall into the local optimal situation, and the searched threshold cannot be guaranteed to be the best result.

In order to perform optimal image segmentation, this paper improves the crossover operator and mutation operator in the traditional genetic algorithm, and adopts the method of adaptive adjustment of the crossover probability value and the mutation probability value. In the early stage of the algorithm iteration, there may be too many individuals eliminated, causing the search to enter the local optimal situation. At this time, the crossover probability value can be set higher. In the middle and late stages of the algorithm iteration, in order to enhance the convergence rate and keep the outstanding individuals at the same time, the crossover probability value should be appropriately small at this time. Setting the crossover probability value into a segment function can speed up the convergence speed, search for the best fitness value faster, and achieve a better image segmentation effect. In the same way, the mutation probability value is improved. In the early stage of the iteration, in order to make the individual easier to be selected, the mutation probability value should not be too large. In the middle and late stages of the iteration, in order to ensure the convergence rate and prevent the loss of outstanding individuals, it is necessary to avoid the situation where the mutation rate is 0. The specific formula is as follows:

$$P_c = \begin{cases} P_{c1} \ , gen \leq 20 \\ P_{c2} \ , gen > 20 \end{cases} \tag{8}$$

$$P_m = \begin{cases} P_{m1} \ , gen \leq 20 \\ P_{m2} \ , gen > 20 \& gen \leq 50 \\ P_{m1} \ , gen > 50 \end{cases} \tag{9}$$

Where gen is the number of iterations. $P_{c1} = 0.8$, $P_{c2} = 0.6$, $P_{m1} = 0.02$, $P_{m2} = 0.03$.

3.3 The OTSU Double Threshold Image Segmentation Based on Improved Genetic Algorithm

The OTSU double threshold image segmentation based on improved genetic algorithm used in this paper is as follows:

Step 1: Formulate coding and decoding rules, mainly using the principle of binary coding. Based on the gray value range of the image (0 ~ 255), select a 16-bit binary string to encode the threshold (T_1 T_2). The first 8 bits represent the threshold value T_1, and the last 8 bits are T_2; the value of the fitness is obtained by decoding the 16-bit binary string into a value between 0 and 255.

Step 2: Initialize the population, the initial value of the population has a direct impact on the execution efficiency and results of the genetic algorithm. If the value is too large, the complexity of the algorithm will increase; if the value is too small, it is difficult to obtain the optimal segmentation threshold due to the limited space. After experimental analysis, the population size in this experiment is set to 20, which is optimal.

Step 3: Select the fitness function, and find the best fitness value by taking formula (7) as the fitness function. For selection, crossover and mutation, this paper uses the roulette selection method as the selection operator, The crossover probability value is the above formula (8). At the same time, start the exchange from the position halfway behind the gene, so that it is convenient to maintain the current optimal value and set the mutation value as the probability value of the change. For each individual, a one-bit mutation is generated at any position, but the mutation probability should not be designed to be large. If the mutation probability is set to be too large, the entire population will converge too slowly, which will affect the calculation time. In this experiment, the mutation probability is set as the above formula (9), the proportion of retained parents in the elite is 0.1, and the maximum number of iterations is set to 60.

Step 4: Termination condition, when the maximum value of the population iteration is met, the algorithm stops running. At this time, the individual with the highest fitness value is the optimal threshold. Output the best threshold value T_1, T_2,calculation time, and the image result after segmentation.

4 Weld Centerline Coordinate Extraction

Using the above method, the weld can be separated from the background. In order to determine the position of the weld, the coordinates of the center of the weld must be calculated, and then the center line of the weld can be fitted with a curve. Image refinement is a key step in image pattern recognition. The idea of the thinning algorithm is to optimize the process of eliminating black pixels layer by layer from the boundary in the original thinning algorithm to improve the efficiency of thinning. The skeleton extraction of the image is to refine the target based on the center of the target, and extract a part of the image to take the center line, which can remove unnecessary information and facilitate recognition.

In this paper, the Zhang-Suen parallel iterative algorithm [11] is used to refine the image. Each iteration of the algorithm is to corrode the target pixels that meet specific conditions. The effect is that the target becomes more and more refined. Iterate continuously until the target after the last corrosion is in this round of operation, and no new pixels are corroded, and the algorithm ends. This paper uses the gray barycentric method to calculate the weld feature image after image processing, so as to obtain the center coordinates of the weld.

$$\bar{x}_i = i \qquad \bar{y}_i = \frac{\sum_{jm}^{jm+w_v} g(i,j)*y_i}{\sum_{jm}^{jm+w_v} g(i,j)} \tag{10}$$

Among them, $\bar{x}_i$ and $\bar{y}_i$ are the abscissa and ordinate of the weld center in the i row; y_i is the ordinate of the i column, j_m is the starting point of the weld in the i row, and $j_m + w_v$ is the end of the weld in the i row. $g(i, j)$ is the gray value of the point in the image coordinates.

5 NURBS Curve Fitting of Weld

After extracting the coordinates of the weld center line above, the weld center coordinate set is obtained, and the cubic NURBS curve fitting is performed on 200 discrete value points obtained by the gray barycentric method. The premise of accurate curve fitting is to calculate the parameters of the curve. These include: the nodal vectors, boundary conditions, control vertices and weight factor.

Determination of the nodal vectors: to calculate the nodal vectors [12], the type value points to be fitted need to be parameterized first. For the cubic NURBS curve, when inverse calculation, the first and last nodes are all set to a repeat degree of 4. In this paper, it uses the cumulative chord length parameterization method.

Determination of the first and end boundary conditions: the boundary conditions usually adopt the endpoint tangent vector condition, which has a fixed tangent direction. When the endpoint tangent vector condition is used, the first and the end satisfy the following equations:

$$d_1 - d_0 = \frac{\Delta_3}{3} q'_0 \quad d_{n+2} - d_{n+1} = \frac{\Delta_{n+2}}{3} q'_n \tag{11}$$

Among them, $d_i(0, 1, ..., n)$ is the vertex of each control, $\Delta_3 = u_4 - u_3$, $\Delta_{n+2} = u_{n+3} - u_{n+2}$, q'_0 is the leading end tangent vector; q'_n is the end tangent vector.

Determination of control vertices: the first and last control vertices of the cubic NURBS curve are the first and last value points of the curve. $d_0 = Q_0$, $d_{n+2} = Q_n$ can be obtained according to the above equation.

$$a_i = \frac{(\Delta_{i+2})^2}{\Delta_i + \Delta_{i+1} + \Delta_{i+2}}, i = 2, 3, \ldots, n \quad b_i = \frac{\Delta_{i+2}(\Delta_i + \Delta_{i+1})}{\Delta_i + \Delta_{i+1} + \Delta_{i+2}} + \frac{\Delta_{i+1}(\Delta_{i+2} + \Delta_{i+3})}{\Delta_{i+1} + \Delta_{i+2} + \Delta_{i+3}}, i = 2, 3, \ldots, n \tag{12}$$

$$c_i = \frac{(\Delta_{i+2})^2}{\Delta_{i+1} + \Delta_{i+2} + \Delta_{i+3}} \quad e_1 = Q_0 - \frac{\Delta_3}{3} q'_0 \quad e_{n+1} = Q_n - \frac{\Delta_{n+2}}{3} q'_n \quad e_i = (\Delta_{i+1} + \Delta_{i+2}) Q_{i-1}, i = 0, 1 \ldots, n \tag{13}$$

Solving the above linear equations can solve all the unknown control vertices.

6 Experimental Results and Analysis

This paper takes the butt weld image collected by the Hikvision MV-CA020-10GM industrial camera as an example to carry out the experiment. The experimental hardware platform is Intel(R) Core (TM) i5–6200 CPU 2.3 GHz processor, 8 GB memory, and the software platform is MATLAB R2018a. Experimental flowchart see Fig. 2.

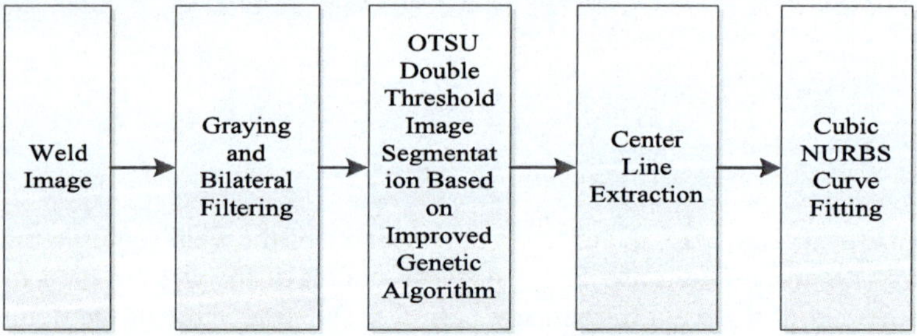

Fig. 2. Flow chart of weld recognition and curve fitting experiment.

6.1 Image Grayscale and Denoising Experiment

In this paper, the weighted average method with higher usage and better effect is used to grayscale the image. (see Fig. 3).

By comparison, as shown in Fig. 3(c), Gaussian filtering cannot protect image details very well. Bilateral filtering as shown in Fig. 3(d) can obviously filter out the noise in the image and effectively reduce the noise interference. It can also achieve a good filtering effect while preserving the image detail information, making the weld contour clearer.

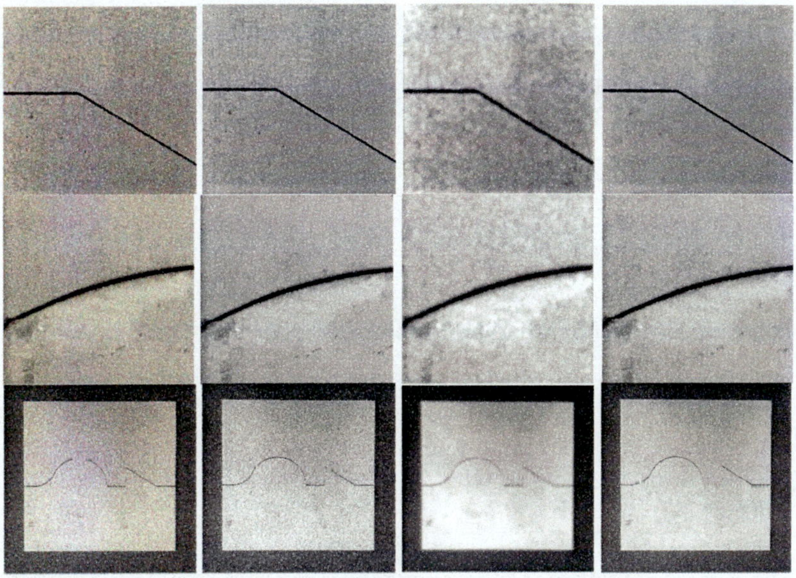

Fig. 3. Grayscale and denoising of weld image.

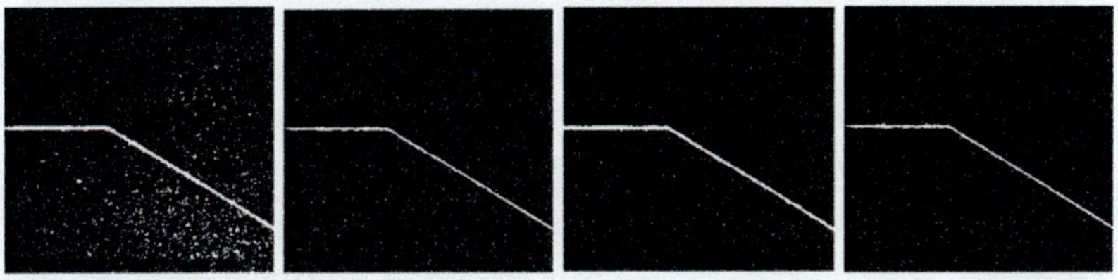

Fig. 4. Four algorithms for threshold segmentation of the poly line weld.

6.2 Weld Image Threshold Segmentation Experiment

This paper selects traditional OTSU algorithm, improved OTSU double threshold segmentation algorithm, traditional genetic algorithm and the algorithm of this paper for comparative experiments. The segmentation results of different algorithms are shown in Fig. 4, Fig. 6 and Fig. 8. The convergence process of the genetic algorithm before and after the improvement is shown in Fig. 5, Fig. 7 and Fig. 9. The evaluation indicators of different algorithms are shown in Table 1, Table 2 and Table 3.

As shown in Fig. 4(a), the weld information has low clarity and cannot eliminate the influence of Gaussian white noise on the segmentation results. In the improved OTSU algorithm, as shown in Fig. 4(b), when a $\in$ (0.5, 0.6), it can be clearly observed that the noise is reduced, part of the noise effect is eliminated, and the clarity of the weld information is improved. Using the convergence process is shown in Fig. 5(a). After 50 iterations, the optimal fitness value is reached, and the convergence process is

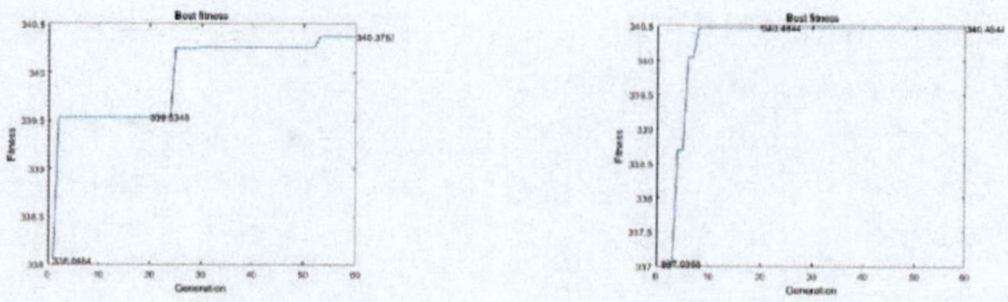

(a) Convergence process of traditional GA (b) Convergence process of improved

Fig. 5. Convergence process before and after genetic algorithm improvement.

Table 1. Comparison of experimental data of different algorithms (the poly line weld)

Algorithm	Threshold T1	Threshold T2	Time(s)
Traditional OTSU	152	85	1.2611
Improved OTSU	157	97	0.9978
Traditional GA	158	102	0.5070
Ours	160	96	0.3060

slow. The result of using the algorithm in this paper is shown in Fig. 4(d). When $T_2 = 0.6*T_1$, the algorithm has good noise resistance, and the segmentation result is clear. As shown in Fig. 5(b). After 20 iterations, the best fitness value is reached. This reduces the time-consuming calculation and improves the real-time performance of segmentation.

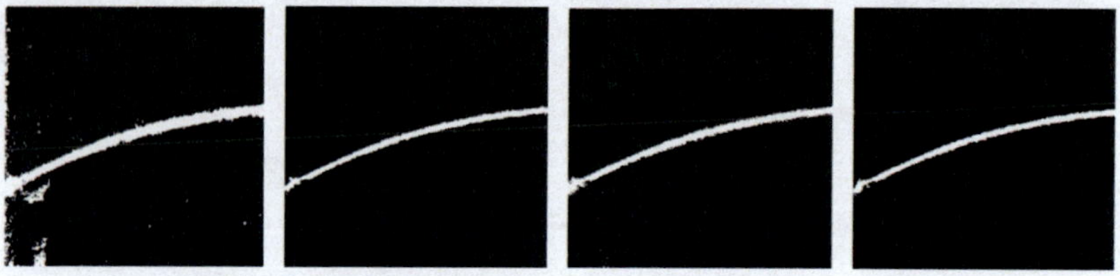

(a) Traditional OTSU (b) Improved OTSU (c) Traditional GA (d) Ours

Fig. 6. Four algorithms for threshold segmentation of the arc weld.

In Fig. 6(a), the traditional OTSU algorithm segmentation result has noise, and the weld has false edges, which affects the definition of the final weld curve. In Fig. 6(b) and Fig. 6(c), there is still a small amount of image information that has nothing to do with the weld. However, the poor performance in the early stage of the algorithm iteration is shown in Fig. 7(a), The algorithm proposed in this paper, as shown in Fig. 6(d). It has good noise resistance, and completely separates the weld information from the background. And the improved genetic algorithm convergence process is shown in Fig. 7(b). When iterates 20 times, the optimal fitness value is reached. The calculation time of the algorithm is greatly reduced.

(a) Convergence process of traditional GA (b) Convergence process of improved GA

Fig. 7. Convergence process before and after genetic algorithm improvement.

Table 2. Comparison of experimental data of different algorithms (the arc line weld)

Algorithm	Threshold T1	Threshold T2	Time(s)
Traditional OTSU	187	93	1.8542
Improved OTSU	182	106	0.8008
Traditional GA	183	107	0.4890
Ours	180	104	0.1840

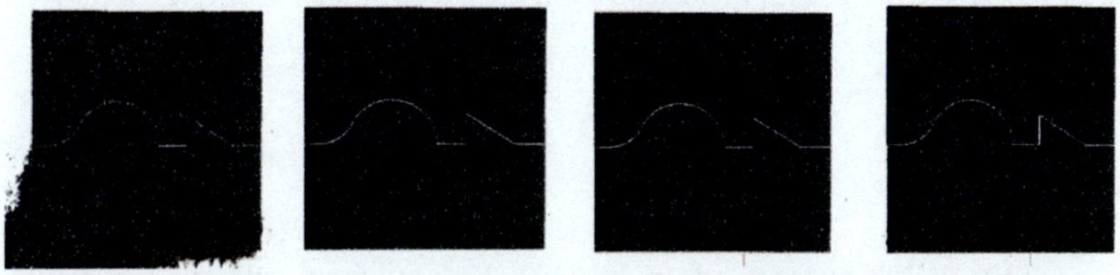

(a) Traditional OTSU (b) Improved OTSU (c) Traditional GA (d) Ours

Fig. 8. Four algorithms for threshold segmentation of the compound weld.

(a) Convergence process of traditional GA (b) Convergence process of improved GA

Fig. 9. Convergence process before and after genetic algorithm improvement.

Table 3. Comparison of experimental data of different algorithms

Algorithm	Threshold T1	Threshold T2	Time(s)
Traditional OTSU	202	100	3.2407
Improved OTSU	203	132	2.1162
Traditional GA	204	127	1.2170
Ours	204	123	0.4660

The algorithm in this paper is shown in Fig. 8(d), the weld curve is clear, and the time is significantly shorter than the traditional GA algorithm. After 30 iterations, the best fitness value is reached as shown in Fig. 9(b), which speeds up the convergence process.

6.3 Extraction of Weld Center Line and NURBS Curve Fitting

Using the above image refinement to extract the center line of the weld, the 200 discrete value points obtained by the gray barycentric method are used to perform the cubic NURBS curve fitting using the relevant parameters obtained in the paper (see Fig. 10).

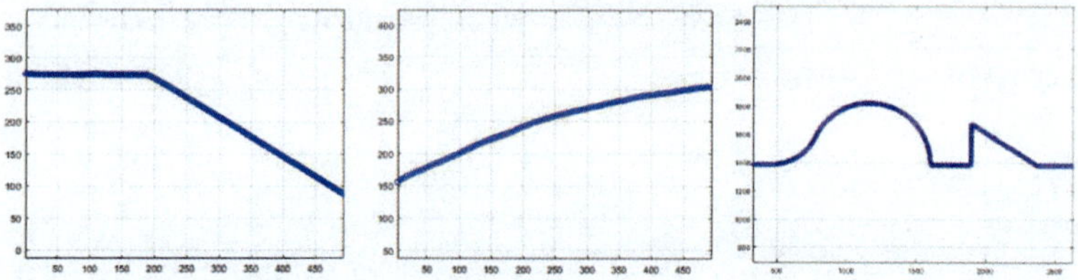

Fig. 10. Cubic NURBS curve fitting for three kinds of weld center line coordinates.

7 Concluding

This paper aims at the uneven illumination of the collected images, the influence of noise, and the long time-consuming calculation of traditional algorithms and poor real-time performance. This paper proposes an OTSU double threshold image segmentation algorithm based on an improved genetic algorithm. The noise resistance, segmentation accuracy and real-time performance have been significantly improved; Compared with the time-consuming solution of traditional OTSU algorithm, the improved algorithm proposed in this paper takes less time and can better meet real-time requirements. Then, the segmented weld image is refined, the center line coordinates are extracted, and so on, and finally the cubic NURBS fitting curve is obtained.

References

1. Tang, B., Kong, J.Y.: Image enhancement and segmentation algorithm for low-contrast small defects on steel late. J. Image Graph. **25**(1), 0081–0090 (2020)
2. Peng, G.: Design and realization of sorting robots system for plug-in components. Huazhong Univ. Sci. Technol. (Nat. Sci. Ed.) **48**(1), 108–114 (2020)
3. Xiang, R.: Recognition of overlapping tomatoes based on overlapping edges at night. Huazhong Univ. Sci. Technol. (Nat. Sci. Ed.) **47**(5), 68–72 (2019)
4. Tang, M.A., Wang, C.Y.: Extraction of the edge of track images based on improved canny algorithm and hough transform. Railway Stand. Des. **65**(8), 1–6 (2021)
5. Chen, Z.W., Li, S.F.: Chip recognition and classification system based on improved Otsu algorithm. Microeltronic. Comput. **38**(4), 6–10 (2021)

6. Song, R.J.: Adaptive Canny edge detection algorithm. J. Nanjing Univ. Posts Telecommun. (Nat. Sci. Ed.) **38**(3), 72–77i (2018)
7. Qian, S.H., Li, J.: An image segmentation algorithm based on power line aerial images with foreign matters. Comput. Dig. Eng. **48**(8), 1986–1997 (2020)
8. Zhang, J.S.: Improved wolf group optimization two-dimensional OTSU threshold segmentation algorithm. J. Electric Power **35**(01), 40–45 (2020)
9. Hong, C.C.: Leather defect detection based on improved bilateral filtering. Laser Technol. **45**(3), 373–377 (2021)
10. Lin, F.T., Hu, W.H., Yang, Y.: Application of 3D surface reconstruction technology based on bilateral filtering in rail inspection. Mach. Tool Hydraul. **49**(6), 6–11 (2021)
11. Hannah, J.C., Gladis, D.: Feature extraction with thinning algorithms for precise cretos copy. Indian J. Sci. Technol. **8**(29), 127–133 (2015)
12. Huo, Y.G.: Effect of different parameterization methods on the cubic NURBS curve fitting errors. Res. Dev. **48**(04), 54–58 (2019)

Modeling and Simulation of Non-common Field Position and Pose Measurement of Dual Cameras

Mujun Xie, Aoshuang Li(✉), Changhong Jiang, and Shaoshi Qin

College of Electrical and Electronic Engineering, Changchun University of Technology, Changchun 130000, China

Abstract. In this paper, in order to solve the automatic alignment problem of large targets loaded in close range by cantilever, a non-common field of view position and pose measurement method with two cameras is proposed based on two cooperative targets in the same plane at the bottom of the vehicle box. The image of the cooperative target on the bottom of the vehicle box is obtained by the camera on the cantilever hanger. Based on the position relationship between the cooperative target and the camera, a position and pose measurement model is established when the bottom of the loading box is parallel to the cantilever spreader. The position and pose relationship between the spreader and the bottom of the loading box is solved. The simulation results show that the method is feasible to measure the position and pose of large target in close range with non-common field of view.

Keywords: Automatic alignment · Machine vision · Non-common field of view · Pose measurement · Cooperative target

1 Introduction

When loading goods on a cantilever crane, workers need to manipulate the cantilever crane to load the goods onto the truck, but the worker's operational sight will be affected by the truck's warehouse grid, which increases the loading time. It is very important to realize the automatic pairing of cargo box and truck [1]. In most vision measurement systems, the multi-camera system [2] has a larger field of view than a single camera [3], and is usually used to obtain the geometric information of the scene for pose measurement [4]. The existing visual pose measurement methods are based on non-cooperative targets [5] and cooperative targets [6]. The measurement method based on the cooperative target requires artificially set mark points [7], which can improve the detection accuracy of the target.

This article is mainly based on the working environment when the cantilever crane is loaded with goods automatically aligned, and proposes a dual-camera non-common field of view pose measurement method. By setting the cooperation target on the vehicle box, install the camera on the cantilever sling to build the pose measure the model and calculate the posture relations between the spreader and the vehicle. This method can

S.-C. Chu et al. (Eds.): ICGEC 2021, LNEE 833, pp. 328–337, 2022.
https://doi.org/10.1007/978-981-16-8430-2_30

save the time of loading goods by the cantilever crane, and is of great significance for improving the loading efficiency.

2 Dual Camera Pose Measurement Model

2.1 Measuring Coordinate System

In order to help the analysis of the pose measurement problem, the following coordinate system is built: take the center of the bottom of the carrier box as the center of the coordinate system to build the measurement reference coordinate system O−XYZ (right-handed coordinate system), as shown in Fig. 1 [8]. O is the center of the bottom of the carrier box Point; A (X_A, Y_A, 0), B (X_B, Y_B, 0) are the cooperation target points. $O'-X'Y'Z'$ is the coordinate system of the cantilever sling. O' is the center point of the sling and the measuring cameras are installed at $A'(X_a, Y_a, 0)$, $B'(X_b, Y_b, 0)$; cooperation the coordinates of targets A and B in the spreader coordinate system are (X'_A, Y'_A, Z'_A), (X'_B, Y'_B, Z'_B); $O_a-x_ay_az_a$ and $O_b-x_by_bz_b$ are the image planes coordinate system; the imaging positions of A and B on the image plane are a (x_a, y_a), b (x_b, y_b).

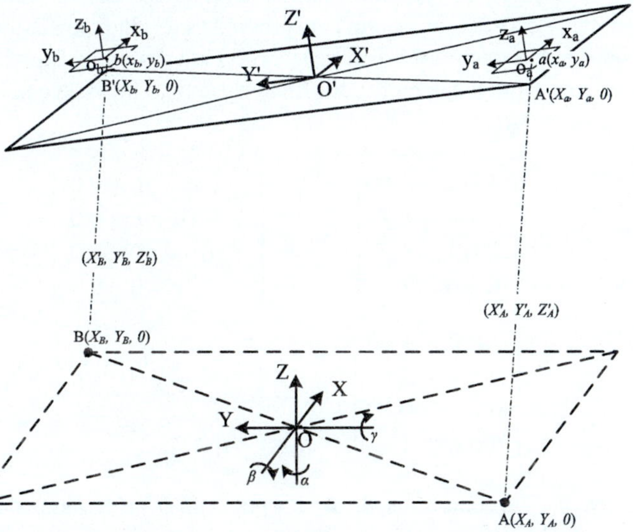

Fig. 1. The model of the measurement coordinate system

In this paper, both the vehicle body and sling are in horizontal state; the XOY plane is parallel to the X′O′Y′ plane, thus ignoring coordinate rotation around X axis and Y axis. Only consider the coordinate system translation (ΔX, ΔY, ΔZ) and α angle around Z axis (The direction of rotation of Fig. 1 is positive).

2.2 Coordinate Transformation

Since the car body and the spreader are horizontal, the coordinate transformation process from the reference coordinate system to the image plane coordinate system is shown in Fig. 2.

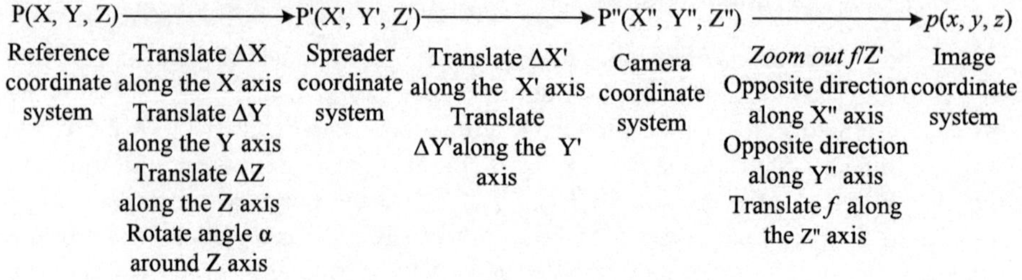

Fig. 2. Horizontal coordinate transformation process

Among them: the matrix from the reference coordinate system to the spreader coordinate system is expressed as R_T; the rotation matrix from the reference coordinate system to the spreader coordinate system is expressed as R_α; the translation matrix from the spreader coordinate system to the camera coordinate system is expressed as R_t; the transformation matrix from the camera coordinate system to the image plane coordinate system is expressed as R_p.

$$R_T = \begin{bmatrix} 1 & 0 & 0 & -\Delta X \\ 0 & 1 & 0 & -\Delta Y \\ 0 & 0 & 1 & -\Delta Z \\ 0 & 0 & 0 & 1 \end{bmatrix}, \; R_\alpha = \begin{bmatrix} cos\alpha & \sin\alpha & 0 & 0 \\ -\sin\alpha & cos\alpha & 0 & 0 \\ 0 & 0 & 1 & 0 \\ 0 & 0 & 0 & 1 \end{bmatrix}$$

$$R_t = \begin{bmatrix} 1 & 0 & 0 & -\Delta X' \\ 0 & 1 & 0 & -\Delta Y' \\ 0 & 0 & 1 & 0 \\ 0 & 0 & 0 & 1 \end{bmatrix}, \; R_p = \begin{bmatrix} -\frac{f}{Z'} & 0 & 0 & 0 \\ 0 & -\frac{f}{Z'} & 0 & 0 \\ 0 & 0 & \frac{f}{Z'} & -f \\ 0 & 0 & 0 & 1 \end{bmatrix}$$

From the above known conditions, the image plane position equation can be obtained as:

$$p = R_P \cdot R_t \cdot R_\alpha \cdot R_T \cdot P$$

Which is:

$$\begin{bmatrix} x \\ y \\ z \\ 1 \end{bmatrix} = \begin{bmatrix} -\frac{f}{Z'} & 0 & 0 & 0 \\ 0 & -\frac{f}{Z'} & 0 & 0 \\ 0 & 0 & \frac{f}{Z'} & -f \\ 0 & 0 & 0 & 1 \end{bmatrix} \begin{bmatrix} 1\,0\,0 & -\Delta X' \\ 0\,1\,0 & -\Delta Y' \\ 0\,0\,1 & 0 \\ 0\,0\,0 & 1 \end{bmatrix}$$

$$\begin{bmatrix} cos\alpha & \sin\alpha & 0\,0 \\ -\sin\alpha & cos\alpha & 0\,0 \\ 0 & 0 & 1\,0 \\ 0 & 0 & 0\,1 \end{bmatrix} \begin{bmatrix} 1\,0\,0 & -\Delta X \\ 0\,1\,0 & -\Delta Y \\ 0\,0\,1 & -\Delta Z \\ 0\,0\,0 & 1 \end{bmatrix} \begin{bmatrix} X \\ Y \\ Z \\ 1 \end{bmatrix}$$

2.3 Establishment of Pose Measurement Model

Transforming from the reference coordinate system O-XYZ to the image coordinate system $O_{a-}x_ay_az_a$, it is known: $\Delta X^{'} = X_a$, $\Delta Y^{'} = Y_a$, $Z^{'} = -\Delta Z$. According to the obtained image position, the image coordinate of point A can be obtained:

$$\begin{aligned} x_a &= \frac{f_a}{\Delta Z}[cos\alpha \cdot (X_A - \Delta X) + sin\alpha \cdot (Y_A - \Delta Y) - X_a] \\ y_a &= \frac{f_a}{\Delta Z}[-\sin\alpha \cdot (X_A - \Delta X) + cos\alpha \cdot (Y_A - \Delta Y) - Y_a] \end{aligned} \tag{1}$$

Transforming from the reference coordinate system O-XYZ to the image plane coordinate system $O_{b-}x_by_bz_b$, it is known: $\Delta X^{'} = X_b$, $\Delta Y^{'} = Y_b$, $Z^{'} = -\Delta Z$, and the image plane coordinates of point B are obtained according to the obtained phase plane position equation:

$$\begin{aligned} x_b &= \frac{f_b}{\Delta Z}[cos\alpha \cdot (X_B - \Delta X) + sin\alpha \cdot (Y_B - \Delta Y) - X_b] \\ y_b &= \frac{f_b}{\Delta Z}[-\sin\alpha \cdot (X_B - \Delta X) + cos\alpha \cdot (Y_B - \Delta Y) - Y_b] \end{aligned} \tag{2}$$

Simultaneous Eqs. (1), (2) can get the pose measurement model about ΔX, ΔY, ΔZ and α:

$$X_A = cos\alpha \cdot (\frac{x_a}{f_a} \cdot \Delta Z + X_a) - sin\alpha \cdot (\frac{y_a}{f_a} \cdot \Delta Z + Y_a) + \Delta X \tag{3}$$

$$Y_A = \sin\alpha \cdot (\frac{x_a}{f_a} \cdot \Delta Z + X_a) + cos\alpha \cdot (\frac{y_a}{f_a} \cdot \Delta Z + Y_a) + \Delta Y \tag{4}$$

$$X_B = cos\alpha \cdot (\frac{x_b}{f_b} \cdot \Delta Z + X_b) - sin\alpha \cdot (\frac{y_b}{f_b} \cdot \Delta Z + Y_b) + \Delta X \tag{5}$$

$$Y_B = \sin\alpha \cdot (\frac{x_b}{f_b} \cdot \Delta Z + X_b) + cos\alpha \cdot (\frac{y_b}{f_b} \cdot \Delta Z + Y_b) + \Delta Y \tag{6}$$

3 Pose Measurement Model Calculation

In the model, X_A, Y_A, X_B, Y_B, X_a, Y_a, X_b, Y_b, x_a, y_a, f_a, x_b, y_b, f_b are known quantities, then ΔX, ΔY, ΔZ and α are the positions to be solved.

3.1 Alpha Calculation

Perform the following operations according to the obtained pose measurement model: (3)–(5), (4)–(6):

Available:

$$\begin{cases} X_A - X_B = cos\alpha \cdot (\frac{x_a}{f_a} - \frac{x_b}{f_b}) \cdot \Delta Z - sin\alpha \cdot (\frac{y_a}{f_a} - \frac{y_b}{f_b}) \cdot \Delta Z \\ \quad + cos\alpha \cdot (X_a - X_b) - \sin\alpha \cdot (Y_a - Y_b) \\ Y_A - Y_B = \sin\alpha \cdot (\frac{x_a}{f_a} - \frac{x_b}{f_b}) \cdot \Delta Z + cos\alpha \cdot (\frac{y_a}{f_a} - \frac{y_b}{f_b}) \cdot \Delta Z \\ \quad + sin\alpha \cdot (X_a - X_b) + cos\alpha \cdot (Y_a - Y_b) \end{cases} \tag{7}$$

Make $C_1 = \sqrt{(\frac{x_a}{f_a} - \frac{x_b}{f_b})^2 + (\frac{y_a}{f_a} - \frac{y_b}{f_b})^2}$, $cos\theta_1 = \frac{\frac{x_a}{f_a} - \frac{x_b}{f_b}}{C_1}$, $\sin\theta_1 = \frac{\frac{y_a}{f_a} - \frac{y_b}{f_b}}{C_1}$.
Make $C_2 = \sqrt{(X_a - X_b)^2 + (Y_a - Y_b)^2}$, $cos\theta_2 = \frac{X_a - X_b}{C_2}$, $sin\theta_2 = \frac{Y_a - Y_b}{C_2}$.
When $cos\theta_1 \geq 0$, $\theta_1 = arcsin(\frac{\frac{y_a}{f_a} - \frac{y_b}{f_b}}{C_1})$, otherwise $\theta_1 = \pi - arcsin(\frac{\frac{y_a}{f_a} - \frac{y_b}{f_b}}{C_1})$.
When $cos\theta_1 \geq 0$, $\theta_2 = arcsin(\frac{Y_a - Y_b}{C_2})$, otherwise $\theta_2 = \pi - arcsin(\frac{Y_a - Y_b}{C_2})$.
After the above transformation, the equations are as follows:

$$\begin{cases} X_A - X_B = C_1 \cdot \Delta Z \cdot cos\alpha \cdot cos\theta_1 - C_1 \cdot \Delta Z \cdot sin\alpha \cdot sin\theta_1 \\ \quad + C_2 \cdot cos\alpha \cdot cos\theta_2 - C_2 \cdot \sin\alpha \cdot sin\theta_2 \\ Y_A - Y_B = C_1 \cdot \Delta Z \cdot \sin\alpha \cdot cos\theta_1 + C_1 \cdot \Delta Z \cdot cos\alpha \cdot sin\theta_1 \\ \quad + C_2 \cdot sin\alpha \cdot cos\theta_2 + C_2 \cdot cos\alpha \cdot sin\theta_2 \end{cases} \tag{8}$$

According to the Eq. 8, we can solve the expression of α

$$\alpha = \pi - 2 \cdot (\theta_1 - \theta_2) \tag{9}$$

3.2 ΔX, ΔY, ΔZ Calculations

We will get the α angle rotated horizontally around the Z axis, θ_1, θ_2 into formula (3) to formula (6), and then ΔX, ΔY, and ΔZ can be calculated.

$$\Delta Z = \frac{(X_A - X_B) - C_2 \cdot cos(\alpha + \theta_2)}{C_1 \cdot cos(\alpha + \theta_1)} = \frac{(Y_A - Y_B) - C_2 \cdot sin(\alpha + \theta_2)}{C_1 \cdot \sin(\alpha + \theta_1)} \tag{10}$$

$$\begin{aligned} \Delta X &= X_A - cos\alpha \cdot (\frac{x_a}{f_a} \cdot \Delta Z + X_a) + \sin\alpha \cdot (\frac{y_a}{f_a} \cdot \Delta Z + Y_a) \\ &= X_B - cos\alpha \cdot (\frac{x_b}{f_b} \cdot \Delta Z + X_b) + \sin\alpha \cdot (\frac{y_b}{f_b} \cdot \Delta Z + Y_b) \end{aligned} \tag{11}$$

$$\Delta Y = Y_A - \sin\alpha \cdot (\frac{x_a}{f_a} \cdot \Delta Z + X_a) - cos\alpha \cdot (\frac{y_a}{f_a} \cdot \Delta Z + Y_a)$$
$$= Y_B - \sin\alpha \cdot (\frac{x_b}{f_b} \cdot \Delta Z + X_b) - cos\alpha \cdot (\frac{y_b}{f_b} \cdot \Delta Z + Y_b) \quad (12)$$

Based on the above calculations, we obtain a computational model for the ΔX, ΔY, ΔZ, α.

4 Pose Measurement Simulation

In order to verify the feasibility of the model, MATLAB software was used for simulation based on the following initial conditions. Suppose the following conditions are known: $X_a = -200$ mm, $Y_a = -600$ mm; $X_b = 200$ mm, $Y_b = 600$ mm; $X_A = -200$ mm, $Y_A = -600$ mm; $X_B = 200$ mm, $Y_B = 600$ mm; $f_a = 25$ mm, $f_b = 25$ mm; Z = 1200 mm;

According to Fig. 1, the rotation matrix R_β around the X axis, the rotation matrix R_γ around the Y axis and the rotation angle α around the Z axis are given.

Among them:

$$R_\beta = \begin{bmatrix} 1 & 0 & 0 & 0 \\ 0 & cos\beta & -\sin\beta & 0 \\ 0 & \sin\beta & cos\beta & 0 \\ 0 & 0 & 0 & 1 \end{bmatrix}, R_\gamma = \begin{bmatrix} cos\gamma & 0 & \sin\gamma & 0 \\ 0 & 1 & 0 & 0 \\ -\sin\gamma & 0 & \cos\gamma & 0 \\ 0 & 0 & 0 & 1 \end{bmatrix}$$

When the x-axis, y-axis, and z-axis all have rotation angles, the image plane equation can be expressed as:

$$p = R_P \cdot R_t \cdot R_\alpha \cdot R_\beta \cdot R_\gamma \cdot R_T \cdot R_P \quad (13)$$

Calculate the A and B image plane coordinates (x_a, y_a) and (x_b, y_b) based on the above conditions. Then according to the measurement model in the article, the position and posture translation ΔX, ΔY, ΔZ and α of the spreader relative to the bottom of the car box are calculated. The measurement error of the model built in the article is analyzed by setting the different inclination angles β and γ of the spreader and the vehicle and different translation values. Let the values of the translational distance from the spreader coordinate system to the base coordinate system of the bottom of the carriage be ΔX, ΔY, ΔZ, and the horizontal rotation angle around the Z axis is α.

1. When the spreader is parallel to the bottom of the carriage, $\beta = \gamma = 0°$, and the partial measurement error changes of ΔX, ΔY, ΔZ, and α are shown in Table 1:

Table 1. The error changes of ΔX(mm), ΔY(mm), ΔZ(mm), α (°)

Position offset variable	Truth value	Measurd value	Error	Truth value	Measurd value	Error	Truth value	Measurd value	Error
ΔX	10	10	0	50	50	0	100	100	0
ΔY	10	10	0	50	50	0	100	100	0
ΔZ	1200	1200	0	1200	1200	0	1200	1200	0
α	1	1	0	1	1	0	1	1	0
ΔX	−10	−10	0	−50	−50	0	−100	−100	0
ΔY	10	10	0	50	50	0	100	100	0
ΔZ	1200	1200	0	1200	1200	0	1200	1200	0
α	1	1	0	1	1	0	1	1	0
ΔX	−10	−10	0	−50	−50	0	−100	−100	0
ΔY	−10	−10	0	−50	−50	0	−100	−100	0
ΔZ	1200	1200	0	1200	1200	0	1200	1200	0
α	1	1	0	1	1	0	1	1	0
ΔX	10	10	0	50	50	0	100	100	0
ΔY	−10	−10	0	−50	−50	0	−100	−100	0
ΔZ	1200	1200	0	1200	1200	0	1200	1200	0
α	1	1	0	1	1	0	1	1	0

It can be seen from Table 1 that when $\beta = 0°$, $\gamma = 0°$, when the spreader is parallel to the bottom of the vehicle box, the error of the center point coordinate translation amount ΔX, ΔY, ΔZ and α is zero, which verifies the pose,the accuracy of the measurement model and settlement.

2. When position offset variable ΔX, ΔY, ΔZ and α are constant, change the inclination angles β and γ of the spreader and the bottom of the carrier box. The measured data are shown in Table 2.

Based on the data, the following error curve is shown in Fig. 3. In order to make the error change more obvious, the absolute value of the error is taken.

From Fig. 3 we can see that when the coordinate position offset variable ΔX, ΔY is constant, the larger the spreader coordinate rotation angle β and γ, the greater the tilt of the spreader and the bottom of the carriage, ΔX, ΔY, ΔZ and the greater the α error, the lower the measurement accuracy.

3. When the inclination angle β and γ are constant, take the different quadrant position offset variable ΔX, ΔY and the partial error changes are shown in Table 3.

Based on the above data, we can get the curve of the error of ΔX, ΔY, ΔZ, and α with the amount of translation when the translation amounts ΔX and ΔY are taken from the four quadrants, as shown in Fig. 4. In order to make the error change more obvious, the absolute value of the error is taken.

Table 2. The error changes of ΔX(mm), ΔY(mm), ΔZ(mm), α(°).

		β = 0° γ = 0°		β = 0.02 γ = 0 .02		β = 0.04 γ = 0.04	
Position offset variable	Truth value	Measured value	Error	Measured value	Error	Measured value	Error
ΔX	10	10	0	10.57	−0.57	11.15	−1.15
ΔY	10	10	0	9.72	0.27	9.44	0.55
ΔZ	1200	1200	0	1218.1	−18.1	1235.9	−35.9
α	1	1	0	1.01	−0.01	1.03	−0.03
		β = 0.08 γ = 0 .08		β = 0.10 γ = 0.10		β = 0.12 γ = 0.12	
Position offset variable	Truth value	Measured value	Error	Measured value	Error	Measured value	Error
ΔX	10	12.34	−2.34	12.95	−2.95	13.56	−3.56
ΔY	10	8.82	1.17	8.49	1.51	8.15	1.84
ΔZ	1200	1270.3	−70.3	1286.9	−86.9	1303.2	−103.2
α	1	1.06	−0.06	1.07	−0.07	1.08	−0.08

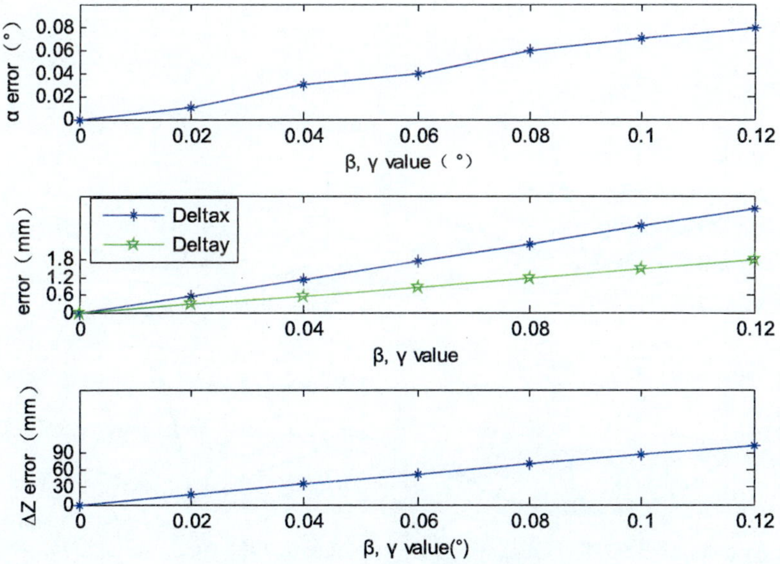

Fig. 3. Error curves of ΔX, ΔY, ΔZ, α

Table 3. The error changes of ΔX(mm), ΔY(mm), ΔZ(mm), α(°).

Position offset variable	Truth value	Measued value	Error	Truth value	Measured value	Error	Truth value	Measured value	Error
ΔX	10	10.86	−0.86	50	56.30	−6.30	100	123.2	−23.2
ΔY	10	9.58	0.42	50	55.13	−5.13	100	122.6	−22.6
ΔZ	1200	1227	−27.1	1200	1337	−137	1200	1475	−275
α	1	1.02	−0.02	1	1.11	−0.11	1	1.22	−0.22
ΔX	−10	−9.47	−0.52	−50	−52.2	2.23	−100	−110	10
ΔY	10	9.47	0.52	50	52.12	−2.12	100	110.4	−10.4
ΔZ	1200	1213	−13	1200	1268	−68	1200	1336	−136
α	1	1.01	−0.01	1	1.05	−0.05	1	1.11	−0.11
ΔX	−10	−9.71	−0.28	−50	−43.7	−6.3	−100	−76.9	−23.1
ΔY	−10	−10.12	0.12	−50	−44.7	−5.28	−100	−77.2	−22.7
ΔZ	1200	1190	9.2	1200	1062	138	1200	925.1	274.9
α	1	0.99	0.01	1	0.88	0.11	1	0.77	0.23
ΔX	10	10.49	−0.49	50	47.65	2.34	100	88.82	11.17
ΔY	−10	−10.50	0.50	−50	−47.7	−2.25	−100	−89.2	−10.8
ΔZ	1200	1185	14	1200	1131	69	1200	1061	139
α	1	0.98	0.02	1	0.94	0.06	1	0.88	0.12

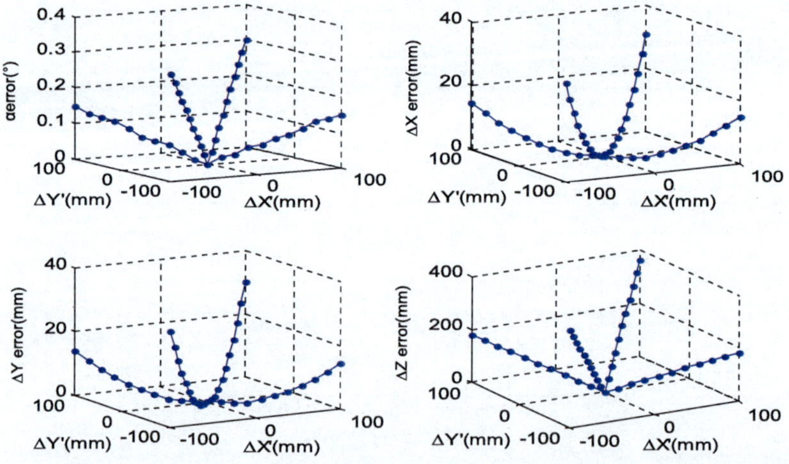

Fig. 4. Errors curves of ΔX, ΔY, ΔZ, and α with the offset variable

From the curve in Fig. 4, we can see that when the inclination angle β and γ are constant, the greater the offset variable between the spreader coordinate and the bottom coordinate of the carriage, the greater the measurement error of ΔX, ΔY, ΔZ and α, and the lower the measurement accuracy. When the accuracy requirement is constant, the measurement range can be determined.

5 Conclusion

In the automatic alignment process of the cantilever crane loading cargo, aiming at the visual measurement characteristics of the non-common field of view of the short-distance large target, this paper proposes a dual-camera non-common field of visual measurement method. By setting the cooperative target under the carrier box, A camera is installed on the cantilever sling to build a pose measurement model when the cantilever sling is parallel to thc bottom of the box, and calculate the coordinate translation (ΔX, ΔY, ΔZ) and rotation around the Z axis between the sling and the bottom of the box Angle α. A simulation experiment is carried out based on MATLAB, and the results show that the dual-camera non-common field-of-view pose measurement method is feasible for short-range pose measurement of large targets.

References

1. Zhao, L., Liu, E., Zhang, W., et al.: Accuracy analysis of monocular three-point position measurement. Opt. Precis. Eng. **22**(5), 1190–1197 (2014)
2. Song, D., Lu, L.: Non-cooperative circle feature pose measurement method based on multi-camera non-common field of view. Infrared Technol. **42**(01), 93–98 (2020)
3. Miao, X., Zhu, F., Hao, Y.: Non-cooperative aircraft pose measurement method for multi-camera non-common field of view. Infrared Laser Eng. **42**(03), 709–715 (2013)
4. Shu, A., Pei, H., Ding, L., Duan, H., Zhou, S.: Binocular visual pose measurement method for space non-cooperative targets. Acta Optica Sinica **40**(17), 107–117 (2020)
5. Lin, T., Jiang, S., Li, R., Ge, Y., Zhou, Y.: Non-cooperative target visual pose measurement and ground verification method. J. Dalian Jiaotong Univ. **41**(03), 34–40 (2020)
6. Shu, A., Pei, H., Duan, H.: Trinocular stereo vision measurement method for space non-cooperative targets. Acta Optica Sinica **41**(06), 163–171 (2021)
7. Wang, X., Cao, Y., Zhou, K.: Single-camera spatial pose measurement method for two-dimensional cooperative targets. Opt. Precis. Eng. **25**(01), 274–280 (2017)
8. Wang, Y., Yuan, F.: A new camera calibration method based on cooperative target attitude measurement. Acta Opt. **36**(11), 180–188 (2016)

Neural Network and Deep Learning

Transformer Combination Weighting Evaluation Model Based on BP Neural Network

Xiaohuan Wu[1,2](✉), Pengcheng Wang[1,2], Liqun Wang[2], Yinfei Xu[2], and Zhongwei Zhao[3]

[1] State Grid Hangzhou Power Supply Company, Hangzhou 310000, Zhejiang, China
[2] Hangzhou Power Equipment Manufacturing Co., Ltd., Hangzhou 310000, Zhejiang, China
[3] School of Information Technology, Zhejiang Gongshang University, Hangzhou 310000, Zhejiang, China

Abstract. The structure of the power transformer is complex, and there are many indicators that characterize the state. It is necessary to integrate multi-dimensional information indicators to get a comprehensive and scientific evaluation conclusion. The analytic hierarchy process is one of the main methods to solve multi-factor decision-making. It uses a hierarchical way to describe more complex problems, which can effectively mine the information between the transformer's multi-dimensional indicators, and combine quantitative and qualitative analysis to ensure The integrity of the data. The importance of power transformer indexes is quite different. In order to obtain more accurate index weights, the influence of expert experience and data distribution must be considered. This paper establishes an analytic hierarchy model that integrates transformer multi-dimensional information, uses fuzzy analytic hierarchy process to obtain subjective weights, uses improved entropy weight method to calculate objective weights, combines subjective and objective weights, and uses BP neural network to learn the nonlinear mapping of state evaluation Relationship, established a transformer state evaluation model based on the combination weighting method.

Keywords: Transformer · Condition assessment · BP neural network

1 Introduction

Power transformer is the key substation equipment in the power system. The safe and stable operation of the transformer is directly related to the safe and stable operation of the power grid. Once a transformer accident causes a large-scale blackout, it will seriously affect residents' lives and economic and social development. Therefore, it is necessary to grasp the operating status of the transformer in real time and take corresponding measures in time and accurately to ensure the safe and stable operation of the transformer [1–3].

At present, domestic and foreign researchers have carried out a lot of research on the evaluation of transformer health status. Literature [4–7] uses the theory of health index combined with a large number of test indicators to evaluate the status and life of transformers, but the weight of each indicator is not easy to determine in engineering

S.-C. Chu et al. (Eds.): ICGEC 2021, LNEE 833, pp. 341–349, 2022.
https://doi.org/10.1007/978-981-16-8430-2_31

applications; Literature [8] uses random fuzzy theory to predict the operating state of the transformer, but each parameter adopts an accurate membership function, ignoring the randomness of the parameters during the transformer test process; Literature [9–11] uses the cloud model to comprehensively evaluate the state of the transformer based on quantitative test data, and does not consider the adverse effects of indicators that are not easy to quantitatively analyze, such as historical faults and family defects of the transformer. In addition, there are literatures that use matter-element models [12], neural networks [13], deep learning [14] and other intelligent algorithms to establish transformer state assessment or fault diagnosis models in combination with transformer routine test data and operating conditions.

In this paper, the fuzzy analytic hierarchy process is used to obtain the subjective weights, the improved entropy weight method is used to calculate the objective weights, and the weights are combined. By learning the nonlinear mapping relationship of the state evaluation through the BP neural network, a transformer based on the combined weighting method is established State evaluation model.

2 Analysis of Transformer Operation State

2.1 Transformer Working Status Classification

The SCADA system [18] was able to collect real-time operating data, but failed to achieve effective evaluation of the operating status. [15]. The division of the transformer operating status should correspond to the maintenance measures under its operating environment. Between the transformer fault state and the healthy state, there is often a transition state that can maintain the operation of the substation equipment but also has a greater operational risk. For this reason, it is necessary to evaluate the multiple operating status of the equipment, considering the multiple monitoring indicators of the equipment, so as to reasonably arrange the status maintenance cycle according to the accurate operating status of the whole machine, and formulate corresponding maintenance measures. According to the electrical equipment state maintenance guidelines, the operating state of the equipment is divided into four states: healthy, good, attention, and fault [8].

2.2 Evaluation Process and System of Transformer Operation State

Using multi-source data such as the status of key components of the transformer and electrical parameters for in-depth fusion analysis can improve the accuracy of the evaluation of the operating status of substation equipment. Integrate historical data collected by the SCADA system and transformer operation and maintenance experience, and select important index parameters that can directly or indirectly reflect the operating status of substation equipment [4–7]. These index parameters have a large correlation with the operating status, and the sampling period is short, which can reflect the operating status of the substation equipment in a timely and accurate manner. According to the target level-project level-index level 3 evaluation levels, the substation equipment status evaluation index system is divided, as shown in Fig. 1.

As shown in Fig. 1, the evaluation target layer, that is, the overall operating state of the transformer, is marked as X.

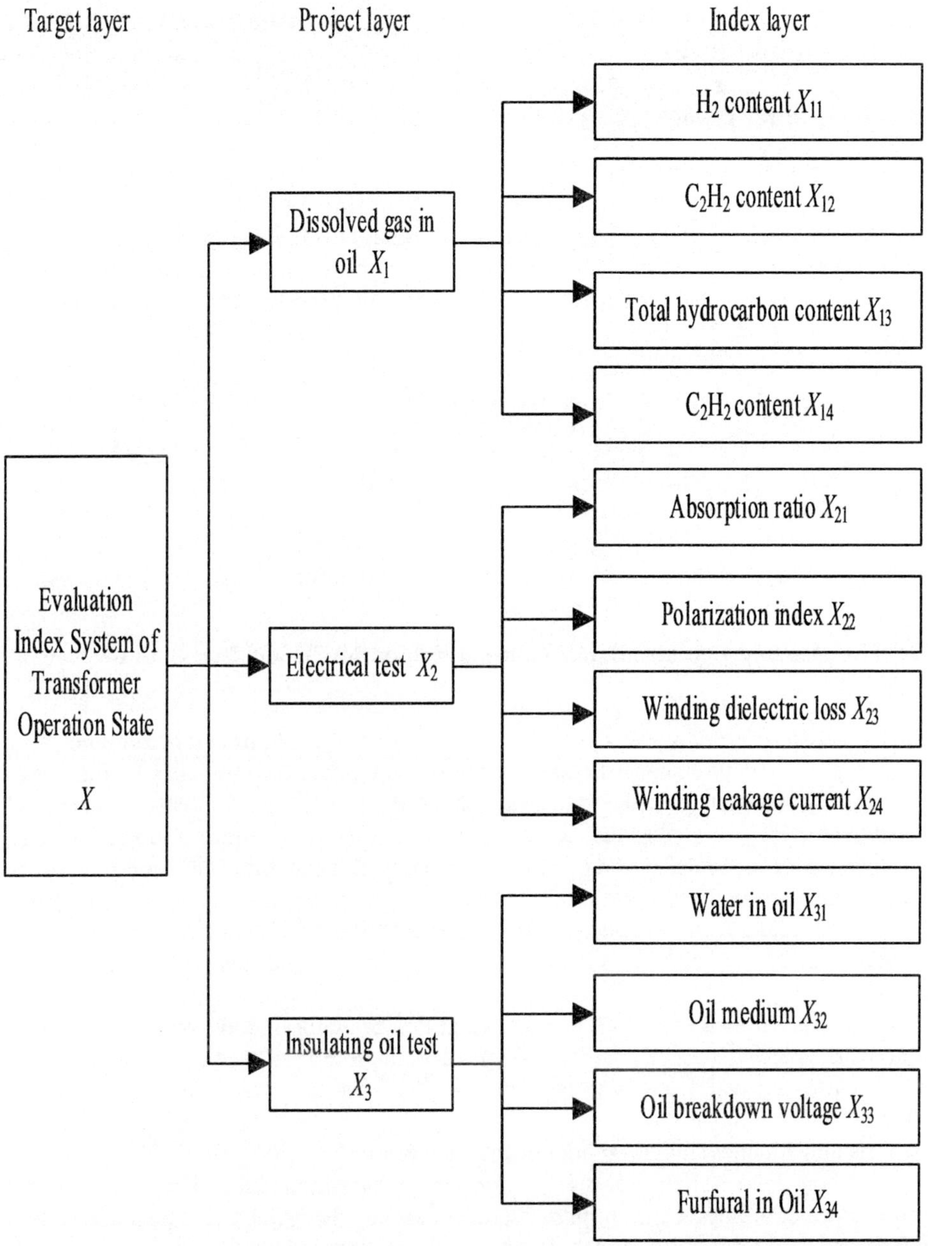

Fig. 1. Evaluation index system of transformer operation status

3 Comprehensive Evaluation Model of Transformer's Overall Operating Status

3.1 BP Neural Network

In the artificial neuron model, the input signal and the corresponding connection weight are first input into the adder for linear accumulation, and the accumulated sum is input into the activation function, and the nonlinear signal is further converted into a linear signal. The neural network can be adjusted by controlling the activation function and the corresponding threshold. The output signal. The function realized by a single neuron can be expressed as:

$$\begin{cases} u_k = \sum_{i=1}^{m} w_{kk} x_i \\ y = f(u_k + b_k) \end{cases} \quad (1)$$

Among them, k represents the number of the neuron, wik represents the weight of the corresponding neuron, xi represents the i-th dimension of the input signal, m represents the dimension of the input signal, uk represents the sum of the product of the input signal and the weight, yk is the output signal, f is the activation function.

3.2 Transformer Combination Weighting Evaluation Model Based on BP Neural Network

Fuzzy Analytic Hierarchy Process can simplify complex problems and recombine them according to the hierarchical structure to effectively improve the efficiency of analyzing problems. However, in the process of establishing the fuzzy judgment matrix, the restriction of the expert level makes the calculated weights very one-sided and subjective. The entropy method calculates based on objective data, which can mine the amount of information contained in the database, but lacks horizontal comparison between indicators, requires high data accuracy, and is easily affected by data distribution, and does not consider expert experience It is also easy to make the weight deviate from the actual situation.

Therefore, a transformer combination weighting evaluation model based on BP neural network is proposed, which not only considers the importance of expert experience, but also fully excavates database information. The specific steps of the model are as follows:

① Establish the evaluation model of the transformer

The status characteristics of the transformer can be mined from many aspects, such as test information, operation monitoring information, environmental status, and so on. Four factors including oil chromatographic analysis electrical test test information, oil chemical test information and operating parameters are selected, including a layered evaluation system composed of 20 bottom-layer indicators.

② Data normalization

Through the overall analysis of the object to be evaluated, the final target layer is determined, and then the influencing factors are decomposed into the criterion layer and the scheme layer, and the hierarchical analysis diagram of the object is established.

Based on the importance of each element in the scheme layer relative to the criterion layer, the judgment matrix R is quantified by fuzzy scale as:

$$R = \begin{bmatrix} r_{11} & r_{12} & \cdots & r_{1n} \\ r_{21} & r_{22} & \cdots & r_{2n} \\ \vdots & \vdots & r_{ij} & \vdots \\ r_{n1} & r_{n2} & \cdots & r_{mn} \end{bmatrix} \tag{2}$$

Among them: where rij represents the fuzzy scale of the i element relative to the j element.Judgment of the consistency of the fuzzy judgment matrix:

$$l_{ij} = r_{ik} - r_{jk} + 0.5 \tag{3}$$

Calculation of indicator weights. There are many ways to calculate weights. The more common ones are the line and normalization method, the square root method, the factor relationship method, etc., because the factor relationship method can clearly distinguish the importance of each index, and the maximum and minimum weights have the largest gap. Therefore, this article chooses the factor relation method to obtain the weight. The calculation formula of subjective weight is:

$$v_i = \frac{1}{n} - \frac{1}{2a} + \frac{1}{na}\sum\nolimits_{k-1}^{n} r_{ik} i = 1, 2, \cdots, n \tag{4}$$

$$\left|v_i - v_j\right| \le \frac{1}{n-1} \tag{5}$$

Among them, n represents the dimension of the indicator; a represents the weight difference factor. The larger the value, the more obvious the difference in the weight of each indicator. This article takes (n−1)/2; vi represents the weight of the i-th indicator.

From this, the subjective weight of the project layer (oil chromatographic analysis, electrical test test information, oil chemical test information, and operating parameters) can be obtained $v = (v_1, v_2, v_3, v_4)$.

③ Improve the entropy weight method to determine the objective weight

The specific steps of entropy method to calculate objective weight are as follows:

$$X = \begin{bmatrix} x_{11} & x_{12} & \cdots & x_{1n} \\ x_{21} & x_{22} & \cdots & x_{2n} \\ \vdots & \vdots & x_{ij} & \vdots \\ x_{m1} & x_{m2} & \cdots & x_{mn} \end{bmatrix} \tag{6}$$

Among them, Xij represents the j index in the i subsystem in the evaluation object matrix X, m represents the number of test objects, and n represents the dimension of the index.

Calculate the characteristic weight of a single indicator:

$$f_{ij} = \frac{x_{ij}}{\sum_{i=1}^{m} x_{ij}} \tag{7}$$

Calculate the entropy value of each indicator:

$$H_j = -\frac{1}{\ln m}\left(\sum\nolimits_{i-1}^{m} f_{ij}\ln f_{ij}\right) \tag{8}$$

Among them, when $f_{ij} = 0$, let $\ln f_{ij} = 0$.

According to the entropy value, the traditional formula for calculating weight is:

$$w_{0j} = \frac{1 - H_j}{n - \sum_{j=1}^{n} H_j} \tag{9}$$

Among them, $0 \leq w_{0j} \leq 1$, $\sum_{j=1}^{n} w_{0j} = 1$.

④ Calculate the combined weight

$$\alpha_{11} = \frac{v_1 w_{11}}{v_1 \sum_{k=1}^{5} w_{1k} + v_2 \sum_{k=1}^{5} w_{2k} + v_3 \sum_{k=1}^{6} w_{3k} + v_4 \sum_{k=1}^{4} w_{4k}} \tag{10}$$

Among them, α_{11} represents the combined weight of the indicators, v_1 represents the weight of the first item in the project layer to the target layer, and w_{11} represents the objective weight of the first indicator of the first project.

⑤ Calculate the overall deterioration degree of the transformer:

$$F = \sum\nolimits_{i=1}^{m} \sum\nolimits_{j=1}^{n} x_{ij}\alpha_{ij} \tag{11}$$

Among them, F represents the overall deterioration of the transformer.

⑥ Transformer status classification and maintenance strategy formulation

This article divides the status of the transformer into four levels: normal, attention, abnormal, and severe. 0–0.2 is the normal state, the indicator is relatively stable and does not exceed the attention value; 0.2–0.4 is; 0.4–0.7 abnormal state, the indicator has approached or slightly exceeded the attention value attention state, and the indicator change is developing in the direction of approaching the attention value; 0.7–1.0 Severe state, the state severely exceeds the attention value.

⑦ Establish a combined weighting evaluation model of BP neural network

In the actual operation and maintenance process, the calculation of the combined weight of the transformer requires many steps. In order to reduce the calculation time and optimize the state evaluation process, the BP neural network is used to deal with nonlinear problems and establish a neural equivalent to the combined weight evaluation method. Network model:

Calculate the subjective weights according to the steps of the fuzzy analytic hierarchy process, use the steps of the improved entropy weight method to calculate the objective weights, calculate the degradation degree, and repeat the labeling of the samples in the sample set many times; Initialize the BP neural network, and input the multiple labeled samples obtained in step 1 into the neural network for training. After multiple iterations, make it reach the specified error. It can be considered that the network can be equivalent to the combined weight calculation process; Input the transformer index to be evaluated into the trained neural network, and according to the output transformer degradation degree, the state level of the transformer can be obtained according to 5), and the corresponding maintenance strategy can be further developed.

4 Analysis on the Evaluation Effect of Transformer's Overall Operation State

In order to verify the effectiveness of the proposed fuzzy comprehensive evaluation method for transformer operating status, an oil-immersed transformer in a 330 kV substation was used for evaluation. The measured data of the 12 key indicators are shown in Table 1, and the data is normalized through the above indicator processing process. The operating state data of the 330 kV transformer from T1 to T7 is verified as a state parameter to be evaluated, during which the transformer develops from a normal state to a fault state. It is known that the transformer will be shut down for maintenance at T7. The specific transformer status index data is shown in Table 1.

Table 1. Transformer state index data

Index parameter	State indicator parameter collection time						
	T_1	T_2	T_3	T_4	T_5	T_6	T_7
$X_{11}/\mu L{\cdot}L^{-1}$	168.80	171.30	184.54	172.46	179.33	169.20	167.80
$X_{12}/\mu L{\cdot}L^{-1}$	0.22	0.22	0.18	0.23	0.18	0.26	0.24
$X_{13}/\mu L{\cdot}L^{-1}$	100.05	125.78	116.90	123.28	123.51	114.80	104.49
$X_{14}/\mu L{\cdot}L^{-1}$	9.04	9.01	8.84	7.56	10.37	9.45	8.84
$X_{21}/\%$	1.40	1.01	1.75	1.41	1.71	1.37	1.45
X_{22}	1.80	1.91	2.25	2.23	2.31	1.87	1.69
$X_{23}/\%$	0.17	0.34	0.16	0.28	0.20	0.35	0.43
X_{24}/A	27.70	24.60	23.50	26.40	31.20	25.30	24.90
$X_{31}/mg \cdot L^{-1}$	9.40	6.90	9.60	15.20	12.00	16.70	21.70
$X_{32}/\%$	1.59	2.40	2.75	2.50	2.72	2.37	2.30
X_{33}/kV	53.00	50.10	49.50	47.40	56.10	53.70	53.20
$X_{34}/mg \cdot L^{-1}$	0.29	0.38	0.37	0.40	0.33	0.29	0.26

Table 2. BP neural network training parameters

Parameter	Value
Maximum number of training	1000
Number of hidden layer (middle layer) neurons	12
Network learning rate	0.001
Training target error	0.5×10^{-3}

The experimental results show that the absolute value of the error between the output of the BP network and the expected degradation degree is within −3.2%, and the correct rate of the state and the expected state predicted by the BP network reaches 100%. It can be seen that the error does not affect the classification of the state level, which is within the acceptable range. Therefore, the BP network model can realize the transformer state evaluation function based on combination weighting, and can achieve a high accuracy rate, and can be used for transformer state evaluation (Table 2).

The deterioration index is input into the trained BP neural network, and the deterioration degree is 0.5355, which is judged as an abnormal state, while the fuzzy analytic hierarchy process and the entropy method obtain the results, which are attention and abnormality respectively. In order to evaluate the status of the transformer in the example, after the operation and maintenance personnel performed on-site power outage overhaul of the equipment, it was found that some windings of the transformer were squeezed out of the wire cake, resulting in uneven distance between the wire cakes and partial winding deformation. Therefore, the transformer should be evaluated as an abnormal level, and maintenance should be carried out in accordance with the corresponding state inspection requirements. The state evaluation method in this article can faithfully reflect the health of the equipment and meet the validity and accuracy of the state evaluation (Table 3).

Table 3. Comparative experiment

Method	Deterioration	State assessment
Fuzzy hierarchical analysis	0.48	Attention
Entropy method	0.55	Abnormal

5 Conclusion

This paper establishes a transformer stratification evaluation model, which considers the three aspects of dissolved gas in oil, electrical test and insulating oil test, but other methods do not consider these three aspects at the same time.Discussed the weight problem in the evaluation process, considering the importance of expert experience, using the fuzzy analytic hierarchy process to obtain the subjective weight, and optimize the solution process. Considering objectivity and reliability, a method of fusing the entropy method theory to obtain the combined weight is designed and improved, so that when the entropy value approaches 1, the entropy weight can still obtain a weight that conforms to the actual situation. Finally, an evaluation model based on BP neural network is established, which simplifies the calculation process and improves the practicability. Through the case analysis, the transformer state evaluation based on the combined weight designed in this paper can effectively evaluate the state of the equipment.

References

1. Zhou, L., Hu, T.: Multifactorial condition assessment for power transformers. IET Gener. Transm. Distrib. **14**(9), 1607–1615 (2020)

2. Eke, S., Clerc, G., Aka-Ngnui, T., et al.: Transformer condition assessment using fuzzy C-means clustering techniques. IEEE Electr. Insul. Mag. **35**(2), 47–55 (2019)
3. Velasquez, R.M.A., Lara, J.V.M.: Health index for transformer condition assessment. IEEE Lat. Am. Trans. **16**(12), 2843–2849 (2018)
4. Wang, N., Zhao, F.: An assessment of the condition of distribution network equipment based on large data fuzzy decision-making. Energies **13**(1), 197 (2020)
5. Qi, B., Zhang, P., Rong, Z., et al.: Differentiated warning rule of power transformer health status based on big data mining. Int. J. Electr. Power Energy Syst. **121**, 106150 (2020)
6. He, D., Wang, R.: State assessment of oil-paper insulation based on fuzzy rough sets. In: 2019 15th International Conference on Computational Intelligence and Security (CIS), pp. 438–441. IEEE (2019)
7. Liu, Q., Zhang, D., Shang, Y., et al.: Evaluation of transformer state based on the life cycle. In: 2018 International Conference on Power System Technology (POWERCON), pp. 4603–4610. IEEE (2018)
8. Cheng, L., Yu, T.: Dissolved gas analysis principle-based intelligent approaches to fault diagnosis and decision making for large oil-immersed power transformers: a survey. Energies **11**(4), 913 (2018)
9. Gong, R., Li, S., Peng, W.: research on multi-attribute decision-making in condition-based maintenance for power transformers based on cloud and kernel vector space models. Energies **13**(22), 5948 (2020)
10. Bagheri, M., Zollanvari, A., Nezhivenko, S.: Transformer fault condition prognosis using vibration signals over cloud environment. IEEE Access **6**, 9862–9874 (2018)
11. Bagheri, M., Nezhivenko, S., Naderi, M.S., et al.: A new vibration analysis approach for transformer fault prognosis over cloud environment. Int. J. Electr. Power Energy Syst. **100**, 104–116 (2018)
12. Chen, Q., Cai, J., Ye, R.: comprehensive assessment of transformer insulation condition based on extension theory. IOP Conf. Series Earth Environ. Sci. **218**(1), 012069 (2019)
13. Kadim, E.J., Azis, N., Jasni, J., et al.: Transformers health index assessment based on neural-fuzzy network. Energies **11**(4), 710 (2018)
14. Hong, K., Jin, M., Huang, H.: Transformer winding fault diagnosis using vibration image and deep learning. IEEE Trans. Power Delivery **36**(2), 376–685 (2020)
15. Zhang, H., Deng, Y.: Entropy measure for orderable sets. Inf. Sci. **561**, 141–151 (2021)
16. Lyu, H.M., Sun, W.J., Shen, S.L., et al.: Risk assessment using a new consulting process in fuzzy AHP. J. Constr. Eng. Manag. **146**(3), 04019112 (2020)
17. Yang, Z., Chang, J.: Interval-valued Pythagorean normal fuzzy information aggregation operators for multi-attribute decision making. IEEE Access **8**, 51295–51314 (2020)
18. Jin, X., Xu, Z., Sun, Y., Shan, J.: Wind turbine online operation status monitoring based on SCADA data analysis and sparse self-encoding neural network. Acta Solar Energy **42**(06), 321–328 (2021)

Research on Target Recognition Algorithm Based on Improved Faster-RCNN

Fangxiao Cheng, Borui Yin, and Ming Chu(✉)

College of Electrical and Electronic Engineering, Changchun University of Technology, Changchun 130000, China

Abstract. Aiming at the problem of missed detection in target detection under special conditions such as occlusion, an improved Faster-RCNN target detection algorithm is proposed. First replace the VGG16 network with Resnet101 and optimize it. Secondly, improve the RPN module, extract features from feature layers with richer semantics, and use two 3×3 convolutions to obtain 5×5 convolution effects, while reducing training parameters and improving the detection accuracy of occluded targets; finally Improve the RoI pooling to RoI Align, cancel the quantization operation, and use the bilinear interpolation method to obtain the image values with the coordinates as floating-point numbers, thereby transforming the feature aggregation process into a continuous operation. The actual measurement results show that the algorithm is suitable for detecting occluded objects in different scenes and has good robustness.

Keywords: Target recognition · Faster-RCNN network · RPN · RoI align

1 Introduction

Target detection is a kind of image segmentation based on target geometric and statistical characteristics, and it is one of the core issues in image understanding. It combines target segmentation and recognition into one, and recognizes the target and locates the coordinate information of the target in the video or static image. Its accuracy and real-time are an important ability of the entire system. Target detection integrates image processing, pattern recognition, artificial intelligence and other technologies. In recent years, it has a wide range of applications in intelligent transportation systems, intelligent monitoring systems, military target detection, and surgical instrument positioning in medical navigation operations. Especially in complex scenes, factors such as illumination, occlusion, and angle changes will cause image changes. When the target is processed in real time, automatic target extraction and recognition are particularly important.

Traditional target detection algorithms generally require artificial design of sample features. In recent years, with the development of computer performance and hardware, the rapid rise of deep learning, especially represented by the target detection framework based on convolutional neural networks, combines computer vision and deep learning to improve the accuracy of target detection. The existing target detection algorithms can be divided into two categories: two-stage detection algorithms and single-stage detection

S.-C. Chu et al. (Eds.): ICGEC 2021, LNEE 833, pp. 350–359, 2022.
https://doi.org/10.1007/978-981-16-8430-2_32

algorithms. One is RCNN [1, 2], SPPnet [3], Fast-RCNN [4] and Faster-RCNN [5] is a candidate box-based method; the other is a regression-based method represented by YOLO [6] and SSD [7].

In the early research [4, 8], the object proposal was generated by an external module [9]. Later, after development, the two-level detection algorithm became more accurate and efficient. R-CNN extracts each candidate region and then inputs it into the CNN classifier to achieve feature extraction, which requires a lot of repetitive work. SPP-NET [3] and Fast-RCNN introduced the idea of regional feature extraction, which significantly accelerated the speed of the entire detector, and at the same time adjusted the input of the candidate box to be large or small, thereby avoiding the problem of RCNN target deformation. Proposed in 2016 The Faster-RCNN detection algorithm introduced the region proposal network RPN (region proposal network) [5] for the problem of large amount of calculation in the candidate region. It uses a fully convolutional network to map the anchor of each sliding window to a low-dimensional feature. This improvement solves the time-consuming problem of CNN series algorithms in regional recommendation and is widely used in two-level detection algorithms. Existing deep learning algorithms cannot effectively distinguish the target from the background, and are prone to miss detection and false detection. In order to solve the existing problems and better realize the target detection, this paper proposes an improved Faster-RCNN occluded object detection method. Replace the VGG16 network with Resnet101, improve the RPN module, improve RoI pooling to RoI Align, and transform the feature aggregation process into a continuous operation. This makes the algorithm in this paper suitable for the detection of occluded objects in different scenarios, and has good robustness.

2 Faster-RCNN Algorithm Principle

The RCNN [1, 2] method combines the selective search area suggestion algorithm with the neural network. Firstly, the selective search is used on the input image to generate the area suggestion, and extract the target feature vector from the area suggestion. The extracted feature vectors are classified by SVM. Compared with traditional methods, the accuracy of RCNN is significantly improved. However, the RCNN algorithm scales the region proposal to a fixed scale, destroys the image information structure, and causes local detail loss. To solve the problem of RCNN, He [3] et al. proposed the Spatial Pyramid Pooling (SPP) method. The SPP can produce a fixed-dimensional output for any dimensional input. The SPP layer divides the input features into local spatial regions of several scales. Pooling dimensionality reduction in these regions can make the output features of fixed dimensions. Although SPPnet can effectively deal with the problem of loss of details, the training of the SPPnet detection algorithm is carried out in stages and requires feature extraction for each candidate region, which increases network parameters and complicates the training process. So that in the feature extraction stage, the cost of time and space is relatively large.

Aiming at the shortcomings of SPPnet, Girshick [4] et al. proposed an end-to-end training Fast-RCNN target detection algorithm. This algorithm integrates target classification and bounding box regression instead of the SVM classifier. This network does not need feature dumping work, realizes the convolution feature sharing of classification and regression, and saves storage overhead.

However, Fast-RCNN must perform a selective search on each input image to obtain a region suggestion, and this step is time-consuming. For this reason, Ren [5] and others creatively proposed the Region Proposal Network (RPN), integrating RPN and the basic network Fast R-CNN into a unified detection structure to share convolutional features, forming the Faster-RCNN target The detection framework and network structure are shown in Fig. 1.

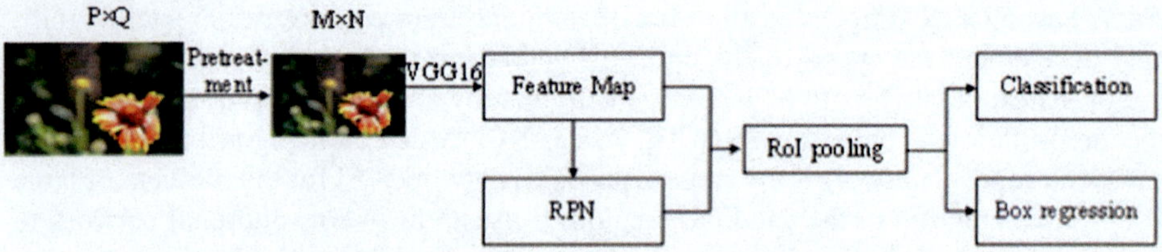

Fig. 1. Faster RCNN network structure

In terms of feature extraction, Faster-RCNN directly uses the convolution part of the VGG16 network framework; in terms of target classification and location refinement, Faster-RCNN trains a classification and location refinement network, and used RPN to perform the selection area. For detailed target classification and location refinement, the algorithm first performs RoI pooling of the candidate area, unifies the candidate area to the same size and connects to the fully connected network, and trains two fully connected branches to perform target classification (Classification) and location refinement (Box regression).

The RPN network is used to replace the time-consuming selective search algorithm to generate high-quality candidate regions. While improving the detection performance of the entire network, the target detection speed is significantly improved. For the input, convs and pooling are first performed to extract the feature map of the sample, and then the detection network and the RPN network share the picture of the full image, which effectively solves the problem of a large amount of calculation of the candidate area and time-consuming. In fact, Faster-RCNN only uses the last layer depth feature, which causes the detection framework to not use the all information. Under special conditions such as occlusion, the detection of targets is prone to missed detection. This affects the performance of the entire detection framework.

3 Improved Faster-RCNN Network

Since Faster-RCNN only uses the last layer depth feature as the shared feature information with RPN, the detection framework cannot use all feature information. Therefore, under special conditions such as occlusion, the detection of the target is prone to missed detection, which reduces the detection performance of the entire framework. Aiming at the missed detection, this paper proposes an improved Faster-RCNN target detection algorithm. Replace the VGG16 network with Resnet101, improve the RPN module, improve RoI pooling to RoI Aglign, and transform the process into a serial operation. This makes the algorithm in this paper suitable for object detection under different occlusion conditions, and has good robustness, as shown in Fig. 2.

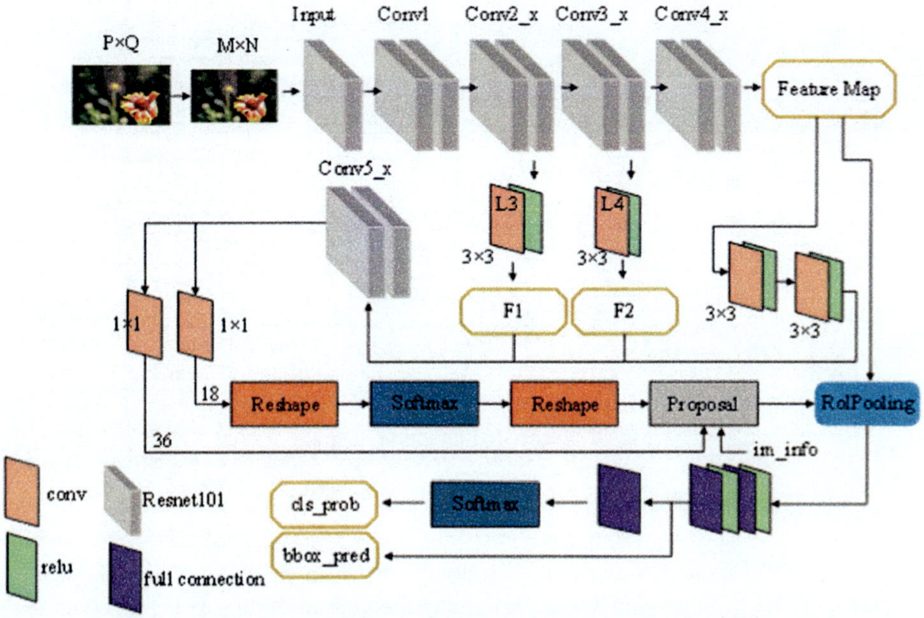

Fig. 2. Improve faster-RCNN network structure diagram

3.1 Resnet 101 Backbone Network

The VGG16 network is all composed of a 3 × 3 convolution kernel and a 2 × 2 pooling layer. Through multiple Conv operations in each block, the amount of parameters is reduced while retaining the same receptive field. By deepening the network structure, the network can extract more abstract high-order features, which is of great help to the improvement of accuracy. However, since most of the parameters in VGG16 come from the first fully connected layer, the disadvantage is that as the network structure deepens, more parameters are used, which leads to more memory usage and consumes a lot of computing resources.

In this paper, the Resnet 101 network is used instead of the VGG16 network. The convolutional layer reduces the training parameters of the network and the time required for network training through weight sharing. Resnet 101 is mainly composed of convolutional neural network. By using the multi-layer convolutional layer and the pooling layer to make the depth feature more extracted of the input image, the output is used for the classification and regression of the subsequent RPN network and Faster RCNN target detection network. Its structure is shown in Fig. 3 Show.

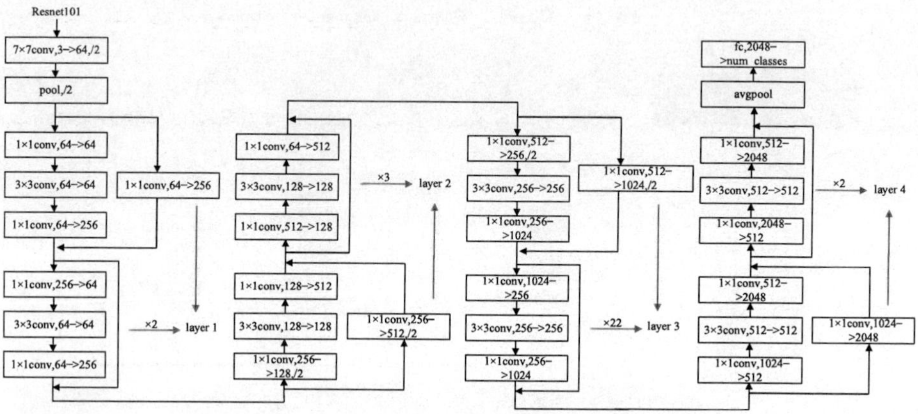

Fig. 3. Resnet 101 structure diagram

3.2 Improved RPN Module

Traditional RPN uses a 3 × 3 sliding window to traverse the feature map output by the convolutional layer to generate anchors, and then cut and filter them, and then use the fully connected layer to make a binary classification for each anchor to determine whether it is foreground or Background, meanwhile, bounding box regression corrects anchors for preliminary regression, and finally outputs more accurate proposals (rois). However, the traditional RPN only uses the deepest feature layer as the candidate frame extraction layer, which will cause a lot of loss of image detail information in the deepest feature layer after multiple downsampling, and a large loss of target feature information, which will bring a certain target positioning error.. The structure of the RPN is shown in Fig. 4(a).

Because in practical applications, the target object has different scale changes. The traditional RPN network only uses a 3 × 3 sliding window to traverse the feature map, and the receptive field is not enough, which limits the ability of model target detection. In order to solve this problem, in the original RPN network, the L2 and L3 feature layers with richer details are used for further feature extraction and L2 regularization with 3 × 3 convolution kernels, and two new feature layers L^c 2 and L^c 3 are obtained. Then, the feature layer L5 is convolved by the 3 × 3 convolution kernel to obtain the feature layer L^c 5. The new feature layer is obtained by information fusion of L^c 5 and the feature layers L^c 3 and L^c 4 as the final candidate frame extraction layer. The depth is the same as that of the feature layer L^c 5. This method can use the image detail information in the lower feature layer and combine the abstract feature information in the higher feature layer without increasing the thickness of the candidate frame extraction layer to improve target recognition And the reliability of positioning. A 3 × 3 sliding window is added, which is connected in series with the original 3 × 3 convolution kernel to convolve the feature map, which is equivalent to the effect of a 5 × 5 convolution kernel receptive field. Through this series connection, the depth and performance of the network can be improved. Two sliding windows are concatenated to obtain a shared feature layer. The shared layer performs two 1 × 1 convolution operations, one for RPN classification loss operation, and the other for bounding box loss operation. The improved structure is shown in Fig. 4 Show.

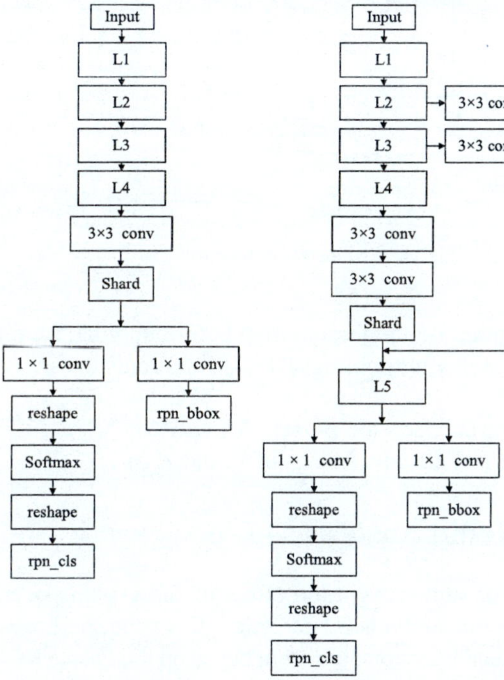

(a) Traditional RPN module structure (b) Improved RPN module structure

Fig. 4. Comparison of RPN module structure before and after improvement

3.3 Improved RoI

This article uses RoI Align instead of RoI Pooling. By canceling the quantization operation, the bilinear interpolation method is used to get the image value whose coordinates are floating-point numbers. First, traverse each candidate area, keeping the floating-point number boundary without quantization, and then divide the candidate area into k × k units. Calculate and fix four coordinate positions in each unit, calculate the values of these four positions by bilinear interpolation, and then perform the maximum pooling operation. The process is shown in Fig. 5:

Using RoI Align instead of RoI Pooling, the number of traversal sampling points is less than RoI Pooling, but better performance can be obtained. The back propagation formula of RoI Align is

$$\frac{\partial L}{\partial x_i} = \sum_r \sum_j \left[d\left(i, i^*(r,j)\right) < 1\right](1 - \Delta h)(1 - \Delta w)\frac{\partial L}{\partial y_{rj}} \tag{1}$$

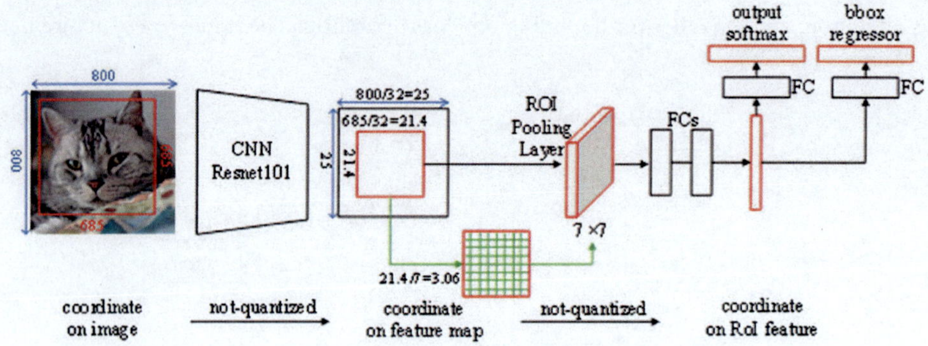

Fig. 5. RoI Align flowchart

x_i represents the pixel point on the feature map before pooling; y_{rj} represents the j point of the r-th candidate area after pooling; $i^*(r,j)$ represents the source of the pixel value of point y_{rj};$x_i^*(r,j)$ is the coordinate position of a floating point number; $d(\cdot)$ represents two The distance between the points, Δh and Δw represent the difference between the horizontal and vertical coordinates of x_i and $x_i^*(r,j)$.

4 Experiment and Result Analysis

Aiming at the poor detection and recognition effect of the existing occluded objects, which are prone to misdetection and missed detection, this paper proposes an improved Faster-RCNN occluded object detection algorithm based on the Faster-RCNN. To verify the performance, firstly, experimental results under several basic networks are compared. Secondly, the improved part of the Faster-RCNN network is compared. Finally, several common obstacle categories are tested and the experimental results are obtained.

4.1 Data Set

This article uses the MSCOCO data set for experimental comparative analysis. This data set aims at scene understanding and is mainly intercepted from complex daily scenes. The target in the image is calibrated by precise segmentation. The image includes 91 types of targets, 328,000 images and 2,500,000 labels.

4.2 Evaluation Index

The article will use single class average accuracy and frame rate per second to assess indicators for target detection methods, and draw the Recall Precision (recall rate-precision) curve to evaluate the performance of the detection method, surrounded by RP curve This area is the single class average accuracy value. The larger the area, the greater the single class average accuracy value and the higher the detection accuracy. The calculation formulas for single class average accuracy, Recall, and Precision are as follows:

$$Precision = \frac{TP_K}{TP_K + FP_K} \quad (2)$$

$$Recall_k = \frac{TP_K}{TP_K + FN_K} \quad (3)$$

Among them, TP is the possibility of detecting a positive specimen, FP is the possibility of detecting a wrong sample, FN is the possibility of not detecting a correct sample, and TN is the possibility of detecting a negative specimen.

4.3 Training Parameters

The number of iterations is set to 100,000,the momentum is set to 0.9, the initial learning rate is set to 0.001, the number of iterations is set to 100,000. When the iterations is 30000 and 60000, the learning rate is resized to 0.0001, 0.00001. The loss convergence of the algorithm in this paper on the COCO data set is shown in Fig. 6.

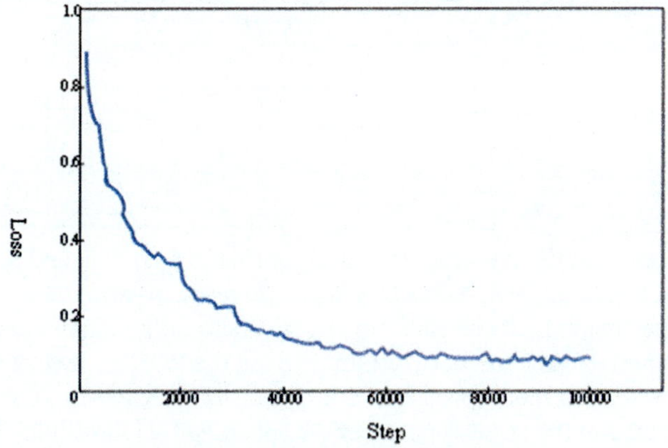

Fig. 6. Loss function graph

4.4 Analysis of Results

Select the target detection method: Faster-RCNN, YOLO v3, SSD, R-FCND, Cascade RCNN and the algorithm of this paper for experimental comparison. YOLO v3 is a single-stage based on FPN, because of its detection characteristics, it has a faster detection speed; SSD introduces a single-stage detector, and does not use RPN and Pooling Operation, so it is more accurate and faster than the previous algorithm YOLO.

Experiment 1 compared whether the skeleton network in Faster-RCNN replaces VGG16 with Resnet101, whether to use the improved RPN module, and whether to replace RoI Pooling with RoI Align. The experimental results are shown in Table 1.

The results are the use of the Resnet101 skeleton improves the detection accuracy by 3.23%. The improved RPN module increased the detection accuracy by 2.34%; the improved RoI Align increased the detection accuracy by 2.67%.

It shows comparative experiments on the COCO set to verify the effectiveness of the proposed innovations. The detection results on COCO are shown in Table 2.

Table 1. Influence of different modules on experimental results.

	mAP%	MR%
Faster-RCNN	84.17	11.53
Faster-RCNN + Resnet101	86.89	10.94
Faster-RCNN + improved RPN	86.14	10.78
Faster-RCNN + RoI Align	86.42	11.13
Ours	89.12	10.31

Table 2. Performance comparison of algorithms

	mAP%	MR%	FPS/s
Faster-RCNN	58.85	52.91	32
YOLOv3	59.31	43.82	41
SSD	61.64	39.19	22
Ours	63.94	32.71	33

In the COCO data set, the AP value of this algorithm is 8.64, 7.8, and 3.7 percentage points higher than Faster-RCNN, YOLO v3 and SSD respectively.

The above experimental data verifies the effectiveness of the innovations in this article. Using the Resnet101 skeleton network, improving the RPN module, and improving RoI Align may strengthen the impact. The PR (Precision-Recall) curve can reflect the accuracy and coverage of the target type under the specified IoU threshold. The PR curve is shown in Fig. 7.

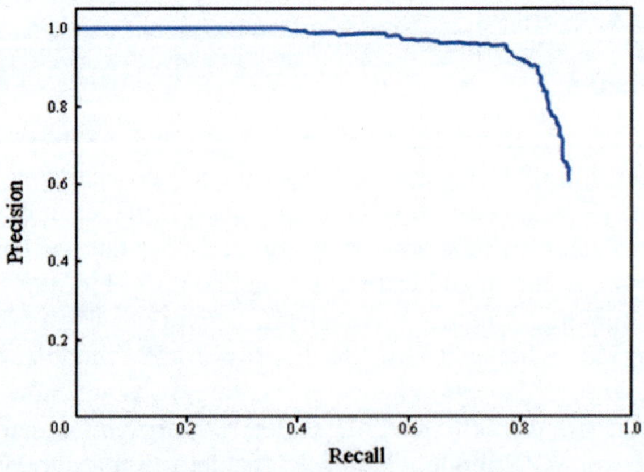

Fig. 7. COCO data set PR curve graph

5 Concluding

This paper proposes an improved Faster-RCNN target detection algorithm. The skeleton network is optimized by replacing the VGG16 network with Resnet101. Improve the RPN module, extract features from feature layers with richer semantics, and use two 3×3 convolutions to obtain 5×5 convolution effects, while reducing training parameters and improving the detection accuracy of occluded targets. In order to cancel the quantization operation, the RoI pooling is improved to RoI Align, and the image value with the coordinates of floating point numbers is obtained using the bilinear interpolation method, thereby transforming the feature aggregation process into a continuous operation. This paper improves the Faster-RCNN algorithm to improve the accuracy of target detection under conditions such as occlusion.

References

1. Girshick, R., Donahue, J., Darrell, T., et al.: Rich feature hierarchies for accurate object detection and semantic segmentation. In: Proceedings of the IEEE Conference on Computer Vision and Pattern Recognition, pp. 580–587(2014)
2. Li, J., Wong, H.C., Lo, S.L., et al.: Multiple object detection by deformable part- based model and R- CNN. IEEE Sig. Process. Lett. **25**(99), 288–292 (2018)
3. He, K.M., Zhang, X.Y., Ren, S.Q., et al.: Spatial pyramid pooling in deep convolutional networks for visual recognition. IEEE Trans. Patt. Anal. Mach. Intell. **37**(9), 1904–1916 (2015)
4. Girshick, R.: Fast R-CNN. In: Proceedings of the 2015 IEEE International Conference on Computer Vision, Santiago, Chile, pp. 1440–1448. IEEE (2015)
5. Ren, S.Q., He, K.M., Girshick, R., et al.: Faster R- CNN: towards real-time object detection with region proposal networks. IEEE Trans. Patt. Anal. Mach. Intell. **39**(6), 1137–1149 (2017)
6. Redmon, J., Divvala, S., Girshick, R., et al.: You only look once: unified, real-time object detection. In: IEEE Computer Vision and Pattern Recognition, Las Vegas, Nevada, pp. 779–788 (2016)
7. Liu, W., Anguelov, D., Erhan, D., et al.: SSD: single shot multibox detector. In: European Conference on Computer Vision, Amsterdam, The Netherlands, pp. 21–37 (2016)
8. Girshick, R., Donahue, J., Darrell, T., Malik, J.: Rich feature hierarchies for accurate object detection and semantic segmentation. In: 2014 IEEE Conference on Computer Vision and Pattern Recognition, Columbus, pp. 580–587 (2014)
9. Uijlings, J.R.R., vande, S., et al.: Selective search for object recognition. Int. J. Comput. Vis. **104**(2), 154–171 (2013)

Identification Method of Dust Accumulation of Air Cooled Condenser Based on Convolution Neural Learning

Bo Zhao(✉), Hongda Lin, Zhihui Wang, and Jiayue Xu

Northeast Electric Power University, Jilin 132012, China
zhaobo@neepu.edu.cn

Abstract. Compared with the traditional wet cooling system, the air cooling system of power station has the characteristics of water saving, which is widely used in water shortage and dust areas. The surface ash of air cooler reduces the heat exchange efficiency and increases the operation cost of power station. However, the structure of ash deposition detection method based on heat balance is complex and its real-time performance is poor. Aiming at the color and texture features of the air cooler image in the state of ash deposition, a recognition method based on neural network learning is proposed. An image recognition experimental system for ash deposition of air cooler in coal-fired power plant is established, and the thermal image data set is obtained. The image of air cooler with different degree of ash deposition is identified by using neural network, and the corresponding relationship between the image of ash deposition and the loss rate of heat transfer coefficient is analyzed. The results show that the accuracy of the thermal imaging image test set of Alexnet, resnet-50, resnet-101, mobilenetv3(large) models are 59.67%, 83.67%, 63.33%, 73% respectively. This method provides a reference and new idea for the intelligent detection and maintenance technology of air cooling system.

Keywords: Air cooled condenser · Convolution neural network · Ash identification · Heat transfer coefficient

1 Introduction

With the continuous development of society, power resources and water resources have been paid more and more attention, and become the problems that human beings must deal with and solve [1]. The use of fresh water for thermal power generation is second only to irrigation, becoming the second largest user of fresh water in many countries [2]. China's thermal power stations are mainly built in places where coal resources are rich and water resources are scarce [3]. Direct air cooling system has many advantages such as large water saving, but there are still some problems [4]. Because the air cooling system is often arranged in the open air environment, it is easy to deposit ash on the surface of the air-cooled condenser. Ash deposition will not only reduce the heat transfer coefficient of the system, reduce the economy of the unit operation, but also reduce the

S.-C. Chu et al. (Eds.): ICGEC 2021, LNEE 833, pp. 360–369, 2022.
https://doi.org/10.1007/978-981-16-8430-2_33

air flow area in the system, which will increase the back pressure, and cause shutdown in serious cases [5].

At present, the monitoring of ash deposition degree of air cooler mostly depends on the operator's visual judgment, with poor real-time performance and large error [6]. Through the artificial neural network, its excellent calculation ability is suitable for the modeling process of various complex system input-output relations. By using correlation analysis, data fitting and artificial intelligence algorithm, the on-line soft sensor model of ash settlement state is established, and the on-line monitoring method of ash settlement state is developed, which is of great significance to improve air cooling efficiency and realize energy saving and emission reduction [7]. Hamzacebi C. [8] proposed an artificial neural network (ANN) structure for seasonal time series forecasting. The proposed structure considers the seasonal period in time series in order to determine the number of input and output neurons. The model was tested for four real-world time series. The results show that the proposed model is especially convenient when the seasonality in time series is strong. Csiszar O., et al. [9] combined neural network with continuous logic and multi criteria decision-making tools to reduce the black box nature of neural models. Shrikrishna K., et al. [10] use three different methods namely 3-dimensional (3-D) CNN, CNN with convolutional long short-term memory (ConvLSTM) layers and vision transformer that use temporal information along with spatial information for plant accession classification. Time series color images of four different plant accessions of Arabidopsis are given as input to the neural network model which in turn identifies plant accession. Sahin S. O., et al. [11] investigate classification and regression for nonuniformly sampled variable length sequential data and introduce a novel long short-term memory (LSTM) architecture. Chen Q., et al. [12] proposes a new network, LD-GAMP, that integrates a data-driven model, ResNet, into a hand-designed algorithm, GAMP, following an unrolling methodology that unfolds the iterative sparse-signal-recovery algorithm to form an interpretable deep network. Zhang Y., et al. [13] presented a novel spatial-temporal graph inception residual network (STGI-ResNet) for network-based traffic prediction. Li Z., et al. [14] used ResNet's residual module and identity mapping, it can effectively process sequential features after which the outer concatenation function is used for sequential and pairwise features. Han C., et al. [15] propose a method of feature fusion of three block residual neural network (ML RESNET) based on multi lead residual neural network (ML RESNET) and 12 lead ECG feature fusion.

In this paper, Alexnet, resnet50, resnet101, mobilenetv3(large) are used to train the thermal imaging data set of air cooler, and the thermal balance method [16] is used to identify the ash deposition state. The method can improve the real-time performance of ash identification and improve the economic benefits of power plant.

2 Experimental and Methods

2.1 Experimental System

The schematic diagram of ash deposit identification system is shown in Fig. 1.

The heat exchange efficiency decline degree of the air cooler is calculated based on the heat balance method, and the data collected by the sensor is uploaded to the industrial computer for processing. The air cooler model includes five parts: heat exchanger, electric

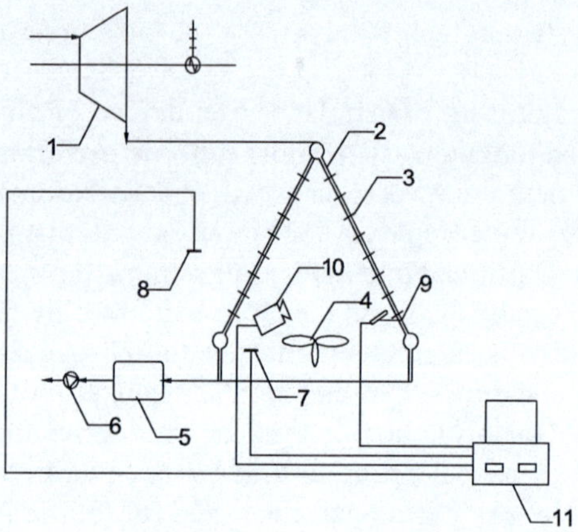

1-turbine; 2-heat exchange tube bundle; 3-fin channel; 4-air cooling fan; 5-condensation tank; 6-condensate pump; 7-entrance air temperature; 8-export air temperature; 9-anemograph; 10-infrared thermal imager; 11-industrial personal computer;

Fig. 1. Schematic diagram of ash deposit identification system

boiler, fan, water pump and water tank.The actual site drawing of the air cooler is shown in Fig. 2(a). The fin tube length of the air cooler is 200 mm, and the base tube width is 19 mm. The schematic diagram of finned tube is shown in Fig. 2(b).

The gray image is shown in Fig. 3. Figure 3(a) shows the finned tube bundle with deep ash deposition, and the heat transfer efficiency is seriously affected. Figure 3(b) shows the finned tube bundle with light ash deposition. The texture and surface temperature of the ash image with different ash deposition degree are obviously different, which can be used as the classification basis.

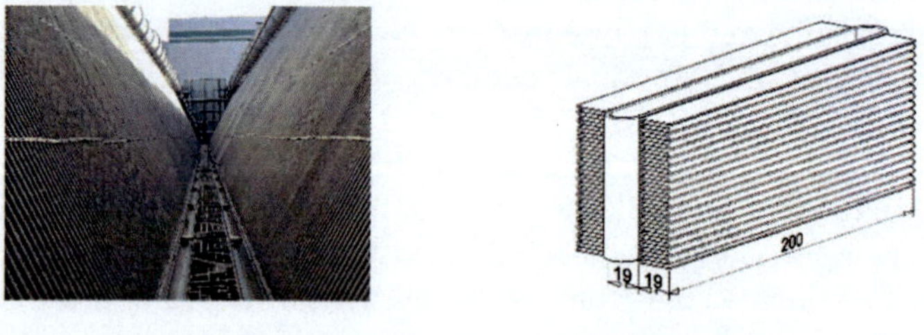

(a) Physical drawing of air cooler (b) Schematic diagram of finned tube

Fig. 2. Structural drawing of air cooler

Based on the principle of heat balance, the heat loss of steam turbine exhaust after passing through the air cooler is equal to the heat taken away by the cooling air of the

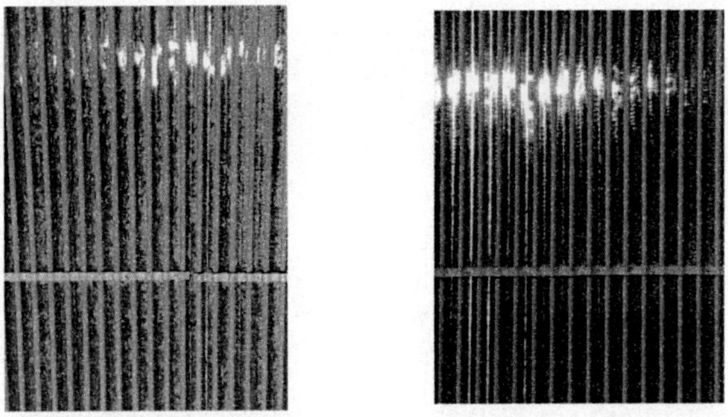

(a) deep ash accumulation (b) light ash accumulation

Fig. 3. Tube bundles with different ash deposition degrees

cooling fan.

$$Q = U \times A \times \Delta T_{\text{lm}} \tag{1}$$

Where Q is the heat load, U is total heat transfer coefficient of air cooler, A is Air side heat transfer area of air cooler, ΔT_{lm} is log mean temperature difference.

The heat load of air cooler is as follows:

$$Q = m_{\text{s}}(h_{\text{s,i}} - h_{\text{c,o}}) = m_{\text{a}} c_{\text{p,a}}(T_{\text{a,i}} - T_{\text{a,o}}) \tag{2}$$

Where m_{s} is steam mass flow rate, $h_{\text{s,i}}$ and $h_{\text{c,o}}$ are Enthalpy of steam at the inlet and condensed steam, m_{a} is Mass flow rate of cold fluid, $c_{\text{p,a}}$ is Specific heat capacity of air, T_{s} is Condensing steam temperature, $T_{\text{a,i}}$ and $T_{\text{a,o}}$ are the temperature of windward and leeward side of heat exchanger.

The logarithmic temperature difference of air cooler is as follows:

$$\Delta T_{\text{lm}} = (T_{\text{a,o}} - T_{\text{a,i}}) / \ln \frac{(T_{\text{s}} - T_{\text{a,i}})}{(T_{\text{s}} - T_{\text{a,o}})} \tag{3}$$

According to Formula (1)–(3), the total heat exchange coefficient U of air cooler under different load and air volume can be determined by obtaining the above temperature and other data. The degree of ash deposition can be measured by the difference of thermal resistance between the fouling state and the clean state of the air cooler. The heat transfer coefficient is used to express the heat transfer ability of fluid and air. The larger the heat transfer coefficient is, the stronger the heat transfer ability is. The total heat exchanger coefficient is as follows:

$$U = \frac{Q}{A \times \Delta T_{\text{lm}}} \tag{4}$$

The total heat transfer coefficient U_0 of fin tube of air cooler under clean condition can be expressed as follows:

$$U_0 = \frac{1}{\frac{1}{h_{\text{i}}} \frac{A_0}{A_{\text{i}}} + \frac{\delta_{\text{w}}}{\lambda} \frac{A_0}{A_{\text{i}}} + \frac{1}{h_0 \eta_0}} \tag{5}$$

Where h_i and h_o are heat transfer coefficient of cold end surface and hot end surface, A_i and A_o are the total heat transfer area of smooth tube without fin and fin side including fin heat transfer area, δ_w is wall thickness of finned tube, λ is wall thermal conductivity, η_0 is total efficiency of ribs.

After a certain time of ash deposition, the heat transfer coefficient U_a of the air cooler is as follows:

$$U_a = \frac{1}{\frac{1}{h_i}\frac{A_0}{A_i} + \frac{\delta_w}{\lambda}\frac{A_0}{A_i} + \frac{1}{h_0\eta_0} + \frac{\delta_f}{\lambda_f}} = \frac{1}{\frac{1}{U_0} + R_f} \tag{6}$$

Where δ_f is thickness of ash scale, λ_f is thermal conductivity of ash scale.

As shown in the following formula, R_f is the thermal resistance of ash scale:

$$R_f = \frac{1}{U_a} - \frac{1}{U_0} = R_a - R_0 \tag{7}$$

Where R_a and R_0 are the heat transfer resistance under ash deposition and clean conditions respectively.

$$K_f = \frac{1}{R_f} = \frac{1}{R_a} - \frac{1}{R_0} \tag{8}$$

η_{loss} is selected to describe the weakening degree of fouling on heat transfer performance:

$$\eta_{loss} = \frac{K_0 - K_f}{K_0} \tag{9}$$

Where K_0 is the heat transfer coefficient under clean conditions. The greater the thermal resistance of ash deposition, the greater the η_{loss}.

2.2 Data Set Acquisition

The experimental data were obtained from the infrared images collected from the field finned tube heat exchanger of a 2 × 350MW unit air cooling island of a power plant in Xinjiang. Based on the heat transfer coefficient of 29.5 W/(m^2·K) under clean condition, the average value of the decline rate of heat transfer coefficient of six grades of finned tube was obtained by converting the decrease rate of heat transfer. The infrared image sets of six grades of gray deposition on finned tubes are shown in Fig. 4.

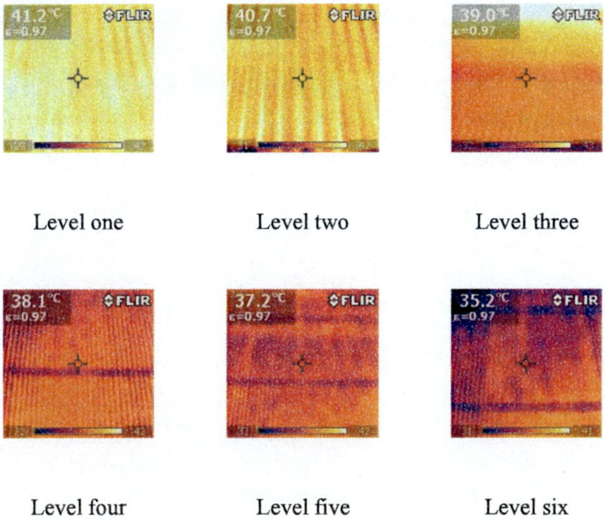

Fig. 4. Infrared image of ash deposition state of finned tube of air cooler and the decrease rate of heat transfer coefficient

The infrared images in the processed images are classified according to the range of heat transfer coefficient decline rate. Finally, the infrared images are 1200, and the ratio of model training set to test set is 3:1. The decreasing rate of heat transfer coefficient can be obtained by identifying the image, so as to directly judge the ash deposition state.

2.3 Identification Method

The local connection, weight sharing and pooling operation of the convolutional neural network can effectively reduce the complexity of the network, reduce the number of training parameters, make the model invariant to translation, distortion and scaling to a certain extent, and has strong robustness and fault tolerance.

The main feature of RESNET model is that residual structure is added. Because identity of identity mapping layer is added, gradient dissipation problem is well solved, the "residual learning module" solves the problem of model degradation, breaks through the limitation of network depth, and further improves the accuracy of the model [17]. The typical RESNET model structure and its residual module are shown in Fig. 5. The principle is that the input is recorded as X, and the output of the residual module is recorded as H(x), at this time $h(x) = f(x) + X$ with the deepening of the model layers, the characteristic difference between the network output f(x) and the lower network output H(x) gradually decreases. The module learns the residual error H(x)-X in the process of model backward conduction, H(x) is used to calculate the partial derivative of X. Because it is greater than 1, it solves the problem that the ratio of output difference of deep network is too small and the gradient of model disappears. The integrity of information is protected to a certain extent by transmitting the information directly from the input to the output. At the same time, because the goal of learning is residual, the difficulty of learning is simplified. The activation function f of the model uses a Rectified

Linear Unit(Relu) to enhance the nonlinear characteristic ability of the model. Compared with tanh, sigmoid and other activation functions, using Relu as activation function, the model converges faster and the parameter configuration is more concise.

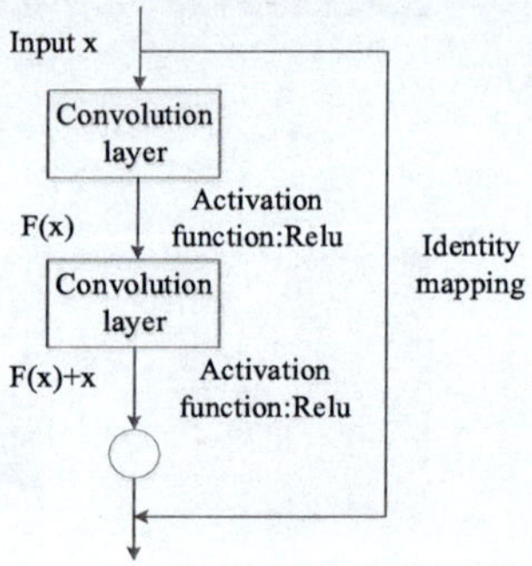

Fig. 5. Structure of residual block

3 Results and Discussion

The server hardware configuration for image recognition is Intel (R) core (TM) i7-7700hq processor, 16G memory and nvdia gtx1070 video card. A deep learning software environment based on Python is built on Windows platform to train the model based on convolutional neural network. Alexnet, resnet-50, resnet-101, mobilenetv3 models (the suffix is the number of layers of the model) were selected as the experimental group.

After 50 epochs of training, the train accuracy of the three models is shown in Fig. 6, the test accuracy of the three models is shown in Fig. 7. The model train loss is shown in Fig. 8 and test loss is shown in Fig. 9.

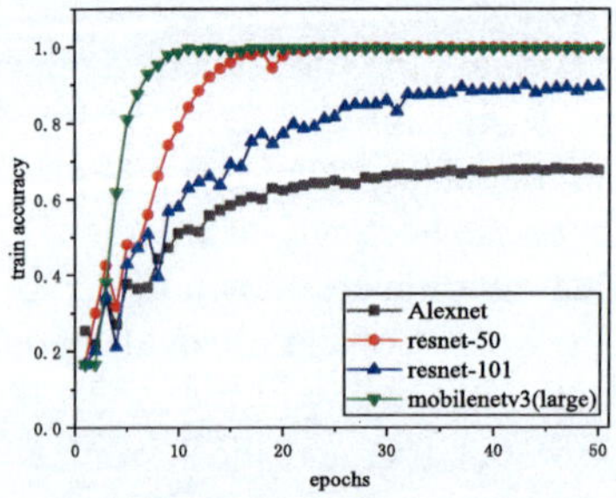

Fig. 6. Train accuracy

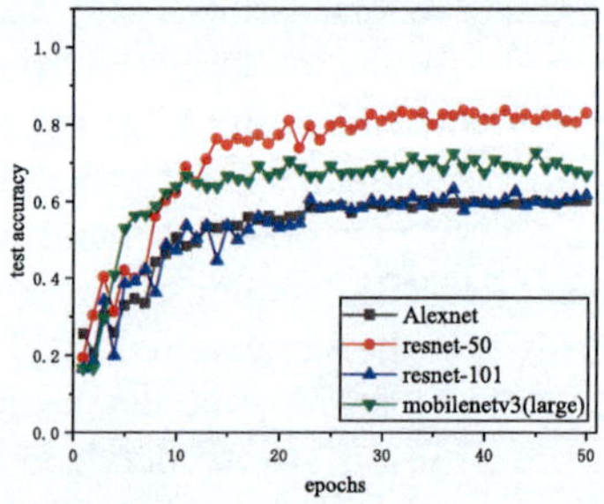

Fig. 7. Test accuracy

According to the recognition accuracy test results of three models in Fig. 6, Fig. 7, Fig. 8, Fig. 9. Three hundred pictures are selected to test with four models, the precision, recall and F1 score of each classification under each training model are calculated, the recognition effect of resnet-50 test set is slightly better than that of Alexnet, resnet-101 models and mobilenetv3(large).

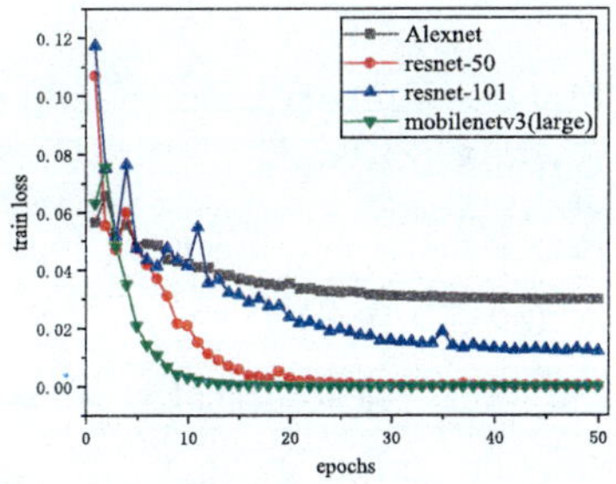

Fig. 8. Train loss

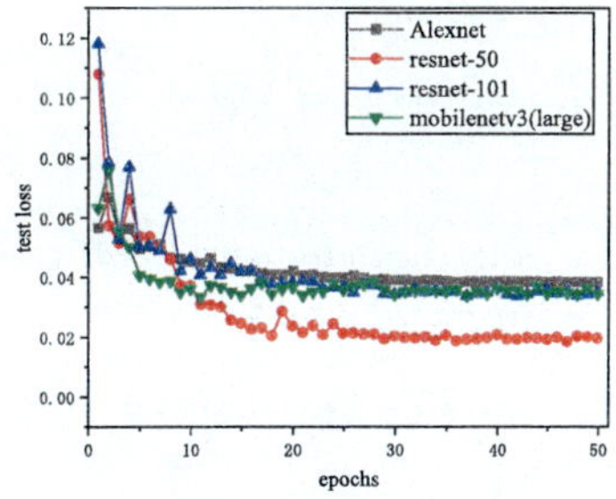

Fig. 9. Test loss

The traditional way to calculate the ash deposition degree requires a large number of sensors, which is difficult to be arranged in large area on site. Too many data acquisition devices are greatly affected by environmental factors, and the calculation method is complex and the degree of automation is low. The image acquisition device is simple to install and needs less equipment. The images collected by infrared cameras are identified based on RESNET model. By establishing the relationship between the image of ash deposition and the degree of ash deposition, the requirement of intelligent identification of ash deposition state of air cooler is realized.

4 Conclusion

1 In order to overcome the shortcomings of traditional ash detection methods for power plant air condensers, such as large detection error and slow detection speed, this paper proposes an air cooler ash state recognition and analysis method based on convolution neural learning. The experimental system of ash accumulation state image recognition is constructed in the laboratory and field, and infrared images collected in the laboratory are screened to prepare data sets, and the training model is constructed the accuracy rate and error size of the identification of the ash deposition state by the type and test model.

2 For the collected infrared image of air cooler ash accumulation; the average accuracy of resnet-50 for six categories is 0.83, average recall is 0.82, and average F1 score is 0.84. Among the four models of Alexnet, resnet-50, resnet-101, and mobilenetv3(large), resnet-50 is the best.

Nomenclature

Q	the heat load, kW	h_i	heat transfer coefficient of cold end surface, W/(m^2·K)
U	total heat transfer coefficient, W/(m^2·K)	h_o	heat transfer coefficient of hot end surface, W/(m^2·K)
A	Air side heat transfer area of air cooler, m^2	A_i	the total heat transfer area of smooth tube without fin heat transfer area, m^2
ΔT_{lm}	log mean temperature difference, ℃	A_o	total heat transfer area of fin side including fin heat transfer area, m^2
m_s	steam mass flow rate, kg·s^{-1}	δ_w	wall thickness of finned tube, mm
$h_{s,i}$	enthalpy of steam at the inlet, kJ·kg^{-1}	λ	wall thermal conductivity, W/(m·K)
$h_{c,o}$	enthalpy of condensed steam, kJ·kg^{-1}	η_0	total efficiency of ribs, dimensionless
m_a	mass flow rate of cold fluid, kg·s^{-1}	R_f	the thermal resistance of ash scale, (m^2·K)/W
$c_{p,a}$	specific heat capacity of air, J /(kg·K)	K_f	the heat transfer coefficient under the condition of ash deposition, W/(m^2·K)
$T_{a,i}$	temperature of windward side of heat exchanger, ℃	K_0	the heat transfer coefficient under clean condition, W/(m^2·K)
$T_{a,o}$	temperature of leeward side of heat exchanger, ℃	η_{loss}	decrease rate of heat transfer coefficient, dimensionless

Acknowledgement. This work was supported by the Science and Technology Development Plan of Jilin Province (20190201098JC); the National Natural Science Foundation of China (51606035).

References

1. Han, W., Chen, X., Zhang, Z.: Research on relationship between urban water use and economic growth based on decoupling analysis theory. Bull. Soil Water Conserv. **10**, 141–146 (2017)
2. Zhang, C., Zhong, L.J., Wang, J.: Decoupling between water use and thermoelectric power generation growth in China. Nat. Energy **3**(9), 792–799 (2018)

3. Zhang, C., Zhong, L.J., Fu, X.T., et al.: Managing scarce water resources in China's coal power industry. Environ. Manage. **57**(6), 1188–1203 (2016)
4. Kröger, D.G.: Air-cooled heat exchangers and cooling towers, Stellenbosch Stellenbosch University (6), p. 96 (2005)
5. Ni, W., Ge, Z., Yang, L., et al.: Piping-main scheme for condensers against the adverse impact of environmental conditions on air-cooled thermal power units. Energies **13**(1), 170 (2019)
6. Ge, Z.H., Du, X.Z., Yang, L.J., et al.: Performance monitoring of direct air-cooled power generating unit with infrared thermography. Appl. Therm. Eng. **31**(4), 418–424 (2010)
7. Loyola-Fuentes, J., Smith, R.: Classification and estimation of unmeasured process variables in crude oil pre-heat trains subject to fouling deposition. Comput. Chem. Eng. **137**, 106779 (2020)
8. Hamzacebi, C.: Improving artificial neural networks' performance in seasonal time series forecasting. Inf. Sci. **178**(23), 4550–4559 (2008)
9. Csiszar, O., Csiszar, G., Dombi, J.: Interpretable neural networks based on continuous-valued logic and multicriteria decision operators. Knowl.-Based Syst. **128** (2020)
10. Shrikrishna, K., Jayant, J.: Spatio-temporal deep neural networks for accession classification of Arabidopsis plants using image sequences. Ecol. Inf. **64** (2021)
11. Sahin, S.O., Kozat, S.S.: Nonuniformly sampled data processing using LSTM networks. IEEE Trans. Neural Netw. Learn. Syst. **30**(5), 1452–1461 (2019)
12. Chen, Q., Zhang, H.C.: Hybrid model for compressive image recovery: integrating ResNet-based denoiser into GAMP. Signal Process. **173**, 107583 (2020)
13. Zhang, Y., Cheng, T., Ren, Y.B.: A graph deep learning method for short-term traffic forecasting on large road networks. Comput.-Aided Civil Infrastruct. Eng. **34**(10), 877–896 (2019)
14. Li, Z., Lin, Y.L., Elofsson, A., et al.: Protein contact map prediction based on ResNet and DenseNet. Biomed Res. Int. 7584968 (2020)
15. Han, C., Shi, L.: ML-ResNet: a novel network to detect and locate myocardial infarction using 12 leads ECG. Comput. Meth. Programs Biomed. **185**, 105138 (2020)
16. Jeon, E., Heo, G., Kim, I., et al.: Conceptual design for combined ocean thermal energy conversion using computational fluid dynamics and heat balance analysis. Int. J. Energy Res. **44**(9), 7477–7494 (2020)
17. Shi, J., Liu, Q.P., Wang, C.F., et al.: Super-resolution reconstruction of MR image with a novel residual learning network algorithm. Phys. Med. Biol. **63**(8), 085011 (2018)

A Pedestrian Intrusion Detection Method Based on Improved Mask R-CNN Model

Xiaoyu Chen(✉), Jianyou Yu, Dezhi Zhao, Jian Sun, Xianghong Chen, Mingli Yi, and Lin Xia

Baishan Power Supply Company of State Grid Jilin Electric Power Company Limited, Baishan, China

Abstract. In order to solve the problem of pedestrian intrusion detection in dangerous areas in power construction sites, a target detection method based on the improved Mask R-CNN model is proposed. First, initialize the parameters of the Mask R-CNN backbone network according to the migration learning strategy to extract the basic features of the image; then introduce the feature pyramid structure for bottom-up feature map extraction to complete multi-scale feature fusion; then, through the multi-scale transformation method to adjust the regional recommendation network, obtain anchor points for regression calculation to complete the detection experiment; finally analyze and evaluate the results.

Keywords: Mask R-CNN model · Electric power construction site · Target detection · Transfer learning

1 Introduction

The surrounding area of power facilities, especially the unattended substation area, is extremely dangerous for pedestrians. At the power facility emergency repair work site, due to the negligence of the staff and failing to dress in accordance with the standard safety, it will cause an electric shock accident and cause unpredictable consequences. This is not negligible and needs to be detected in time and early warning to avoid danger.

Pedestrian detection is a very hot topic in the field of computer vision. However, due to factors such as illumination, various colors of clothing, and changes in pedestrian postures, pedestrian detection is a more difficult problem. And regarding the problem of pedestrian detection, there are also many methods, such as the Haar feature proposed by Paul Viola combined with the Adaboost method to train a classifier for face detection. Leibe et al. established a top-down and bottom-up method based on key point detection, Hough voting and Chamfer distance template matching. Ronfard et al. used a classifier trained by a Support Vector Machine (SVM) to complete pedestrian detection. In this classifier, each part of the human body is described by "graphic structure"; the external features of each part are described by first-order and second-order steps.

Traditional video analysis methods mostly rely on manual design features, there are a large number of false detections, missed detections, and poor generalization capabilities. With the rapid development of convolutional neural networks, target detection models

S.-C. Chu et al. (Eds.): ICGEC 2021, LNEE 833, pp. 370–380, 2022.
https://doi.org/10.1007/978-981-16-8430-2_34

based on deep learning gradually replace traditional manual algorithms and become the mainstream trend in the field of image detection. Bi Lin et al. built convolutional neural networks (CNNs) to detect the wearing of helmets, but the effect is not ideal when the background is complex, the lighting and the video quality of the camera is poor. Based on the YOLO V3 model, Shi Hui and others can realize end-to-end helmet detection and recognition, and the detection speed is significantly improved, but the YOLO V3 model directly divides the image, resulting in rough positioning information. Xu Shoukun et al. proposed a multi-component combined detection method based on the improved Faster R-CNN model, which eliminates the false detection target and improves the accuracy, but the target confidence of the experimental results is low, which leads to the problem of false detection. The instance segmentation framework Mask R-CNN adds mask branches on the basis of the Faster R-CNN model. The positioning information is accurate, and high-quality segmentation results can be obtained. At the same time, the confidence level is above 90%, and the target detection accuracy is further improved. In the target feature extraction Stronger expression skills.

Due to the advantages of Mask R-CNN in target detection, this paper proposes a target detection method based on the improved Mask R-CNN model for the problem of pedestrians in dangerous areas in power construction scenarios. In the first stage, the power construction scene data set (person and helmet date, PHDate) is established by collecting samples, and the weights of the Microsoft Common Objects Data Set (COCO Data Set) are called to train the Mask R-CNN model in combination with the migration learning strategy; In the second stage, small targets cannot be detected due to the fact that the actual collected video data has a long distance and large target size differences. Adjust the area and suggest network parameters to achieve multi-scale transformation operations; the third stage introduces Laplacian for low-quality surveillance video The convolutional layer is sharpened to enhance the edge contrast of the image and highlight the target contour, which is used to improve the detection effect of low-quality video.

2 Mask R-CNN Algorithm

Mask R-CNN is a general target detection algorithm for objects. It can be used to build an instance segmentation framework for research on "target detection", "target instance segmentation", and "target key point detection". It has excellent target detection performance and the overall structure of the algorithm. As shown in Fig. 1. First, residual neural networks (ResNet) combined with feature pyramid networks (FPN) are used to extract deep convolutional feature maps of the input image. Then the feature map is sent to the region proposal networks (region proposal networks, RPN) to generate a suggestion window to obtain a high-quality candidate frame, that is, to find out the possible position of the target in the map in advance to correct the frame, and map the suggestion window to the convolutional feature On the map. Then use the RoiAlign region alignment layer to generate a fixed size for each region of interest (ROI), and expand the dimension of the output feature map through the network head to achieve precise alignment of the output and input regions of interest, so that the target positioning information is more accurate. Finally, through the full convolutional network (full convolutional network, FCN) output high-quality binary segmentation mask, fully connected layers (fully connected layers, FC) output prediction box and class.

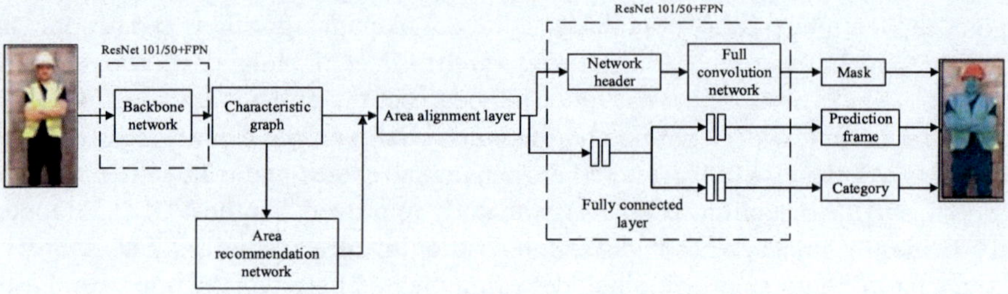

Fig. 1. Mask R-CNN model network structure

2.1 ResNet Combined with FPN Detection Algorithm

The residual network ResNet is used as the feature extraction network, combined with the FPN algorithm to build a multi-scale feature pyramid model, which mainly solves the multi-scale fusion problem. The architecture is shown in Fig. 2. The input image is extracted from the bottom-up feature map by (Cl, C2, C3, C4, C5) 5 layers of convolutional layers, and then the 1×1 convolution operation is performed to change the dimension of the feature map, and the feature map is changed from top to bottom. 2 times of upsampling, and the previous layer network is superimposed and fused, and 3×3 convolution is used to convolve each fusion result to eliminate the aliasing effect of upsampling. The fusion feature layer is (P2, P3, P4, P5), which is combined with the top-level feature map P6 and input into the RPN network to extract candidate regions. The feature pyramid architecture enhances high-level features, and the feature maps used in each layer of prediction incorporate features of different resolutions and different semantic strengths.

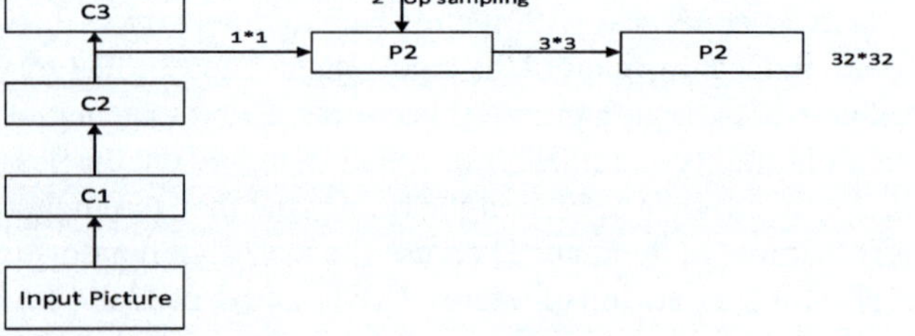

Fig. 2. ResNet + FPN structure

2.2 Regional Proposal Network

The function of the regional suggestion network is to extract a certain number of suggested regions with target confidence on the multi-layer feature map output by FPN, and its principle is shown in Fig. 3. RPN takes the convolution feature matrix as input and outputs a series of rectangular candidate boxes and probability values. The sliding window mechanism is adopted, and sliding windows are added to each layer of the feature map (P2, P3, P4, P5, P6), and each sliding window corresponds to k initial suggestion regions, which are called anchors, which are anchor points.

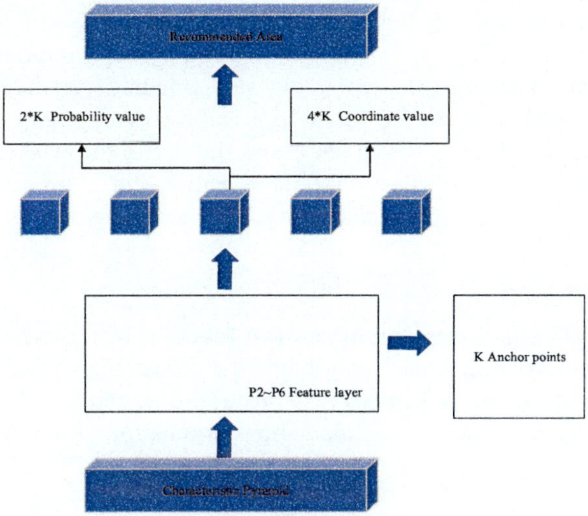

Fig. 3. RPN network

Let n be the side length of the sliding window. When a sliding window of $n \times n$ size is traversed on the feature pyramid layer, each position of the sliding corresponds to k different anchor points on the original image, and a fully connected layer outputs $2 \times k$-dimensional vector, corresponding to the probability value of k anchor points; another fully connected layer outputs a $4 \times k$-dimensional vector, which represents the coordinate values of k anchor points. The Mask R-CNN algorithm presets 5 scale sizes (32, 64, 128, 256, 512), and sets the anchor point K width ratio to (1:1, 1:2, 2:1), so each position anchor The point is 15.

Because the anchor points often overlap, the suggestion frame will overlap multiple times on the same target. In order to solve the problem of repeated suggestions, a non-maximum suppression method is used for the generated candidate frames (non-maximum suppression, NMS). The NMS algorithm generates a list of suggestions sorted by probability value. If the probability value is less than 0.3, it is judged as the background, and if the probability value is greater than 0.7, it is judged as the foreground. The sorted list is iterated, and then the overlap rate IoU between the target window and the original marked window is used as Measure and filter out suggestion boxes with higher overlap rate scores. The calculation formula of IoU is

$$IoU = \frac{(DR \cap GT)}{(DR \cup GT)} \tag{1}$$

In formula (1): DR (detection result) is the detection result, that is, the final target frame; GT (ground truth) is the true value.

2.3 ROIAlign Layer

In order to solve the problem of misalignment caused by feature misalignment, an ROIAlign layer is proposed. The misalignment problem has little impact on the classification task, but it will have a very large negative impact when predicting the pixel-level precision mask. The ROIAlign layer process is: using bilinear interpolation to calculate the precise value of the input feature at two sampling positions in each ROI block, avoiding any quantification of the ROI boundary, and finally aggregate the results. The output dimension of each ROI is, which represents the size of the mask and represents a category. The ROIAlign layer will not only improve the accuracy of detection, but also facilitate instance segmentation and achieve precise positioning.

2.4 Loss Function

The Mask R-CNN algorithm uses a multi-task loss function, which is the sum of the loss of classification, regression, and mask prediction. It can be used as a basis for measuring the effect of model detection. When the loss function Loss reaches the minimum, the model detection effect is the best. The calculation formula for

$$L = L_{\mathrm{cls}} + L_{blox} + L_{mask} \tag{2}$$

In formula (2): L_{cls} is the classification loss; L_{blox} is the prediction box regression loss; L_{mask} is the mask regression loss.

For each region of interest, the dimensional matrix defined $i * m * m$ by the mask branches represents i the regions corresponding to $m * m$ different categories. Each pixel is a sigmod activation function to calculate the relative entropy, so as to obtain the L_{mask} average relative entropy error.

The loss function allows the network to generate a mask for each category without competing with other categories, and can separate the prediction of the mask and the category.

3 Model Establishment and Detection Based on Mask R-CNN

The model building process is divided into three parts: the preprocessing phase, the model training phase, and the detection target phase. The process is shown in Fig. 4.

3.1 Data Processing

Collecting pictures of power construction sites, perform data enhancement processing such as segmentation, rotation by 90°, mirroring, and translation, and establish training data sets and test data sets for power construction scenes.

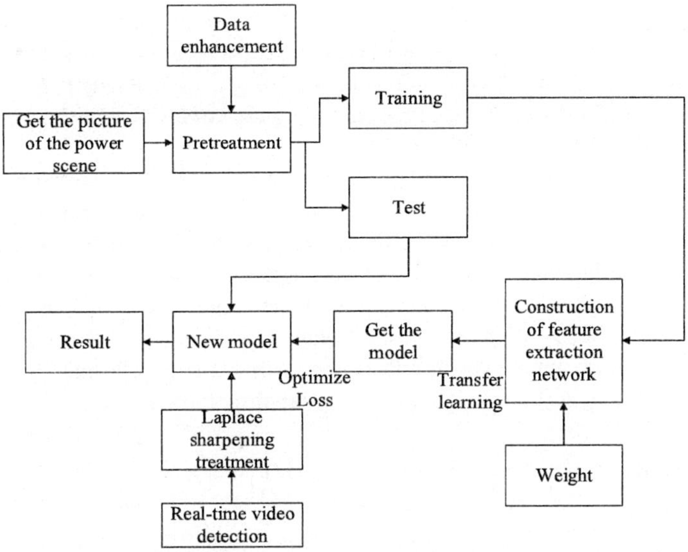

Fig. 4. Training and testing process based on Mask R-CNN model

3.2 Model Construction

First, building a complete feature extraction network, and input the training data set into the ResNet + FPN network architecture. Then use the migration learning method to perform model training on the basis of the pre-trained model that has been trained on the COCO data set, and use the randomly generated weights as the starting point of training to solve the problem of inaccurate labeling mask coverage. Finally, the loss function Loss is calculated, and the Mask R-CNN model is adjusted and then repeatedly trained until the minimum loss function is obtained. At this time, the model detection effect is the best.

3.3 Target Detection

The test data set is retrieved and send it to the optimized model to perform target detection experiments, and obtain the information contained in the detected target in the image: the prediction frame, the category of the target, and the mask.

Video detection is expanded on the work of still image detection. Obtain the video stream of the electric power construction site, convert it into an image frame, and sharpen it if the pixel is too low. For the task of detecting pedestrian intrusion targets in dangerous areas, when the input image frame detects pedestrians, an alarm is issued to remind the pedestrians to evacuate the dangerous area in time.

4 Experimental Results and Analysis

Since the public data set resources for pedestrian detection experiments are not disclosed at this stage, model training based on convolutional neural networks relies on a large number of samples. In order to make up for the lack of training samples, manually collect pictures of the construction site, add multiple substation videos, and use the multimedia video processing tool FFmpeg to obtain image frames from the video. The main body of the frame operation is the select function as a filter statement, and the input video is traversed to obtain a fixed time. Spaced pictures.

In multimedia processing, FFmpeg provides a variety of built-in filters, which can be used in combination with complex commands to direct decoded frames from one filter to another. This processing reduces the decoding and encoding operations, the overall training picture quality will not be reduced, and the problem of data damage is prevented.

The samples are preprocessed, and the final PHDate training data set is expanded to 5000 sheets, and the test data set is 1000 sheets. Use deep learning image annotation tool VIA (VGG image annotator) to annotate pictures. Using polygon format to label the target data set is better than rectangular box labeling to obtain a mask.

4.1 Model Performance Analysis Based on Mask R-CNN

The COCO data set used by the original Mask R-CNN model has about 80,000 pictures and 81 categories, while the power scene PHDate data set contains 5000 pictures, two categories. Compared with the COCO data set, the PHDate background is simpler and single. In the feature extraction stage, considering the greater impact of different networks on the experimental results, the ResNet50 + FPN and ResNet101 + FPN feature extraction networks are selected for experiments. The experimental structure parameters are shown in Table 1.

Table 1. ResNet network parameter setting

Stage	Parameter settings
C1	7 * 7,64, Step 2 3 * 3, maximum value pooling, step length 2
C2	$\begin{bmatrix} 1*1, 64 \\ 3*3, 64 \\ 1*1, 256 \end{bmatrix} *3$
C3	$\begin{bmatrix} 1*1, 128 \\ 3*3, 128 \\ 1*1, 512 \end{bmatrix} *4$
C4	$\begin{bmatrix} 1*1, 256 \\ 3*3, 256 \\ 1*1, 1024 \end{bmatrix} *6 \quad \begin{bmatrix} 1*1, 256 \\ 3*3, 256 \\ 1*1, 1024 \end{bmatrix} *23$

(continued)

Table 1. *(continued)*

Stage	Parameter settings
C5	$\begin{bmatrix} 1*1, 512 \\ 3*3, 512 \\ 1*1, 2046 \end{bmatrix} * 3$

The two network convergence curves of the training model process are shown in Fig. 5, the final number of iterations is 300,000 times, the curve smoothness is set to 0.6, the final loss rate of ResNet50 + FPN residual extraction network is 7.654%, and ResNet101 + FPN is 6.147%. With the increase in the number of network layers and the greater the amount of target detection calculations, the ResNet50 structure is lighter and simpler, occupies less GPU memory, and the training time is shorter, so ResNet50 + FPN is selected as the reference network structure.

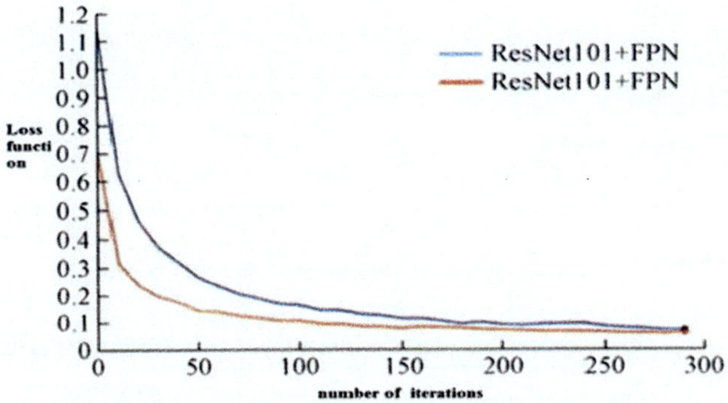

Fig. 5. The relationship between the loss rate of different networks and the number of iterations

The COCO data set pre-trained model is used to accelerate the convergence. The weight parameters of the pre-trained model are imported into the ResNet50 + FPN structure to initialize the network, and the PHDate data set is used for fine-tuning. The weight attenuation coefficient is set to 0.0001 during the tuning process, and the learning rate is 0.001, the learning impulse is 0.9. Analyze the feature extraction effect of the model convolutional layer, the four stages Cl–C4 and the mask generate weight distribution. The initial weight is a small random number, and the overall output of each layer satisfies the Gaussian distribution with a mean value of 0. This can attenuate the weight to a smaller value and reduce the problem of model overfitting to a certain extent.

4.2 Model Performance Evaluation

In order to evaluate the effectiveness of the algorithm in detecting targets, a multi-target average accuracy (MAP) index is used to measure the detection performance of the model. During the training phase of the RPN network, the scale and aspect ratio are adjusted to complete the multi-scale transformation. The parameter setting is shown in Table 2. In the same environment, the original Mask R-CNN model and the model obtained after changing the anchor point scale are performed on the test set (Table 3). It can be seen that when the anchor point size is reduced by half, the size is (16, 32, 64, 128, 256) The detection effect is the most ideal. After obtaining the best size, adjust the aspect ratio of the anchor point and set up multiple sets of data for experiments. The results are shown in Table 4. The results are compared with the experimental results of literature [2] and literature [4], and the results are shown in Table 5.

The target area classification confidence represents the quality of the model measurement. From the experimental results, it can be seen that the improved Mask R-CNN model proposed in this paper can achieve the highest average accuracy of multiple targets under the condition of ensuring the confidence.

Table 2. RPN network parameters

Parameter name	SetPoint
Anchor step	1
Anchor points number	256
R OI Number	2000
N MS threshold value	0.7

Table 3. Comparison of scale transformation experiment

Model	Size of the anchor points of each layer	Average accuracy rate
Original model	(32,64,128,256,512)	93.88%
Double size	(64,128,256,512,1024)	50.83%
Reduce by 50%	(16,32,64,128,256)	94.62%
Reduce by 75%	(8,16,32,64,128)	76.88%

Table 4. Comparison of the aspect ratio experiment

Horizontal aspect ratio	Accuracy:
(1,2,4)	85.74%
(0.5,1,2)	94.62%
(0.25,1,2)	94.22%
(0.25,1,4)	95.22%

Table 5. Comparison of the experimental results

Method	Reliability	Average accuracy rate
Literature: [2]	–	92.13%
Literature: [4]	0.2	91.42%
M ask R-CNN	0.9	93.87%
This algorithm of this algorithm	0.9	95.22%

5 The Conclusion

Aiming at the problem of pedestrian intrusion detection in dangerous areas of electric power construction sites, a target detection method based on the improved Mask R-CNN model is proposed. The electrical scene detection model is obtained through the migration learning strategy training data set, and the multi-scale transformation method is used to optimize the system to solve the problem of small target misdetection and missed detection, so that the average accuracy of multi-day standards is improved, and the Laplace algorithm is used for sharpening. Modified preprocessing improves the accuracy of the model under low-quality surveillance video. The method in this paper is quick to deploy and easy to operate, and can be extended to other application scenarios. In future work, we will focus on solving the problem of model compression to adapt to the embedded environment with limited processing space.

Acknowledgment. This work was funded by the "Research and Application of Key Technologies for the Whole Process Management and Control of Substation Operations" (Project Number: SGJLBS00AJJS2000262).

References

1. Lin, B., Wei, X., Jun, C.: Research on miner's hard hat wearing recognition based on convolutional neural network. Gold Sci. Technol. **25**(4) 2017
2. Shi, H., Chen, X., Yang, Y.: Improve the detection method of YOLO v3 helmet wearing. Comput. Eng. Appl. **55**(011), 213–220 (2019)
3. Viola, P., Jones, M.J., Snow, D.: Detecting pedestrians using patterns of motion and appearance. Int. J. Comput. Vision **63**(2), 153–161 (2005)
4. Xu, S., Wang, Y., Gu, Y., et al.: Research on helmet wearing detection based on improved Faster RCNN. Appl. Res. Comput. **37**(341(03)), 267–271 (2020)
5. Leibe, B., Seemann, E., Schiele, B.: Pedestrian detection in crowded scenes. In: IEEE Computer Society Conference on Computer Vision & Pattern Recognition. IEEE (2005)
6. Ronfard, R., Schmid, C., Triggs, B.: Learning to parse pictures of people. In: Heyden, A., Sparr, G., Nielsen, M., Johansen, P. (eds.) Computer Vision — ECCV 2002: 7th European Conference on Computer Vision Copenhagen, Denmark, May 28–31, 2002 Proceedings, Part IV, pp. 700–714. Springer Berlin Heidelberg, Berlin, Heidelberg (2002). https://doi.org/10.1007/3-540-47979-1_47

7. Ren, S., He, K., Girshick, R., et al.: Faster R-CNN: towards real-time object detection with region proposal networks. IEEE Trans. Pattern Anal. Mach. Intell. **39**(6), 1137–1149 (2017)
8. He, K., Gkioxari, G., Dollár, P., et al.: Mask R-CNN. In: IEEE Transactions on Pattern Analysis & Machine Intelligence. IEEE (2017)
9. Yao, H., Jiang, J., Qi, M., et al.: Image de-motion blur algorithm based on Laplacian and bilateral filtering. Sens. Microsyst. **36**(001), 139–142 (2017)

A Flight Simulation Model Architecture for Reinforcement Learning

KaiXuan Wang[1,2], YuTing Shen[3,4,5], FuQuan Zhang[6](✉), Nan Zhao[1], and Lijie Yang[1]

[1] China Academy of Launch Vehicle Technology, Beijing 100076, People's Republic of China
[2] School of Automation Science and Electrical Engineering, Beihang University, Beijing 100191, People's Republic of China
[3] National Space Science Center, Chinese Academy of Sciences, Beijing 100190, China
[4] University of Chinese Academy of Sciences, Beijing 100039, People's Republic of China
[5] The 54th Research Institute of China Electronics Technology Group Corporation, Shijiazhuang 050000, People's Republic of China
[6] Fujian Provincial Key Laboratory of Information Processing and Intelligent Control, (Minjiang University), Fuzhou 350121, People's Republic of China

Abstract. For the reinforcement learning of agents in virtual environment, a flight simulation model architecture is designed by combining the integration method, flight algorithm, external force calculation, earth model and other factors in flight simulation. This architecture realizes the separation of data and algorithm in the process of flight computation at the software level, and adopts inheritance and pointer registration mechanism to achieve unified and hierarchical management of data related to flight computation. The state transfer model of aircraft is designed to realize the smooth connection of flight states between different stages, which can realize efficient large-scale parallel computation in reinforcement learning. The architecture proposed in this paper provides a flexible and efficient interaction mechanism for enhancing reinforcement learning of aircraft agents.

Keywords: Reinforcement learning · Agent · Flight calculation · Software architecture design

1 Preface

It is an effective way to evaluate the performance of aircraft through agent reinforcement learning in virtual environment [1], but the traditional simulation methods can not optimize the performance on the basis of the evaluation. With the development of artificial intelligence, reinforcement learning through interaction with environment is an effective way to optimize flight parameters of aircraft and achieve flight strategy optimization. It is necessary to establish accurate, efficient and flexible flight computational model and software architecture for the realization of reinforcement learning for aircraft agents. This research is of great significance for aircraft performance optimization and flight strategy optimization.

S.-C. Chu et al. (Eds.): ICGEC 2021, LNEE 833, pp. 381–391, 2022.
https://doi.org/10.1007/978-981-16-8430-2_35

2 Research Tracking

At present, most of the research related to flight computation focuses on flight control algorithms [2], aircraft simulation [3] and software implementation [4]. Researchers generally do not design and study the high-level architecture of flight computing. In the aspect of flight simulation, most models are realized by the principle of process-oriented software design [5]. There is no case of using the object-oriented method for reinforcement learning and parallelization, nor is there any case of analyzing and designing the architecture based on the state transition model. In the aspect of reinforcement learning for flight performance optimization, the characteristics of aircraft or high confidence flight simulation are seldom considered. And the conventional path search [6] or key point tracking [7] methods are generally used to generate flight constraints and carry out flight calculation. For reinforcement learning training, there are few researches on large-scale parallel optimization methods based on state transfer of aircraft.

3 Basic Architecture of Aircraft Agent Reinforcement Learning

Reinforcement learning for aircraft performance optimization is a sequential decision problem, in which an aircraft agent interacts with the environment through actions, obtains interaction data with reward value, and completes the training of the agent [8]. The value function of state s under strategy π is $v_\pi(s)$, and the Behrman equation of v_π is:

$$v_\pi(s) = \sum_a \pi(a|s) \sum_{s',r} p(s', r|s, a)\big[r + \gamma v_\pi(s')\big] \tag{1}$$

Where, γ is the constant coefficient, s' is the next transition state, and r is the reward value. The Behrman equation of v_π expresses the relationship between state value and subsequent state value with equations. The task of reinforcement learning is to find an optimal strategy. For example, for the finite MDP problem, the optimal state value function is:

$$v_*(s) = \max_\pi v_\pi(s) = E_\pi\big[R_{t+1} + \gamma G_{t+1}|S_t = s\big] \tag{2}$$

The optimal action value function is:

$$q_*(s, a) = \max_\pi q_\pi(s, a) = E\big[R_{t+1} + \gamma v_*(S_{t+1})|S_t = s, A_t = a\big] \tag{3}$$

Calculating either $v_*(s)$ or $q_*(s, a)$ requires interaction between the agent and the environment at a series of discrete moments to maximize the target benefit over time. In the course of flight performance optimization, the expectation of value function must be solved. Therefore, multiple simulations are needed for the same state. Moreover, due to the characteristics of sequential decision problems, restarting flight calculation every time will lead to a large number of repeated calculations, leading to a decrease in learning efficiency. To solve the reinforcement learning problem of flight performance optimization, it is necessary to research the flight computation model or software architecture which supports state transfer and parallel computation.

4 Discrete Flight Computing Architecture

The flight of the aircraft is a continuous process of motion in accordance with time t, and transfers sequentially in the direction of t. When the influence of the subsequent states on the current state is not considered, the flight process of the aircraft meets the non-after-effect, and the MDP model can be used to describe the discrete flight calculation problem. The aircraft agent changes its state dynamically at time interval Δt, and the state transition equation is obtained from the aircraft kinematics model. In order to control the motion state of the aircraft, it is assumed that the control variable related to the changes between different states S is α, and the kinematic parameter related to the aircraft itself is Λ^{T}. The state transfer process of the flight calculation model built according to the kinematics principle of the aircraft is as follows:

$$S_t = f(S_{t-1}, \alpha, \Lambda^{\mathrm{T}}) \tag{4}$$

Therefore, the flight process of the aircraft can be regarded as the MDP problem given a complete environment model, and the flight performance can be optimized by using reinforcement learning method.

The flight calculation process is based on the kinematics method of the aircraft, the integral iteration is used to complete the whole flight path calculation of the aircraft, and the continuous flight process of the aircraft is simulated. In order to make the flight calculation process adapt to the discrete state, it is necessary to design the specific architecture to meet the state transition between different states. In reinforcement learning training, the optimization process of value network or value decision network needs to involve a lot of computation, so the flight computation model should be able to adapt to the parallel computation requirements starting from any state.

5 Software Infrastructure Design

According to the object-oriented software design method, the flight calculation model is analyzed. This model separates the calculation process from the calculation data. The classes supporting model calculation and the relationships among them were extracted, and the basic integral model class, flight integral model class, flight calculation model class and aircraft model class were constructed. Each class is responsible for its own core computing data and core computing capabilities, and its architecture is shown (see Fig. 1). This architecture lays a foundation for flight data management, flight computation management, flight state transfer mechanism design, breadth-first parallel computation design and implementation.

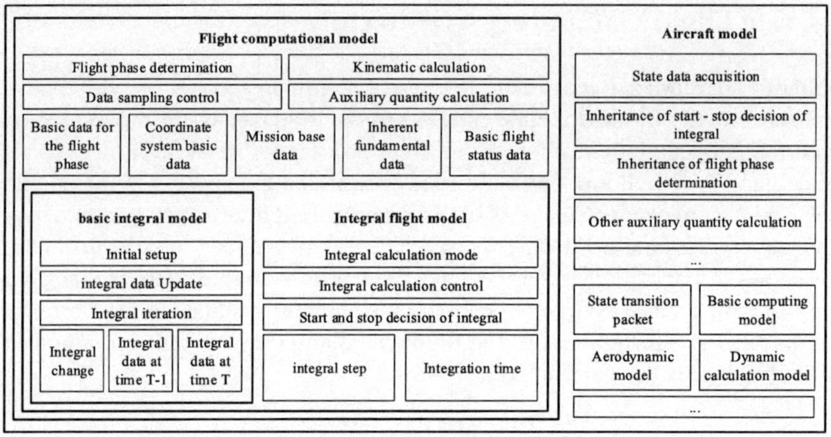

Fig. 1. Software infrastructure

5.1 Basic Integration Model

The basic integration model is the basis of the integration iteration, and it implements the specific integration algorithm and integration mechanism internally. The specific functions are as follows:

A) Initialization. Initialize the variables used for integration and use array to manage the integration variables.

B) Update the integral data. Update the integral variable and integrand component used for the integration iteration.

C) Integral iteration. Calculate the product component according to the specified step size and the specified calculation model.

5.2 Flight Integration Model

The flight integration model includes the basic integration model, which is the basis for the calculation of relevant integral data and flight control algorithm in flight calculation. Specific functions are as follows:

A) Definition of integral calculation model. The virtual function is defined in the integral calculation model, and the class interface of flight calculation model is reserved.

B) Decision of start and stop of flight points. An interface virtual function is provided to determine whether to stop integration, which is used as the start and stop condition of the flight path integration control.

C) Integral calculation control. According to the start and stop conditions of the integral in the track calculation, the calculation model is called and the integral model is updated.

5.3 Flight Computation Model

The flight computation model class inherits from the flight integral model class, defines the basic data required for flight computation, and realizes the functions and process control related to flight computation. The specific functions are as follows:

A) Flight phase determination. According to the current state of the flight stage to judge, and according to the flight stage of the relevant solution.

B) Kinematics calculation. According to the flight operation method, complete the related basic calculation.

C) Data sampling control. According to a certain sampling period, complete the sampling of all kinds of basic data.

D) Calculation of auxiliary quantity. According to the standard flight calculation model class, complete the calculation of aerodynamic, geographic data, coordinate transformation matrix, etc.

5.4 Aircraft Model

The aircraft model is inherited from the flight calculation model, and the virtual function in the flight integral model and flight calculation model is realized. The aircraft model contains the calculation models related to aerodynamics, power and overload, which is the concrete realization of the flight calculation model. Aircraft model class can construct various types of aircraft according to different functions and different flight stages. The internal attributes and methods in the aircraft model are inherited from the flight calculation model, so the abstract principle in the object-oriented method can be used to realize the unified control and scheduling of different aircraft models.

6 Data Management Architecture

6.1 Data Architecture

In the algorithm module of the flight calculation model, the data structure is formed according to the principle of low coupling and high reuse, and at the same time, the principle of calculating the invariable data only once in each iteration is taken into account. The flight stage data, initial coordinate system data, flight task data, aircraft inherent data and flight state data used for calculation were sorted out to determine their contents and initialization sequence, as shown in Fig. 2.

The types of data are organized as basic data and derived data. Basic data is applied to the flight calculation model, which belongs to the common data of flight calculation. Derived data is applied to the aircraft model and inherited from the basic data.

6.2 Basic Data

Basic Data of Flight Stage. Fundamental data of flight stages are used to control the application of different computational models at each flight stage. Enumeration variables are used to determine flight stages such as preparation, engine start, take-off, flight and landing.

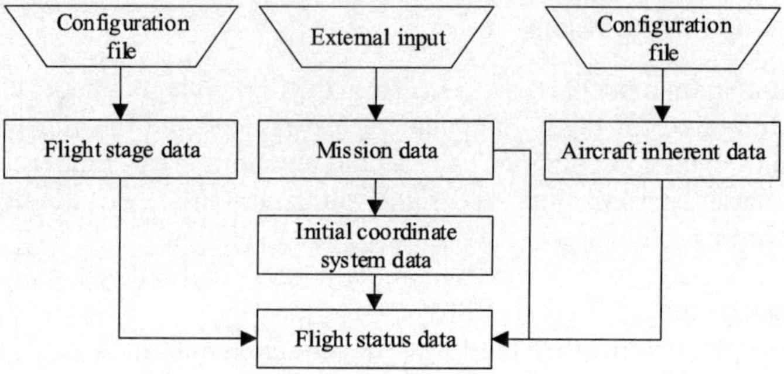

Fig. 2. Data model system

Basic Data of the Task. Mission basic data is used to clarify the current state of the aircraft and determine the mission objectives. At the same time, the reusable data can be determined only from the mission data, including the take-off location and landing location, and the derived data such as flight direction and theoretical range can be obtained.

The Inherent Basic Data of the Aircraft. The inherent basic data of the aircraft are constant data, including the mass, length, area and other basic data of each component of the aircraft.

Basic Data of the Initial Coordinate System. For long-haul aircraft, the basic data of the initial coordinate system is used to calculate the inherent reusable data in the ground Cartesian coordinate system according to the take-off position, landing point position and the earth model. The ground Cartesian coordinate system data include the components of the geocentric vector diameter of the take-off point in the initial coordinate system, the components of the geocentric vector diameter of the landing point in the initial coordinate system, the angular velocity of the earth rotation, the gravitational acceleration of the take-off point, and all kinds of transformation matrices independent of the flight position, such as the Gothic acceleration matrix, etc. For small flight aircraft, directly specify the Gaussian Cartesian coordinate system.

Basic Data of Flight Status. Flight state basic data is used to record each time point of flying state, including flight time, basic product component (including quality, position, velocity and acceleration), the amount of space position (altitude, latitude, longitude and altitude), the attitude data volume (pitching Angle and yaw Angle, roll Angle) and other auxiliary materials (the current voyage, pressure), etc.

6.3 Derived Data

Derived Data Definition. According to different aircraft model classes, relevant basic data can be inherited for expansion. For example, navigation point constraint can be added to flight mission data, and overload can be added to flight status data. The corresponding relationship of specific expansion is shown in Fig. 3.

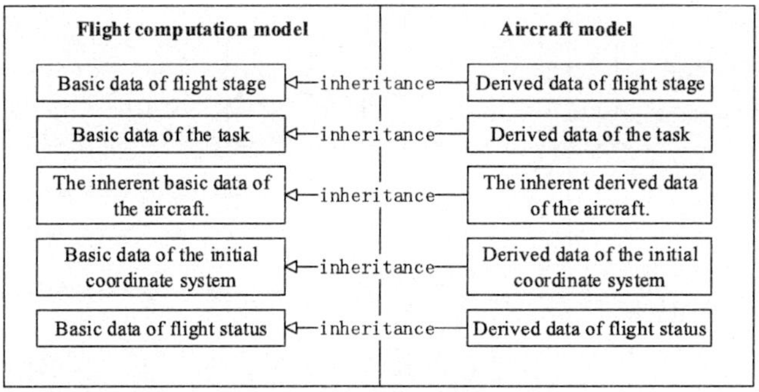

Fig. 3. Data expansion system

Derived Data Registration. In the flight computing model, the pointer pointing to the basic data is declared and the dynamic initialization and registration of the basic data are completed. In the aircraft model, the derived data can be re-registered and replaced according to the basic data superclass type. The flight computation model is responsible for the assignment and operation of the base data part of the derived data. The aircraft model class completes the assignment and operation of the new part in the derived data. This registration and invocation mechanism enables custom data expansion without changing the flight computation model class structure.

7 Computational Process Architecture

According to the divide-and-conquer cooperation of flight integration model, flight calculation model and aircraft model, the flight calculation is completed. The flight integral model is responsible for the initialization of the basic integral model and the start and stop control of the integral. The flight calculation model class is responsible for the identification of flight stages and the solution and control of each flight stage. The virtual function is realized in the aircraft model class, the specific calculation model is defined and the solution of auxiliary data is completed. The calculation process is shown in Fig. 4, where "(v)" represents a virtual function and "call" represents a method call.

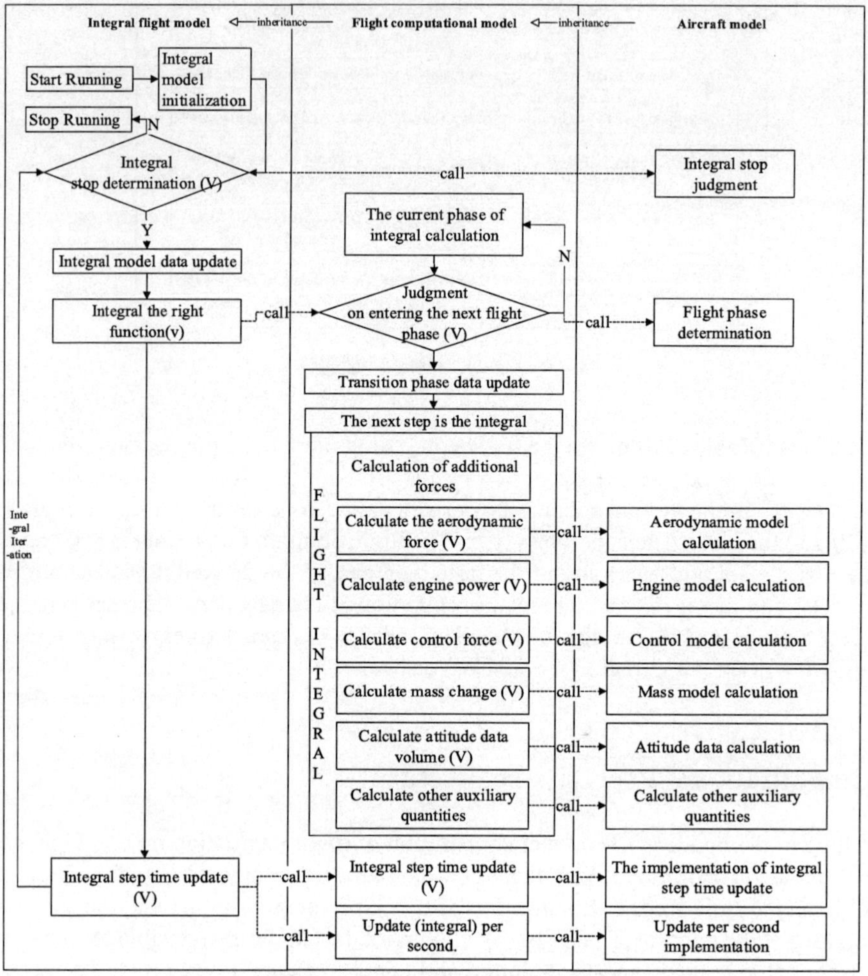

Fig. 4. Flight calculation process

8 Flight Status Transfer Architecture

The aircraft model inherits from the flight calculation model and is completely compatible with the basic data of flight calculation. Therefore, under the condition of ensuring the compatibility of derived data, the smooth state transfer of different aircraft model objects and the same aircraft model object from the previous state to the next state can be realized. This architecture uses the state transition data package to implement the transition between different states. The state transfer packet is based on flight stage data, initial coordinate system data, flight mission data, aircraft inherent data and flight state data, as well as the integral variables in the flight integration model. The state transfer packet can update the data of the basic integral model, the flight integral model, the flight computation model and the aircraft model in one operation. This architecture guarantees the same state of the two aircraft models before and after the transfer, and realizes the transfer of the flight state of the entity calculation model with different flight states. The flight state transfer mechanism is shown in Fig. 5.

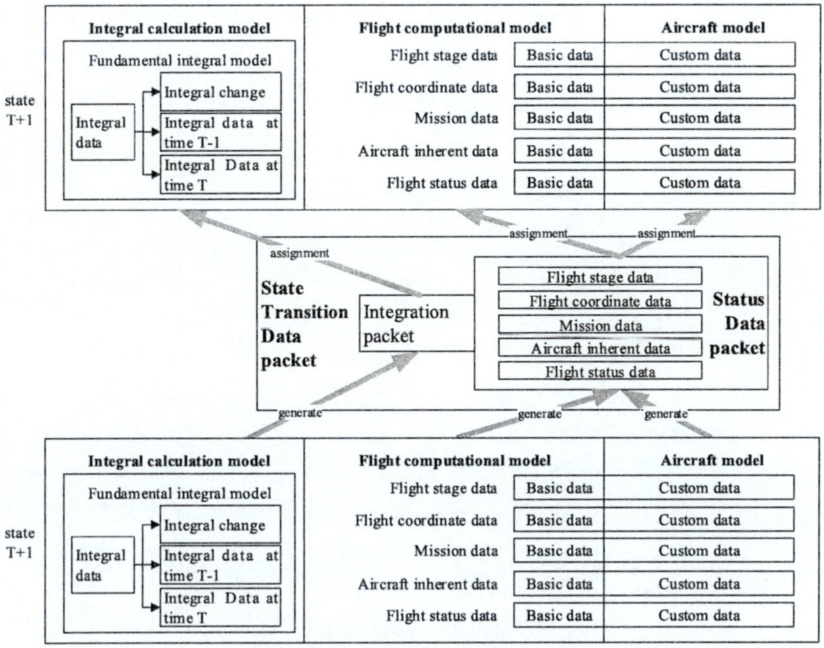

Fig. 5. Flight state transfer mechanism

9 Reinforcement Learning Computing Support

The traditional flight calculation method is depth-first, and all flight paths can be traversed by the parallel method of one-by-one calculation. Therefore, reinforcement learning training using traditional parallel mode will produce a large number of repeated calculations, and it is impossible to carry out phase iteration (see Fig. 6).

The aircraft model is designed by object-oriented method, which can record the current state and decision parameters in any state S, and can also be calculated from any state as the starting point. This architecture can solve the problem of multi-stage iteration and avoid a lot of repeated calculations. In order to meet the need of parallelization, this architecture supports the use of thread pool or distributed technology to build multiple aircraft model object containers M. It is controlled by scheduling model to carry out parallel computation. The aircraft model object used to calculate procedure P can be run in container M, as shown in Fig. 7. The switching of model objects is carried out through state transfer packets and state transfer mechanism.

Due to the object-oriented software design method and considering the state transition of multiple flight stages, the propellant state transition model and the feedback

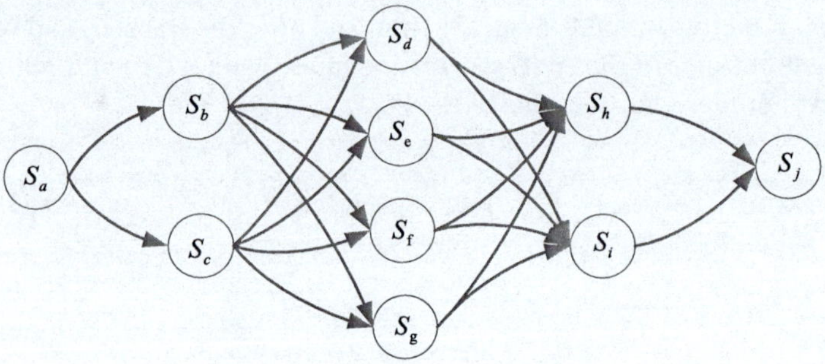

Fig. 6. State transition diagram.

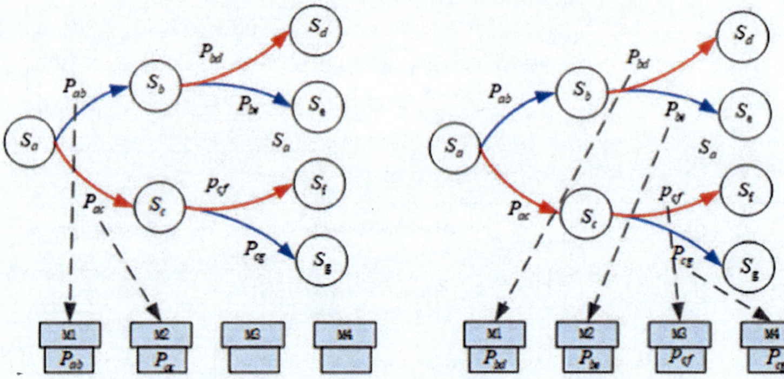

Fig. 7. Parallel calculation process

state transition model can realize parallel computation for different flight stages. This framework has significant advantages in large-scale reinforcement learning training.

10 Conclusion

In this paper, an object-oriented flight computing architecture based on state transition model is constructed to meet reinforcement learning requirements. In this architecture, data and calculation methods can be separated, and parallelization and breadth-first flight computation can be realized through state transfer mechanism. On this basis, the state transfer types related to flight computation are analyzed. This paper provides a good idea for using reinforcement learning method to complete flight calculation and aircraft performance optimization.

References

1. Xian, G., Yongchun, F.: Introduction to Reinforcement Learning Principles. Publishing House of Electronics Industry, Beijing (2018)

2. Kondrat'ev, A.I., Tyumentsev, Y.V. Application of neural networks for synthesizing flight control algorithms. I neural network inverse dynamics method for aircraft flight control. Russ. Aeronaut. **56**, 135–144 (2013)
3. Been, Y., Kang, Y., Kim, S., Shin, S., Jang, S.: Development of the snapshot method for six degree-of-freedom flight dynamics simulation of a high aspect ratio wing aircraft. Int. J. Aeronaut. Space Sci. **19**(4), 904–919 (2018). https://doi.org/10.1007/s42405-018-0097-0
4. Vogeltanz, T.: A survey of free software for the design, analysis, modelling, and simulation of an unmanned aerial vehicle. Arch. Comput. Meth. Eng. **23**(3), 449–514 (2015). https://doi.org/10.1007/s11831-015-9147-y
5. Belokon', S.A., Zolotukhin, Y.N., Filippov, M.N.: Architecture of a platform for hardware-in-the-loop simulation of flying vehicle control systems. Optoelectron. Instrument. Proc. **53**, 345–350 (2017)
6. Yakovlev, K.S., Makarov, D.A., Baskin, E.S.: Automatic path planning for an unmanned drone with constrained flight dynamics. Sci. Tech. Inf. Process. **42**(5), 347–358 (2015). https://doi.org/10.3103/S0147688215050093
7. Izuta, S., Takahashi, M.: Path planning to improve reachability in a forced landing. J. Intell. Rob. Syst. **86**(2), 291–307 (2016). https://doi.org/10.1007/s10846-016-0431-3
8. Rodriguez-Ramos, A., Sampedro, C., Bavle, H., de la Puente, P., Campoy, P.: A deep reinforcement learning strategy for UAV autonomous landing on a moving platform. J. Intell. Rob. Syst. **93**(1–2), 351–366 (2018). https://doi.org/10.1007/s10846-018-0891-8

NLP and Clustering

HDMacBERT-FGM-CRF Model-Based Chinese Event Element Extraction

Peng Chen[1], Baohua Qiang[1](✉), Yufeng Wang[2], Chen Wang[1], Ruidong Chen[1], Baolian Li[2], Bo Peng[2], and Xiangyu Zhou[1]

[1] Guangxi Key Laboratory of Image and Graphic Intelligent Processing, Guilin University of Electronic Technology, Guilin 541004, Guangxi, China
qiangbh@guet.edu.cn

[2] The 54th Research Institute of China Electronics Technology Group Corporation, Shijiazhuang 050002, China

Abstract. Because of the inaccuracy of event element extraction caused by long text, the errors of pre-training and fine-tuning stages, and the single sample of semantic features in the current Chinese event element extraction model, Hierarchical Decomposition MacBERT and Fast Gradient Method-based Chinese event element extraction algorithm by jointing Conditional Random Field (HDMacBERT-FGM-CRF) is proposed. This model extends absolute location encoding based on hierarchical decomposition so that the HDMacBERT model can directly process more than 512 Token texts. The HDMacBERT model based on the Mask strategy of similar words is used to mitigate errors. The FGM-based approach implements adversarial networks and increases the diversity of semantic feature samples. The P, R and F_1 of the HDMacBERT-FGM-CRF model proposed in this paper are 2.52%, 4.03% and 3.26% higher than those of the classical Chinese event element extraction model BERT-CRF on the DuEE dataset. The experimental results show that the proposed model can learn more comprehensive deep-level semantic information of text and improve Chinese event element extraction.

Keywords: Chinese event element extraction · HDMacBERT · Fast gradient method · Conditional random field

1 Introduction

Event extraction (EE) is a core task of text data mining. Its purpose is to automatically identify event information from free text and present it in a structured form [1]. It is generally divided into event trigger word extraction and event element extraction [2]. The event element is more challenging to identify and extract than the event trigger word, but it is also more critical in describing event information.

Chinese event element extraction is applying event extraction tasks in the Chinese field, which is mainly divided into two parts: feature selection and classification models. In feature selection, unlike English event element extraction, Chinese text has the characteristics of semantic fuzziness, complex word boundary, and high-dimensional sparsity [3]. Therefore, it needs to design an effective method to learn text semantic features

S.-C. Chu et al. (Eds.): ICGEC 2021, LNEE 833, pp. 395–406, 2022.
https://doi.org/10.1007/978-981-16-8430-2_36

and context correspondence and usually uses a word embedding model or dynamic pre-training model to learn text context semantic features. For classification models, Support Vector Machine (SVM), Maximum Entropy Markov Model (MEMM), and Conditional Random Field (CRF) are usually used to label candidate sequences. Existing studies show that using a pre-training model can learn deep semantic features and significantly improve the effect of event element extraction.

Pre-training models are generally divided into two categories, static pre-training model and dynamic pre-training model. Static pre-training models such as Word2Vec [4], GloVe [5], and dynamic pre-training models such as BERT [6], MacBERT [7]. Static pre-training models are faster and can learn semantic information between each word. However, there is a problem of polysemy, which was alleviated by Google's release of dynamic pre-training model BERT in 2019. However, there are some problems in the BERT pre-training model, such as the error in pre-training and fine-tuning stages. The pre-training model cannot process long text directly, and fewer semantic feature samples are generated according to the vocabulary.

This paper uses the HDMacBERT pre-training model to make feature selection, alleviating the errors of pre-training and fine-tuning stages. By extending the range of absolute position encoding through hierarchical decomposition [8], the HDMacBERT model is derived, enabling the model to process long text directly and guarantees the integrity of sentence semantic features. In the Embedding layer, a Fast Gradient Method (FGM) is used to add perturbation [9], which increases the diversity of semantic feature samples and improves the effect of event element extraction. Finally, CRF converts the state feature matrix, fusing the semantic features to get a good labelling result.

The contributions of this study are as follows:

(1) Using the HDMacBERT model to learn the semantic features of text, alleviate the errors of pre-training and fine-tuning stages based on meeting the characteristics of Chinese text.
(2) Hierarchical decomposition is used to expand the absolute location encoding range of MacBERT, resulting in the HDMacBERT model, which can directly process more than 512 Token texts.
(3) Using FGM to implement antagonism training, adding disturbance to Embedding layer to increase the diversity of semantic feature samples and improve the effect of Chinese event element extraction.

The remainder of this paper is organized as follows: Related research is reviewed in Sect. 2. The proposed method is illustrated in Sect. 3. Results of three comparative experiments are presented in Sect. 4. Finally, conclusions are drawn in Sect. 5.

2 Related Work

The methods of extracting event elements can be divided into two categories: pattern matching-based method and machine learning-based method [1]. The pattern matching-based method identifies and extracts certain types of event elements under the guidance of some templates, which shows good performance in specific areas. Based on machine learning, event element extraction is modelled as a multi-class task, and event element extraction is completed by extracting features for classification.

The method based on pattern matching is that researchers extract event elements for building templates for professional domain data. Existing work, such as Jiang Jifa [10], has proposed an event pattern matching method in the field, IEPAM, which achieves excellent results in the extraction of flight accident event elements from MUC-7 corpus. Xingsi Xue et al. [11] presents an innovative interactive ontology matching technique based on Partial Reference Alignment (PRA) to better balance between the large workload posed on users and the demand of improving the quality of ontology alignment. Arendarenko et al. [5] established a knowledge base through the pre-processing process of sequential labelling, sentence morphology analysis, and text semantics analysis of enterprise business text, followed by rule-based event element extraction, which achieved good results in enterprise business text event element extraction. Xingsi Xue et al. [12] presents an effective technology to solve the ontology heterogeneous problem based on ontology matching to identify the correspondences between the entities of heterogeneous ontologies.However, it takes much human time to make templates. Moreover, the templates are limited to the domain background, so it is difficult to apply them in the generic domain event element extraction tasks.

Machine learning-based methods extract and model event elements into multiple classified tasks. Machine learning enables models to automatically learn text features from the corpus without spending much time making templates. It is the leading research direction at present. Work has been done, such as Jacobs et al. [13], to convert ten economic news text categories into multiple binary categories, learn semantic text features through LSTM, and finally use SVM to get good classification results. In 2016, Feng et al. [14] proposed to use a two-way long-term and short-term memory network for event element detection, followed by the second classification of detected event elements to get a reasonable label. Zheng et al. [15] proposed a sequence-to-sequence model, Doc2EDAG, which uses multiple Transformer models [16] to detect entities to form directed acyclic diagrams, and then extracts text-level event elements similar to the classification of named entity recognition. Because the machine learning method works well in extracting event elements, this paper also uses the machine learning method to extract event elements in the Chinese field.

Feature selection is essential in machine learning-based event element extraction, and the pre-training model can better learn the deep semantic information of the text. Existing work, such as Ein-Dor et al. [17] converts event element extraction tasks into categorical tasks. First, the pre-training model BERT is used to train on the news corpus. Then the generated model is used to detect common news event elements, resulting in a better news event element extraction model. Zhao et al. [18] used the pre-training model RoBERTa to conduct text effective analysis, multi-scale entity selection, and entity analysis on financial text data to improve the effect of event element extraction.

In summary, previous event element extraction is mainly based on pattern matching. However, template making takes much manual time, and its generalization ability is weak, so it is gradually discarded. Using the BERT and its improved pre-training model to extract event elements has achieved good results in recent years. Machine learning method based on the BERT model has gradually become a mainstream direction of event element extraction. However, the current method based on the BERT model only uses the learning ability of the pre-training model, without considering the errors of pre-training and fine-tuning stages, the problem that longer text cannot be directly processed, and the problem of fewer semantic feature samples. Therefore, this paper uses the MacBERT to learn the profound text semantic features to alleviate the errors of pre-training and fine-tuning stages. Hierarchical decomposition is used to extend the absolute location encoding range of the MacBERT so that longer text can be processed directly. Adding perturbations to the Embedding layer using FGM increases the diversity of the semantic feature samples. It makes the model more context-aware and improves the performance of event element extraction.

3 HDMacBERT-FGM-CRF Model

3.1 Model Framework

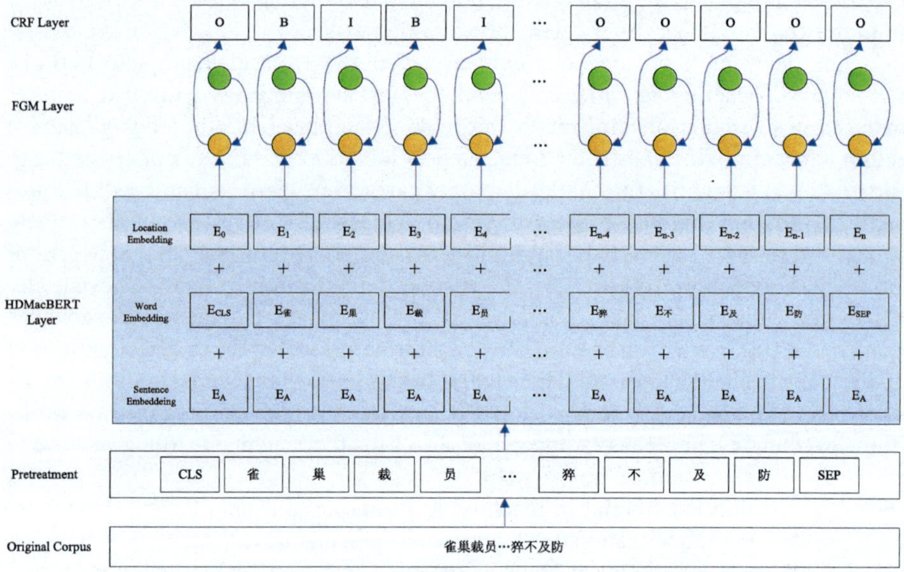

Fig. 1. Diagram of HDMacBERT-FGM-CRF model

Figure 1 shows the overall framework of this model. It mainly includes the HDMacBERT layer, FGM disturbance layer, and CRF layer. The specific functions and implementations of each layer in the model are detailed in Sects. 3.2 to 3.4.

This paper presents a Chinese event element extraction model based on the MacBERT pre-training model, the HDMacBERT-FGM-CRF model, to extract relevant elements from news events. The original corpus is first divided into word granularity, and input into the HDMacBERT (Hierarchical Decomposition MacBERT) layer to avoid problems such as unregistered words and ambiguous word boundaries. Next, the initial Embedding of the text is obtained by adding word, sentence, and position vectors together. Then disturbances are added to the initial Embedding by FGM to obtain the final Embedding. Finally, CRF is used to obtain a good labelling result.

3.2 HDMacBERT Model

Hierarchical decomposition of the absolute location coding range of the MacBERT pre-training model yields the HDMacBERT model, which allows the pre-training model to process long text directly without changing the original model structure.

MacBERT Model

The MacBERT is a pre-training model that continuously optimizes training based on the BERT model. Unlike BERT's All Mask strategy for random words, MacBERT uses the Whole Mask and Ngram Mask strategies to select Mask Token for training, where the unigram to 4-g columns were 40%, 30%, 20%, and 10%, respectively [7]. To alleviate the error of pre-training and fine-tuning stages, Mask Token is replaced by similar words based on Wordvector calculation instead of Mask strategy in the fine-tuning stage. In rare cases, no similar words are found, and random Mask replacement is used.

Hierarchical Decomposition MacBERT Model

Using the MacBERT pre-training model to extract Chinese event elements, although a similar word Mask strategy reduces the error of pre-training and fine-tuning stages, MacBERT is an absolute position encoding and cannot process long text directly, which results in text semantic incompleteness. This paper uses hierarchical decomposition to extend the absolute location encoding range of MacBERT, which can process n Tokens to process n^2 Tokens directly if the graphics card resources allow.

Hierarchical decomposition assumes that the trained absolute location coding vector is $p_1, p_2, \ldots, p_n$ and the extended coding vector is $q_1, q_2, \ldots, q_m$, where $m > n$. The extended function of the absolute position-encoding is defined as Eq. (1), where $\alpha \in (0, 1)$ is a hyperparameter, $u_1, u_2, \ldots, u_n$ is the "base" of absolute location coding. The "base" refers to the representation of each position vector in the original absolute position code. The position codes αu_i and $(1-\alpha)u_j$ corresponding to i and j are superimposed to represent the position codes of $(i-1)\times n + j$, extending from n to m.

$$q_{(i-1)\times n+j} = \alpha u_i + (1-\alpha)u_j \tag{1}$$

in order not to change MacBERT's original absolute position encoding, keep the position vector constant when the number of Tokens does not exceed n, that is, satisfy $q_1 = p_1, q_2 = p_2, \cdots, q_n = p_n$, so each u_i can be reversed. The "base" function of the absolute position-encoding is defined as follows:

$$u_i = \frac{p_i - \alpha p_1}{1-\alpha}, i = 1, 2, \cdots, n \tag{2}$$

the HDMacBERT parameter is still $p_1, p_2, \ldots, p_n$, but it can represent n^2 Tokens, and the position vectors of the first n Tokens are the same as before. Thus, although the training time is increased in Chinese event element extraction, it can avoid the situation that more than 512 Token texts cannot be processed directly and ensure the semantic integrity of the text.

3.3 FGM Model-Based Countermeasure Training

In deep learning and machine learning, antagonistic training generally has two meanings: one is to generate an antagonistic network, increase the difficulty of model training to improve model learning and care about the robustness of the model under minor disturbances. For example, there are one-to-one, one-to-many, many-to-one, many-to-many correspondences in Chinese event element extraction. Therefore, finding the corresponding relationship between each element and event type can improve the extraction effect. This paper uses FGM-based antagonistic networks to improve model learning ability and find the corresponding relationship between elements and types.

FGM-based antagonistic training adds disturbance to the Embedding layer of text semantic feature learning to increase the diversity of semantic feature samples. However, no matter what kind of antagonistic training, the antagonistic function of the model training is defined as follows:

$$\underset{\theta}{min}\mathbb{E}_{(x,y)\sim T}\left[\underset{\Delta x\in\Omega}{max}L(x+\Delta x, y;\theta)\right] \tag{3}$$

where T represents the training set extracted from event elements, x represents the input, y represents the element label, θ is the model parameter, $L(x + \Delta x, y; \theta)$ is the loss of a single sample, Δx is the added disturbance, Ω is the perturbation space. For each sample, adding perturbation Δx to x improves the learning ability of the model by making the loss of a single sample as large as possible and making model predictions as incorrect as possible.

The standard way to reduce loss is the gradient descent algorithm. This paper uses the ascending gradient method to calculate Δx, which interferes with loss reduction and increases the difficulty of model training, thus increasing the ability of the model to learn deep-level semantic features of text. The disturbance function of the model training is defined as follows:

$$\Delta x = \nabla_x L(x, y; \theta) \tag{4}$$

if Δx too large, the semantic features of model learning are mostly erroneous information, thus under-fitting reduces the effect of Chinese element extraction, to prevent Δx from being too large for standardization. The standardized function of the model training is defined as follows:

$$\Delta x = \text{sign}(\nabla_x L(x, y; \theta)) \tag{5}$$

Equation (3), Eq. (4), and Eq. (5) are sorted out to get the definition of FGM-based antagonistic training function as follows:

$$\underset{\theta}{min}\mathbb{E}_{(x,y)\sim T}[L(x+\Delta x, y;\theta)] \tag{6}$$

3.4 CRF Model-Based Predicting Probability Value of Sequence Tags

This paper uses a BIO mode based on sequence labelling, where "B" denotes the beginning of an event element, "I" denotes the content of this event element, and "O" denotes not an event element. Therefore, there is a strong dependency between tags and adjacent tags in the text. For example, if the current tag is "B-Company", the following tag cannot be "I-People". It can only be "I-Company" or "O". CRF can capture the dependency characteristics between context tags and infer a reasonable sequence of tags. Therefore, after the HDMacBERT passes through the FGM output vector $p_1, p_2, \ldots, p_n$, the CRF layer is used to calculate the probability value of the tag sequence $q_1, q_2, \ldots, q_n$. The probability function of the HDMacBERT-FGM-CRF model is defined as follows:

$$p(y \mid s) = \frac{\exp\left(\sum_i \left(W_{CRF}^{q_i} p_i + b_{CRF}^{(q_{i-1}, q_i)}\right)\right)}{\sum_{y'} \exp\left(\sum_i \left(W_{CRF}^{q_i'} p_i + b_{CRF}^{\left(q_{i-1}', q_i'\right)}\right)\right)} \tag{7}$$

where y' represents any of the tag sequences, $W_{CRF}^{q_i}$ represents the model parameters of the tag q_i, and $b_{CRF}^{(q_{i-1}, q_i)}$ represents the deviation between q_{i-1} and q_i.

During the training process, the maximum likelihood function of (p_i, q_i) is optimized by regularization. The regular function of the HDMacBERT-FGM-CRF model is defined as follows:

$$L = \sum_{i=1}^{N} \log(p(y_i \mid s_i)) + \frac{\lambda}{2} \|\theta\|^2 \tag{8}$$

where $p(y_i \mid s_i)$ is the probability value from the original sequence to the model prediction sequence, λ and θ are the regularization parameters. CRF derives tags with reasonable sequence based on the maximum likelihood function.

4 Experiments

4.1 Experimental Datasets

This paper crawls nearly 10,000 military category news texts from major news websites such as Xinhua, People's Net, and NetEase. Finally, it collates some data to get 16 event element categories in 2 event categories, totalling 300 military category data. Based on military data, add the largest Chinese event extraction dataset in the industry, Baidu DuEE. The DuEE contains 65 event categories totalling 213 event element categories, totalling 13375 data. The total data is 13675. The training set and test set are divided into 8:2 scales. There are 10938 data in the training set and 2737 data in the test set. During the training, 20% of 10938 data in the training set is used as the verification set. The framework of event elements for event categories is shown in Table 1. There are 67 event categories in total, and three are shown here. Each event type corresponds to several event elements. For example, event type "Military-Multinational Joint Exercise" has event elements such as "Trigger", "Start" and so on.

Table 1. Event framework

Event types	Military-multinational joint exercise
Military-multinational joint exercise	Trigger, start, end, country, purpose, exercise name, equipment, military, location
Finance/transaction-sale/acquisition	Trigger, time, seller, transaction, selling price, buyer
Finance/Transaction-financing	Trigger, time, follow investor, lead investor, financing rounds, financing amount, financing party

4.2 Evaluation Criteria and Environmental Configuration

In this paper, the accuracy (Precision, P), recall (Recall, R), F value (F_1) are used as the criteria to evaluate the extraction of Chinese event elements. The calculations for the three are shown in Eq. (9), Eq. (10), and Eq. (11).

$$P = \frac{TP}{TP + FP} \tag{9}$$

$$R = \frac{TP}{TP + FN} \tag{10}$$

$$\mathrm{F}_1 = \frac{2 \times P \times R}{P + R} \tag{11}$$

among them, TP is the number of word levels that the model predicts correctly for both event types and event elements, FP is the number of word levels that the model predicts incorrectly for event types or event elements, and FN is the number of word levels that the model does not extract correctly for event elements. The experimental environment is shown in Table 2.

Table 2. Experimental environment

Experimental environment	Environment configuration
Operating system	Ubuntu 16.04.5
CPU	Intel(R) Xeon(R) CPU E5–2630 v4 @ 2.20 GHz
GPU	Tesla V100-PCIE-16 GB
CU_DRV	10.2, 440.64.00
Python	3.6
In-depth learning framework	Tensorflow 1.14, Keras 2.3.1, Bert4Keras 0.9.3

4.3 Setup of Experimental Parameters

This paper uses the HDMacBERT pre-training model to learn the deep semantic features of text, disturbs the Embedding layer by FGM, and obtains good labelling by CRF. The

HDMacBERT-FGM-CRF parameters of the in-depth learning model are shown in Table 3. This parameter is mainly based on the model parameters proposed by Chen Xingyue [19].

Table 3. Model parameters

Parameters	Parameters values
Learning rate	0.00002
Maximum text length	600
batch size	8
Epoch	20
CRF learning rate	100
Perturbation parameters	0.5

4.4 Contrast Experiment Settings

To verify the validity of the proposed HDMacBERT-FGM-CRF model, the comparison experiments with the following models are carried out:

BERT-CRF model [20]: Uses the BERT pre-training model published by Google in 2019 to learn the deep semantic characteristics of the text. CRF captures the dependency characteristics between context tags. BERT solves the problem of polysemy in static pre-training models such as Word2Vec.

MacBERT-CRF model: Use the MacBERT model published by SCIR Lab of the Harbin University of Technology in 2020 to learn the deep semantic features of a text. MacBERT effectively mitigates errors in the pre-training and fine-tuning phases.

MacBERT-FGM-CRF model: FGM mode is used to generate the anti-network, adding disturbance in the Embedding layer of the model to increase the learning ability of the model.

The comparison test results are shown in Table 4.

Table 4. Benchmark algorithm comparison

Models	$P(\%)$	$R(\%)$	$F_1(\%)$
BERT-CRF	76.29	73.78	75.05
MacBERT-CRF	76.05	76.71	76.38
MacBERT-FGM-CRF	78.62	76.12	77.35
HDMacBERT-FGM-CRF	78.81	77.81	78.31

From Table 4, the following conclusions can be drawn:

(1) Comparing the performance of the BERT-CRF model and the MacBERT-CRF model in the experiment, it is shown that although the BERT pre-training model in the BERT-CRF model can learn the deep semantic features of the text, due to the adoption of random word Mask strategy, the pre-training and fine-tuning stages have enormous differences, resulting in a lower recall rate R-value of Chinese event element extraction. The MacBERT similar word Mask strategy effectively alleviates this problem and improves the extraction performance without reducing the accuracy. Compared with model BERT-CRF, model MacBERT-CRF increased *R*-value by 2.93% and F_1-value by 1.33%. Additionally, *P*-value decreased by 0.24% because MacBERT model masking strategy always uses similar words and can't learn word meanings more extensively than BERT model using random words.
(2) Comparing the performance of the MacBERT-CRF model and the MacBERT-FGM-CRF model in the experiment, it is shown that adding disturbance to the Embedding layer by FGM to model MacBERT-FGM-CRF can effectively improve the extraction of Chinese event elements by increasing the sample diversity of semantic features. Compared with the MacBERT-CRF model, the MacBERT-FGM-CRF model increased *P*-value by 2.57%, and F_1-value by 0.97%. Additionally, adding disturbance in Embedding layer by FGM in MacBERT-FGM-CRF model greatly increases the *P*-value of event extraction, but adding negative samples to antagonistic training results in a slight decrease in recall rate R-value, which is 0.59%.
(3) Compared with the MacBERT-FGM-CRF model, the HDMacBERT-FGM-CRF model decomposes the absolute position-coding of MacBERT pre-training model hierarchically, which enables HDMacBERT to process more than 512 Token texts directly without changing the initial structure of MacBERT, which ensures the integrity of text semantic information and improves the overall effect of Chinese event element extraction. Compared with the MacBERT-FGM-CRF model, the HDMacBERT-FGM-CRF model increased *P*-value by 0.19%, *R*-value by 1.69%, F_1-value by 0.96%. Compared with the BERT-CRF model, the HDMacBERT-FGM-CRF model increased *P*-value by 2.52%, *R*-value by 4.03%, and F_1-value by 3.26%.

5 Conclusions

This paper presents an HDMacBERT-FGM-CRF model for antagonistic training based on the MacBERT pre-training model and FGM method. The model first extends the absolute location encoding range of MacBERT by hierarchical decomposition, enabling the HDMacBERT model to directly learn more than 512 profound text semantic features of Token while maintaining the hierarchical structure of the pre-training model. Secondly, disturbances are added to the Embedding layer of the model by FGM to increase the sample diversity of the semantic features. Finally, CRF is used to capture the dependency characteristics between context tags to infer a reasonable sequence of tags. The HDMacBERT-FGM-CRF model proposed in this paper has better Chinese event element extraction performance and verifies the effectiveness of using hierarchical decomposition MacBERT model to enhance text semantic feature learning ability through three sets of comparative experiments and improve model learning ability through FGM. However, direct processing of extra-long text requires more hardware. In the future, we can consider slicing and stitching text to learn profound text semantic information. In addition, due to the large pre-training model and slow training speed, using model compression technology without reducing the model effect is also one of the directions of future research.

Acknowledgments. This work was supported in part by the National Natural Science Foundation of China under Grant 61762025; in part by the Development Foundation of the 54th Research Institute of China Electronics Technology Group Corporation under Grant SXX18138X017; in part by the Natural Science Foundation of Guangxi under Grants 2019GXNSFDA185007, and 2019GXNSFDA185006; in part by the Guangxi Key Research and Development Project under Grants AB18126053, AB18126063, AD18281002, and AD19110137; in part by the Guilin Science and Technology Development Program under Grants 20190211–17; in part by the Guangxi Key Science and Technology Planning Project under Grants AA18118031 and AA18242028; in part by the Innovation Project of GUET Graduate Education under Grants 2020YCXS052.

References

1. Xiang, W., Wang, B.: Survey of Chinese event extraction research. Comput. Technol. Dev. **30**(2), 1–6 (2020). https://doi.org/10.3969/j.issn.1673-629X.2020.02.001
2. Chen, Y., Liu, S., Zhang, X., et al.: Automatically labeled data generation for large scale event extraction. In: Proceedings of the 55th Annual Meeting of the Association for Computational Linguistics, vol. 1: Long Papers), pp. 409–419 (2017). https://doi.org/10.18653/v1/P17-1038
3. Wollmer, M., Weninger, F., Knaup, T., et al.: YouTube movie reviews: sentiment analysis in an audio-visual context. IEEE Intell. Syst. **28**(3), 46–53 (2013)
4. Feldman, R., Rosenfeld, B., Bar-Haim, R., et al.: The stock sonar-sentiment analysis of stocks basedon a hybrid approach. In: Conference on Innovative Applications of Artificial Intelligence, pp. 1642–1647. DBLP (2011)
5. Arendarenko, E., Kakkonen, T.: Ontology-based information and event extraction for business intelligence. In: Ramsay, A., Agre, G. (eds.) AIMSA 2012. LNCS (LNAI), vol. 7557, pp. 89–102. Springer, Heidelberg (2012). https://doi.org/10.1007/978-3-642-33185-5_10
6. Devlin, J., Chang, M., Lee, K., et al.: Bert: pre-training of deep bi-directional transformers for language understanding. In: Proceedings of Conference of the North American Chapter of the Association for Computational Linguistics: Human Language Technologies. Stroudsburg, pp. 4171–4186. ACL (2019). arXiv:1810.04805v1 [cs.CL]
7. Cui, Y., Che, W., Liu, T., et al.: MacBERT: Revisiting Pre-Trained Models for Chinese Natural Language Processing (2020). arXiv preprint arXiv:2004.13922, https://doi.org/10.18653/v1/2020.findings-emnlp.58
8. Su, J.: Hierarchical decomposition and position coding, let Bert deal with super long text (2020). https://kexue.fm/archives/7947
9. Goodfellow, L.J., Shlens, J., Szegedy, C., et al.: FGM:Explaining and Harnessing Adversarial Examples (2015). arXiv preprint arXiv:1412.6572v3
10. Jiang, J.: An event IE pattern acquisition method. Comput. Eng. **15**, 96–98 (2005). https://doi.org/10.3969/j.issn.1000-3428.2005.15.035
11. Xue, X., Yao, X.: Interactive ontology matching based on partial reference alignment. Appl. Soft Comput. **72**, 355–370 (2018). https://doi.org/10.1016/j.asoc.2018.08.003

12. Xue, X., Yao, X.: Collaborative ontology matching based on compact interactive evolutionary algorithm. Knowl.-Based Syst. **137**, 94–103 (2017). https://doi.org/10.1016/j.knosys.2017.09.017
13. Jacobs, G., Lefever, E., Véronique, H.: Economic event detection in company-specific news text. In: Proceedings of the First Workshop on Economics and Natural Language Processing, pp. 1–10 (2018). https://doi.org/10.18653/v1/W18-3101
14. Feng, X., Qin, B., Liu, T.: A language-independent neural network for event detection. Sci. China Inf. Sci. **61**(9), 1–12 (2018). https://doi.org/10.1007/s11432-017-9359-x
15. Zheng, S., Cao, W., Xu, W., et al.: Doc2EDAG: An end-to-end document-level framework for Chinese financial event extraction. In: Proceedings of the 2019 Conference on Empirical Methods in Natural Language Processing and the 9th International Joint Conference on Natural Language Processing (EMNLP-IJCNLP),pp. 337–346 (2019). https://doi.org/10.18653/v1/D19-1032
16. Vaswani A, Shazeer N, Parmar N, et al. Attention is all you need[C]//Proceedings of the31st Conference on Neural Information Processing Systems (NIPS 2017),2017:5998–6008.
17. Ein-Dor, L., Gera, A., Toledo-Ronen, O., et al.: Financial event extraction using wikipedia-based weak supervision. In: Proceedings of the 57th Annual Meeting of the Association for Computational Linguistics, pp. 10–15 (2019)
18. Zhao, L., Li, L., Zheng, X.: A BERT based Sentiment Analysis and Key Entity Detection Approach for Online Financial Texts (2020). arXiv preprint arXiv:2001.05326
19. Chen, X., Ni, L., Ni, Z.: Financial event extraction based on ELECTRA and part-of-speech. Data Anal. Knowl. Disc., 1–18 (2021). 1–18. https://doi.org/10.11925/infotech.2096-3467.2020.1296
20. Peiying, L.I., Jian, Y.A.N.G.: Prosodic unit boundary prediction of MYANMAR based on BERT-CRF model. Comput. Sci. Appl. **11**(03), 505–514 (2021). https://doi.org/10.12677/CSA.2021.113051

Abstractive Sentences Summarization Model Combining Entity Information and Bidirectional Decoding

Chen Wang[1], Baohua Qiang[1](✉), Bo Peng[2], Peng Chen[1], Ruidong Chen[1], Hong Zheng[1], and Yufeng Wang[2]

[1] Guangxi Key Laboratory of Image and Graphic Intelligent Processing, (Guilin University of Electronic Technology), Guilin Guangxi 541004, China
qiangbh@guet.edu.cn

[2] The 54th Research Institute of China Electronics Technology Group Corporation, Shijiazhuang 050002, China

Abstract. Text summarization is a valuable research task in the field of Natural Language Processing (NLP). The Sequence-to-Sequence model is currently the mainstream model of abstractive summarization. However, most abstractive summarization models do not consider named entity information separately. Named entity incompleteness often occurs when long text generates a summary. Based on the characteristics of generating summaries, error accumulation is also the main problem causing summary discontinuity. In this paper, we proposed an abstractive summarization model NER_Bideocder, which combines entity information with bidirectional decoding based on a sequence-to-sequence model. Named entity recognition tags are used for text, and the contents of the tags are used as entity information when encoding; meanwhile, decoding uses a bidirectional LSTM structure, which combines the decoding results in both directions with attention mechanism to generate summarization. Experiments on NLPCC2017 task3: Single Document Summarization dataset show that the performance of this model is better than other models.

Keywords: Abstractive summarization · Bidirectional decoding · Named entity · Sequence to sequence model

1 Introduction

Text data generated by the Internet is increasing. Text summarization is a solution to the problem of text information overload. In 1958, Luhn [1] et al. proposed a method to sort text sentences based on high-frequency keywords to get a summary. Thereafter, based on Luhn's ideas, researchers began to focus on text summarization techniques. News texts now contain more content, and short text-only automatic summarization no longer meet current needs.

Based on the method of generation summaries, there are two main text summarization methods: extractive summarization and abstractive summarization. Extractive text summarization makes up summaries by extracting sentences from the original text based on

S.-C. Chu et al. (Eds.): ICGEC 2021, LNEE 833, pp. 407–418, 2022.
https://doi.org/10.1007/978-981-16-8430-2_37

word frequency, keywords and key sentences. At present, the extractive summarization method is mature, but the generated summaries are not consistent and readable.

Abstractive text summarization uses a neural network training model to generate summaries word by word, and the process of generating summaries is very close to that of humans reading articles and rewriting summaries. In recent years, abstractive text summarization has become a research hotspot.

During the generation of text summary, the length of the original text and the generated summary are extremely inconsistent. For the traditional seq2seq (Sequence to Sequence) model, there are some problems that need to be refined from the full-text information of the encoder to the digest of the decoder (1) OOV problem. The number of words in the source document corpus is usually large, but the number of commonly used words is relatively fixed. Therefore, some words are usually filtered out to form a glossary based on their frequency. This will result in the word UNK being encountered when generating the summary. (2) Readability of the summary. Decoding is usually done using greedy algorithms or beam-search methods. These methods accumulate errors and sometimes produce sentences that are not smooth. (3) It is difficult to generate a summary from the long text. Long text as an encoder input will be more different from the length of the summary, requiring the encoder to highly condense the contents of the document, and the decoder can fully understand the contents to generate the summary, making it more difficult. In the face of the complex semantics of Chinese vocabulary, the generation of Chinese text summarization often faces the problem of incomplete entity recognition, such as the incomplete names of people and places in the summary. Due to the complexity of Chinese vocabulary semantics, the generation of Chinese text summarization often faces the problem of incomplete entity recognition, such as the incomplete names of people and places in the summary. In addition, the generation of summary based on the seq2seq model is word by word, which results in error accumulation. In this paper, we proposed an abstractive text summarization model that combines entity information and bidirectional decoding. Entities in the text are tagged as entity information and fused into the network structure, and the result of bidirectional decoding is combined to generate the summary, which reduces the error accumulation in generating the summary verbatim by word.

The main contributions of this study are as follows:

- Named entity recognition tags as entity information to solve incomplete entity generation problem;
- A bidirectional decoding mechanism is proposed to solve the error accumulation problem in a summary generation;
- Experiments show that this model is superior to existing baseline models on NLPCC2017 task3: Single Document Summarization dataset.

2 Related Work

Text summarization technology starts with extractive summarization and is divided into supervised and unsupervised methods. Supervised methods mainly include traditional machine learning and deep learning. Unsupervised methods are mainly based on word frequency statistics, clustering, graph algorithms and combinatorial informatics. Beginning with a summary method based on word frequency statistics proposed by Luhn et al. to generate a summary based on word frequency to score sentences. Split the single document into a list of sentences, break up the list of sentences into words, rate the list of sentences according to the most frequently occurring part of the keywords, and take out the top three to five sentences to form a summary. Since then, this frequency-based approach has become the basis of text summarization research, which incorporates various features of text into the information contained in words. Baxendale [2] found through statistics that the author usually makes a conclusive description at the beginning and end of the article, most of the main sentences are at the beginning and some at the end, so he proposed adding the feature of sentence position, which provided new research ideas for researchers.

TextRank [3] is an unsupervised extractive summarization method based on graph algorithm, and its idea comes from Google's PageRank. By dividing the text into several components (sentences), a node connection graph is constructed, the similarity between sentences is used as the weight of the edges, the TextRank value of the sentences is calculated by cyclic iteration, then the top-ranked sentences are extracted and combined into text summaries.

Early text summarization research mainly focused on extractive summarization method. With the deepening of machine learning and neural network research, extractive summarization methods are not limited to statistical word frequency or graph-based algorithms. SummaRuNNer [4] is an RNN-based extractive summarization model. In addition to RNN encoding at the word level, RNN encoding at the sentence level is also performed. The last trained target is whether each sentence belongs to a summary sentence or not.

According to the characteristics of extractive summarization, the resulting summary is extracted from the original text, there will be no semantic bias or sentence incompleteness, but most of them have poor coherence, and it is difficult to refine more information from the limited summary text. Compared to extractive summarization, most abstractive summarizations use neural network models, and the process of abstractive summarization is more in line with the process of human summary writing. Since abstractive summarization tasks are similar to machine translation tasks, researchers have started their research in the field of abstractive summarization based on the idea of machine translation. In recent years, deep learning has been deeply studied, seq2seq (Sequence-to-Sequence) model is the mainstream of generative abstracts.

Rush et al. [5] applied deep learning to text summarization, proposed an ABS (Attention-Based Summarization) model, and used it on the DUC-2004 dataset. Chopra et al. [6] introduced a convolution neural network to encode the original information and decoded it with a cyclic neural network to generate a more coherent summary. Nallapati et al. [7] used bidirectional looping neural networks in the encoding phase and also processed low-frequency words. These methods are all based on English datasets and are not suitable for the grammatical and semantic features of Chinese. Chen et al. [8] made the Chinese text summarization dataset LCSTS public, which facilitates the study of Chinese text summarization.

Nallapati [9] et al. used the attentional encoder-decoder RNN model of machine translation directly for text summarization and proposed improvements on the model based on the characteristics of summary generation. Limit decoder-vocabulary to words in the current mini-batch text, and add high-frequency words as well which purpose is to solve the computing bottleneck of the softmax layer caused by the large vocabulary of the decoder. A key issue in generating a summary is to identify key concepts and entities in the original text. Therefore, in addition to word vectors, additional part-of-speech labels, entity labels, TF-IDF statistics are added. In addition, a generator-pointer switch is proposed, and when the switch is on, the decoder generates a word from the target dictionary; when the switch is off, the decoder generates a pointer to the position of the original word and copies it to the summary.

Xue [10] et al. proposed a large-scale biomedical ontology partitioning and matching framework, which ensured the semantic integrity of the ontology and the appropriate size of each segment. Then, the matching technique based on Adaptive Compact Evolutionary Algorithm (ACEA) is used to determine the alignment of ontology segments, which can effectively match similar ontology segments.

Intra-attention is used in Deep Reinforced Model [11], which contains two attention mechanisms called Intra-temporal attention and Intra-decoder attention, to solve the problem of phrase duplication. The former is used in encoders to calculate weights for each word in the input text so that the generated content information can overwrite the original text. The latter works on the decoder and calculates weights on the words that have already been generated so that duplicate content can be avoided. Then they are stitched together to predict the next word.

3 NER_Bidecoder Model

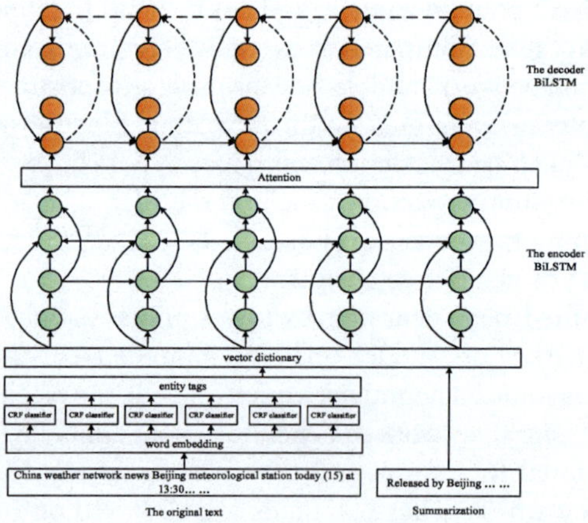

Fig. 1. The overall diagram of NER_Bidecoder

The overall framework of the model is shown in Fig. 1. For the original text entered, we mark the named entity. Our model is based on the sequence-to-sequence model.

For the encoder, we use a common bidirectional LSTM structure. For the decoder, we no longer use unidirectional network decoding but improve to bidirectional LSTM decoding; moreover, we introduce attention mechanism in the decoding process, and beam-search method is used to get the predicted results. Next, we will describe named entity tags, bidirectional decoding, and beam-search in our model.

3.1 Named Entity Tags

According to the current abstractive summarization research, the summarization effect works well on short text datasets. On long text datasets, the problem of summary inconsistency is common, most of which is caused by the incomplete named entity when generating a summary. Different types of named entities (person names, place names, organization names, etc.) are in different locations when generating a summary, which requires that the model not only summarize the contents of the original text but also generate the named entities accurately according to the grammar [12]. The Chinese language is very flexible, and new named entities often appear. Therefore, abstractive summarization requires special attention to named entities to improve the overall quality of the summary. Without context information for named entities, it is difficult to know the meaning of words or entities with different meanings when ambiguous. "他 说的 话 里 有 很大的 水分。" The phrase "水分" can be interpreted as material water or implies an untruthful ingredient in a situation. Numbers in the original text may be used to represent time, quantity, percentage, etc., and it is difficult to accurately express them in summary.

Named entity recognition is widely used in question and answer systems, machine translation and reading comprehension in the field of natural language processing (NLP). Li [13] et al. introduced named entities into the field of machine translation and found that adding features with named entities to the source language can produce better machine translation results. Considering that the machine translation task is similar to the abstractive summarization task, we tag the original text with named entities before the abstractive summarization model so that the model encoder can retain the entity information and increase the decoder's recognition of named entities when generating the summary.

The research on named entity recognition is mature at present, and there are many models for reference. Entity sequence labelling in named entity recognition mainly uses BIO and BIOES. In BIOES, B, Begin, means start; I, Intermediate, means intermediate; E, End, means the end; S, or Single, stands for a single character; O, Other, for other, used to mark unrelated characters. An example of BIOES labelling is shown in Fig. 2.

Original text:	小	明	在	北	京	大	学	的	燕	园	看	了	中	国	男	篮	的	一	场	比	赛
BIOES labelling:	B-PER	E-PER	0	B-ORG	I-ORG	I-ORG	E-ORG	0	B-LOC	E-LOC	0	0	B-ORG	I-ORG	I-ORG	E-ORG	0	0	0	0	0

Fig. 2. The example of BIOES labelling

This named entity labelling method, if used directly in abstractive text summarization, can increase the loss of source information by the encoder. In addition, named entities usually contain specific nouns, numbers, or abbreviations, etc. Named entity models are trained specifically for text summary datasets, have little data available, and require a lot of manual labelling. Therefore, the named entity labels used in this article are NER labelling tool, StanfordNER [14]. The labeled entities are divided into four categories, PERSON, ORG, GPE and MISC, indicates person name, organization name, place name and number respectively; the boundaries of labeled entities are represented as <PERSON> </PERSON>, <ORG> </ORG>, <GPE> </GPE>, <MISC> </MISC>. For text data, we label entities based on words. An example of labelling is shown in Fig. 3.

Original text:	新华网	杭州	1月16日	电（记者	岳德亮）
StanfordNER labelling:	<ORGANIZATION>新华网</ORGANIZATION>	<GPE>杭州</GPE>	<MISC>1月1 6日</MISC>	电（记者	<PERSON>岳德亮</PERSON>）

Fig. 3. The example of StanfordNER labelling

3.2 Encoder

In our model, the encoder uses a bidirectional LSTM structure, forward LSTM is called Forward Layer, and reverse LSTM is called Backward Layer, which jointly connects the output layer. Forward layers are calculated forward from one moment to t, and the output of forward hidden layers is obtained and saved at each moment. The Backward layer is inverted from time t to time 1 to get and save the output of the backward hidden layer at each time. Finally, the final output is obtained at each time by combining the output from the Forward and Backward layers at the corresponding time, as shown in Eq. (1)–(3):

$$h_t = f(w_1 x_t + w_2 h_{t-1}) \tag{1}$$

$$h_t^{'} = f(w_3 x_t + w_5 h_{t-1}^{'}) \tag{2}$$

$$o_t = g(w_4 h_t + w_6 h_t^{'}) \tag{3}$$

Where h_t denotes the hidden layer state of the Forward layer at time t, $h_t^{'}$ denotes the hidden layer state of the Backward layer at time t, and o_t denotes the output of the combination of the Forward and Backward layers at time t. $w_1 \sim w_6$ represents six weights, which are respectively: the weights of the input layer to the Forward and Backward layers, w_1, w_3; the weights between the Forward layer and the Backward layer, w_2, w_5; the weights from the Forward and Backward layers to the output layer, w_4, w_6.

The bidirectional LSTM network structure is shown in Fig. 4.

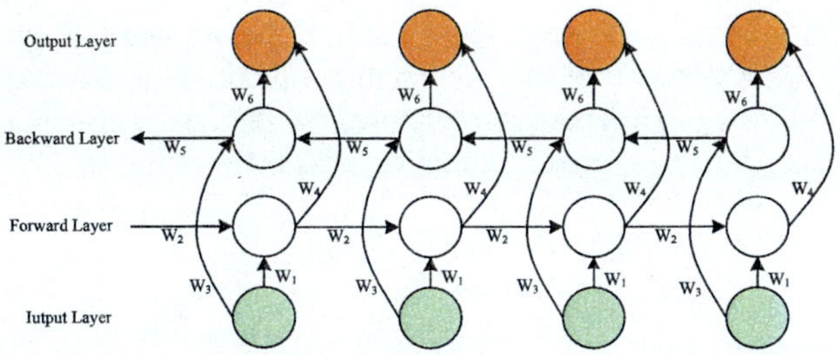

Fig. 4. Bidirectional LSTM network structure diagram

3.3 Decoder

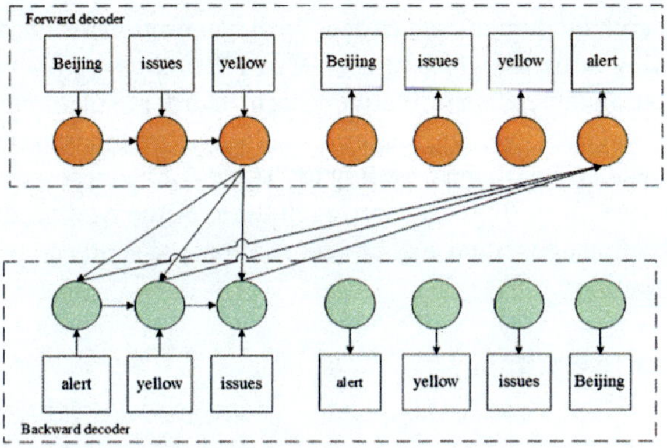

Fig. 5. Bidirectional decoding structure diagram

The general sequence-to-sequence model decoding process is to generate a summary word by word from left to right. We found that such a word-by-character summary will cause a direction tilt problem, that is, error accumulation. The accuracy of the first few words of the generated summary is generally high. If the summary results are predicted incorrectly, errors will occur in the subsequently generated summary, resulting in the lower accuracy of the part after the generated summary. This model uses a bidirectional decoding structure to alleviate this directional skew and then generates a summary through the attention mechanism. The bidirectional decoding structure is shown in Fig. 5.

Bidirectional decoding can be seen as the coexistence of Forward decoder and Backward decoder. Each decoder predicts the result of time t by combining the internal information of the decoder with the result sequence already decoded by another decoder. The result sequence of the Forward and Backward decoders predicting the nth word is as follows Eq. (4), (5):

$$H^{forward} = [h_1^{forward}, h_2^{forward}, \ldots, h_n^{forward}] \tag{4}$$

$$H^{backward} = [h_1^{backward}, h_2^{backward}, \ldots, h_n^{backward}] \tag{5}$$

$h_n^{forward}$ in $H^{forward}$ represents the predicted result of the nth word of Forward decoder, $h_n^{backward}$ in $H^{backward}$ represents the predicted result of the nth word of Backward decoder. In the bidirectional decoding structure, we also need to combine the result of attention mechanism. The output of attention is as follows Eq. (6):

$$Attention(Q, K, T) = softmax(\frac{QK^T}{\sqrt{d_k}})V \tag{6}$$

In Forward decoder, $Q = H^{forward}$, $K = V = H^{backward}$; in Backward decoder, $Q = H^{backward}$, $K = V = H^{forward}$. Attention is used as the feature to predict the nth word combined with $H^{forward}$ and $H^{backward}$.

3.4 Beam-Search

Generally, beam-search method is used to decode the prediction results. Considering the bidirectional decoding structure, *topk* results of two decoders are obtained at the same time, and the calculation dimension is relatively high. In order to facilitate the calculation, we make a simple weighted average fusion of the top^3 combination results of the two decoders into top^2 results, and then according to the idea of beam search method, we get the *topk* combination with the highest score each time for the next prediction. Until the decoder in both directions generates a complete summary, a sentence with high score is taken as the final summary.

4 Experimental Results

4.1 Datasets

This experiment uses nlpcc2017 task3: single document summary data set, which is provided by CCF International Conference on natural language processing and Chinese computing (nlpcc2017) in 2017. Among them, 49500 news texts are used as training sets, and 500 news texts are used as test sets. The contents of the texts cover a wide range, including sports, entertainment, politics, science and technology, finance and so on. The average length of the dataset text is 1036 words; The average length of summary is 45 words, and the longest is no more than 60 words; There are 11493 longer texts with a length of 640 words.

4.2 Evaluation Metrics

The generated summarization and reference summarization may have similar meanings but different word order, for example, if the generated summarization is "There is a cat on a mat." and the reference summary is "A cat is sitting on a mat", accuracy cannot be used directly as an evaluation metrics of the text summarization. Rouge (Recall-Oriented Understudy for Gisting Evaluation) [15] method commonly used in text summarization research is used for evaluation metrics, which is mainly calculated based on Recall rate. In our experiment, rouge-1, rouge-2 and rouge-l were used as evaluation metrics. ROUGE-N is the main statistic for recall rates on N-gram. L in ROUGE-L refers to the longest common subsequence (LCS), and ROUGE-L is calculated using the longest common subsequence of model-generated summary C and reference summary S.

4.3 Experimental Steps

In the data preprocessing stage, the original text is segmented using the Jieba open-source word segmentation tool. The segmented data is labelled with named entities using the StanfordNER labelling tool. Given that entity tags have been added, a dictionary is built to vectorize text words in the order they appear.

Build an abstractive text summarization model based on the sequence-to-sequence model. First, build an Embedding layer, set the vector dimension to 512, decoder and encoder use bidirectional LSTM, decoder decode with Attention results, and finally use beam-search method to generate the summary. The algorithm flowchart is shown in Fig. 6.

The model training is set to 20 epochs, and cross-entropy is used as the loss function. After ten epochs of training, the loss value decreases to less than 1 and stabilizes to 20 epochs.

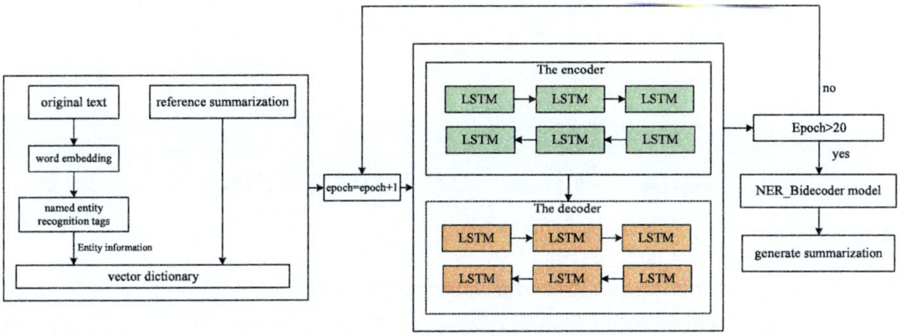

Fig. 6. The algorithm flowchart

4.4 Baseline Models

The comparative experiment of this experiment adopts six text summarization models, which are respectively Lexpagerank, MEAD, SubModular, Uni Attention, NLP_ONE, Pointer - Generator. The first three are extractive summarization methods, and the last three are abstractive summarization methods.

The LexPageRank [16] model constructs a sentence connectivity matrix based on cosine similarity. If the cosine similarity between two sentences exceeds a specific pre-defined threshold, the corresponding edges are added to the join matrix. MEAD [17] scores sentence by taking into account four common features of the sentence, including the centroid, the position of the sentence, the common subsequence, and keywords. The Submodular [18] model uses the diminishing returns characteristic of the submodel function to select important sentences to generate a summary. The UniAttention [19] model is the basic attention sequence to sequence model, which enables attention mechanism consideration and summary generation of input information from the original text. NLP_ONE [20] is the model that won first place in the NLPCC 2017 Chinese single-document summary evaluation task. Pointer-generator is the latest text summarization model published by ACL 2017, which uses a pointer mechanism to solve the problem of errors in output information and rare words.

4.5 Analysis of Experimental Results

Based on the above comparison model and our model, the experimental results are shown in Table 1:

Table 1. Experimental result

Method	ROUGE-1	ROUGE-2	ROUGE-L
LexPageRank	0.23634	0.10884	0.17578
MEAD	0.28674	0.14872	0.22365
Submodular	0.29704	0.15283	0.21668
Uni attention	0.33752	0.20667	0.29462
NLP_ONE	0.34983	0.21181	0.30686
Point-generator	0.36002	0.21978	0.29888
Bidecoder	**0.37638**	**0.22281**	**0.36845**
NER_Bidecoder	**0.38146**	**0.22901**	**0.37421**

Where Bidecoder is a bidirectional decoding model experiment without adding entity information, NER_Bidecoder is a model experiment for combining entity information and bidirectional decoding.

From the table above, we can see that the ROUGE-1, ROUGE-2 and ROUGE-L of the abstractive summarization model are 4 to 10 percentage points higher than that of extractive summarization model. Abstractive summarization does work better than extractive summarization.

The ROUGE-1 and ROUGE-2 of the experiment Bidecoder in this paper are about 1 percentage point higher than those of the baseline model and 7 percentage points higher than those of the current one. ROUGE-L can represent the longest common subsequence for generating summaries and reference summaries. It can be seen that the Bidecoder method is superior to other models in summarizing results.

The ROUGE-1, ROUGE-2, ROUGE-L of NER_Bidecoder with entity information added are higher than those of Bidecoder and other existing models, with ROUGE-L increased by about 8 percentage points. It is shown that the model combining entity information and bidirectional decoding can alleviate the problems of incoherence and error accumulation to a certain extent.

5 Conclusion

In this paper, we proposed an abstractive summarization model that combines entity information with bidirectional decoding. It tags the original text with named entity, preserves entity information, and reduces the incomplete named entity in the summary generation. The decoder uses a bidirectional decoding structure to alleviate error accumulation in unidirectional decoding. Experiments show that our model performs better in generating summaries on NLPCC datasets. However, the vectorization method for text containing entity information is too simple to confuse semantics in some cases, and the length of the sequence increases dramatically after vectorization for long text, resulting in a long text summary that is far less effective than longer text.

Acknowledgments. This work was supported in part by the National Natural Science Foundation of China under Grant 61762025; in part by the Development Foundation of the 54th Research Institute of China Electronics Technology Group Corporation under Grant SXX18138X017; in part by the Natural Science Foundation of Guangxi under Grants 2019GXNSFDA185007, and 2019GXNSFDA185006; in part by the Guangxi Key Research and Development Project under Grants AB18126053, AB18126063, AD18281002, and AD19110137; in part by the Guilin Science and Technology Development Program under Grants 20190211–17; in part by the Guangxi Key Science and Technology Planning Project under Grants AA18118031 and AA18242028; in part by the Innovation Project of GUET Graduate Education under Grants 2020YCXS052; in part by the Guangxi Colleges and Universities Key Laboratory of Intelligent Processing of Computer Image and Graphics under Grants GIIP201603 and GIIP1806; in part by the Guangxi Key Laboratory of Cryptography and Information Security under Grant GCIS201906.

References

1. Luhn, H.P.: The automatic creation of literature abstracts. IBM J. Res. Dev. **2**(2), 159–165 (1958). https://doi.org/10.1147/rd.22.0159
2. Baxendale, P.B.: Machine-made index for technical literature—an experiment. IBM J. Res. Dev. **2**(4), 354–361 (2010). https://doi.org/10.1147/rd.24.0354
3. Mihalcea, R., Tarau, P.: Textrank: bringing order into text. In: Proceedings of the 2004 Conference on Empirical Methods in Natural Language Processing, pp. 404–411 (2004). http://dx.doi.org/
4. Nallapati, R., Zhai, F., Zhou, B.: SummaRuNNer: a recurrent neural network based sequence model for extractive summarization of documents. In: The Thirty-First AAAI Conference on Artificial Intelligence (AAAI-2017), Computer Science, arXiv:1611.04230 [cs.CL] (2016)
5. Rush, A.M, Chopra, S, Weston, J.: A neural attention model for abstractive sentence summarization. Computer Science. In: Proceedings of EMNLP 2015 (2015). https://doi.org/10.18653/v1/D15-1044
6. Chopra, S., Auli, M., Rush, A.M.: Abstractive sentence summarization with attentive recurrent neural networks. In: Proceedings of the 2016 Conference of the North American Chapter of the Association for Computational Linguistics: Human Language Technologies, pp. 93–98 (2016). https://doi.org/10.18653/v1/N16-1012
7. Nallapati, R., Xiang, B., Zhou, B.: Sequence-to-Sequence RNNs for Text Summarization. ArXiv, abs/1602.06023 (2016)
8. Chen, Q., Hu, B., Zhu, F.: LCSTS: a large scale chinese short text summarization dataset, Emnlp2015. In: Proceedings of the 2015 Conference on Empirical Methods in Natural Language Processing, Lisbon, Portugal, pp. 1967–1972 (2015). https://doi.org/10.18653/v1/D15-1229

9. Nallapati, R., Zhou, B., dos Santos, C.: Abstractive text summarization using sequence-to-sequence RNNs and Beyond. In: Proceedings of The 20th SIGNLL Conference on Computational Natural Language Learning, Berlin, German, pp. 280–290 (2016). https://doi.org/10.18653/v1/K16-1028
10. Xue, X., Yao, X.: Interactive ontology matching based on partial reference alignment. Appl. Soft Comput. **72**, 355–370 (2018). https://doi.org/10.1016/j.asoc.2021.107343
11. Paulus, R., Xiong, C., Socher, R.: A Deep Reinforced Model for Abstractive Summarization. Computation and Language, arXiv:1705.04304 [cs.CL] (2017)
12. Xue, X., Liu, J.: Collaborative ontology matching based on compact interactive evolutionary algorithm. Knowl. Based Syst. **137**, 94–103 (2017). https://doi.org/10.1016/j.knosys.2017.09.017
13. Li, Z., Wang, X., Aw, A., Chng, E.S., Li, H.: Named-entity tagging and domain adaptation for better customized translation. In: Proceedings of the Seventh Named Entities Workshop, Melbourne, Australia, pp 41–46 (2018). https://doi.org/10.18653/v1/W18-2407
14. Finkel, J.R., Grenager, T., Manning, C.: Incorporating non-local information into information extraction systems by Gibbs sampling. In: Proceedings of the 43nd Annual Meeting of the Association for Computational Linguistics, Ann Arbor, Michigan, pp. 363–370 (2005). https://doi.org/10.3115/1219840.1219885
15. Lin, C.-Y.: ROUGE: a package for automatic evaluation of summaries. Text summarization branches out, Barcelona, Spain, pp. 74–81 (2004)
16. Erkan, G., Radev, D.R.: LexPageRank: prestige in multi-document text summarization. In: Proceedings of the 2004 Conference on Empirical Methods in Natural Language Processing, Barcelona, Spain, pp. 365–371 (2004)
17. Radev, D., Allison, T., Blair-Goldensohn, S., Blitzer, J., Zhang, Z.: MEAD - A Platform for Multidocument Multilingual Text Summarization. LREC 2004, Lisbon, Portugal (2004)
18. Lin, H., Bilmes, J.: Multi-document summarization via budgeted maximization of submodular functions. In: Human Language Technologies: Conference of the North American Chapter of the Association of Computational Linguistics, Proceedings, Los Angeles, California, USA (2010)
19. Chopra, S., Auli, M., Rush, A.M.: Abstractive sentence summarization with attentive recurrent neural networks. In: Proceedings of the 2016 Conference of the North American Chapter of the Association for Computational Linguistics: Human Language Technologies, San Diego, California, pp 93–98 (2016). https://doi.org/10.18653/v1/N16-1012.
20. Hou, L., Hu, P., Chao, B.: Abstractive document summarization via neural model with joint attention. Springer, Cham (2017). https://doi.org/10.1007/978-3-319-73618-1_28

Multi-group Discrete Symbiotic Organisms Search Algorithm Based Clustering Method for Large-Scale Traveling Salesman Problem

Shu-Chuan Chu[1,2], Zhi-Gang Du[1], and Jeng-Shyang Pan[1,3](✉)

[1] College of Computer Science and Engineering, Shandong University of Science and Technology, Qingdao 266590, China
[2] College of Science and Engineering, Flinders University, 1284 South Road, Clovelly Park, SA 5042, Australia
[3] College of Information Science and Technology, Dalian Maritime University, Dalian 116026, China

Abstract. In recent years, the Traveling Salesman Problem (TSP) has become an indispensable part of the current supply chain system. The solution and optimization of this problem, especially the Large-Scale Traveling Salesman Problem (LSTSP), attract a large number of scholars and experts. The swarm intelligent algorithm, also known as the meta-heuristic evolutionary algorithm, is also developing rapidly in modern times. Especially the outstanding performance in solving complex problems has led many scholars to use the idea of swarm intelligence to solve or optimize some complex problems. The Symbiotic Organisms Search(SOS) algorithm is a swarm intelligent algorithm that has emerged in recent years. Its excellent global optimization and convergence capabilities have won the favor of many experts and researchers. This paper proposes a Multi-group Discrete SOS based on the K-Means clustering (MDSOSKMC) algorithm for solving LTSP. In order to test the performance of our proposed algorithm, we use some large-scale instance set in TSPLIB for verification.

Keywords: Discrete symbiotic organism search · Traveling salesman problem · Meta-heuristic evolutionary algorithm

1 Introduction

The traveling salesman problem is one of the well-known problems in the field of mathematics [1]. TSP has a long history. The earliest description is the knight travel problem studied by Euler in 1759, that is, for the 64 squares in the chessboard, visit the 64 squares once and only once, and finally return to the starting point. TSP was introduced by the American RAND company in 1948, and the emergence of the new method of linear planning [2]. The company's influence has made TSP a well-known and popular problem. The TSP problem can be

S.-C. Chu et al. (Eds.): ICGEC 2021, LNEE 833, pp. 419–428, 2022.
https://doi.org/10.1007/978-981-16-8430-2_38

described in general terms: suppose a traveling businessman wants to visit n cities, he must choose the path to be taken, the limit of the path is that each city can only visit once, and finally return to the original departure city. The ultimate goal of this problem is to solve that the length of the selected path is the smallest value among all paths [3].

TSP problems can be mathematically modeled using undirected weighted graphs in graph theory of discrete mathematics. Each city can be used as a vertex in the graph, and the distance between cities can be expressed as the weight of two vertices in the graph. Then the TSP problem becomes a traversal problem of graph $G = (V, E)$.Where $V = \{c_1, c_2, ..., c_n\}$represents the set of cities, and $V = \{d(c_i, c_j)|i = j\}$ represents the weight between two cities, that is, the distance. When we need to traverse the minimum path of the graph, we need to find a permutation π of all cities and try to minimize it according to the following Eq. 1.

$$f(\pi) = d(c_{\pi_1}, c_{\pi_n}) + (\sum_{i=1}^{n-1} d(c_{\pi_i}, c_{\pi_{i+1}})) \tag{1}$$

TSP is a Non-deterministic Polynomial Complete (NPC) problem. When the problem size is small, we can directly exhaust the problem space. When the problem size increases, the problem space will also increase exponentially. Generally, when the scale of the problem expands to a certain degree, we can only obtain an approximate optimal solution by using some intelligent algorithms using a greedy strategy. The greedy strategy sacrifices the accuracy of the understanding in some cases (the obtained solution may not be the best solution), but it saves a lot of time (the time complexity is directly reduced to the degree of polynomial).

In recent years, Swarm intelligent algorithms [4,5], also known as metaheuristic evolutionary algorithms [6], have developed rapidly [7–9]. Particle Swarm Optimization (PSO) is a classic algorithm inspired by bird foraging behavior [10–12]. The Cat Swarm Optimization (CSO) [13,14] is inspired by the behavior between cat groups. The QUasi-Affine TRansformation Evolutionary (QUATRE) algorithm (2020b) is generated by simulating the process of affine-transformation in geometry [15,16]. There are also Differential Evolution (DE) [17,18], Artificial Bee Colony (ABC), Cuckoo Search (CS) [19–21] and so on. The above algorithms are all proposed to solve continuous problems. In order to solve the discrete problems in real life and retain the advantages of the global optimization ability of the clustering intelligent algorithm [22,23], the researchers proposed the following two strategies: First, they proposed special discrete intelligent algorithms[22], such as ant colony optimization (ACO) [24,25], Genetic Algorithms (GA) [26,27] and so on. The other is discretization based on the original continuous intelligent algorithm, such as Discrete PSO (DPSO) [28].

The Symbiotic Organisms Search (SOS) [29,30], was proposed by Cheng and Prayogo first proposed the algorithm in 2014, is inspired by the three symbiotic relationships of mutual benefit, symbiosis, and parasitism in biological systems. Because of its excellent global optimization ability in solving continuous problems, the algorithm has become a very promising meta-heuristic evolutionary algorithm. In 2017, Absalom El-Shamir Ezugwu and Aderem Oluyinka Adewumi jointly proposed the Discrete SOS (DSOS) algorithm to solve the TSP problem [31,32]. In order to further solve the large-scale TSP problem in real life, a multi-group discrete SOS (MDSOSKC) algorithm based on the K-Means clustering method is proposed in this paper.

2 Related Work

2.1 SOS Algorithm

The original SOS algorithm is an outstanding meta-heuristic evolutionary algorithm that has evolved by simulating the three symbiotic relationships of organisms in nature, namely Mutualism, Commensalism, and Parasitism.

During the Mutualism phase, creatures coexist in the biological world through a mutually beneficial relationship, such as rhinos and rhino birds in nature. We describe this relationship in the biological world through Eqs. 2–3.

$$\begin{cases} \mathbf{X}_{inew} = \mathbf{X}_i + r_1 \times (\mathbf{X}_{gbest} - \overline{\mathbf{X}} \times b_1) \\ \mathbf{X}_{jnew} = \mathbf{X}_i + r_2 \times (\mathbf{X}_{gbest} - \overline{\mathbf{X}} \times b_2) \end{cases}, \tag{2}$$

$$\begin{cases} \overline{\mathbf{X}} = (\mathbf{X}_i + \mathbf{X}_j)/2 \\ b_1 = round(1 + random(0,1)) \\ b_2 = round(1 + random(0,1)) \end{cases}, \tag{3}$$

where $\mathbf{X}_i$ and $\mathbf{X}_j$ represent the positions of the two creatures, respectively, and they update their positions to $\mathbf{X}_{inew}$ and $\mathbf{X}_{jnew}$ through the mutually beneficial stage, respectively. r_1 and r_2 are random variables between 0 and 1. b_1 and b_2 are the mutually beneficial factors of organisms $\mathbf{X}_i$ and $\mathbf{X}_j$, respectively. $\overline{\mathbf{X}}$ is in the center transition state. $\mathbf{X}_{Gbest}$ is the global optimal position to which the algorithm has recently converged.

During the Commensalism phase, one creature gains all its benefits by interacting with another creature, but the effects of their interaction with one another. Just like the interaction between cattle egrets (birds living near grazing cows) and cattle, foods such as insects needed by cattle egrets are produced from the feces of cattle, but the living conditions of cattle are not affected by the existence of cattle egrets, only Cattle egrets can benefit from the relationship between the creatures. By simulating this relationship in the biological world, the Commensalism phase of the SOS algorithm is formed, as shown in Eq. 4.

$$\mathbf{X}_{inew} = \mathbf{X}_i + r_3 \times (\mathbf{X}_{gbest} - \mathbf{X}_j), \quad r_3 \in (-1, 1) \tag{4}$$

In the parasitic stage, the relationship between two organisms is mainly to benefit themselves by harming the interests of the other party. The relationship between virus hosts can be described as Eqs. 5–6.

$$\begin{cases} \mathbf{X}_{parasite} = \mathbf{X}_i, \quad P > 0.5 \\ \mathbf{X}_{parasite} = random(0,1) \times (UP - LP) + LP, \quad others \end{cases}, \tag{5}$$

$$\mathbf{X}_j = \mathbf{X}_{parasite}, \quad if \quad f(\mathbf{X}_{parasite}) < f(\mathbf{X}_j) \quad , \tag{6}$$

where $\mathbf{X}_i$ and $\mathbf{X}_j$ represent the state of the virus and the human host, $\mathbf{X}_{parasite}$ represents the state of the intermediate host, and UP and LP represent the range of the biological activity interval, respectively.

3 Our Proposed SOS Algorithm for Multi-group Communication Strategy Based on K-Means Clustering Method

In order to improve the performance of the algorithm in solving large-scale TSP problems, in this section we propose a discrete SOS algorithm with a clustering function and an inter-group communication strategy. The specific idea of the algorithm is to the first cluster by the Euclidean distance between cities, and divide a relatively large city size group into k (K-Means cluster number) smaller city size groups. Then each smaller city group independently performs path optimization of the intelligent algorithm. Finally, after the optimization of each small-scale city group is completed, the multi-group discrete SOS algorithm with a communication strategy is used to connect the paths between the small-scale cities.

3.1 Small-Scale TSP Problem After K-Means Clustering Optimization

In this section, we propose a multi-group discrete SOS algorithm with communication strategy and Gene-Transfer(GT) mutation operation. We introduce a Gene Transfer mutation operation and inter-group Communication Strategy to improve the convergence ability and speed of the algorithm.

GT Mutation Operation. is inspired by biological genetic modification technology. We can use the location $\mathbf{X}_i$ of an organism as a recipient organism chromosome, and it's neighbor organism $\mathbf{X}_j$ as a donor organism chromosome. By transferring some of the excellent genes in the donor chromosome $\mathbf{X}_j$ to the recipient chromosome $\mathbf{X}_i$, the position state (gene performance) of the organism $\mathbf{X}_i$ is better. Suppose $\mathbf{X}_i$ = (3, 7, 4, 5, 2, 6, 1), $\mathbf{X}_j$ = (2, 4, 7, 6, 1, 3, 5), and the Hybrid $B = \{3, 5\}$ between $\mathbf{X}_i$ and $\mathbf{X}_j$. The specific process of GT Mutation Operation is shown in Fig. 1.

After we added GT mutation operation, we improved the original discrete SOS algorithm into four mutation operations, namely swap mutation operation, inverse mutation operation, insert mutation operation and GT mutation operation. In the multi-group discrete SOS algorithm, the biological update strategy is shown in Eq. 7.

$$\mathbf{X}_i = min(Swap(\mathbf{X}_i, \mathbf{X}_j), Inverse(\mathbf{X}_i, \mathbf{X}_j), Inverse(\mathbf{X}_i, \mathbf{X}_j), GT(\mathbf{X}_i, \mathbf{X}_j)) \tag{7}$$

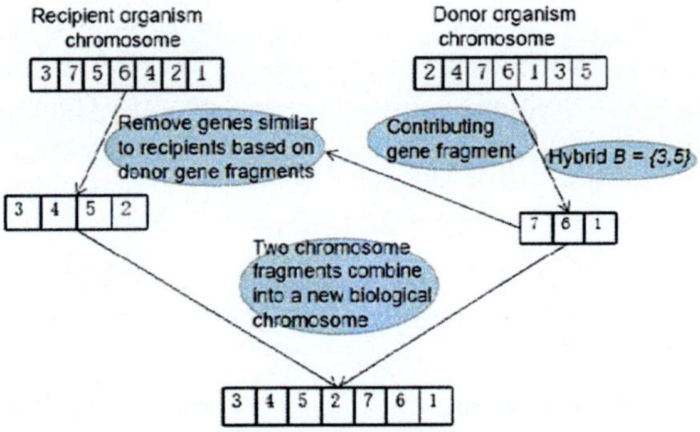

Fig. 1. GT mutation operation.

Intergroup Communication Strategy. While improving the mutation strategy, we have also improved the shortcoming of the algorithm's single optimization direction. Through the multi-group parallel thought, the convergence performance of the algorithm is improved and more spatial regions are explored. Through the communication behaviors and strategies between biological groups, the groups are not isolated for optimization. Through the communication strategies between groups, the groups learn from each other, thereby improving the performance of the algorithm and avoiding the algorithm from falling into a local optimum. The communication method uses a full communication strategy, that is, each subgroup communicates with other subgroups.

The use of inter-group communication strategies is to use the corresponding learning strategies to communicate between groups after each group of biological search iterates to a certain number (or step size). The learning strategy is as shown in Eq. 8.

$$\begin{cases} \mathbf{X}_i^{gbest} = \mathbf{X}_i^{gbest} + p \times (\mathbf{X}_i^{gbest} - \mathbf{X}_j^{gbest}) \\ p = \frac{2f(\mathbf{X}_i^{gbest})}{f(\mathbf{X}_i^{gbest}) + f(\mathbf{X}_i^{gbest})} \end{cases}, \quad (8)$$

where $\mathbf{X}_i^{gbest}$ represents the intra-group optimal solution of the i-th subgroup, p represents the inter-group learning factor, and the size of the p value is related to the optimal optimum value of the current group within the i-th group and the j-th group.

3.2 Path Connectivity Between Multiple Small-Scale Urban Agglomerations

After the algorithm optimizes the path optimization among all the small-scale cities divided by the K-Means clustering algorithm, it is necessary to connect the paths between the small-scale city groups. In this section, we propose a connection rule that is based on the locations of the first visited city and the last visited city of each divided small-scale city group. The specific steps are shown below.

Step 1. According to Eq. 9, the first visited city and last visited the city of each small-scale city group optimized by the intelligent algorithm, and calculate the center position between the two cities.

$$t_k = (x, y) = c^k_{\pi_1} + c^k_{\pi_n}/2 \quad , \tag{9}$$

where t_k represents the base points obtained by the k-th small-scale city group, and $c^k_{\pi_1}$ and $c^n_{\pi_1}$ represent the first visited city and the n-th (last) visited city obtained after optimizing the k-th small-scale city, respectively. Step 2. The multi-group discrete SOS algorithm is used to optimize the base point t of all small-scale cities. Step 3. According to the arrangement order of the optimal solutions of the small-scale cities obtained in step 2, directly splice k (the number of K-Means clusters) small-scale cities. If the order of the optimal solution of the small-scale cities obtained in step 2 is (3, 2, 1, 4), then k small-scale city clusters can be connected into a large-scale city cluster $[\pi_3, \pi_2, \pi_1, \pi_4]$.

4 Experimental Results

In order to verify the performance of our proposed MDSOSKMC algorithm, we compared the K-Means clustering method to four, six, eight, ten, and fifteen types of multi-group discrete SOS algorithms and the original DSOS algorithm in seven based on TSPLIB. Large-scale data sets were compared. This section is mainly divided into two sections. Section 4.1 mainly introduces some evaluation parameters and the settings of parameters used in the algorithm. Section 4.2 mainly gives some final comparison results after setting them based on various parameters.

4.1 Setting of Parameters

The parameters used by the MDSOSKCM algorithm and the original DSOS algorithm to solve large-scale TSP problems (including the size of the population, the number of the entire population divided into subgroups, the communication step between groups, and various control parameters) are set as Table 1 shows.

4.2 Performance Comparison Between MDSOSKMC and DSOS Algorithm

The following experimental results are compared based on the TSPLIB test set. We compare the original DSOS and MDSOSKMC algorithms to verify the performance of our proposed new algorithm. In this simulation experiment, the results of clustering of 4, 6, 8, 10, and 15 clusters using the Comparison of

Table 1. Parameter settings for our and other algorithms

Name	d1291	d1655	dsj1000	fl1577	pcb1173	pcb3038	pr1002
DSOS	1.17E+06	1.51E+06	3.41E+08	9.75E+05	9.16E+05	4.34E+06	3.93E+06
MDSOSKMC_4	1.41E+05	1.64E+05	3.68E+07	8.54E+04	1.25E+05	4.41E+05	5.17E+05
MDSOSKMC_6	1.94E+05	1.35E+05	3.04E+07	8.47E+04	1.02E+05	3.82E+05	4.12E+05
MDSOSKMC_8	9.77E+04	1.14E+05	2.63E+07	1.02E+05	8.99E+04	3.27E+05	3.74E+05
MDSOSKMC_10	9.02E+04	1.11E+05	2.48E+07	1.23E+05	8.18E+04	3.03E+05	3.53E+05
MDSOSKMC_15	7.20E+04	1.15E+05	2.37E+07	8.83E+04	7.51E+04	2.49E+05	3.32E+05

MDSOSKMC and DSOS algorithms. The algorithm was compared with the DSOS algorithm to show the performance of the improved algorithm. Tables 2 show the mean "*Mean*", the best value "*Best*" and the variance "*std.*" obtained by performing 10 iterations of each algorithm according to the seven large-scale TSP problems of TSPLIB.

Table 2. Comparison results

Name		d1291	d1655	dsj1000	fl1577	pcb1173	pcb3038
DSOS	Mean	1.19E+06	1.55E+06	3.45E+08	9.83E+05	9.35E+05	4.36E+06
	Best	1.17E+06	1.51E+06	3.41E+08	9.75E+05	9.16E+05	4.34E+06
	std	1.63E+04	2.25E+04	2.82E+06	6.76E+03	1.34E+04	1.68E+04
MDSOS_4	Mean	1.98E+05	1.70E+05	3.94E+07	1.63E+05	1.32E+05	4.50E+05
	Best	1.41E+05	1.64E+05	3.68E+07	8.54E+04	1.25E+05	4.41E+05
	std	7.66E+04	4.31E+03	2.67E+06	9.62E+04	5.48E+03	8.38E+03
MDSOS_6	Mean	2.21E+05	1.45E+05	3.30E+07	1.18E+05	1.07E+05	3.91E+05
	Best	1.94E+05	1.35E+05	3.04E+07	8.47E+04	1.02E+05	3.82E+05
	std	4.17E+04	7.59E+03	2.48E+06	1.75E+04	3.78E+03	7.28E+03
MDSOS_8	Mean	1.83E+05	1.35E+05	2.79E+07	1.25E+05	9.26E+04	3.46E+05
	Best	9.77E+04	1.14E+05	2.63E+07	1.02E+05	8.99E+04	3.27E+05
	std	3.80E+04	2.39E+04	1.71E+06	1.36E+04	2.18E+03	1.23E+04
MDSOS_10	Mean	1.38E+05	1.45E+05	2.57E+07	1.57E+05	8.41E+04	3.14E+05
	Best	9.02E+04	1.11E+05	2.48E+07	1.23E+05	8.18E+04	3.03E+05
	std	2.68E+04	2.04E+04	7.16E+05	2.13E+04	1.77E+03	8.03E+03
MDSOS_15	Mean	7.59E+04	1.37E+05	2.43E+07	1.10E+05	7.65E+04	2.54E+05
	Best	7.20E+04	1.15E+05	2.37E+07	8.83E+04	7.51E+04	2.49E+05
	std.	2.43E+03	1.75E+04	3.19E+05	1.40E+04	7.53E+02	4.95E+03

Where i of $MDSOSKMC\_i$ in Table 2 represents the number of K-Means clusters. The data in the table are blacked out to indicate the best data value.

4.3 Analysis of Results

According to the experimental results described above, we can clearly find that compared with the DSOS algorithm, the performance of our newly proposed MDSOSKMC algorithm is significantly more outstanding. And we can get from the experimental results that the performance of the algorithm is good or bad, but it has a lot to do with the number of K-Means clusters set for the large-scale TSP problem. Suppose a large-scale TSP problem contains N city locations without classification. Under the condition, the size of its solution space is $N!$. If using K-Means clustering into x clusters, the size of the solution space under ideal conditions is shown in Eq. 10.

$$O(N|x) = x! \times (\frac{N}{x})! \tag{10}$$

Under ideal conditions (large-scale TSP problem is more uniform in clustering), we can use the above formula to find a smaller solution space. In other words, if the solution space $O(N|x)$ is minimized, then $x = \sqrt{N}$. Because the number of classifications is a positive integer greater than 2, $x = \lfloor\sqrt{N}\rfloor$ or $\lceil\sqrt{N}\rceil$.

5 Conclusion

In this paper, we propose a Multi-group Discrete SOS based on K-Means Clustering (MDSOSKMC) algorithm in order to solve the large-scale TSP problem that occurs in practical problems. This algorithm uses a K-Means clustering algorithm to divide a larger TSP problem into multiple smaller TSP problems for solving. Then use our proposed multi-group discrete SOS algorithm with a communication strategy to solve it. By improving the mutation operation mode, multi-group parallel strategy, and optimal solution path cross elimination strategy to improve the algorithm's convergence speed and convergence ability, avoid the algorithm Premature fall into local optimum. The use of inter-group communication strategies to improve the ability of organisms to learn between different groups and the ability to optimize group life has been improved, which overcomes the shortcomings of the original DSOS algorithm that is easy to fall into a local optimum when resolving large-scale TSP. In order to further verify the performance of our proposed MDSOSKMC algorithm in solving large-scale TSP problems, we used 7 large-scale test sets in TSPLIB and analyzed the use of K-Means clustering to set multiple classification numbers. final result. The experimental results show that the MDSOSKMC algorithm proposed in this paper has a better ability to solve large-scale TSP problems, and if it can be combined with some efficient methods, it can further benefit.

In future research, we will further study and optimize smart algorithms with better performance, as well as parallel group communication strategies, initial strategies, and sampling strategies, so that more and more intelligent algorithms will shine in a wide range of practical applications.

References

1. Hoffman, K.L., Padberg, M., Rinaldi, G., et al.: Traveling salesman problem. Encycl. Oper. Res. Manag. Sci. **1**, 1573–1578 (2013)
2. Campuzano, G., Obreque, C., Aguayo, M.M.: Accelerating the miller-tucker-zemlin model for the asymmetric traveling salesman problem. Expert Syst. Appl. **148**, 113229 (2020)
3. Gao, W.: New ant colony optimization algorithm for the traveling salesman problem. Int. J. Comput. Intell. Syst. **13**(1), 44–55 (2020)
4. Chu, S.-C., Dy, Z.-G., Peng, Y.-J., Pan, J.-S.: Fuzzy hierarchical surrogate assists probabilistic particle swarm optimization for expensive high dimensional problem. Knowl.-Based Syst. **220**, 106939 (2021)
5. Pei, H., Pan, J.-S., Chu, S.-C.: Improved binary grey wolf optimizer and its application for feature selection. Knowl.-Based Syst. **195**, 10576 (2020)
6. Tan, F.-G., et al.: Detection of wrong components in patch component based on transfer learning. J. Netw. Intell. **5**, 1–9 (2020)
7. Du, Z.G., Pan, J.S., Chu, S.C., Luo, H.J., Hu, P.: Quasi-affine transformation evolutionary algorithm with communication schemes for application of RSSI in wireless sensor networks. IEEE Access **8**, 8583–8594 (2020)
8. Tian, A.-Q., Chu, S.-C., Pan, J.-S., Cui, H., Zheng, W.-M.: A compact pigeon-inspired optimization for maximum short-term generation mode in cascade hydroelectric power station. Sustainability **12**(3), 767 (2020)
9. Nassrullah, H.A., Flayyih, W.N., Nasrullah, M.A.: Enhancement of LSB audio steganography based on carrier and message characteristics. J. Inf. Hiding Multimedia Signal Process. **11**(3), 126–137 (2020)
10. Eberhart, R., Kennedy, J.: Particle swarm optimization, vol. 4, pp. 1942–1948 (1995)
11. Pei, H., Pan, J.-S., Chu, S.-C., Chai, Q.-W., Liu, T., Li, Z.-C.: New hybrid algorithms for prediction of daily load of power network. Appl. Sci. **9**(21), 4514 (2019)
12. Sun, C., Jin, Y., Cheng, R., Ding, J., Zeng, J.: Surrogate-assisted cooperative swarm optimization of high-dimensional expensive problems. IEEE Trans. Evol. Comput. **21**(4), 644–660 (2017)
13. Chu, S.-C., Tsai, P., Pan, J.-S.: Cat swarm optimization. In: Yang, Q., Webb, G. (eds.) PRICAI 2006. LNCS (LNAI), vol. 4099, pp. 854–858. Springer, Heidelberg (2006). https://doi.org/10.1007/978-3-540-36668-3_94
14. Bahrami, M., Bozorg-Haddad, O., Chu, X.: Cat swarm optimization (CSO) algorithm. In: Bozorg-Haddad, O. (ed.) Advanced Optimization by Nature-Inspired Algorithms. SCI, vol. 720, pp. 9–18. Springer, Singapore (2018). https://doi.org/10.1007/978-981-10-5221-7_2
15. Sun, X.-X., Pan, J.-S., Chu, S.-C., Pei, H., Tian, A.-Q.: A novel pigeon-inspired optimization with quasi-affine transformation evolutionary algorithm for dv-hop in wireless sensor networks. Int. J. Distrib. Sensor Netw. **16**(6), 1550147720932749 (2020)
16. Meng, Z., Pan, J.-S., Xu, H.: Quasi-affine transformation evolutionary (QUATRE) algorithm: a cooperative swarm based algorithm for global optimization. Knowl.-Based Syst. **109**, 104–121 (2016)
17. Deng, W., Xu, J., Song, Y., Zhao, H.: Differential evolution algorithm with wavelet basis function and optimal mutation strategy for complex optimization problem. Appl. Soft Comput. **100**, 106724 (2021)

18. Deng, W., Liu, H., Junjie, X., Zhao, H., Song, Y.: An improved quantum-inspired differential evolution algorithm for deep belief network. IEEE Trans. Instrument. Meas. **69**(10), 7319–7327 (2020)
19. Pan, J.-S., Song, P.-C., Chu, S.-C., Peng, Y.-J.: Improved compact cuckoo search algorithm applied to location of drone logistics hub. Mathematics **8**(3), 333 (2020)
20. Sun, Y., Yin, S., Li, H.: A new wavelet threshold function based on gaussian kernel function for image de-noising. J. Inf. Hiding Multimedia Signal Process. **10**(1), 91–101 (2019)
21. Song, P.-C., Pan, J.-S., Chu, S.-C.: A parallel compact cuckoo search algorithm for three-dimensional path planning. Appl. Soft Comput. **94**, 106443 (2020)
22. Li, W., Yan, W., Ding, Q., Zhang, R., Chen, Y.-C.: Discrete synchronization method for continuous chaotic systems and its application in secure communication. J. Netw. Intell. **5**(2), 62–76 (2020)
23. Chu, S.-C., Roddick, J.F., Su, C.-J., Pan, J.-S.: Constrained ant colony optimization for data clustering. In: Zhang, C., W. Guesgen, H., Yeap, W.-K. (eds.) PRICAI 2004. LNCS (LNAI), vol. 3157, pp. 534–543. Springer, Heidelberg (2004). https://doi.org/10.1007/978-3-540-28633-2_57
24. Dorigo, M., Birattari, M., Stutzle, T.: Ant colony optimization. IEEE Comput. Intell. Mag. **1**(4), 28–39 (2006)
25. Wu, J.M.T., Zhan, J., Lin, J.C.W.: Ant colony system sanitization approach to hiding sensitive itemsets. IEEE Access **5**, 10024–10039 (2017)
26. Sivanandam, S.N., Deepa, S.N.: Genetic algorithms, pp. 15–37 (2008)
27. Corus, D., Dang, D.C., Eremeev, A.V., Lehre, P.K.: Level-based analysis of genetic algorithms and other search processes. IEEE Trans. Evol. Comput. **22**(5), 707–719 (2017)
28. Phung, M.D., Quach, C.H., Dinh, T.H., Ha, Q.: Enhanced discrete particle swarm optimization path planning for UAV vision-based surface inspection. Autom. Constr. **81**, 25–33 (2017)
29. Chu, S.-C., Du, Z.-G., Pan, J.-S.: Symbiotic organism search algorithm with multi-group quantum-behavior communication scheme applied in wireless sensor networks. Appl. Sci. **10**(3), 930 (2020)
30. Du, A.-G., Pan, J.-S., Chu, S.-C., Chiu, Y.-J.: Improved binary symbiotic organism search algorithm with transfer functions for feature selection. IEEE Access **8**, 225730–225744 (2020)
31. Ezugwu, A.E.S., Adewumi, A.O.: Discrete symbiotic organisms search algorithm for travelling salesman problem. Expert Syst. Appl. **87**, 70–78 (2017)
32. Wang, Y., Wu, Y.W., Xu, N.: Discrete symbiotic organism search with excellence coefficients and self-escape for traveling salesman problem. Comput. Ind. Eng. **131**, 269–281 (2019)

Cluster Analysis of Smart Meter Load Based on Electricity Behavior Characteristics

Dachi Zhang[1](✉), Xuanyuan Zhao[2], Yunfeng Guo[3], Jingying Yang[1], Peng Liu[1], and Yu Lu[1]

[1] Economic and Technological Research Institute of State Grid Jilin Electric Power Co., Ltd, Jilin, China
[2] Key Laboratory of Modern Power System Simulation and Control and Renewable Energy Technology, Ministry of Education (Northeast Electric Power University), Jilin 132012, China
[3] State Grid Jilin Electric Power Co. Ltd, Changchun, China

Abstract. Smart meters are installed at different locations in the power distribution area according to the distribution of users. Due to the geographical environment differences in the power distribution area, the differences in economic levels between individual residents, the differences in lifestyles, and the differences in business models and working hours among other commercial users, the data collected by smart meters varies greatly. Therefore, under the framework of MapReduce, the user's electricity consumption behavior characteristics are extracted from the electricity consumption big data; then the Canopy algorithm is used to determine the number of user clusters to avoid unreasonable user classification affecting the clustering results; finally, K-Means is used in the clustering algorithm realizes the cluster analysis of users.

Keywords: Smart meter · Big data · Feature extraction · User clustering

1 Data Acquisition of Smart Meter Data

Before the smart grid was proposed, the informatization of the power grid was driven by business and was limited by the high cost of data collection and data storage. Generally, only a small amount of electricity data was collected to meet business needs, while ignoring the use of other information. value. With the introduction of smart grids to large-scale construction, the reduction of production costs of data acquisition equipment and data storage equipment, and the advent of the era of big data, data has now become a valuable asset in various fields.

To alleviate the dependence of the feature Self Extraction Model on voltage sag sample data and improve the ability of feature capture, a voltage sag event type identification algorithm based on improved auxiliary classifier generative adversarial networks (AC-GAN) is propose [1]. A comprehensive classification method of user power consumption behavior considering user load characteristics and adjustable potential is proposed [2]. An adaptive S-transform (AST) directly controls the standard deviation of window function to control resolution [3]. A multi-dimensional typical scenario generation algorithm for distribution networks based on load clustering and network equivalence [4].

S.-C. Chu et al. (Eds.): ICGEC 2021, LNEE 833, pp. 429–437, 2022.
https://doi.org/10.1007/978-981-16-8430-2_39

The load data collected by smart meters of more than 5000 power users in a certain power distribution area in Ireland are used as the experimental data of this subject, and the sampling period of load data is 30 min [5]. Among them, the users participating in this topic can be divided into three categories: ordinary title users, small and medium enterprise users (commercial), and other users.

2 Different Types of Load Analysis

2.1 Analysis of Resident Load Characteristics

Figure 1 shows the load curve of a certain resident in different seasons. It can be seen from the figure that the user has similar electricity consumption patterns in different seasons, that is, at a low load level from 0:00 to 12:00, and after 12:00, the load rises sharply; and the load in different seasons It can be found from the measurement time that due to the influence of climate as the dominant factor, the load level of this user in summer and winter is significantly higher than the load in spring and autumn.

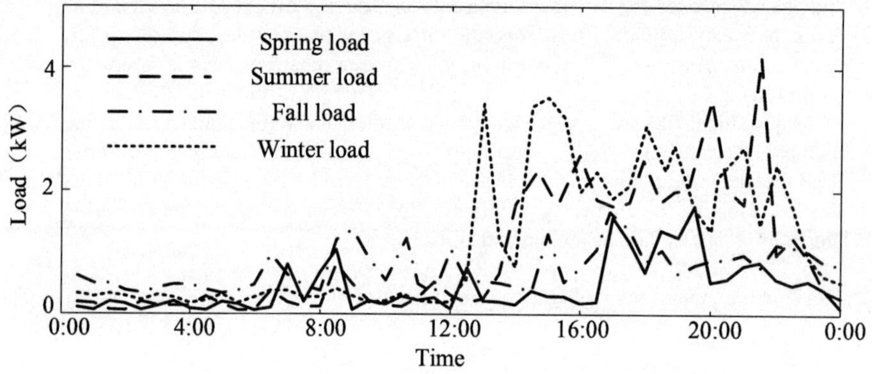

Fig. 1. Load curve of a certain resident in different seasons.

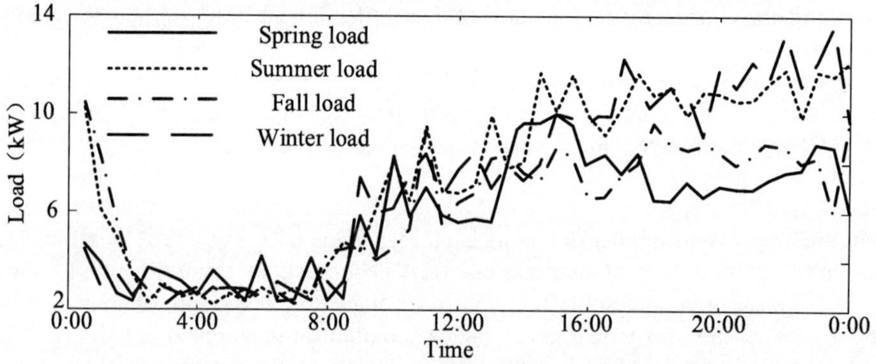

Fig. 2. Load curve of a small and medium-sized user in different seasons.

2.2 Analysis of Load Characteristics of SMEs

Figure 2 shows the load curve of a small and medium-sized user in different seasons. Similar to the change in the load of residential users, the load of this small and medium-sized user in the four seasons of the year also shows a trend of less in spring and autumn, and more in summer and winter. Different from the residential electricity consumption situation shown in Fig. 1, the user's load gradually increases from 8:00, which means that people start to work and various electrical equipment of the household start to operate.

2.3 Analysis of Regional Total Load Characteristics

The total load is obtained by accumulating the load of all users in the power distribution area. Figure 3 shows a total load of users and the load curve of a single user. The user load shown in the figure is at a level before 12:00 every day. The state of low load, and after 12:00, the load has increased significantly.

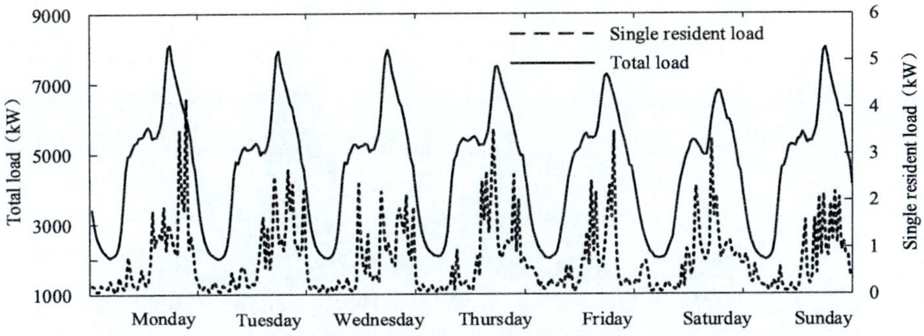

Fig. 3. Load curve between total load and single user

It can be seen from the above analysis that there is a significant difference between the load characteristics of the distribution network level and the load characteristics of the power user level, and the load curve between the power user levels also has a certain difference.

3 Feature Extraction of Electricity Consumption Based on MapReduce

During the execution of MapReduce, MapReduce can be decomposed into Map tasks and Reduce tasks, that is, first use the Map program to decompose the data into independent data blocks, then use the computer to implement parallel calculations, and finally use the Reduce program to summarize the results. The MapReduce program only contains the Map function and the Reduce function, and the input and output of both are key-value pairs in a user-defined format [6–8].

Figure 4 respectively show the number of users with peak load of all residents in Ireland at different times in different seasons. It can be seen from the figure that no matter

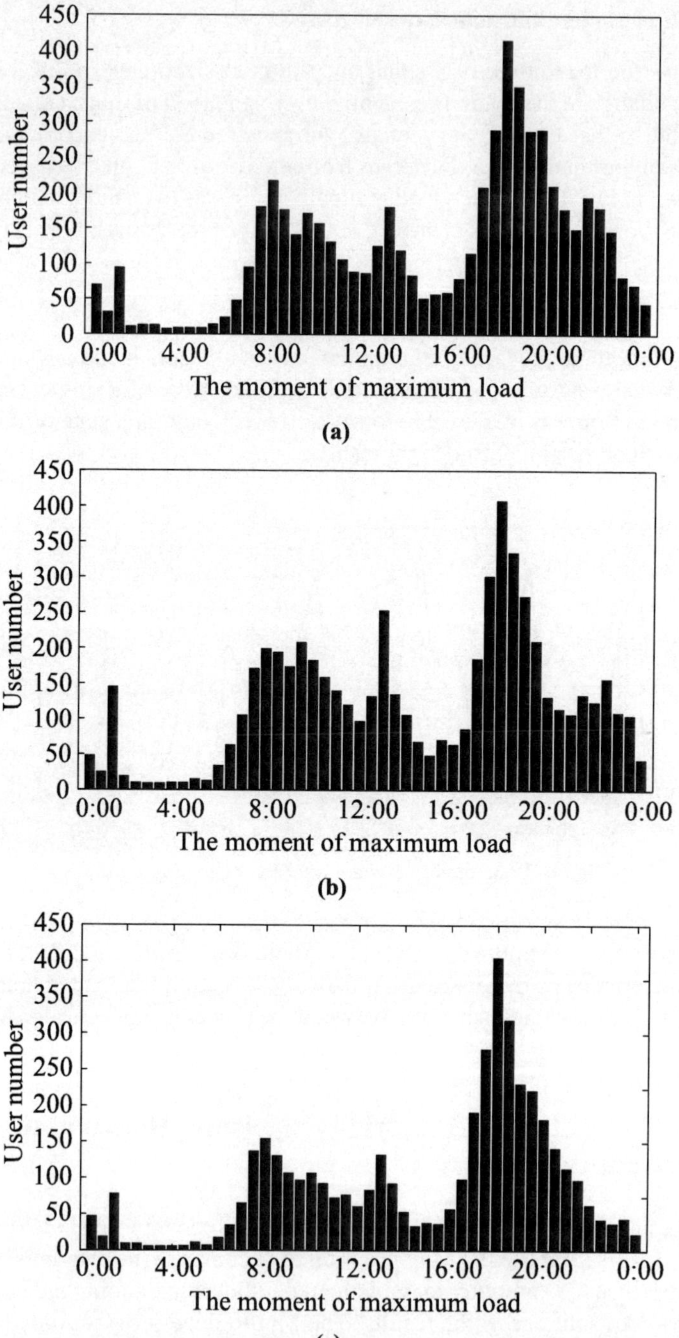

Fig. 4. Time distribution of peak load of each user in each season

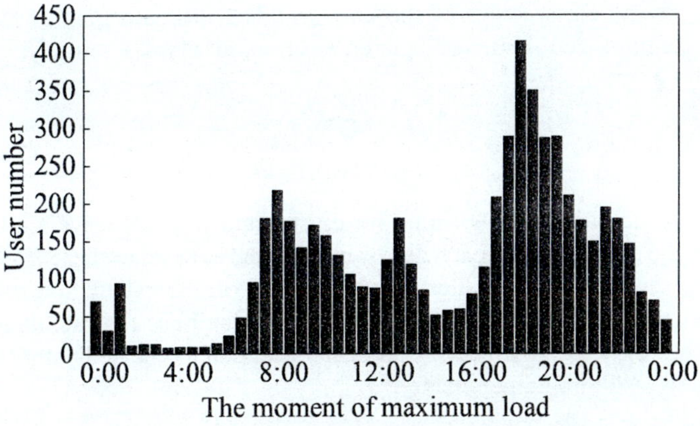

(d)

Fig. 4. continued

in that season, the number of users distributed at different times is similar, indicating that the power consumption patterns of users in the distribution area remain unchanged. In addition, according to the distribution of the number of users, 24 h a day can be divided into 4 time periods: 0:00–5:00, 5:00–12:00, 12:00–17:00 and 17:00–0:00, and extract the maximum, minimum, and average loads in the four periods as features for cluster analysis.

4 User Clustering of Smart Meter Big Data Based on K-Means

4.1 K-Means Clustering Based on Canopy

The measurement methods based on Euclidean distance, Manhattan distance and Minkowski distance are commonly used standards to measure the effect of clustering. The expressions of these three measurement methods are shown in Eq. (1–3), respectively.

$$d(i,j) = \sqrt{|x_1 - y_1|^2 + |x_2 - y_2|^2 + \ldots + |x_n - y_n|^2} \tag{1}$$

$$d(i,j) = |x_1 - y_1| + |x_2 - y_2| + \ldots + |x_n - y_n| \tag{2}$$

$$d(i,j) = \sqrt[m]{|x_1 - y_1|^m + |x_2 - y_2|^m + \ldots + |x_n - y_n|^m} \tag{3}$$

Where $(x_1, x_2, \ldots, x_n)$ and $(y_1, y_2, \ldots, y_n)$ are the data of two dimensions of n in the data set.

Suppose $X = \{x_i\}, i = 1, 2, \ldots, n$ is a data set that contains n data objects to be clustered. The K-Means algorithm divides the data set X so that the square of the error between the center of the class and each object in the class is the smallest [9].

$$J(c_k) = \sum_{x_i \in c_k} \|x_i - a_k\|^2 \tag{4}$$

Where a_k is the class center of the class c_k. The ultimate goal of the K-Means algorithm is to minimize the sum of squared errors of all classes, namely:

$$J(C) = \sum_{k=1}^{K} J(c_k) \tag{5}$$

When using the K-Means algorithm for cluster analysis, first give K class centers, then calculate the distance from each data sample in the data set to these K class centers, and assign the sample to the smallest distance based on the calculation results In that category. Then recalculate the center of this category, and use the calculation result as the new center of this category. The cluster center is updated repeatedly until the samples in the clustering result no longer change.

4.2 Analysis of User Clustering Results

Affected by the climate throughout the year, the electricity consumption of different users in different seasons will vary significantly. Therefore, according to the different seasons, all users are clustered separately. The number of clusters in different seasons and the number of users in each category are shown in Table 1.

Table 1. Clustering results of users in different seasons.

Season	Category	User number
Spring	The first category	78
	The second category	1423
	The third category	4208
Summer	The first category	60
	The second category	1712
	The third category	3937
Fall	The first category	111
	The second category	5598
Winter	The first category	4096
	The second category	1508
	The third category	105

Figure 5 shows the load curves obtained by summing loads of different types of users in different quarters. Figures (a–d) correspond to the four seasons of spring, summer, autumn and winter. It can be seen from the figure that there are obvious differences between different types of total loads in the same quarter, which also shows that before using smart meter big data for load forecasting, it is necessary to cluster users based on their electricity consumption behavior.

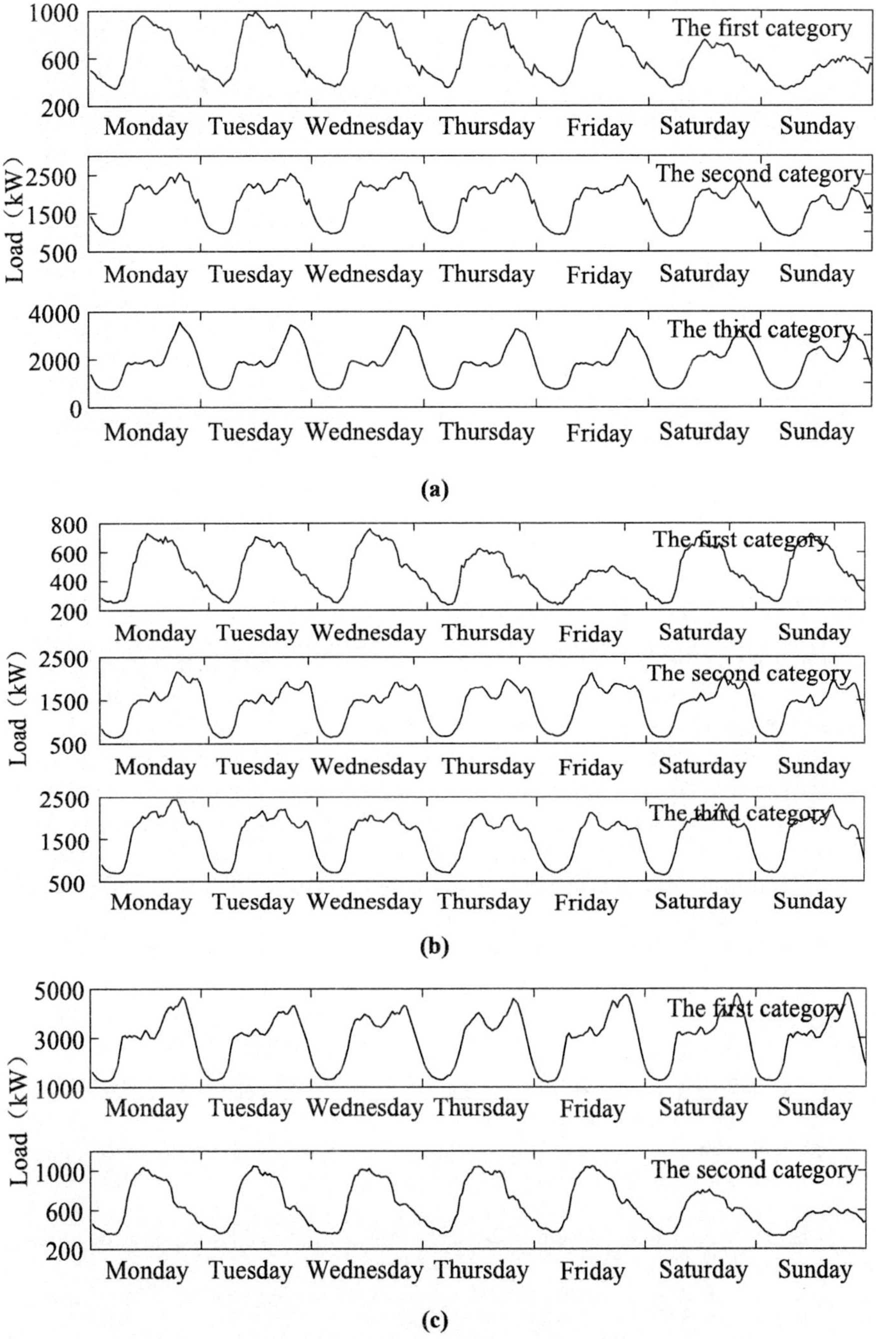

Fig. 5. Clustering results of users in different seasons

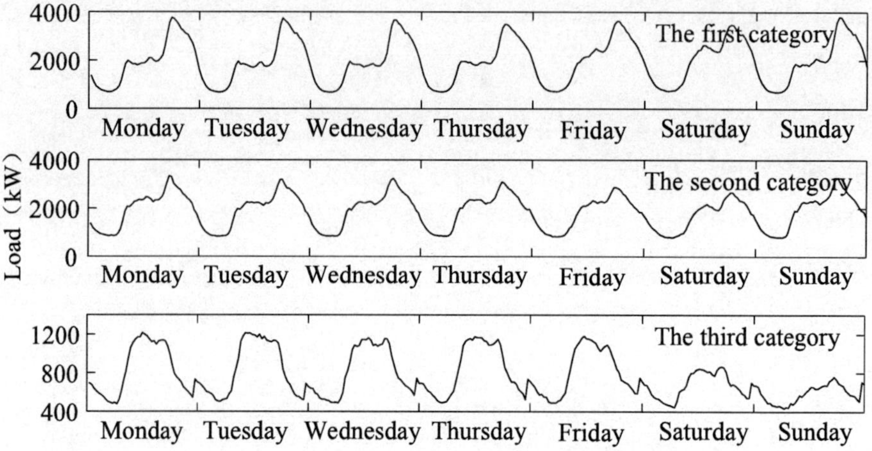

(d)

Fig. 5. continued

5 Conclusion

This article makes a brief analysis of Irish user-level load and distribution network-level load. The difference in load between different users and different types of users is analyzed.First of all, considering that the load of a single household electricity user collected by a smart meter is extremely small and has high volatility, the use of 19 representative users The characteristics of electricity usage behavior are used as the premise of user clustering for clustering; then, the MapReduce algorithm is introduced for feature extraction. It effectively reduces the time required for feature extraction; finally, the Canopy algorithm is used to obtain the initial clustering centers of electricity users in different quarters, which avoids the problem that the K-Means algorithm is greatly affected by the initial clustering centers.

References

1. Haoyuan, S., et al.: Research on voltage sag event type identification based on improved generative adversarial networks. In: Proceedings of the CSEE, 20 Mar 2021, pp. 1–13 (2021)
2. Yi, S., et al.: A comprehensive clustering method of user load characteristics and adjustable potential based on power big data. In: Proceedings of the CSEE, 20 Marh 2021, pp. 1–14 (2021)
3. Pan, L., et al.: Application of adaptive S-transform in power quality feature extraction. In: Proceedings of the CSEE, 20 Marh 2021, pp. 1–9 (2021)
4. Bo, L., et al.: A multi-dimensional typical scenarios generation algorithm for distribution network based on load clustering and network structure equivalence. In: Proceedings of the CSEE, 20 Mar 2021, pp. 1–12 (2021)
5. Quilumba, F.L., et al.: Using smart meter data to improve the accuracy of intraday load forecasting considering customer behavior similarities. IEEE Trans. Smart Grid **6**(2), 911–918 (2015)

6. Martins, R., Manquinho, V., Lynce, I.: Improving linear search algorithms with model-based approaches for MaxSAT solving. J. Exp. Theor. Artif. Intell. **27**(5), 673–701 (2015)
7. Ding, X.O., et al.: Association relationships study of multi-dimensional data quality. J. Softw. **27**(7), 1626–1644 (2016)
8. Yang, D.H., et al.: The optimization of the big data cleaning based on task merging. Chinese J. Comput. **39**(1), 97–108 (2016)
9. Zhao, L., Wen-hao, Y., Chong-Guang, R.: Population initialization based on search space partition and Canopy Kmeans clustering. Control Decis. **35**(11), 2767–2772 (2020)

Improved Affinity Propagation Clustering Based on K-Nearest Neighbors and Canopy Algorithm

Zhihe Wang, Gang Zhang, Hui Du(✉), and Yiyang Ni

School of Computer Science and Engineering, Northwest Normal University, Lanzhou 730070, China
duhuiywy@nwnu.edu.cn

Abstract. For large-scale data sets, the traditional AP clustering algorithm processes all data samples through the idea of iteration and three metrics calculation. This kind of processing will lead to poor clustering effect and high time consumption. Based on this problem, an algorithm which is called KCAP based on Affinity Propagation clustering and Canopy clustering algorithm was proposed. Firstly, KNN distance was used to calculate the local density for each data. Secondly, we used the Canopy algorithm to roughly cluster the data sets, and the center points were formed into a new data set Y. Then, we applied affinity propagation clustering algorithm to cluster the data set Y. Finally, we assigned the data to the nearest center. Compared with the traditional AP algorithm, K-means algorithm and FCM algorithm, the experimental results on synthetic data sets and UCI data sets show that algorithm KCAP not only improves the clustering accuracy, but also have higher efficiency.

Keywords: K-Nearest Neighbors (KNN) · Initial center · Canopy algorithm · Affinity Propagation (AP) · Rough clustering

1 Introduction

Cluster analysis is a process of separating several subsets from data set, which we call the result of clustering. It can be seen that the similarity of the samples in the same clusters is higher, and that of the samples in the different clusters is lower. Cluster analysis is an important technology in the field of data mining [1]. It has been widely used in various fields, including gene expression [2], image processing [3], text mining [4], pattern recognition [5] and so on.

The traditional AP clustering algorithm [6] was proposed by Frey et al. in 2007 and published in Science. The advantage of AP algorithm is that it needn't setting the number of clusters in advance. Lu et al. [7] proposed an algorithm using AP. Firstly, they used the local distance to construct the similarity matrix, and applied AP clustering algorithm to cluster, which could obtain preliminary clustering, then used the global distance as the similarity measure, and repeatedly applied AP clustering algorithm to cluster on the preliminary clustering data sets. By using this method, they got the good performance in large-scale and complex data sets. Fan et al. [8] used a density-adaptive

S.-C. Chu et al. (Eds.): ICGEC 2021, LNEE 833, pp. 438–448, 2022.
https://doi.org/10.1007/978-981-16-8430-2_40

preference estimation and Nature Nearest Neighbor searching to measure the similarity matrix, which obtain a better performance than traditional AP. Li et al. [9] proposed an adjustable preference value method, which constantly updates preference value in operation. Comparing with the traditional AP clustering algorithm, it not only improves the accuracy of the algorithm, but also reduces the number of iterations. Refianti et al. [10] proposed a new calculation method of preference value, which is based on the similarity distribution of preference value calculation model, and applied this method for the adaptive AP algorithm(AAP), which achieves good results. However, large-scale data sets will also lead to low clustering efficiency in this method. Through the study of reference [11], it is found that the Canopy algorithm can be used for fast rough clustering of data through threshold selection, and avoid low clustering efficiency problem in large-scale data sets. Thus, we use the Canopy clustering algorithm to improve performance of AP algorithm.

In the experiments, we use synthetic data sets and UCI data sets to evaluate the results, and compare with the traditional AP algorithm, K-means algorithm and FCM algorithm. Experimental results show that our algorithm not only improved the accuracy, but also greatly improved the clustering efficiency in large-scale data sets.

2 Related Work

2.1 AP Clustering Algorithm

The advantage of traditional AP clustering algorithm is that it doesn't need to determine the cluster center and the number of clusters in advance, each points have the same probability to become the exemplar. It is clustering by continuous iterative message passing through the similarity between points. The pseudo code and formula of the traditional AP algorithm are as follows [12].

$$s(x_i, x_k) = -||x_i - x_k||^2 \tag{1}$$

$$r_0(x_i, x_k) = s(x_i, x_k) - \max_{x_i \neq x_k}\left\{s(x_i, x_j)\right\} \tag{2}$$

$$a_0(x_i, x_k) = 0 \tag{3}$$

$$r_t(x_i, x_k) = \begin{cases} s(x_i, x_k) - \underset{x_j \neq x_k}{max}\{a_{t-1}(x_i, x_j) + s(x_i, x_j)\}(x_i \neq x_k) \\ s(x_i, x_k) - \max\limits_{x_j \neq x_k}\{s(x_i, x_j)\}(x_i = x_k) \end{cases} \tag{4}$$

$$a_t(x_i, x_k) = \begin{cases} \min\{0, r_{t-1}(x_i, x_k) + \sum\limits_{j \neq i,k} \max\{0, r_{t-1}(x_i, x_k)\}\}(x_i \neq x_k) \\ \sum\limits_{x_j \neq x_k} \max\left\{0, r_{t-1}(x_j, x_k)\right\}(x_i = x_k) \end{cases} \tag{5}$$

$$r_t(x_i, x_k) = (1 - \lambda)r_t(x_i, x_k) + \lambda r_{t-1}(x_i, x_k) \tag{6}$$

$$a_t(x_i, x_k) = (1 - \lambda)a_t(x_i, x_k) + \lambda a_{t-1}(x_i, x_k) \tag{7}$$

$$r_t(x_k, x_k) + a_t(x_k, x_k) \tag{8}$$

where, x_i and x_j denote positions of point i and point j, respectively. x_k is to be the exemplar for x_i, $s(x_i, x_k)$ denote the similarity between x_i and x_k, when $x_i = x_k$, we called the value of $s(x_k, x_k)$ as preference value which could affect the number of clusters. Generally, preference value was set as $s(x_k, x_k) = \text{median}(S)$. The responsibility $r_t(x_i, x_k)$, denotes how appropriate x_k is become the exemplar for x_i. The availability $a_t(x_i, x_k)$, it means how suitable it is for x_i to choose x_k as the exemplar. $r_0(x_i, x_k)$ and $a_0(x_i, x_k)$ denotes the responsibility and the availability when $t = 0$, respectively. Parameters λ mainly plays the role of convergence, default value is 0.5. If $r_t(x_k, x_k) + a_t(x_k, x_k) > 0$, we can identify point x_i as the exemplar. Then we add x_i to C, C is the set of cluster centers.

Algorithm 1. AP Clustering Algorithm.

Input: Data set X ; λ ; max_iter;

Initialization:

Initialized the similarity of data set X by (1) and $P = median(S)$

Initialized the $r_0(x_i, x_k)$ and $a_0(x_i, x_k)$ by using (2)(3)

Clustering:

While in max_iter:

  Update $r_t(x_i, x_k)$ and $a_t(x_i, x_k)$ by (4)(5)(6)(7)

  Terminated if (8) is not changed or (8) is below some threshold

If (8) >0:

  Add x_k to C

Output: $C = \{c_1, c_2, \ldots, c_k\}$

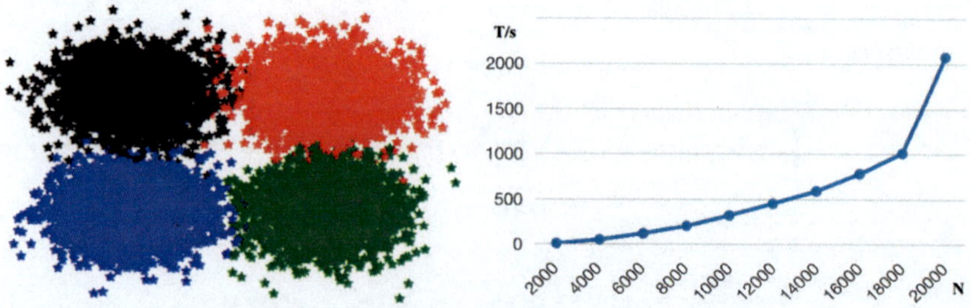

Fig. 1. Time consumption in synthetic data sets

But the traditional AP algorithm still has some defects, for large-scale data sets, the traditional AP clustering algorithm processes all data samples through the idea of iteration and three metrics (the similarity metric, the responsibility metric, and the availability metric) calculation. With the increase of the amount of samples, this kind of processing will lead to poor clustering effect and a sharp increase in time consumption. As show in Fig. 1.

3 The Proposed Algorithm KCAP

To solve above problem, an Improved Affinity Propagation Clustering Based on KNN and Canopy Algorithm was proposed which we called the KCAP (KNN, Canopy and Affinity Propagation) algorithm.

3.1 KCAP Algorithm

The Canopy algorithm usually used as the pre-clustering algorithm to initial the cluster center and the number of the clusters by choosing the appropriate thresholds $T1$ and $T2$, and $T1 > T2$ [13]. However, through experiments, a small number of clusters can be find when the thresholds are larger. And it's easier to find the center at this time. The small thresholds we choose the more clusters we can find. Through this idea, we can use the Canopy algorithm for rough clustering of data sets (The selection of thresholds value would be introduced in more detail in Sect. 4.1.2 and Sect. 4.2.2). Further research, to avoid local optimum of the traditional Canopy algorithm, K-Nearest Neighbors algorithm [14, 15] be used as local density calculation between the samples. The dense sample is selected as the initial center of the Canopy algorithm. Then, we obtain the subset Y which constructed by the centers of the Canopy algorithm's result. Finally, we applied AP clustering algorithm to cluster the data set Y and assigned all data points to the nearest center. The proposed algorithm improved accuracy and efficiency of clustering results, and the method fully considered the influence of different density points on the selection of center points.

The point $x_i \in X$, $dist(\cdot, \cdot)$ is Euclidean Distance, $KNN(x_i)$ denotes a set of the k samples of x_i, then we take K-Nearest Neighbors distance ($D_{knn}(x_i)$) as Eq. (9).

$$D_{knn}(x_i) = \frac{1}{k} \sum_{x_j \in KNN(x_i)} dist(x_i, x_j) \tag{9}$$

From the Eq. (9), if the samples are located in the dense area, the value of $D_{knn}(x_i)$ is small; otherwise, the value of $D_{knn}(x_i)$ is large. Therefore, the density of the point x_i, denote as ρ, is calculated as Eq. (10).

$$\rho = \frac{1}{D_{knn}(x_i)} \tag{10}$$

The proposed algorithm is summarized as Algorithm 2.

Algorithm 2. KCAP Clustering Algorithm

Input: Data set X ; $T1$; $T2$; k ; λ ; max_iter;

Rough Clustering:

Calculate the density of the samples by using (9)(10), and sorted by density

While X is not empty:

  Select the max density point x_i from X to initialize center c of Canopy

  Remove x_i from X

  Loop through remaining samples in X :

    If $dist(x_j, c) < T1$: Add x_j to c;

    If $dist(x_j, c) < T2$: Remove x_j from X ;

  Add c to Cl

Construct the new data set Y by Cl

Initialization:

Initialized the similarity of data set Y by (1) and $P = median(S)$

Initialized the $r_0(y_i, y_k)$ and $a_0(y_i, y_k)$ by using (2)(3)

Clustering:

While in max_iter:

  Update $r_t(y_i, y_k)$ and $a_t(y_i, y_k)$ by (4)(5)(6)(7)

  Terminated if (8) is not changed or (8) is below some thresholds

If (8) >0:

  Add y_k to C

Output: $C = \{c_1, c_2, \ldots, c_k\}$

The flow chart of the proposed algorithm can be shown as Fig. 2.

Time complexity analysis: For the given data set $X = (x_1, x_2, x_3, \ldots, x_n) \subset R_s$, we introduced K-D Tree [16] to calculate the K-Nearest Neighbors, the computation complexity is $O(n \log n)$. The maximum complexity of the Canopy algorithm is $O(n^2)$. The second step is clustering by AP algorithm, data set $Y = (y_1, y_2, y_3, \ldots, y_N) \subset R_d$, $N << n$, so the computation complexity of AP is $O(N^2T)$, T is the number of iterative of KCAP algorithm, the total time consumption of KCAP algorithm is $O(n \log n + n^2 + N^2T) < O(n^2)$. But for the traditional AP algorithm, the complexity is $O(n^2t)$, t is the

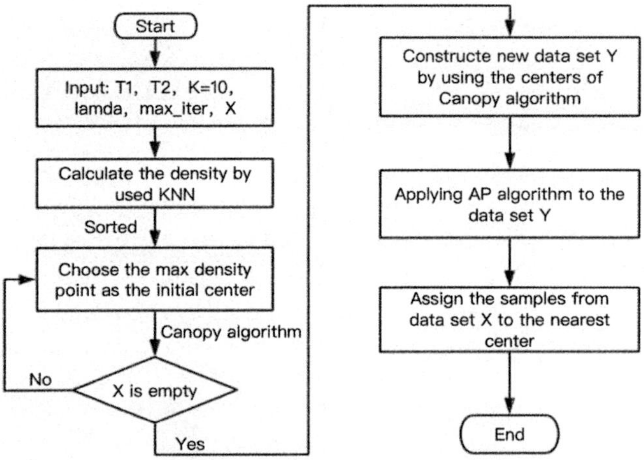

Fig. 2. The flow chart of KCAP

number of iterative of AP. To sum up, KCAP algorithm is more efficient than traditional AP algorithm.

4 Experiments and Analysis

In this section, we will verify the performance of KCAP algorithm on synthetic data sets and UCI data sets, and comparing the clustering results with K-means algorithm, FCM algorithm and traditional AP algorithm. The experiments work with Intel Core i5, 1.4 GHz processor and 16 GB RAM running python 3.7.4.

4.1 Experiments on Synthetic Data Sets

4.1.1 Synthetic Data Sets

The algorithms are tested on two-dimensional synthetic data sets, and we use the data sets to visualize the experimental results. Square 1 has simple data samples distribution, which includes 1,000 points in 4 clusters. S1 has complex data samples distribution, which consists by 15 clusters, including 5,000 points.

4.1.2 Parameter Analysis

In the above two-dimensional data sets, we set $T1$ and $T2$ at (0, 0.2]. Firstly, we normalizing the data sets and all samples are distributed between 0 to 1. In a large number of experiments, we found that if we set the larger thresholds, many points would be filtered, and we couldn't restore the original distribution of the data. Based on this, we set the thresholds from 0 to 0.2. Besides, for the parameters k from KNN algorithm, after excluding the influence of some extremely values, the effect of k value on density calculate is not obvious. To obtain the local density in a small range, we set $k = 10$, and all the results in this paper are calculated by this. In AP and KCAP algorithms, we selected $P = median(S)$ following the paper [6] to test the evaluation of clustering results in synthetic data sets.

4.1.3 The Results in Synthetic Data Sets

As show in above, the color of center points is black. Figure 3 shows that K-means, FCM and KCAP algorithms have a good clustering effect, but the traditional AP algorithm clustering effect is not well. Figure 4 shows that KCAP has better performance than three others, and it solves the problem that the traditional AP algorithm takes a long time on large-scale data sets.

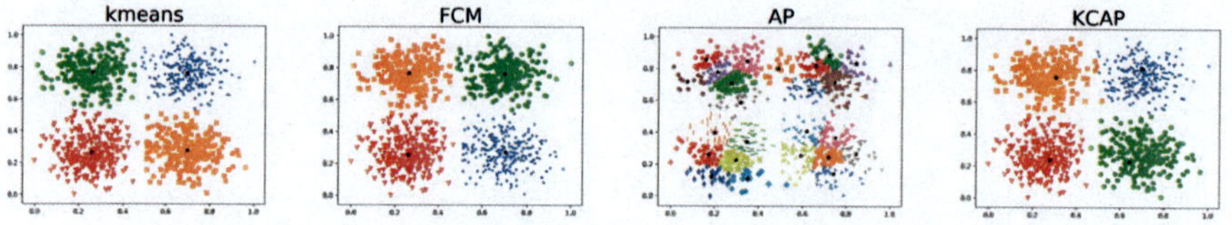

Fig. 3. Clustering results of the Square1 data set on the four

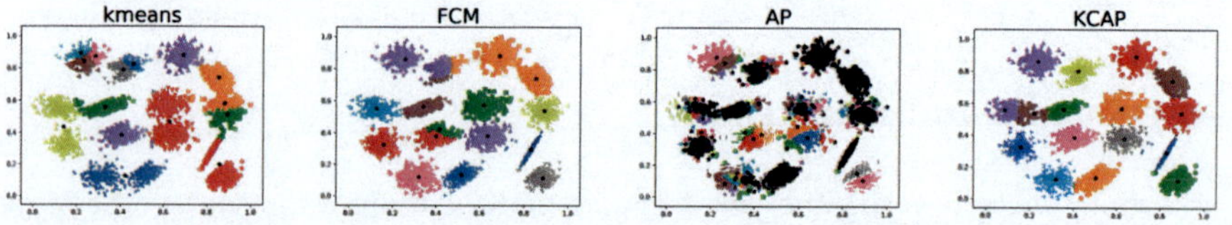

Fig. 4. Clustering results of the S1 data set on the four

4.2 Experiments on UCI Data Sets

4.2.1 UCI Data Sets

The following experiments will verify the effectiveness of the algorithm through UCI data sets. There are Iris, Wine, Seeds, Glass, HTRU_2, Ecoli, Banknotes, Haberman, Thyroid, Blood. The information of the data sets is show in Table 1.

Table 1. The basic information of the UCI data sets

Data sets	Instances	Attributes	Classes
Iris	150	4	3
Wine	178	13	3
Seeds	210	7	3
Glass	214	9	6

(continued)

Table 1. (*continued*)

Data sets	Instances	Attributes	Classes
HTRU_2	17898	8	2
Ecoli	336	7	8
Banknotes	1372	4	2
Haberman	306	3	2
Thyroid	215	5	3
Blood	748	4	2

4.2.2 Parameter Analysis

For UCI data sets, we set $T1$ and $T2$ at [0.2, 0.6]. The k value of KNN algorithm is set as 10. Table 3 shows the results by adjusting the parameters setting. All the parameters and results are shown in Table 3.

4.2.3 The Results in UCI Data Sets

As shown in the following tables, we run the algorithms for 10 times and calculate the means of clustering results to avoid local optimum results of the random center selection of K-means and FCM algorithms. Then, Normalized Mutual Information (NMI) [17], Adjusted Rand Index (ARI) [18] and Silhouette Coefficient (Silhouette) [19] are used to evaluate the quality of the clustering results, the larger values meaning the better performance of the clustering.

In Table 2, we evaluate the time consumption of the traditional AP algorithm and KCAP algorithm. Rough clustering not only reduces the size of the data set, but also selects some more representative data points for clustering, moreover, it also reduces the number of iterations. As show in Table 2, KCAP algorithm has greatly reduces time consumption.

Table 2. The time consumption of KCAP and AP algorithm (s)

	AP	KCAP
Iris	0.054	**0.031**
Wine	**0.051**	0.11
Seeds	0.07	**0.055**
Glass	0.073	**0.03**
HTRU_2	747.8	**0.552**
Ecoli	0.097	**0.035**

(*continued*)

Table 2. (*continued*)

	AP	KCAP
Banknotes	3.092	**0.267**
Haberman	0.159	**0.019**
Thyroid	0.071	**0.019**
Blood	1.218	**0.042**

In Table 3, we got results by adjusted the parameters. Except Iris data set, KCAP algorithm has favorable performance compared to the above three algorithms, especially when the data set is large and complex, the accuracy of KCAP algorithm is improved more obviously. Due to the small scale and uncomplicated structure of Iris data set, the performance of KCAP algorithm is slightly worse than that of the traditional AP. In consequence, KCAP algorithm outperforms the traditional AP algorithm.

Table 3. The acceptable results of KCAP and AP with optimized parameters

		K-means	FCM	AP	KCAP	Parameters	
Iris	NMI	0.716	0.717	**0.758**	0.742	AP	P = −1.8
	ARI	0.682	0.705	**0.743**	0.716	KCAP	P = −1
	Sil	0.681	0.65	0.658	**0.932**		T1 = 0.3
							T2 = 0.2
Wine	NMI	0.828	0.559	**0.861**	**0.861**	AP	P = −5
	ARI	0.845	0.481	**0.88**	**0.88**	KCAP	P = −5
	Sil	0.475	0.336	0.474	**0.75**		T1 = 0.4
							T2 = 0.34
Seeds	NMI	0.67	0.644	0.689	**0.75**	AP	P = −4
	ARI	0.699	0.653	0.715	**0.748**	KCAP	P = −1.8
	Sil	0.621	0.578	0.62	**0.893**		T1 = 0.3
							T2 = 0.2
Glass	NMI	0.331	0.323	0.319	**0.477**	AP	P = −2.3
	ARI	0.179	0.191	0.166	**0.298**	KCAP	P = −1.2
	Sil	0.483	0.35	0.515	**0.881**		T1 = 0.4
							T2 = 0.3
HTRU_2	NMI	0.342	0.268	0.195	**0.572**	AP	P = −1
	ARI	0.532	0.365	0.007	**0.711**	KCAP	P = −3
	Sil	0.806	0.606	0.282	**0.908**		T1 = 0.6
							T2 = 0.5

(*continued*)

Table 3. (*continued*)

		K-means	FCM	AP	KCAP	Parameters	
Ecoli	NMI	0.586	0.527	0.61	**0.646**	AP	P = −1.4
	ARI	0.87	0.491	0.417	**0.654**	KCAP	P = −0.41
	Sil	0.364	0.086	0.403	**0.91**		T1 = 0.4
							T2 = 0.3
Banknotes	NMI	0.017	0.097	0.359	**0.455**	AP	P = −4.21
	ARI	0.022	0.13	0.161	**0.169**	KCAP	P = −0.2
	Sil	0.5	0.435	0.439	**0.968**		T1 = 0.2
							T2 = 0.12
Haberman	NMI	0.001	0.002	0.001	**0.003**	AP	P = −7
	ARI	−0.003	−0.001	−0.004	**0.021**	KCAP	P = −0.6
	Sil	0.563	0.433	0.565	**0.821**		T1 = 0.5
							T2 = 0.4
Thyroid	NMI	0.59	0.327	0.587	**0.61**	AP	P = −5
	ARI	0.605	0.197	0.612	**0.62**	KCAP	P = −0.6
	Sil	0.732	0.141	0.754	**0.943**		T1 = 0.4
							T2 = 0.3
Blood	NMI	0	0.002	0.018	**0.035**	AP	P = −5
	ARI	−0.006	−0.003	0.033	**0.031**	KCAP	P = −1.5
	Sil	0.67	0.639	0.8	**0.924**		T1 = 0.3
							T2 = 0.2

5 Conclusions

To solve the problem of low clustering effect and large time consumption of the traditional AP clustering algorithm in large-scale data sets, we proposed KCAP clustering algorithm. The algorithm calculates the local density of the samples by using KNN distance and selects high-density points as the initial center for the Canopy algorithm which avoids that the Canopy algorithm convergence to local optimum from randomly center selection. Then, we introduced AP algorithm to cluster the new data sets Y which constructed by the cluster centers of improved Canopy algorithm. Comparing with the K-means algorithm, FCM algorithm and traditional AP algorithm, experimental result on synthetic data sets and UCI data sets shows that the proposed algorithm can achieves better effect.

However, the Euclidean distance is still used to measure the similarity of samples. The algorithm does not perform well when dealing with non-convex data sets, which is the direction of future research.

Acknowledgments. This work was supported by the [Northwest Normal University] under Grant [Research on Retina Image Segmentation and Aided Diagnosis Technology based on Deep Learning] [61962054].

References

1. Han, J.M., Kamber, M., Pei, J.: Data Mining: Concepts and Techniques. 3rd edn. Machine Press (2012)
2. Zeng, Y., Xu, Z., He, Y., et al.: Fuzzy entropy clustering by searching local border points for the analysis of gene expression data. Knowl.-Based Syst. **190**, 105309 (2019)
3. Zhou, S.B., Xu, Z.: Automatic grayscale image segmentation based on affinity propagation clustering. Patt. Anal. Appl. **23**(5), 331–348 (2019)
4. Lakshmi, R., Baskar, S.: Efficient text document clustering with new similarity measures. Int. J. Bus. Intell. Data Mining **18**(1), 49–72 (2021)
5. Gao, M., Shi, G.Y.: Ship-handling behavior pattern recognition using AIS sub-trajectory clustering analysis based on the T-SNE and spectral clustering algorithms. Ocean Eng. **205**, 106919 (2020)
6. Frey, B.J., Dueck, D.: Clustering by passing messages between data points. Science **315**(5814), 972–976 (2007)
7. Lu, P.L., Wang, Z.D.: Multilevel affinity propagation clustering based on density-sensitive distance. J. Lanzhou Univ. Technol. **39**(6), 85–89 (2013)
8. Fan, Z.Y., Jiang, J., Weng, S.Q., et al.: Adaptive density distribution inspired affinity propagation clustering. Neural Comput. Appl. **111**, 27–35 (2017)
9. Li, P., Ji, H.F., Wang, B.L., et al.: Adjustable preference affinity propagation clustering. Patt. Recogn. Lett. **85**, 72–78 (2017)
10. Refianti, R., Mutiara, A.B., Juarna, A., et al.: A preference model on adaptive affinity propagation. Int. J. Electr. Comput. Eng. (IJECE) **8**(3), 1805–1813 (2018)
11. Mccallum, A., Nigam, K., Ungar, L.H.: Efficient clustering of high-dimensional data sets with application to reference matching. In: Proceedings of the Sixth ACM SIGKDD International Conference on Knowledge Discovery and Data Mining, pp.169–178. ACM, New York, Boston, Massachusetts (2000)
12. Liu, Y.C., Liu, J., Jin, Y., et al.: An affinity propagation clustering based particle swarm optimizer for dynamic optimization. Knowl.-Based Syst. **195**, 105711 (2020)
13. Xia, D.L., Ning, F.F., He, W.N.: Research on parallel adaptive canopy-k-means clustering algorithm for big data mining based on cloud platform. J. Grid Comput. **18**(2), 263–273 (2020)
14. Sun, L., Qin, X.Y., Ding, W.P., et al.: Density peaks clustering based on k-nearest neighbors and self-recommendation. Int. J. Mach. Learn. Cybern. **12**(7), 1913–1938 (2021). https://doi.org/10.1007/s13042-021-01284-x
15. Liu, Y.H., Ma, Z., Fang, Y.: Adaptive density peak clustering based on K-nearest neighbors with aggregating strategy. Knowl. Based Syst. **133**, 208–220 (2017)
16. Bentley, J.L.: Multi-dimensional binary search trees used for associative searching. Commun. ACM **18**(9), 509–517 (1975)
17. Strehl, A., Ghosh, J.: Cluster ensembles - a knowledge reuse framework for combining multiple partitions. J. Mach. Learn. Res. **3**(3), 583–617 (2002)
18. Xuan, V.N., Epps, J., Bailey, J.: Bibliometrics: Information theoretic measures for clusterings comparison. In: Proceedings of the International Conference on Machine Learning, vol. 11, No. 1, pp. 2837–2854. ACM, New York (2010)
19. Rousseeuw, P.J.: Silhouettes: a graphical aid to the interpretation and validation of cluster analysis. Comput. Appl. Math. **20**, 53–65 (1987)

Research on Multi-dimensional Load Characteristics Based on Vertical and Horizontal Hybrid Clustering

Xiaotong Zhang[1], Zhijun Yuan[2], Yicheng Jin[2], Shunjiang Wang[3(✉)], Feng Sun[1], Bing Xie[1], and Xingtong Ma[1]

[1] Institute of Electric Power Science of State Grid Liaoning Electric Power Co., Ltd., Shenyang 110000, Liaoning Province, China
[2] College of Electrical Engineering, Northeast Electric Power University, Jilin City 132012, Jilin Province, China
[3] State Grid Liaoning Electric Power Co., Ltd., Shenyang 110000, Liaoning Province, China

Abstract. In- depth and multi-dimensional research on load characteristics is an urgent need for the refined development of power systems under the background of power big data. This paper proposes a multi-dimensional load characteristic research method based on vertical and horizontal hybrid clustering. First, starting from the vertical and horizontal aspects, the K-means calculation method is used twice to cluster the load. Secondly, in order to analyze the load characteristics of the system in depth, this paper considers the mathematical characteristics of the load data and analyzes it. Finally, by selecting 8760 load data measured in a certain area power grid throughout the year, the validity and practicability of the research method proposed in this paper are verified, and at the same time, it provides a reference for the follow-up related work.

Keywords: Electric power big data · Vertical and horizontal mixed clustering · Multi-dimensional · Mathematical characteristics · Load characteristics

1 Introduction

With the gradual emphasis on customer-side interaction and services by China's power companies, power users have gradually actively participated in the various businesses of power grid companies, and the source-grid-load interaction adjustment has gradually increased. Among them, the multi-dimensional research of user load characteristics is particularly important. Based on advanced communication, measurement, and data management technologies, a large number of power users' data has been widely collected by power grid companies, and the user behavior characteristics reflected by the data have gradually become prominent, and users generally exhibit complex and diverse power consumption behaviors [1–4].

At present, many scientific researchers have made great efforts in the study of load characteristics [5] to select for a typical daily load characteristic of the conventional large error method, proposed based on the calculated load characteristics typical daily

S.-C. Chu et al. (Eds.): ICGEC 2021, LNEE 833, pp. 449–457, 2022.
https://doi.org/10.1007/978-981-16-8430-2_41

K-means method, herein proposed method can be effectively selected typical daily load curve. Literature [6] introduces user power consumption characteristics into the load forecasting model, puts forward the concept of cluster load characteristic curve, and then proposes a "cluster-regression" short-term load forecasting method for large power users based on cluster load characteristic curve [7]. By analyzing the Northwest Power Grid 876 nearly 10 years 0 tracking load data, summarized in the latest daily load characteristics and load characteristics Northwest Power Grid, and summarize the history of changes of load characteristics, combined with the economic development, Chanyejiegou, environmental protection policy and other factors, look for internal load characteristics and laws change, the Northwest power grid 202 predicted 5-year load characteristics.

In recent years, load clustering methods based on neural networks and deep learning have also been widely used. The literature [8] effectively improves the clustering efficiency of high-dimensional load data by using the clustering ensemble method based on the support vector machine of the convolutional neural network; the literature [9] proposes a residential electricity consumption based on clustering and deep confidence network Load pattern recognition method, which provides reliable distribution network maintenance for residential electricity load on the basis of classification and recognition; literature [10] uses large-scale information as the input of SOM, and performs cluster analysis on different characteristic data. The ideal clustering effect is obtained.

In this paper, clustering and the study of load characteristics are linked together, using the k-means algorithm as a calculation tool, starting from the vertical and horizontal aspects of the matrix, and considering the mathematical characteristics of the load data, multi-dimensional and deep mining of electricity The actual physical meaning contained in big data, and the clustering of load data to analyze the characteristics of the load, providing for subsequent load forecasting, power and electricity balance, new energy consumption, peak shaving analysis, demand-side response and other related work Reference.

2 The Overall Thinking Process of This Article

2.1 Basic Process

This paper uses the k-means algorithm as the calculation tool, based on 8760 load data of a certain place in a year. The data can be regarded as a matrix of 24 * 365, starting from the vertical and horizontal aspects of the matrix, and considering the load data Mathematical and physical characteristics, multi-dimensional and deep mining of the actual physical meaning contained in power big data, and he characteristics of the load, for subsequent load forecasting, power and electricity balance, new energy consumption, and adjustment Peak analysis, demand-side response and other related work provide reference basis.

2.2 Features of This Article

This article has the following characteristics:

(1) This paper uses the K-means algorithm to perform cluster analysis on the load many times, deeply excavates the effective information behind the large power data, and then performs in-depth cluster analysis on the load characteristics of the system.

(2) The data in this paper is the load data of a certain power grid for 365 days a year. The data sampling period is 1 h, which can be regarded as a 24 * 365 matrix. The behavior of the matrix is horizontal and the columns are vertical. This article analyzes the overall change trend of the daily load curve throughout the year from the vertical and horizontal perspectives.
(3) This article analyzes the large power data from the perspective of mathematical statistics, and solves the mathematical characteristics of the load data for daily load, and obtains the related change trend of the load characteristic curve, which is used for subsequent load forecasting, power balance analysis, and new energy consumption. Peak shaving analysis, demand-side response and other related work provide a certain reference basis.

3 Model Solving Method and Process

3.1 Preprocessing of Grid Raw Load Data

With the continuous advancement of science and technology and the continuous improvement of the automation level of the power grid, although the reliability of the power grid load information collection system is constantly improving, there may be abnormal daily load data in the actual load information collection system. In view of this situation, interpolation methods can be used to reduce the impact on the entire system analysis.

3.1.1 Identification of Abnormal Data

Due to the variability of power system operation, system abnormal data can be further divided into situations such as missing data, constant data, and large changes in data amplitude. Different methods are adopted to identify abnormalities in different situations.

1) For the two situations of missing and constant data, the screening method can be used for direct identification, so I won't repeat them too much.
2) In the case of large changes in the data amplitude, the Laida criterion can be used to judge.

The specific principles are as follows: Raida criterion (3δ principle):

$$X_{i\max} = X_{ie} + 3 \times \delta \tag{1}$$

$$X_{i\min} = X_{ie} - 3 \times \delta \tag{2}$$

3.1.2 Correction of Abnormal Data

After the above-mentioned different methods are used to identify the abnormal data in different situations, the abnormal data can be further processed, that is, the data can be corrected, and the original data can be processed by the method of supplementing the difference. The purpose of data correction is to reduce the influence of abnormal data on the analysis of the entire system.

3.2 K-means Algorithm

The efficiency of the K-means algorithm makes this algorithm the preferred algorithm for researchers to process user load data. Based on the superiority of the algorithm in terms of time and accuracy, the K-means clustering algorithm was selected to carry out the cluster analysis of load data, and then to study the load characteristics of the area, and to provide relevant basis for subsequent load forecasting work.

3.2.1 The Basic Flow of the Algorithm

The K-means algorithm first randomly selects K initial clustering center algorithms, and the input is the number of clusters K and the load data group containing N objects. Secondly, it is necessary to classify all load data sample points. By calculating the distance D(C) from each sample point to K cluster centers, the cluster centers are continuously adjusted; finally, load groups with the same load characteristics category have the greatest similarity The load groups with inconsistent categories have the smallest similarity. When the cluster center does not change, the K- type load pattern is output. Specific steps are as follows:

Step 1 Select the cluster center. Randomly select K sample points of load data as the initial clustering center.

Step 2 Sample classification. The D(C) value calculated first has the meaning: the sum of the squared distances of each user's load data to K load cluster centers, and then divide each load data sample into the cluster center closest to it.

$$D(C) = \min \sum_{m=1}^{K} \sum_{X \in C_m} d^2(C_m, X_n) \tag{3}$$

(Among them, u_{mn} means that the nth load sample data belongs to the m category, C_m is the cluster center and $d(C_m, X_n)$ is the distance from the nth load to the m category.)

Step 3 Update cluster centers. According to the above division results, and based on the CH index, iteratively update the centers of the K classes.

Step 4 Convergence judgment. Judge whether the convergence condition is met, and output the result if it is met, and return to the previous step if it is not met. Until the conditions are met.

3.2.2 Determination of the Number of Clusters K

When the daily load data set is clustered, when the number of clusters is greater than or equal to 6, the percentage value of the daily load clustering results belonging to certain classes is relatively small. After increasing the number of clustering classes, they belong to the cluster. The daily load data of the class category is very small, about 1%–3%, all less than 5%, and the clustering result for the entire system is small. Therefore, these small probability categories may not be considered when clustering. When performing daily load clustering, when the number of clusters K is 5, the proportion of various types of load is shown in Fig. 1 below, and the minimum proportion of load categories is 5%. When the number of clusters K is 6, the proportion of various types of loads is shown in Fig. 2 below, and the minimum proportion of load categories is 2%. According to the above principles and the actual situation, the number of clusters K is set to 5 in this study. (Similarly, the number of annual hourly load clusters can be found to be 4.)

4 Multi-Dimensional Analysis of Load Characteristics by Vertical and Horizontal Hybrid Clustering

In order to verify the effectiveness of the research method proposed in this article, the algorithm is written by MATLAB R2018b, and the operating system is win10. This article takes the data group of a 10 kV power grid in a certain area as an example. The data collected by the data group is 365 days a year in that place. The data collection period is one hour. The data can be regarded as a 24 * 365 matrix. In-depth analysis of the load data of the region from the vertical and horizontal of the matrix provides a certain reference basis for subsequent load forecasting, power and electricity balance, new energy consumption, peak shaving analysis, demand-side response and other related work.

4.1 Longitudinal Clustering and Their Multi-dimensional Analysis of Load

The data collected by the collection system in this article can be a 24 * 365 matrix. Starting from the longitudinal direction of the matrix, perform longitudinal cluster analysis on the original daily load data group of the region for one year, and consider the mathematical and physical characteristics of its load. Analyze the load characteristics.

4.1.1 Load Characteristics of Longitudinal Cluster Analysis

Starting from a longitudinal matrix, K-means algorithm is to calculate the tool, collected on the grid to load the original cluster, obtain the clustering results are in Table below (Table 1):

Table 1. Daily load classification results

	First	Second	Third	Fourth	Fifth
Days	1012	99	76	70	18
Percentage	28%	27%	21%	19%	5%

From the results of daily load classification, draw similar daily load curves. Figure 3 shows the original load curve of the power grid. Figure 3 is a first category to the second five load the summary graph in FIG. From the load curve in FIG can to be seen, the same load of between similar variation, not with the variation of the load type having a difference.

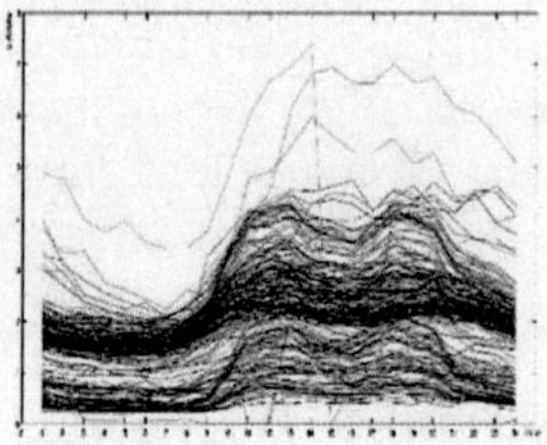

Fig. 1. Summary of the original load data curve of the power grid

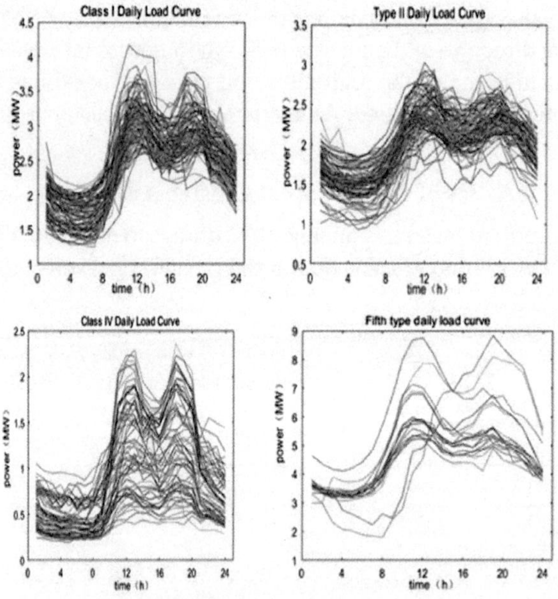

Fig. 2. Various daily load curves

4.2 Multi-Dimensional Analysis of Load Characteristics Considering Mathematical and Physical Characteristics

With the continuous improvement of smart meter functions and the continuous improvement of the reliability of the collected electricity big data, it is particularly important to analyze the mathematical and statistical characteristics of big data from the mathematical nature, and to deeply explore the hidden physical information behind the big electricity data.

It is the minimum value of the load at a certain moment. Might as well draw the cluster center curve and its extreme value curve of various load categories for analysis,

1. It can be seen from the above figure that the change trend of the load cluster center load curve and the load extreme value curve of the same category is basically the same, which again verifies the similarity of the load within the category, and provides strong evidence for the subsequent load forecasting work.
2. According to the above figure, From this, we can know the change margin of the load at a certain moment, and this load margin can be regarded as a controllable load, and reasonable adjustments can be made to it to increase the rate of new energy consumption.

4.3 Horizontal Cluster Analysis of Load Characteristics

The data collected by the collection system in this article can be a 24 * 365 matrix. Starting from the longitudinal direction of the matrix, the relevant characteristics of the daily load curve in this area have been studied above. Starting from the horizontal direction of the matrix, horizontal cluster analysis is performed on the original annual hourly load data group of the region for a year, and the load characteristics are analyzed in depth in multiple dimensions.

Starting from the longitudinal direction of the matrix, the K-means algorithm is used as the calculation tool to cluster the original loads collected by the power grid. The clustering results are shown in the following table (Table 2):

Table 2. Year load classification results

	First	Second	Third	Fourth
Days	8	2	5	9
Percentage	33%	8%	21%	38%

From the results of the annual hourly load classification, draw similar annual hourly load curves. Figure 3 is a first category to the second four load the summary graph in FIG. From the load curve in FIG can to be seen, the same load of between similar variation, not with the variation of the load type having a difference.

From the analysis of the above figure, the following conclusions can be drawn:

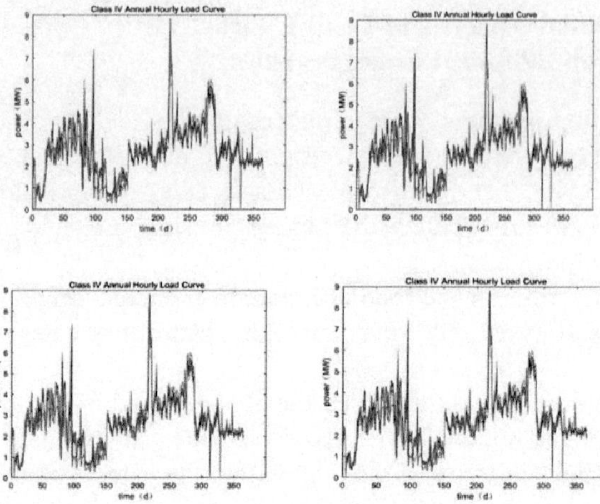

Fig. 3. Various types of annual hourly load curves

(1) Starting from the longitudinal direction, the load curve for the same hour on a certain day of the year has a certain similarity, which shows that people's production and life have a certain regularity and periodicity, which has a reference basis for subsequent load forecasting.

(2) The magnitude of different loads at the same time point is different, and the load value reflects the magnitude of the load level in the area.

5 Conclusion

Based on the historical data collected by the power grid collection system, this paper uses the K-means algorithm twice to cluster analysis of the power grid load in a certain area from the vertical and horizontal aspects and considers the mathematical characteristics of the data, and multi-dimensional and deep mining of the big data. The physical significance verifies the validity and practicability of the research method proposed in this article, and at the same time provides a reference for the follow-up related work.

References

1. Weihong, Y., et al.: Research on the clustering analysis algorithm of consumer electricity consumption behavior based on regulation potential index. Electr. Power Constr. **39**(6), 96–104 (2018)
2. Chunyan, L., et al.: Power consumption behavior analysis of AP clustering users based on optimized SAX and weighted load characteristic indexes. J. Electr. Eng. Technol. **34**(S1), 368–377 (2019)
3. Jun, L., Yanping, Z., Wenhao, P.: An interactive demand response method for smart communities that takes into account the clustering of electricity consumption behavior. Autom. Electr. Power Syst. **41**(17), 113–120 (2017)
4. Gangjun, G., et al.: Cluster optimization strategy for behavior analysis of smart electricity users. Autom. Electr. Power Syst. **42**(2), 58–63 (2018)

5. Ying, C., et al.: Analysis and prediction of load characteristics of Northwest Power Grid. Qinghai Electr. Power **39**(4), 06–11 (2020)
6. Sun, Y., Mao, Y., Li, Z., Zhang, X., Li, F.: Comprehensive clustering method of user load characteristics and adjustable potential for electric power big data. Chinese Soc. Electr. Eng.
7. Huang, N., et al.: Short term electrical load forecasting using mutual information based feature selection with generalized minimum-redundancy and maxi-mum-relevance criteria. Entropy **18**(9), 330 (2016)

Big Data Analysis and Ontology System

An Improved Fireworks Algorithm for Ontology Matching

Huanting Shi[1](✉) and Jinyuan Shi[2]

[1] North University of China, Taiyuan 030051, Shanxi, China
[2] Taiyuan University of Technology, Taiyuan 030051, Shanxi, China

Abstract. Ontology, presentation of the domain clearly and accurately, is established by knowledge extraction or hand. However, due to differences in construction methods or understanding of domain knowledge, it leads to ontology heterogencity. This problem interferes with the flow of information between machines. Ontology matching is designed to address this challenge by using different similarity measures. As a branch of ontology matching, ontology meta-matching has the advantages of high efficiency and accuracy in solving ontology heterogeneity. It aims to acquire the matching pairs with the highest confidence by using the best aggregation weights to integrate the similarity measures. Fireworks algorithm (FA) is proposed to solve complex optimization problems, and the experimental results show that it has good performance in optimization accuracy and search ability. However, using roulette to select gaussian variant sparks has greater uncertainty. In addition, the alignment given by experts is impractical in the process of ontology matching, it not only consumes a lot of resources but also is prone to errors. In order to improve these tasks, two novel strategies (Maximum Confidence Model (MCM) and Elite Solution Priority Mutation Strategy (ESPMS)) are proposed in this paper to increase population diversity and evaluate the validity of matching pairs. In order to verify the validity of the proposed method, the benchmark of OAEI (Ontology Alignment Evaluation Initiative) is used.

Keywords: Ontology matching · Fireworks algorithm · MCM · ESPMS

1 Introduction

The development of the Semantic Web enables machines to provide intelligent ser-vices [1]. However, the current level of intelligence is still in a lower plane, that is, information cannot be effectively shared between machines. The reason is that the data comes from various organizations and is constructed by different methods [2].

Knowledge extraction and ontology matching are proposed to improve this state. Knowledge extraction aims to analyze and mine the structure and relation-ship of data which includes link data, text data, etc. However, due to the particularity of data obtained by knowledge extraction, a better data structure, ontology, is proposed to store data compared than database.

S.-C. Chu et al. (Eds.): ICGEC 2021, LNEE 833, pp. 461–470, 2022.
https://doi.org/10.1007/978-981-16-8430-2_42

Ontology, as one of the forms of structured data, can not only accurately represent the structural relationship between entities, but also has the characteristics of simplicity and clarity when describing various attributes inside entities [3]. The structure of data in ontology is divided into two levels: schema layer and instance layer. The first layer focuses on the properties, concepts, axioms, and constraints of domain knowledge, the second layer is individuals. Because of its special structure and the ability of accurately describe the domain knowledge, ontology is widely used in biomedicine, business management, and other fields.

However, due to every organization has its own needs for domain knowledge or have different representations of the same knowledge, it leads to ontology heterogeneity [4, 5]. In order to address this task, several ontology matchings were proposed, entity matching and ontology meta-matching, respectively. The former is defined as a discrete optimization problem while the latter is continuous.

Entity matching has focused on internal structures that includes a large number of restricted attributes. A new model [6] with four layers structure is proposed to address entities and obtains the fuzzy relationship of matching pairs.

However, matching pairs are achieved by ontology meta-matching that is the task of this paper. EA based on selection pressure is proposed in this paper [7], but slow convergence is still a challenge [8]. This paper proposes a method to solve ontology meta-matching based on multiple targets, but its memory is costly. A freely available framework for scientific experimentation is proposed to address ontology meta-matching [9]. The Simulated Annealing Particle Swarm Optimization (SAPSO) method is proposed in to address sensor ontology, and the experimental results show the effectiveness of it [10]. However, the performance of SAPSO is dependent on the initial value, and its parameters sensitive. A greedy algorithm [11] is proposed to solve ontology meta-matching, which aims to obtain a set of reasonable results with fewer iterations by introducing the concept of granularity. However, the experimental results demonstrate that the performance of the proposed algorithm is inferior to EA. One conclusion can be drawn from the above is that the ability of the optimization algorithm will determine the quality of the matching pairs when solving ontology meta-matching.

The fireworks algorithm [12] is proposed by observing the phenomenon of fireworks explosions. Ying Tan and Yuanchun Zhu verify the effectiveness of the fireworks algorithm by nine basic test functions, and then compare it with the PSO algorithm and clone PSO algorithm. The experiment results demonstrate that the proposed algorithm is superior to PSO and clone PSO in optimization accuracy and convergence speed. The fireworks algorithm designs two kinds of fireworks to increase the diversity of the population, namely, normal explosions and gaussian explosions. However, since gaussian fireworks are selected by roulette, more iterations are needed to jump out of the local optimal solution. Therefore, the elite solution priority mutation strategy is proposed to address this challenge.

In addition, some semi-automatic systems [13] adjust the parameters manually, and reference alignment, which is obtained manually, is used to evaluate the matching pairs. However, due to reference alignment is established with time-consuming and error-prone when the number of entities in the source and target ontologies is too large, it leads to do not work well in actual projects [14, 15]. Therefore, the model of maximum confidence is proposed to evaluate the effectiveness of ontology matching results.

The following discussion is used to clarify the contribution of this paper more clearly:

1. An elite solution priority mutation strategy (ESPMS) is proposed to improve the performance of FA.
2. The model of maximum confidence (MCM) is proposed to evaluate the validity of matching pairs.
3. The OAEI standard data set (benchmark) is used to evaluate the effective-ness of the proposed method.

The rest of this paper is organized as follows: Sect. 2 presents the concept of ontology matching. The implementation detail of the model of confidence maximum is shown in Sect. 3. Section 4 is the concrete realization process of elite solution priority mutation strategy. Section 5 shows the experiment results and analysis. The conclusions and future research directions are presented in Sect. 6.

2 Relevant Concepts

Ontology: The domain knowledge [16] is represented by the most accurate semantic specification techniques and the most standard structuring techniques. where "the most accurate semantic specification techniques" aims at expressing exact domain knowledge by experts or machines, and "the most standard structuring techniques" aims at showing relationships between entities by structured approach.

Ontology Matching: For the given source ontology and target ontology, the relationship between the two ontologies is determined by setting relevant parameters and using externally available resources [17, 17]. Figure 1 shows the course of ontology matching, where O_s and O_t are source ontology and target ontology respectively. p represents related parameters. r represents relationships between entities. s represents external resources available for selection, for instance, reference alignment. A is the set of matching pairs.

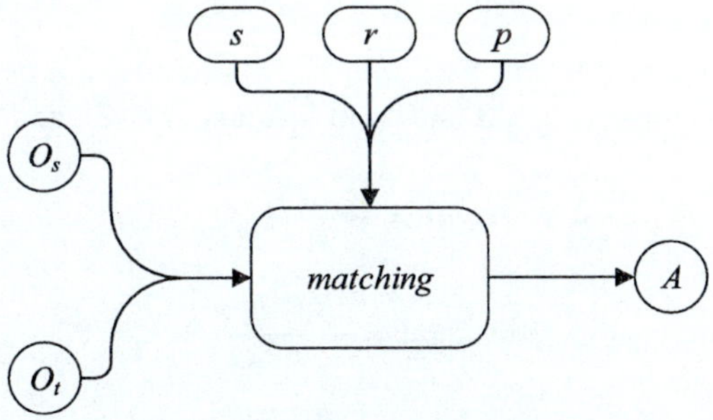

Fig. 1. Ontology matching

Ontology Meta-matching: [19] A set of matching pairs is obtained with the highest confidence by using a variety of similarity measures and then assigning weights to them. Figure 2 shows the course of ontology meta-matching. where O_s and O_t are source ontology and target ontology respectively. SM_i represents a kind of similarity measure. M_i represents the similarity matrix calculated by using the ith similarity measure. w_i represents the weights assigned to the ith similarity measure. n represents the number of similarity measures used by the matching course. M represents the final aggregation matrix. *Matching pairs* is the set of entities which has achieved processing.

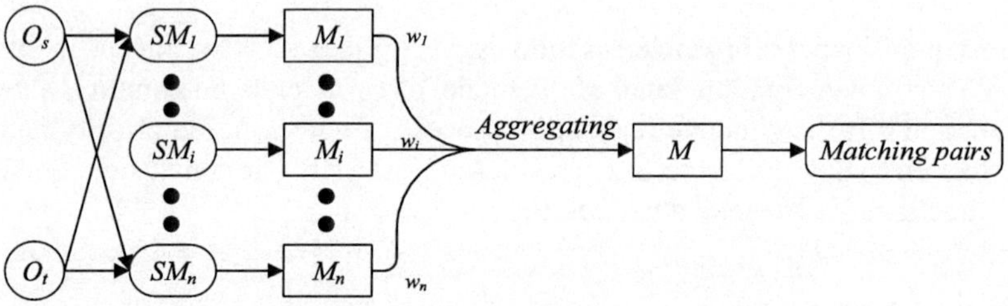

Fig. 2. The process of ontology meta-matching

3 The Confidence Maximization Model

The steps of FA based ontology meta-matching are as follows:

Step 1: All the entities related information of the concept layer in the ontology is analyzed.
Step 2: The corresponding matrices are calculated by using the similarity measures with different characteristics.
Step 3: A set of the most satisfactory weight vectors is obtained by the FA.
Step 4: The aggregation matrix $\boldsymbol{M}$ is generated by processing the weight vector in Step 3 and the similarity matrices in Step 2.
Step 5: The evaluation function is used to evaluate the matching result.

Three indicators, namely *recall*, *precision* and *f-measure*, are used to evaluate the results of ontology matching [11]. As shown in formula (1), (2), and (3):

$$recall = \frac{|R \cap S|}{R} \tag{1}$$

$$precision = \frac{|R \cap A|}{A} \tag{2}$$

$$f - measure = \frac{recall \times precision}{\alpha \times recall + (1 - \alpha) \times precision} \tag{3}$$

where R represents reference alignment, A is a set of matching pairs. In order to balance the influence of *recall* and *precision*, the value of α is generally set to 0.5 [19]. Therefore, the formula (3) has the following change:

$$f - measure = \frac{2 \times recall \times precision}{recall + precision} \tag{4}$$

However, when the scale of the entities in the ontology is too large, building reference alignment manually is time-consuming and error-prone. Therefore, in order to overcome this challenge, we propose a maximum confidence model (MCM) to evaluate the validity of ontology matching results. This model obtains a set of matching pairs with maximum similarity by the maximum confidence function (MCF). MCF is defined as follows:

$$MCF = \sum\nolimits_{i=0}^{n} smv_{\max} \tag{5}$$

where n is the number of rows in the aggregation matrix M. $smv_{\max}$ represents the similarity value of a matching pair with the maximum confidence in the matrix. Note that only if the value of $smv_{\max}$ is greater than the threshold will it be considered is vaild.

In addition, three evaluation functions are proposed by the MCM, which are used to evaluate the quality of results of the fireworks algorithm based on ontology matching, namely $recall_{MCF}$, $precison_{MCF}$, and $f\text{-}measure_{MCF}$. They are defined as follows:

$$recall_{MCF} = \frac{n_M}{N} \tag{6}$$

A set of matching pairs with the maximum confidence value, which is called ***S***, is produced by MCF. where n_M represents the cardinality of the ***S***.N has the following limit:

$$N = Min(num(O_s), num(O_t)) \tag{7}$$

where num(O_s) is the number of entities in O_s.

$$precision_{MCF} = \frac{MCF}{N} \tag{8}$$

The average similarity value of matching pairs in set ***S*** is considered as $precison_{MCF}$. $f\text{-}measure_{MCF}$ is calculated as follows:

$$f - measure_{MCF} = \frac{2 \times recall_{MCF} \times precision_{MCF}}{recall_{MCF} + precision_{MCF}} \tag{9}$$

4 Elite Solution Priority Mutation Strategy

4.1 Fireworks Algorithm

Inspired by the explosion of fireworks, Ying Tan and Yuanchun Zhu propose fireworks algorithm. They find that the higher-quality fireworks exploded with more sparks closer to the center of the explosion, while the lower-quality fire-works execute the strategy oppositely. The number of sparks and the amplitude of the explosion are generated by this mechanism. Among them, fire-works with higher fitness values are used for local search, while the global search is realized by the fireworks with lower fitness values.

Formula (10) and (11) are used to generate sparks and calculate explosion amplitudes. These formulas are designed to address the minimization problems.

$$S_i = m \frac{y_{\max} - f(x_i) + \xi}{\sum_{i=1}^{n} (y_{\max} - f(x_i)) + \xi} \tag{10}$$

$$A_i = \widehat{A} \frac{f(x_i) - y_{\min} + \xi}{\sum_{i=1}^{n} (f(x_i) - y_{\min}) + \xi} \tag{11}$$

where y_{max} represents the firework with the highest (worst) fitness value. y_{min} represents the firework with the lowest (best) fitness value. $f(x_i)$ is the fitness value of ith firework. n is the number of populations. m and $\widehat{A}$ are constants that limit the number of sparks and the size of amplitude, respectively.

What's more, in order to prevent the condition that the explosion is too spectacular, Formula (12) will be used to limit the number of sparks.

$$\widehat{S_i} = \begin{cases} round(a \cdot m), \; S_i < am \\ round(b \cdot m), \; S_i > bm, a < b < 1 \\ round(S_i), \quad otherwise \end{cases} \tag{12}$$

The dimension of fireworks explosion be calculated by formula (13).

$$z = round(d \cdot rand(0, 1)) \tag{13}$$

In addition, gaussian fireworks are designed to increase the diversity of the population. The Gaussian coefficient is generated by a gaussian distribution which mean and standard deviation are 1, respectively. Implementation details are shown in formula (14).

$$x_{g-i}^{t+1} = g \times x_i^t \tag{14}$$

where, t represents the t-th iteration, and i represents the ith firework in the population with gaussian variation.

The Manhattan distance or Euclidean distance will be used to calculate the fireworks of next generation population. Implementation details are shown in formula (15) and (16).

$$R(x_i) = \sum_{j \in N_{all}} d(x_i, x_j) = \sum_{j \in N_{all}} \left\| x_i - x_j \right\| \tag{15}$$

$$p(x_i) = \frac{R(x_i)}{\sum_{j \in N_{all}} R(x_j)} \tag{16}$$

where the sum of the number of fireworks and sparks in the current iteration is represented N_{all}, and they will be stored in set S. $p(x_i)$ is the probability of the *ith* particle (firework or spark) being selected in set S.

4.2 Implementation Details of Elite Solution Priority Mutation Strategy

Due to the fast convergence speed of the fireworks algorithm, it is not recommended when solving ontology element matching. If the population converges too fast, the algorithm will reach the local optimal solution instead of the global optimal solution. On the other hand, the fireworks algorithm uses roulette to select *k* fireworks for gaussian explosion to improve the diversity of the population. However, this approach is not conducive to the better individuals in the population to shift out of the local optimal solution.

In addition, the elite strategy of the fireworks algorithm is used to compare the spark with the elite fireworks in the population, and the best spark is selected by the tournament mechanism. However, when the population is trapped in the local optimum, the deadlock can only be broken by the gaussian mutation strategy of roulette with greater randomness. Therefore, the elite solution priority mutation strategy is proposed to overcome this drawback.

The Elite Solution Priority Mutation Strategy (ESPMS): In addition to the normal explosion, elite fireworks are also processed by gaussian explosion in each generation. The probability of gaussian variation of the remaining fireworks in the population is calculated by the following formula (17):

$$p(x_i) = \frac{f(x_i)}{\sum_{i=1}^{n-1} f(x_i)} \tag{17}$$

where $f(x_i)$ represents the fitness value of the *i*th firework, and *p* represents the gaussian variation probability of the *i*th firework.

5 Experimental Results and Analysis

Benchmark is a test set containing a variety of heterogeneous conditions, whose final matching can well verify the availability of the system. Where the ontologies 101–104 have different lexical relation; 201–210 have same structure relation; 221–247 have same semantic relation; 248–266 have differences in semantic, structure and lexical relation.

In order to verify the effectiveness of the method in this paper, the benchmark track of OAEI is used. The average value of the 30 times test results is considered the final experimental result. Table 1 shows the experimental results of the fireworks algorithm and the improved fireworks algorithm. The first column of the table represents the ordinal number of the case, the second column represents the results of the FA, and the results based on the improved FA (IFA) are presented in the third column. where f is f-measure.

The experimental results show that the improved fireworks algorithm is superior to the traditional fireworks algorithm. Especially in the case 221–247, the $f-measure_{MCF}$

Table 1. The result of FA and IFA.

ID	f_{FA}	f_{IFA}	ID	f_{FA}	f_{IFA}	ID	f_{FA}	f_{IFA}
101	1.0	1.0	103	1.0	1.0	104	1.0	1.0
201	0.97	1.0	201-2	0.96	0.97	201-4	0.95	0.95
201-6	0.92	0.94	201-8	0.93	0.93	202	0.00	0.02
202-2	0.89	0.89	202-4	0.65	0.68	202-6	0.60	0.61
202-8	0.34	0.34	203	1.0	1.0	204	1.0	1.0
205	0.90	0.92	206	0.91	0.93	207	0.89	0.91
208	0.92	0.94	209	0.60	0.61	210	0.64	0.65
221	0.98	1.0	222	0.99	1.0	223	0.97	1.0
224	1.0	1.0	225	0.98	1.0	228	0.96	1.0
232	0.99	1.0	233	0.98	0.99	236	1.0	1.0
237	0.97	1.0	238	0.99	1.0	239	1.0	1.0
240	1.0	1.0	241	0.98	1.0	246	1.0	1.0
247	0.99	1.0	248	0.02	0.04	248-2	0.85	0.89
248-4	0.98	1.0	248-6	0.50	0.51	248-8	0.37	0.38
249	0.22	0.23	249-2	0.81	0.84	249-4	0.72	0.73
249-6	0.55	0.57	249-8	0.40	0.41	250	0.29	0.30
250-2	0.81	0.82	250-4	0.78	0.79	250-6	0.41	0.43
250-8	0.32	0.33	251	0.01	0.01	251-2	0.80	0.82
251-4	0.71	0.73	251-6	0.57	0.57	251-8	0.40	0.41
252	0.03	0.03	252-2	0.80	0.81	252-4	0.78	0.79
252-6	0.76	0.77	252-8	0.85	0.85	253	0.08	0.09
253-2	0.87	0.87	253-4	0.71	0.72	253-6	0.51	0.53
253-8	0.35	0.37	254	0.24	0.25	254-2	0.84	0.85
254-4	0.71	0.75	254-6	0.50	0.51	254-8	0.19	0.21
257	0.33	0.33	257-2	0.86	0.86	257-4	0.76	0.80
257-6	0.54	0.55	257-8	0.34	0.35	258	0.17	0.19
258-2	0.85	0.87	258-4	0.80	0.80	258-6	0.61	0.59
258-8	0.40	0.42	259	0.21	0.21	259-2	0.79	0.81
259-4	0.75	0.78	259-6	0.80	0.80	259-8	0.83	0.85
260	0.05	0.05	260-2	0.60	0.61	260-4	0.47	0.48
260-6	0.37	0.38	260-8	0.31	0.32	261	0.07	0.07
261-2	0.71	0.72	261-4	0.78	0.80	261-6	0.76	0.77

(*continued*)

Table 1. (*continued*)

ID	f_{FA}	f_{IFA}	ID	f_{FA}	f_{IFA}	ID	f_{FA}	f_{IFA}
261-8	0.81	0.82	262	0.17	0.18	262-2	0.31	0.32
262-4	0.34	0.35	262-6	0.41	0.42	262-8	0.34	0.34
265	0.03	0.07	266	0.20	0.23			

of the improved fireworks algorithm almost all reached 1.0, while the experimental results of the traditional fireworks algorithm only reached 1.0 in 232, 241, 246. The reason is that the elite solution priority mutation strategy makes the elite solution in the population better complete the global search.

6 Conclusions and Future Works

With the development of ontology matching, efficient information exchange between machines can be realized. As a branch of ontology matching, ontology meta-matching can effectively address ontology heterogeneity. Therefore, the improved fireworks algorithm is proposed to overcome the task that the algorithm falls into the local optimal solution. In addition, in order to solve the limitations of ontology matching with reference alignment, the maxi-mum confidence model is proposed to evaluate the validity of matching pairs. Experimental results demonstrate the effectiveness of the pro-posed method by testing the benchmark track. In the future, we will focus on solving large-scale ontology as well as complex ontology by IFA.

In addition, it is also a good research direction to realize the multi-objective optimization of IFA, which aims to optimize multiple objectives at the same time and gives users multi-level choices to achieve the purpose of intelligent interaction.

References

1. Zhdanova, A.V., Shvaiko, P.: Community-driven ontology matching. In: European Semantic Web Conference, vol. 4011, pp. 34–49. Springer, Heidelberg (2006)
2. Shvaiko, P., Euzenat, J.: Ontology matching: state of the art and future challenges. IEEE Trans. Knowl. Data Eng. **25**(1), 158–176 (2013)
3. Otero-Cerdeira, L., Rodríguez-Martínez, F.J., Gómez-Rodríguez, A.: Ontology matching: a literature review. Expert Syst. Appl. **42**(2), 949–971 (2015)
4. Xue, X., Chen, J.: Optimizing ontology alignment through hybrid population-based incremental learning algorithm. Memetic Comput. **11**(2), 209–217 (2018). https://doi.org/10.1007/s12293-018-0255-8
5. Xue, X., Chen, J.: Optimizing sensor ontology alignment through compact co-firefly algorithm. Sensors **20**(7), 2056 (2020)
6. Gherbi, S., Khadir, M.T.: ONTMAT1: ontology matching using a reasoner and property restriction. Int. J. Web Eng. Technol. **15**(2), 119 (2020)
7. Lv, Q., Jiang, C., Li, H.: Solving ontology meta-matching problem through an evolutionary algorithm with approximate evaluation indicators and adaptive selection pressure. IEEE Access **9**, 3046–3064 (2020)

8. Lv, Q., Jiang, C., Li, H.: An interactive multi-objective ontology matching technique. In: Hassanien, A.-E., Chang, K.-C., Mincong, T. (eds.) AMLTA 2021. AISC, vol. 1339, pp. 955–964. Springer, Cham (2021). https://doi.org/10.1007/978-3-030-69717-4_89
9. Ferranti, N., Mouro, J., Mendonça, F., de Souza, J., Soares, S.: A framework for evaluating ontology meta-matching approaches. J. Intell. Inf. Syst. **56**(2), 207–231 (2020). https://doi.org/10.1007/s10844-020-00615-8
10. Zhu, H., Xue, X., Geng, A., Ren, H.: Matching sensor ontologies with simulated annealing particle swarm optimization. Mobile Inf. Syst. **2021**(1), 1–11 (2021)
11. Martinez-Gil, J., Aldana-Montes, J.F.: Evaluation of two heuristic approaches to solve the ontology meta-matching problem. Knowl. Inf. Syst. **26**(2), 225–247 (2011). https://doi.org/10.1007/s10115-009-0277-0
12. Xue, X.S., Chen, J.F.: Using compact evolutionary tabu search algorithm for matching sensor ontologies. Swarm Evol. Comput. **48**, 25–30 (2019)
13. Tan, Y., Zhu, Y.: Fireworks algorithm for optimization. In: Tan, Y., Shi, Y., Tan, K.C. (eds.) ICSI 2010. LNCS, vol. 6145, pp. 355–364. Springer, Heidelberg (2010). https://doi.org/10.1007/978-3-642-13495-1_44
14. Xue, X.S., Liu, J.H.: Collaborative ontology matching based on compact interactive evolutionary algorithm. Knowl.-Based Syst. **137**, 94–103 (2017)
15. Xue, X.S., Yao, X.: Interactive ontology matching based on partial reference alignment. Appl. Soft Comput. **72**, 355–370 (2018)
16. Guarino, N., Oberle, D., Staab, S.: What is an ontology? In: Staab, S., Studer, R. (eds.) Handbook on Ontologies. IHIS, pp. 1–17. Springer, Heidelberg (2009). https://doi.org/10.1007/978-3-540-92673-3_0
17. Jiang, C., Xue, X.: A uniform compact genetic algorithm for matching bibliographic ontologies. Appl. Intell. **51**(10), 7517–7532 (2021). https://doi.org/10.1007/s10489-021-02208-6
18. Xue, X., Jiang, C., Wang, H., Tsai, P.-W., Mao, G., Zhu, H.: An improved multi-objective evolutionary optimization algorithm with inverse model for matching sensor ontologies. Soft. Comput. **25**(18), 12227–12240 (2021). https://doi.org/10.1007/s00500-021-05895-y
19. Xue, X.S., Zhang, J.: Matching large-scale biomedical ontologies with central concept based partitioning algorithm and adaptive compact evolutionary algorithm. Appl. Soft Comput. **106**, 1–11 (2021)

Revealing Top-k Dominant Individuals in Incomplete Data Based on Spark Environment

Ke Wang[1], Jimmy Ming-Tai Wu[1(✉)], Binge Cui[1], and Jerry Chun-Wei Lin[2]

[1] College of Computer Science and Engineering, Shandong University of Science and Technology, Shandong, China
wmt@wmt35.idv.tw, cuibinge@sdust.edu.cn

[2] Department of Computer Science, Electrical Engineering and Mathematical Sciences, Western Norway University of Applied Sciences, Bergen, Norway
jerrylin@ieee.org

Abstract. A large number of incomplete datasets exist in our life, which may be caused by missing data transmission, privacy information protection, and sudden device failure. It is more difficult to mine the key information in incomplete datasets because of missing dimension values in some uncertain dimensions. top-k dominated (TKD) query can be better applied to the information mining of this dataset. Compared with the traditional query algorithm, this algorithm does not need to set the threshold before the beginning, and the top-k key information can be obtained by simply setting in the algorithm. At present, there are some algorithms to perform TKD query on incomplete datasets, but most of them are stand-alone algorithms applied to small datasets and parallel algorithms based on MapReduce applied to large datasets. For large datasets, the algorithm based on MapReduce framework is defective due to the distributed computing framework of MapReduce, for example, when multiple MapReduce jobs are serially calculated, some framework start-up time and communication time between nodes are often wasted, leading to the performance of the algorithm to be improved. In this paper, a TKD query algorithm based on Spark computing framework is proposed. The data interaction between multiple jobs can be realized based on memory, which avoids a large amount of disk interaction when the algorithm runs. In addition, we also propose a new method to calculate the dominant number of objects based on bitmap index table, which simplifies the calculation process. Therefore, the performance of the algorithm has been further improved by using the new computing method and the new computing framework.

1 Introduction

In real life, there are many examples of incomplete datasets. For example, the movie rating dataset is a classic case of incomplete datasets. Since it is impossible for the audience to see all the movies in the dataset, only the movies they

S.-C. Chu et al. (Eds.): ICGEC 2021, LNEE 833, pp. 471–480, 2022.
https://doi.org/10.1007/978-981-16-8430-2_43

have seen will be rated when they are rated, thus resulting in data missing. For example, the score record of a movie exists in the form of (–,–,1,3) in the dataset, which indicates that the score of audience one and audience two is missing, and only audience three and four have rated the movie. The movie rating dataset is used to construct the movie recommendation system, for example, there are two movies o= (–,–,1,3), o'= (–,1,2,3), according to the definition of the dominant relationship between the two objects [9]: When the values of all dimensions of object one are not worse than those of the corresponding dimensions of object two, and the values of at least one dimension are better than those of the corresponding dimensions of object two, we can get that object one dominates object two. According to the dominant relationship, we compare the two movies. In the third dimension, the dimension value of movie o is better than that of o', and in the fourth dimension, the dimension value of movie o is equal to that of o'. Therefore, movie o dominates o' is represented as $o \prec o'$, which can be regarded as movie o is better than o'. According to the dominant relationship, we can find the top-k most popular movies. It can be concluded from the above examples and references [1,3,8] that due to the particularity of incomplete datasets, its processing and calculation methods are different from those of complete datasets.

Based on the verification in [9,10,14], it is possible to apply Skyline to handle top-k dominated query in incomplete datasets. At present, some algorithms have been proposed to implement TKD query on incomplete datasets. For example, the Bitmap Indexing Guided (BIG) algorithm based on bitmap index tables [12] implements score calculation through bitmap pruning strategy and fast bit-by-bit operations. When the amount of data becomes larger, the requirements for the feasibility and performance of the algorithm will also become higher. Therefore, in order to adapt to TKD query in large-scale datasets, an MapReduced Enhanced Bitmap Index Guided Algorithm (MRBIG) is proposed [6]. In the early stage of calculating the dominant number of objects, MRBIG algorithm uses the MapReduce framework to assign the computing task to multiple nodes for parallel computation, which makes it possible to conduct TKD query in large-scale datasets.

Hadoop's MapReduce framework is not designed to deal with cyclic and iterative data streams [5], so there are many problems such as computational efficiency in the scenarios of data reuse in multiple parallel runs. Therefore, Spark framework arises at the right moment. Based on the MapReduce computing framework, Spark improves its computing process, thus greatly speeding up the running and reading and writing speed of data analysis and mining, and introduces a new RDD computing model for data calculation and storage. Based on these advantages of Spark computing framework, an efficient algorithm is proposed in this paper. The algorithm also optimizes the calculation formula of the dominant number of objects. The main contributions of this paper are as follows:

1. We propose a new algorithm, which is implemented on the basis of a new computing framework Spark. Therefore, this algorithm has the general advantages of Spark framework, such as high computational efficiency, convenient use, strong versatility and high compatibility.

2. Based on the bitmap index table, we propose a new formula for calculating the dominant number of objects, which simplifies the complicated calculation process, reduces the amount of calculation and improves the operating efficiency.

The remaining structure of this article is as follows: Sect. 2 reviews the contents of TKD query, MapReduce and Spark. Section 3 introduces the related problems and definitions. Section 4 introduces the bitmap index table and our algorithm. The introduction and results of the experiment are shown in Sect. 5. Section 6 summarizes the paper.

2 Related Work

In this section, we review the content and research of TKD query, MapReduce computing framework and Spark computing framework.

2.1 Top-k Dominance

The original intention of TKD query design is to find the key information in the complete dataset. With the migration of time and the prevalence of incomplete datasets, this query method is used in incomplete datasets. The concept of top-k dominated query was first proposed in [14], and a TKD dominated query algorithm based on Skyline was proposed. Then the top-k dominated query is discussed and studied in more depth in [13], so that the information of interest can be found in any uncertain data set. More details and concepts about top-k query and skyline query of uncertain datasets are shown in article [15], so that people can better understand the basic concepts.

2.2 MapReduce

MapRduce is a core component of Hadoop distributed system architecture, which is used to solve the problem of massive data analysis and calculation. A MapReduce program can be divided into two stages: Map and Reduce [4]. In the Map stage, the data in the disk is read in blocks. After reading, each block of data is assigned to a MapTask for calculation and processing. MapTasks are executed in parallel and do not affect each other. In the Reduce stage, the number of ReduceTasks is determined according to the number of partitions of the data. In the ReduceTask, the data with the same key value is summarized and the result is output. In principle, a MapReduce programming model can only be composed of a Map stage and a Reduce stage, and the Reduce stage can only follow the Map stage to process the data after the Map stage. However, when the algorithm logic is very complex and a MapReduce program can not effectively solve the problem, multiple MapReduce programs need to be executed serially. When multiple MapReduce serial execution is involved, there will be problems such as Hadoop multiple startup and shutdown and data communication between multiple jobs based on disk [2], which will lead to longer execution time of the algorithm.

2.3 Spark

Spark is a big data analysis and computing framework written by Scala language [11]. It was written in 2009 in AMPLab Laboratory of Berkeley University [16], and was set as an open source project in 2010. Since 2015, Spark has been used by a large number of domestic companies due to its open source, fast, universal, scalable and other characteristics. The core modules of Spark framework include Spark Core, Spark SQL, Spark Streaming, Spark MLlib and Spark GraphX. Spark Core module contains the most basic technology of Spark framework, and other modules are extended by this module. As the main goal of Spark framework is to analyze and calculate large-scale data, it can be said that it is a computing framework, so other data storage and resource scheduling problems also need the support of HDFS, HBase and Yarn modules. Compared with MapReduce computing framework, Spark introduces a memory based elastic distributed data set (RDD) for data storage and calculation [17]. It does not need to store the intermediate results to disk, but to memory, which reduces the I/O operation with disk and improves the time efficiency. It can be said that Spark computing framework is a new framework for computing and processing large-scale data [18].

3 Preliminary

In this section, we will formally show the domination relationship between objects, and introduce the problem definition of TKD query in incomplete datasets. Each object in incomplete dataset S contains d dimensions. When a dimension value of an object is "–", it means that the dimension value of the dimension is missing. As shown in Table 1, we give an example of a small incomplete dataset. The dataset contains ten objects, each of which contains four dimensions. We use four bit vectors to represent the values of the four dimensions. For example, the object o_1= (–,1,2,–) has missing observations on d_1 and d_4.

Definition 1. ***In incomplete datasets, the dominance relationship*** *[9]* ***between two objects is defined as follows:***
There are two objects o and o′, compare the dimension j which contains values in common between the two objects. If o dominates o′, it is denoted as $o \prec o'$, then the following two conditions must be satisfied: (i) $\forall$ j with $o.[j] \leq o'.[j]$; and (ii) $\exists$ j with $o.[j] < o'.[j]$.

For example. In Table 1, o_1= (-,1,2,-) and o_8= (-,3,2,-). $o_1.[2] < o_8.[2]$ and $o_1.[3] = o_8.[3]$, so we can get the object $o_1 \prec o_8$.

Table 1. An sample incomplete dataset

Object ID	Multidimensional data
o_1	(– , 1 , 2 , –)
o_2	(1 , – , 3 , 2)
o_3	(3 , 1 , – , –)
o_4	(– , – , – , 1)
o_5	(– , 2 , 1 , –)
o_6	(– , 2 , – , 3)
o_7	(1 , 1 , – , –)
o_8	(– , 3 , 2 , –)
o_9	(2 , – , 2 , 2)
o_{10}	(3 , 3 , – , –)

Definition 2. ***Object Score***
In an incomplete dataset S, the Score of a certain object o can be expressed as $Score(o)$, which is equal to the number of objects dominated by object o in dataset S. i.e., Score(o) = $|\{o'|o \prec o' \wedge o' \in S - \{o\}\}|$.

For example, for the o_1 in Table 1, $o_1 \prec \{o_2, o_6, o_8, o_{10}\}$ can be obtained according to the dominant relation in Definition 1, so Score(o_1) = $|\{o_2, o_6, o_8, o_{10}\}|$ = 4.

Definition 3. ***TKD query in an incomplete dataset***
After TKD query in incomplete dataset S, the result data set S_k is obtained. The objects in the dataset S_k must be the k objects with the highest score in S. i.e.,$|S_k|=k$, and $\forall o \in S_k, \forall o' \in S$, Score(o) $\geq$ Score(o').

4 TKD Query on Large Incomplete Datasets

4.1 The Bitmap Index Table

First of all, our algorithm is based on the data in the bitmap index table. Therefore, the bitmap index table is very important to our algorithm. In order to facilitate the establishment of the bitmap index table, we first establish the dimension table of the dataset S, as shown in Table 2. The bitmap index table of incomplete datasets should be established according to the following steps: (i) First, we need to determine which values are composed of each dimension in the dataset, A vector v_i is used to represent the values of the i-th dimension of the dataset. For example, the first dimension in dataset S consists of "–", "1", "2", "3". so v_1= {–,1,2,3}. Each value in the vector v_i constructs a column in the i-th dimension of the bit graph indexing table. The length of each column is the number of objects in the dataset + 1. The starting value of each column is the value in the vector v_i, and the rest is completed by 1. (*ii*) After the first step, we need to compare the starting value of each column with all the values of the corresponding dimension, and then update the value of the column. When the initial value of the column is "–", we do not need to modify the data of the corresponding column. Otherwise, we need to compare the initial value with the dimension value of the corresponding dimension. When the dimension value is equal to or better than the initial value, we change the 1 of the corresponding position to 0.

Table 2. A sample incomplete dataset table

Object	d_1	d_2	d_3	d_4
o_1	–	1	2	–
o_2	1	–	3	2
o_3	3	1	–	–
o_4	–	–	–	1
o_5	–	2	1	–
o_6	–	2	–	3
o_7	1	1	–	–
o_8	–	3	2	–
o_9	2	–	2	2
o_{10}	3	3	–	–

According to the data in Table 2 and the above process of establishing bitmap index table, we have established the bitmap index table of the dataset S, as shown in Table 3.

Table 3. A sample bitmap index table

Items	d_1	–	1	2	3	d_2	–	1	2	3	d_3	–	1	2	3	d_4	–	1	2	3
o_1	–	1	1	1	1	1	1	0	0	0	2	1	1	0	0	–	1	1	1	1
o_2	1	1	0	0	0	–	1	1	1	1	3	1	1	1	0	2	1	1	0	0
o_3	3	1	1	1	0	1	1	0	0	0	–	1	1	1	1	–	1	1	1	1
o_4	–	1	1	1	1	–	1	1	1	1	–	1	1	1	1	1	1	0	0	0
o_5	–	1	1	1	1	2	1	1	0	0	1	1	0	0	0	–	1	1	1	1
o_6	–	1	1	1	1	2	1	1	0	0	–	1	1	1	1	3	1	1	1	0
o_7	1	1	0	0	0	1	1	0	0	0	–	1	1	1	1	–	1	1	1	1
o_8	–	1	1	1	1	3	1	1	1	0	2	1	1	0	0	–	1	1	1	1
o_9	2	1	1	0	0	–	1	1	1	1	2	1	1	0	0	2	1	1	0	0
o_{10}	3	1	1	1	0	3	1	1	1	0	–	1	1	1	1	–	1	1	1	1

4.2 Proposed Algorith

After the bitmap index table is created, we can use the data to calculate the dominant number of objects. Here, the proposed algorithm introduces two sets Q and T to calculate the dominating number of objects. It is no longer like BIG, MRBIG algorithms, which need to calculate Q, P, ϕ , and $nonD$. the proposed algorithm simplifies the calculation process and improves the efficiency of the algorithm. The set Q of object o represents the object whose value is missing or not better than o in the comparable dimension. From the description, we can get that the set Q of object o contains the following objects: (i) Objects dominated by object o. (ii) Objects that cannot be compared with object o. (iii) Objects that are not dominated by object o. The set T of object o contains the object whose values are missing or equal to the values of the corresponding dimensions of object o. There are two kinds of objects in set T: (i) Objects that cannot be compared with object o. (ii) Objects that are not dominated by object o. So the inclusion relation between sets Q and T is shown in Fig. 1.

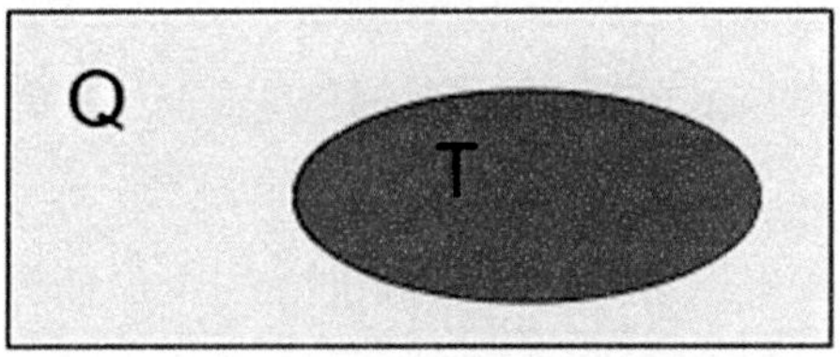

Fig. 1. Inclusion relationship between Q and T.

According to this relation, we can get $Score(o) = |Q(o) - T(o)|$. By analyzing the bitmap index table, we can see that for an object o, the left column of the first column with 0 in the corresponding dimension of the corresponding row in the bitmap index table is the Q set of the object in the dimension, which is represented by the vector $[Q_i]$. The set Q is obtained by taking the intersection of all vectors $[Q_i]$ of object o, expressed as:

$$Q = \bigcap_{i=1}^{d} [Q_i]$$

Then, we need to rescan the database, compare the dimension values to get the vector $[T_i]$ of each dimension of the object, and all vectors $[T_i]$ take intersection, and get the object set T, which is expressed as:

$$T = \bigcap_{i=1}^{d} [T_i]$$

The proposed algorithm proposed in this paper is to use Spark Program to realize the above process, which solves the problem of low execution time efficiency of the original algorithm based on MapReduce framework. First, we use a NoSQL database HBase which is similar to relational database but supports distributed and large-scale data storage to store bitmap index table. the proposed algorithm reads the bitmap index table in HBase into memory, obtains the score of each object through several times of data processing model RDD analysis and calculation, and then takes the top-k objects with the highest score to output. Note that after many RDDs data processing and calculation, the intermediate result is stored in memory, which reduces the operation of the algorithm and disk I/O, and improves the time efficiency of the algorithm.

5 Experimental Results

We use different sizes of real data sets movielens to experiment. The three data sets include 1682 users and 943 movies, 6040 users and 3706 movies, 71567 users and 10677 movies. Movielens dataset is used in many papers, such as [7,9]. The operating system of the computer used in the experiment is Linux Ubuntu 16.04 LTS, and the memory size is 32 GB. The Framework version used is Hadoop 2.8.5, HBase 1.3.1, Spark 3.0.0. The cluster environment built in the experiment has one master node and five data nodes, and the number of data nodes can be adjusted according to the requirements. We compare the time efficiency of BIG, MRBIG and the proposed algorithms for TKD query in the prepared dataset. By comparison, we can conclude that the proposed algorithm has better performance in large data sets (Fig. 2).

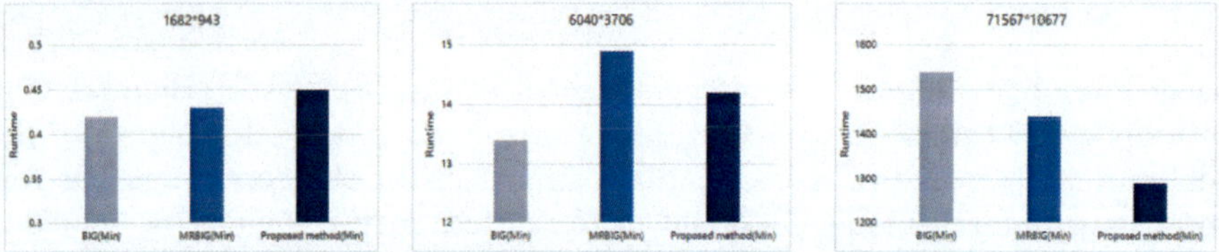

Fig. 2. Runtime under different database sizes

6 Conclusion and Future Work

In this paper, we propose an efficient TKD query algorithm for incomplete datasets based on Spark computing framework. The algorithm takes advantage of the advantages of fast start-up time of Spark task and memory based communication among multiple Spark jobs to improve the time efficiency of TKD query for incomplete datasets. However, because Spark computing framework stores intermediate results in memory when analyzing and calculating data, in real life, when processing large-scale incomplete datasets, Spark jobs may fail due to memory constraints. In this case, MapReduce computing framework is a better choice, Therefore, Spark framework can not completely replace MapReduce framework. Every framework has a suitable application scenario. Because the algorithm based on Spark may be limited by memory, so in the future work, how to optimize our algorithm and study the algorithm with better space complexity is our urgent task.

Acknowledgment. This research is supported by Shandong Provincial Natural Science Foundation (ZR201911150391).

References

1. Antova, L., Koch, C., Olteanu, D.: From complete to incomplete information and back. In: Proceedings of the 2007 ACM SIGMOD International Conference on Management of Data, pp. 713–724 (2007)
2. Bu, Y., Howe, B., Balazinska, M., Ernst, M.D.: Haloop: efficient iterative data processing on large clusters. Proc. VLDB Endowment **3**(1–2), 285–296 (2010)
3. Cheng, W., Jin, X., Sun, J.T., Lin, X., Zhang, X., Wang, W.: Searching dimension incomplete databases. IEEE Trans. Knowl. Data Eng. **26**, 725–738 (2013)
4. Dean, J., Ghemawat, S.: Mapreduce: simplified data processing on large clusters. In: Proceedings of the 6th Conference on Symposium on Operating Systems Design Implementation, vol. 6 (2004)
5. Dean, J., Ghemawat, S.: Mapreduce: simplified data processing on large clusters. Commun. ACM **51**(1), 107–113 (2008)
6. Ezatpoor, P., Zhan, J., Wu, J.M.T., Chiu, C.: Finding top-k dominance on incomplete big data using mapreduce framework. IEEE Access **6**, 7872–7887 (2018)
7. Gao, Y., Miao, X., Cui, H., Chen, G., Li, Q.: Processing k-skyband, constrained skyline, and group-by skyline queries on incomplete data. Expert Syst. Appl. **41**(10), 4959–4974 (2014)
8. Imieliński, T., Lipski Jr, W.: Incomplete information in relational databases. In: Readings in Artificial Intelligence and Databases, pp. 342–360. Elsevier (1989)
9. Khalefa, M.E., Mokbel, M.F., Levandoski, J.J.: Skyline query processing for incomplete data. In: 2008 IEEE 24th International Conference on Data Engineering, pp. 556–565. IEEE (2008)
10. Lofi, C., El Maarry, K., Balke, W.T.: Skyline queries in crowd-enabled databases. In: Proceedings of the 16th International Conference on Extending Database Technology, pp. 465–476 (2013)
11. Meng, X., et al.: Mllib: machine learning in apache spark. J. Mach. Learn. Res. **17**(1), 1235–1241 (2016)
12. Miao, X., Gao, Y., Zheng, B., Chen, G., Cui, H.: Top-k dominating queries on incomplete data. IEEE Trans. Knowl. Data Eng. **28**(1), 252–266 (2015)
13. Nguyen, H.T.H., Cao, J.: Top-k dominance range-based uncertain queries. In: Cheema, M.A., Zhang, W., Chang, L. (eds.) ADC 2016. LNCS, vol. 9877, pp. 191–203. Springer, Cham (2016). https://doi.org/10.1007/978-3-319-46922-5_15
14. Papadias, D., Tao, Y., Fu, G., Seeger, B.: Progressive skyline computation in database systems. ACM Trans. Database Syst. (TODS) **30**(1), 41–82 (2005)
15. Wang, Y., Li, X., Li, X., Wang, Y.: A survey of queries over uncertain data. Knowl. Inf. Syst. **37**(3), 485–530 (2013). https://doi.org/10.1007/s10115-013-0638-6
16. Zaharia, M., et al.: Resilient distributed datasets: A fault-tolerant abstraction for in-memory cluster computing. In: 9th {USENIX} Symposium on Networked Systems Design and Implementation ({NSDI} 12), pp. 15–28 (2012)

17. Zaharia, M., Chowdhury, M., Franklin, M.J., Shenker, S., Stoica, I., et al.: Spark: cluster computing with working sets. HotCloud **10**(10–10), 95 (2010)
18. Zaharia, M., et al.: Apache spark: a unified engine for big data processing. Commun. ACM **59**(11), 56–65 (2016)

Optimizing Task Processing in Big Data with Federated Learning

Chunyi Wu(✉) and Ya Li

Chongqing College of Electronic Engineering, Chongqing 401331, China

Abstract. With the rapid development of cloud computing and big data technology, task processing based on edge computing and fog computing in big data has become a research hotspot nowadays. Most related research work is based on modularized program and existing intelligent algorithms. However, there are two bottleneck problems: firstly, many existing task processing frameworks are always based on optimization algorithm, which requires multiple iterations to achieve the optimum target searching, which is often a time-consuming process; secondly, there always be data island problems among task processing clusters, which becomes the bottleneck of data sharing during big data analysis. To solve this problem, based on the idea of fast task migration, this work proposes a task processing approach based on federated learning and deep feature learning in big data to achieve accurate task classification and rapid task allocation named FLOM (Federated Learning based Optimization Methodology for large-scale tasks processing). We employ federated learning framework to realize data aggregation and establish personalized models. Deep feature learning is exploited to train task classification models based on deep neural network. Experiments based on Spark show that FLOM can effectively achieve accurate task classification and improve the performance in terms of overall load balancing effect.

Keywords: Task pocessing · Big data · Federated learning · Deep feature learning

1 Introduction

Recently, edge computing [1–3] technology is in the rapid development. The large-scale tasks processing based on edge environment for big data [4–6] has become one of the research hotspots. In order to realize the parallel processing of large-scale tasks, we can exploit the combination mode of cloud data center and multiple edge computing centers to realize data sharing and multi-category tasks processing. An efficient task processing approach can achieve the parallel processing capability of multiple clusters, including cloud data center and edge computing center. On the contrary, the system may encounter the increased computing cost and load imbalance when the task processing approach based on traditional framework and optimization model is in the face of the problem of multiple computing centers and large-scale multi-category tasks. Therefore, it is necessary to explore an efficient task processing approach in the process of big data processing.

S.-C. Chu et al. (Eds.): ICGEC 2021, LNEE 833, pp. 481–492, 2022.
https://doi.org/10.1007/978-981-16-8430-2_44

In recent years, some new computing paradigms have attracted wide attention and developed rapidly. In the comprehensive technology survey [7] on the federated machine learning, researchers indicate that the optimization method based on federated learning framework is firstly presented by Google [8] in 2016. In [8], researchers have introduced a new setting for the distributed optimization called *federated optimization*. The motivation of this idea is to keep the training data locally on users' mobile devices instead of logging it to the uniform data center for the training. It achieves the distributed mobile phone applications in a high-quality centralized model with user data protection. After that, Georgios A. Kaissis [9] et al. have presented a secure, privacy-preserving and federated machine learning in medical imaging. It aims to improve patient care and address the demands for data protection while the utilization is mandatory. An authoritative overview [10] on the recent advances in deep learning indicates that Deep Learning (DL) was first introduced into Machine Learning (ML) in 1986 and then used in Artificial Neural Networks (ANN) in 2000. Deep learning is composed of multiple hidden layers to learn data features with multiple abstraction layers [11]. Weisong Shi [12], one of the proposers of Edge Computing (EC) published an authoritative review on edge computing technology in 2016, which indicates that edge computing is a kind of enabling technology allowing computation to be performed in the edge of networks. Herein the upstream data represents the IoT services and the downstream data represents the cloud services. Since edge computing was proposed, it has been customized and used in various fields.

Nowadays, with the increasing number of task requests from users and the diversification of task types, the combination of different domains has become a trend to overcome this issue. The purpose of this work is to combine the idea of distributed computing with the cutting-edge artificial intelligence [13–15] to realize the technology integration of multiple fields, so as to complete the reasonable allocation of large-scale tasks in the edge computing environment of big data [16, 17]. Our goal is to design and implement an efficient task processing approach to achieve accurate task classification and global load balancing, so as to reduce the cost of task processing.

Based on this motivation, we propose a large-scale task processing approach for big data processing based on the idea of federated learning in the edge computing environment. It employs the federated learning framework to realize the data interaction and model parameter updating of multiple edge computing centers and the cloud center without exposing private data, and realizes multi-category tasks processing and global load balancing from the whole system level. At the same time, the deep feature learning method is used to fully analyze the features of the substrate network resources and the current task requests, construct the feature matrix as the input of the federated learning model based on the deep convolutional network, and obtain the probability distribution of a certain task assigned to each node in the edge computing centers and the cloud as the output of the model, so as to realize the large-scale tasks allocation.

2 Model

Figure 1 describes the framework details of the proposed approach in the edge computing environment. It intuitively shows the relationship between the cloud server and the local edge computing nodes. Firstly, the user data from different kinds of task requests in the substrate network is uploaded to the cloud data center and integrated with feature data of task processing nodes in the cloud data center and edge computing centers as the public dataset for initial model training. Then each edge computing center downloads the initial model to the local server, and updates the edge computing model with the real-time updated personalized user data. We upload the updated user model (edge computing model) to the cloud data center to update DE_{fed} to complete the encryption parameter sharing among the edge computing nodes that train and process different types of task data without affecting the privacy of user data, and then realize the federated model training by the way of interaction between the federated model and the edge computing model. In the model application stage, the federated model classifies and identifies real-time task requests according to the federated model trained by historical dataset. It assigns task requests onto the optimal edge task computing node or directly sends them to the cloud data center for efficient processing. It can be seen that a "Lotus shaped" federated model for big data processing in the federal cloud-edge data center is established, which is named as "Lotus model".

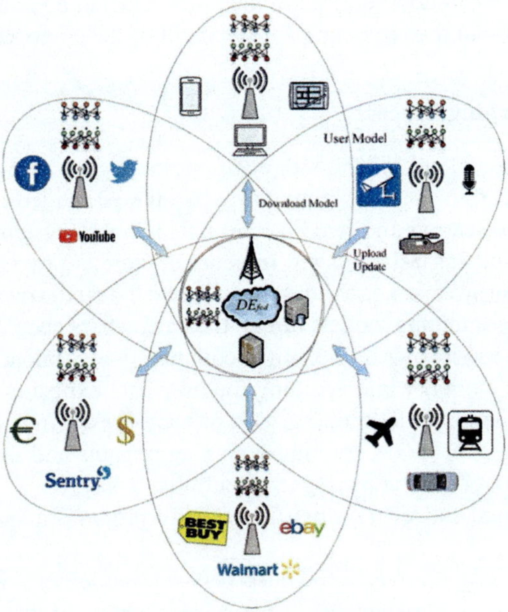

Fig. 1. The view of FLOM's architecture "Lotus model".

3 Methodology

3.1 Feature Extraction

We extract features of task requests and physical hosts in the federated cloud center with edge computing to construct feature matrixes as the input of our deep network model.

It is worth noting that there exist many features can describe the cluster state and task requirements, and the inaccurate feature extraction can't achieve accurate description of the system. If the number of features is too small, the task can't be described comprehensively. Whereas, if the number of features is too large, the load of data acquisition tools and the complexity of deep network model will be increased. Hereupon, according to resource management pattern in Alibaba Cloud Service and Google Cloud Service [18], we select more than twenty key features from the attributes of each task request and host, including CPU capacity provisioned (MHZ), CPU usage (MHZ), memory capacity provisioned [KB], memory usage [KB], disk read throughput [KB/s], disk write throughput [KB/s], network received throughput [KB/s], network transmitted throughput [KB/s], CPU capacity required (MHZ), memory capacity required (KB), network resource request (KB/s), disk resource request (KB/s), the ratio of the available computing resource of the physical node to the amount demanded of computing capability of a task request, as well as the ratio of available CPU amount of the physical node to the total amount of available CPU resource in the cloud data center or edge computing center, etc. By extracting the features from current n task requests and $m + l$ hosts during a time window Δt, we can normalize and integrate the obtained features to construct a $(m + l) * 23$-dimensional feature matrix for the input of deep network model.

3.2 Federated Learning Process

Our designed federated cloud model with edge computing adopting the typical architecture and federal learning process aims to realize the encrypted model training and sharing for task classification and overall load balancing of system using large-scale masking data from kinds of task requests. In general, there are two main entities in the typical federal learning system, i.e., the data owners and the model owner. Herein, we can assume that the edge computing models run as the data owners and the cloud data center model operates as the model owner. We aim to collaboratively train a shared model (viz. the cloud data center model) while avoiding the user data exposing to the other users, which achieve the federal model training with privacy-preserving. The whole process differs from the traditional model training which aggregates and integrates individual resource to complete model establishing in a centralized way.

As for the federated model training, the specific process of model training is as follows:

Step 1: Since we have employed the deep network model to train the task classification model in this work, thus we need to construct feature matrixes for input of the model. We elaborately select twenty-five key features from the attributes of the received task requests and physical hosts that include CPU capacity provisioned, CPU usage and memory capacity provisioned, etc. via the distributed data acquisition tool Ganglia [19] during tons of tasks processing. Herein, aiming at reflecting the interaction between the current task requirement and available resources in the historical data set, as well as the interaction between the current resource surplus of a certain task processing node and the total resource surplus in the cloud data center or edge computing center, we design and calculate the following indicators in terms of CPU as two important elements in the input features, which are also applicable to memory, network and disk resources.

$$ratio_{cpu}^{q/p} = \frac{R_c^q}{L_c^p}. \tag{1}$$

$$ratio_{cpu}^{p/total} = \frac{L_c^p}{total_c}. \tag{2}$$

After extracting the features of physical hosts and task requests, we need to carry out normalization to accelerate the training process and enable the algorithm to converge quickly. In order to clearly reflect the probability distribution of each task assigned to each physical host in the output of deep network model, we use the normalized values to construct a feature matrix M_q composed of features from a certain task and all the available hosts which can be allocated onto in each feature input process.

$$M_q = \begin{bmatrix} CPU\ cores_1 & CPU\ capacity\ provisioned_1 & CPU\ usage_1 & \cdots & ratio_{cpu}^{q/1} & \cdots & ratio_{dsk(w)}^{1/total} \\ CPU\ cores_2 & CPU\ capacity\ provisioned_2 & CPU\ usage_2 & \cdots & ratio_{cpu}^{q/2} & \cdots & ratio_{dsk(w)}^{2/total} \\ \vdots & \vdots & \vdots & \vdots & \vdots & \cdots & \vdots \\ CPU\ cores_p & CPU\ capacity\ provisioned_p & CPU\ usage_p & \cdots & ratio_{cpu}^{q/p} & \cdots & ratio_{dsk(w)}^{p/total} \\ \vdots & \vdots & \vdots & \vdots & \vdots & \vdots & \vdots \\ CPU\ cores_{m+l} & CPU\ capacity\ provisioned_{m+l} & CPU\ usage_{m+l} & \cdots & ratio_{cpu}^{q/(m+l)} & \cdots & ratio_{dsk(w)}^{(m+l)/total} \end{bmatrix}. \tag{3}$$

Step 2: As the public data set from task requests and physical hosts has been integrated, the cloud data center gets ready for the parameter initialization and training of the cloud data center model. Herein, cloud data center employs the deep network model constructed by CNN and it specifies the hyperparameters of the federated cloud center with edge computing model DE_{fed}, i.e., learning rate, batch size and dropout rate, etc. The cloud date center model is trained based on the public datasets. Subsequently, the cloud data center will distribute the initialized DE_{fed} among the edge computing centers to lead the edge computing center models' training for local models establishing.

Step 3: Taking advantage of the obtained cloud data center model, each edge computing center model can exploit its local data from users' task requests and physical hosts in its own cluster to update the parameters of the deep network model. In the model DE_{fed}, we adopt deep neural model to learn both the cloud data center model and edge computing center models. Herein, the deep networks are employed to achieve feature learning and the task requests classification in right of inputting feature matrixes from the raw task request data. The goal of the model learning is to find the optimal parameters o_w^z that can minimize the loss function $L(o_w^z)$:

$$o_w^{z*} = \arg\min_{o_w^z} L(o_w^z). \tag{4}$$

where w is the serial number of the models in the clusters including v edge computing centers and the cloud data center. Hereupon o_w^z represents parameters of the wth model in the iteration z. And then the updated parameters of edge computing models will be sent to the cloud data center.

Step 4: In this step, we combine cloud data center model with the edge computing model to establish the federated model DE_{fed}. The trained cloud data center model aggregates the edge computing center models from clusters and then broadcasts updated federated model parameters to the v clusters. Herein, the cloud data center aims to minimize the global loss function $L(o_G{}^z)$ for the designed federated model.

$$L(o_G^z) = \frac{1}{v+1} \sum_{w=1}^{v+1} L(o_w^z). \tag{5}$$

Specifically, this process can be used to minimize the loss for our deep neural networks. We assume that $f_{DE}()$ is the federated model to be learned. It also can be treated as a prediction function using deep network in the task classification. Thus the learning objective is as follows:

$$\arg\min_{o_w^z} L = \sum_{w=1}^{v+1} l(y_w, f_{DE}(x_w)). \tag{6}$$

where $l()$ is the loss for the deep neural network. $\{x_w, y_w\}_{w=1}{}^{v+1}$ are the task data samples from the training datasets. In this work, we employ the cross-entropy loss for the task requests classification.

$$\arg\min_{o_w^z} L = \sum_{w=1}^{v+1} l(y_w^z, f_{DE}(x_w^z)). \tag{7}$$

In the overall learning process, Step 3–4 will be repeated until the global loss function $L(o_G{}^z)$ converges to an acceptable range or a desired task classification accuracy is obtained.

Note that we have employed CNN based deep neural model to train both cloud data center model and edge computing center model. We can take the feature matrix as the input of the deep model and the probabilities of mapping task requests onto physical hosts as the output. The basic components of the designed network for our proposal are shown in Fig. 2. It consists of an input layer, convolutional layers, pooling layers, a fully connected layer and a softmax layer. We determine the number of hidden layers via sets of experimentations since the optimal number of hidden layers of deep neural network belongs to an open research question. Herein, we explore the performance of the designed model by changing the number of hidden layers from 1 to 5.

Generally, the result from full connection layer is transmitted to the output layer for the probability distribution of each task request that indicates the probability of obtaining a desired performance benefit after allocating the task request onto a certain physical host in the cluster. The output layer contains a softmax function which is a generalization of the logistic regression. The probability Pr_p can be expressed as:

$$\Pr_p = \frac{e^{v_p^l}}{\sum_K e^{v_K^l}}. \tag{8}$$

where ${v_p}^l$ is the *pth* output of the *lth* convolutional layer. Using the softmax function, the multi-dimensional vector can be transformed into real values from 0 to 1 and their sum is 1.

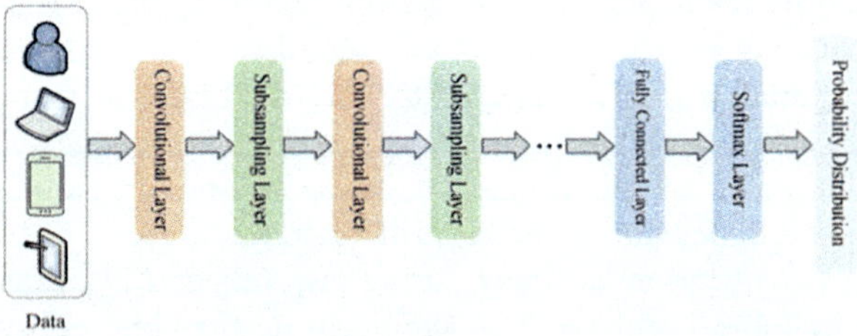

Fig. 2. The deep learning process of FLOM.

As for the over-fitting due to the lack of public available labeled training samples that limits the performance for big data feature learning, an adaptive distribution function is established to set the activating rate in each hidden layer with a probability ξ. It is always in accordance with the law of human cognition of things when we use the normal distribution to explain the law of nature within the widely used distribution functions. Herein, we can't directly employ the normal distribution to set the activating rate of each hidden layer since it is a monotonically increasing function. Thus we introduce a distribution model ξ in the lth layer in the deep neural network as follows:

$$\xi = \begin{cases} 1 - \frac{1}{\delta\sqrt{2\pi}} \int_{-\infty}^{l} \exp(-\frac{(l-\frac{L}{2})^2}{2\delta^2}) dl, & L = 2k; \\ 1 - \frac{1}{\delta\sqrt{2\pi}} \int_{-\infty}^{l} \exp(-\frac{(l-\frac{L+1}{2})^2}{2\delta^2}) dl, & L = 2k + 1. \end{cases} \tag{9}$$

where L presents the total number of the layers in the deep neural network, and δ is the parameter for decay velocity control. The parameter of the deep network model can be trained by the combination of higher-order backpropagation and the designed distribution function. The specific process can be found in [20].

4 Experimental Analysis

We evaluate the proposed approach on the dataset composed by the task requests simulated by Whetstone [21]. Our approach is compared with the existing approach FIFO, Fair Scheduler and Capacity Scheduler [22] based on two indicators: the effectiveness of task requests classification and the load balancing effect.

We exploit the simulation tools to generate the certain kinds of task requests in parallel with different resource requirements continuously getting to cloud data center and edge computing centers to compose the synthetic task dataset. The task requests are divided into the training set and testing set herein. It is worth noting that they can be used to establish labeled dataset during the model training while the partial labeled dataset during semi-supervised learning for the model testing. The hyperparameters of all the comparison approaches are tuned by using the cross-validation. The deep learning framework TensorFlow is exploited to achieve the deep network modeling. We employ the Paillier additively homomorphic encryption [23] for the privacy security in python. The proposed federated model fetches the resource information and state of cloud data center and edge computing centers regularly. OpenStack [24] is employed to create virtualization scenarios for deployment of the big data processing framework Spark [25]. All the experiments are carried out in the same Linux workstation with an Intel Xeon with the configuration of three 3.4 GHz CPUs and 512 GB memory. In this work, 32 processing nodes with different configuration are created as the edge computing nodes and resource pool in the cloud data center. The latest version of Spark 2.4.6 is used in this work, which needs to be built on the basis of Hadoop.

The indexes of classification accuracy during task requests identification for each subject are shown in Table 1 and Table 2. The mean results obtained by multiple sets of experiments are shown in Table 3. The results indicate that our proposed federated model can achieve decent classification performance in terms of precision, recall and F-measure on all the task requests from users. Compared to the other approaches, the proposed FLOM approach significantly improves the accuracy in case of 600, 1200, 1800, 2400, 3600 and 4500 task requests.

As for the precision, the proposed FLOM approach achieves 0.8857 in case of 3600 task requests that is the best result among all the datasets in different scales. We receive the best result of recall (0.8891) in the case of 600 task requests while the best results of F-measure (0.8805) in the case of 4500 task requests. Although there is no result obtained in the different data scales that possesses the absolute advantage in all three indicators, on the whole, we can still get a relatively decent result (0.8762, 0.8848, 0.8805) in the case of 4500 task requests. It significantly demonstrates the effectiveness of our proposed federated model for task processing in big data.

Table 1. Four types of task request classification results.

	Positive	Negative
The task has been correctly classified	*tp*	*fp*
The task hasn't been correctly classified	*fn*	*tn*

Table 2. Measuring accuracy of task request classification.

Indexes	Calculation formulas
Precision	$p = \frac{Num(tp)}{Num(tp)+Num(fp)}$
Recall	$r = \frac{Num(tp)}{Num(tp)+Num(fn)}$
F-measure	$F = 2 \times \frac{p \times r}{p+r}$
Accuracy	$Accuracy = \frac{Num(tp)+Num(tn)}{Total}$

The results in Fig. 3 shows the comparison of four kinds of approaches in accuracy with 3-layer, 4-layer and 5-layer deep neural network respectively. On the whole, the proposed FLOM approach shows its advantage of accuracy for task request classification in three kinds of deep network architectures herein except for the result in the 5-layer neural network with 2400 task requests. Specifically, as shown in Fig. 3(a), in the 3-layer architecture, the accuracy of these approaches fluctuates with the change of task request

Table 3. Comparison in effectiveness of task requests classification.

	Precision	Recall	F-measure
600 task requests	0.8692	0.8891	0.8790
1200 task requests	0.8592	0.8841	0.8715
1800 task requests	0.8694	0.8750	0.8722
2400 task requests	0.8665	0.8831	0.8747
3600 task requests	0.8857	0.8692	0.8773
4500 task requests	0.8762	0.8848	0.8805

scale. The results of Fair Scheduler and Capacity Scheduler are similar, but they are always better than FIFO. As for FLOM, its results are always better than the other three approaches. The maximum value of accuracy is obtained when the task scale is 1800 and the minimum value is obtained when the task scale is 4500.

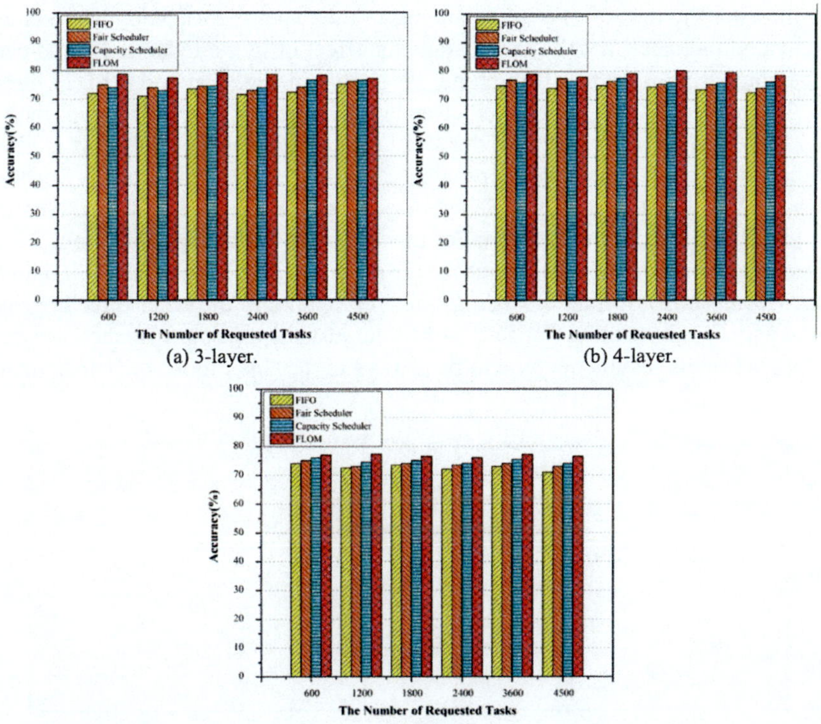

(a) 3-layer. (b) 4-layer. (c) 5-layer.

Fig. 3. Comparison of FLOM with FIFO, Fair Scheduler and Capacity Scheduler in task allocation effectiveness under different neural network structures.

In Fig. 3(b), on the whole, the results in terms of accuracy of all algorithms are almost stable from the beginning to the end in the 4-layer architecture deep network. In the initial stage, the results of Fair Scheduler are better than those of FIFO and Capacity Scheduler. However, the results of Capacity Scheduler begin to be better than Fair Scheduler from the task scale of 1800. The results of Fair Scheduler and Capacity Scheduler are always better than FIFO in the whole process. For FLOM, it shows the performance advantages over the other three approaches in the overall process. The maximum value of accuracy is obtained when the task scale is 2400, and the minimum value is obtained when the task scale is 1200. Figure 3(c) has shown the specific classification results of four approaches in the case of 5-layer neural network. FLOM still maintains the performance advantage in task request classification except for the task scale of 2400. Its result is only a little lower than Capacity Scheduler, but still higher than the other two approaches. Therefore, in general, it can be observed that the proposed FLOM approach has certain performance advantages in tasks classification in big data.

We analyze the load balancing effect of FLOM with the other task processing approaches. We compare their performance within 4-layer neural network since it possesses the potential to obtain relatively favorable results. And in the following experiments, we all employ this network structure by default. We employ the standard deviation value to measure the degree of load balancing in the experiments. The lower standard deviation value indicates that the load balancing effect of in the federated cloud center with edge computing is better at this time. As shown in Fig. 4(a) and 4(b), in the two kinds of task request scales, the results have shown a trend of decreasing with the time changing. For the task scale of 1200, the standard deviation value of FLOM is always smaller than those of the other three approaches. While the results of Capacity Scheduler is always smaller than those of Fair Scheduler except for the time at 1750s. And their standard deviation values are both smaller than those of FIFO approach from 250s to 2750s. As for the scale of 3600 task requests, the proposed FLOM approach still maintain its advantages in terms of load balancing effect compared to the other three approach. Compared with Fair Scheduler, Capacity Scheduler just keeps its performance advantage before 750s while its results become to be always higher than those of Fair Scheduler

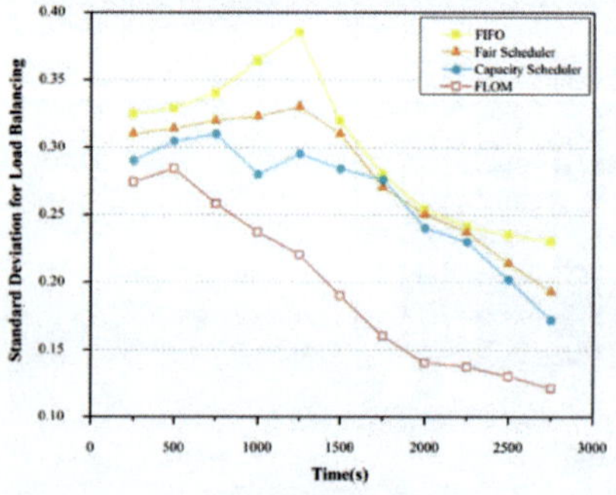

(a) 1200 task requests.

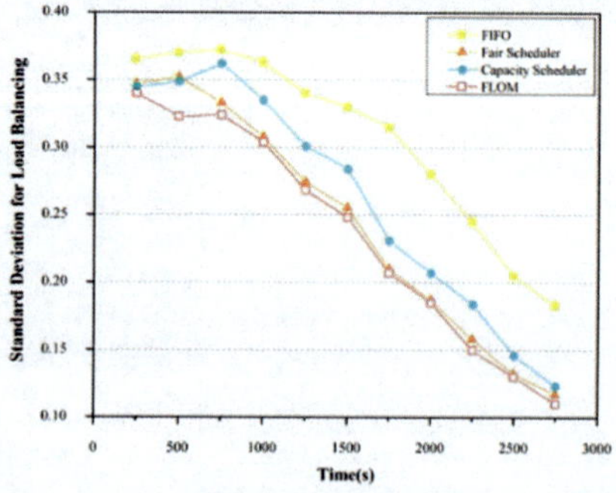

(b) 3600 task requests.

Fig. 4. Comparison of FIFO, Fair Scheduler, Capacity Scheduler and FLOM in load balancing with the task scale of 1200 and 3600 task requests.

in the later stage. FIFO always has the higher standard deviation value compared to the other three approaches. The results of federated learning model significantly decrease the other three approaches by 5.98% at most. From the above, we can observe that FLOM can achieve the desired load balancing of the whole cluster of the federated data center in the process of large-scale task processing in 4-layer neural network structure.

5 Conclusion

This work studies the combination of federated learning and deep feature learning in the edge computing environment that supports large-scale tasks processing for big data. Herein, 1) an optimized large-scale task processing methodology based on federated learning in the edge computing environment for big data is proposed; 2) a federated model for task processing based on the federated learning framework and deep feature learning is designed. We achieve accurate task classification and overall load balancing by using the federated model trained through parameter updating among the cloud data center model and the edge computing models as well as the deep convolutional network with dropout; 3) the superiority of FLOM for large-scale tasks processing is evaluated in comparison with state-of-the-art approaches in the edge computing environment. Experimental results show that FLOM can accurately classify tasks, realize overall load balancing and decrease consumption of task execution simultaneously.

Acknowledgements. This work is supported in part by the Science and Technology Research Program of Chongqing Municipal Education Commission (Grant No. KJQN202103108), the Doctoral Initiation Project of Chongqing College Of Electronic Engineering (120724), the Chongqing University Innovation Research Group Project (CXQT21031), and the School Project of Chongqing College Of Electronic Engineering (XJZK202104).

References

1. Ren, J., Yu, G., He, Y., et al.: Collaborative cloud and edge computing for latency minimization. IEEE Trans. Veh. Technol. **68**(5), 5031–5044 (2019)
2. Luo, Z., Liwang, M., Huang, L., et al.: Caching mechanism for mobile edge computing in V2I networks. IEEE Trans. Emerg. Telecommun. Technol. **30**(8), e3689 (2019)
3. Chang, Z., Guo, W., Guo, X., et al.: Incentive mechanism for edge computing-based blockchain. IEEE Trans. Industr. Inf. **16**(11), 7105–7114 (2020)
4. Hu, D., Feng, D., Xie, Y., et al.: Efficient provenance management via clustering and hybrid storage in big data environments. IEEE Trans. Big Data **6**(4), 792–803 (2019)
5. Liu, X., Zhu, Q., Pramanik, S., et al.: VA-store: a virtual approximate store approach to supporting repetitive big data in genome sequence analyses. IEEE Trans. Knowl. Data Eng. **32**(3), 602–616 (2020)
6. Bersani, M.M., Marconi, F., Tamburri, D.A., et al.: Verifying big data topologies by-design: a semi-automated approach. J. Big Data **6**(1), 1–23 (2019)
7. Yang, Q., Liu, Y., Chen, T., Tong, Y.: Federated machine learning: concept and applications. ACM Trans. Intell. Syst. Technol. (TIST) **10**(2), 12 (2019)
8. Konen, J., Mcmahan, H.B., Ramage, D., et al.: Federated optimization: distributed machine learning for on-device intelligence (2016)

9. Kaissis, G.A., Makowski, M.R., Daniel, R., et al.: Secure, privacy-preserving and federated machine learning in medical imaging. Nat. Mach. Intell. **2**, 305–311 (2020)
10. Minar, M.R., Naher, J.: Recent advances in deep learning: an overview (2018)
11. Lecun, Y., Bengio, Y., Hinton, G.: Deep learning. Nature **521**(7553), 436–444 (2015)
12. Shi, W., Cao, J., Zhang, Q., et al.: Edge computing: vision and challenges. IEEE Internet Things J. **3**(5), 637–646 (2016)
13. Miller, T.: Explanation in artificial intelligence: Insights from the social sciences. Artif. Intell. **267**, 1–38 (2019)
14. Williams, T., Szafir, D., Chakraborti, T., et al.: Report on the first international workshop on virtual, augmented, and mixed reality for human-robot interaction. AI Mag. **39**(4), 64–66 (2018)
15. Hutson, M.: Artificial intelligence faces reproducibility crisis. Science **359**(6377), 725–726 (2018)
16. Adhikari, M., Amgoth, T.: An enhanced dynamic load balancing mechanism for task deployment in IaaS Cloud. In: 2018 IEEE International Conference on Computing, Power and Communication Technologies (GUCON) (2019)
17. Liu, R., Marakkalage, S.H., Padmal, M., et al.: Collaborative SLAM based on WiFi fingerprint similarity and motion information. IEEE Internet Things J. **7**(3), 1826–1840 (2020)
18. Xue, Y., Zhang, H., Ma, H.: Performance evaluation of image and video cloud services. In: 2018 IEEE 20th International Conference on High Performance Computing and Communications (2018)
19. Massie, M.L., Chun, B.N., Culler, D.E.: The ganglia distributed monitoring system: design, implementation, and experience. Parallel Comput. **30**(7), 817–840 (2004)
20. Zhang, Q., Yang, L.T., Chen, Z., et al.: An adaptive dropout deep computation model for industrial IoT big data learning with crowdsourcing to cloud computing. IEEE Trans. Ind. Inform. **15**, 2330–2337 (2018)
21. Raju, P.S., Govindarajulu, P.: Performance improvement of multi-core architecture using whetstone application in Linux. Int. J. Comput. Sci. Netw. Secur. **13**, 47 (2013)
22. Han, R., Liu, C.H., Zong, Z., et al.: Workload-adaptive configuration tuning for hierarchical cloud schedulers. IEEE Trans. Parallel Distrib. Syst. **30**(12), 2879–2895 (2019)
23. Le, T.P., Aono, Y., Hayashi, T., et al.: Privacy-preserving deep learning via additively homomorphic encryption. IEEE Trans. Inf. Forensics Secur. **13**(5), 1333–1345 (2017)
24. Cotroneo, D., Natella, R., Rosiello, S.: Overload control for virtual network functions under CPU contention. Future Gener. Comput. Syst. **99**, 164–176 (2019)
25. Lunga, D., Gerrand, J., Yang, L., et al.: Apache spark accelerated deep learning inference for large scale satellite image analytics. IEEE J. Sel. Top. Appl. Earth Obs. Remote Sens. **13**, 271–283 (2020)

Privacy of Web Browsers: A Challenge in Digital Forensics

Adnan Ahmed[1], Abdul Rehman Javed[2], Zunera Jalil[2], Gautam Srivastava[3](✉), and Thippa Reddy Gadekallu[4]

[1] Institute of Avionics and Aeronautics Air University, Islamabad, Pakistan
200106@students.au.edu.pk

[2] Department of Cyber Security, Air University Islamabad, Islamabad, Pakistan
abdulrehman.cs@au.edu.pk, zunera.jalil@mail.au.edu.pk

[3] Department of Mathematics and Computer Science, Brandon University, Brandon, Canada
srivastavag@brandonu.ca

[4] School of Information Technology and Engineering, Vellore Institute of Technology, Vellore, India
thippareddy.g@vit.ac.in

Abstract. Web browsers are ubiquitous applications to access public and private applications over the Internet, Intranet, and Extranet. The increased demand for cybersecurity, including data privacy, secrecy, and anonymity, becomes the reason for enhanced privacy and anonymity in common web browsers and specialized web browsers to achieve such purposes. These features are great challenges and obstacles for forensic investigators. In this paper, a collection of research will be analyzed, that have been carried out to identify and collect artifacts of web browsers having secrecy features for examination, validation, and find out potential ways to use the collected information during active investigations. As a result, live forensics can become more relevant and dependable for collecting reasonable artifacts from private browsers. From common browsers using private browsing facilities, even removing web browsers after committing criminal activities can also be identified by analyzing the registry, supporting factual evidence gathering in any Digital Forensic investigation.

Keywords: Privacy browsing · Web forensics · Private browsing · Anonymity

1 Introduction

In recent times, the Internet has become pervasive for daily tasks associated with desktop and mobile computer devices [1,2,6,15,20,24,25]. Web browsers are used to connect and access Internet websites, search engines, e-mail correspondence, social networking, etc. [5]. The Internet has revolutionized banking with e-commerce, which resulted in growing financial fraud by extracting the users' critical data after their browsing sessions [8,10,23]. As the Web continues

S.-C. Chu et al. (Eds.): ICGEC 2021, LNEE 833, pp. 493–504, 2022.
https://doi.org/10.1007/978-981-16-8430-2_45

to evolve, browsers have become complex application platforms that mediate a significant part of our online activities [26]. The connectivity of web browser and web-server establishes various data/metadata have been exchanged between them, and various traces are developed on computing machines. Such traces are in the forms of history, cache, temporary files, cookies, etc., which are useful for tracking web user's digital footprints in any Digital Forensic investigation. These artifacts facilitate investigators to identify and collect the required digital footprints. The fact that tracking can so readily be performed using browser fingerprinting is potentially a major threat to web users' privacy, which users currently have no control [3,7,42]. These artifacts have been shown to be useful for service providers to understand the user behavior, region, expectations to promote the sale, get a competitive edge, and customize offers. Unfortunately, criminals also use these artifacts for malicious purposes, including hacking, unauthorized access, and password cracking, to name just a few.

The increased number of cyber-crimes in this era are witnesses to the misuse of browser metadata and have made end-users conscious about their security [28,40,43], privacy/secrecy, and anonymity, which also started privacy competitions among various web browser vendors. Each of them is now offering to protect its customers' private information, which has undoubtedly enhanced its customers' privacy and anonymity. This secrecy has introduced many challenges and obstacles globally, especially to digital forensic investigators, because privacy and anonymity features are equally utilized for legitimate and illegitimate purposes. Cyber-criminals have taken advantage of private browsing modes to clear any digital traces leftover on machines used and leave forensic examiners empty-handed [32]. Criminals use private browsers for malicious activities such as smuggling, gambling, and human trafficking, to name just a few. We know that the most famous browser to ensure anonymity is TOR (The Onion Routing)[1]. Onion routing is a simple concept in that the end-user, or initiator of network traffic, encrypts traffic with multiple layers [45]. A layer of encryption exists for each hop inside the TOR network. The encrypted traffic moves through the Tor network, and each node removes one layer of encryption, analogous to removing an onion layer. At the last TOR network node, the final layer of encryption is removed, and the traffic proceeds out onto the Internet [45]. Tor Browser is also configured and performs some filtering to resist browser fingerprinting, and first-party isolation to resist sharing state or linking of identifiers across origins [12].

Apart from TOR, it is known that all major vendors started providing a new feature known as Private Browsing Mode [8]. Private browsing is defined as a "web browser mode" in which information about visited websites is not saved [14]. As a result, the challenge is simplified where there is no need to capture potentially vast quantities of private browsing history records, only the activation of the function itself [16]. To validate vendor's assertions, we have studied comprehensive research articles that claim that most web browsers do not provide or offer private browsing facilities in the true sense and leave obvious artifacts on storage to detect and investigate the browsing history along with

[1] https://www.torproject.org/.

necessary timestamps. Gabet *et al.* expected enhanced privacy browsers would produce fewer recoverable artifacts than the common web browsers in private browsing mode [14]. However, both produced almost the same number of recoverable artifacts, with the common web browsers producing one more artifact than the enhanced privacy web browsers. Few limitations were observed by authors of the aforementioned papers, which are the use of limited forensic tools for analysis due to licensing issues. Furthermore, the labs were not made on Windows 10 platforms which are expected to capture more artifacts than any other older Windows Operating System. Finally, researchers' focus was to collect artifacts from permanent storage like history, cache, temporary files, etc.

This paper makes the following contributions:

- Reasonable artifacts can be collected in case of the use of common browsing as well as private browsing.
- Presentable artifacts can be collected from static forensics/storage using the Windows 10 OS platform.
- How can an investigator determine that browser has been deleted/removed from the machine after committing a crime?
- We also proposed a solution that equally safeguards the interest of its user and investigator. It ensures the secrecy, privacy, and anonymity of customers and protects society's interest by availability and provisioning of reasonable artifacts to the investigator in case of a need or as much as permissible by the Law.

The paper is organized as follows: Sect. 2 provides the related work to private browser and private mode. Section 3 presents the overview of the proposed approach. Section 4 provides our results and findings. Finally, Sect. 5 provides the Discussion, and lastly Sect. 6 concludes the paper and leads toward the Future Works.

2 Related Work

Private browsing is very significant issue and lot of research has been carried out in recent years, such as Horsman *et al.* who documented the examination of 30 web browsers to determine the presence of 'private mode' and where available, the 'privateness' of the mode in question [18]. The authors created 30 VM (virtual machines) and installed 30 of the most famous as well as common web browsers globally. Synopsis of their work is given in Table 1:

The analysis of this research shows that 17 web browsers offer private mode with actual privacy. The rest of the browsers leave sufficient artifacts for collection by investigators in their respective examination of browser history, cookies, downloads, and form auto-fill information from storage. Furthermore, the examination of memory through live forensics is beneficial to conclude any digital forensic investigation successfully.

Gabet *et al.* performed a comparative study over six web browsers, including Dooble, Comodo Dragon, Epic, Chrome, Edge, and Firefox [14]. Uniform

Table 1. Summary of web browsers

Sr.	Browser	Private function	Is it private
1	Avant	Yes	No
2	Brave	Yes	Yes
3	Chrome	Yes	Yes
4	Chromium	Yes	No
5	Comodo Ice Dragon	Yes	Yes
6	Comodo Dragon	Yes	No
7	Dooble	Yes	Yes
8	Edge	Yes	No
9	Epic	Private when proxy is on	No
10	Falkon	N/A	N/A
11	Fire Fox	Yes	Yes
12	Green Browser	No	N/A
13	Internet Explorer	Yes	No
14	Konqueror	N/A	N/A
15	Links	N/A	N/A
16	Lynx	N/A	N/A
17	Maxthon	Yes	Yes
18	Midori	Yes	Yes
19	Netsurf	No	N/A
20	Opera	Yes	Yes
21	Pale Moon	Yes	Yes
22	Puffin	N/A	N/A
23	Seamonkey	Yes	Yes
24	Sleipnir	No	N/A
25	Slim Jet	Yes	Yes
26	Tor Browser	Yes	Yes
27	Torch	Yes	Yes
28	UC Browser	Yes	Yes
29	Vivaldi	Yes	Yes
30	Water Fox	Yes	Yes

traffic was produced, and captured artifacts of all mentioned browsers analyzed through Forensic Toolkit (FTK) and Autopsy. Ten rounds of trials were carried out to ensure the reliability of the test. Finally, the study did not produce sufficient evidence to conclude that enhanced privacy browsers provide better privacy than common web browsers with privacy mode. The authors' research also showed that Firefox and Dooble provide the best privacy, but artifacts can also be recovered from both web browsers, although they are constrained but could not establish relevance with forensic investigation. In our opinion, Gabet *et al.* did not collect artifacts from live forensics and directly from Operating Systems that were in use, which might help investigate traces more effectively.

The authors in [22] collected TOR artifacts from the registry, memory, and storage. The authors showed that TOR leaves its artifacts on installation in the registry even after deletion. Few registry values could facilitate investigators to know that TOR was installed or used for web browsing. Memory analysis with respect to process showed that two processes in memory, `Firefox.exe` and `tor.exe` leave artifacts. Memory analysis with respect to browsing provides a number of artifacts to ensure which websites were accessed, links of social media account profiles, watched videos, keyword searched, etc., all through string search. Live Forensics is an investigative development model for obtaining evidence of computer usage [44]. The authors identified one problem during the conclusion phase: the whole research was conducted over Windows 7, an older operating system (OS)(Windows 8.1 and/or Windows 10 may capture more artifacts. Therefore, we initiate our research from the point left in this research.

3 Proposed Approach

Our research work starts by collecting presentable artifacts from most secure browsers in privacy mode, i.e., Firefox, Dooble, Edge, and TOR using Windows 10 64-bits. The browsing artifacts are collected from memory and storage before closing the browsers and after closing the browsers and analyzing their effectiveness to complete an investigation. Further, we analyzed the registry of the OS to verify the artifacts presented even removal of the browsers [22]. We suggest that every browser, whether private or capable of private browsing, should record sufficient and accurate user activities, but the storage of such digital footprints should be protected in terms of confidentiality as well as integrity. Asynchronous encryption may facilitate implementing this solution, as one key can be installed with the installation of a browser to record data in ciphertext that cannot be retrievable for anyone except vendor (owner) due to possession of the corresponding private key [33]. The artifacts can be deciphered at the request of any legitimate authority. Legitimate authorities can be defined in a licensed agreement considering the laws and regulations.

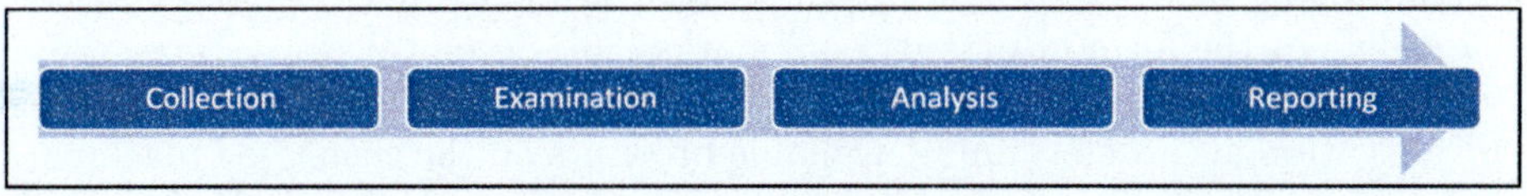

Fig. 1. Proceeding method for investigation

A physical machine with Microsoft Windows 10 64-bit was used. Compatible Web browsers, including Firefox, Dooble, Microsoft Edge, and TOR, were installed. The research used digital forensic-based steps from the NIST method to obtain data for analysis (see Fig. 1). All browsers executed in their private mode except Microsoft Edge, which was run in common mode to identify, distinguish and collect artifacts. Algorithm 1 demonstrates the activities were performed sequentially through each browser.

Algorithm 1. Pseudocode for Activities performed through each browser

1: Create the registry backup before installation of browsers for future comparison.
2: Create the registry backup after installing browsers and comparing key values to identify newly created hives and keys.
3: Restart the computer and create a memory image through FTK imager including page file before starting any browsing activity. Restart process cleans the memory from every temporary and permanent file and fresh load all the Operating system, drives, utilities, etc. files, and services.
4: Executed each browser in private mode except Microsoft Edge and accessed Google.com.pk. Searched Air university through Google, accessed Facebook.com, accessed YouTube.com, access Wikipedia.org and hec.gov.pk sites in sequences.
5: Capture memory dump and create an image of storage partitioned without closing the browsers after accessing mentioned sites from each browser to capture the artifacts created by each browser.
6: Capture another memory dump and create an image of similar storage after closing the web browsers.
7: Restart the computer and create another image of similar storage fresh boot-up.
8: Remove all browsers and create a copy of the registry at the end.
9: Analyze pre-installed, post-installed, and post-removal data of browser registry.
10: Analyze the memory dumps (before closing the browser and after the browser) and storage images (before closing the browser, after closing of the browser, and after a restart of the machine) for existence/availability and recording of the artifacts.

4 Findings

Registry analysis reflected that all browsers leave traceable artifacts after installation to detect which browser was installed on a suspect's machine, and even the un-installation does not remove all of the artifacts from the registry. First, artifacts proving installation/use of the browser are generated in memory and on hard disk in the form of default bookmarks [36]. Secondly, every browser is not private or offers a private browsing facility. Hence, the investigator can plan his investigation after determining available browsers on the suspected machine. Thirdly, the analysis of memory before closing the browsers and the browser leaves clear and presentable browsing artifacts. The analysis is given in Table 2.

Lastly, artifacts cannot be collected from storage or after a hard restart of a computer in case of private browsing. Findings are given in Table 3.

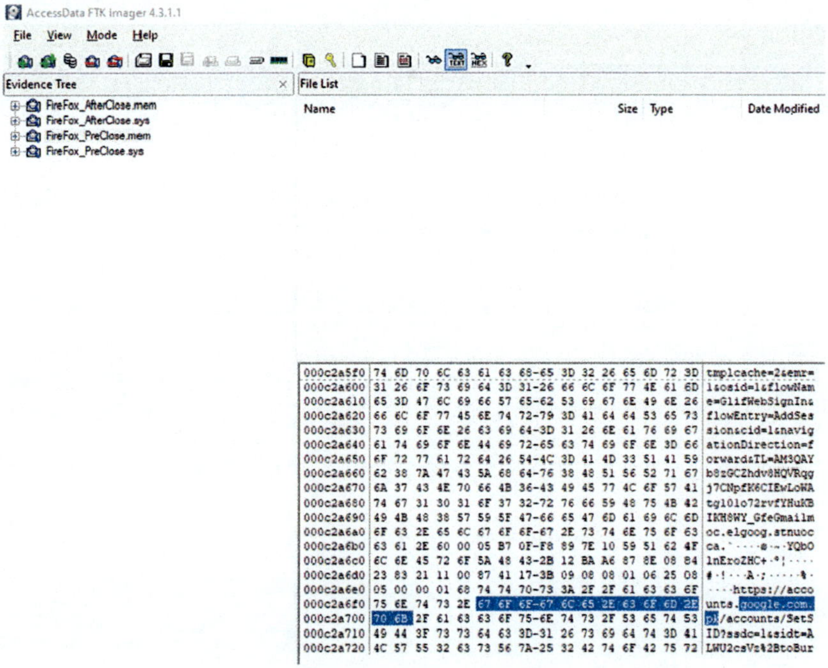

Fig. 2. Artifacts of firefox

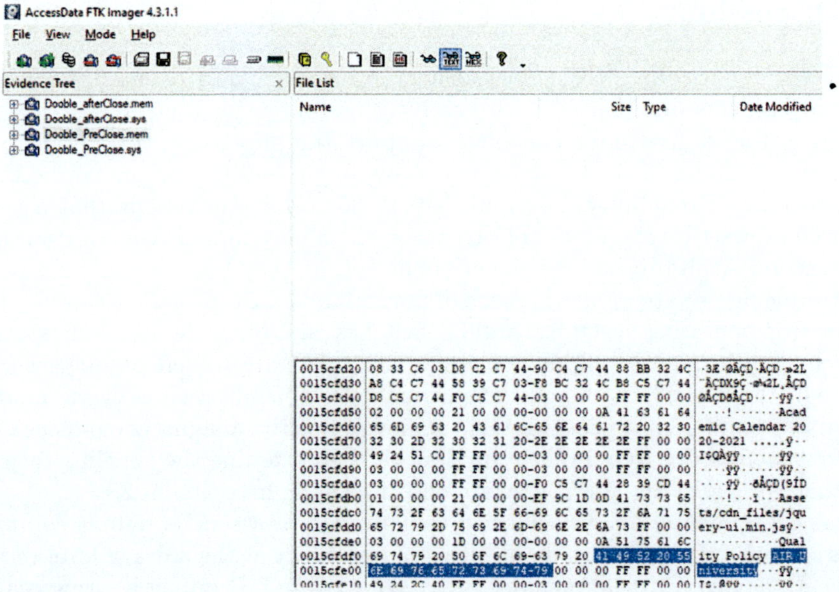

Fig. 3. Artifacts of dooble

Table 2. Traceable artifacts through live forensic (before and after closing the web browsers)

	Fire fox	Dooble	TOR	Edge
Google	Yes	Yes	Yes	Yes
Air University	Yes	Yes	Yes	Yes
Facebook	Yes	Yes	Yes	Yes
YouTube	Yes	Yes	Yes	Yes
Wikipedia	Yes	Yes	Yes	Yes
HEC	Yes	Yes	Yes	Yes

Table 3. Traceable artifacts through storage (cache, history and temporary files) before and after closing the web browsers

	Fire fox	Dooble	TOR	Edge
Google	No	No	No	Yes
Air University	No	No	No	Yes
Facebook	No	No	No	Yes
Youtube	No	No	No	Yes
Wikipedia	No	No	No	Yes
Hec	No	No	No	Yes

5 Discussion

We are firmly within a technology-driven era with digital devices documenting many aspects of our lives both passively and intentionally [17,19,21]. Consequently, this digital data can often support the effective investigation of a criminal act should one take place, and providing this information can be both accessed and its reliability determined [9,17,35]. This study finds that a common web browser leaves reasonable artifacts which may support carrying out any investigation. Web browser developers believe that they have enhanced privacy by introducing private modes in place of normal mode. Despite their claims, they cannot stop operating systems, routers, switches, and other devices from storing artifacts [39]. It is now well known that artifacts are always left on devices [39]. Although these artifacts' magnitude varies with different web browser modes, analysis with different aspects can help achieve a healthy amount of evidence [39]. The first step of the web forensic framework is to examine the timeline for any clue that will assist the forensic examiner in forming an opinion [38].

Our experimental analysis showed that private browsers as well as common browsers with private modes do not leave any artifact at the storage level unless Hibernate mode or virtual memory has been activated. Running a string search of all URLs and keywords used during the test on the entire hard disk did not reveal any trace other than the one found in the `pagefile.sys` [41]. So, we

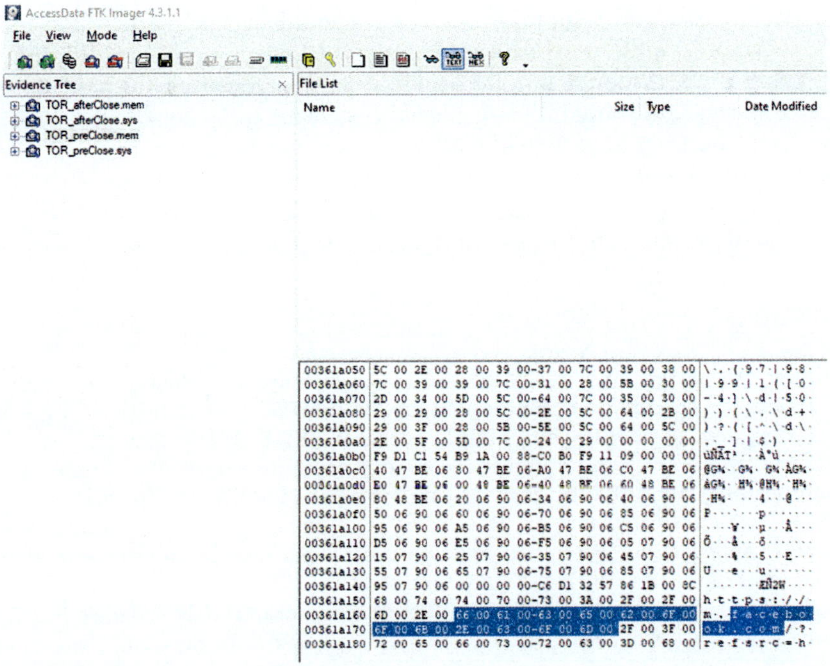

Fig. 4. Artifacts of TOR

can conclude that artifacts on a suspect's machine can be collected from live forensics as most secure browsers with the most privacy and anonymity features can be used to commit a crime (Figs. 2, 3 and 4). The removal/uninstallation of the browser after committing a crime cannot be untraceable as its artifacts can be collected from live forensics and registry values. However, if a criminal has managed to shutdown/power off the machine, the artifacts created through secure browsing cannot be retrieved in normal cases, according to this study. In such cases, Hibernate, memory paging, and memory slack may facilitate an investigator to capture required artifacts if available. Static forensics can also support in the manner in which an investigator may analyze the registry to identify used browsers and rely on the evidence/data provided if provided by 3rd parties, e.g., accessed servers/services, ISPs (Internet Service Providers), CSP (Could Service Providers), and other agencies.

6 Conclusion and Future Work

In this work, we explored common web browsers that tend to leave reasonable artifacts which may support carrying out any digital investigation. Web browser developers believe that they have enhanced privacy by offering private modes in place of general modes. Despite their claims, they cannot stop operating systems, routers, switches, and other devices from storing artifacts on hard drives. Future enhancement of this research may include the recovery of data from storage, i.e., cache, history, and temporary files that should be imaged through bit-by-bit. Furthermore, using more advanced analysis tools may produce differing and better results, which could not be achieved in our research due to time constraints and the non-availability of licensed software. Blockchain [11,29,30,37] and machine learning algorithms [4,13,27,31,34] can be integrated with digital forensics to provide privacy and security. Finally, network forensics should also be included in the study's scope, which we believe can significantly improve the results of artifacts collected from criminal investigations.

References

1. Abbasi, A., Javed, A.R., Chakraborty, C., Nebhen, J., Zehra, W., Jalil, Z.: Elstream: an ensemble learning approach for concept drift detection in dynamic social big data stream learning. IEEE Access **9**, 66408–66419 (2021)
2. Adautin, E.D., Ar, N.M.: Forensic reconstruction and analysis of residual artifacts from portable web browser. Int. J. Comput. Appl **128**(18), 19–24 (2015)
3. Afzal, S., Asim, M., Javed, A.R., Beg, M.O., Baker, T.: Urldeepdetect: a deep learning approach for detecting malicious URLs using semantic vector models. J. Netw. Syst. Manag. **29**(3), 1–27 (2021)
4. Agrawal, S., Sarkar, S., Srivastava, G., Maddikunta, P.K.R., Gadekallu, T.R.: Genetically optimized prediction of remaining useful life. Sustain. Comput. Inf. Syst. **31**, 100565 (2021)
5. Ahmed, W., et al.: Security in next generation mobile payment systems: a comprehensive survey (2021). arXiv preprint arXiv:2105.12097
6. Ahmed, W., Shahzad, F., Javed, A.R., Iqbal, F., Ali, L.: Whatsapp network forensics: discovering the IP addresses of suspects. In: 2021 11th IFIP International Conference on New Technologies, Mobility and Security (NTMS), pp. 1–7. IEEE (2021)
7. Al-Fannah, N.M., Mitchell, C.: Too little too late: can we control browser fingerprinting? J. Intell. Capital (2020)
8. Ar, N.M., et al.: Forensic evidence collection by reconstruction of artifacts in portable web browser. Int. J. Comput. Appl. **91**(4), 32–35 (2014)
9. Basit, A., Zafar, M., Javed, A.R., Jalil, Z.: A novel ensemble machine learning method to detect phishing attack. In: 2020 IEEE 23rd International Multitopic Conference (INMIC), pp. 1–5. IEEE (2020)
10. Basit, A., Zafar, M., Liu, X., Javed, A.R., Jalil, Z., Kifayat, K.: A comprehensive survey of AI-enabled phishing attacks detection techniques. Telecommun. Syst. **76**, 1–16 (2020)
11. Ch, R., Srivastava, G., Gadekallu, T.R., Maddikunta, P.K.R., Bhattacharya, S.: Security and privacy of UAV data using blockchain technology. J. Inf. Secur. Appl. **55**, 102670 (2020)
12. Dahlberg, R., Pulls, T., Ritter, T., Syverson, P.: Privacy-preserving & incrementally-deployable support for certificate transparency in tor. Proc. Priv. Enhanc. Technol. **2**, 194–213 (2021)
13. Deepa, N., Prabadevi, B., Srivastava, G.: Integrated ranking algorithm for efficient decision making. Int. J. Inf. Technol. Decis. Mak. **20**(02), 597–618 (2021)
14. Gabet, R.M., Seigfried-Spellar, K.C., Rogers, M.K.: A comparative forensic analysis of privacy enhanced web browsers and private browsing modes of common web browsers. Int. J. Electron. Secur. Dig. For. **10**(4), 356–371 (2018)

15. Hina, M., Ali, M., Javed, A.R., Gabban, F., Khan, L.A., Jalil, Z.: Sefaced: semantic-based forensic analysis and classification of e-mail data using deep learning. IEEE Access **9**, 98398–98411 (2021)
16. Horsman, G.: The challenge of identifying historic 'private browsing' sessions on suspect devices. For. Sci. Int. Dig. Invest. **34**, 300980 (2020)
17. Horsman, G., Errickson, D.: When finding nothing may be evidence of something: anti-forensics and digital tool marks. Sci. Just. **59**(5), 565–572 (2019)
18. Horsman, G., et al.: A forensic examination of web browser privacy-modes. For. Sci. Int. Rep. **1**, 100036 (2019)
19. Iqbal, F., Batool, R., Fung, B.C., Aleem, S., Abbasi, A., Javed, A.R.: Tweet-to-act: towards tweet-mining framework for extracting terrorist attack-related information and reporting. IEEE Access **9**, 115535–115547 (2021)
20. Iwendi, C., et al.: Keysplitwatermark: zero watermarking algorithm for software protection against cyber-attacks. IEEE Access **8**, 72650–72660 (2020)
21. Iwendi, C., Rehman, S.U., Javed, A.R., Khan, S., Srivastava, G.: Sustainable security for the internet of things using artificial intelligence architectures. ACM Trans. Internet Technol. (TOIT) **21**(3), 1–22 (2021)
22. Jadoon, A.K., Iqbal, W., Amjad, M.F., Afzal, H., Bangash, Y.A.: Forensic analysis of tor browser: a case study for privacy and anonymity on the web. For. Sci. Int. **299**, 59–73 (2019)
23. Javed, A.R., Baker, T., Asim, M., Beg, M., Al-Bayatti, A.H.: Alphalogger: detecting motion-based side-channel attack using smartphone keystrokes (2020)
24. Javed, A.R., Jalil, Z.: Byte-level object identification for forensic investigation of digital images. In: 2020 International Conference on Cyber Warfare and Security (ICCWS), pp. 1–4. IEEE (2020)
25. Javed, A.R., Rehman, S.U., Khan, M.U., Alazab, M., Khan, H.U.: Betalogger: Smartphone sensor-based side-channel attack detection and text inference using language modeling and dense multilayer neural network. Trans. Asian Low-Res. Lang. Inf. Process. **20**(5), 1–17 (2021)
26. Karami, S., Ilia, P., Polakis, J.: Awakening the web's sleeper agents: misusing service workers for privacy leakage. In: Network and Distributed System Security Symposium (NDSS) (2021)
27. Khan, H., et al.: Fake review classification using supervised machine learning. In: Del Bimbo, A., et al. (eds.) ICPR 2021. LNCS, vol. 12664, pp. 269–288. Springer, Cham (2021). https://doi.org/10.1007/978-3-030-68799-1_19
28. Kumar, P., et al.: PPSF: a privacy-preserving and secure framework using blockchain-based machine-learning for IoT-driven smart cities. IEEE Trans. Netw. Sci. Eng. **8**, 2326–2341 (2021)
29. Kumar, R., Kumar, P., Tripathi, R., Gupta, G.P., Gadekallu, T.R., Srivastava, G.: Sp2f: a secured privacy-preserving framework for smart agricultural unmanned aerial vehicles. Comput. Netw. **187**, 1078919 (2021)
30. Kumar, R., Tripathi, R., Marchang, N., Srivastava, G., Gadekallu, T.R., Xiong, N.N.: A secured distributed detection system based on IPFS and blockchain for industrial image and video data security. J. Para. Distrib. Comput. **152**, 128–143 (2021)
31. Maddikunta, P.K.R., Srivastava, G., Gadekallu, T.R., Deepa, N., Boopathy, P.: Predictive model for battery life in IoT networks. IET Intell. Transp. Syst. **14**(11), 1388–1395 (2020)
32. Mahlous, A.R., Mahlous, H.: Private browsing forensic analysis: a case study of privacy preservation in the brave browser (2020)

33. Malina, L., Srivastava, G., Dzurenda, P., Hajny, J., Fujdiak, R.: A secure publish/subscribe protocol for internet of things. In: Proceedings of the 14th International Conference on Availability, Reliability and Security, pp. 1–10 (2019)
34. Manoj, M., et al.: An incentive based approach for covid-19 planning using blockchain technology. In: 2020 IEEE Globecom Workshops (GC Wkshps), pp. 1–6. IEEE (2020)
35. Muhammad, A., Asad, M., Javed, A.R.: Robust early stage botnet detection using machine learning. In: 2020 International Conference on Cyber Warfare and Security (ICCWS), pp. 1–6. IEEE (2020)
36. Muir, M., Leimich, P., Buchanan, W.J.: A forensic audit of the tor browser bundle. Dig. Invest. **29**, 118–128 (2019)
37. Prabadevi, B., et al.: Toward blockchain for edge-of-things: a new paradigm, opportunities, and future directions. IEEE Internet Things Mag. **4**, 102–108 (2021)
38. Rahman, R.U., Tomar, D.S.: A new web forensic framework for bot crime investigation. For. Sci. Int. Dig. Invest. **33**, 300943 (2020)
39. Rasool, A., Jalil, Z.: A review of web browser forensic analysis tools and techniques. Researchpedia J. Comput. **1**, 15–21 (2020)
40. Rupa, Ch., Srivastava, G., Gadekallu, T.R., Maddikunta, P.K.R., Bhattacharya, S.: A blockchain based cloud integrated IoT architecture using a hybrid design. In: Gao, H., Wang, X., Iqbal, M., Yin, Y., Yin, J., Gu, N. (eds.) CollaborateCom 2020. LNICST, vol. 350, pp. 550–559. Springer, Cham (2021). https://doi.org/10.1007/978-3-030-67540-0_36
41. Said, H., Al Mutawa, N., Al Awadhi, I., Guimaraes, M.: Forensic analysis of private browsing artifacts. In: 2011 International Conference on Innovations in Information Technology, pp. 197–202. IEEE (2011)
42. Shabbir, M., et al.: Enhancing security of health information using modular encryption standard in mobile cloud computing. IEEE Access **9**, 8820–8834 (2021)
43. Srivastava, G., Deepa, N., Prabadevi, B., Reddy, M.P.K.: An ensemble model for intrusion detection in the internet of softwarized things. In: Adjunct proceedings of the 2021 International Conference on Distributed Computing and Networking, pp. 25–30 (2021)
44. Umar, R., Yudhana, A., Faiz, M.N.: Experimental analysis of web browser sessions using live forensics method. Int. J. Electr. Comput. Eng. **8**(5), 2951–2958 (2018)
45. Warren, A.: Tor browser artifacts in windows 10. SANS Inst. InfoSec Read, Room (2017)

Mining High Utility-probability Sequential Patterns in Bigdata Environments

Jimmy Ming-Tai Wu[1(✉)], Shuo Liu[1], Mu-En Wu[2], and Jerry Chun-Wei Lin[3]

[1] College of Computer Science and Engineering, Shandong University of Science and Technology, Shandong, China
wmt@wmt35.idv.tw

[2] Department of Information and Finance Management (IFM), National Taipei University of Technology, Taipei, Taiwan

[3] Department of Computer Science, Electrical Engineering and Mathematical Sciences, Western Norway University of Applied Sciences, Bergen, Norway
jerrylin@ieee.org

Abstract. High utility-probability sequential pattern mining (HUP-SPM) has become a novel research errand realm in data mining. In practical application, most of the data are collected by a variety of sensors, forming an uncertain database. Therefore, the data collected in real life can be expressed in terms of probability. In previous research work, for the sake of heightening the performance of mining in small data sets and single machine, a great deal of algorithm have been posed. Unfortunately, these algorithms have loopholes in mining large databases with mickle records. Therefore, it is very urgent to construct an efficient model to deal with large uncertain dateabas data sets. In this article, a MapReduce framework is moved to mine effective information in multitudinous uncertain data sets and reduce the number of database scans. Experiments on a large number of uncertain data sets show that the posed Hadoop algorithm has a strong ability to mine the high utility-probability sequential pattern.

1 Introduction

Valuable and useful information is obtained by knowledge discovery from databases (KDD) [2,3,5]. Firstly, association rule mining is proposed, which is called ARM [10]. Then, frequent itemsets mining(FIM), is proposed. Both of them have become an momentous part of KDD. Apriori algorithm's core is to quarry the frequently occurring data set in the data value. The frequent itemsets are mined through two stages: candidate set generation and downward closing detection of plots. And the algorithm has been wide-ranging exploit in commerce, network security and other signorys. Fp-Growth algorithm finds Frequent item

S.-C. Chu et al. (Eds.): ICGEC 2021, LNEE 833, pp. 505–514, 2022.
https://doi.org/10.1007/978-981-16-8430-2_46

sets by storing data sets in FP (frequent pattern) trees [10]. Compared with Apriori algorithm, Fp-Growth algorithm only needs to traversal the database twice to find Frequent item sets efficiently. The process of finding frequent itemsets by the algorithm is as follows: First, construct FP tree; Then mining frequent itemsets from FP Tree. However, the above-mentioned algorithms not consider the number of item sets, unit profit, weight and other factors.

For the purpose of settling the mishap of appeal, Chen proposed high-utility itemset mining (HUIM) [8,15,22,23] in combination with the number of itemsets and unit profit, which can mine more valuable information and is called high-utility itemset. Due to the HUIM structure has no downward closure property, it needs to pay a huge price to search valuable information in a large database. The high transaction weighted utility item set (HTWUI) can be found by cyphering the upper bound of item set. Through this method, you only need to search the database twice to find the HUI. IHUP [6], Up-Growth [20], HUP-growth [11], EFIM [14], HUIM-Miner [14] and FHM [27] have been proposed to mine efficient item sets in succession. However, this algorithm also has a limitation that it does not take into account the order of itemsets. Consider the order pattern $< (chocolate), (banana), (yogurt) >$, which indicates that the customer purchased chocolate, banana, and yogurt products in this order. That is to say, the HUIM model does not consider the order of itemsets. Sequential pattern mining (SPM) is proposed to mine the information between itemsets and develop business models better. However, the above contents are in a certain database, and the occurrence value of each itemset is accurate. In practical application, most of the data are collected by various intelligent sensors, forming an uncertain database. Therefore, the data gathered may be expressed by probability.

In daily behavior, large data sets collected by sensors and intelligent wireless devices are often uncertain. Lin et al. [12] proposed an efficient model to mine the high utility item sets of uncertain databases. Zhang et al. [25] added utility and sequence patterns to the uncertain database and resulted in high utility probability sequential pattern. Other mining algorithms [1,13,19,21] for uncertain applications in uncertain databases are constantly developing. Despite the above mentioned method in the practical application of the Internet of things joined the utility and uncertainty factors, but these methods only consider running a small data set on a single machine, as in real life, the data set collected by sensors and intelligent wireless devices is huge, will spend a lot of time to run on a single or even cannot be conducted on single machine processing. Hadoop is designed to process large data sets. For it can be executed in parallel, so it saves running time. Hadoop is composed of HDFS, MapReduce, YARN, Common and other components. HDFS is a distributed file storage system that provides data storage services for various distributed computing frameworks. When running a MapReduce computational task, the task process is divided into two phases: the Map phase and the Reduce phase, with each phase using key-value pairs as input and output. Based on MapReduce platform, the BIGHUSP algorithm is proposed to mine HUSP in large data sets. Experiments on a large number of uncertain data sets show that the algorithm has a strong ability to mine highly utility probability sequential pattern. The main contributions of this paper are as follows.

1. The method proposed in this paper runs on the Hadoop platform and mines HUPSPs from the distributed database, so it has the advantages of Hadoop in implementation. For example, Hadoop distributes data through available computer clusters, completes storage and computing tasks,with high scalability.
2. Since traditional algorithms are not suitable for treating large data sets, the method proposed in this paper performs multiple MapReduce tasks to search the original database, and proposed model will not carry out any repeated operations or modifications on the original database.
3. The proposed method divides the original database into several independent parts, thus ensuring the parallel execution of tasks on the large data set and reducing the database run time.

The remaining structure of this paper is as follows: Sect. 2 reviews the related work of proposed method. Section 3 introduces the definition of mining HUPSPs. Experimental results and summary are presented in Sect. 4 and Sect. 5, respectively.

2 Related Work

In this section, we will review some of the research related to proposed method.

2.1 Proposed Method

Agrawal et al. designed Sequence Pattern Mining (SPM) [4] and proposed some algorithms to mine Sequence Pattern Sets (SPs) from Sequence Database. Srikant et al. [18] proposed the GSP algorithm. Zaki et al. [24] proposed the Spade algorithm to find SP more effectively. However, these algorithms have some disadvantages, resulting in a large number of candidate sets. To deal with this limitation, Pei et al. designed the PrefixSpan [17] algorithm. Owing to the huge amount of uncertain data in life submissions, mining frequent itemsets (UFIM) in uncertain databases has turn into a split-new research direction in the signory of data mining. At present, it mainly consists of two models, including probability frequency model [7] and expectation support model [9]. In the first model, if the probability of an item set is not less than a predetermined minimum probability, it is called UFI. In the expectation support model, frequent itemsets mean that the expected support is higher than the pre-specified minimum support threshold.

Berbecker et al. [7] designed another model similar to the expectation support model for mining probability frequency patterns in uncertain databases. Tong et al. [19] contrasted the modi of the expected support model and the probabilistic frequency model. Muzammal. [16] et al. designed three algorithms related to expected support. Zhao et al. [26] proposed the U-PrefixSpan algorithm and designed two uncertain sequence models. The problem of efficient sequential pattern mining (HUSPM) has been studied extensively, and a sequential weighted utility (SWU) model has been proposed to maintain the properties of downward closures (DCs) during data mining. The SWU of the itemset is specified as the sum of all the sequence utilities that contain the itemset. To improve the mining speed of HUPSP, sequence weighted downclosure (SWDC) was used to prune inefficient sequences.

3 Preliminary and Problem Statement

Table 1. An uncertain quantitative sequence database

SID	Sequence	Probability
S_1	$< (a,4),(b,5),[(a,2),(c,2),(e,3)] >$	0.7
S_2	$< [(b,2),(c,2)],[(a,2),(d,2)],[(a,3),(b,3)] >$	0.8
S_3	$< (f,2),[(f,2),(d,2)],[(b,4),(c,2)] >$	0.5
S_4	$< (b,2),(c,2),[(a,2),(b,3)],[(a,5),(b,2)] >$	0.9

Table 2. A profit table

$Item$	profit
a	3
b	2
c	4
d	5
e	2
f	3

3.1 Preliminaries

Suppose $I = \{i_1, i_2, ..., i_m\}$ represents m different records, $USD = \{S_1, S_2, ..., S_n\}$ means that the uncertain quantitative sequence database contains n sequences, and each sequence can be uniquely represented and denoted as SID. Each sequence S_q in USD consists of several different sequences such that $S_q = \{I_1, I_2, ..., I_k\}$. Thus, we can have that $i_j \in I$, and $S_q \subseteq USD$. Table 1 shows the running examples in the original database, it contains the internal utility (purchase quantity,probability) of the sequence I_j in each sequences S_q.This example consists of six items and four sequences. Table 2 shows the unit profit $pr(i_j)$ of each item i_j. In the running example,The minimum utility threshold is denoted as ε, and the minimum expected support threshold is expressed as μ.

Definition 1. *Presuming the sequence embodies r items,it is a r-sequence.*

Like below, the sequence $\{<(a),(f),(b,f)>\}$ is a 4-sequence since it embodies 4 items.

Definition 2. *The utility of an item i_j in a itemsets I is denoted as $iu(i_j, I_k, S_q)$ and defined as:*

$$iu(i_j, I_k, S_q) = q(i_j, I_k, S_q) \times pr(i_j). \tag{1}$$

For example, in sequence S_4 can be represented as $\{I_1, I_2, I_3, I_4\}$. The internal utility of (a) in $\{I_3\}$ in $\{S_4\}$ are separately counted as: $\{iu(a, I_3, S_4)\} = 2 \times 3 = 6$.

Definition 3. *If $S \subseteq S_q$, The occurrence probability of sequence S is the same as sequence Sq.*

In Table 1, $<(b)> \subseteq S_1$ and $<(a),(d)> \subseteq S_3$. Moreover, sp$(S_1) = 0.7$, sp$(S_3) = 0.5$; sp$(<(b)>, S_1) = 0.7$ and sp $<(a),(d)>, S_3) = 0.5$.

Definition 4. *In an uncertain quantitative sequence database USD, the utility of a sequence S are defined as:*

$$su(S) = \sum_{S \subseteq S_q \wedge S_q \in USD} su(S, S_q). \tag{2}$$

Definition 5. *$SU(S_q)$ (sequence utility) expressed as:*

$$SU(S_q) = \sum_{i_j \in I_k \wedge I_k \in S_q} SU(i_j, I_k, S_q). \tag{3}$$

Definition 6. *The utility of a transaction in transaction database is denoted as $tu(T_q)$ and expressed as:*

$$tu(T_q) = \sum_{i_j \subseteq T_q} u(i_j, T_q). \tag{4}$$

Definition 7. *In a database USD, the total of utilities of all sequences incorporating S expressed as sequence weighted utility (SWU), that is:*

$$SWU(S) = \sum_{S \subseteq S_q \wedge S_q \in USD} SU(S_q). \tag{5}$$

For the running example, the *twu* of a sequence S= $<(b),(c)>$ can be ciphered as $SWU(S) = \text{SU}(S_2) + \text{SU}(S_3) + \text{SU}(S_4) = 43 + 38 + 43 = 124$.

Definition 8. *The total of the utilities of all sequences in USD expressed as TSU, scilicet the sum of the utility in sequence database USD:*

$$TSU = \sum_{S_q \in USD} SU(S_q). \tag{6}$$

In order to consider both the probability and utility of the sequences, High Utility-Probability Sequential Patterns (HUPSP) as follows:

Definition 9. *HUPSPs* $\leftarrow \{S \mid sp(S) \geq \mid USD \mid \times \mu \wedge su(S) \geq TSU \times \epsilon\}$

3.2 Problem Statement

Given a large-scale database and a profit table, according to the above circumscriptions, our trouble is to rummage high utility-probability sequential patterns (HUPSP) from a MapReduce-based framework.

4 Proposed a MapReduce Framework

Pertaining to this section, first proposed a MapReduce framework based on Hadoop platform. An algorithm is designed, which consists of three tasks that perform the MapReduce framework. The first task is called the initialization process, the second is called the Updating process, and the last task is called the mining process. After the processing of calculation and mining, the high utility probability sequence pattern is finally output. The proposed framework is shown in Fig. 1. The stages are described as follows:

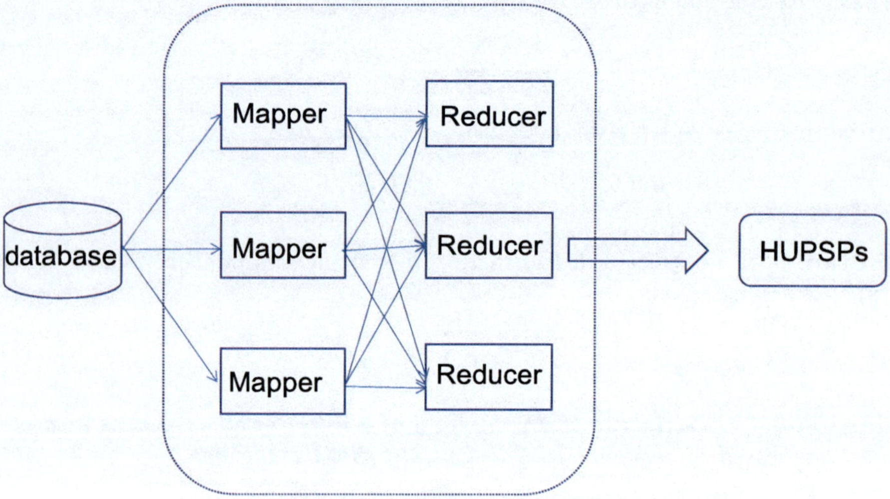

Fig. 1. The MapReduce framework

4.1 Initialization Phase

The original database as the input data are distributed several independent partition sub-databases and sent to each Mapper for processing, and a pair in the form of <key, value> is generated. The Reducer merges and calculates the pairs with the same key, and obtains the occurrence probability $sp(S)$ of each sequence S and its sequence weighted utility $SWU(S)$.

4.2 Updating Phase

In the previous stage, the *SWU* of each sequence is obtained. We update the sequence according to the *SWU* of each sequence, prune the 1-sequence that does not fulfil the minimum value of the weighted utility and probability of the sequence, and update the original database, so as to reduce the search space.

4.3 Mining Phase

At this stage, the algorithm finds the high utility probability sequence in parallel through multiple iterations. According to the results obtained in the previous stage, each sequence in the partitioned sub database is pruned. The sub databases of each partition are processed by MapReduce framework. Mapper continues to iterate until no candidate set is generated, and then finds the high utility probability sequence we need.

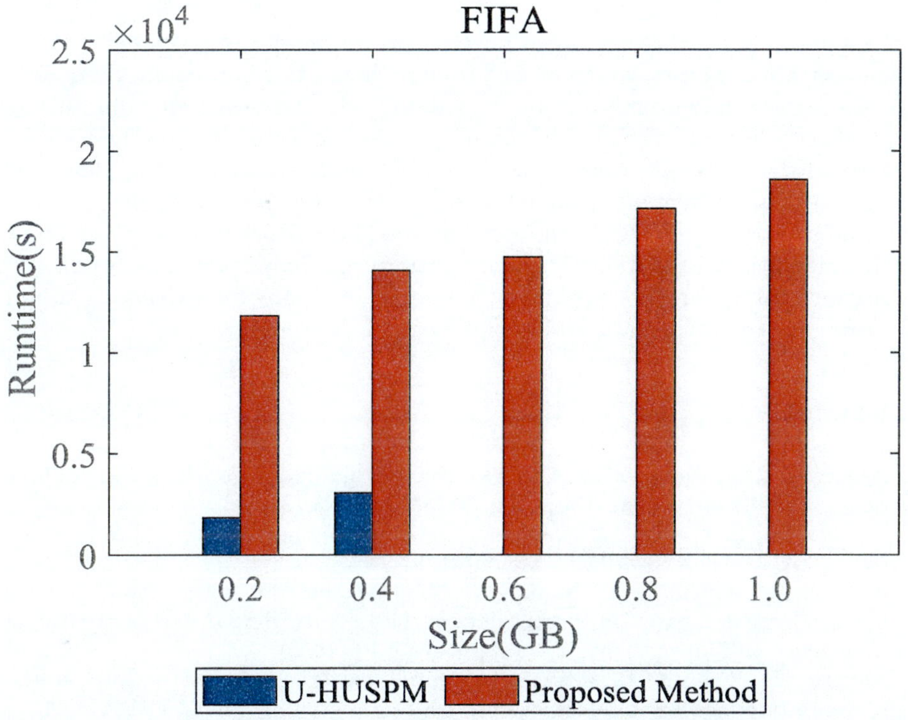

Fig. 2. Runtimes under different database sizes

5 Experimental Results

In this section, our experiment evaluates the running time of U-HUSPM algorithm under data sets of different extents. The experimental data set is replicated the basis of the authentic data set FIFA, and the data sets of 0.2, 0.4, 0.6, 0.8, 1.0GB are respectively formed. The experimental result is exhibited in the Fig. 2. Through pictures we can acquire that, as the extent of the data set is mounting, U-HUSPM algorithm on a single machine failed to get experimental results due to memory leakage, while the designed algorithm can still be executed.

From the experimental result, it can see that the algorithm posed in this thesis outshine the most advanced algorithms, especially in processing with large data sets. This is because when the data volume is large, the algorithm is executed on a Hadoop-based platform and the data set is divided into multiple partitions to process the data in parallel.

6 Conclusion and Future Work

In this paper, this algorithm based on Hadoop framework is proposed to treat large uncertain data sets, which is used to mine high utility probability sequence patterns, reduce the running time on large databases, and prune the unpromising candidate sets in time. The algorithm designed in this paper runs on the Hadoop platform, distributes data through the available computer cluster, completes the storage and computing tasks, and has high scalability. In the future, we will continue to optimize the algorithm based on the MapReduce framework and how to mine high utility probability sequence patterns on Spark platform. Spark and MapReduce framework are significant research directions for extracting useful info from a large extent data sets in the scope of data mining.

References

1. Aggarwal, C.C., Philip, S.Y.: A survey of uncertain data algorithms and applications. IEEE Trans. Knowl. Data Eng. **21**(5), 609–623 (2008)
2. Agrawal, R., Imielinski, T., Swami, A.: Database mining: a performance perspective. IEEE Trans. Knowl. Data Eng. **5**(6), 914–925 (1993)
3. Agrawal, R., Imieliński, T., Swami, A.: Mining association rules between sets of items in large databases. In: Proceedings of the 1993 ACM SIGMOD International Conference on Management of Data, pp. 207–216 (1993)
4. Agrawal, R., Srikant, R.: Mining sequential patterns. In: Proceedings of the Eleventh International Conference on Data Engineering, pp. 3–14. IEEE (1995)
5. Agrawal, R., Srikant, R., et al.: Fast algorithms for mining association rules. In: Proceedings of 20th International Conference Very Large Data Bases, VLDB, vol. 1215, pp. 487–499. Citeseer (1994)
6. Ahmed, C.F., Tanbeer, S.K., Jeong, B.S., Lee, Y.K.: Efficient tree structures for high utility pattern mining in incremental databases. IEEE Trans. Knowl. Data Eng. **21**(12), 1708–1721 (2009)

7. Bernecker, T., Kriegel, H.P., Renz, M., Verhein, F., Zuefle, A.: Probabilistic frequent itemset mining in uncertain databases. In: Proceedings of the 15th ACM SIGKDD International Conference on Knowledge Discovery and Data Mining, pp. 119–128 (2009)
8. Chan, R., Yang, Q., Shen, Y.D.: Mining high utility itemsets. In: Third IEEE International Conference on Data Mining, pp. 19–19. IEEE Computer Society (2003)
9. Chui, C.-K., Kao, B., Hung, E.: Mining frequent itemsets from uncertain data. In: Zhou, Z.-H., Li, H., Yang, Q. (eds.) PAKDD 2007. LNCS (LNAI), vol. 4426, pp. 47–58. Springer, Heidelberg (2007). https://doi.org/10.1007/978-3-540-71701-0_8
10. Han, J., Pei, J., Yin, Y., Mao, R.: Mining frequent patterns without candidate generation: a frequent-pattern tree approach. Data Min. Knowl. Disc. **8**(1), 53–87 (2004)
11. Lin, C.W., Hong, T.P., Lu, W.H.: An effective tree structure for mining high utility itemsets. Expert Syst. Appl. **38**(6), 7419–7424 (2011)
12. Lin, J.C.-W., Gan, W., Fournier-Viger, P., Hong, T.-P., Tseng, V.S.: Efficiently mining uncertain high-utility itemsets. Soft Comput. **21**(11), 2801–2820 (2016). https://doi.org/10.1007/s00500-016-2159-1
13. Liu, C., Chen, L., Zhang, C.: Summarizing probabilistic frequent patterns: a fast approach. In: Proceedings of the 19th ACM SIGKDD International Conference on Knowledge Discovery and Data Mining, pp. 527–535 (2013)
14. Liu, M., Qu, J.: Mining high utility itemsets without candidate generation. In: Proceedings of the 21st ACM International Conference on Information and Knowledge Management, pp. 55–64 (2012)
15. Liu, Y., Liao, W., Choudhary, A.: A two-phase algorithm for fast discovery of high utility itemsets. In: Ho, T.B., Cheung, D., Liu, H. (eds.) PAKDD 2005. LNCS (LNAI), vol. 3518, pp. 689–695. Springer, Heidelberg (2005). https://doi.org/10.1007/11430919_79
16. Muzammal, M., Raman, R.: Mining sequential patterns from probabilistic databases. In: Huang, J.Z., Cao, L., Srivastava, J. (eds.) PAKDD 2011. LNCS (LNAI), vol. 6635, pp. 210–221. Springer, Heidelberg (2011). https://doi.org/10.1007/978-3-642-20847-8_18
17. Pei, J., et al.: Mining sequential patterns by pattern-growth: the prefixspan approach. IEEE Trans. Knowl. Data Eng. **16**(11), 1424–1440 (2004)
18. Srikant, R., Agrawal, R.: Mining sequential patterns: generalizations and performance improvements. In: Apers, P., Bouzeghoub, M., Gardarin, G. (eds.) EDBT 1996. LNCS, vol. 1057, pp. 1–17. Springer, Heidelberg (1996). https://doi.org/10.1007/BFb0014140
19. Tong, Y., Chen, L., Cheng, Y., Yu, P.S.: Mining frequent itemsets over uncertain databases. arXiv preprint arXiv:1208.0292 (2012)
20. Tseng, V.S., Wu, C.W., Shie, B.E., Yu, P.S.: Up-growth: an efficient algorithm for high utility itemset mining. In: Proceedings of the 16th ACM SIGKDD International Conference on Knowledge Discovery and Data Mining, pp. 253–262 (2010)
21. Wang, L., Cheung, D.W.L., Cheng, R., Lee, S.D., Yang, X.S.: Efficient mining of frequent item sets on large uncertain databases. IEEE Trans. Knowl. Data Eng. **24**(12), 2170–2183 (2011)
22. Yao, H., Hamilton, H.J.: Mining itemset utilities from transaction databases. Data Knowl. Eng. **59**(3), 603–626 (2006)
23. Yao, H., Hamilton, H.J., Butz, C.J.: A foundational approach to mining itemset utilities from databases. In: Proceedings of the 2004 SIAM International Conference on Data Mining, pp. 482–486. SIAM (2004)

24. Zaki, M.J.: An efficient algorithm for mining frequent sequences. In: Machine Learning, Citeseer (2001)
25. Zhang, B., Lin, J.C.W., Fournier-Viger, P., Li, T.: Mining of high utility-probability sequential patterns from uncertain databases. PLoS ONE **12**(7), e0180931 (2017)
26. Zhao, Z., Yan, D., Ng, W.: Mining probabilistically frequent sequential patterns in large uncertain databases. IEEE Trans. Knowl. Data Eng. **26**(5), 1171–1184 (2013)
27. Zida, S., Fournier-Viger, P., Lin, J.C.-W., Wu, C.-W., Tseng, V.S.: EFIM: a highly efficient algorithm for high-utility itemset mining. In: Sidorov, G., Galicia-Haro, S.N. (eds.) MICAI 2015. LNCS (LNAI), vol. 9413, pp. 530–546. Springer, Cham (2015). https://doi.org/10.1007/978-3-319-27060-9_44

The Timeliness Position Recommendation Based on Geographical Impacts and Social Impacts

Baolian Li[1], Jianghua Zhu[1], Yufeng Wang[1(✉)], and Haiyao Xie[2]

[1] The 54th Research Institute of China Electronics Technology Group Corporation, Shijiazhuang 050002, China

[2] Beijing Institute of Technology, Beijing 100081, China

Abstract. The popularity of mobile phones and the rapid development of location-based social networks (LBSN) provide the rich data support for position recommendation. This paper presents a study of timeliness position recommendation by mining the time cycle of human geographical and social activities, and proposes to re-divide the geographical circle and social circle according to the time state. The geographical factors and social factors are quantified reasonably according to the space aggregation effect on the users' check-in and the user's behaviors. Finally, the recommendation model GSTS (Geographical, Social, Temporal, Spatial) is proposed, which integrates multi-geographical factors and multi-social factors by matrix decomposition technique. The experimental results show that the proposed model can achieve the best recommendation performance.

Keywords: Timeliness position recommendation · Geographic impacts · Social impacts · GSTS model

1 Introduction

The explosive growth of information makes it difficult for users to quickly find the relevant information. The collaborative filtering recommendation techniques usually have lacks of information about user's behavior which limits their learning and predictive ability for generating quality recommendations [1–3]. According to the spatial aggregation effect of user check-in and user behavior, the geographical factors and social factors can be reasonably quantified. In addition, the location based social network (LBSN) has emerged. Benefitting from the location based services, LBSN has successfully attracted billions of users. The massive social activity information provides rich data support for mining the user behavior patterns. To better support location-based services, it is significant to mine the user social behavior and give personalized location recommendation algorithm.

LBSN contains rich social and check-in information. Geographical information reveals the intrinsic relationships between positions, and social information reveals the intrinsic relationships between users, while check-in information reflects the real movement traces in the offline world. Intuitively, analysis and mining of geographical, social and check-in information can help recommender systems to recognize user

S.-C. Chu et al. (Eds.): ICGEC 2021, LNEE 833, pp. 515–526, 2022.
https://doi.org/10.1007/978-981-16-8430-2_47

behavior patterns better and provide more effective location recommendations. The main contributions are as follows:

(1) In order to introduce the time cycle characteristics in the recommendation model, the geographical position and social friends are divided into new circles according to the time state.
(2) A variety of geographical factors and social factors are researched, such as *Geographical preference, Similarity of subject, Authority, and Proximity of intimacy*. At the same time, we put forward reasonable calculation methods to quantify the above factors, which are injected into the recommendation model to reflect the impacts of geographical factors and social factors to users.
(3) A new personalized position recommendation model GSTS is presented, which incorporates geographical and social characteristics, and to some extent reduces the limitations of the matrix's sparsity and cold start problem on the performance of recommendation [2];
(4) The matrix decomposition technique is used to study the hidden features of the user and the position, and obtain the prediction matrix by the matrix multiplication. The experimental results show that the GSTS model can improve the performance of recommendation on real data sets.

2 Related Work

In this section, we highlight related work about location recommendations in LBSNs. The Model-based recommendation methods [4–10] construct a recommendation model with user preference data and learn model parameters from training, in which ontology can be used to learn the relationship among parameters. As the main time spent in the training phase, once the model parameters are determined, they can be quickly used to recommend during the application phase.

Jamali et al. [11] studied the widespread communication mechanism in sociology and proposed the SocialMF recommendation model. The model considered the social factors, and believed that the user preference is influenced by the trust of the friends. Therefore, the eigenvector of the direct social friends is used to calculate the hidden vector. The experiment carried out on the real data set proved the validity of the method.

In social network, user's score can clearly reflect the user's preference [12]. Ye, Yin et al. [13] studied user enrollment data to explore user preferences, social impacts, and geographic impacts. In order to overcome data sparsity, they used a collaborative filtering mechanism based on user and project-based hybridization.

Cho, Cheng, Gao et al. [14–16] analyzed the user history attendance data, and found that the user's activity is usually around the vicinity of the residence. In history, about 20% signing positions distance distribution is less than 1 km, 60% between 1 km and 10 km, which indicated that users always prefer to access position near the residence. The time information of behavior is an important clue to infer the rules of user's activity, and it is helpful to improve the accuracy of recommendation by using the time state [17–19]. Literature [18] is used to analyze the characteristics of twenty million positions of more than twenty thousand users, and it is found that the user mobility pattern has obvious time periodicity.

However, the present methods are not very good to consider geographic impact, social impact, and time period effects together. This paper will present a timeliness position recommendation model based on the geographical impacts and social impacts.

3 Divide Circle Based on Time State

3.1 Division of the Geographical Circle

Based on the analysis of user history attendance data [18], this paper proposes a geographical distance variable D and $C(l_j)$ which represents the adjacent position of the position l_j. If the recommended position is not in the user's current position, the position l_j will not be considered and $C(l_j)$ will be ignored.

3.2 Construct Geographical Circle Based on Time State

This paper divides the day into 24 time states, and uses the geography information of check-in data as the input data to construct the geographical circle based on the time state. The core idea is that the geographical position shows different frequency of checking in different time states. Therefore, the original geographical circles are divided into multiple sub-geosphere, and each sub-geosphere corresponds to a separate time state.

Definition 1. Social circle based on time state (LC_l^t): A circle satisfies all of the following two conditions in a given time state t:

(1) There is a direct geographical relationship between the position x and y in the original geography circle, that is $LN_{x,y} = 1$;
(2) $|L_x^t| > 0$ && $|L_y^t| > 0$, which means that the position x and y have been checked in the time state t, and the double vertical line symbol refer to the cardinal number.

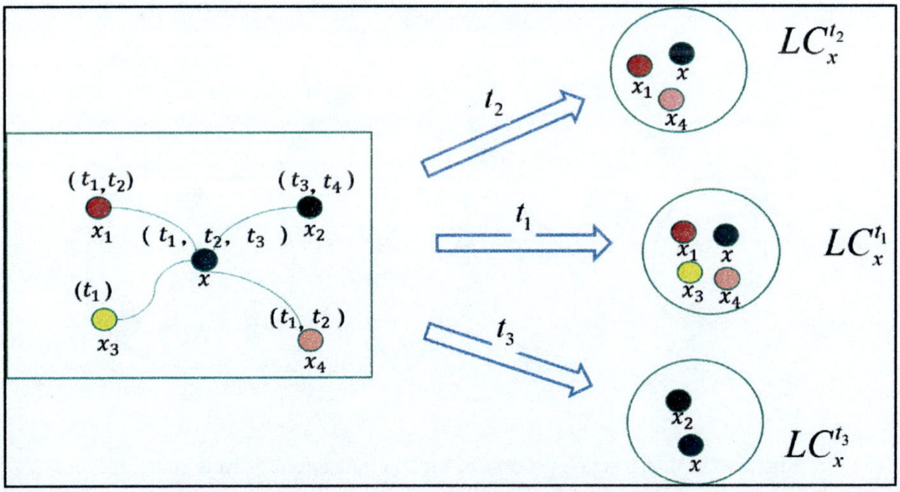

Fig. 1. The construction process of geographical circle based on time state

If y satisfies the above conditions in a certain time state t, it is recorded as $y \in LC_x^t$, otherwise $y \notin LC_x^t$.

In summary, this paper proposes a time-based geographic circle by dividing the original position network using the time-period characteristics. It should be noted that the following discussion of geographical factors are limited to the geographical circle that in specific time state. Figure 1 shows the construction process of geographical circle based on time state, of which x and t indicate position and time state respectively.

3.3 Construct Social Circle Based on Time State

The original social network SN which contains all social relations is divided into multiple sub-social networks. Each sub-social network corresponds to a separate time state, denoted as SN^t. The following is the definition of time-based social circle:

Definition 2. Time-based social circles (SC_u^t): All users who meet the following two conditions constitute the social circle of user u in a given time state.

(1) There is a direct social relationship between users u and v in the original social network, that is $SN_{u,v} = 1$;
(2) $|L_u^t| > 0$ && $|L_v^t| > 0$, which means that the user u and user v have been checked in the time state t, and the double vertical line symbol refer to the cardinal number.

Figure 2 shows the construction process of social circle based on time state.

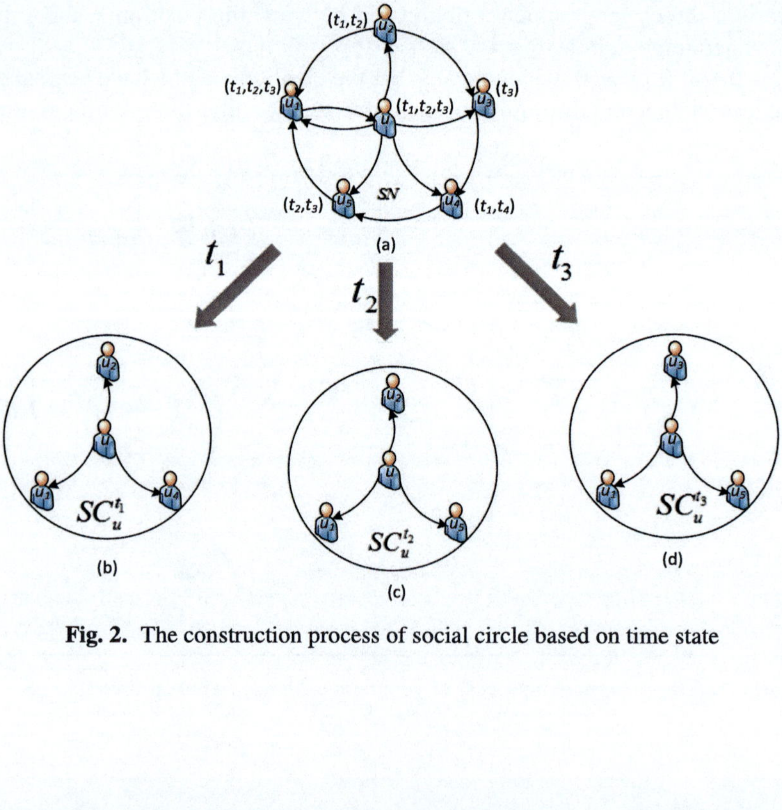

Fig. 2. The construction process of social circle based on time state

4 Geographical and Social Influence

4.1 Geographical Preference

Definition 3 (Q_l^t): The geographical preference is the preference distribution of the user u, denoted as D_u^t, to the subject of position at a certain time t.

Calculation:

Step 1: In this paper, we use the symbol D_l to represent the subject distribution vector for position l:

$$D_l = [I_{tp_1}, I_{tp_2}, \ldots, I_{tp_n}] \tag{1}$$

Where I_{tp_n} represents the subject identifier, and I_{tp_n}=1 if the position l belongs to the subject tp_n, otherwise 0. n is the number of position subject.

Step 2: In order to obtain the personal preference distribution, the check-in frequency matrix CI^t is calculated and normalized firstly. Check in frequency matrix CI^t, which refers to *check-in frequency* matrix for some certain subject, can be obtained by querying the value of the user's check-in on the corresponding position, and then the CI^t is normalized as the weight of the subject preference over the *check-in frequency matrix* for all subjects, denoted as $CI_{u,l}^{t*}$.

$$CI_{u,l}^{t*} = \frac{CI_{u,l}^t}{\sum\limits_{l \in L_u^t} CI_{u,l}^t} \tag{2}$$

We get the personal preference distribution vector D_u^t by the weighted average of position subjects that are checked and the normalized CI^t, the specific formula is as follows:

$$D_u^t = \sum_{l \in L_u^t} CI_{u,l}^{t*} D_l \tag{3}$$

Finally, geographical preference $Q_{u,l}$ that between the user preference distribution and the subject distribution of the position is calculated as Possibility of potential user u's accessing to position l:

$$Q_{u,l} = \cos in(D_u, D_l) \tag{4}$$

4.2 Similarity of Subject

Definition 4 (T_l^t): The similarity of subject is obtained by calculating Cosine similarity between two positions, denoted as T_l^t.

Calculation: A cosine similarity between two positions is calculated by the subject distribution vector D_l.

$$T_{x,y} = \cos in(D_x, D_y)$$

4.3 Authority

Definition 5 **($S_{u,v}$):** It is the degree of recognition that user u to user v in a given environment.

Calculation: Trust based on expert level is employed to calculate the authority, and the expert level of the target users can be measured by reverence that other users to him in a given time. This assumption is based on the PageRank algorithm proposed for calculating the importance of web pages by Google. Specifically, the symbol EL_u^t is used to represent the expert level of the user u in the temporal state t and it is calculated using the following formula:

$$EL_u^t = (1-d)N_u^t + d\sum_{v \in SC_u^t} \frac{EL_v^t}{M_v^t} \tag{5}$$

Where N_u^t is the check-in times of the user u in the time state t, M_v^t is the number of friends in the social circle of the user v in the time state t, and d is the attenuation factor.

The calculation of the expert level of the user is carried out under the premise of determining the time state, and expert level of the user needs to be calculated separately. Once the user expert level is calculated at different time, the authority can be assigned to each friend in the circle, and the assignment of the authority value is proportional to the expert level:

$$S_{u,v}^t = \begin{cases} EL_v^t, & \text{if } v \in SC_u^t \\ 0, & \text{otherwise} \end{cases} \tag{6}$$

4.4 Proximity of Intimacy

Definition 6 **($F_{u,v}^t$):** The degree of proximity of intimacy represents the degree of intimacy between the user's social links.

Calculation: In this paper, the sociology theory was borrowed to measure the proximity of intimacy, and the number of public friends that in the social circle is taken to calculate the degree of intimacy $M_{u,v}$ between two friends, that is:

$$M_{u,v} = \begin{cases} 1 + |SN_u \cap SN_v|, & \text{if } SN_{u,v} = 1 \\ 0, & \text{otherwise} \end{cases} \tag{7}$$

Where $|SN_u \cap SN_v|$ represents the number of public friends between user u and user v in the original social network. The degree of intimacy is equal to $1 + |SN_u \cap SN_v|$ when the two users have a direct social relationship.

Based on the above analysis, the following formula is used to measure the proximity of intimacy between users:

$$F_{u,v}^t = \begin{cases} \frac{M_{u,v}}{d_{u,v}^t}, & \text{if } v \in SC_u^t \\ 0, & \text{otherwise} \end{cases} \tag{8}$$

$d_{u,v}^t$ is the spatial distance between users' center position under time t.

4.5 Position Recommendation Model Based on Matrix Decomposition

We propose a GSTS model with geographical factors and social factors according to the spatial data, temporal data and social networks in LBSN. This model injects geographical factors and social factors, such as *Geographical preference, Similarity of subject, Authority, and Proximity of intimacy.* The user and the position of the hidden feature matrix are learned by the matrix decomposition technique, and potential positions that are checked by users possibly are recommended to users by the matrix multiplication.

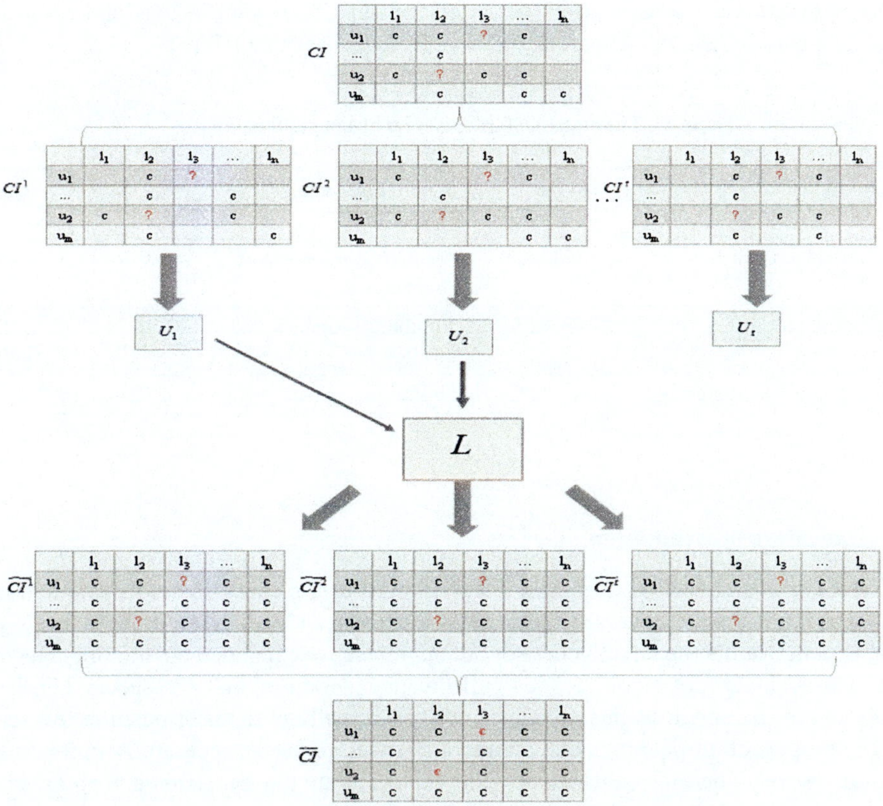

Fig. 3. The framework of position recommendation system

Figure 3 shows the framework of the position recommendation model, where "c" is the user's check-in on the corresponding position, and "?" is the user preference that needs to be predicted. First, the recommendation model takes the users' check-in frequency matrix *CI* as the input. Second, the hidden matrix of the user and the hidden matrix of position are obtained through the learning algorithm. Specifically, the users' check-in matrix is decomposed into the user hidden feature matrix $U^t \in \mathbb{R}^{n \times k}$ and the position hidden feature matrix $L \in \mathbb{R}^{m \times k}$ (k is the vector dimension of the hidden space, and inverted "!" refers to a general matrix with n rows and k columns or m rows and k columns.) by matrix decomposition technique that take geographical factors and social factors as constraints. Finally, the users' check-in preference to the position are predicted by the matrix multiplication, and the recommended result $\widetilde{CI}$ can be obtained.

5 Experiment

5.1 Data Set

In this paper, we use Foursquare data set that is released by Gao, Tang et al. [18]. The Foursquare data set contains user social relationships, user sign-in logs and position topic information, which includes 4163 users from California (Table 1).

Table 1. The statistical information of Foursquare dataset

User number	4163	Number of social connections	32512
Number of positions	121142	Average number of social connections	8
Number of records	483813	Number of topics in the position	36
Average number of signings	116	Sparse degree of signings	5.34×10^{-4}

5.2 Indicators of Evaluation

We select 80% and 60% as a training set from each user's check-in records randomly in different time states. Generally, 80%/20% of training and validation data are used. In order to demonstrate the effectiveness of our approach, we expanded the training data to 60%. The evaluation of recommended performance is focused on two aspects: (1) The proportion of the positions that are checked by user really in the user position test set; (2) The proportion of positions that are recommended to users successfully in the user position test set. Therefore, this paper will adopt accurate rate $precision@N$ and recall rate $recall@N$ as the evaluation criteria of the above two aspects:

$$precision@N = \frac{\sum_{u_i \in U} |TopN(u_i) \cap L(u_i)|}{\sum_{u_i \in U} |TopN(u_i)|} \tag{9}$$

$$recall@N = \frac{\sum_{u_i \in U} |TopN(u_i) \cap L(u_i)|}{\sum_{u_i \in U} |L(u_i)|} \tag{10}$$

Where $L(u_i)$ represents the set of positions where the user u_i has checked in the test set, and the number of positions in the set varies with the user. $TopN(u_i)$ is the set of top N positions that are generated by the GSTS model. In this paper, we designed $N = 5$ and $N = 10$ respectively, and compared the recommended performance under the two values. The geographical factors and social factors are assigned the same weight respectively.

We compared our proposed GSTS model with other three recommended models in accuracy and recall rates. It is important to note that the accuracy and recall rates that are obtained using the recommendation model are usually not high in absolute values because the check-in data in the location-based social network is usually very sparse, which is normal. Figure 4 shows the performance of each position recommendation model in different conditions.

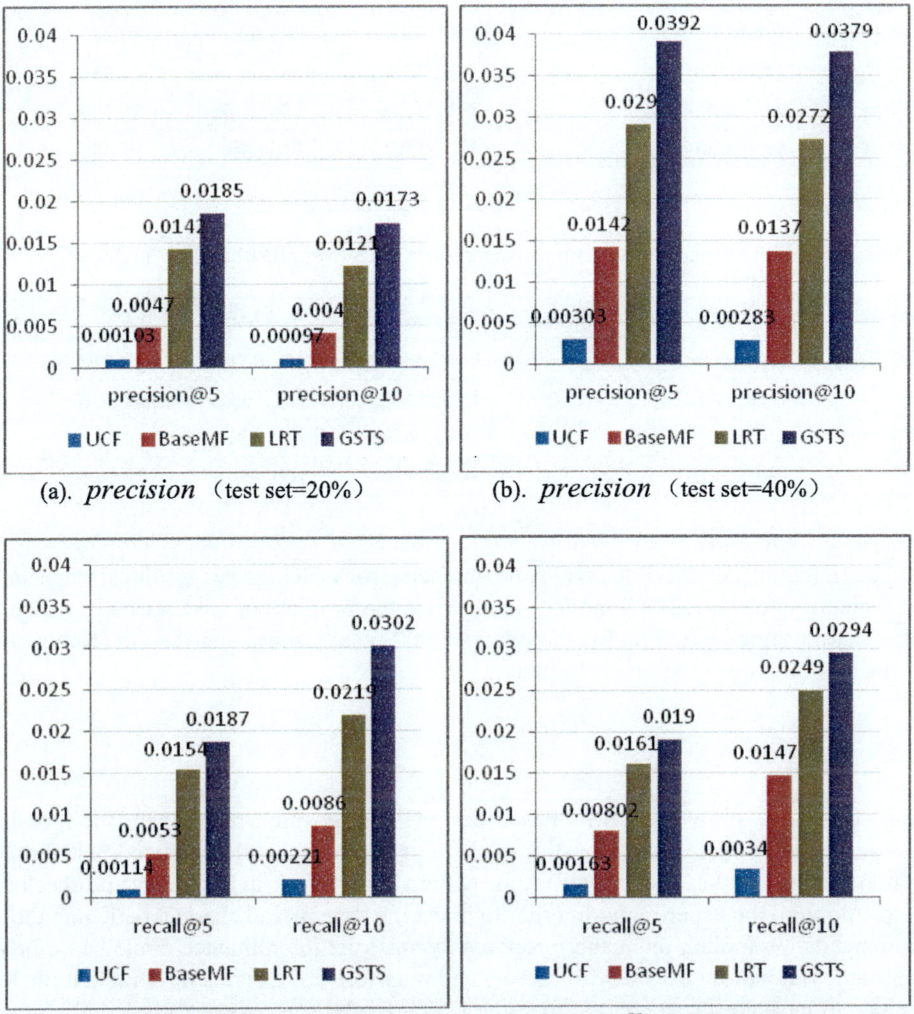

(a). *precision* (test set=20%) (b). *precision* (test set=40%)

(c). *recall* (test set = 20%) (d). *recall* (test set = 40%)

Fig. 4. Experimental results of our method and other algorithms

It can be seen from the experimental results that our proposed GSTS recommendation model can achieve the highest accuracy and recall rate in each condition, proving the effectiveness of the GSTS model in position recommendation. Tables 2 and 3 show the relative precision improvements and the relative recall improvements of GSTS over other baseline models respectively.

Table 2. The relative precision improvements of GSTS over other baseline models

Test set	Indicators of evaluation	UserCF	BaseMF	LRT
20%	precision@5	1696.12%	293.62%	30.28%
	precision@10	1683.51%	332.5%	42.98%
40%	precision@5	1193.73%	176.06%	35.17%
	precision@10	1252.5%	176.64%	39.33%

Table 3. The relative recall improvements of GSTS over other baseline models

Test set	Indicators of evaluation	UserCF	BaseMF	LRT
20%	recall@5	1540.35%	252.83%	21.42%
	recall@10	1266.51%	251.16%	37.90%
40%	recall@5	1065.64%	139.40%	19.25%
	recall@10	764.71%	100%	18.07%

Due to the introduction of time-cycle characteristics such as geographical impacts and social impacts, the GSTS model can alleviate the problem of cold start and sparse of the matrix effectively. Finally, the experimental results show that the GSTS model has the best recommendation performance.

6 Conclusion

In this paper, we study the LBSN-based personalized position recommendation, and focus on the time cycle characteristics of the user's geographical and social behavior. We find that the user has shown a clear time pattern in the check-in frequency and check-in topic through the experiment. In order to inject the time period characteristic into the recommendation model, this paper proposes to construct the influence circle based on time state. We explore a variety of factors and puts forward a series of computational methods to measure these factors, then the GSTS model was proposed by combining the above factors. The final experimental results show that the GSTS model can provide effective position recommendation and improve the performance obviously.

Acknowledgments. This work was funded in part by the Development Foundation of the 54th Research Institute of China Electronics Technology Group Corporation under Grant SXX18138X017.

References

1. Koren, Y.: Collaborative filtering with temporal dynamics. In: Proceedings of KDD 2009, vol. 53, no. 4, pp. 89–97 (2009)
2. Natarajan, S., Vairavasundaram, S., Natarajan, S., Gandomi, A.H.: Resolving data sparsity and cold start problem in collaborative filtering recommender system using Linked Open Data. Expert Syst. Appl. **149**, 113248 (2020)
3. Singh, P.K., Pramanik, P.K.D., Choudhury, P.: An improved similarity calculation method for collaborative filtering-based recommendation considering neighbor's liking and disliking of categorical attributes of items. J. Inf. Optim. Sci. **40**(2), 397–412 (2019)
4. Sitkrongwong, P., Maneeroj, S., Takasu, A.: Latent probabilistic model for context-aware recommendations. In: 2013 IEEE/WIC/ACM International Joint Conferences on Web Intelligence (WI) and Intelligent Agent Technologies (IAT), pp. 95–100. IEEE Computer Society (2013)
5. Qi, L., Enhong, C., Hui, X., et al.: Enhancing collaborative filtering by user interest expansion via personalized ranking. IEEE Trans. Syst. Man Cybern. Part B Cybern. **42**(1), 218–233 (2012). A Publication of the IEEE Systems Man & Cybernetics Society
6. Chen, Y., Canny, J.F.: Recommending ephemeral items at web scale. In: International ACM SIGIR Conference on Research & Development in Information Retrieval, pp. 1013–1022. ACM (2011)
7. Liu, H., Amin, M., Yan, B., et al.: Generating supplemental content information using virtual profiles. In: Proceedings of the 7th ACM Conference on Recommender Systems, pp. 295–302. ACM (2013)
8. Agarwal, D., Gurevich, M.: Fast top-k retrieval for model based recommendation. In: Proceedings of the fifth ACM International Conference on Web Search and Data Mining, pp. 483–492. ACM (2012)
9. Xue, X., Chen, J.: Optimizing ontology alignment through hybrid population-based incremental learning algorithm. Memetic Comput. **11**(2), 209–217 (2019)
10. Xue, X., Jiawei, L.: A compact brain storm algorithm for matching ontologies. IEEE Access **8**, 43898–43907 (2020)
11. Jamali, M., Ester, M.: A matrix factorization technique with trust propagation for recommendation in social networks. In: RecSys, pp. 135–142 (2010)
12. Yu, Y., Chen, X.: A survey of point-of-interest recommendation in location-based social networks. In: Workshops at the Twenty-Ninth AAAI Conference on Artificial Intelligence (2015)
13. Ye, M., Yin, P., Lee, W.C., et al.: Exploiting geographical influence for collaborative point-of-interest recommendation. In: Proceedings of the 34th International ACM SIGIR Conference on Research and Development in Information Retrieval, pp. 325–334. ACM (2011)
14. Cho, E., Myers, S.A., Leskovec, J.: Friendship and mobility: user movement in location-based social networks. In: Proceedings of the 17th ACM SIGKDD International Conference on Knowledge Discovery and Data Mining, pp. 1082–1090. ACM (2011)
15. Cheng, C., Yang, H., King, I., et al.: Fused matrix factorization with geographical and social influence in location-based social networks. In: Twenty-Sixth AAAI Conference on Artificial Intelligence (2012)

16. Gao, H., Tang, J., Liu, H.: Exploring social-historical ties on location-based social networks. In: ICWSM (2012)
17. Cheng, Z., Caverlee, J., Lee, K., et al.: Exploring millions of footprints in location sharing services. In: ICWSM 2011, pp. 81–88 (2011)
18. Gao, H., Tang, J., Hu, X., et al.: Modeling temporal effects of human mobile behavior on location-based social networks. In: Proceedings of the 22nd ACM International Conference on Information & Knowledge Management, pp. 1673–1678. ACM (2013)
19. Gao, H., Tang, J., Hu, X., et al.: Exploring temporal effects for location recommendation on location-based social networks. In: Proceedings of the 7th ACM Conference on Recommender Systems, pp. 93–100. ACM (2013)

Smart Grid and Energy Internet

Division of Power Grid Dispatching Domain and Analysis of Operation Characteristics

Shunjiang Wang[1](✉), Yifan Wang[2], Tianhang Wang[2], Jiancheng Hao[3], Weichun Ge[1], and Aihua Wang[1]

[1] State Grid Liaoning Electric Power Co., Ltd., Shenyang 110000, LN, China
[2] School of Electrical Engineering, Northeast Electric Power University, Jilin 132012, China
[3] Electric Power Research Institute of State Grid Liaoning Electric Power Co., Ltd., Shenyang 110000, LN, China

Abstract. In order to alleviate the problems of unstable operation and complex dispatching caused by some regions have large precentage of renewables, this paper proposes a voltage stability L index and operation domain division index on top of random power current calculation. On the basis of ensuring the voltage stability of grid with large precentage of recycled energy, the power grid is divided into three dispatching domains: general, exception and urgent domains. Finally, a simulation example is carried out on the basis of real operated figures of regional network in Liaoning province to simulate the operated features of different network dispatching domains, and verify the effectiveness and correctness of the proposed index.

Keywords: Renewable energy · Stochastic power flow calculation · Voltage stability L index · Network dispatching domain

1 Introduction

With the convening of the "fourteenth five year plan", the share of electric power generated by renewable energy sources in the network will rise rapidly [1]. However, because of the episodic and stochastic quality of the renewal energy, which has a significant influence on the voltage reliability on the power grid, and brings a series of challenges to the coordinated dispatching of power, load and storage in the power grid [2, 3], which greatly increases the complexity of dispatching. Therefore, how to realize the reasonable division of network scheduling domains have already been amongst the topical issues for specialists and academics in terms of ensuring the voltage stability of power grid.

The influence of uncertainty factors on network currents has been intensively studied by numerous authors. A probability model of stationary voltage stability of electricity transmission systems on the basis of random currents and static domain of voltage stability region is presented in document [4]. Voltage possibilities instability was evaluated. The paper [5] improves successive tide algorithm for wind farms, then analyzes the static voltage stability of the system accurately. The paper [6] the uncertainty factor of the static voltage stability margin for wind power systems is evaluated. In this paper,

S.-C. Chu et al. (Eds.): ICGEC 2021, LNEE 833, pp. 529–537, 2022.
https://doi.org/10.1007/978-981-16-8430-2_48

an improved algorithm for continuous current has been taken into account [7]. A load clustering based stochastic power flow model has been developed to provide an analysis of static voltage stability. At the same time, with the wide application of high proportion renewable energy, the source load memory will also serve as a resource for system scheduling. Thus, the features of all kinds of dispatching resources shall be taken into full consideration, as well as optimal dispatching scheme of power system is achieved from a holistic point of view in order to achieve coordinated interaction between source and load storage, and maximize the utilization of system resources [8].

Aiming at the uncertain factors of renewable energy, this paper proposes a voltage stability L index and power grid operation domain division index based on stochastic power flow calculation. To ensure the voltage stability of network, the grid scheduling domain is divided and the operation characteristics of different dispatching domains are analyzed.

2 Stochastic Power Flow Model

2.1 Stochastic Power Flow Simulation for Wind Power Systems

High percentage of wind grid access brings more stochasticity to the system, and the system state is more complex. Stochastic power flow was first studied in the 1970s, which was proposed by B. Borkowska [9]. It is a special power flow problem considering a variety of random variables. It can analyze the probability distribution information of system state variables, so it can more accurately and comprehensively analyze the system operation characteristics.

Wind power generally obeys the Weibull distribution to output its conditional conditional odds ratio development function:

$$f(v) = (\frac{k}{c})(\frac{v}{c})^{k-1} \exp(-\frac{v}{c})^k \tag{1}$$

In the above equation, v means wind speed, c means scale factor, k means shapefactor.

The distribution is shown in Fig. 1 for the range of wind energy export.

$$P_{WT} = \begin{cases} 0 & v \leq v_{ci} \\ k_1 v + k_2 & v_{ci} < v \leq v_r \\ P_N & v_r < v \leq v_{co} \\ 0 & v_{co} < v \end{cases} \tag{2}$$

In the above equation, P_N is the nominal power for the fan, v_r is the nominal air speed, v_{ci} is the access air speed for the fan, and v_{co} is the cut-off air speed, $k_1 = P_N/(v_r - v_{ci})$; $k_2 = -k_1 v_{ci}$.

In the study of stochastic power flow with wind power, the linearized AC power flow model is generally considered:

$$\begin{cases} \Delta X = S_0 \Delta W \\ \Delta Z = T_0 \Delta W \end{cases} \tag{3}$$

Where $\triangle X$, $\triangle Z$ and $\triangle W$ are the random fluctuations of system node voltage, branch power flow and system node injection power relative to the reference point respectively, and S_0 and T_0 represent the matrix of nodal village voltage and branch tide sensitivity to the changes of each node injection power respectively.

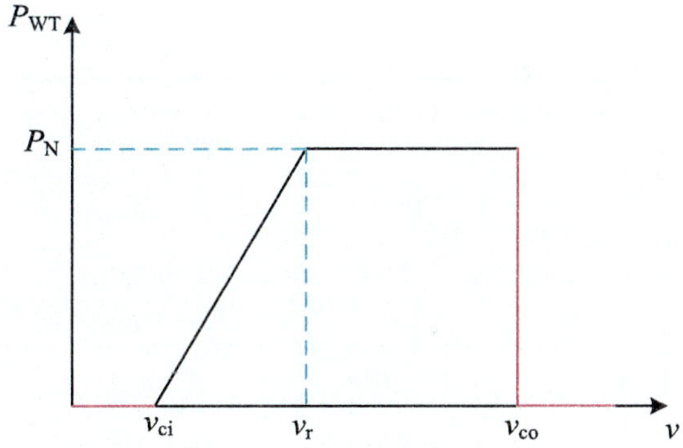

Fig. 1. Wind power curve

2.2 Stochastic Power Flow Method for Power System with Wind Power

So far, stochastic power flow problem has been put forward, and its solving methods are divided into three categories: approximate method, simulation method and analytical method [10]. This article uses the MC method and semi invariant method are used to calculate the stochastic power flow, which not only improves the calculation efficiency, but also obtains the calculation accuracy. The algorithm flow is shown in the figure.

3 Stochastic Voltage Stability Assessment Under Uncertainty

3.1 Random Sampling of Uncertainty of Load Composition

The load is one of the essential units of the electrical system. Different load models have different effects on the outcome of voltage steady phase method of analysis. The commonly used comprehensive load type used in voltage steady phase method of analysis is load type:

$$\begin{cases} P = a_p V^2 + b_p V + c_p \\ Q = a_Q V^2 + b_Q V + c_Q \end{cases} \tag{4}$$

Where a, b and c are power coefficients, which respectively represent the proportion of fixed electrical load, fixed current load and fixed impedance load respectively.

The power coefficient is sampled and analyzed by Monte Carlo method. Firstly, it is possible to determine the initial values of active and reactive loads from the initial operating values of the system, then power coefficient in the comprehensive load model is determined by sampling. Although the composition of load will change with the uncertainty factors such as weather, temperature and user, some coefficients will only fluctuate within 0–1 range The power coefficient is distributed evenly in 0–1 to obtain different load ratio models.

3.2 Random Sampling of Load Level Inaccuracy

Affected by weather and temperature, power system load forecasts are also very uncertain. The probability density functions of load active and reactive power meet the normal distribution as follows:

$$\begin{cases} f = (P) = \dfrac{1}{\sqrt{2\pi}\sigma_{BP,i}} \exp(-\dfrac{(P - \mu_{BP,i})^2}{2\sigma_{BP,i}^2} \\ f = (Q) = \dfrac{1}{\sqrt{2\pi}\sigma_{BQ,i}} \exp(-\dfrac{(P - \mu_{BQ,i})^2}{2\sigma_{BQ,i}^2} \end{cases} \tag{5}$$

Where, $\mu_{BP,i}$, $\mu_{BQ,i}$, ${\sigma^2}_{BP,\ i}{\sigma^2}_{BQ,\ i}$ can be obtained from the previous power system experience.

4 Voltage Stability L Index Calculated on the Basis of Stochastic Power Flow

Power system with high stochasticity and therefore large errors in the deterministic assessment results. This artical considers unknown elements and introduces stochastic currents, presenting a voltage stability L index based on random current calculations.

This equation in the power of the knots in polar co-ordinates can be composed of the matrix:

$$W = F(X, Y) \tag{6}$$

In the above equation, W represents injected power, X means state variable, Y means network structure parameter.

The above formula is expanded by Taylor series in the ground state:

$$\begin{aligned} W &= W_0 + \Delta W \\ &= F(X_0, Y_0) + F_x'(X_0, Y_0)\Delta X + F_y'(X_0, Y_0)\Delta Y \end{aligned} \tag{7}$$

When calculating the L index of mean voltage stability, X is expressed as the node voltage.

Assuming that the network structure parameter Y is constant, there are:

$$\Delta W = F_x'(X_0, Y_0)\Delta X \tag{8}$$

In the rectangular coordinate system, it is:

$$W = [P_1\ Q_1 \ldots P_{(n-1)}\ Q_{n-1}]^T \tag{9}$$

$$V = \left[e_1 f_1 \ldots e_{n-1} f_{n-1}\right]^T \tag{10}$$

In the formula, W is obtained by Monte Carlo sampling calculation, and all uncertainties are transformed into random disturbance of node injection power. Deterministic power flow calculation is carried out for each system data obtained by Monte Carlo sampling, and then according to the distribution of input variables, the distribution functions of node voltages and branch currents are acquired using the semi-invariant method combined with the gram Charlie series, based on which the L-index of voltage stability based on random current computation is proposed.

In formula (6), injection power factor λ is obtained as follows:

$$\frac{dW}{d\lambda} = J\frac{dV}{d\lambda} \tag{11}$$

The results are as follows:

$$\frac{dV}{d\lambda} = J^{-1}\frac{dW}{d\lambda} = S_0\frac{dW}{d\lambda} \tag{12}$$

Can be recorded as:

$$\Delta V = S_0 \Delta W \tag{13}$$

By the semi-invariant property, the r-level semi-invariance $\Delta W^{(r)}$ of the jet power at every node in the system is given by the r-level semi-invariance for node load, there are:

$$\begin{cases} \Delta P^{(r)} = \Delta P_G^{(r)} + \Delta P_L^{(r)} \\ \Delta Q^{(r)} = \Delta Q_G^{(r)} + \Delta Q_L^{(r)} \end{cases} \tag{14}$$

Thus:

$$\Delta V^{(r)} = S_0^{(r)} \Delta W^{(r)} \tag{15}$$

Based on the semi-invariant additivity, the r-level semi invariant of the voltage stability L index are calculated as follows:

$$\Delta L_k^{(r)} = \left| 1 - \frac{\sum\limits_{i \in \alpha_G} F_{ji} \cdot V_i^r}{V_j^r} \right| \tag{16}$$

The probability density and distribution function of voltage correlation margin are obtained by Gram-Charlie series:

$$f(L) = \varphi(L) + \frac{\Delta(L)^1 \varphi^1(L)}{1!} + \ldots + \frac{\Delta(L)^n \varphi^n(L)}{n!} \tag{17}$$

$$F(L) = \Phi(L) + \frac{\Delta(L)^1 \Phi^1(L)}{1!} + \ldots + \frac{\Delta(L)^n \Phi^n(L)}{n!} \tag{18}$$

From the probability distribution curve, we can clearly and intuitively understand the node random voltage stability and the probability of the maximum injected disturbance power.

5 Division Index Among Grid Dispatching Domain

Depending on the different operation states among power grid, power grid dispatching can be divided into three different domains (in normal operation stage, called normal dispatching domain; in abnormal operation stage, called abnormal dispatching domain; in emergency operation state, called emergency dispatching domain). Figure 2 shows the grid dispatch domain division.

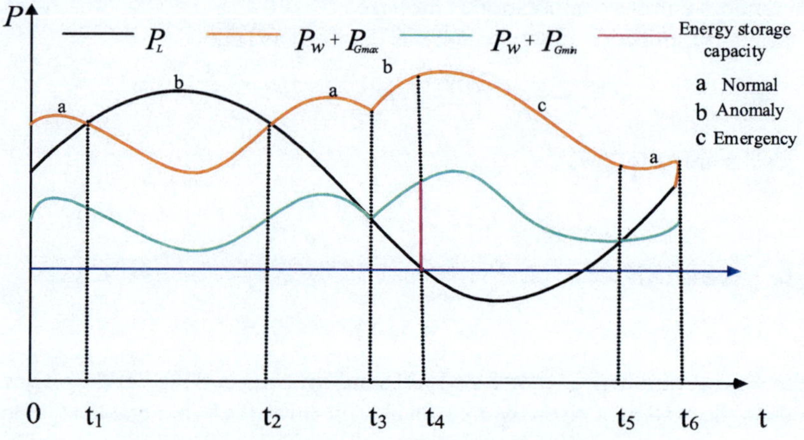

Fig. 2. Division diagram of power grid dispatching domain

As can be seen from Fig. 1, as the grid operates to a tipping point in each dispatch territory, the weakest node can be found by observing the voltage of the whole network, the breakdown voltage can be calculated by continuous power flow.

Changes in weak node voltage with power. The key dot defines the exponential of power grid dispatching domain is to observe the voltage of weak node.

In this paper, the index of power grid dispatching domain is defined as:

$$K_p = \frac{V_i}{V_{im}} \tag{19}$$

Where V_i represent the voltage of the weak point, V_{im} represent its lapse voltage.

Based on the grid dispatch domain indicator K_p, the power grid dispatching domain can be divided according to the following formula:

Normal scheduling domain:

$$\frac{V_{i\max}}{V_{im}} < K_p < \frac{V_{i\min}}{V_{im}} \tag{20}$$

Exception scheduling domain:

$$\frac{V_{i\min}}{V_{im}} < K_p < \frac{V_{is}}{V_{im}} \tag{21}$$

$$1 < K_p < \frac{V_{i\max}}{V_{im}} \tag{22}$$

Emergency scheduling domain:

$$1 < K_p < \frac{V_{i\max}}{V_{im}} \tag{23}$$

In the formula, V_{imin} is the voltage at the weaker node when the output of the conventional cell is at its minimum, V_{imax} is the voltage at the weaker node of the conventional cell when its output is at its maximum, and V_{is} is the weaker power voltage storage device with full capacity.

6 Case Simulation and Operation Characteristics Analysis of Power Grid Dispatching Domain

6.1 Example Simulation

Analysis of operational data from a regional power grid in Liaoning Province, 200 MW wind power is incorporated at y, J and s, and the wind power accounts for 35% of the total installed capacity. Figure 2 shows the system connection drawing.

By using L index to analyze the actual power grid, M area is the weak node of the whole network. According to K_p method and relevant data, the actual power grid K_{pmin} value is approximately 1.12, K_{pmax} value is approximately 1.06, K_{ps} value is approximately 1.19.

By calculating K_p index, the operation domain of local 24-h power grid is divided, Fig. 3 shows the simulation outcome.

Figure 3 and Table 1 shows that in a 24-h operation state of a power grid in Liaoning province, the wind speed is small in the daytime, the wind generation is small, and with high load. In the period of 6:15–16:30 and 17:30–20:30, the power grid generally operates in the normal region with high wind speeds, high wind power generation and low load at night. In the period of 00:00–6:15, 16:30–17:30 and 20:30–23:00, the power grid operates in the abnormal region, and in the period of 23:00–24:0 During this period, the power grid was operating in the emergency domain.

6.2 Analysis of the Network Dispatch Domain Operating Characteristics

Combined with the example simulation, the grid dispatch domain is analysed for its operative features. When the power grid is in the normal dispatching area, the system has the ability to fully absorb wind power, and the system has sufficient flexibility. When the system operates in the normal dispatching domain and the boundary state of the normal dispatching domain, there is no wind abandonment; when the system operates in the normal dispatching domain, the conventional unit output, wind power output and load power jointly affect wind energy consumption.

Even when the grid is in an abnormal dispatch domain, the system is still able to fully absorb the wind. When the system is running in the boundary state of the abnormal

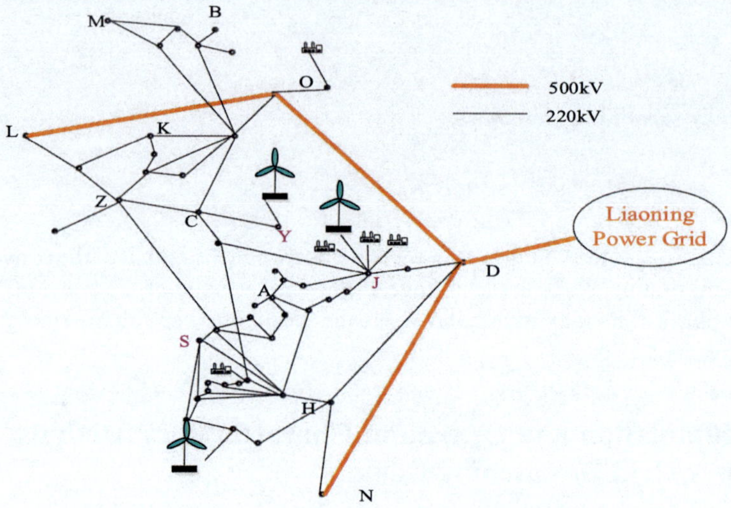

Fig. 3. Actual system one-line diagram

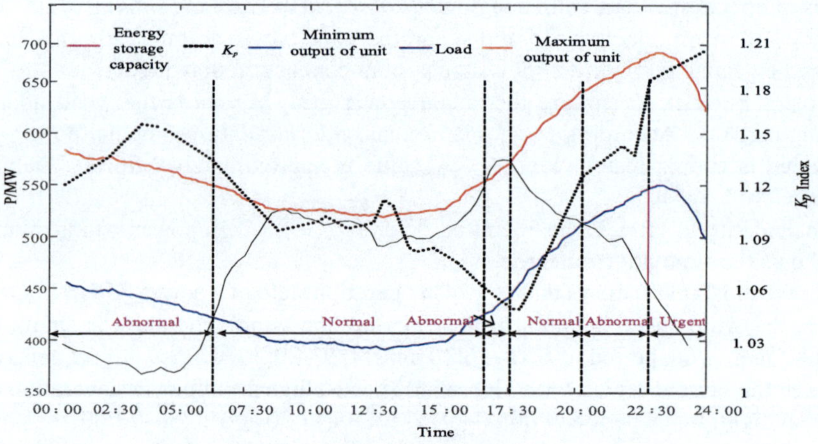

Fig. 4. Actual grid operation domain division diagram

Table 1. Division of actual power grid operation area

Time	Division of domain
00:00–6:15	Anomaly domain
6:15–16:30	Normal domain
16:30–17:30	Anomaly domain
17:30–20:30	Normal domain
20:30–23:00	Anomaly domain
23:00–24:00	Emergency domain

dispatching domain, its movement trend faces the interior of the abnormal dispatching domain. If the system has no access to other flexible resources, the flexibility of the system is close to the lower limit. If the system keeps on absorbing wind to satisfy the load and wind fluctuations, on the basis of regulating the output of conventional units, the electric energy storage units will be connected to the system as a system flexibility resource to stabilize the wind power fluctuation, so as to increasing the security for system operation.

When the network is in the emergency dispatching area, wind power is not fully assimilated, and the system is not flexible enough. To guarantee safe, wind power, as a special flexible resource, can only provide the system with downstream adjustable capacity by abandoning wind.

7 Conclusion

This paper combines the voltage stability L index to address the grid operation instability and scheduling complexity caused by the high-ratio connection of renewable energy to the grid, this paper puts forward the L index of voltage stability based on random power flow calculation and the division index of power grid operation domain, and divides the power grid dispatching domain based on the overall voltage reliability of the network, so as to provide certain auxiliary decision-making function for power grid dispatching personnel.

References

1. Li, J., Guo, B., et al.: Optimal allocation strategy for energy storage capacity of wind solar energy storage system. Acta Electrotechnics Sinica **33**(06), 1189–1196 (2018)
2. Ai, X., Tayerjiang, B., et al.: Spinning reserve optimization of power system with wind power based on scenario set. Power Grid Technol. **42**(03), 835–841 (2018)
3. Xu, B., Jin, L., et al.: Study on evaluation model of intermittent energy priority dispatch. J. Electr. Eng. **12**(12), 45–51 (2017)
4. Wang, D., Yu, Y.: Probabilistic static voltage stability security assessment for power transmission system. China Acad. J. Electron. Publishing House **43**(03), 10–14 (2010)
5. Xiong, C., Zhang, X., Meng, Y., et al.: Analysis on static voltage stability of system with large-scale wind power integration. Power Syst. Prot. Control **40**(21), 132–137 (2012)
6. Zhu, X., Liu, W., Zhang, J.: Probabilistic load flow method considering large-scale wind power integration. Proc. CSEE **33**(07), 77–85+16 (2013)
7. Duan, G., Qin, W., Lu, R., et al.: Static voltage stability analysis considering the wind power and uncertainty of load. Power Syst. Prot. Control **46**(12), 108–114 (2018)
8. Xu, H., Li, Y., Miao, Y., et al.: Optimization dispatch strategy considering renewable energy consumptive benefits based on "source-load-energy storage" coordination in power system. Power Syst. Prot. Control **45**(17), 18–25 (2017)
9. Borkowska, B.: Probabilistic load flow. IEEE Trans. Power Apparatus Syst. **93**(3), 752–759 (1974)
10. Xie, X., Yuan, Y., Zou, W.: Reliability evaluation of microgrid power supply in emergency non-scheduled isolated mode. Autom. Eletr. Power Syst. **36**(09), 18–23 (2012)

Distribution Network Load Forecasting Based on Smart Meter User Behavior Clustering

Shunjiang Wang[1](✉), Qianbin Dai[2], Guiping Zhou[1], Yangyang Ge[3], Peng Jin[1], and Yifan Wang[2]

[1] State Grid Liaoning Electric Power Co., Ltd., Shenyang 110000, LN, China
[2] School of Electrical Engineering, Northeast Electric Power University, Jilin 132012, China
[3] Electric Power Research Institute of State Grid Liaoning Electric Power Co., Ltd., Shenyang 110000, LN, China

Abstract. Aiming at the demand of distribution network load forecasting under the massive user power consumption data, this paper analyzes the user's power consumption behavior characteristics to study the distribution network user clustering and load forecasting. Firstly, by extracting 6 features that characterize users' electricity consumption behaviors, clustering of users based on Gaussian mixture model is carried out, and original feature sets are established respectively according to the clustering results, and conditional mutual information is used as a feature selection method to perform feature selection for different types of loads. Determine the optimal feature subset, and finally establish the optimal prediction model with long and short-term memory network as the prediction method. The experimental results show that the new method proposed in this paper has a significant effect on the improvement of the distribution network prediction accuracy.

Keywords: User clustering · Long short-term memory · Distribution network load forecast

1 Introduction

The construction and development of smart grid is a huge historical project, and the complex smart grid construction project is still in progress. In the smart grid, as its main support, power users are an indispensable part of realizing the intelligent power grid. The intelligent meters installed by power users make the power department obtain not only the energy data recorded by traditional meters, but also the information of voltage, active power and reactive power, weather and geography collected by various sensors [1]. The access of smart meter provides the basis for analyzing the user's electricity consumption behavior. With the increase of data accumulation time, the user side data increases and the complexity increases constantly, which brings opportunities and challenges to the power industry [2].

Analyzing the massive data collected by smart meters can enable the power service department to have a deeper understanding of users' power demand, and can formulate targeted services according to the power demand of different users. Reference [3] uses

S.-C. Chu et al. (Eds.): ICGEC 2021, LNEE 833, pp. 538–547, 2022.
https://doi.org/10.1007/978-981-16-8430-2_49

the data collected by smart meters to analyze the electricity consumption behavior of different users, clusters the users in the area according to 18 kinds of electricity consumption behavior characteristics, and finally carries out load forecasting for different types of loads. The forecast results shows that it is a feasible method to improve the accuracy of prediction when the user load is clustered and then predicted. Reference [4] extracts features that can characterize users' electricity consumption behaviors, such as daily minimum load rate, daily peak-valley difference, and daily peak-valley difference rate, and then clusters users through these features. Reference [5] uses massive smart meter data to compare and analyze the advantages and disadvantages of three clustering methods: Gaussian Mixture Model (GMM), K-Means and K-Medoid mapping, and analyzes a variety of factors that may affect users' electricity consumption behavior. Reference [6] considers the differences in electricity consumption patterns of residential users, clusters user load data as samples, establishes load forecasting models for different user groups, and uses a variety of classic clustering and forecasting methods to combine. The above studies have made great contributions to the use of smart meter data and smart meter data mining. At the same time, it also shows that the effective use of smart meter data is of great value.

The bottom users of distribution network are greatly affected by the electricity consumption behavior, and the load fluctuation is strong, which leads to the traditional clustering method cannot cluster the bottom users of distribution network accurately. Therefore, this paper uses the electricity consumption behavior clustering of smart meter users based on Gaussian mixture model to cluster different types of users reasonably according to the characteristics of electricity consumption behavior, and carries out targeted modeling; considering that the traditional neural network cannot deal with the problem, in order to improve the accuracy of distribution network load forecasting model, a load forecasting model based on long-term and short-term memory neural network is selected.

2 Clustering of Electricity Consumption Behavior of Smart Meter Users Based on Gaussian Mixture Model

2.1 GMM Clustering Based on Gap Statistics

The Gaussian mixture model is weighted by M Gaussian distribution models, which is shown in Eq. (1).

$$p(x|\lambda) = \sum_{i=1}^{K} \omega_i g(x|\mu_i, \Sigma_i) \tag{1}$$

Where x is the sample data (input feature), ω_i, $i = 1, ..., K$, which represents the mixed weight. $g(x|\mu_i, \Sigma_i)$ represents the probability that sample x belongs to class k electricity consumption behavior.

When the probability density expression of sample data x,g and the clustering number K are known, the maximum likelihood function can be used to solve the super parameters μ, Σ and ω through the maximum expectation (EM) algorithm [7]. All the undetermined parameters in the Gaussian mixture density function are denoted as λ, that is $\lambda = \{\omega_i, \mu_i, \Sigma_i\}, i = 1, ..., K$.

EM algorithm uses the initialization parameter $\lambda = \{\omega_i, \mu_i, \Sigma_i\}$ to get the new parameter $\overline{\lambda}$ through the calculation of a posteriori probability. The new model parameters will become the initial parameters in the next iteration, and continue to calculate the new model parameters until the model parameters reach the convergence threshold and satisfy $p(X|\overline{\lambda}) \geq p(X|\lambda)$.

Gap Statistics (GS) is used to determine the optimal number of GMM clusters [8]. For a given data set, $\{x_{ij}\}, i = 1, 2, ..., n, j = 1, 2, ...p$, contains the p-dimensional features of n independent observation samples. Suppose that the data has been clustered into k classes $C_1, C_2, ..., C_k$, $n_r = |C_r|$, and the Euclidean distance $D_{ii'}$ is used to represent the distance between the observation value *i* and *i'*, $\Sigma_j (x_{ij} - x_{i'j})^2$. *Dr* is the Euclidean distance between all the points in class *r*.

$$W_k = \sum_{r=1}^{k} \frac{1}{2n_r} D_r \tag{2}$$

W is the sum of *k*-class deviations.

The Gap statistic is defined as:

$$Gap_n(k) = E_n^*\{\log(W_k)\} - \log(W_k) \tag{3}$$

Where *En* is the expectation under the reference distribution.

B reference data sets are generated in a given data set, and these reference data are grouped into *k*-class respectively. According to formula (2), W_{kb} and gap statistics are obtained.

$$Gap(k) = (1/B) \sum_b \log(W_{kb}^*) - \log(W_k) \tag{4}$$

$$s_k = sd_k \sqrt{(1 + 1/B)} \tag{5}$$

The minimum *K* value of the formula $Gap(k) \geq Gap(k + 1) - s_{k+1}$ is the optimal cluster number *K*.

2.2 Clustering Feature Selection

This paper uses six kinds of electricity behavior characteristics to cluster the load data. According to the distribution of users at different times, a day can be roughly divided into four periods: 22:30–06:30, 06:30–09:00, 09:00–15:30, 15:30–22:30, and extracting average power for each period, and calculating the standard deviation of each period according to the average power. Through the analysis of the load data of smart meter users, it is found that the load curve trend of different users in the same day is different, the arrival time of peak value and valley value is different, and the load fluctuation is different. Therefore, the paper also uses the difference scores of annual average relative standard deviation, working day and non-working day as the characteristics of user power consumption behavior [9, 10].

3 Long Short-Term Memory

Long short term memory neural network is a special form of RNN. It takes the hidden layer of the neural network as a storage unit to deal with the relationship between long-term and short-term information in time series. Among them, neurons represent the selected activation function, which is obviously the traditional RNN. The internal structure of the hidden layer is very simple, and there is no choice between the current time input and the previous time state input, which may cause the recent non critical information to occupy a lot of storage space, and the long-term key information cannot be saved and transmitted, so that the final output has a large deviation [11].

The structure of LSTM unit is shown in Fig. 1. At any time t, $\boldsymbol{i}_t$, $\boldsymbol{f}_t$, $\boldsymbol{o}_t$ and $\tilde{\boldsymbol{c}}_t$ represent input gate, forgetting gate, output gate and candidate value respectively. $\boldsymbol{x}_t$ is the input of the current time, $\boldsymbol{h}_{t-1}$ represents the output of LSTM at the previous time, i.e. t − 1, and $\boldsymbol{\sigma}$ represents the sigmoid activation function. In the internal structure of the cell, the forgetting gate is used to determine the proportion of the previous memory value removing the current cell state. Similarly, the input gate is used to determine the input proportion entering the current cell state. c_t is the long-term memory value at time t, $\boldsymbol{h}_t$ is the output value at time t, and $\boldsymbol{x}_t$ is the prediction output value.

Figure 2 shows the network structure of LSTM expanded by time, and j represents the characteristic input variable at time t. The LSTM unit in the current state will receive feedback information $\boldsymbol{h}_{t-1}$ from the former LSTM unit to capture the time dependence. The training and learning of long-term and short-term memory network aims to reduce the square difference function f by using the back propagation method of gradient descent. In the training process, the weights and deviations are adjusted iteratively through their gradients. When a batch of input training sets are learned by back propagation optimization algorithm, an iteration is completed [12, 13].

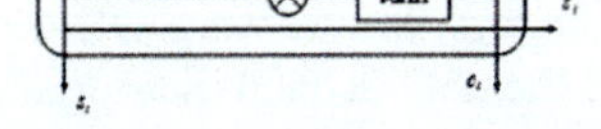

Fig. 1. Internal topology of LSTM hidden layer

4 Classified Load Forecasting Model Based on LSTM

According to the characteristics of large load fluctuation of bottom users in distribution network, firstly, the characteristics of power consumption behavior are used to carry out reasonable clustering analysis for users; secondly, in order to reduce the feature quantity

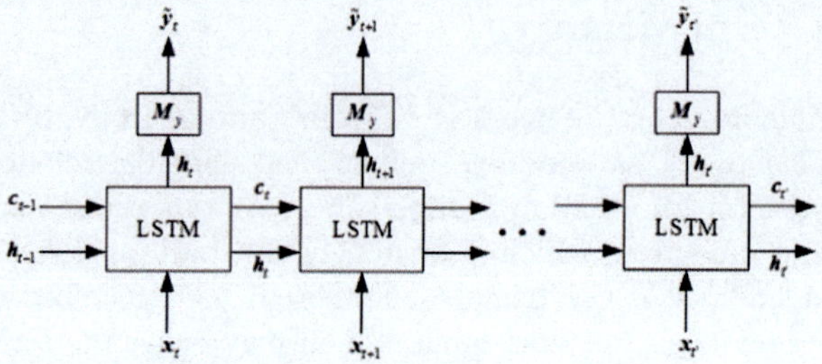

Fig. 2. LSTM network structure diagram

used in modeling, according to the feature with maximum mutual information with the target variable, the remaining feature which can provide maximum information gain for the target variable is selected, and the CMI value is calculated and sorted [14]; the LSTM model is taken as an example The predictor takes the average absolute percentage error of the prediction results as the decision variable, and uses the sequence forward selection method to determine various optimal feature subsets according to the ranking order of characteristic CMI values. Finally, it trains the LSTM prediction model with various optimal feature subsets as the input, and inputs the test set data to complete the prediction [15]. The specific process is shown in Fig. 3.

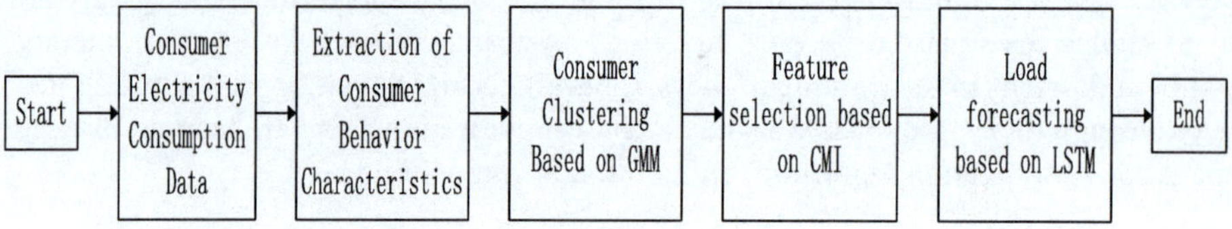

Fig. 3. Flow chart of classified load forecasting based on LSTM

5 Experiments

The power load data is extracted from more than 5000 smart meter users in Ireland in 2010. Considering seasonal factors, the training set, verification set and test set are divided for each quarter. 90% of the data in each quarter are included in the training set, and the remaining 5% are used as the verification set and the test set.

After clustering different types of users in each quarter by using electricity consumption behavior, LSTM is trained by using training set data In the prediction model, the verification set data is used as input to determine the optimal feature subset obtained by using CMI feature selection method, and the final prediction result is obtained by using the most characteristic subset in the test set as input. The average absolute percentage error is used as the evaluation index to verify the effectiveness of the proposed method [16].

5.1 Feature Selection Results

In this paper, the verification set data is used to carry out hourly load forecasting to determine the optimal feature subset. CMI feature selection method combined with LSTM forecasting model is used for the three types of load in the first quarter, and the forecasting error curve is obtained when different dimension features are input. With the increase of the number of features, the prediction error decreases obviously. However, when the number of features increases to a certain number, the error does not decrease. According to the calculation, the optimal feature subset dimension of the first type of users in the first quarter through the verification set experiment is 21, corresponding to the MAPE value of 1.801; the optimal feature subset dimension of the second type of users through the verification set experiment is 19, corresponding to the MAPE value of 1.834; the optimal feature subset dimension of the third type of users through the verification set experiment is 35, corresponding to the MAPE value of 2.346.

Table 1 gives the optimal feature subset dimension and corresponding MAPE value of each type of smart meter users in Ireland who carry out CMI feature selection combined with sequence forward search in four quarters.

Table 1. Optimal feature subset dimension and MAPE of different users in four seasons

	Class 1 User		Class 2 User		Class 3 User	
	Optimal feature subset dimension	MAPE	Optimal feature subset dimension	MAPE	Optimal feature subset dimension	MAPE
First season	21	1.801	19	1.834	35	2.346
Second season	18	1.925	25	1.755	19	1.823
Third season	22	2.875	30	2.634	–	–
Fourth season	29	1.786	22	1.985	27	2.052

5.2 Analysis of Prediction Results

By constructing the optimal feature subsets of different types of load, the optimal feature subsets of different types of load are combined with the long-term and short-term memory network modeling to carry out load forecasting. In addition, the same quarterly experimental data set but without clustering is used, the feature selection is directly carried out, and the short-term and long-term memory network is combined to carry out load forecasting as a comparative test The necessity and effectiveness of clustering algorithm.

Figure 4 shows the forecast results of the two methods in each quarter and compares them with the measured load. In Fig. 4, the LSTM model is used for forecasting, and the forecasting results are well matched with the measured load curve whether clustering or not. However, it is still obvious that the curve results obtained by clustering analysis of load data are better than those obtained by direct clustering analysis. Careful observation of Fig. 4 shows that the fitting degree of the load curve in the third quarter is slightly worse than that in the other three quarters. However, it can be seen that the fitting effect of the load curve after clustering analysis and the measured load in the third quarter is significantly better than that of the non-clustering method, which effectively verifies the necessity and effectiveness of clustering again.

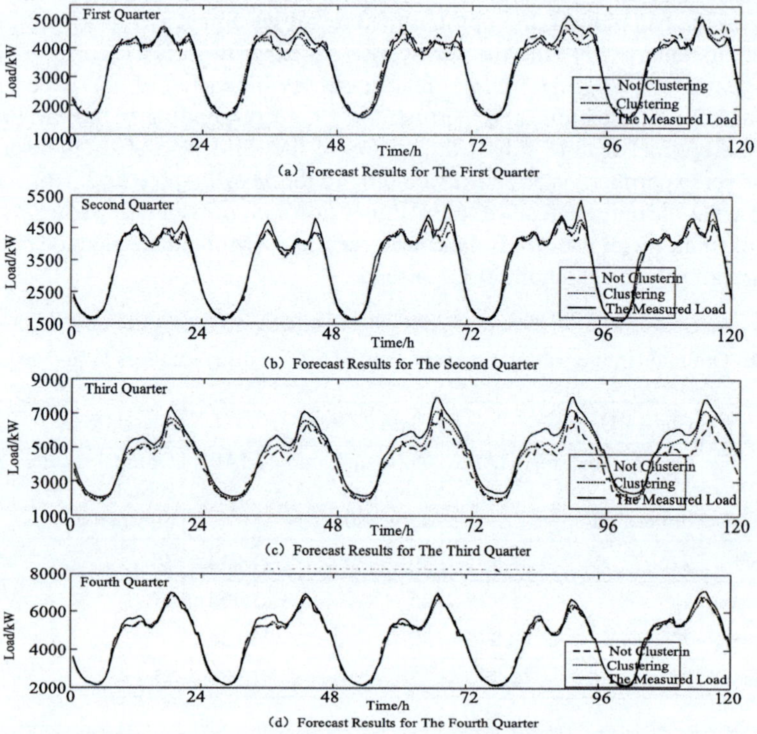

Fig. 4. Prediction curve and measured load curve of different methods

In order to verify the effectiveness of the proposed predictor long-term and short-term memory network, LSTM, SVR and back propagation neural network (BPNN) combined with CMI feature selection method are used to carry out comparative experiments on user clustering results. In the experiment, cross validation method is used to determine the optimal parameters of SVR when different features are input. The number of hidden layer neurons of BPNN is $p = 2n + 1$, where n is the number of input features of BPNN.

Figure 5 shows the optimal feature subset obtained by feature selection after cluster analysis, and the curves of predicted load and measured load obtained by combining the three predictors.

In Fig. 5, the optimal feature subset selected by clustering analysis combined with CMI and the three predictors fit well with the measured load curve, but it is still obvious

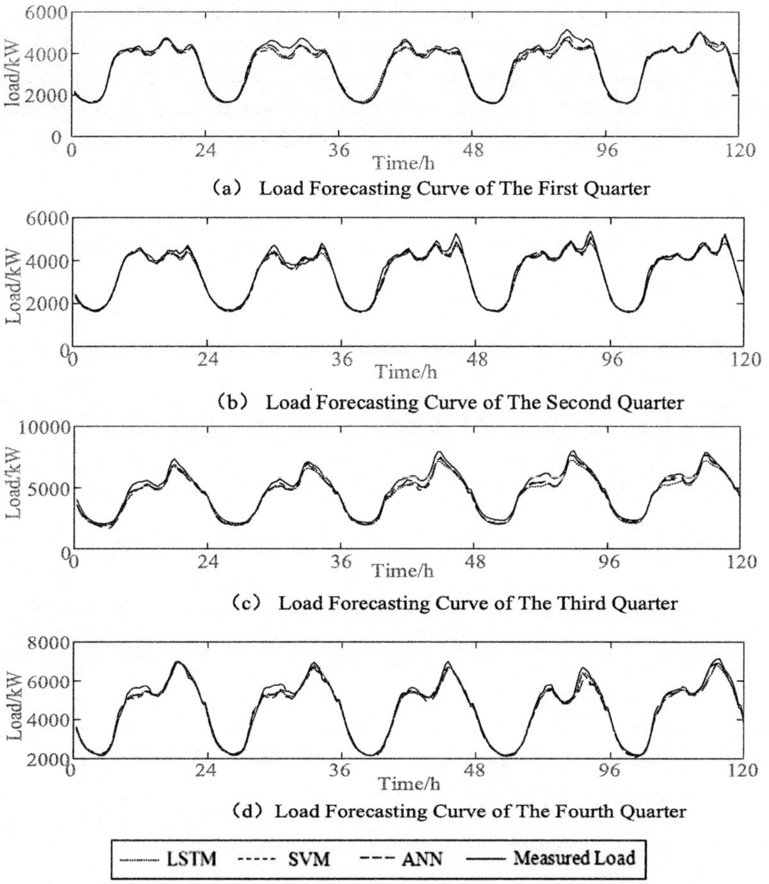

Fig. 5. Prediction curve and measured load curve of different prediction models

that the curve result obtained by using LSTM as predictor is better than that of the other two predictors.

6 Conclusion

In this paper, Irish smart meter data is used to predict the bottom users of low-voltage distribution network, and conditional mutual information is used as the feature selection method. Because the bottom users of distribution network are greatly affected by the electricity consumption behavior and the load fluctuation is strong, the traditional clustering method cannot cluster the bottom users of distribution network accurately. This paper makes reasonable clustering for different types of users according to the characteristics of electricity consumption behavior Compared with the traditional direct prediction at the distribution network level, the targeted modeling can fully consider the diversity and difference of different users, and reduce the impact of the large load fluctuation of the bottom users of the distribution network. Considering that the traditional neural network cannot process the correlation information between time series and cannot autonomously memorize the long-term key information, this paper adopts the negative prediction method based on long-term and short-term memory neural network based on the prediction model, LSTM can automatically remember and forget the input features, effectively judge and screen the features, and thus increase the accuracy of prediction.

References

1. Dongxia, Z., Xin, M., Liping, L., et al.: Research on big data technology development of smart grid. Proc. CSEE **1**, 2–12 (2015)
2. Khan, A.R., Mahmood, A., Safdar, A., et al.: Load forecasting, dynamic pricing and DSM in smart grid: a review. Renew. Sustain. Energy Rev. **54**, 1311–1322 (2016)
3. Beckel, C., Sadamori, L., Staake, T., et al.: Revealing household characteristics from smart meter data. Energy **78**, 397–410 (2014)
4. Jundong, F., Yao, Y., Shanjiang, L.: Characteristic weight analysis of residential power load in Intelligent Community. Prot. Control Power Syst. **44**(18), 41–45 (2016)
5. Mcloughlin, F., Duffy, A., Conlon, M.: A clustering approach to domestic electricity load profile characterisation using smart metering data. Appl. Energy **141**, 190–199 (2015)
6. Hang, H.H., Lin, L.S., Chen, N., et al.: Particle swarm optimization based non-intrusive demand monitoring and load identification in smart meters. In: Industry Applications Society Meeting, pp. 1–8. IEEE (2012)
7. Huang, N., Hu, Z., Cai, G., et al.: Short term electrical load forecasting using mutual information based feature selection with generalized minimum-redundancy and maximum-relevance criteria. Entropy **18**(9), 330 (2016)
8. Quilumba, F.L., Lee, W.J., Huang, H., et al.: Using smart meter data to improve the accuracy of intraday load forecasting considering customer behavior similarities. IEEE Trans. Smart Grid **6**(2), 911–918 (2015)
9. Jin, Y., Honggang, G., Jianzhou, W., et al.: A hybrid system based on LSTM for short-term power load forecasting. Energies **13**(23), 6241 (2020)
10. Wi, Y.M., Joo, S.K., Song, K.B.: Holiday load Forecasting using fuzzy polynomial regression with weather feature selection and adjustment. IEEE Trans. Power Syst. **27**(2), 596–603 (2012)
11. Sun, G., Chuanwen, J., Xu, W., et al.: Short-term building load forecast based on a data-mining feature selection and LSTM-RNN method. IEEJ Trans. Electr. Electron. Eng. **15**(7), 1002–1010 (2020)
12. Muzaffar, S., Afshari, A.: Short-term load forecasts using LSTM networks. Energy Procedia **158**, 2922–2927 (2019)
13. Lin, X., Nantian, H., Shuye, Z., Panpan, W.: Feature selection of low redundancy short-term load forecasting based on conditional mutual information. J. Northeast Electr. Power Univ. **39**(02), 30–38 (2019)
14. Teng, Z., Lintong, W., Yan, Z., et al.: Identification of user power consumption related factors and prediction method of power consumption based on mutual information and random forest algorithm. Proc. CSEE **36**(3), 604–614 (2016)
15. Cecati, C., Kolbusz, J., Różycki, P., et al.: A novel RBF training algorithm for short-term electric load forecasting and comparative studies. IEEE Trans. Industr. Electron. **62**(10), 6519–6529 (2015)

16. Jurado, S., Nebot, À., Mugica, F., et al.: Hybrid methodologies for electricity load forecasting: entropy-based feature selection with machine learning and soft computing techniques. Energy **86**, 276–291 (2015)

Regulation Strategy of Electric Vehicle Charging Station Taking into Account New Energy Consumption

Yuxing Li[1](✉), Dongyuan Liu[2], Jingying Yang[1], Qunying Li[3], Lei Song[1], and Xuefeng Gao[1]

[1] Economic and Technological Research Institute of State Grid Jilin Electric Power Co., LTD., Jilin 130000, China
liyx01@jl.sgcc.com.cn
[2] Northeast Electric Power University, Jilin 132000, China
[3] State Grid Jilin Electric Power Co., LTD., Jilin 130000, China

Abstract. The safety and economics of the grid will be considerably impacted as large-scale electric vehicles and deeply random distributed renewable energy sources are connected to the grid. On the basis of considering the uncertainties of wind power, photovoltaic power output and charging and discharging of electric vehicles, and ensuring the normal operation of electric vehicles, this paper proposes a control strategy for centralized charging stations of electric vehicles taking into account the consumption of renewable energy accommodation. GA algorithm is used to calculate the 24 h charge and discharge power of the charge and discharge system and the cascade battery energy storage system. Through the analysis of the calculation examples, the load of the regulated centralized EV charging station can effectively reduce the peak and valley of the urban daily equivalent load curve connected with wind power and photovoltaic power generation. Thus, the impact on the power grid caused by the grid-connected operation of distributed energy resources is reduced, and the purpose of peak clipping and valley filling and renewable energy accommodation is realized. In addition, the advantages of GA algorithm in overall optimization performance are also verified.

Keywords: Wind power and photovoltaic power · Charging stations for electric vehicles · The GA algorithm

1 Introduction

In recent years, in order to deal with the energy crisis, environmental pollution, climate change and other issues, the proportion of distributed renewable energy sources installed in the power grid has been increasing. However, due to the intermittency and volatility of wind energy and solar energy, a large part of renewable energy can not be absorbed, resulting in energy waste. Therefore, controllable units and energy storage equipment need to set aside large rotating reserves to cooperate with distributed new

S.-C. Chu et al. (Eds.): ICGEC 2021, LNEE 833, pp. 548–559, 2022.
https://doi.org/10.1007/978-981-16-8430-2_50

energy power generation online [1, 2]. On the other hand, electric vehicles perform adequately in environmental protection and resource conservation and are widely promoted and applied. However, large-scale disorderly connection of electric vehicles to the power grid for charging will increase the peak-valley difference of the power grid. As a kind of distributed flexible energy storage device, making full use of the coordination and complementary characteristics of electric vehicles and distributed generation is of great significance to reduce the peak valley difference on the demand side, enhance the balance of power supply and demand, and upgrade the utilization rate of renewable energy [3].

Although wind power prediction has made great progress through research and practice, the prediction error is still substantial [4]. For example, [5] suggested that the mean absolute percentage error (MAPE) of wind power prediction software actually put into commercial operation around the world is about 14%–20%. Literature [6] pro-posed to use a combination of discrete distribution and continuous function to de-scribe the prediction error. Literature [7] establishes a probability statistical model of forecast error, which takes the measured wind speed in the current period and the predicted wind speed in the next period as the joint condition, and adopts the representation method of discrete probability. In literature [8], ARMA was used to generate landscape output scenes, and Kantorovich distance method was applied to re-duce and screen the generated scene sets.

The research on charging load and scheduling control strategy of electric vehicles has obtained a lot of research results at home and abroad. Existing charging load determination methods can be divided into determination analysis methods based on travel demand [9], Monte Carlo simulation method [10] and probability analysis method of charging station charging load [11, 12]. The charging control of valley filling introduced in literature [13] increases the number of electric vehicles that can be accommodated by the system from 20% to 50% of the total number of vehicles. Literature [14] takes the Danish power system as an example to analyze the role of charging control in promoting wind power absorption and reducing greenhouse gas emissions.

However, the randomness and volatility of wind power and photovoltaic power generation are affected by uncertain factors such as season, climate, altitude, weather, temperature and so on, which is problematic to be compensated by prediction error. The modeling process of the above model is cumbersome, and there are still some errors compared with the actual output of wind power photovoltaic. The above research on electric vehicles regards electric vehicles as a whole to predict their charging load, but not according to the types of electric vehicles; In addition, the load of electric vehicles is merely added, without considering the potential of load regulation of centralized charging and changing stations for electric vehicles and the role of promoting wind power and photovoltaic absorption. Therefore, the adjustable characteristics of electric vehicle charging station coincide with the demand of suppressing the random fluctuation of distributed renewable energy.

In view of this, this paper first builds an EV charging model based on the Monte Carlo simulation method based on the probability density function of the initial charging time, charging duration, daily driving distance and SOC of four types of electric bus, taxi, business car and private car. Secondly, the Weibull distribution and Beta distribution, which have the highest applicability, are used to describe the uncertainty of wind speed and irradiance respectively, and the Monte Carlo simulation method is used to simulate the output of distributed renewable energy. Finally, taking the maximum peak clipping and valley filling and renewable energy consumption of EV charging stations as the objective function, the power balance, capacity limit and reserve margin of the charging stations are considered in the constraint conditions. Based on this, a regulatory strategy model of EV charging stations taking into account the consumption of new energy is established. The charging and discharging power between the charging and discharging system and the grid, the energy storage system and the grid, and the charging and discharging power between the charging and discharging system and the energy storage system in the centralized charging station of electric vehicles were solved by GA algorithm.

2 Distributed Power Model

2.1 Electric Vehicle Model

The large-scale electric vehicle access to the power grid will cause serious impact on the operation and planning of the power system. Based on the analysis of the operation characteristics of different types of electric vehicles on the simulation of electric vehicle load, which can provide a reference for the coordinated and optimal operation of the power grid and the distribution of power demand, achieve the purpose of peak load reduction and valley filling, and ensure the stable operation of the power grid.

Operating Characteristics of Electric Vehicles. Under the influence of operation system and work and rest rules, the operating characteristics of different types of electric vehicles are linked to the initial charging time, charging duration, daily driving mileage, SOC and other factors. The following is the analysis of each influencing factor and the establishment of its mathematical model:

The initial charging time of EV obeys normal distribution, and its probability density function is as follows:

$$f_x(t) = \frac{1}{\sigma_x\sqrt{2\pi}} exp\left[-\frac{(t-\mu_x)^2}{2\sigma_x^2}\right] \tag{1}$$

Where, μ_x, σ_x^2 are the expectation and standard deviation of time t.

The charging time of electric vehicles depends on the electric energy consumed and charging power by electric vehicles. The consumed electric energy is determined by the state of charge before and after charging and is related to daily driving distance, as shown in Eq. (2):

$$T_c = \frac{(SOC_2 - SOC_1)E}{P_c\eta_c} \tag{2}$$

Where, η_c is the charging efficiency. SOC_1 and SOC_2 represent the state of charge in the initial and end state of charging. E is the battery capacity of electric vehicles.

The daily mileage of electric vehicles is determined by the user's occupation and travel purpose. According to the law given in a survey data of DOT, the daily mileage is found to obey lognormal distribution through statistical data, and its probability density function is:

$$f_d(x) = \frac{1}{x\sigma_d\sqrt{2\pi}} exp\left[-\frac{(lnx - \mu_d)^2}{2\sigma_d^2} \right] \tag{3}$$

Where, x is the mileage. μ_d, σ_d^2 are expectation and standard deviation of lnx.

The state of charge (SOC) at the beginning and end of electric vehicle charging determines the power demand during the charging period:

$$SOC = \left(1 - \frac{x}{x_m}\right) \times 100\% \tag{4}$$

Where, x is daily mileage. x_m is maximum daily mileage of the vehicle.

Load Characteristics of Electric Vehicles. Monte Carlo simulation method is a random simulation method based on probability and statistics methods. It uses random numbers to give a probability model to a complex problem, so that the solution of the complex problem corresponds to some characteristics of the random variables in the model to achieve the purpose of solving. In this paper, four types of electric buses, taxis, business cars and private cars are considered, and the discharge behavior of electric vehicles is considered. The Monte Carlo random sampling method is adopted to predict the 24-h load of electric vehicles. The process is as follows:

Monte Carlo simulation is a kind of stochastic simulation method based on probability and statistics method, using the random number given as probability model to a complex problem, the solution of the complex problem is matched with some characteristics of the random variables in the model to achieve the purpose of solving. In this paper, four types of electric buses, taxis, business cars and private cars are considered, and the discharge behavior of electric vehicles is considered. The Monte Carlo random sampling method is adopted to predict the 24-h load of electric vehicles. The process is as follows:

1 According to the operation characteristics of different types of electric vehicles, the corresponding parameter ranges of initial charging time, charging duration, charging power, daily mileage and SOC are set;
2 Setting the sampling proportion of different types of electric vehicles;
3 To extract each of four types of the electric vehicles in turn;
4 Extract the car start charging time, charging, electric power, daily mileage and charged state;
5 Calculate the charging time and get the daily charging load of the car from the charging power;
6 Repeat steps 3, 4, and 5 to add up the charging load of all vehicles.

To extract the samples in each period will accumulate the get all types of electric vehicle charging load in one day, the calculation formula is as follows (5):

$$P_{ev}^t = \sum_{t=1}^{1440} \left(\sum_{i=1}^{N_{gi}} P_{gi}^i + \sum_{i=1}^{N_{cz}} P_{cz}^i + \sum_{i=1}^{N_{gw}} P_{gw}^i + \sum_{i=1}^{N_{sj}} P_{sj}^i \right) \tag{5}$$

Where, N_{gi}, N_{cz}, N_{gw} and N_{sj} are the numbers of electric buses, taxis, official cars and private cars respectively. P^i_{gi}, P^i_{cz}, P^i_{gw} and P^i_{sj} are respectively the charge-discharge power of the i-th electric bus, taxi, business car and private car at time t.

2.2 Uncertainty of Renewable Energy

Uncertainty of Wind Power. Wind power converts the mechanical energy of natural wind into electrical energy by wind turbine, and the wind speed is determined by season, climate, altitude and other factors, so it can be considered that the mathematical expectation of wind speed changes with time. Through a large number of specific research and comparisons, it is generally considered that the wind speed obeys the distribution law of positive skewness. Two parameter Weibull curve is the most suitable probability density function among many functions used to simulate wind speed distribution [15]. The expression is shown in (6):

$$\begin{Bmatrix} f\left(V_{WT}\right) = \frac{k_t}{c_t}\left(\frac{V_{Wt}}{c_t}\right)^{k_t-1}\exp\left(-\left(\frac{V_{Wt}}{c_t}\right)^{k_t}\right) \\ k_t = \left(\frac{\overline{\sigma_t}}{V_{Wt}}\right)^{-1.086} \quad c_t = \frac{V_{Wt}}{\Gamma(1+1/k)} \end{Bmatrix} \tag{6}$$

Where, V_{Wt} is the wind speed at time t. k_t and c_t are the shape parameters and scale parameters of Weibull distribution at time t respectively. The relationship between wind speed and wind power output can be described as a piecewise function, as shown in Eq. (7):

$$P_{Wt}(V_{Wt}) = \begin{cases} 0 \quad V_{W_t} < V_{W_{in}}, V_{W_t} > V_{W_{off}} \\ P_{W_R}\frac{V_{Wt}-V_{Win}}{V_{W_R}-V_{W_{in}}} \quad V_{W_{in}} \le V_{W_t} \le V_{W_R} \\ P_{W_R} \quad V_{W_R} \le V_{W_t} \le V_{W_{off}} \end{cases} \tag{7}$$

Where, $V_{W_{in}}$, $V_{W_{off}}$ and V_{W_R} are respectively the connection, cut-off and the rated wind speed of the wind turbine. P_{W_R} is the rated output power of wind turbine.

Uncertainty of Photovoltaic Power. The Photovoltaic power generation is a kind of make use of the photovoltaic effect, converting solar energy to electrical energy form of clean power generation. Photovoltaic power generation is affected by many factors, such as weather, light intensity, temperature and so on. Among them, light intensity has the greatest influence on photovoltaic power generation, while light intensity is affected by regional environment and latitude. It can be considered that light intensity in a certain region is a random variable. According to statistical data, the illumination intensity distribution can be approximately described by Beta distribution [15], and its probability density function is shown as (8):

$$\begin{Bmatrix} f\left(R_{a_t}\right) = \frac{\Gamma(a_t+b_t)}{\Gamma(a_t)\Gamma(b_t)}\left(\frac{R_{a_t}}{R_{a_{max}}}\right)^{a_t-1}\left(1-\frac{R_{a_t}}{R_{a_{max}}}\right)^{b_t-1} \\ a_t = \overline{R_{a_t}}^2\left(\frac{1-\overline{R_{a_t}}}{\sigma_t^2} - \frac{1}{\overline{R_{a_t}}}\right) \quad b_t = a_t\left(\frac{1}{\overline{R_{a_t}}} - 1\right) \end{Bmatrix} \tag{8}$$

Where, R_{a_t} is the solar radiation at time t. a_t and b_t are the shape parameters of the beta distribution at time t. Since photovoltaic output power varies linearly with illumination intensity, the energy conversion relationship between photovoltaic unit output and solar radiance is shown in Eq. (9):

$$P_{S_t}(R_{a_t}) = \begin{cases} P_{SR}\left(\frac{R_{a_t}^2}{R_{amax}R_{ac}}\right) & 0 < R_{a_t} < R_{a_c} \\ P_{SR}\left(\frac{R_{a_t}}{R_{amax}}\right) & R_{a_t} \geq R_{a_c} \end{cases} \quad (9)$$

Where, $R_{a_{max}}$ is the maximum solar radiation. R_{a_c} is the fixed-point radiation. P_{S_t} is the rated output power of the photovoltaic unit.

3 Regulation Strategy Model of Electric Vehicle Charging Station

Wind power and photovoltaic power generation with randomness and volatility will increase the peak-valley difference of the power grid after grid connection, resulting in the waste of renewable energy. In order to reduce the waste of wind power and photovoltaic energy and relieve the pressure of peak load on thermal power units, the orderly charge and discharge control of electric vehicle charging stations can be con-trolled to absorb renewable energy and achieve peak clipping and valley filling.

3.1 The Objective Function

With peak and valley reduction and maximum consumption of renewable energy as the goal, the expression is as follows:

$$minf_1(P_{bess}) = \sqrt{\sum\nolimits_{t=1}^{T}(P_{load}^t + P_{bess}^t - P_{wind}^t - P_{solar}^t - P_{av})^2/T} \quad (10)$$

$$P_{bess}^t = P_{ch}^t + P_{tc}^t \quad (11)$$

Where, P_{load}^t is daily equivalent load demand. P_{bess}^t is charging and discharging power of the charging station. Positive when charging and negative when discharging. P_{wind}^t is wind power in time period t. P_{solar}^t is photovoltaic power generation in period t. P_{av} is average load demand. T is scheduling cycle. P_{ch}^t is charging and discharging power of the charging and changing system. P_{tc}^t is charging power of the cascade battery energy storage system.

3.2 The Constraint

The electricity demand of an EV charging station is mainly determined by the charge and discharge system and the energy storage system. The charge and discharge pow-er constraints of the charge and discharge system and the energy storage system are shown as follows:

$$-P_{ch\_max} \leq P_{ch}^t \leq P_{ch\_max} \quad (12)$$

$$-P_{tc\_max} \leq P_{tc}^{t} \leq P_{tc\_max} \tag{13}$$

Where, P_{ch_max} is maximum dispatchable charge and discharge power of a charge-discharge system. P_{tc_max} is maximum dispatchable charge and discharge power of cascade battery energy storage system.

The charge-discharge power of the charge-discharge system also needs to meet the charging demand of electric vehicles, and has a certain reserve margin.

$$Q_{ch}^{0} + \sum\nolimits_{1}^{t} P_{ch}^{t} \Delta t + P_{e}^{t} \Delta t \geq (1 + k_{ch}) Q_{car}^{t} \tag{14}$$

Where, Q_{ch}^{0}, k_{ch} are total initial battery charge and reserve capacity factor of the charge-discharge system.

At the same time, the system has the relationship of power balance:

$$P_{load}^{t} + P_{ch}^{t} + P_{tc}^{t} = P_{th}^{t} + P_{wind}^{t} + P_{solar}^{t} \tag{15}$$

Where, P_{th}^{t} is generating power of thermal power plant in time period t.

3.3 Intelligent Optimization Algorithm

With the rapid development of computer technology, intelligent calculation method in the application of the smart grid scheduling more and more widely. Centralized charging stations in order to charge and discharge within the scheduling strategy in each period should not only consider the charging and discharging system, storage system and network between charge and discharge power, but also in electricity de-mand, charging station capacity and many other constraints. The solution process is relatively complex. Therefore, this paper uses GA algorithm to solve the problem.

GA algorithm based on population coding simulation of the chromosome of nature, through the selection, crossover and mutation to change the information in the individual chromosomes. Calculate the fitness value of each individual in a population and the evolution law of retaining high fitness individuals. By setting the algorithm to simulate the natural evolution algebra iteration, populations evolve finally met the objective function of optimal group. Finally, the optimal individual in the group is obtained through global search.

The application of GA algorithm in this paper is divided into the following steps:

Coding. In order to facilitate the calculation of crossover variation, genetic algorithm encodes the original data into gene sequence. In this paper, the charging and discharging pow-er between the charging and discharging system and the power grid, the energy storage system and the power grid, and the charging and discharging system and the energy storage system in each period of time are coded by binary.

Initial Group Setting. Based on the 2.2 section as the constraint conditions for random initialization of population.

The Hybrid. The process of genetic evolution is divided into selection, crossover and variation. Firstly, the fitness of N generation individuals is used as reference index to select two N generation individuals with higher fitness as crossover objects. Second, the partial structure of two N-generation individuals is replaced and recombined in a certain way to generate the N + 1 generation individuals inheriting some genetic characteristics of the N-generation individuals. Finally, a new gene sequence will be generated by variation with a certain probability during the crossover process.

Fitness Function. The fitness function is introduced to simulate the degree of adaptation of the organ-ism in the environment. The higher the fitness value of the individual is in the process of group iteration, the more conducive the individual is to the current environment. At the time of iteration to the next generation, the higher the strain value of the gene sequence has a high probability of being inherited to the next generation population.

4 Simulation Results and Analysis

According to the probability characteristics of wind speed, irradiance and driving characteristics of electric vehicles, this paper adopts the Monte Carlo simulation method for simulation of distributed power supply. In order to ensure continuity, the variance coefficients of adjacent simulation data of wind power and photovoltaic power generation output are both less than the set value, and the maximum value is set as 0.05%.Monte Carlo simulation extraction times for 1000 times, wind power, photovoltaic power generation and electric discharge planning simulation parameters as shown in Table 1. 24 h of wind power and photovoltaic power output as shown in the figure below.

Table 1. Simulation parameters of wind power, photovoltaic power and EVs

Wind power	Solar power	EV	
Access wind speed: 3 m/s	PV panel area: 19.81 m^2	Discharging Aspiration	
Rated wind speed: 15 m/s	Solar irradiance: 1000 w/m^2		
Excision wind speed: 25 m/s	Conversion efficiency: 0.15	[0.75, 1]	100%
Weibull-scale: 9.2	Beta-scale:0.95	[0.5, 0.75]	45%
Weibull-shape: 2.6	Beta-shape:0.95	[0, 0.5]	0
Rated power:100 Mw		Number of EVs:800	

According to the daily load curve of a region and the output of wind power generation and photovoltaic power generation, the forecast curve of daily equivalent load after the integration of wind power and photovoltaic power generation into the grid and the load curve of electric vehicles simulated by Monte Carlo are obtained as shown in Fig. 2:

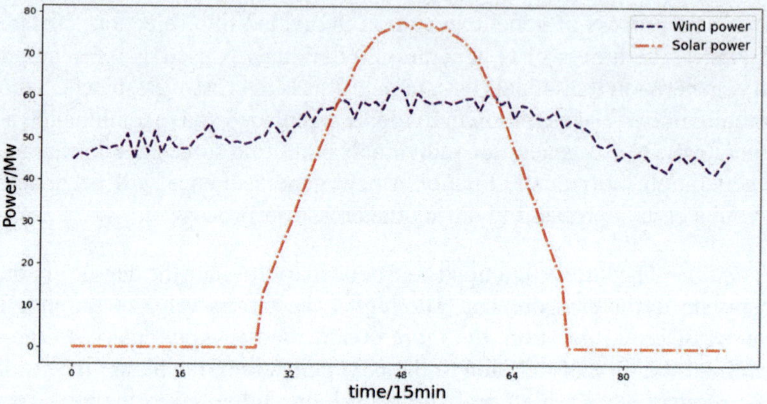

Fig. 1. Wind power generation and photovoltaic power generation curve.

The following conclusions can be drawn from the picture above:

1) 0:00 to 5:00 is the trough period of municipal power load. Wind power can relieve peak load regulation pressure of some thermal power units. The peak period of urban power load is from 14:00 to 0:00, while the peak shaving pressure of thermal power units has not been alleviated due to small wind power output and no photovoltaic power generation at night.
2) There is overlap between peak and valley periods of electric vehicle charging load and peak and valley periods of equivalent daily load in the city. If the charging and discharging of EV are not managed in an orderly way, superposition of EV charging

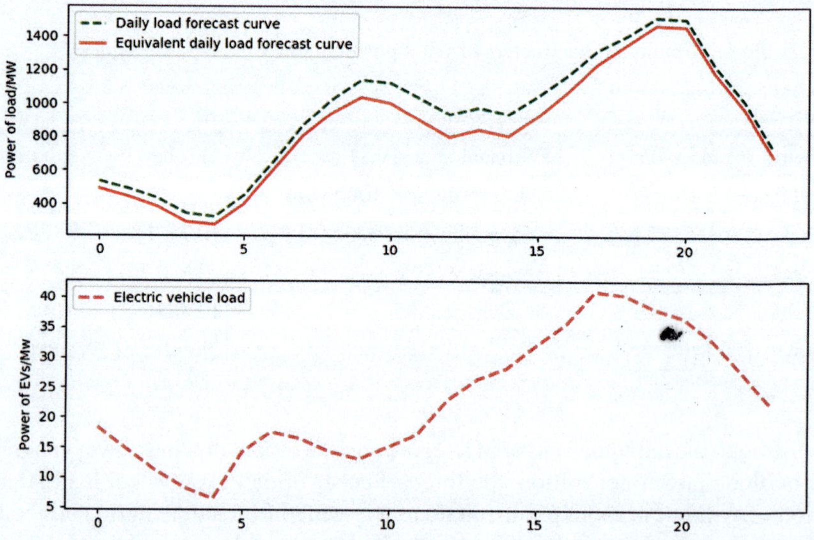

Fig. 2. Average daily equivalent load and electric vehicle load.

load will lead to peak increase of equivalent load peak, which will aggravate peak-valley difference of power grid and bring adverse effects on stable operation of power grid.

A centralized EV charging station with a cascade battery energy storage system is established according to the charging and battery changing needs of electric vehicles in a certain area. The charging station is regulated on the basis of giving priority to the consumption of wind power generation and photovoltaic power generation resources. Known charging stations have some extra full battery charge and discharge system, the number is set to the user, in 10% of the electricity demand. According to the superimposed equivalent daily load curve of wind power and photovoltaic power generation and load curve of electric vehicles. GA algorithm is adopted to develop the scheduling plan of charging stations, so that the charging stations work in different periods of time under the operation state of maximizing consumption of wind power and photovoltaic power generation and peak clipping and valley filling. The optimized running state of the charging station obtained through simulation is shown in Fig. 3.

The charging station regulated can adjust the charging and discharging operating states in different periods according to the daily equivalent load curve. In the trough period of power consumption, both the charging and discharging system and the energy storage system are charged, which can promote the consumption of wind power on the one hand, and serve as a reserve for electric vehicles to change electricity in the daytime on the other hand. In the peak of power consumption, the charging and discharging system and the energy storage system can reduce the charging and discharging to the power grid to reduce the peak-regulating pressure of thermal power units and reduce the peak-valley difference of the power grid.

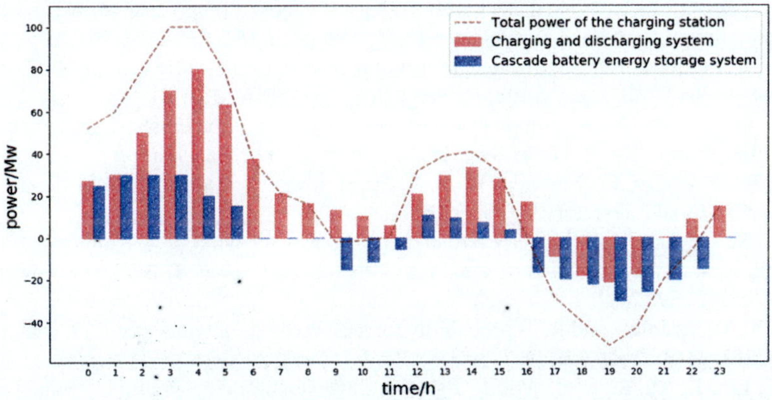

Fig. 3. Optimized running state of charging station

5 Conclusion

In this paper, the operating characteristics of electric vehicles and the influencing factors of wind power generation and photovoltaic power generation are analyzed in detail. The charging and discharging model of electric vehicles, wind power generation and photovoltaic power generation are modeled by the Monte Carlo simulation method. Secondly, on the basis of considering peak clipping and valley filling and absorbing renewable energy, the regulation strategy model of EV charging station is modeled, and the operating state of charging station is solved by the GA algorithm. Finally, put forward by this paper is verified by an example of the model can effectively ease the load curve, given to promote wind power photovoltaic and thermal power unit load pressure reduction.

Although the larger the capacity of the charging station's energy storage system is, the better the suppressing effect of the charging station on the net load curve will be. However, high energy storage capacity will increase the initial investment. So the optimal allocation of electric vehicle charging station energy storage system capacity issues for further research. In addition, the genetic algorithm for the selection of the initial population is dependent, the potential ability of parallel mechanism has not been fully utilized. The future will be combined with some heuristic algorithm improved genetic algorithm, to explore more efficient solutions.

References

1. Chen, R., Sun, H., Guo, Q., et al.: Reducing generation uncertainty by integrating CSP with wind power: an adaptive robust optimization-based analysis. IEEE Trans. Sustain. Energy **6**(2), 583–594 (2015)
2. Mu, G., Guo, G., Li, J., et al.: The control strategy of energy storage externality for reducing wind curtailment from wind farm cluster. Energy Procedia **152**, 233–239 (2018)
3. Darabi, Z., Ferdowsi, M.: Aggregated impact of plug-in hybrid electric vehicles on electricity demand profile. IEEE Trans. Sustain. Energy **2**(4), 501–508 (2011)
4. Xu, M., Qiao, Y., Lu, Z.: A comprehensive error evaluation method for short-term wind power prediction. Autom. Electr. Power Syst. **35**(12), 20–26 (2011). (in Chinese)
5. Bessa, R., Monteiro, C., Miranda, V., et al.: Wind power forecasting state-of-the-art 2009. Argonne National Laboratory, Argonne, Illinois (2009)
6. Tewari, S., Geyer, C.J., Mohan, N.: A statistical model for wind power forecast error and its application to the estimation of penalties in liberalized markets. IEEE Trans. Power Syst. **26**(4), 2031–2039 (2011)
7. Wang, S., Yu, J.: Joint conditions probability forecast method for wind speed and wind power. Proc. CSEE **31**(7), 7–15 (2011)
8. Shi, Y., Liu, J., Xu, S., et al.: Wind - light - storage stochastic programming model of distribution network considering time series characteristics. Power Syst. Prot. Control **47**(10), 23–32 (2019)
9. Michael, K.M., Kevin, S., Robert, P.: Impacts assessment of plug-in hybrid vehicles on electric utilities and regional U.S. power grids part1: technical analysis. Pacific Northwest National Laboratory, Washington (2007)
10. Claire, W.: Plug-in hybrid electric vehicle impacts on hourly electricity demand in the United States. Energy Policy **39**(6), 3766–3778 (2011)

11. Gan, L., Xiao-Ping, Z.: Modeling of plug-in hybrid electric vehicle charging demand in probabilistic power flow calculations. IEEE Trans. Smart Grid **3**(1), 492–499 (2012)
12. Bae, S., Kwasinski, A.: Spatial and temporal model of electric vehicle charging demand. IEEE Trans. Smart Grid **3**(1), 394–403 (2012)
13. Denholm, P., Short, W.: An evaluation of utility system impacts and benefits of optimally dispatched plug-in hybrid electric vehicles. National Renewable Energy Laboratory, Colorado (2006)
14. Henrik, L., Willett, K.: Integration of renewable energy into the transport and electricity sectors through V2G. Energy Policy **36**(9), 3578–3587 (2008)
15. Biswas, P., Suganthan, P.N., Qu, B.Y., et al.: Multi objective economic-environmental power dispatch with stochastic wind-solar-small hydro power. Energy **150**(2), 1039–1057 (2018)

Evaluation Model of Power Staff Skill State Based on Improved Analytic Hierarchy Process_Fuzzy Comprehensive Evaluation

Xiuming Wang[1](✉), Zhaoyang Qu[2], Lichao Wang[1], Miao Li[1], Chunhui Liu[1], Yaowei Liu[3], Zhongfeng Kan[4], and Huashun Li[4]

[1] Training Center of Jilin Province, State Grid Corporation of China, Changchun, China
[2] School of Computer, Northeast Electric Power University, Jilin City, Jilin, China
[3] State Grid Jilin Electric Power Co., Ltd., Changchun, China
[4] State Grid Jilin Electric Power Co., Ltd., Jilin Power Supply Company, Jilin, China

Abstract. In view of the problem that the State Grid Corporation cannot visually reflect the comprehensive level of power employees' skills in skill evaluation, this paper proposes a power employee skill status evaluation model based on improved analytic hierarchy process_fuzzy comprehensive evaluation. Researchers establish a power employee skill evaluation index system based on the research situation and the construction principles of the index system. Aiming at the problem that the traditional analytic hierarchy process adopts a single weight calculation method, it is easy to cause a large difference in weighting due to different method selections, and an improved analytic hierarchy process is proposed to assign weights to indicators. The improved analytic hierarchy process uses geometric average method, arithmetic average method, eigenvector method and least square method to calculate weights respectively and takes average value instead of single method to calculate weights. The grading standard calculates the membership degree of the index. The researcher obtains the evaluation result of the skill status of the electric power staff through the corresponding fuzzy evaluation calculation. The researcher applies the improved analytic hierarchy process_fuzzy comprehensive evaluation to the comprehensive evaluation of the skills of electric power employees. The effectiveness of the model is verified by calculating the actual data of employees in a training center of a certain province of the State Grid.

Keywords: State Grid · Index weight · Evaluation index system · Improved analytic hierarchy process · Fuzzy comprehensive evaluation

1 Introduction

With the comprehensive advancement of the construction of the Energy Internet system for power grid companies, many applications and management systems continue to go online. As an important part of the power system, the State Grid Training Center's safe and reliable operation plays a very important role in maintaining the stability of the power system. Having a group of high-quality electric power employees who adapt to

S.-C. Chu et al. (Eds.): ICGEC 2021, LNEE 833, pp. 560–572, 2022.
https://doi.org/10.1007/978-981-16-8430-2_51

the new situation and have strong skills is the basis for ensuring the safe and reliable operation of the State Grid training system. Staff skill training is an effective way to improve the professional skills of electric power employees. Although many electric power companies have formulated relevant assessment and evaluation systems for the development and effects of training, it is still difficult to make effective conclusions about the training effects. There is still no breakthrough in the research on the comprehensive evaluation of the skills of electric power employees. The research of training evaluation is still in the exploratory period, and the evaluation methods and evaluation modes need to be improved. Literature [1, 2] uses the analytic hierarchy process to determine the index weight. It calculates the membership degree of the index by the grade division standard. it passes the corresponding fuzzy evaluation Calculate to obtain the evaluation result. The subjective weight of the index parameters in the literature [3, 4] is determined by the analytic hierarchy process. The objective weight of the index parameters is determined according to the entropy weight method. The optimal combination weight of the model is selected through the minimum variance theory, which effectively overcomes the weight of the evaluation model. Subjective influence; Finally, the factor set, evaluation set and membership function of the model are established through the fuzzy comprehensive analysis method, and D-S evidence theory is introduced to realize the integration of various types of index parameters. Literature [5] proposed the effectiveness evaluation based on analytic hierarchy process and fuzzy comprehensive evaluation. Based on the analytic hierarchy process and fuzzy comprehensive evaluation, establish an evaluation index system to determine the weight of each index. Quantitative and specific evaluation based on real data as an example provides a quantitative basis for the development and use of accurate assurance models.

However, the above methods lack a scientific and reasonable index system. The selection of the evaluation index is mainly based on the importance of the index in the evaluation model. Evaluation indicators should select important indicators and discard unimportant indicators. Therefore, important indexes should be selected and unimportant indexes should be abandoned. Based on the repeated investigation of a provincial power company, this paper follows the construction principle of the index system to establish the power staff skill evaluation index system [6, 7]. Therefore, this paper proposes an assessment model of power employees skill status based on improved analytic hierarchy process_fuzzy comprehensive evaluation. The target layer, criterion layer, index layer and each index weight are determined by improved analytic hierarchy process. The multi-level fuzzy comprehensive evaluation method is used to comprehensively evaluate the skill status of power employees.

2 Skills Evaluation Index System of Electric Power Employees

For a complex system, if you want to establish a scientific reliability evaluation index system, you need to comprehensively understand the reliability of the system from different aspects, understand the nature of the index system, and the principles for establishing the index system. The indicators of the scientific system can only be established after the above steps [8]. According to the analysis results of the influencing factors of electric power employees' skills, this paper designs an index screening method. The method flow includes the following 6 steps. As shown in Fig. 1.

1) Step 1: Primary selection of evaluation indicators. The preliminary selection of evaluation indicators refers to the collection of relevant data to obtain a large number of existing relevant evaluation indicator sets. On this basis, the evaluation indicators are fully sorted out, the connotation of the indicators is clarified, and the concept of the indicators is clearly described.
2) Step 2: Determination of indicator attributes. Index attributes refer to the principles that should be followed in the screening process. For example, the rationality, necessity, and collectability of indicators represent various screening principles for evaluating indicator performance.
3) Step 3: Assignment of indicator attributes. Regarding the index attribute as the evaluation target, assign a value to each primary selected index. The value range of the attribute value is [0, 1]. The larger the attribute value, the higher the degree of satisfaction of the evaluation index to the evaluation criterion.
4) Step 4: Aggregation of indicator attribute values. After processing steps 2 and 3, each evaluation index corresponds to multiple attribute values. The screening process needs to use aggregation functions to aggregate multiple attribute values into a single value.
5) Step 5: Sort the evaluation indicators. Sort the evaluation indicators according to the aggregation results. The larger the aggregation result value, the better the performance of the evaluation index.
6) Step 6: Screening of evaluation indicators. On the basis of the evaluation index ranking, according to the operation quality evaluation target and the evaluation environment, the threshold value of the index ranking value is set, and the index set used to establish the evaluation index system is determined and screened. In the indicator screening method, steps 1, 2, 3, and 4 are the key links, and are the most important for indicator screening.

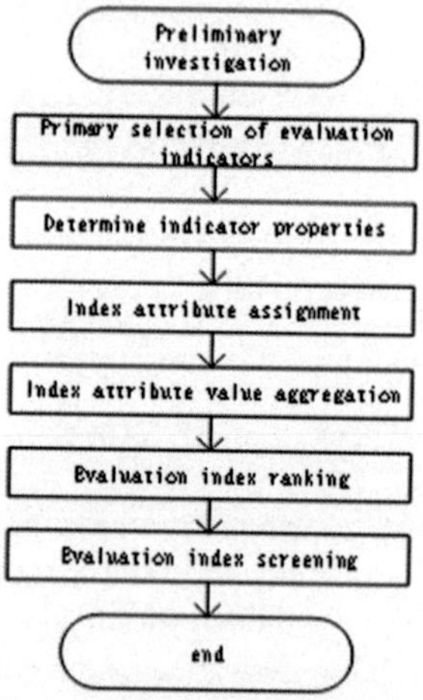

Fig. 1. Process of selecting index system

In order to achieve an accurate assessment of the skill level of the employees of the State Grid, this paper established a three-tiered evaluation system for the skill index of electric power employees. Respectively from the basic knowledge of professional ethics, basic theory knowledge, quality management knowledge and basic knowledge of safety production four angles to evaluate the skills of power employees.

Based on the repeated investigation of a provincial power limited company, following the construction principle of index system, the power staff skill evaluation index system is established as follows. In the first-level subsystem, there are four first-level indicators, namely, basic knowledge of professional ethics, basic theoretical knowledge, quality management knowledge and basic knowledge of safety production. In the secondary subsystem, the basic knowledge of professional ethics can be subdivided into five secondary indicators, the basic theoretical knowledge can be subdivided into five secondary indicators, the quality management knowledge can be subdivided into four secondary indicators, and the basic knowledge of safety production can be subdivided into four secondary indicators, a total of 18 indicators. The skill level of electric power employees is evaluated from four aspects. Basic knowledge of professional ethics, basic theoretical knowledge, quality management knowledge and basic knowledge of safety production, which comprehensively reflects the professional skill level of electric power employees. The power staff skill evaluation index system is shown in Fig. 2.

3 Based on Improved Analytic Hierarchy Process_Fuzzy Comprehensive Evaluation Method

The skill evaluation of electric power employees is a comprehensive evaluation problem involving multiple factors. Scientific index evaluation system is the basis for correct evaluation of electric power employees skill status. Reasonable evaluation method is the key to effective evaluation of employees' skills. This study compares the existing staff evaluation criteria of different methods, considers the comprehensive principle, establishes a reasonable index system, and combines the improved analytic hierarchy process [9] and fuzzy comprehensive evaluation method [10, 11] to construct an improved analytic hierarchy process-fuzzy comprehensive evaluation model to evaluate the skill status of electric power employees, reduce the subjectivity and fuzziness of the evaluation results caused by the influence of subjective evaluation indexes, and improve the accuracy. The basic steps are as follows.

3.1 Calculation of the Comprehensive Weight of Index

Analytic hierarchy process generally calculates index weight according to four steps. Introduce as follows.

Step 1. Construct hierarchical structure model. By analyzing the overall object, the various influencing factors are divided into different levels according to different attributes and classification. Low level factors affect high level, high level factors dominate low level.

Step 2. Construct all judgment matrices in each level. For all elements in each layer, its importance needs to be quantified. The importance difference between the two elements

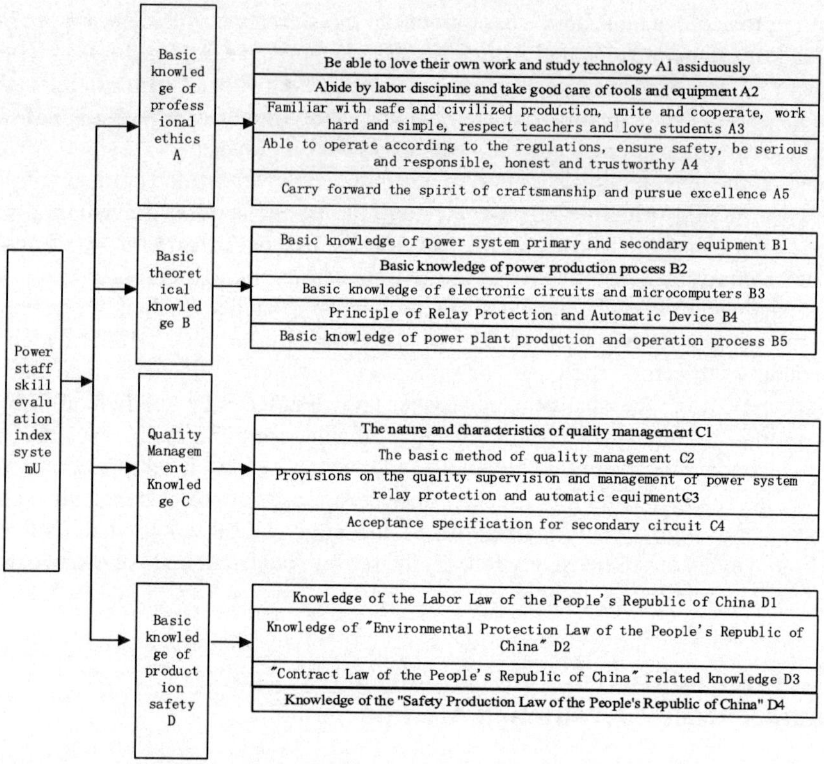

Fig. 2. Evaluation index system of power staff skills

is quantified by number 1–9. Construction of judgment matrix

$$A = \left(a_{ij}\right)_{n*n} \tag{1}$$

Step 3. Calculate the weight vector and do consistency check. The maximum eigenvalue of the judgment matrix is calculated, and then the consistency index CI is calculated. $\lambda_{\max}$ is the maximum eigenvalue of the judgment matrix.

$$CI = \frac{\lambda_{max} - n}{n - 1} \tag{2}$$

The consistency is tested by testing the consistency ratio, as follows.

$$CR = \frac{CI}{RI} \tag{3}$$

Step 4. Calculate the combination weight vector and do consistency check. After step 3, the combination weight of each element to the target is calculated, and the consistency is tested by testing the consistency ratio.

The traditional analytic hierarchy process only uses a single method to calculate the weight, and usually only uses the eigenvector method. Although the calculation process of single method is simple, it is easy to cause inaccurate weighting due to the lack of stability of single method. Therefore, this paper uses the improved method to calculate the weight vector W. The improved method is a combination algorithm including geometric average method, arithmetic average method, eigenvector method and least square method.

(1) Weight W calculation by geometric average method.

$$W_i = \frac{\left(\prod_{j=1}^{n} a_{ij}\right)^{\frac{1}{n}}}{\sum_{i=1}^{n}\left(\prod_{j=1}^{n} a_{ij}\right)^{\frac{1}{n}}}, \quad i = 1, 2, \cdots, n \tag{4}$$

In the above formula, a_{ij} is the element in the constructed judgment matrix. The elements in the judgment matrix are multiplied by rows and then the n-th power is opened. Finally, the obtained vector is normalized to obtain the weight vector.

(2) The arithmetic average weight $\boldsymbol{W}$.

$$W_i = \frac{1}{n}\sum_{j=1}^{n}\frac{a_{ij}}{\sum_{k=1}^{n} a_{kj}}, \quad i = 1, 2, \cdots, n \tag{5}$$

(3) Weight $\boldsymbol{W}$ calculated by eigenvector method.

$$\boldsymbol{AW} = \lambda_{\max}\boldsymbol{W} \tag{6}$$

(4) Least square method to calculate weight $\boldsymbol{W}$.

$$\min Z = \sum_{i=1}^{n}\sum_{j=1}^{n}\left(a_{ij}w_j - w_i\right)^2 \tag{7}$$

$$\text{s.t.} \sum_{i=1}^{n} w_i = 1 \tag{8}$$

Type in W_i in the above formula is the element in the weight vector, representing the weight of each index.

When calculating weights for the same model, even if the same judgment matrix is constructed, the weight values obtained by different weight calculation methods may still be very different, and the weight obtained is not reliable. Moreover, the calculation difference may be larger for different models, and the applicability of the algorithm is poor. Therefore, this paper uses the above four methods to calculate the weights respectively, and takes the average value of the weights obtained by the four methods as the final weight of each index. Using the improved analytic hierarchy process to calculate the weight comprehensively considers the advantages of various methods, which can make up for the shortcomings of single method in reliability and applicability.

3.2 Determination of Membership Matrix

1) Positive indicators

The positive index refers to the higher the index value, The better the skill status of power employees, such as promoting the craftsman spirit, pursuing excellence and basic knowledge of electric power production process.

When $x_{im} > v_{im1}$,the value of the index x is greater than the grade standard "excellent", the membership degree corresponding to this index is 1, and the membership degrees of other grades are 0.

$$r_{im1} = 1, r_{im2} = r_{im3} = r_{im4} = 0 \tag{9}$$

When the value of the index x_{im} is between the level criteria v_{imj} and $v_{im(j+1)}$, the degree of membership of the $(j + 1)$ level is as follows.

$$r_{im(j+1)} = \frac{V_{imj} - x_{imj}}{V_{imj} - V_{im(j+1)}}, j = 1, 2, 3 \tag{10}$$

The degree of membership of the j level is as follows.

$$r_{imj} = 1 - r_{im(j+1)}, j = 1, 2, 3 \tag{11}$$

The membership degree of other levels is 0.

When the value of the index x_{im} is less than the grade standard *difference*, the membership degree corresponding to this index is 1, and the membership degree of other grades is 0.

$$r_{im1} = r_{im2} = r_{im3} = 0, r_{im4} = 1 \tag{12}$$

2) Reverse indicators

Reverse indicators refer to indicators such as the basic methods of quality management and the acceptance of secondary loops when the value of the indicator is larger, the skills of electric power employees become worse.

When $x_{im} < v_{im1}$, the value of the index x_{im} is less than that excellent, the membership degree corresponding to this index is 1, and the membership degree of other grades is 0.

$$r_{im1} = r_{im2} = r_{im3} = 0, r_{im4} = 1 \tag{13}$$

When the value of the index x_{im} is between the level criteria v_{imj} and $v_{im(j+1)}$, the degree of membership of the $(j + 1)$ level is as follows.

$$r_{im(j+1)} = \frac{x_{imj} - V_{imj}}{V_{im(j+1)} - V_{imj}}, j = 1, 2, 3 \tag{14}$$

The degree of membership of the j level is as follows.

$$r_{imj} = 1 - r_{im(j+1)}, j = 1, 2, 3 \tag{15}$$

The membership degree of other levels is 0.

When $x_{im} < v_{im4}$, the value of the index x_{im} is greater than difference, the membership degree corresponding to this index is 1, and the membership degrees of other grades are 0.

$$r_{im1} = r_{im2} = r_{im3} = 0, r_{im4} = 1 \tag{16}$$

In formulas (1)–(8) x_{im} is the actual measured value of the m-th index in the i-th quasi-measurement layer, r_{imn} is the degree of membership of the n-th level of the m-th index in the i-th quasi-measurement layer, v_{imj} is the j-th grade standard.

4 Fuzzy Comprehensive Evaluation Model of Electric Power Staff Skills

Fuzzy evaluation is divided into the first-level fuzzy evaluation of the index layer and the criterion layer and the second-level fuzzy evaluation of the criterion layer and the target layer. The first-level fuzzy evaluation adopts the formula.

$$B_i = W_{ci} \circ R_i = (w_{ci1}, w_{ci2}, \cdots, w_{cim}) \circ \begin{bmatrix} r_{i11} & r_{i12} & r_{i13} & r_{i14} \\ r_{i21} & r_{i22} & r_{i23} & r_{i24} \\ \wedge & \wedge & \wedge & \wedge \\ r_{im1} & r_{im2} & r_{im3} & r_{im4} \end{bmatrix} = (B_{i1}, B_{i2}, B_{i3}, B_{i4}) \tag{17}$$

In the above formula, B_i is the fuzzy evaluation of the i-th criterion layer, w_{ci} is the weight set of the corresponding i-th criterion layer in the index layer,R_i is the membership degree matrix of the v-th criterion layer, and w_{cim} is the m-th in the i-th quasi-measurement layer The weight of the index to the criterion layer, r_{imn} is the degree of membership of the m-th index in the i-th quasi-measurement layer, B_{in} is the degree of membership of the nth level of the i-th criterion layer, $\circ$ is the fuzzy operator, and is calculated by $M(\bullet, +)$.

The second-level fuzzy evaluation adopts the formula.

$$S = W_B \circ B = (w_{B1}, w_{B2}, w_{B3}, w_{B4}) \circ \begin{bmatrix} W_{C1} \circ R_1 \\ W_{C2} \circ R_2 \\ W_{C3} \circ R_3 \\ W_{C4} \circ R_4 \end{bmatrix} = (S_1, S_2, S_3, S_4) \tag{18}$$

In the above formula, S is the fuzzy evaluation of the target, W_B is the weight set of the criterion layer to the target layer, B is the first-level fuzzy evaluation, W_{Bi} is the weight of the i-th criterion layer, w_{ci} is the weight set of the i-th criterion layer corresponding to the index, R_i is the membership degree matrix of the i-th criterion layer, and S_n is the n-th degree of membership of the target, $\circ$ is the fuzzy operator, and is calculated by $M(\bullet, +)$.

5 Experimental Verification

In order to verify the reliability of the improved analytic hierarchy process_fuzzy comprehensive evaluation model proposed in this paper, six groups of skill assessment data of electric power employees are randomly selected from the professional skills training platform of the power grid enterprise training center personnel. The experimental data is shown in Table 1.

Table 1. Staff skill assessment data

Serianumber	A	B	C	D
1	79	81	85	93
2	80	70	96	78
3	75	86	78	82
4	70	78	83	80
5	90	81	74	80
6	84	89	91	70

According to the actual value of each indicator of the assessment data, formulas (1)–(8) are used to calculate the membership degree of each indicator. Refer to Table 2 for the evaluation index data of the total power staff's skill status. The fuzzy evaluation formulas (9) and (10) are used to obtain the evaluation results of the skills of the total power staff, as shown in Table 3.

From the first-level fuzzy evaluation results of the criterion level, it can be seen that the evaluation of the skill status of the total power staff is more polarized, because there is a big difference between "Being able to love their own work and diligently studying technology" and "Promoting the spirit of craftsmanship and pursuing excellence"; The quality management knowledge evaluation level is "Medium", the power system relay protection mastery has a greater impact on the results, the basic theoretical knowledge evaluation level is "excellent", the mastery of the overall skill.

level is good, and the professional ethics evaluation level is Medium, according to the results of the first-level fuzzy evaluation, corresponding measures can be taken to improve the skill level of employees in response to specific problems in the skill level. Finally, the total skill level evaluation set D = [0.1693, 0.1451, 0.4505, 0.2350] is obtained. According to the principle of maximum membership degree, the total skill level evaluation status is "Medium", and the comprehensive inspection of the training center department and other departments to assess the mastery status of the total skill level The results are similar, indicating that the evaluation system more objectively reflects the actual status of the power staff's skills and knowledge in the past year (Fig. 3).

Table 2. Evaluation index weights and membership degrees of total skill level status

Criterion layer	Index layer	Weights	Index value	Membership
Basic knowledge of professional ethics A	Be able to love their job and study technology assiduously	0.3333	2.85	[0, 0.8500, 0.1500, 0]
	Abide by labor discipline and take good care of tools and equipment	0.3333	1.61	[0, 0.6100, 0.3900, 0]
	Familiar with safe and civilized production, unite and cooperate, work hard and simple, respect teachers and love apprentices	0.3333	0.93	[0, 0, 0, 1]
	Be able to operate according to the rules, ensure safety, be serious and responsible, honest and trustworthy	0.3333	1.65	[0, 0, 0, 1]
	Carry forward the spirit of craftsmanship and pursue excellence	0.3333	2.67	[0, 0.6100, 0.1500, 0]
Basic theoretical knowledge	Basic knowledge of power system primary and secondary equipment	0.3936	1	[0, 0, 1, 0]
	Basic knowledge of power production process	0.1375	0.97	[1, 0, 0, 0]
	Basic knowledge of electronic circuits and microcomputers	0.0753	0.31	[1, 0, 0, 0]
	Principles of Relay Protection and Automatic Devices	0.3936	0.46	[0, 0, 0, 1]
	Basic knowledge of power plant production and operation process	0.1375	0.51	[0, 0, 0, 1]

(continued)

Table 2. (*continued*)

Criterion layer	Index layer	Weights	Index value	Membership
Quality management knowledge	The nature and characteristics of quality management	0.4236	7	[0, 0, 0.500, 0.500]
	Basic methods of quality management	0.2270	0.95	[1, 0, 0, 0]
	Provisions on the quality supervision and management of power system relay protection and automatic equipment	0.0753	0.01	[1, 0, 0, 0]
	Acceptance standard of secondary circuit	0.3936	1	[1, 0, 0, 0]
Basic knowledge of production safety	Knowledge of the "Labor Law of the People's Republic of China"	0.5252	41.7	[0, 0, 0.5850, 0.4150]
	Knowledge of the Environmental Protection Law of the People's Republic of China	0.2518	52.14	[0, 0, 0.6430, 0.3570]
	Knowledge of the "Contract Law of the People's Republic of China"	0.2229	75	[0, 0.500, 0.500, 0]
	Knowledge of the "Safety Production Law of the People's Republic of China"	0.2516	62	[0, 0.400, 0.400, 0]

Table 3. Fuzzy evaluation results of total staff skill level

Target layer	Criterion layer	First-level evaluation	Secondary evaluation
Staff skill level	Basic knowledge of professional ethics A	[0, 0.4866, 0.1800, 0.3333]	[0.1693, 0.1451, 0.4505, 0.2350]
	Basic theoretical knowledge B	[0.2128, 0, 0.3936, 0.3936]	
	Quality Management Knowledge C	[0.5763, 0, 0.2118, 0.2118]	
	Basic knowledge of production safety D	[0, 0.1114, 0.5807, 0.3079]	

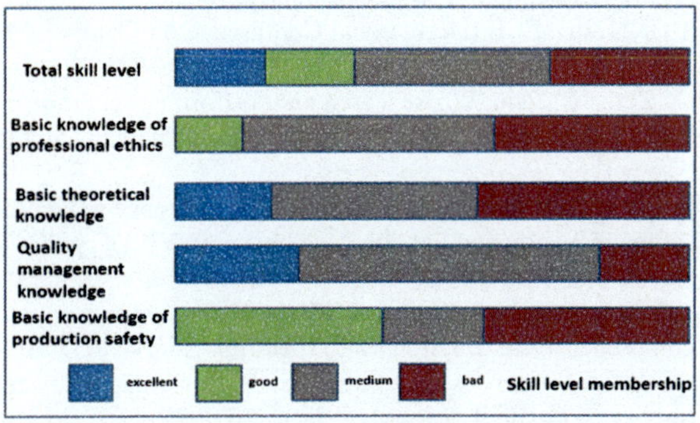

Fig. 3. Fuzzy evaluation result of total staff skill level

6 Conclusion

Aiming at the problem that the skill evaluation of electric power employees cannot visually and intuitively reflect the comprehensive skill level of electric power employees, this paper proposes an improved analytic hierarchy process_fuzzy comprehensive evaluation of electric power employees' skill status evaluation model. First, the general idea of comprehensive evaluation of power employees' skills is given; secondly, the power employee skills evaluation index system is established based on the research situation and the principles of index system construction; Then propose a method based on improved analytic hierarchy process_fuzzy comprehensive evaluation, The improved analytic hierarchy process is adopted, and the original single calculation method is replaced with four weight calculation methods, which weakens the difference in weight calculation caused by different weight calculation methods. This method makes the weighting result more reliable. Calculate the corresponding membership function according to the grade standard, and carry out two fuzzy evaluations, and determine the skill status of electric power employees according to the principle of maximum membership; Finally, the improved analytic hierarchy process_fuzzy comprehensive evaluation method is applied to the comprehensive evaluation of electric power employees' skills. The experimental results show that the comprehensive evaluation model can vividly and intuitively reflect the comprehensive skill level of electric power employees, and has practical application value.

Acknowledgment. This work was funded by the "Visualization Research on Intelligent Analysis of Power Staff Skill State" (Project Number: SGJLPX00PGJS2000067).

References

1. Wang, C., Fang, H., Zhu, Y., Pan, G.: Evaluation of river ecological conditions in the northern Jiangsu Plain based on analytic hierarchy process-fuzzy comprehensive evaluation. Rural water conservancy and hydropower in China, pp. 1–9, 16 June 2021. http://kns.cnki.net/kcms/detail/42.1419.TV.20210602.1805.020.html.GIR
2. Shick, R.: Fast R-CNN. In: IEEE International Conference on Computer Vision, pp. 1440–1448. IEEE, Piscataway (2015)
3. Zhang, K., Guo, J., Zeng, B., Zhu, W.: Research on transformer condition evaluation method based on optimal weight and fuzzy comprehensive analysis. J. Wuhan Univ. (Eng. Ed.) **54**(06), 563–570 (2021)
4. He, K., Zhang, X., Ren, S., et al.: Deep residual learning for image recognition. In: The IEEE Conference on Computer Vision and Pattern Recognition (CVPR), pp. 770–778. IEEE Press (2016)
5. Wang, H., Mou, Z., Guo, Y.: Evaluation of UAV precision support effectiveness based on AHP and fuzzy comprehensive evaluation. Ship Electron. Eng. **41**(05), 109–112 (2021)
6. Zhang, L., Wang, X.: Comprehensively evaluate the information system by using the hierarchical analysis method. Intell. Rep. Mag. **1**, 83–85 (2005)
7. Wei, P., Guo, L.: Fuzzy comprehensive evaluation of power supply capacity of distribution network based on demand-side response. Hydropower Energy Sci. **37**(10), 181–185, 208 (2019)
8. Zhang, D., Lin, G., Song, Q., et al.: State evaluation of electric energy metering devices based on grey theory and fuzzy analytic hierarchy process. Electr. Meas. Instrum. **56**(11), 134–139, 152 (2019)
9. Zhang, Y., Tian, M., Pen, H.: Selection of smart meters based on analytic hierarchy process. Electr. Meas. Instrum. **57**(12), 131–136 (2020)
10. Li, X., Liu, Y., Zhu, X.: Research on the evaluation index system and its application of the construction scale of high-voltage distribution network. Proc. Chin. Soc. Electr. Eng. **26**(17), 18–24 (2016)
11. Lu, Y., Wei, Y., Zhao, Q., Li, S.: Power emergency robot path planning based on analytic hierarchy process and improved A~* algorithm. Power Syst. Prot. Control **49**(09), 82–89 (2021)

Information Recommendation Method of Network Blended Skill Learning Based on Personalized Adaptive of Electric Power Staff

Yan Gao[1](✉), Lei Wang[2], Miao Li[1], Chunhui Liu[1], and Lichao Wang[1]

[1] Training Center of Jilin Province, State Grid Corporation of China, Changchun, China
[2] School of Computer, Northeast Electric Power University, Jilin City, Jilin, China

Abstract. In question of low satisfaction and accuracy of traditional staff skill information recommendation learning course, this paper comprehensively uses the relevant raw data generated by various skills training. Modularize the sub-factor sequence using the modular matrix, and calculate the mean value of the correlation coefficient corresponding to each sub-factor sequence. This paper uses the coupling degree function to study the coupling relationship between the network hybrid information, and completes the modular processing of the network hybrid information. This article judges whether the weight value of the network hybrid information is higher than the threshold set in advance, and calculates the weight of the network hybrid information. Define the similarity between the network hybrid information, and obtain the similarity distribution of the network hybrid information in the information concentration. Personalized adaptive learning improves the learning demand orientation of network hybrid information. The purpose of calculating the skill similarity of electric power employees is to design a network hybrid information recommendation algorithm to realize the comprehensive analysis and processing of the skill status of electric power employees. Training managers and trainers can optimize training programs based on the feedback analysis results, and accurately train, consolidate and improve the technical capabilities of employees. Experimental results show that the recommendation algorithm has greatly improved the convergence speed and accuracy.

Keywords: Power staff · Skill learning · Information recommendation

1 Introduction

With the advancement of the construction of the power Internet of Things, various power equipment is becoming more and more intelligent. The electric power industry puts forward higher requirements on the skill level of electric power employees, and the skill status of electric power employees directly affects the smooth development of daily work of the enterprise. However, the existing training system lacks learning and recommending methods for different skill states. How to train the skill level of employees quickly and efficiently has become an important part of the electric power training work.

S.-C. Chu et al. (Eds.): ICGEC 2021, LNEE 833, pp. 573–583, 2022.
https://doi.org/10.1007/978-981-16-8430-2_52

Foreign researchers in order to promote the research of information recommendation system, a large number of information recommendation algorithms and electric power staff behavior learning algorithms have been proposed, which has further promoted the academic research on information recommendation systems. The United States [1] has also applied the information recommendation system to social networks, bringing new challenges and impetus to the long-term development of the information recommendation system. Many domestic scholars [2] have also conducted a lot of in-depth research on information recommendation systems. At the beginning, they introduced [3] personalized service technology into the information recommendation system. By comparing the execution procedures of the existing information recommendation system, they designed a personalized recommendation algorithm for information. Achieved very good results. In reference to the problems of personalized information recommendation, the literature [4] introduced cognitive computing and situational awareness in analyzing the concepts, methods and adaptation aspects of personalized information recommendation, and proposed a personalized information recommendation model and framework. Deepen the connotation of personalized information adaptive recommendation, through the cognitive computing and situational awareness viewpoints and methods, the cognitive computing and situational awareness are integrated, and the skill characteristics of electric power employees are extracted. Literature [5] constructed a framework of personalized information adaptive recommendation model based on cognitive computing and context awareness. In view of the problem that the traditional collaborative filtering algorithm has too large granularity in the description of the learning needs of electric power employees in the hybrid recommendation, it is proposed to apply incremental learning to the design of the hybrid recommendation algorithm. Incremental learning establishes a content-based electric power employee model, combined with a collaborative filtering recommendation mechanism to achieve mixed-information recommendation. However, the information recommendation ability in the existing research still has shortcomings in terms of accuracy, recall and coverage. Based on the above background, this article applies personalized adaptation to the network hybrid information recommendation, thereby improving the ability of the network hybrid information recommendation. It better solves the personalized learning plan recommendation for electric power employees' mastery of different skills during the training process.

2 Design of Network Hybrid Information Recommendation Method

2.1 Modular Processing of Network Hybrid Information

If the network hybrid information recommendation is regarded as a system, modular processing refers to distinguish the key factors of the network hybrid information recommendation, and the sub-factors in the module are reasonably analyzed. If the network hybrid information recommendation is regarded as a system, the modular processing refers to distinguish the key factors of network hybrid information recommendation, and reasonably analyze the sub-factors in the module.

In order to construct and modularize the matrix of personalized adaptive learning based on power employees' skills, we use n subsequences to deal with the factors of network mixed information recommendation.

$$(X_0, X_1, \cdots, X_n) = \begin{pmatrix} x_0(1) & x_1(1) & \cdots & x_n(1) \\ x_0(2) & x_1(2) & \cdots & x_n(2) \\ \vdots & \vdots & \vdots & \vdots \\ x_0(m) & x_1(m) & \cdots & x_n(m) \end{pmatrix} \quad (1)$$

In the formula, m represents the sub-factor of personalized adaptive learning. Using personalized adaptive learning to modularize the sequence of sub-factors.

$$(X_1', X_2' \cdots, X_n') = \begin{pmatrix} x_1'(1) & x_2'(1) & \cdots & x_{nt}'(1) \\ x_1'(2) & x_2'(2) & \cdots & x_{n-1}'(2) \\ \vdots & \vdots & \vdots & \vdots \\ x_1'(m) & x_2'(m) & \cdots & x_n'(m) \end{pmatrix} \quad (2)$$

In order to better and more clearly show the relationship between network mixed information factors and sequences, we calculate the average value of each sub factor sequence and its corresponding correlation coefficient, which is called correlation sequence.

$$r_{i0} = \frac{1}{m} \sum\nolimits_{k=1}^{m} S_i(k) \quad (3)$$

$S_i(k)$ represents the k-th factor in the network hybrid information module i, and the recommended utility of calculating the sub-factor u_{ij} for the network hybrid information.

$$u_{ij} = \begin{cases} \frac{(arcmaxS_i(j) - r_{i0})}{S_i(j)}, & \text{the recommended effect of } u_{ij} \text{ is negative} \\ \frac{(r_n - arcmaxS_i(j))}{S_i(j)}, & \text{the recommended effect of } u_{ij} \text{ is positive} \end{cases} \quad (4)$$

In the formula, arcmax $S_i(j)$ represents the recommended maximum value of network mixed information, $S_i(j)$ represents the maximum coefficient, and r_{i0}.

Represents the statistical number of network mixed information divided from the network mixed information collection.

$$T_m = \left\{ \frac{(u_1 \times u_2 \times \ldots u_m)}{\left[\prod(u_i + u_j)\right]} \right\}^{\frac{1}{m}}, i \neq j \quad (5)$$

The range of T_m value is from 0 to 1, u_m represents the overall propagation effect of network mixed information.

$$u_i = \lambda_{ij} \times u_{ij} \quad (6)$$

In the formula, λ_{ij} represents the hybrid information coupling factor. Although Eq. (6) can be used to calculate the degree of coupling between network hybrid information. For network mixed information recommendation, generally, it cannot well reflect the specific situation of personalized adaptation level. For the above reasons, we set up a function about the degree of coupling coordination to calculate the degree of coupling between systems the mixed information of the network, and the coordination degree function constructed is as follows.

$$\begin{cases} F = \alpha u_1 + \ldots + \beta u_m \\ A = \sqrt{T_m} \times F \end{cases} \tag{7}$$

In the formula, A is used as a function of the degree of coordination of the coupling, T_m represents the coupling degree of the network hybrid information, F represents the coordination index, and α and β represent the overall recommendation weight of the network hybrid information. The sub factor sequence is modularized through the modularized matrix. The correlation coefficient corresponding to each sub factor sequence is calculated and averaged. Finally, the coupling relationship between network mixed information is discussed by using the results of coupling function, and a series of specific modular processing of network mixed information is completed.

2.2 Similarity Calculation of Network Hybrid Information

When solving the similarity of the network hybrid information, the vector space method is mainly used to determine whether the weight value of the network hybrid information is higher than the threshold set in advance, if the weight of the network hybrid information is higher than the threshold value, then extract the information as recommended information. The weight calculation formula for network hybrid information is as follows.

$$\text{Weight}(t)_d = [\alpha(t)_d + \beta(t)_d] \times \text{TFIDF}(t)_d \tag{8}$$

In the formula, Weight(t)d represents the weight of the term t in the network hybrid information d, $\alpha(t)_d$ and $\beta(t)_d$.

Represents the weighting coefficient of the network hybrid information, respectively representing the title weight and ending weight of the word t in the network hybrid information d. For the TFIDF value of the word t in the network hybrid information d, it can be calculated by Eq. (9).

$$TFIDF(t)_d = \frac{TF(t)_d \times \log(N_k/n_i + 0.01)}{\sqrt{\sum_{ted}[TF(t)_d \times \log(N_k/n_i + 0.01)]^2}} \tag{9}$$

In the formula, N_k represents the amount of information in the network hybrid information set, $TF(t)_d$ represents the word frequency of the word t in the network hybrid information d, and n_i represents all the network hybrid information in the network hybrid information set that contains the word t quantity.

Through personalized adaptive learning, the similarity between network hybrid information is defined as follows.

$$Sim(s_i, s_j) = \frac{|C_i \cap C_j|}{|C_i \cup C_i|} \tag{10}$$

In the formula, C represents the set of keywords in the network hybrid information. By calculating the ratio between the number of the same keywords and the total number of keywords between the two network hybrid information s_i and s_j, the value of the two network hybrid information is determined similarity. From Eq. (10), it can be seen that the value range of the network hybrid information similarity is between 0 and 1, where 0 means that the two network hybrid information are completely dissimilar, and 1 means that the two network hybrid information are completely similar. By calculating the similarity of any two network hybrid information in the entire network hybrid information set, the similarity distribution of the information centralized network hybrid information can be obtained, which is represented by the following matrix.

$$S = \begin{bmatrix} s_{11} & s_{12} & \cdots & s_{1n} \\ s_{21} & s_{22} & \cdots & s_{2n} \\ \vdots & \vdots & \ddots & \vdots \\ s_{n1} & s_{n2} & \cdots & s_{nm} \end{bmatrix} \tag{11}$$

In the formula, s_{nm} represents the similarity between network hybrid information s_n and s_m.

By judging whether the weight value of the network hybrid information is exceeding the preset threshold, and weight of the network hybrid information is calculated, and by defining the similarity between the network hybrid information, the information concentration of the network hybrid information is obtained. According to the distribution of similarity, the similarity value of network mixed information is calculated. And from this to design the network mixed information recommendation algorithm, and finally achieve more accurate network mixed information rationalization recommendation.

2.3 Network Hybrid Information Recommendation Algorithm

Each piece of information released by design electric power employees will involve one or more topics, and the characteristics of information released by electric power employees match the topic model of network hybrid information. Use keywords and topics to judge the distribution of search or release information of power employees, and preliminarily judge the knowledge needs and learning orientation of power employees. The definition $C = \{C1, C2, ..., CT\}$ represents the topic set of network hybrid information. For a piece of network hybrid information t, the posterior probability is $p(Ci|t)$, so, The main part vector of the network hybrid information formed by the posterior probability is $(p(C1|t), p(C2|t), ..., p(CT|t))$, T represents the topic set of the network hybrid information Number.. For the orientation of the learning needs of electric power employees, suppose $\{t1, t2, ..., td\}$ represents the network hybrid information set released by the power employees, then the T-dimensional vector $(v1, v2, ..., vT)$ represents the power employees' Orientation of learning needs of formula information.

$$v_i = \frac{1}{d}\sum_{j=1}^{d} p(C_i|t_j) \tag{12}$$

According to the weight of learning needs of electric power employees, we use personalized adaptive learning to improve v_i, and get the following formula.

$$v_i = \frac{1}{d}\sum_{j=1}^{d} \alpha_j p\left(C_i|t_j\right) \tag{13}$$

In the above formula, $\sum d_j = 1, \alpha_j = 1$, α_j represents the weight value of power employees' learning demand for network hybrid information, which can reflect the skill set of power employees for network hybrid information. When the value of αj becomes larger, the greater the power employees' ability to network hybrid information. The higher the preference of the type information.

$$P\left(z_{u_i} = j|Z_{t,\neg i}, t, \varphi, \alpha\right) \propto \frac{P\left(z_{t_i} = j, Z_{t,\neg i}, t|\varphi, \alpha\right)}{P\left(Z_{t,-i}, t|\varphi, \alpha\right)} \tag{14}$$

The topic distribution of the word wn is $Vwn = (v1, v2, ..., vT)$, then the standardized probability of vi is as follows.

$$v_j = \frac{P\left(z_{x_i} = j|Z_{t,\neg i}, t, \varphi \cdot \alpha\right)}{\sum_{j=1}^{T} P\left(z_{u_i} = j|Z_{t,\neg}, t, \varphi, \alpha\right)} \tag{15}$$

The probability formula for network hybrid information t belonging to the j-th topic is as follows.

$$\theta_{t,j} = \frac{n(j,t) + \alpha}{n(t) + T\alpha} \tag{16}$$

According to θ_t, j can define the average skill similarity of power employees in M cycles.

$$sim(U_{k-1}, U_k) = \frac{u_{k-1}}{\mu_{k-1} \times \mu_k} \tag{17}$$

U_{k-1} represents the weight of personalized adaptive learning processing, and U_k represents the weight of personalized adaptive learning skills. In summary, the use of personalized adaptive learning improves the learning demand orientation of network hybrid information. The skill similarity of power employees is calculated and counted, and the recommendation algorithm of network mixed information is designed to meet the reasonable recommendation of network mixed information and make the recommendation results more accurate and efficient.

3 Experimental Results and Analysis

3.1 Establish Experimental Data Set

The information released by 1,000 users and 3,000 users comes from partial data of a training website. The experimental data set includes network hybrid information type, launch time, network demand, number of likes, number of comments, number of fans, and number of views. Experimental data is divided into training set and test set. The number of network hybrid information recommended for users is set to 20. Evaluate the network hybrid information according to the evaluation method, and display the experimental result data. The simulation environment is shown in Fig. 1.

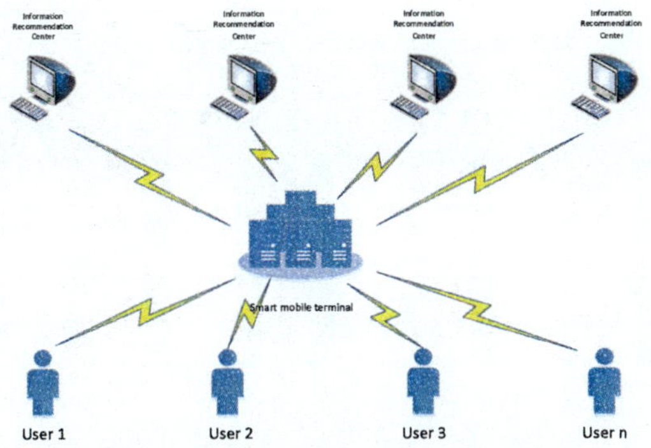

Fig. 1. Simulation environment diagram

3.2 Comparison of Network Hybrid Information Recommendation Methods

This paper applies personalized adaptive learning to the network hybrid information recommendation algorithm, and compares it with literature [4] network hybrid information recommendation algorithm and literature [5] network hybrid information recommendation algorithm. Calculate the hybrid data value to compare the performance of the information recommendation algorithm. The lower the hybrid data value, the better the performance of the network hybrid information recommendation algorithm. When the number of network mixed information is 10, 20, 30, 40, 50, the mixed data Perplexity value of the literature [4] network mixed information recommendation algorithm is 2877.1, 2882.6, 2967.5, 3132.9, 3419.5, respectively. Literature [5] The mixed data Perplexity value of the network hybrid information recommendation algorithm is 2722.9, 2725.1, 2807.5, 2970.8, 3315.9, respectively. The Perplexity values of the hybrid data based on the personalized adaptive network hybrid information are 2671.2, 2688.3, 2762.2, 2891.6, and 3030.8, respectively. The experimental data results are shown in Fig. 2.

It is inferred from the experimental data in Fig. 2 that the Perplexity value of the personalized and adaptive network hybrid information recommendation algorithm is smaller than the two information recommendation algorithms, indicating that the algorithm has better performance.

3.3 Offline Calculation Recommended Experiment

The recommended network hybrid information is recorded as the information required for user sense learning as N. The total recommended network hybrid information is M. There is no recommended, but the number of network hybrid information that users feel learning needs is N. The calculation formula for accuracy and recall is as follows.

$$p = \frac{N}{M} \tag{18}$$

$$p' = \frac{N}{N + N'} \tag{19}$$

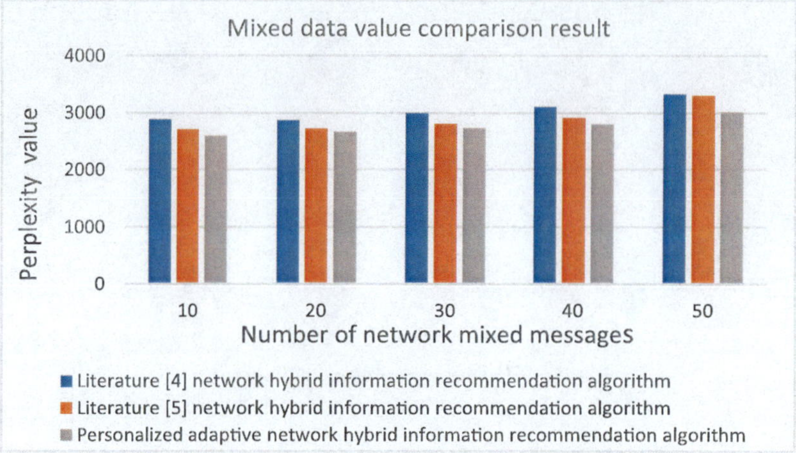

Fig. 2. Comparison results of mixed data values

Use K to represent the number of similar users, and experiment by taking different K values. When the K value is 10, 20, 40, 80, and 90, the accuracy rates are 15.88%, 19.48%, 21.38%, 23.68%, 22.69%, respectively. The corresponding recall rates were 7.11%, 8.84%, 10.01%, 11.12%, and 11.05%. The experimental results are shown in Fig. 3.

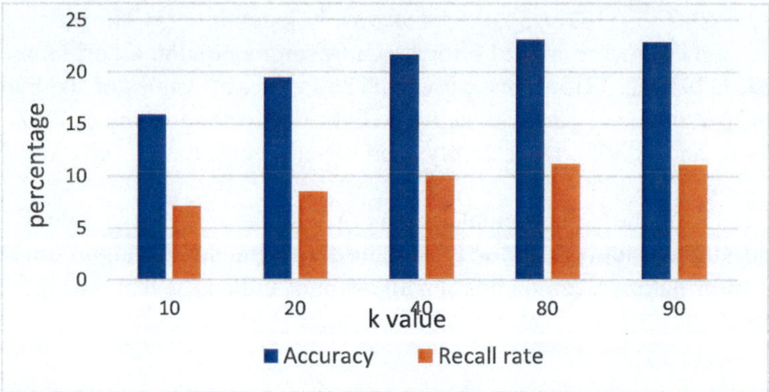

Fig. 3. Performance comparison of network hybrid information recommendation methods under different K values.

From the results in Fig. 3 that when the number of users with high similarity is 80, the accuracy and recall rate of the network hybrid information recommendation are both the highest. In order to verify the advantages of the network hybrid information recommendation method based on personalized self-adaptation, a comparative test was carried out with the literature [4] network hybrid information recommendation method and literature [5] network hybrid information recommendation method. In terms of accuracy, the network hybrid information recommendation method based on personalized self-adaptation and literature [4] network hybrid information recommendation method and literature [5] network hybrid information recommendation method are 21.32%, 13.41%, and 13.88%, respectively. In terms of recall rate, the network hybrid information recommendation method based on personalized self-adaptation and the literature [4] network hybrid information recommendation method and literature [5] the network hybrid information recommendation method are 10.08%, 6.56%, and 6.99%, respectively.

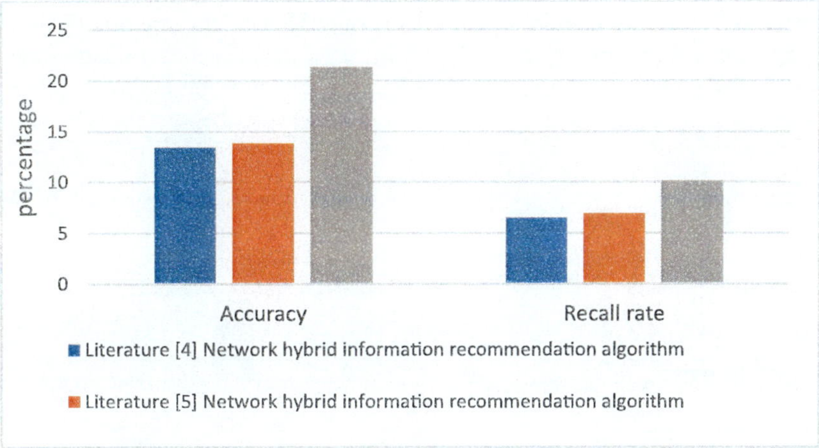

Fig. 4. Performance comparison of information recommendation methods

From the results in Fig. 4, it can be seen that the performance of the network hybrid information recommendation method based on personalized adaptation is better than the two information recommendation methods, and the accuracy and recall rate of the network hybrid information recommendation have been improved.

3.4 Online Personalized Recommendation Experiment

The online personalized recommendation experiment data set selects 1,000 registered users of a training website, uses data preprocessing technology to convert likes, reposts, and comments into users' learning needs and preferences as the test set of the experiment, and the rest of the data is used as the training set. The coverage rate is introduced as an indicator of information recommendation performance.

$$Q = \frac{N_d}{N_v} \tag{20}$$

N_d represents the number of network hybrid information used for recommendation, and N_v represents the total number of network hybrid information. Calculate the average accuracy, recall and coverage of the network hybrid information of 1000 users, and compare it with the literature [4] network hybrid information recommendation method and literature [5] network hybrid information recommendation method. In terms of accuracy, the network hybrid information recommendation method based on personalized self-adaptation and literature [4] network hybrid information recommendation method and literature [5] network hybrid information recommendation method are 13.66%, 0.704%, and 9.77%, respectively. In terms of recall rate, the network hybrid information recommendation method based on personalized self-adaptation and literature [4] network hybrid information recommendation method and literature [5] network hybrid information recommendation method are 8.44%, 0.55%, and 5.33%, respectively. In terms of coverage, the network hybrid information recommendation method based on personalized self-adaptation and literature [4] network hybrid information recommendation method and literature [5] network hybrid information recommendation method are 7.94%, 5.94%, and 3.16%, respectively. The comparison result is shown in Fig. 5.

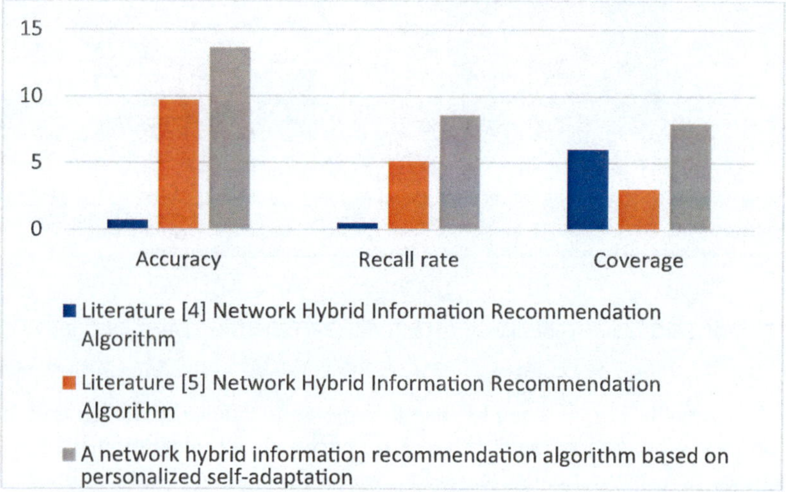

Fig. 5. Performance comparison of online recommendation methods

From the results in Fig. 5 that the network hybrid information recommendation method based on personalized self-adaptation is significantly better than the two recommendation methods in terms of recommendation performance. In summary, the network hybrid information recommendation method based on personalized self-adaptation has higher advantages in user satisfaction, and also has obvious advantages in terms of accuracy, recall and coverage, indicating that it is based on personalized self-adaptation The performance of the network hybrid information recommendation method is better.

4 Conclusion

This paper proposes a network hybrid information recommendation based on individualized self-adaptation of electric power employees. This recommendation method has better recommendation performance, and is superior to other methods in terms of recommendation accuracy and personalized adaptive learning of network hybrid information, and has higher accuracy, recall and coverage. In the power training industry, it effectively meets the company's needs for employee personalized learning information recommendation. This research still has some shortcomings. Although the information recommendation accuracy, recall rate and coverage rate studied in this article have been improved to a certain extent, the overall level is still not up to a high level. Therefore, it is necessary to further improve the recommendation efficiency of network hybrid information, so as to meet the needs of most users in all walks of life.

Acknowledgments. This work was funded by the "Visualization Research on Intelligent Analysis of Power Staff Skill State" (Project Number: SGJLPX00PGJS2000067).

References

1. Sun, Y., Zhu, B., Zhu, L.: Research progress of domestic information recommendation based on knowledge graph. Inf. Theor. Pract. **42**(12), 163–169, 149 (2019)
2. Sun, Y., Li, Y., Zhu, J.: Domestic research progress of information recommendation based on big data: architecture system. Comput. Digit. Eng. **48**(12), 2806–2814 (2020)
3. Sun, Y., Li, C.: Domestic IT planning research progress: core content. Comput. Digit. Eng. **48**(5), 1036–1044 (2020)
4. Huijuan, W., Sun, H.: Research on the framework of personalized information adaptive recommendation model based on cognitive computing and situational awareness. Inf. Sci. **36**(5), 114–118 (2018)
5. Zhang, C., Liang, Y., Fang, H.: Social network information recommendation method supporting privacy protection. J. Shandong Univ. (Sci. Edn.) **55**(3), 13–22 (2020)
6. Sun, Y., Dong, H.: Research on personalized recommendation of digital library based on semantic grid—system structure and overall framework. Inf. Theor. Pract. **32**(6), 63–66, 62 (2009)
7. Qin, C., Zhu, H., Zhuang, F., et al.: Research review of recommendation systems based on knowledge graphs. Sci. Chin. Inf. Sci. **50**(7), 937–956 (2020)
8. Pan, C., He, L., Zhijie, X., et al.: Bilinear graph attention network recommendation algorithm fused with knowledge graph. Comput. Eng. Appl. **57**(1), 29–37 (2021)
9. Liu, Q., Chen, S., Huo, H.: Entity recommendation model based on user preference dissemination of knowledge graph. Comput. Appl. Res. **37**(10), 2926–2931 (2020)
10. Zhu, R., Zhang, J., Zhu, J., et al.: A two-stage deep learning multi-task recommendation model based on KCNN and MKR. J. Shaanxi Normal Univ. Nat. Sci. Edn. **48**(6), 82–89 (2020)

A Differential Evolution Algorithm for Wind Farm Layout Optimization Using a New Bilevel Programming Model

Lijuan Chen[1], Hecheng Li[1,2](✉), and Yu Shen[1]

[1] School of Mathematics and Statistics, Qinghai Normal University, Xining 810016, People's Republic of China
lihecheng@qhnu.edu.cn

[2] Academy of Plateau Science and Sustainability, Xining 810016, People's Republic of China

Abstract. The wind farm layout optimization (WFLO) is still regarded as an NP-hard problem involving a lot of constraints owing to the multiple wake effects. In order to efficiently cope with the problem, both a new optimization model and an improved differential evolution algorithm are developed. First of all, a new bilevel programming model is presented for the problems of this kind. Different from other existing models, the bilevel programming problem aims at the maximum profit at upper level as well as the maximum power output at lower level. Then, in the wind farm (WF) areas given, wind turbines (WTs) are placed at different locations and the wake effect among different wind turbines is taken into account. In addition, by designing a hybrid mutation operator, an improved differential evolution algorithm (IDE) is proposed for dealing with the bilevel programming model. Finally, the comparative analysis of maximal power output among the proposed algorithm and other four advanced methods is conducted under two wind conditions collected from commercial wind farms, and the results demonstrate that the proposed method is feasible and effective.

Keywords: Wind farm layout optimization · Bilevel programming model · Differential evolution · Profit

1 Introduction

The GWEC [1] concluded that 2020 was the best year in history because the global wind industry shows a greatly growth of new installations. And it calculated that global wind power installations will need to increase threefold over the next decade to meet the 2050 goal of net zero emissions and avoid the severe effects of climate change.

Wind resources are widely distributed in China. WT is a device that can converts the kinetic energy in the air into electrical energy, and its life span is generally 20 years. In a WF, each WT is located at different position. The wind speed at downstream WT will be lower than that of upstream WT due to the wake effect on the wind direction [2]. The wake effect will not only lead to the loss of wind energy, but also lead to WTs damaged and reduce service life in the wake area. In reality, the area of the WF is limited,

S.-C. Chu et al. (Eds.): ICGEC 2021, LNEE 833, pp. 584–596, 2022.
https://doi.org/10.1007/978-981-16-8430-2_53

and the price of the WT is very expensive. If the distance of two WTs is too close, the downstream WT is always in the wake region of the upstream WT. On the contrary, if the distance of two WTs is too far, land and wind resources can't be fully utilized. Therefore, it is very necessary to study how to locate the WT in the limited land.

There are two ways to solve the problem, grid method and coordinate method. Mosetti *et al.* [3] divide the area of the WF into some square grids, where the WT is put in the center of each grid. 1 or 0 mean that the turbine is located or not located at the center of the grid. Then, the GA with binary-coding is used for dealing with the optimization problem firstly. The authors tested the performance of the presented algorithm under three kinds of wind conditions. Sun *et al.* [4] argued that the omnidirectional restriction based on experience needs to be taken into account, and the minimum distance between any two WTs is set to 5D (D represents the rotor diameter of WTs). In [5], the minimum distance between any two WTs is set to 5R, where R represents the rotor radius of WTs. Song [6] concluded that the energy output using GA with the real code is about 5%–7% higher than that using the GA with binary code.

From the hierarchy, such problems can be solved by single level and bi-level models. In the single level model, a common way to optimize a WF's layout is to minimize the cost per unit of power. In view of the shortcomings of the scheme given in [3], Grady [7] optimized the approach in [3] by expanding the population size and increasing iterations for the goal of minimizing per unit of power. Chen [8] researched the WFLO with different hub heights by using a genetic algorithm, and found that the output power of WF with different hub heights is better than that using the fixed hub heights of WTs when other conditions keep constant. Finally, the authors took a large WF as an example to analyze the benefits of WFs with different hub heights. Chen [9] investigated the WFLO using a greedy algorithm when multiple hub heights of WTs are involved. The numerical cases in the flat terrain and complex terrain were used for demonstrating the proposed greedy algorithm's efficiency. The results showed that the proposed algorithm has a lower computational cost than GA. However, just as other greedy search techniques, its global search ability is limited. As an often adopted model, the shortcoming of minimizing cost per unit of power is obvious. The minimum objective function value does not mean that both $1/p$ and cost can be optimized for common variables at same time, and it does not also take the preference of decision makers into account. Another way is to maximize the expected power output. Long [10] proposed a data-driven evolutionary algorithm to resolve the problem by saving the high computational cost. Through the model, the bad candidate solutions are filtered. Finally, the efficiency of algorithm is greatly improved. Wang [5] proposed a new encoding scheme, where the variables to be optimized are the positions of the WTs. The authors maximized the objective function when the scale of WTs is from small to large. This method only considers the output capacity of WF, but it does not take the profit into account that the producer cares.

The bilevel programming model is also presented and become a common method for dealing with this kind of WT location optimization. Zhang [11] proposed a bilevel programming model for optimizing the layout of offshore WFs. However, the model is established simply for offshore WFs. Long [12] proposed a two-echelon model. In the first echelon model, the location of turbines is at the center of grid which is obtained by RKGA. In the second echelon model, the optimal location in the first-echelon model was taken as the initial layout, and the final optimal positions of WTs were obtained by PSO. Similarly, long only consider the power output, the cost and profit are also not involved.

For the cost of WF, most of wind farm layouts adopt the simplified model proposed by [3]. However, this cost model cannot provide the profit value for decision makers. Tian [13] applied a new cost model where the total investment cost of WF mainly includes the construction cost and operation cost which is reasonable in real-world application.

In the research on WFLO, dozens of heuristic algorithms and optimization methods have been adopted, like Monte Carlo simulation [14], simulated annealing algorithm [15], random search algorithm [16], PSO [17], ACO [18], and so on.

Differential evolution (DE) is one of evolutionary algorithm family. Since the approach was proposed, many variants have been developed [19–21]. Because of its simple structure, easy implementation, strong robustness and global optimization ability, DE has been widely used in chemical industry, electric power, artificial neural network, signal processing, wind farm layout optimization and so on [5, 22–24].

Shakoor [25] made a comparative analysis of commonly used wake models and concluded that Jesen's wake model is easy to operate and has high accuracy.

In this manuscript, in order to balance the power output and the cost in the WF operation, a bilevel programming model is presented and solved by a newly designed differential evolution algorithm. In the model, Jesen's wake model is adopted.

2 Wind Farm Model

2.1 Wind Speed Probability Density Function [12]

Wind characteristics are usually represented by two factors: wind speed and direction. In real-world applications, both the wind speed and the direction involve randomness. However, through dozens of experiential observations, it is found that the probability density function of the wind speed could be expressed by Weibull distribution with two parameters as follows.

$$f(v, c_i(\theta), k_i(\theta)) = \frac{k_i(\theta)}{c_i(\theta)}(\frac{v}{c_i(\theta)})^{k_i(\theta)-1} \times e^{-(\frac{v}{c_i(\theta)})^{k_i(\theta)}} \tag{1}$$

In Eq. (1), v denotes wind speed and $k_i(\theta)$ represents the shape parameter which is not influenced by the wake effect. Generally, $k_i(\theta)$ is always taken as 2, $c_i(\theta)$ is the scale parameter updated by $c_i(\theta) = c_i(\theta) \times (1 - VD_i)$, $i = 1, 2, \cdots N$. Here, VD_i is the whole velocity deficit of the WT i caused by the wakes of all the other WTs, θ is the wind direction and $0° \leq \theta < 360°$ [12].

2.2 Wake Effect Model

Jesen's wake model supposed that the wake region is conical. Assuming WT i and j coordinates are (x_i, y_i) and (x_j, y_j) respectively. When the WT i is influenced by the wake of the WT j, the velocity deficit of i caused by j is VD_{ji}, VD_i is the whole velocity deficit of WT i caused by the wakes of all the other WTs, the details as follows [10].

$$a = 0.5\left(1 - \sqrt{1 - C_T}\right) \tag{2}$$

$$\tilde{k} = 0.5/\ln(z/z_0) \tag{3}$$

$$d_{j,i} = \left|\left(x_j - x_i\right)\cos\theta + \left(y_j - y_i\right)\sin\theta\right| \tag{4}$$

$$VD_{j,i} = 1 - \frac{v_i}{v_j} = \frac{2a}{\left(1 + \frac{\tilde{k}d_{j,i}}{R}\right)^2} \tag{5}$$

$$VD_i = \sqrt{\sum_{j=1, j\neq i}^{N} (VD_{j,i})^2} \tag{6}$$

Where a denotes the axial induction factor, C_T is called the fixed thrust coefficient, $\tilde{k}$ is called the environment constant, z is height of WT, R is the radius of the WT, z_0 is the roughness of the ground surface, N is the amount of WTs, $d_{j,i}$ is the length between the WT i and the WT j on the wind direction θ, v_i is the wind speed at WT i, and v_j represents the wind speed at WT j.

2.3 Cost Model

The cost model of WFs is often adopted which is more consistent with real situation. The whole investment cost C_Z of WF mainly includes the construction cost C_{CP} and the operation cost C_{OM} [13].

$$C_Z = C_{CP} + C_{OM} \tag{7}$$

$$C_{CP} = c_{cp}d_1NP^* + c_FA_F \tag{8}$$

$$d_1 = \frac{2}{3} + \frac{1}{3}e^{-d_2N^2} \tag{9}$$

$$C_{OM} = c_{om}NP^* \times 20 \tag{10}$$

$$c_{om} = b_0 + (b_1 - b_0)(1 - e^{-b_2\delta}) \tag{11}$$

$$b_0 = b_{01}e^{b_{02}N} \tag{12}$$

c_{cp} is the unit investment for the purchase and installation of units in the construction of WFs, P^* represents the rated power of the WT, A_F is the measure of the WF, c_F is the unit land acquisition fee, N is the quantity of WTs, c_{om} is the operation and maintenance cost of per unit of power, b_1, b_2, b_{01} and b_{02} are the constant coefficients based on engineering experience (Table 1).

Table 1. Parameter setting of cost model [13]

Parameter	Value	Parameter	Value
c_{cp}	10000 ¥/kW	b_1	2.57
d_2	1.74×10^{-3}	b_2	4.1×10^{-4}
P^*	1500 kW	b_{01}	1.4×10^{-2}
c_F	1.2 ¥/m^2	b_{02}	1.6×10^{-4}
A_F	4 km^2	δ	4

2.4 Expected Power Model

The type of WT adopted in this manuscript is GE1.5-77, and its parameters are displayed in Table 3. The power curve function of WT i at the wind speed v is listed in (13):

$$p_i(v) = \begin{cases} 0, & v \geq v_{co}, v < v_{ci} \\ \frac{e^v}{\alpha + \beta e^v}, & v_{ci} \leq v < v_r \\ P_r, & v_r \leq v < v_{co} \end{cases} \tag{13}$$

Where v indicates the wind speed, v_{ci} represents the cut-in wind speed, v_{co} denotes the cut-out wind speed, v_r means the rated wind speed, α and β are constant [10].

If the wind resources of the WF contain h wind directions and N WTs, the output power of a WT in the WF can be expressed as Eq. (14), in which $f_i(\theta)$ is the probability that the wind direction pertains to interval i [10].

$$E(P) = \sum_{i=1}^{h} f_i(\theta) \int_0^{\infty} p_i(v) \frac{k_i(\theta)}{c_i(\theta)} \left(\frac{v}{c_i(\theta)}\right)^{k_i(\theta)-1} \times e^{-\left(\frac{v}{c_i(\theta)}\right)^{k_i(\theta)}} dv \tag{14}$$

3 Proposed Method

3.1 Bilevel Programming Model

A new bilevel programming model can be presented as follows, when the type of WT is predetermined, C_Z is only related to the number N, N is the amount of WTs, and M is the price per unit of power. Assuming that the service span of each WT is 20 years, compared with 8760 h a year, the upper-level objective function stands for the net income of N WTs for running 20 years.

$$\begin{cases} \max\limits_{N} P \times M \times 8760 \times 20 - C_Z \\ \max\limits_{(N,(x,y))} P \\ \text{s.t.} \ x_{\min} \leq x \leq x_{\max} \\ y_{\min} \leq y \leq y_{\max} \\ \sqrt{(x_i - x_j)^2 + (y_i - y_j)^2} \geq 5R \end{cases} \tag{15}$$

Lower-level model represents the maximum power output in the given WF, in which the output power P is related to both the values of N and the position of WTs.

In the lower-level model, the position of each WT is regarded as a variable. When N is fixed as a parameter, the maximum power output can be obtained by optimizing the position of each WT. Once the maximum power output is got, one can determine the total net income from the upper level objective. As a result, the maximum net income can be got by selecting the optimal value of N.

3.2 Improved Differential Evolution (IDE)

Step 1. Initialization

A population P using uniform distribution is initialized in the WF area, and the power output values of the overall layout as well as all WTs are calculated.

Step 2. Mutation

At the beginning of evolution, DE/rand/2/bin mutation is adopted to increase the explore ability of the algorithm [19]. In the late stage of evolution, a new mutation, DE/mean/2/bin, is designed and adopted to refine the exploit ability of the algorithm. In the designed mutation, the location information of some high quality individuals is used and expected to generate better mutation offspring. The mutation operation is presented as follows.

$$\text{DE/mean/2/bin} \qquad V_{i,G} = x_{r_1,G} + F(x_{mean,G} - x_{r_2,G}) + F(x_{r_3,G} - x_{r_4,G}) \tag{16}$$

Where $x_{mean,G}$ is the mean location of the top 20% individuals, $x_{r_m,G}$ (m = 2, 3, 4) are individuals selected randomly from population and $x_{r_1,G}$ is chosen by using the roulette wheel selection depending on the power output of each individual, that is to say, the worse individuals can be chosen with larger probability.

Step 3. Crossover

The binomial crossover is adopted to obtain crossover offspring $U_{i,G}$ [19]. After mutation and crossover, the intermediate population Q is obtained.

Step 4. Selection

IDE uses each individual in Q to displace randomly the worse one in P, after doing so, a new layout S can be obtained. The next generation of population is selected according to (17)

$$P = \begin{cases} S, & if \;\; power(P) < power(S) \\ P, & otherwise \end{cases} \tag{17}$$

Step 5. Circulation

Return to Step 2 until the stop criteria is satisfied.

In above steps, if the new individual in S is located outside the WF area, the boundary absorption is carried out. The process of proposed algorithm is displayed in Table 2.

It should be noted that the caching technique is also adopted in the proposed IDE to reduce the computational time. When WT e is moved, its velocity deficit can be calculated by (6). For other WTs which are not moved, one just need to recalculate its velocity deficit influenced by WT e. For common evaluation methods, the velocity deficit between any two WTs needs to be computed and the computational time complexity is $O(N^2)$. When the caching technique is used, only $2(N-1)$ inspects of the velocity deficit are required and the computation complexity is $O(N)$ [5].

Table 2. The framework of the proposed algorithm

Pseudo-code of Proposed Algorithm (IDE)
1.Iteration: $T = 0$, the maximum iteration number: $T_{\max}$;
2. An initial population P is generated which contains N WTs and calculate the power output for each individual according to (14).
3. for N=15:100
4. while $T \le \kappa T_{\max}$
5. Implement the mutation (DE/rand/2), crossover and selection on P to get the next generation p of population and record the power output.
6. T=T+1
7. end while
8. while $\kappa T_{\max} < T \le T_{\max}$
9. Implement the mutation(16), crossover and selection on P to generate the next generation of population and record the power output.
10.T=T+1
11. end while
12. Output the optimal layout for each N.
13. end for
14. Output the optimal layout with the maximum profit.

4 Computational Experiment and Discussions

4.1 Parameter Setting

In this manuscript, for the purpose of the comparison, in the proposed IDE, we use the same parameter setting and wind conditions as those in reference [5]. The power and profit of the corresponding WF are calculated under two popular wind conditions, the data of wind conditions are provided in Tables 4 and 5.

In Tables 4 and 5, ε_n is the frequency of occurrence in the wind direction interval $[\theta_{n-1}, \theta_n]$. It can be found from Table 4 that the concentration trend of wind directions is obvious. In the wind condition 2, the wind direction with higher frequency is mainly concentrate on 75°–105°.

Table 3. Parameter setting [5]

Parameter	Implication	Value
Z	Hub height of WT (m)	80
R	Rotor radius (m)	40
C_T	Thrust coefficient	0.8
P_r	Rated power (kw)	1500
$\tilde{k}$	Environment constant	0.01
V_{ci}	Cut-in speed (m/s)	3.5
V_r	Rated speed (m/s)	14
V_{co}	Cut-out speed (m/s)	25
α	Parameters of logistic function of wind	6.0268
(β)	Power curve	(0.0007)
F	Scaling factor	0.9
CR	Crossover rate	0.9
T_{MAX}	Maximum iterations	150000
κ	Iteration factor	0.8
M	Price of per unit of power (¥/kwh)	0.5

Table 4. Wind condition 1

n	θ_{n-1}	θ_n	k	c	ε_n	n	θ_{n-1}	θ_n	k	c	ε_n
1	0°	15°	2	7	0.0003	13	180°	195°	2	10	0.1909
2	15°	30°	2	5	0.00723	14	195°	210°	2	8.5	0.1162
3	30°	45°	2	5	0.0237	15	210°	225°	2	8.5	0.0793
4	45°	60°	2	5	0.0242	16	225°	240°	2	6.5	0.00802
5	60°	75°	2	5	0.0222	17	240°	255°	2	4.6	0.0041
6	75°	90°	2	4	0.0301	18	255°	270°	2	2.6	0.0008
7	90°	105°	2	5	0.0397	19	270°	285°	2	8	0.001
8	105°	120°	2	6	0.0268	20	270°	300°	2	5	0.0005
9	120°	135°	2	7	0.0626	21	300°	315°	2	6.4	0.0013
10	135°	150°	2	7	0.0801	22	315°	330°	2	5.2	0.0031
11	150°	165°	2	8	0.1025	23	330°	345°	2	4.5	0.0085
12	165°	180°	2	9.5	0.1445	24	345°	360°	2	3.9	0.0222

Table 5. Wind condition 2

n	θ_{n-1}	θ_n	k	c	ε_n	n	θ_{n-1}	θ_n	k	c	ε_n
1	0°	15°	2	13	0.01	13	180°	195°	2	13	0.01
2	15°	30°	2	13	0.01	14	195°	210°	2	13	0.01
3	30°	45°	2	13	0.01	15	210°	225°	2	13	0.01
4	45°	60°	2	13	0.01	16	225°	240°	2	13	0.01
5	60°	75°	2	13	0.01	17	240°	255°	2	13	0.01
6	75°	90°	2	13	0.2	18	255°	270°	2	13	0.01
7	90°	105°	2	13	0.6	19	270°	285°	2	13	0.01
8	105°	120°	2	13	0.01	20	270°	300°	2	13	0.01
9	120°	135°	2	13	0.01	21	300°	315°	2	13	0.01
10	135°	150°	2	13	0.01	22	315°	330°	2	13	0.01
11	150°	165°	2	13	0.01	23	330°	345°	2	13	0.01
12	165°	180°	2	13	0.01	24	345°	360°	2	13	0.01

Table 6. Sizes of WF with different *N*.

N	Side length (m)
[15, 30)	2000
[30, 35)	2200
35	2400
40	2600
60	3100
80	3600
100	4000

4.2 Computational Results

Table 7 and Table 8 present the means and standard deviations of the power output values provided by IDE in 30 independent runs under two kinds of wind conditions, respectively. The computational results are compared with those provided by other algorithms [5]. In these two tables, the best ones are shown in bold font, AV represents the mean of power output values in 30 independent runs and *std* represents the standard deviation.

From Table 7 and Table 8, one can observe that IDE performed better than other compared algorithms on at least 6 computational examples. However, on 3 instances with bigger *N*, DEEM shows better performance than IDE. Except for DEEM, IDE is still better than other compared algorithms on all tested examples.

Table 7. Experimental results of IDE and four compared algorithms in wind condition 1 (kW)

N	TDA(AV ± std dev)	CLPSO(AV ± std dev)	JADE(AV ± std dev)	DEEM(AV ± std dev)	IDE(AV ± std dev)
15	5923.3 ± 105.26	5923.60 ± 35.84	5954.68 ± 24.7	6183.32 ± 49.90	**6197.71 ± 42.37**
20	7308.67 ± 151.64	7125.12 ± 75.66	7178.9 ± 103.86	7674.78 ± 62.76	**7827.52 ± 90.86**
25	8257.07 ± 178.85	7892.72 ± 71.44	7886.95 ± 155.52	8828.37 ± 156.98	**8852.44 ± 100.98**
30	9487.57 ± 129.37	8902.46 ± 73.32	8996.17 ± 56.71	10085.18 ± 112.36	**10177.32 ± 120.24**
35	10764.48 ± 220.19	9977.47 ± 67.45	9936.61 ± 85.8	11413.48 ± 151.55	**11518.71 ± 95.03**
40	11954.19 ± 133.80	11001.07 ± 77.25	10857.41 ± 81.75	12640.05 ± 225.18	**12745.58 ± 161.85**
60	15512.11 ± 235.16	14024.53 ± 145.60	13647.24 ± 117.31	**16538.61 ± 209.79**	16338.67 ± 152.52
80	19171.57 ± 169.58	16935.41 ± 138.92	16235.83 ± 177.18	**20006.09 ± 150.02**	19916.52 ± 134.59
100	22340.84 ± 309.75	19206.77 ± 126.08	18256.93 ± 246.9	**23142.42 ± 204.74**	22956.98 ± 277.31

Table 8. Experimental results of IDE and four compared algorithms in wind condition 2 (kW)

N	TDA(AV ± std)	CLPSO(AV ± std)	JADE(AV ± std)	DEEM(AV ± std)	IDE(AV ± std)
15	12860.78 ± 96.24	12874.64 ± 27.85	12890.42 ± 22.43	13041.60 ± 21.09	**13048.21 ± 10.29**
20	16413.36 ± 172.64	16304.34 ± 105.90	16357.57 ± 89.68	16957.70 ± 66.12	**16959.33 ± 27.02**
25	19639.62 ± 160.21	18824.44 ± 76.65	18916.37 ± 90.27	19981.99 ± 159.58	**20024.92 ± 105.72**
30	22830.94 ± 157.07	21747.00 ± 138.87	21778.58 ± 114.99	23524.68 ± 219.16	**23641.55 ± 131.72**
35	26088.91 ± 385.63	24737.83 ± 113.95	24610.20 ± 217.57	26814.6 ± 154.97	**26870.19 ± 193.85**
40	29432.24 ± 299.83	27731.31 ± 203.63	27449.53 ± 112.49	**30440.38 ± 191.49**	30358.61 ± 279.53
60	40618.21 ± 618.58	36894.43 ± 171.90	36194.62 ± 281.48	41644.07 ± 425.66	**41724.31 ± 291.30**
80	51563.61 ± 247.88	45930.86 ± 238.22	44113.56 ± 389.51	**52608.78 ± 380.53**	52436.27 ± 164.40
100	60755.42 ± 427.75	53432.41 ± 310.76	50420.59 ± 1241.15	**62332.23 ± 300.90**	62170.39 ± 642.46

4.3 The Presented Optimization Layout

In order to clarify the efficiency of presented model, the area of the WF is limited to the size of 2000 m × 2000 m in this manuscript. Through the newly designed bilevel programming model, one can obtain the maximize profit while N is 28 under wind condition 1 and N is 47 under wind condition 2. The result is new and is not provided in all listed literature.

The optimal layouts of WF is listed in Fig. 1(a) represents the optimal layout in wind condition 1 and (b) stands for that in wind condition 2.

From Fig. 1, one can find that some turbines need to be put near the boundary of the WF. The reason can be analyzed as follows: due to the influence of the wake effect, if the WT is located near the boundary, the loss of the wind speed is less than at other locations, which can increase the power output of WF.

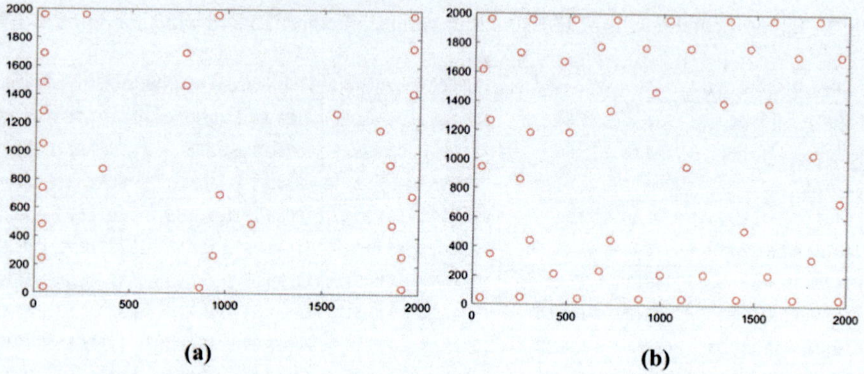

Fig. 1. Optimal layout of wind farm under different condition

5 Conclusion

In this manuscript, first of all, a new bilevel programming model was given for WFLO in flat area, which aims at the maximum profit during the WF operation. The model is new and not presented in existing literature. Secondly, an IDE is developed, in which the information of better individuals is adopted in the mutation operator. Finally, the proposed model and algorithm are utilized to deal with some popular computation examples in WFLO problems. The computational results shown the proposed schemes are feasible and efficient and the optimal value of N is obtained at which the maximum profit can be got during the WF operation.

In the next work, we will research offshore WF problems, such as optimal cable connection layout and so on.

Acknowledgments. This work is partially supported by the National Natural Science Foundation of China under Grant No. 61966030, the Natural Science Foundation of Qinghai Province of China under Grant No. 2018-ZJ-901 and the Key Laboratory of LoT of Qinghai province of china under Grant 2020-ZJ-Y16.

References

1. GWEC. Global wind report (2021). https://gwec.net/global-wind-report-2021
2. Fischetti, M., Pisinger, D.: Mathematical optimization and algorithms for offshore windfarm design: an overview. Bus. Inf. Syst. Eng. **61**(4), 469–485 (2019). https://doi.org/10.1007/s12599-018-0538-0
3. Mosetti, G., Poloni, C., Diviacco, B.: Optimization of wind turbine positioning in large windfarms by means of a genetic algorithm. J. Wind Eng. Ind. Aerodyn. **51**, 105–116 (1994). https://doi.org/10.1016/0167-6105(94)90080-9
4. Sun, H.Y., Yang, H.X., Gao, X.X.: Investigation into spacing restriction and layout optimization of wind farm with multiple types of wind turbines. Energy **168**, 637–650 (2019). https://doi.org/10.1016/j.energy.2018.11.073

5. Wang, Y., Liu, H., Long, H., Zhang, Z.J., Yang, S.X.: Differential evolution with a new encoding mechanism for optimizing wind farm layout. IEEE Trans. Ind. Inf. **14**(3), 1040–1054 (2018). https://doi.org/10.1109/TII.2017.2743761
6. Song, M.X., Chen, K., Wang, J.: Three-dimensional wind turbine positioning using Gaussian particle swarm optimization with differential evolution. J. Wind Eng. Ind. Aerodyn. **172**, 317–324 (2018). https://doi.org/10.1016/j.jweia.2017.10.032
7. Grady, S.A., Hussaini, M.Y., Abdullah, M.M.: Placement of wind turbines using genetic algorithms. Renew. Energy **30**, 259–270 (2005). https://doi.org/10.1016/j.renene.2004.05.007
8. Chen, Y., Li, H., Jin, K., Song, Q.: Wind farm layout optimization using genetic algorithm with different hub height wind turbines. Energy Convers. Manage. **70**, 56–65 (2013). https://doi.org/10.1016/j.enconman.2013.02.007
9. Chen, K., Song, M.X., Zhang, X., Wang, S.F.: Wind turbine layout optimization with multiple hub height wind turbines using greedy algorithm. Renew. Energy **96**, 676–686 (2016). https://doi.org/10.1016/j.renene.2016.05.018
10. Long, H., Li, P., Gu, W.: A data-driven evolutionary algorithm for wind farm layout optimization. Energy **208**, 1–14 (2020). https://doi.org/10.1016/j.energy.2020.118310
11. Zhang, J.H., Jiang, Y.W.: Joint optimization of the number, type and layout of wind turbines for a new offshore wind farm. Renew. Sustain. Energy **12**, 1–13 (2020). https://doi.org/10.1063/5.0020204
12. Long, H., Zhang, Z.J.: A two-echelon wind farm layout planning model. IEEE Trans. Sustain. Energy **6**(3), 863–871 (2015). https://doi.org/10.1109/TSTE.2015.2415037
13. Tian, L.L., Zhao, N., Wu, C.H.: Study on optimization of wind farm layout. Acta Energiae Solaris Sinica **35**(3), 462–468 (2014). https://doi.org/10.3969/j.issn.0254-0096.2014.03.015
14. Marmidis, G., Lazarou, S., Pyrgioti, E.: Optimal placement of wind turbines in a wind park using Monte Carlo simulation. Renew. Energy **33**, 1455–1460 (2008). https://doi.org/10.1016/j.renene.2007.09.004
15. Rivas, R.A., Clausen, G., Hansen, K.S., Jensen, L.E.: Solving the turbine positioning problem for large offshore wind farms by simulated annealing. Wind Eng. **33**(3), 287–298 (2009)
16. Feng, J., Shen, W.Z.: Solving the wind farm layout optimization problem using random search algorithm. Renew. Energy **78**, 182–192 (2015). https://doi.org/10.1016/j.renene.2015.01.005
17. Devibala, P., Rajendran, C.: Optimal micro-sitting of wind turbines in a wind park using particle swarm optimization algorithm. Int. J. Eng. CS & Tech. **2**, 1–6 (2012)
18. Eroglu, Y., Seçkiner, S.U.: Design of wind farm layout using ant colony algorithm. Renew. Energy **44**, 53–62 (2012). https://doi.org/10.1016/j.renene.2011.12.013
19. Qin, A.K., Huang, V.L., Suganthan, P.N.: Differential evolution algorithm with strategy adaptation for global numerical optimization. IEEE Trans. Evol. Comput. **13**(2), 398–417 (2009)
20. Zhang, J.Q., Sanderson, C.: JADE: adaptive differential evolution with optional external archive. IEEE Trans. Evol. Comput. **13**(5), 945–958 (2009)
21. Wang, Y., Cai, Z.X.: Combining multi-objective optimization with differential evolution to solve constrained optimization problems. IEEE Trans. Evol. Comput. **16**(1), 117–134 (2012). https://doi.org/10.1109/TEVC.2010.2093582
22. Zhang, W.Q., Wang, Y., Yang, Y., Gen, M.: Hybrid multi-objective evolutionary algorithm based on differential evolution for flow shop scheduling problems. Comput. Ind. Eng. **130**, 661–670 (2019). https://doi.org/10.1016/j.cie.2019.03.019
23. Zhang, Q., Zou, D.X., Duan, N., Shen, X.: An adaptive differential evolutionary algorithm incorporating multiple mutation strategies for the economic load dispatch problem. Appl. Soft Comput. J. **78**, 641–669 (2019). https://doi.org/10.1016/j.asoc.2019.03.019
24. Adamuthe, A.C., Patil, J.T.: Differential evolution algorithm for optimizing virtual machine placement problem in cloud computing. Int. J. Intel. Syst. Appl. **7**, 58–65 (2018). https://doi.org/10.5815/ijisa.2018.07.06

25. Shakoor, R., Hassan, M.Y., Raheem, A., Wu, Y.K.: Wake effect modeling: a review of wind farm layout optimization using Jensen's model. Renew. Sustain. Energy Rev. **58**, 1048–1059 (2016)
26. Wagner, M., Day, J., Neumann, F.: A fast and effective local search algorithm for optimizing the placement of wind turbines. Renew. Energy **51**, 64–70 (2013). https://doi.org/10.1016/j.renene.2012.09.008

Genetic Algorithm and Machine Learning

Test Paper Automatic Generating Method Based on Hybrid Genetic Algorithm

Yan Wang[1](✉) and Xinhua Wang[2]

[1] College of Pharmacy, Jilin University, Changchun 130012, China
shenmary@jlu.edu.cn

[2] School of Computer Science, Northeast Electric Power University, Jilin 132012, China
wangxh@neepu.edu.cn

Abstract. As an important part of computer-assisted instruction, test paper automatic generation system provides data and intelligent information support for the teaching process. In order to solve the problem that the result of test paper generating system is not ideal, a hybrid genetic algorithm is proposed to improve the multi-objective optimization process of test paper automatic generating. The selection operator adopts the $\mu + \lambda$ selection strategy of combinatorial optimization evolutionary algorithm, which improves the global search ability of the algorithm. The crossover operator adopts the method of subsection crossover according to the question type to ensure that the number of questions of each question type remains unchanged after crossover. Mutation operator adopts mutation mechanism based on tabu search, which solves the defect of premature convergence of genetic algorithm. The experimental results show that the improved algorithm improves the efficiency and success rate of test paper automatic generation.

Keywords: Computer assisted instruction · Test paper automatic generation · Hybrid genetic algorithm

1 Introduction

In recent years, with the continuous development of computer-aided education research, computer education resource management system has become an important part of computer-aided testing [1, 2]. In order to comply with the development trend of national education facing modernization, network education resource management system, as an important part of modern education system, has been greatly developed in China. In the education resource management system, the intelligent test paper generation function determines whether the education resource management system can effectively test the true level of the tested, and is an important yardstick to evaluate the quality of the system. Therefore, the research of intelligent test paper generation algorithm has become an important research topic in computer-aided education. Intelligent test paper generation is a test paper automatically generated by computer according to the constraints set by teachers in advance, which meets the requirements of knowledge structure distribution, question type distribution, difficulty distribution, discrimination distribution, cognitive level distribution, time distribution and score distribution [3, 4].

S.-C. Chu et al. (Eds.): ICGEC 2021, LNEE 833, pp. 599–609, 2022.
https://doi.org/10.1007/978-981-16-8430-2_54

The process of intelligent test paper generation involves algorithm and data structure, mathematical model building, constraint optimization control, artificial intelligence and many other technical problems. The key problem of intelligent test paper generation is to meet the constraint optimization conditions [5, 6]. Aiming at the shortcomings of the existing test paper generation algorithms, such as low success rate, long time consuming, and the generated test paper is difficult to meet the actual test requirements, this paper makes a more in-depth study on the application of hybrid genetic algorithm in intelligent test paper generation, a hybrid genetic algorithm is proposed to improve the multi-objective optimization process of test paper automatic generating. The results show that the improved algorithm improves the efficiency and success rate of test paper automatic generation.

2 Problem Description

2.1 Mathematical Models

We need to determine the attributes of each question, including: question type, chapter, knowledge point, question number, score, answer time, difficulty, discrimination, cognitive level classification, subjective and objective question type classification, as shown in formula (1).

$$A = \begin{bmatrix} a_{1,1} & a_{1,2} & \cdots & a_{1,M} \\ a_{2,1} & a_{2,2} & \cdots & a_{2,M} \\ \vdots & \vdots & \vdots & \vdots \\ a_{n,1} & a_{n,2} & \cdots & a_{n,M} \end{bmatrix} \tag{1}$$

In formula (1), Each row of matrix A represents a question, and each column represents an attribute of the question, which is the total number of all the questions extracted from the question bank. M is the number of attributes of the test question.

2.2 Constraint Condition

Each question has a total of 10 attributes. These 10 attributes are: question number, score, time, question type, chapter, knowledge point, difficulty, discrimination, recognition classification and question form. Intelligent test paper generation includes the following 10 constraints:

(1) The total score, the total number of questions, and the time constraints of answering questions.

$$\sum\nolimits_{i=1}^{n} x(i) \times A(i, 2) = P \tag{2}$$

$$\sum\nolimits_{i=1}^{n} x(i) = \mathrm{Q} \tag{3}$$

$$\sum\nolimits_{i=1}^{n} x(i) \times A(i, 3) = T \tag{4}$$

In formula (2), (3) and (4), $x(i) = 1$ means that the i-th question is selected, otherwise it means that it is not selected. $A(i, 2)$ is the reference score of the i-th question, and $A(i, 3)$ is the reference time of the i-th question. P is the total score of questions, Q is the total number of questions, T is the total time to answer questions.

(2) The score of each item was restricted.

$$\sum\nolimits_{i=1}^{n} x(i) \times y_1(j, i) \times A(i, 2) = b_1(j) \tag{5}$$

In formula (5), $y_1(j, i) = 1$ means that the i-th question is the j-th question, otherwise it does not belong to the j-th question type. $b_1(j)$ is the total score of the type j questions entered by the test sample.

(3) Each question type contains the restriction of the number of questions;

$$\sum\nolimits_{i=1}^{n} x(i) \times y_1(j, i) = c_1(j) \tag{6}$$

In formula (6), $y_1(j, i) = 1$ means that the i-th question is the j-th question, otherwise it does not belong to the j-th question type. $c_1(j)$ is the total number of type j questions input by the test sample.

(4) The score of each chapter is restricted.

$$\sum\nolimits_{i=1}^{n} x(i) \times y_2(j, i) \times A(i, 2) = b_2(j) \tag{7}$$

In formula (7), $y_2(j, i) = 1$ means that the first question belongs to chapter j, otherwise it does not belong to chapter j. $b_2(j)$ is the total score of chapter j test questions entered by the test samples.

(5) The first chapter contains the number of questions.

$$\sum\nolimits_{i=1}^{n} x(i) \times y_2(j, i) = c_2(j) \tag{8}$$

In formula (8), $y_2(j, i) = 1$ means that the i-th question belongs to chapter j, otherwise it does not belong to chapter j. $c_2(j)$ is the total number of questions in Chapter j entered by the test sample.

(6) The paper knowledge point constraints.

$$\sum\nolimits_{j=1}^{m} sign\left(\sum\nolimits_{i=1}^{n} x(i) \times y_3(j, i)\right) = k,\ sign(x) = \begin{cases} 1, x > 0 \\ 0, x \le 0 \end{cases} \tag{9}$$

In formula (9), $y_3(j, i) = 1$ means that the i-th question belongs to the j-th knowledge point, otherwise it means that it does not belong to the j-th knowledge point, m is the total number of knowledge points and k is the number of knowledge points covered.

(7) Test paper difficulty constraints.

$$\sum_{i=1}^{n} x(i) \times y_4(j, i) \times A(i, 2) = b_4(j) \tag{10}$$

In formula (10), $y_4(j, i) = 1$ means that the j-th question belongs to the j-th difficulty, otherwise it does not belong to this difficulty. $b_4(j)$ is the total score of the type j difficulty question input by the test sample.

(8) Test paper discrimination constraints.

$$\sum_{i=1}^{n} x(i) \times y_5(j, i) \times A(i, 2) = b_5(j) \tag{11}$$

In formula (11), $y_5(j, i) = 1$ means that the j-th question belongs to the j-th discrimination, otherwise it does not belong to the j-th discrimination. $b_5(j)$ is the total score of the j-type discrimination test questions entered by the test samples.

(9) Test paper recognition constraints.

$$\sum_{i=1}^{n} x(i) \times y_6(j, i) \times A(i, 2) = b_6(j) \tag{12}$$

In formula (12), A means that the i-th question belongs to the j-th degree of cognition, otherwise it does not. $b_6(j)$ is the total score of the j-th recognition test questions input by the test samples.

(10) Test paper form constraints.

$$\prod_{i=1}^{n} ((x(i) = 1|A(i, 10)) \subset R_m) = 1 \tag{13}$$

In formula (13), $A(i, 10)$ is the formal attribute of the i-th question, R_m is the set of questions, $R_m = \{0, 1\}$, $m = 0$ is the objective question, and $m = 1$ is the subjective question.

2.3 Objective Function

The test questions meeting the constraints are taken out from the question bank and stored in matrix A. The constraints include: total score, total volume, and answer time; The restriction of the score and the number of questions of each question type; The restriction of the score and the number of questions in each chapter. The knowledge points, difficulty, discrimination and form constraints of the test paper.

In general, not all constraint conditions of multi-optimization constrained problems can be satisfied, and the total number of unsatisfied constraint conditions is the minimum [7]. The classic method to solve the multi-objective optimization problem is the weight sum method, that is to give a weight to each standardized objective function, and transform the multi-objective problem into a single objective problem with a title objective function [8], as shown in formula (13).

$$\min z = w_1 z_1'(x) + w_2 z_2'(x) + \cdots + w_k z_k'(x) \tag{14}$$

In formula (14), $z_i'(x)$ is the result of normalizing the objective function $z_i(x)$, which is the absolute value of the error between the i-th constraint and the user's requirement, w_i is the weight given by the test paper maker to the i-th constraint. According to the implementation, some constraints are important, such as the total score of the test, the score of each question type, the answer time and the difficulty. Some constraints are moderate, such as chapter content, distribution of knowledge points and recognition degree. Some constraints are not very important, such as the form of the test paper, that is, subjective questions, objective questions, discrimination. In order to satisfy their constraints, some important constraints should have larger weights, and some unimportant constraints can be sacrificed appropriately [9].

3 Proposed Method

Through the introduction of the theoretical basis, mathematical model and related algorithms of intelligent test paper generation, aiming at the "premature" phenomenon of genetic algorithm, an improved algorithm based on hybrid genetic algorithm is proposed to solve the problem of intelligent test paper generation, which solves the problems existing in the traditional test paper generation method. The algorithm adopts the integer coding method according to the question type, It improves the ability to find the best permutation and combination of constraint items. The three genetic algorithms adopt the following strategies: the niche technology of deterministic selection mechanism of combinatorial optimization evolutionary algorithm $\mu + \lambda$ is used in the selection operator to improve the global search ability of the algorithm [10]; The crossover algorithm adopts the strategy of segmented crossover based on similar individuals, which ensures that the number of selected questions remains unchanged after the crossover of each question type; The mutation operator adopts the mutation mechanism based on tabu search, which is helpful to improve the global search performance of the algorithm [11]. In addition, in order to solve the problem of the loss of excellent individuals due to random factors in the process of selection, crossover and constrained optimization, this paper adopts the non-inferior sorting mechanism of elite retention [12].

3.1 Chromosome Coding

According to the characteristics of the same type of questions in the question bank with the same score, this paper adopts integer coding as the coding scheme. For example, there are N questions, 7 types of questions the number of each type is $(N_1, N_2, \ldots, N_7)$, chromosome code $(J_{1,1}, J_{2,1}, \ldots, J_{Nm,1}), \ldots, (J_{1,7}, J_{2,7}, \ldots, J_{Nq,7})$, where $J_{i,j} = k$ means that the *j-th* question in the *i*-type question is selected, k means the question number, m means that m questions in the 1-type question is selected, and q means that q questions in the 7-type question is selected (see Fig. 1).

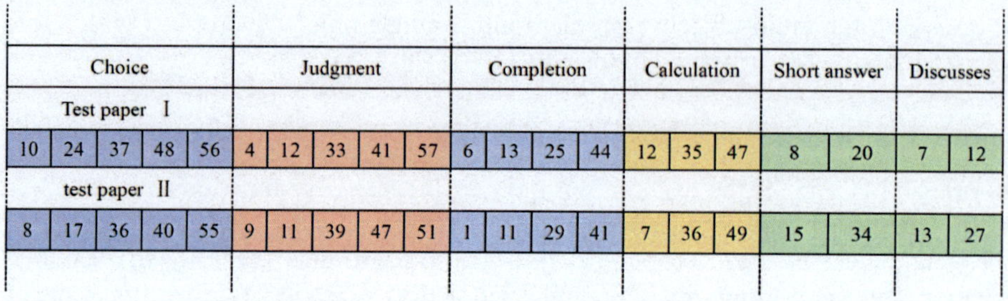

Fig. 1. Examples of chromosome coding design for different types of test paper database

This coding strategy can effectively reduce the search space and simplify the problem solving. Considering the efficiency and quality of the search, we use the population entropy heuristic method to quickly generate high-quality initial population. In general, population initialization is random, but the efficiency and quality of search should be considered. On the one hand, the initial population should be dispersed in the solution space as much as possible [13]. On the other hand, the estimation method of population entropy can quickly produce some excellent solutions as the initial individuals [14].

3.2 Fitness Function

In genetic algorithm, the fitness function is used to distinguish the individuals in the population. In general, fitness function is transformed from objective function. The design of fitness function should mainly meet the principles of normative, functional and conventional design. The fitness function is set as follows:

$$Fit(x) = \frac{1}{1 + c + f(x)} \tag{15}$$

In formula (15), f (x) represents the objective function and c is a constant. According to the basic principle of generating test paper, the smaller the objective function is, the better. It means that the deviation is small, and the larger the fitness function is, the better. In addition, in order to ensure that the fitness function value is non-negative, the value here is -0.99, which can ensure that the fitness function value is large enough. The construction method of fitness function is called "lower bound construction method", which can maintain the diversity of the population and improve the competitiveness of the test paper when it is used in the paper generation problem.

3.3 Operator Design

(1) Selection operator

The selection operator adopts $\mu + \lambda$ selection strategy, which is *a* deterministic selection process. The basic idea of selection strategy is: μ parent individual in the population produces λ offspring through genetic operation, and μ better individual is selected from the set composed of μ parent and λ offspring by sorting fitness function value to enter the next generation population. For every individual $x \in P$, the niche count is calculated as follows:

$$nc(x) = \sum\nolimits_{x \in P} max\left(\frac{\sigma_{share} - dz(x, y)}{\sigma_{share}}, 0\right) \tag{16}$$

In formula (16), $dz(x, y)$ is the Euclidean geometric distance between any two individuals, and σ_{share} is the size of niche count. Each σ_{share} is composed of a group of individuals with similar fitness function values, representing the neighborhood of an individual in the objective function space. The fitness function of every individual is adjust calculated as follows:

$$Fit'(x) = \frac{Fit(x)}{nc(x)} \tag{17}$$

In formula (17), by adjusting the fitness function, individuals in the same neighborhood can contribute to the number of microhabitats, and individuals in the crowded neighborhood have a relatively large number of microhabitats, thus reducing the probability of being selected as parents. This method can produce the fastest local convergence speed, ensure the diversity of individual population, and improve the evolution effect of crossover operation.

(2) Crossover operator

The crossover operator adopts the strategy of crossing in sections according to the question types, so as to ensure that the number of selected questions remains unchanged after each question type is crossed (see Fig. 2).

	Choice					Judgment					Completion				Calculation			Short answer		Discusses	
Paternal chromosome I	10	24	37	48	56	4	12	33	41	57	6	13	25	44	12	35	47	8	20	7	12
Paternal chromosome II	8	17	36	40	55	9	11	39	47	51	1	11	29	41	7	36	49	15	34	13	27
Offspring chromosome I	10	24	37	40	55	4	12	39	47	51	6	13	29	41	12	35	49	8	34	7	27
Offspring chromosome II	8	17	36	48	56	9	11	33	41	57	1	11	25	44	7	36	47	15	20	13	12

Fig. 2. Examples of cross section by the same question type of the test paper database

There may be illegal phenomenon of repeated item number in the offspring chromosome generated after crossing. In this case, the item number that appears should be replaced by the one that does not appear in the paragraph to get a new offspring.

(3) Mutation operator

When the fitness function curve of offspring generated by crossover operation remains horizontal and does not reach the optimal value, it means that the algorithm converges prematurely. The reason for this phenomenon lies in the defect of effective genes. Mutation operation overcomes this situation to a certain extent and is conducive to increasing the diversity of the population.

The mutation operator adopts the local search mechanism based on tabu list, which is beneficial to improve the global search performance of the algorithm. In this paper, the graph coloring algorithm based on ant colony is applied, and it is improved and applied to the local search strategy of mutation algorithm. The algorithm stores a tabu list for each type of question to realize tabu search. In the early stage of the search, random search is carried out with high probability, in order to improve the global search performance of the algorithm and avoid the "premature" of the algorithm; In the later stage of the search, greedy search is carried out with high probability, in order to speed up the convergence process of local search. In addition, the algorithm also realizes the random and relevance search of the same type of test questions in the test question database, mainly through the correlation of test question P and two randomly selected test questions P_1 and P_2, and participates in tabu search as a whole, so as to improve the overall search performance of the algorithm.

3.4 Algorithm Framework

Therefore, the proposed algorithm framework is described as follows:

***Step* 1:** parameter initialization (including the maximum number of iterations T, population size N, crossover probability P_c, mutation probability P_m);
***Step* 2:** the design of coding scheme, coding according to the question type;
***Step* 3:** randomly generate N individuals $x_t^1, x_t^2, \cdots, x_t^N$, and evaluate the advantages and disadvantages of the individual population;
***Step* 4:** calculate the fitness $Fit(x_t^i)$, $i \in [1, N]$, average fitness Fit_{avg} and maximum fitness Fit_{max} of every individual in the population, and sort the individuals according to the descending order of fitness function value;
***Step* 5:** according to the selection strategy $\mu + \lambda$, select the individuals in the population;
***Step* 6:** according to the crossover probability P_c, the crossover operation is performed on the individuals in the population;
***Step* 7:** according to the mutation probability P_m, perform mutation operation on individuals in the population.
7.1: calculate the mutation number and local search probability;
7.2: call tabu algorithm search process to generate new individuals and add them into population N;
***Step* 8:** if $t < T$, $t \leftarrow t + 1$, turn to *step* 4; Otherwise, turn to ***step* 9**;
***Step* 9:** stop the operation and output the result.

4 Experimental Test

4.1 Experimental Environment

The hardware and software environment of the computer are as follows: CPU Intel Core i9-9900k, RAM 64G, operating system Windows 10 (64 bit), program development environment Matlab R2019b.

4.2 Scheme Design

In the aspect of algorithm performance evaluation, the test database of pharmacy course in Jilin University is used as the test object. Under the same software and hardware environment, the algorithm proposed in this paper and the algorithm in reference [15] are run independently for 20 times, and the average value of the calculation results is taken for comparison.

(1) Parameter setting
The initialization parameters of the two algorithms are the same, in which the population size is 50 and the iterations number is 500. The range of crossover probability P_C is $P_c \in [0.40, 0.99]$, and the range of mutation probability P_m is $P_m \in [0.001, 0.1]$.
(2) Performance testing
The average value of fitness function after running is shown in Table 1. It can be seen the average value of fitness function of this algorithm is less than that of compared algorithm. It can be seen the running result of the algorithm in this paper is better than that of the comparison algorithm when the number of iterations is less than that of the comparison algorithm.

Table 1. Comparison of fitness function average value

Algorithm	Proposed	Compared
Average value	0.008242	0.008889

The comparison of constraint satisfaction between the proposed algorithm and the comparison algorithm is given (see Fig. 3), and 10 types of constraints are considered, including 63 constraints. The abscissa is the constraint number, and the ordinate is the absolute value of the difference between the constraint function value obtained by the algorithm and the user input value. It can be seen in most cases, the constraint function value of the proposed algorithm is closer to the user input value. The results of the algorithm proposed in this paper are better to satisfy the constraints, and its practicability should be stronger. The results show that the proposed algorithm is better than the comparison algorithm for 29 constraints, the same as the comparison algorithm for 17 constraints, and worse than the comparison algorithm for the other 17 constraints.

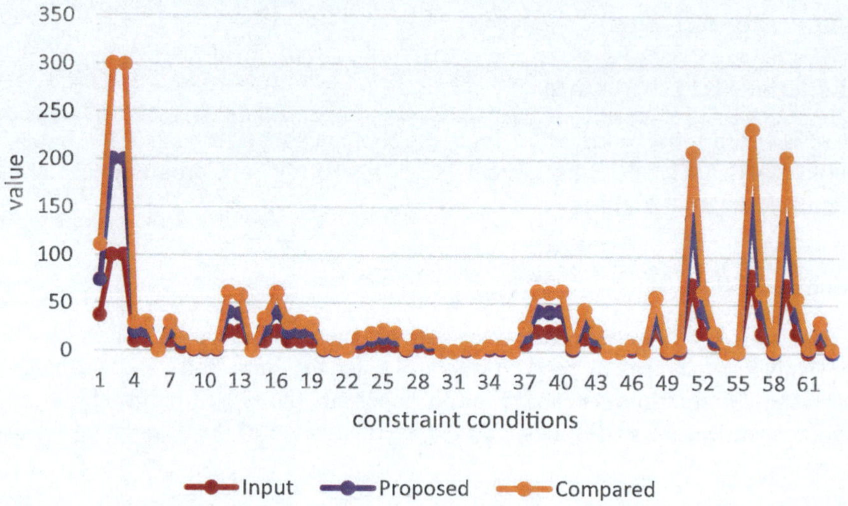

Fig. 3. The comparison between the proposed algorithm and the comparison algorithm on the satisfaction of constraint conditions

Therefore, the results obtained in this paper are more superior to the constraint, and the practical application should be stronger.

Acknowledgements. The study is supported by the Education and reform research project of Jilin University (2021XYB186); Fund for doctoral research of Northeast Electric Power University (BSJXM-2020219); Science and technology research program of Jilin Provincial Department of Education (JJKH20210115KJ).

References

1. Xiong, K., Huang, X.: Research on auto-generating test paper system based on LDA and genetic algorithm. In: 9th IEEE International Conference on Software Engineering and Service Science (ICSESS), Beijing, China, pp. 416–421 (2018)
2. Liu, C., Hu, Y.: Research on intelligent test paper generating strategy based on genetic algorithm and BP neural network. In: 2nd Joint International Information Technology, Mechanical and Electronic Engineering Conference (JIMEC), Chongqing, China, pp. 157–160 (2017)
3. Jun, N.: An improved genetic algorithm for Intelligent test paper generation. In: 7th International Conference on Intelligent Computation Technology and Automation (ICICTA), Changsha, China, pp. 72–75 (2014)
4. Xue, J.: Intelligent generating test paper scheme for English class based on genetic algorithm. In: 6th International Conference on Measuring Technology and Mechatronics Automation (ICMTMA), Zhangjiajie, China, pp. 358–362 (2014)
5. Wang, L.: Research on computer intelligent test paper composition based on genetic algorithm. In: 3rd International Conference on Machinery, Materials and Information Technology Applications (ICMMITA), Qingdao, China, pp.1029–1034 (2015)

6. Ma, Q., Pan, H.: Researching of intelligence test paper generation software based on improved genetic algorithm in Java program design teaching. In: 2nd International Conference on Mechatronics Engineering and Information Technology (ICMEIT), Dalian, China, pp. 232–235 (2017)
7. Zhang, K., Chen, M., Xu, X., Yen, G.: Multi-objective evolution strategy for multimodal multi-objective optimization. Appl. Soft Comput. **101**, 107004 (2021)
8. Luo, J., Yang, Y., Liu, Q., Li, X., Chen, M.: A new hybrid memetic multi-objective optimization algorithm for multi-objective optimization. Inf. Sci. **448**, 164–186 (2018)
9. Zhang, S., Wang, H., Yang, D., Huang, M.: Hybrid multi-objective genetic algorithm for multi-objective optimization problems. In: 27th Chinese Control and Decision Conference (CCDC), Qingdao, China, pp.1970–1974 (2015)
10. Wang, H., Olhofer, M., Jin, Y.: A mini-review on preference modeling and articulation in multi-objective optimization: current status and challenges. Complex Intel. Syst. **3**(4), 233–245 (2017)
11. Ge, Y., Wang, A., Zhao, Z., Ye, J.: A Tabu-genetic hybrid search algorithm for job-shop scheduling problem. In: 3rd International Conference on Power, Energy and Mechanical Engineering (ICPEME), Prague, Czech Republic, p. 04007 (2019)
12. Chen, X.: A diesel-wind-solar-battery micro grid's optimal configuration model based on the modified non-inferior sorting genetic algorithm. In: International Conference on Advances in Energy, Environment and Chemical Science (AEECS), Changsha, China, pp. 176–181 (2016)
13. Kazimipour, B., Li, X., Qin, A.: A review of population initialization techniques for evolutionary algorithms. In: IEEE Congress on Evolutionary Computation (CEC), Beijing, China, pp. 2585–2592 (2014)
14. Poikolainen, I., Neri, F., Caraffini, F.: Cluster-based population initialization for differential evolution frameworks. Inf. Sci. **297**, 216–235 (2015)
15. Zhang, K., Zhu, L.: Application of improved genetic algorithm in automatic test paper generation. In: Chinese Automation Congress (CAC), Wuhan, China, pp. 495–498 (2015)

Improved Multi-objective Genetic Algorithm for Unit Economic Operating Load Distribution

Minghao Guo(✉), Wei Wang, and Renhui Chen

Changchun University of Technology, Yan'an Street 2055, Changchun 130012, Jilin, China
lyh_899@qq.com, {wangwei_dianqi,chenrenhui}@ccut.edu.cn

Abstract. Genetic Algorithm is an optimization technique to simulate the evolution of biological populations in nature, but standard genetic algorithms often fail to obtain the optimization of multi-target models. Based on the study of standard multi-objective genetic algorithm, this paper puts forward a new mechanism for dealing with constraints, and uses it to solve the problem of economic operation load distribution of power system units through improved multi-objective genetic algorithm, so that the power system achieves maximum economical (i.e. the power generation cost is the minimum), and the paper obtains the ideal results through the calculation.

Keywords: Multi-objective genetic algorithm · Power distribution · Pareto · Correction mechanism

1 Introduction

Thermal power generation is the main way to obtain electric energy in various countries. How to ensure the power quality and production efficiency while making the power energy consumption such as coal the least, that is, the lowest power generation cost, has become a problem that many scholars are committed to solving. At present, there are many methods to calculate the minimum cost of load distribution, such as equal consumption micro-increase method, linear programming method, nonlinear programming method, dynamic programming method and so on [1–4]. In recent years, the intelligent algorithm based on artificial intelligence has global convergence in theory, has no strict requirements for initial points and is suitable for programming, and is suitable for solving nonlinear optimization problems of dynamic systems. Genetic algorithm, particle swarm optimization algorithm, other algorithms have been used to solve power generation optimization scheduling problems in wind, light, water and thermal power systems. At the same time, the group characteristics of this algorithm provide an effective way to solve multi-objective optimization problems [5]. However, there are problems in the above intelligent algorithms, precocious phenomenon will occur in the genetic algorithm, this phenomenon causes the excellent properties of the tutor genetic algorithm losing the global search capabilities, and the parameter setting in the particle group algorithm is usually determined by experimental methods, leading to the optimization performance

S.-C. Chu et al. (Eds.): ICGEC 2021, LNEE 833, pp. 610–618, 2022.
https://doi.org/10.1007/978-981-16-8430-2_55

of the method is closely related to human experience, and it is difficult to optimize the algorithm performance.

Therefore, this paper relies on the basis of applying the multi-objective genetic algorithm, and in the process of standard multi-objective optimization, a new mechanism to deal with constraints is proposed to ensure the accuracy of objective optimization and to meet the characteristics of universality and self-adaptability of genetic algorithm.

2 Establishment of Mathematical Models

2.1 Objective Function

The economic distribution of unit load exists in the power generation part of the power system, and its goal is to make the generation cost minimum on the basis that the combined unit meets the operating constraints of the power system [6], $\sum F$ by studying the experimental unit data, the following dynamic equations are established without considering the net loss:

$$\mathrm{F} = \left[\sum_{i=1}^{m} \sum_{t=1}^{n} F_{i(t)}\left(P_{i(t)}\right) \times S\right] + \mathrm{C_i} + \mathrm{H_i} \tag{1}$$

$$F_{i(t)}\left(P_{i(t)}\right) = a_i + b_i P_{i(t)} + c_i P_{i(t)}^2 \tag{2}$$

Type: F is the objective function, that is, the total cost of the demand load of n units in m period of time; $F_{i(t)}\left(P_{i(t)}\right)$ is the fuel consumption for the n unit during the i period; $P_{i(t)}$ is the output for the i unit during the period; S is the cost of one ton of standard coal; $\mathrm{C_i}$ is the cold start-up cost for the i unit; $\mathrm{H_i}$ is the hot start-up cost for the i unit; $a_i b_i c_i$ are fuel cost function coefficients for the i unit.

2.2 Constraints

The inequality constraints in load economic distribution are the same as those in power flow calculation. The active power of any power plant should not exceed its upper and lower limits [6] namely:

$$\mathrm{P_{imin}} \le P_{i(t)} \le \mathrm{P_{imax}}(\mathrm{i} = 1, 2, 3 \ldots \mathrm{m}) \tag{3}$$

The equality constraint condition is that the unit combination satisfies the load demand of the system in a certain period of time, that is:

$$P_{D(t)} = \sum_{i=1}^{m} P_{i(t)}(\mathrm{t} = 1, 2, 3 \ldots \mathrm{n}) \tag{4}$$

Type: $P_{D(t)}$ for the system t period of load demand: $P_{i(t)}$ for the i unit in the t period of the distribution of load; $\mathrm{P_{imin}}$, $\mathrm{P_{imax}}$ for the i unit of the minimum, maximum power output.

3 Multi-objective Genetic Algorithm and Optimization

3.1 Multi-objective Genetic Algorithm

Genetic algorithm is a random global search and optimization method to simulate the evolution mechanism of natural organisms. It draws on Darwin's theory of evolution and Mendel's theory of genetics. Its essence is an efficient and fast intelligent algorithm. The algorithm flow chart is shown in Fig. 1:

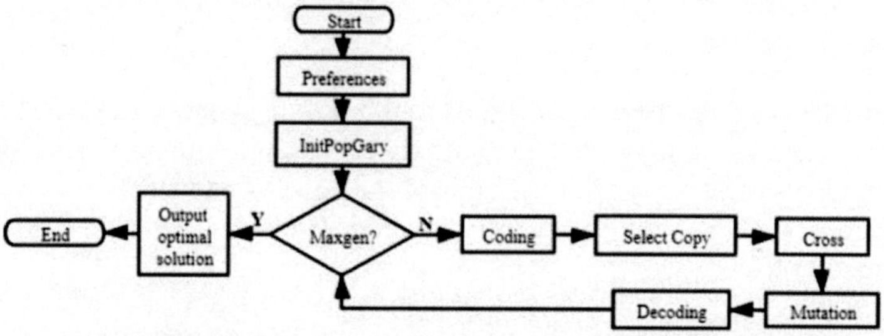

Fig. 1. Flow chart of a multi-objective genetic algorithm

For a multi-objective optimization problem, the optimal solution is not unique, but is deconstructed into a set of optimal solutions, which is called Pareto optimal solution [7], At the feasible solution set, there is no better solution than the individual of the population corresponding to the Pareto optimal solution, and there is no difference between the advantages and disadvantages of the optimal solution set.

3.2 Improved Multi-objective Genetic Algorithm

Based on multi-objective genetic algorithm, a new correction mechanism is proposed for dealing with constraints under reciprocal constraints. At the same time, in order to ensure the high randomness of the progeny population and accelerate the convergence of the output, the generation gap mechanism is introduced into the operation selection replication operator.

Treatment of Constraints

There are many variables in multi-objective optimization problems in complex systems, and inequality and equality constraints are more complex and difficult to deal with. At present, most genetic algorithms use the following methods to deal with equality constraints, but there are some drawbacks.

Restart Method

Random generation of population matrix, will not meet the constraints of the population through judgment to generate again, and this method is not suitable for complex constraints, if the number of evolution, population number, chromosome length will lead to evolution into a cycle, greatly reduce efficiency, so undesirable.

Penalty Function Method
When dealing with equality constraints, penalty function can be introduced into fitness function or objective function to punish randomly generated default items and reduce the fitness of default items. Namely:

$$F(x) = \begin{cases} f(x) & x \ \textit{satisfying the constraints} \\ f(x) + k(x) & x \ \textit{does not satisfy constraints} \end{cases}$$

Type: F(x) new fitness function; $f(x)$ original fitness function; $k(x)$ penalty function.

The difficulty of setting an ideal penalty function is how to consider the degree of breach of contract. Once the complexity of constraints increases, the difficulty will increase.

Considering the above problems, this paper simulates the principle of 'Apical Advantage' in the plant world, regards the objective function as the 'Trunk' of the plant, regards the randomly generated population as the 'Top Bud', and regards the randomly generated population that does not meet the constraints as the 'Lateral Bud'. In order to ensure the growth of the trunk, the lateral bud, which affects the supply of auxin to the bud, at the same time achieve the purpose of fast processing constraints, further accurate search area extremum.

During selecting, crossing and mutation will get new populations. If the equality constraints are not met, after do the following:

$$NP_{i(t)} = \frac{P_{i(t)}}{\sum_{i=1}^{n} \sum_{t=1}^{m} P_{i(t)}} \times D_{(t)} \tag{5}$$

Type: $NP_{i(t)}$—output allocated by the corrected t unit at the time of i;
$P_{i(t)}$—output of the i unit at moment t;
$D_{(t)}$—output required at moment t;
n—number of units;
m—total time period requiring output allocation.

When the new output population matrix is obtained, because this method controls the output matrix in the inequality constraint in the first iteration, it accelerates the convergence speed while ensuring the optimal solution.

Introduction of Generation Gap Mechanism
After calculating the fitness by roulette, random ergodic sampling and other operators, the number of all the previous generations is calculated. In this paper, the generation gap mechanism is introduced in the process of selecting replication, that is, the number of next generation individuals is different from that of the previous generation. Then the progeny population is reinserted into the objective function instead of the parent population by operation to ensure the diversity of the genetic algorithm and to accelerate the convergence and reduce the operation time.

Therefore, the improved multi-objective genetic algorithm flow in Fig. 2:

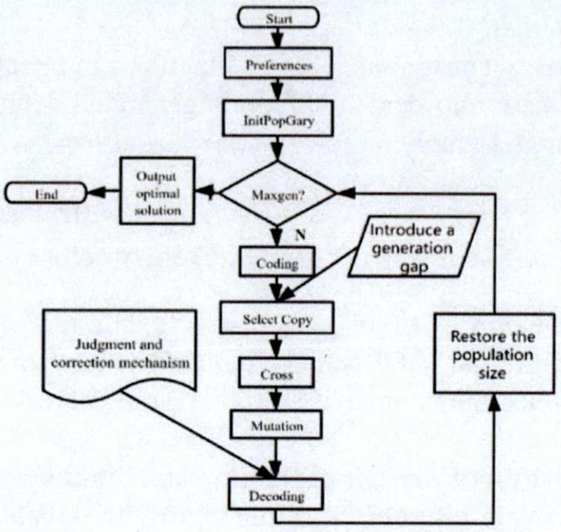

Fig. 2. Flow chart of improved genetic algorithm

4 Example Analysis

4.1 Example Description

Based on literature [8] in the first hour of the experimental unit data, 5 units parallel and regardless of network loss and climbing time as an example (see Table 1 for specific data):

Table 1. Data for five units

	Unit1	Unit2	Unit3	Unit4	Unit5
P_{max} (MW)	455	130	130	80	55
P_{min} (MW)	150	20	20	20	55
a ($/h)	1000	700	680	370	660
b ($/MWh)	16.19	16.60	16.50	22.26	25.92
c ($/MW^{2}h)	0.00048	0.002	0.00211	0.00712	0.00413
H-C	4500	550	560	170	30
C-C	9000	1100	1120	340	60
CH	5	4	4	2	0
IS	8	−5	−5	−3	−1

Note: $P_{max,min}$ unit output upper and lower sections (MW); H-C($): cold start-up costs; H-C ($): hot start-up costs; CH(h): cold start-up duration; IS(h): initial state.

According to formula (1), (2), (3), (4), the following objective functions can be established:

$$
\begin{aligned}
F_1 &= 1000 + 16.19P_1 + 0.00048P_1^2 \\
F_2 &= 700 + 16.60P_2 + 0.002P_2^2 \\
F_3 &= 680 + 16.50P_3 + 0.00211P_3^2 \\
F_4 &= 370 + 22.26P_4 + 0.00712P_4^2 \\
F_5 &= 660 + 25.92 \times 55 + 0.00413 \times 55^2 \\
\mathrm{F} &= F_1 + F_2 + F_3 + F_4 + F_5 + 1100 + 1120 + 340 + 60
\end{aligned}
$$

The constraints are:

$$
\begin{aligned}
150\text{ MW} &\le P_1 \le 455\text{ MW} \\
20\text{ MW} &\le P_2 \le 130\text{ MW} \\
20\text{ MW} &\le P_3 \le 130\text{ MW} \\
20\text{ MW} &\le P_4 \le 80\text{ MW} \\
P_1 + P_2 + P_3 + P_4 &= 500 - P_5 = 345\text{ MW}
\end{aligned}
$$

(1) The process of calculating the distribution of active power using the equal consumption method is as follows. According to the consumption characteristic function, each unit $F_1F_2F_3F_4$ is:

$$
\begin{aligned}
\lambda_1 &= \frac{dF_1}{dP_1} = 16.19 + 0.00096P_1 \\
\lambda_2 &= \frac{dF_2}{dP_2} = 16.60 + 0.004P_2 \\
\lambda_3 &= \frac{dF_3}{dP_3} = 16.50 + 0.00422P_3 \\
\lambda_4 &= \frac{dF_4}{dP_4} = 22.26 + 0.01424P_4
\end{aligned}
$$

Order $\lambda_1 = \lambda_2 = \lambda_3 = \lambda_4$, it can be concluded:

$$
\begin{aligned}
P_1 &= 427.08 + 4.17P_2 \\
P_3 &= 23.70 + 0.95P_2 \\
P_4 &= -397.47 + 0.28P_2
\end{aligned}
$$

Based on the above constraints are known:

$$
\begin{aligned}
P_1 + P_2 + P_3 + P_4 = 427.08 + 4.17P_2 + P_2 + 23.70 + 0.95P_2 \\
-397.47 + 0.28P_2 = 345
\end{aligned}
$$

Solvable $P_2 = 45.576$ MW; $P_1 = 617.057$ MW; $P_3 = 66.997$ MW; $P_4 = -384.708$ MW.

Because P_1, P_4 beyond the limits, here's what makes P_4 least effort $P_4 = 20$ MW, corresponding $P_1 = 345 - P_2 - P_3 - P_4 = 212.425$ MW.

So the object function:

$$
\mathrm{F} = 13470.6156(\$)
$$

(2) The parameters of Matlab multi-target genetic algorithm are as follows: selected chromosome length 20, maximum genetic algebra 200, population 50, cross probability 0.9, variation probability 0.05; population generation rate is 0.95. Using the mixed coding method, the optimization results are as follows (the target function result retains 4 decimal places):

$$P_1 = 243.2273\ \text{MW}$$
$$P_2 = 23.4784\ \text{MW}$$
$$P_3 = 43.5517\ \text{MW}$$
$$P_4 = 34.7425\ \text{MW}$$
$$\text{F} = 13329.7527(\$)$$

(3) The selection of parameters of the improved multi-target genetic algorithm is the same as the standard multi-target algorithm:

$$P_1 = 277.0553\ \text{MW}$$
$$P_2 = 23.9318\ \text{MW}$$
$$P_3 = 23.8039\ \text{MW}$$
$$P_4 = 20.2090\ \text{MW}$$
$$\text{F} = 13235.5966(\$)$$

4.2 Interpretation of Result

Summing up results of three kinds of methods shown in Table 2:

Table 2. Comparison of results

Units	Micro-increment method	Standard GA	Improved GA
P_1 (MW)	212.4256	243.2273	277.0553
P_2 (MW)	45.5766	23.4784	23.9318
P_3 (MW)	66.9978	43.5517	23.8039
P_4 (MW)	20.0000	34.7425	20.2090
P_5 (MW)	55.0000	55.0000	55.0000
F ($)	13470.6156	13329.7527	13235.5966

Table 2 results show that the improved genetic algorithm's accuracy relative to micro-increment method and standard genetic algorithm increase by 1.78% and 0.72%, respectively.

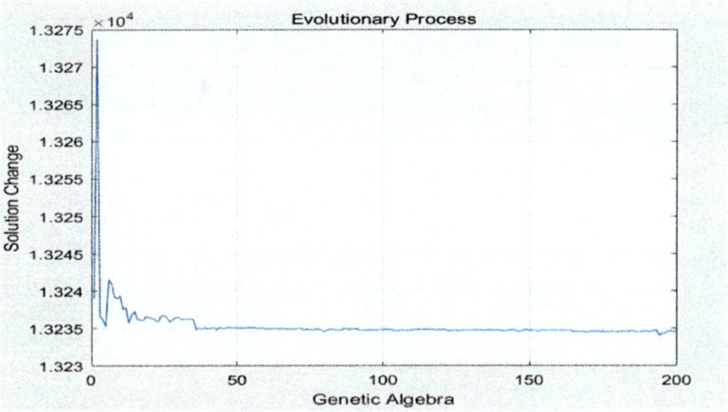

Fig. 3. Target iterative evolution curve

An iterative curve for the improvement of the multi-objective genetic algorithm is shown in Fig. 3:

Figure 3 for the evolution of improved genetic algorithm for curve, evident in the iterative convergence around 39 times the operation, and the early jump is because of the uncertainty of the early generation of random population, and subsequently a sharp decline that is the effect of the correction mechanism to ensure that the extreme value of the search space, speed up convergence to the optimal solution.

5 Conclusion

This paper proposes a power distribution algorithm of multiple thermal power units. The example calculation results shall wait until the following conclusions: comprehensive analysis of the results to verify the feasibility of the proposed optimization algorithm in this article. In a slight increase of rate method in mathematics, as has been seen as an optimal solution, but there's room for promotion.

The model currently considered in this paper is a multi-objective power allocation problem under general conditions, and the author will subsequently consider the multi-objective problem under complex conditions.

References

1. Habibi, Y.Y.: A new optimized power flow model for hydraulic power system. Chin. J. Electr. Eng. **28**(10), 37–45 (2008). (in Chinese)
2. Zhifu, W., Ming, Z., Baohua, L., et al.: Optimal dispatching model and its application in power market. Grid Technol. **30**(15), 45–49 (2006). (in Chinese)
3. Yanling, W., Particle, Z., Yihan, Y.: Electricity rights regulation market based on water-fire electricity replacement. Chin. J. Electr. Eng. **26**(5), 131–136 (2006). (in Chinese)
4. Ma, R., He, R., Yan, H., et al.: A multi-objective optimization subgroup segmental bidding model considering water-fire coordination. Chin. J. Electr. Eng. **24**(11), 53–57 (2004). (in Chinese)

5. Wu, J., Tang, L., Huang, H., Lu, Y.: Multi-objective economic scheduling of power generation in water-fire power system based on genetic algorithm and data envelopment analysis. Grid Technol. **35**(05), 76–81 (2011). (in Chinese)
6. Chen, Y., Wang, J., Wu, J.: An improved genetic algorithm for economic load distribution in power system. Chin. Electr. Power Educ. **2006**(S3), 150–152 (2006). (in Chinese)
7. Deb, K., Pratap, A., Agarwal, S., Meyarivan, T.: A fast and elitist multiobjective genetic algorithm: NSGA-II. IEEE Trans. Evol. Comput. **6**(2), 182–197 (2002). https://doi.org/10.1109/4235.996017
8. Kazarlis, S.A., Bakirtzis, A.G., Petridis, V.: A genetic algorithm solution to the unit commitment problem. IEEE Trans. Power Syst. **11**(1), 83–92 (1996). https://doi.org/10.1109/59.485989

An Improved GA-Based Recommendation System for Soil Fertilization

Usman Ahmed[1], Jerry Chun-Wei Lin[1(✉)], Gautam Srivastava[2,3], and Jimmy Ming-Tai Wu[4]

[1] Western Norway University of Applied Sciences, Bergen, Norway
Usman.Ahmed@hvl.no, jerrylin@ieee.org
[2] Brandon University, Brandon, Canada
SRIVASTAVAG@brandonu.ca
[3] Chinese Medical University, Taichung, Taiwan
[4] Shandong University of Science and Tehcnology, Qingdao, China
wmt@wmt35.idv.tw

Abstract. We present an improved GA-based approach for nutrient recommendation that utilizes time-series sensor data and suggests various settings for different crops in this article. After that, a neighborhood-based method is provided for handling exploration and exploitation in order to optimize the parameters to maximize the production. The final judgment is determined based on the similarity between recommendations of the patterns and real-time sensor data. The experimental findings indicate that the suggested model is capable of recommending optimal patterns and assisting in effectively increasing the annual production.

Keywords: Optimization · Senor network · Soil nutrient · Agriculture

1 Introduction

Precision agriculture is a critical component for the long-term intensity of agriculture, in which information, communication technology and others are necessary but insufficient for sustainable agricultural systems [8]. Technology should be integrated into farmers' practices and managed by their experience-based and useful knowledge in order to contribute to enhancing farming sustainability. Precision agriculture has made significant strides in recent years as a result of the development of novel, sustainable, and low-cost pesticides, as well as fertilizers. The increased use of information and communication technology is one of the causes behind the information and communication technology (ICT). ICT is a network-based data transmission system that communicates information about the condition of production, process control, equipment, operations, and the environment [8]. Through the use of ICT, we can improve the efficiency of farming operations and management while reducing agricultural costs and labor. ICT makes use of model-based techniques such as prediction, statistical learning, and decision support systems.

S.-C. Chu et al. (Eds.): ICGEC 2021, LNEE 833, pp. 619–628, 2022.
https://doi.org/10.1007/978-981-16-8430-2_56

Numerous approaches [5,8] have documented regional soil heterogeneity within various agricultural regions. Additionally, the soil monitoring approach has been refined by exposing different crops to a variety of environmental variables, such as the usage of an important nutrient in response to certain climate conditions. The stated approach aids in the growth of plants by using fertilizer and providing a high yield. It increases productivity and nutrient utilization, both of which are essential in soil engineering. Precision farming using GPS (global positioning system) will provide useful information for managing local ecosystems and agricultural optimally [8]. This assists the decision support system in making decisions and in the deployment of water management and spraying drones.

Soil is the earth's most fundamental organic resource; it aids in food production while also balancing the quality of the local, global, and regional environment. On the other side, the equilibrium between fertility and yield production is disturbed by increased output induced by the use of seed varieties, chemical fertilizers, and pesticides [8]. We offer an enhanced GA technique in this article for recommending optimal nutrients, leading in excellent yield output. The experimental findings indicated that the proposed model enhanced the nutrition recommendation's performance. Our method provides an alternate method for determining the appropriate nutrient to maximize yield output while preserving soil fertility.

2 Related Work

Farm recommendation systems have evolved in the literature as a result of several trials with diverse approaches to agriculture crops [5]. Genetic improvement of seed quality results in higher agricultural yield. To genetically enhance seeds, the combination of genotype and phenotype may be investigated in a unique environment [7]. Additionally, soil quality, weeds, nitrogen levels, and water management may all contribute to increased productivity [4]. Additionally, the learning-based method is critical for increasing agricultural yield and advancing precision agriculture [9]. The author devised a strategy for estimating greenhouse gas emissions in order to estimate potato yields in Iranian fields [3]. The author collected data for many crops personally; the generated model is then evaluated under a variety of situations and by specialists. Electricity, fertilizers, and seed quality all have an effect on predicting production. The model has a predicting accuracy of 98% and a gas emission prediction accuracy of 99%. The authors employ a wireless sensor network to monitor soil moisture and subsequently assist with irrigation system automation [6]. The sensor collects data on soil moisture and pH levels in order to optimize crop output. The created method assists in providing information on moisture levels and soil requirements, as well as fertilizers and pesticides [6].

Additionally, a regression model based on neural networks is presented to forecast the presence of rain in the relevant geographical region. The author obtained the data from a meteorological station in Ahmednagar, India. Temperature, humidity, values, and rainfall data for the previous 10 years are included.

Table 1. A representative sample of the population once it has been initiated.

Instance	*Nitrogen 2001*	*Phosphorous 2001*	*Potassium 2001*	*Nitrogen 2002*	*Phosphorous 2002*	*Potassium 2002*	....	*Nitrogen 2010*	*Phosphorous 2010*	*Potassium 2010*	*Yield*
I1	49	30	15	42	17	22	....	57	22	12	1717.5
I2	59	31	15	46	21	18	....	46	34	17	1878.9
I3	44	31	5	54	10	7	....	54	37	13	1880.2
I4	44	5	7	53	21	17	....	50	26	19	1719.5
I5	40	11	21	40	21	10	....	48	29	18	1961.6
I6	44	23	13	59	33	8	....	50	32	13	1720.1
I7	58	11	17	57	16	23	....	46	20	19	1880.5
I8	40	33	17	47	28	22	....	48	13	13	1720
I9	54	11	15	40	19	10	....	46	13	19	1878.3
I10	57	32	21	50	17	22	....	45	44	18	2039.1

The regression method is able to improving the rainfall forecast for the specified geolocation [1]. Agriculture datasets have also been subjected to various data mining methods. For instance, in cite27, the authors proposed a model based on clustering and including fuzzy k-mean. The soil is grouped according to the characteristics of the sensor values obtained. The article then compares the findings to Google maps and segmentation maps of local streets [2]. According to the authors, the model is an appropriate approach for presenting data to scientists and landowners. The authors of cite28 presented two machine learning models, dubbed PLSR and ANFIS. The model could forecast the amount of water used in the irrigation system by utilizing the soil condition, weather, and crop condition as primary predictors as illustrated in Table 2.

Table 2. A representative sample of year-by-year data gathering for cotton farming.

SID	Time	Nitrogen	Phosphorous	Potassium
s1	2001	108	67	11
s2	2002	182	67	12
s3	2003	177	78	11
.	.	.	.	.
.	.	.	.	.
.	.	.	.	.
.	.	.	.	.
s10	2010	290	70	25

3 Preliminaries and Problem Statement

Let $N = \{N_1, N_2, \ldots, N_m\}$ be a set of nutrients levels and the data a collection of instances such as $I = \{I_1, I_2, \ldots, I_n\}$, where each instance is a collection of time-series data belonging to N and is uniquely identified by the string I_{id}. Each nutrient in an instance has a value, such as $v(N_k, I_c)$ (depending on the crop state and sensor data). Table 1 illustrates a straightforward procedure for

Algorithm 1. Designed approach

INPUT: P, N_i and number of generations max_{gen}, the size of neighbors n_s, crossover probability p_c, mutation probability p_m.
OUTPUT: Optimize parameter with yield values

```
while max_gen do
    for all i ∈ pop do
        P'_i ← Randomly select an individual from N_i.
        P'_i ← Exploration(P)
        child ← CrossMutation(P'_i, P_i).
    end for
    P ← PopulationSelection(P)
    P ← PopulationExploitation(P)
end while
Return Optimize_parameter.
```

initializing the crop time series database. Additionally, Table 1 s a sampling of the cotton crop data collection.

Problem Statement: This study proposes a model for the ICT environment that enables efficient investigation of proper nutrition use in order to build a knowledge-based system. The developed information is then applied directly to the environment, with farmers being recommended a plan for balancing soil fertility and crop output. Additionally, the suggested configuration aids in crop production improvement. Without previous threshold values, the model can determine the optimal nutrient amount and identify patterns from time series data. The model utilizes a genetic algorithm to process the data and propose the most current optimal settings for distant locations.

4 Designed Model

We illustrate the ICT architecture in Fig. 1, where each sensor in the distant area has its own set of nutrient levels (N_i). The sensor values are kept locally and periodically sent to the database (weekly, monthly, yearly). All sensors for each nutrient are gathered and combined over the Internet in order to log the remote area's massive data collection. The developed method is then implemented and optimizes for decision-making sequence nutrients. The framework is described in full below.

The created model's aim is to simultaneously maintain soil fertility (nutrient levels) and improve crop output (yield). The model determines the optimization parameters through the use of the Euclidean distance-based search technique. The suggested exploration and exploitation approach contributes to both local and global optimization. Occasionally, even high-quality soil is incapable of producing a large yield. The reason for this is because crops have been continuously cultivated for a longer period of time. After a few years, the nutrients level decreases, resulting in a decrease in quality and production. As a result, we analyzed a set of data from the same farmed area and then advised the pre-crop

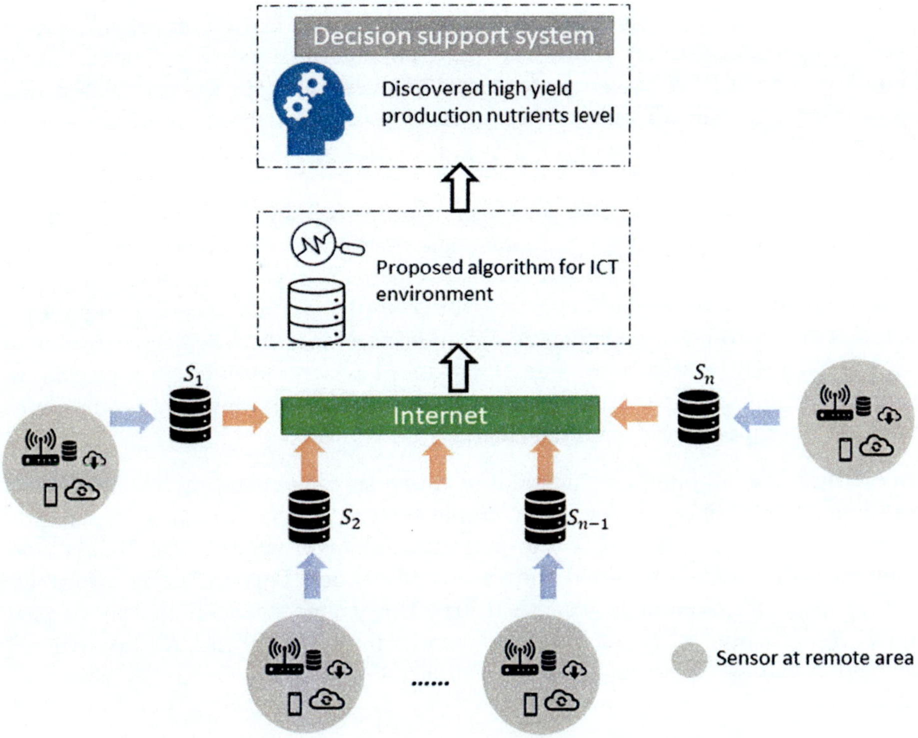

Fig. 1. A framework for ICT development.

nutrient level. In comparison to the conventional technique, which utilizes just the soil's historical or current state. The created model is capable of recommending a set of nutrient concentrations and optimizing yield output as indicated in Eq. 1.

$$\begin{aligned} &F(x) \leftarrow \text{maximize}\,[f_1(X)] \\ &\text{where } \{f_1(X) \leftarrow max(yield(X))\} \end{aligned} \tag{1}$$

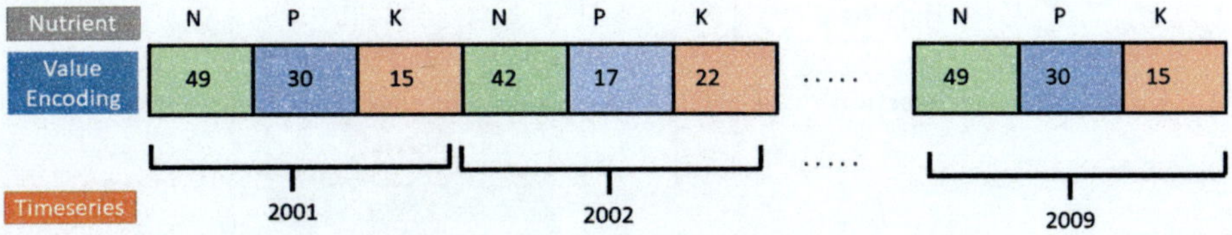

Fig. 2. Example of time-series data encoding.

Initialization: In the first stage, we suggest an initialization technique that is specific to the situation at hand. The value-encoded technique (nutrient values) is employed. We utilize the roulette selection mechanism for the selection mechanism. To begin, the chromosomes are reset to zero. Then a random value of length $len(N_i)$ is chosen. The loop begins with roulette selection, which randomly selects individuals from a population of all chromosomes using a roulette wheel. The size of the roulette selection is proportional to the fitness value of each chromosome - the higher the value, the greater the probability of selection. Following that, we add the fitness values of each chromosome in the population. Then, we used a random number generator to spin the roulette wheel and obtain the index and chromosomal values. We do the same thing for each chromosome in terms of assigning a selection value. This completes the initialization procedure for the *pop* population size. The population matrix contains all chromosomes that have been populated and attached.

Encoding: We employ the nutrient's value as the encoding technique, as described in Algorithm 1. We then employed the two-factor encoding model, i.e., order and values, to aid in the permutation task, whereas the binary and value encoding models retained the values and order. Due to the fact that we must propose nutritional levels, we utilize the values encoded in the created model. As indicated in Eq. 2, the nutritional values (N, P, and K) are utilized as encoded values. The year-based encoding method's time series are shown in Fig. 2.

$$B_{q,j} = \begin{cases} nutrient\_value(), \ i_j \in I_q \\ 0, \qquad\qquad\qquad\qquad \text{otherwise} \end{cases} \tag{2}$$

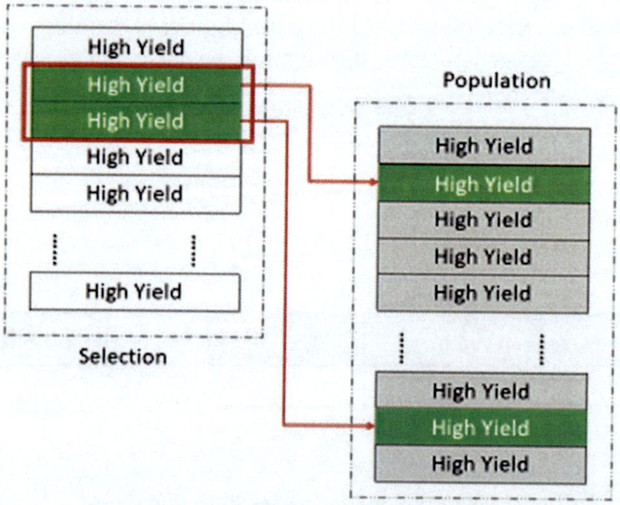

Fig. 3. Population selection mechanism.

Exploration of Neighbors and Population Evolution: For each iteration, we develop a strategy to transfer the data from one generation to another, as mentioned in Fig. 5 and 6. In addition, the designed model proposes a neighbor-based crossover and mutation technique. The initialized population's Euclidean distance is determined using the Eq. 3.

$$d_{ij} = \sqrt{\sum_{k=1}^{n} (x_{ik} - x_{jk})^2} \tag{3}$$

The equation calculates the distance between two individuals by computing the root square and difference between each individual's points. The individual with the highest degree of similarity is chosen, allowing for the exploration of global optimization. We choose each neighbor individually and then do a one-point cross mutation. We create a technique for data transmission between iterations for each iteration. Two successive members of the preceding generation are randomly chosen. We replace it with the current generation, and the random replacement serves to develop individuals in many ways for evolutionary advancement. Additionally, expanding the population's search field through inclusion benefits population variety while avoiding local optimization failure. As seen in Fig. 4, we also suggest the population exploitation technique as a means of identifying more effective solutions. Following that, we combine the prior and new populations and manually eliminate duplicates. Following that, we choose candidates using a rating system.

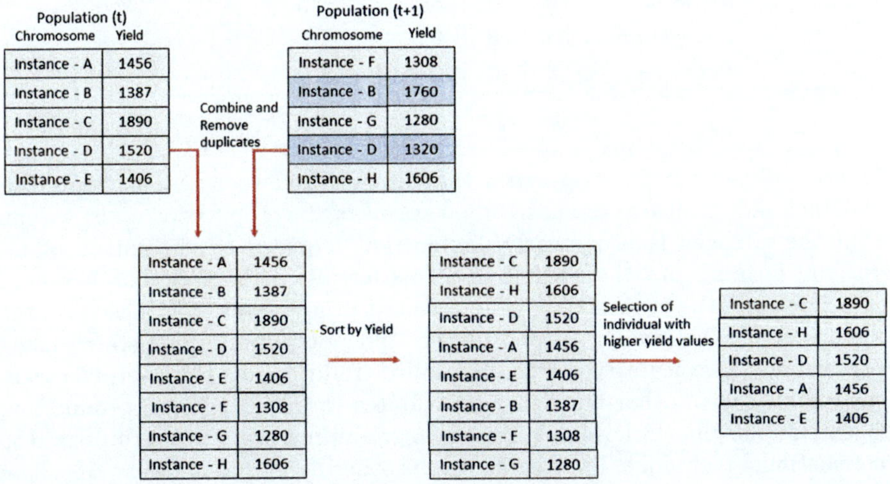

Fig. 4. Population exploitation method.

5 Experiments

This section discusses the quantity of nutrients consumed by the crop during the last decade. The following Table 3 summarizes the features of several crops. The best and worst values are specified. Experiments are conducted using the Linux mint distribution, a core i7 CPU, and 16 GB of RAM. We set the population size to 100, the maximal generation to 100, the neighborhood exploration to 10, and the mutation and crossover probabilities to 0.01 and 1.0, respectively, for the improved GA-based model. We next utilized the 2001–2009 records to estimate the 2010 nutrient levels for cotton, peanuts, maize, and rice.

Table 3. Characteristic of dataset used.

Type	Nutrients	Max	Min	Best max	Best min	Yield	Best yield
Cotton	*Nitrogen*	350	180	290	110	1600	1900
	Phosphorous	110	60	70	60		
	Potassium	30	15	25	10		
Ground nut	*Nitrogen*	90	40	30	15	1100	1600
	Phosphorous	110	60	55	35		
	Potassium	35	15	45	25		
Maize	*Nitrogen*	275	90	170	60	1400	2400
	Phosphorous	70	30	50	15		
	Potassium	90	20	60	10		
Rice	*Nitrogen*	80	70	55	45	110	2200
	Phosphorous	38	20	35	15		
	Potassium	25	10	20	10		

Figure 5 illustrates all crop types that are utilized to assess nutrient loss trends. Each plot indicates the nutritional status of the crop. As a result, we can see that the nitrogen level is nearly double that required by the cotton plant. Traditional farming practices result in a low fertility level and thus low crop output. However, the suggested setting resulted in a greater yield based on the intended model. The amount of phosphorus and potassium administered using IGA is virtually identical to the amount applied traditionally. However, nitrogen, when combined with other nutrients, contributes to increased crop production. Nitrogen and phosphorus levels are greater in groundnut crops when compared to the conventional technique. Potassium, on the other hand, has a lower value. The yield value is nearly same for the conventional technique. The intended model, on the other hand, performs slightly better. For the maize crop, the model need further generation in order to expand knowledge. For maize, IGA recommends higher nutritional levels than the conventional approach.

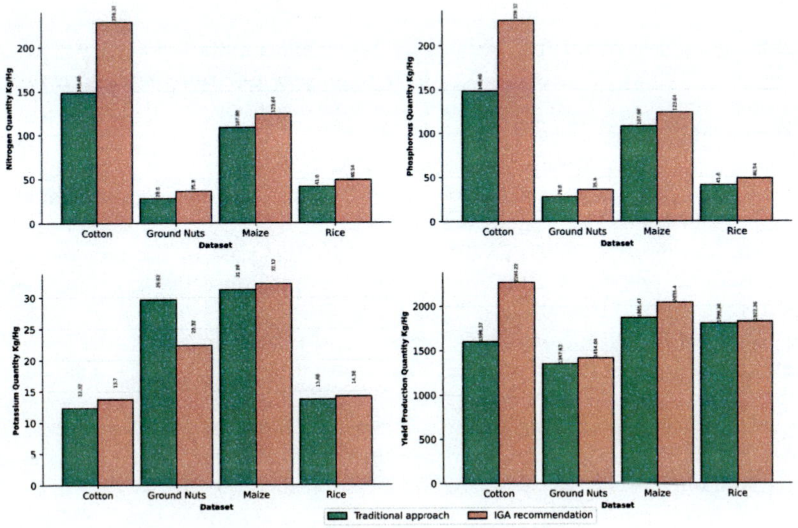

Fig. 5. Nitrogen, Phosphorous, Potassium, and Yield results were compared to the conventional technique.

6 Conclusion and Future Work

In this work, we suggested an IGA technique for recommending the best nutrient settings for various crops. The method employs an optimization strategy that includes techniques for neighborhood exploration and exploitation. The model was able to enhance local optimization in order to prevent premature individuals from being included in the population strategy. Additionally, we employ variety as a means of expanding population knowledge. The findings from real-world datasets demonstrated that the suggested technique outperforms the usual suggestion. As a result, the algorithm is capable of optimizing production while maintaining nutritional levels. We intend to optimize the search strategy and individual repair techniques in the future in order to extract valuable parameters. This contributes to the optimization of computation resources and the suggestion to keep crops for soil fertilization.

References

1. Bendre, M.R., Thool, R.C., Thool, V.R.: Big data in precision agriculture: weather forecasting for future farming. In: The International Conference on Next Generation Computing Technologies (2015)
2. Hot, E., Popovic-Bugarin, V.: Soil data clustering by using k-means and fuzzy k-means algorithm. In: The Telecommunications Forum Telfor (2015)
3. Khoshnevisan, B., Rafiee, S., Omid, M., Mousazadeh, H., Rajaeifar, M.A.: Application of artificial neural networks for prediction of output energy and GHG emissions in potato production in Iran. Agric. Syst. **123**, 120–127 (2014)

4. Khoury, C.K., et al.: Increasing homogeneity in global food supplies and the implications for food security. Proc. Nat. Acad. Sci. **111**(11), 4001–4006 (2014)
5. Morshed, A., Dutta, R., Aryal, J.: Recommending environmental knowledge as linked open data cloud using semantic machine learning. In: IEEE International Conference on Data Engineering Workshops (2013)
6. Muniasamy, A.: Applications of data mining techniques in smart farming for sustainable agriculture. In: Modern Techniques for Agricultural Disease Management and Crop Yield Prediction, pp. 142–178 (2020)
7. Parent, B., Tardieu, F.: Can current crop models be used in the phenotyping era for predicting the genetic variability of yield of plants subjected to drought or high temperature? J. Exp. Bot. **65**(21), 6179–6189 (2014)
8. Priya, R., Ramesh, D.: Adaboost.RT based soil n-p-k prediction model for soil and crop specific data: a predictive modelling approach. In: Big Data Analytics, pp. 322–331 (2018)
9. Rehman, T.U., Mahmud, M.S., Chang, Y.K., Jin, J., Shin, J.: Current and future applications of statistical machine learning algorithms for agricultural machine vision systems. Comput. Electron. Agric. **156**, 585–605 (2019)

Survey on Initialization Methods of Cuckoo Search Algorithm

Wen Xiao[1,2], Maomao Liang[1,2], Lijin Wang[1,2,3](✉), and Yiwen Zhong[1,2]

[1] College of Computer and Information Science, Fujian Agriculture and Forestry University, 350002 Fuzhou, China
lijinwang@fafu.edu.cn

[2] Key Laboratory of Smart Agriculture and Forestry, (Fujian Agriculture and Forestry University) Fujian Province University, 350002 Fuzhou, China

[3] Digital Fujian Tourism Big Data Institute (Wuyi University), 354300 Wuyishan, China

Abstract. We know that initialization has a key point on the performance of evolutionary algorithms. However, there are relatively few studies on the initialization of the cuckoo search algorithm. Therefore, this paper considers a systematic study of the initialization method of the cuckoo search algorithm. Several classic initial methods are used, including chaotic map, uniform design, opposition-based learning, quadratic interpolation, and some superior probability distribution methods. Through comparison on 20 test functions, the experiment result shows that opposition-based learning can perform well in both low and high dimensions among these initial methods. In addition, the initialization method has little effect on low-dimensional functions, but a suitable initialization method can greatly affect the performance of high-dimensional functions. Furthermore, increase population size appropriately can lead better performance while dimension raised.

Keywords: Cuckoo Search · Initialization · high-dimensional · population size

1 Introduction

The Cuckoo Search (CS) [1] algorithm is an efficient optimization algorithm proposed by Yang. It is inspired by the cuckoo's brooding behavior in nature. CS algorithm are few parameters, simple operation, strong optimization ability, easy implementation, etc. so many scholars pay attention to CS algorithm and a large number of CS variants have been designed.

Some variants hope to improve search ability between the processing of Levy flight and biased random walk. Mareli and Twala [2] proposed three methods to dynamically adjust discover probability p_a according to the distribution function. Wang et al. [3] utilized uniform distribution to dynamic generate step size α_0. The literature [4] considered generating the values of p_a and α_0 through chaotic

S.-C. Chu et al. (Eds.): ICGEC 2021, LNEE 833, pp. 629–640, 2022.
https://doi.org/10.1007/978-981-16-8430-2_57

mapping. Wang at el. [5] proposed two CS algorithms that consider dimensional improvement. Peng at el. [6] proposed three different operators and chose different operator in different stage adaptive. Salgotra at el. [7] proposed a adaptive cuckoo search, which not only dynamical generate population size and discover probability but also refine formulas that generate new individual.

Some variants consider hybrid with other algorithms. Wang et al. [8] hybridize cuckoo search algorithm and particle swarm algorithm. Hu et al. [9] combined cooperative collaborative algorithm and CS algorithm. Long et al. [10] hybridize cuckoo search with grey wolf optimizer to deal complicate industry problem. Considering applying the CS to multi-objective problems, Yang et al. [11] proposed a multi-objective cuckoo search algorithm. Cai et al. [12] proposed a hybrid multi-objective cuckoo search algorithm to solve defect prediction problem. Zhang et al. [13,14] proposed two improved multi-objective CS algorithms. Chen et al. [15] proposed a decomposed based CS algorithm to solve multi-objective problem.

The above improvement strategies have improved the performance of CS. But studies on the population initialization of the CS algorithm are relatively few, and there is no comprehensive description of the relevant papers on the initialization of the cuckoo algorithm. According to the research of [16,17], the initialization method have significant impact on CS algorithm. Studies related to population initialization include those based on opposition-based learning [18], uniform design [19], quadratic interpolation [20], chaotic systems [21], etc. have all proved this point. But these are mainly for genetic algorithms, differential evolution algorithms, and there are relatively few studies on population initialization of cuckoo search algorithms. Li et al. [16] analysed the impact of initialization methods on evolutionary algorithms. After research and comparison in the paper, they found that the cuckoo search algorithm has a strong dependence on initialization, especially for large-scale and multi-mode problem, but the initialization methods Li mainly used are various probability distribution functions in mathematics (normal distribution, exponential distribution, beta distribution, etc.). Therefore, it is necessary to consider the influence of traditional initialization methods on the CS algorithm.

Based on this, this paper has conducted a systematic study on the impact of some classic population initial methods on the CS algorithm compared with promising methods acquired in paper [16], hoping to help subsequent researchers in the selection of population initialization methods for CS. A total of 20 test functions are selected, of which 10 are selected for traditional test functions, and 10 are selected for rotation transformation function CEC2005.

2 Cuckoo Search

Cuckoo search algorithm is a promising optimization algorithm. In the initialization phase, the CS algorithm generates an initial population by randomly sampling from D dimensional search space $[x_{j,low}, x_{j,high}]$(j=1,2,...,D).

$$x_{i,j,0} = x_{i,j,\text{low}} + r(x_{i,j,\text{high}} - x_{i,j,\text{low}}) \quad i = 1, 2, \cdots, NP, j = 1, 2, \cdots, D \quad (1)$$

where r represents a random number subjected to uniform distribution in (0,1), and NP is the population size.

When the initialization is completed, the CS algorithm enters the iterative phase, generating new individuals through Levy flight and biased random walk, and greedily select the best individual according to the individual's fitness in each iteration.

The Levy flight is generating new individual rely on best individual of current. The detail can be summarized as

$$X_{i,G+1} = X_{i,G} + \alpha_0 \frac{\phi \times \mu}{|\nu|^{\frac{1}{\beta}}}(X_{i,G} - X_b) \tag{2}$$

where $X_{i,G}X_{i,G+1}$ respectively represent ith individual of G generation and ith individual of G+1 generation. X_b represents the best individual of current generation. β is a constant(generally β=1.5), α_0 is scale factor(generally α_0=0.01), μ and ν are subjected to standard normal distribution, and ϕ is given by

$$\phi = \left(\frac{\Gamma(1+\beta) \times \sin(\frac{\pi \times \beta}{2})}{\Gamma\left(\frac{1+\beta}{2}\right) \times \beta \times 2^{\frac{\beta-1}{2}}} \right)^{\frac{1}{\beta}} \tag{3}$$

where Γ is the gamma function.

The local random walk is to maintain the global search ability of the algorithm by generating new solutions related with current optimal individual. The specific operation can be written as

$$x_{i,j,G+1} = x_{i,j,G} + \alpha s \otimes H(p_a - \epsilon) \otimes (x_{x,j,G} - x_{y,j,G}) \tag{4}$$

where x and y are two random numbers generated from the population size, αs and ϵ are random numbers generated from uniform distribution, p_a is a fraction probability and H is step function.

3 Initialization Methods

Our purpose is investigating the influence of several classic initialization methods on the CS algorithm. These initialization methods respectively are uniform distribution, beta distribution, rayleigh distribution, chaotic map, uniform design, opposition-based learning and quadratic interpolation. The following will give a brief introduction to these methods.

3.1 Uniform Distribution

Uniform distribution is a simple probability distribution which is widely used in initialization. It's probability density function(PDF) is

$$f(x) = \begin{cases} \frac{1}{h-l} & if\ l \le x \le h, \\ 0 & else \end{cases} \tag{5}$$

where l and h are the bounds. And this distribution often written as U(l,h).

3.2 Beta Distribution

Beta distribution is a probability distribution range from 0 to 1. The probability density function(PDF) can be written as

$$f(x;a,b) = \frac{\Gamma(a+b)}{\Gamma(a)\Gamma(b)} x^{a-1}(1-x)^{b-1} \tag{6}$$

where Γ is the gamma function,a and b is shape parameters. Its notation is $Be(a,b)$.

3.3 Rayleigh Distribution

Rayleigh distribution represent non negative variable. It's probability density function(PDF) is

$$f(x;a) = \frac{x}{a^2} e^{-x^2/(2a^2)} \quad x \geq 0 \tag{7}$$

where a is the scale parameter. And the notation is $Rayl(a)$.

3.4 Chaotic Method

Chaos theory is a theory about the study of chaotic dynamic systems [22], which is very dependent on initial conditions. A method of simulating and quantifying a chaotic system is to generate a chaotic sequence through chaotic mapping. In theory, it can traverse every area of the search space. Therefore, the chaotic initialization method [21,23] can obtain a better initialization population . In order to generate an initial population, we need to choose a suitable chaotic map. In this article, we have chosen the Sine map, and its operator is shown below.

$$x_{i+1} = \frac{a}{4} sin(\pi x_i) \quad a = 4 \tag{8}$$

where x_i and x_{i+1} respectively are current value and next generational value. The initial value we choose a random number subjected to U(0,1).

3.5 Uniform Design

Uniform design(UD) is a experiment design method, which is based on space filling and scattered the point uniformly on search space. Literatures [19,24] apply uniform design to population initialization and can enhance the performance of evolution algorithms.

Suppose we need to initialize N individuals in D dimensions. First we need to generate a base sequence $H = (h_1, h_2, ..., h_s)$, whose length is greater than or equal to D and each element should be a random integer less than N. Then generate a uniform matrix based on this base sequence.

$$U_{i,j} = (i * h) \; mod \; N, \quad i = 1, 2, ..., N; j = 1, 2, .., s \tag{9}$$

Then you can use this uniform matrix to generate the initial population $X = (x_1, x_2, ..., x_n)$ as following formula.

$$x_i = l_i + \frac{U_{i,j}}{N} * (u_i - l_i) \tag{10}$$

where x_i is ith individual of population in range $[l_i, u_i]$.

3.6 Opposition-Based Learning

The idea of opposition-based learning(OBL) is not only to consider and evaluate current solution but also its opposite positions at the same time to obtain a more appropriate solution. Rahmanayan et al. [18] applied OBL to differential evolution algorithm(DE), and the result proved that it can improve the performance of DE very well. The formula of opposition-based learning is

$$x^{'} = l + u - x \tag{11}$$

where x is value in the range of [l,u], and $x^{'}$ is its opposite solution.

3.7 Quadratic Interpolation

Quadratic interpolation(QI) is long-standing method for solving optimization problems. In this method, we randomly select three different points (denoted as x, y, z) from the population. According to the defined operator, there will be a parabola to generate a new point r_i, the operator formula is

$$r_i = \frac{1}{2}\frac{(y^2 - z^2) * f(x) + (z^2 - x^2) * f(y) + (x^2 - y^2) * f(z)}{(y - z) * f(x) + (z - x) * f(y) + (x - y) * f(z)} \quad i = 1, 2, ..., NP. \tag{12}$$

where $f(x), f(y), f(z)$ is the fitness of x, y, z.

4 Experiment and Results

In this part, a total of 20 test functions are chosen for in order to compare the effect of these initial methods on the CS algorithm. These test functions can split into two categories. The first is classic test functions. The selection of these functions is based on their function shapes. Among them, F_{sph} and F_{ros} are single-mode functions, and the remaining 8 are multi-mode functions. The second category is the top 10 test functions selected from CEC2005 [25]. In addition, indicator of function error (Error) is selected to assess the performance of the algorithm.

Function error (Error) can be written as: $Error = f(X) - f(X^*)$, where X^* is the known global optimal solution, and X is the optimal solution acquired from current population. In addition, there are Wilcoxon statistical indicators to compare the advantages and disadvantages of the two algorithms. According to the Wilcoxon indicator, the comparison result of the algorithm will be summarized as "+/=/–". "-" means that the first one is better than the second one, "+" reveals that the first is worse than the second, and "=" denotes that the first one is similar to the second.

In our experiment, the population size (NP) is assigned to 30, the parameter p_a is 0.25, α_0 is 0.01, and β is 1.5. The number of runs (runTimes) of each function is set to 25. The maximum number of evaluations (NFS) of the function is 10000*D. The value of these parameters keep constant unless a change mentioned.

4.1 Different Dimensions Comparison

The study here is to show the improvements of other advanced initialization methods compared with uniform distribution method U(0,1), who is commonly used for initialization. Therefore, in all comparisons, the U(0,1) initialization is set as the baseline method. Other initial methods are contrasted with the uniform distribution method. For example, in Table 1, "1/19/0" represent Be(2.5, 2.5) superior U(0,1) in 1 functions, equal in 19 functions and inferior in 0 functions. The Table 1 shows a comparison result of the CS algorithm with different initialization based CS algorithm when D = 30. The following Table 2 is the comparison results when D = 50.

Table 1. Error(Mean) obtained by different initial methods (D = 30)

Fun	U(0,1)	Be(2.5,2.5)	Rayl(0.4)	Chaos	UD	OBL	QI
F_{sph}	1.44E-30	**5.70E-31**	**5.05E-31**	**1.11E-30**	1.65E-30	**7.91E-31**	**6.55E-31**
F_{ros}	1.21E+01	**8.81E+00**	**1.10E+01**	**9.45E+00**	**8.66E+00**	**1.02E+01**	1.22E+01
F_{ack}	**6.68E-14**	4.62E-02	1.62E-13	3.73E-02	8.93E-11	3.73E-02	3.73E-02
F_{grw}	**4.44E-18**	4.75E-16	1.33E-15	1.92E-14	2.66E-17	**0.00E+00**	3.38E-16
F_{ras}	2.42E+01	**2.19E+01**	**2.31E+01**	2.50E+01	**2.37E+01**	**2.29E+01**	**1.97E+01**
F_{sch}	**1.47E+03**	1.59E+03	1.67E+03	**1.45E+03**	**1.22E+03**	1.48E+03	1.59E+03
F_{sal}	3.88E-01	**3.74E-01**	3.92E-01	3.90E-01	**3.60E-01**	**3.88E-01**	**3.77E-01**
F_{wht}	3.65E+02	**3.48E+02**	3.82E+02	3.84E+02	**3.43E+02**	3.77E+02	3.79E+02
F_{pn1}	**4.12E-18**	4.05E-17	4.48E-17	**2.93E-18**	2.59E-17	**5.30E-19**	1.29E-17
F_{pn2}	1.14E-25	**1.39E-28**	**2.70E-26**	**3.00E-28**	9.79E-25	**4.92E-29**	**1.02E-26**
F_1	6.78E-30	**3.00E-30**	**1.48E-30**	9.66E-30	**2.37E-30**	**4.48E-30**	9.47E-30
F_2	**7.23E-03**	9.03E-03	8.23E-03	**6.58E-03**	8.78E-03	8.90E-03	8.17E-03
F_3	**1.99E+06**	2.08E+06	2.34E+06	2.22E+06	2.32E+06	2.00E+06	2.23E+06
F_4	**1.97E+03**	2.29E+03	**1.69E+03**	**1.62E+03**	**1.19E+03**	**1.44E+03**	**1.77E+03**
F_5	3.01E+03	**2.87E+03**	3.17E+03	3.14E+03	**2.93E+03**	3.22E+03	**2.96E+03**
F_6	2.51E+01	**1.55E+01**	**2.40E+01**	**2.47E+01**	2.81E+01	**2.08E+01**	**1.67E+01**
F_7	**1.24E-03**	1.26E-03	**1.10E-03**	**1.03E-03**	**5.59E-04**	**1.00E-03**	**3.88E-04**
F_8	**2.09E+01**	2.10E+01	**2.09E+01**	**2.09E+01**	2.09E+01	**2.09E+01**	2.10E+01
F_9	**2.90E+01**	3.07E+01	**2.78E+01**	**2.53E+01**	2.98E+01	**2.76E+01**	**2.75E+01**
F_{10}	1.63E+02	**1.59E+02**	1.77E+02	1.64E+02	1.77E+02	1.67E+02	1.71E+02
+/=/-		1/19/0	1/17/2	1/19/0	3/17/0	3/17/0	1/19/0

From Table 1 and Table 2, we can see that compared to D = 10, UD and OBL have advantages over U(0,1) in D = 30, which are both better than U(0,1) on 3 functions. When D = 50, Be(2.5, 2.5), UD and OBL have better effects, and have three, two, three functions better than U(0,1) respectively.

Table 2. Error(Mean) obtained by different initial methods (D = 50)

Fun	U(0,1)	Be(2.5,2.5)	Rayl(0.4)	Chaos	UD	OBL	QI
F_{sph}	**1.25E-31**	1.80E-31	1.94E-31	3.47E-31	**1.12E-31**	1.44E-31	1.63E-31
F_{ros}	4.01E+01	**3.38E+01**	**3.12E+01**	**3.22E+01**	**3.36E+01**	**3.48E+01**	**3.39E+01**
F_{ack}	9.18E-01	**7.86E-01**	**7.56E-01**	**8.37E-01**	**6.85E-01**	**6.68E-01**	**8.18E-01**
F_{grw}	3.95E-04	**3.94E-04**	1.68E-03	9.86E-04	1.28E-03	**2.96E-04**	5.92E-04
F_{ras}	4.26E+01	**3.80E+01**	**4.12E+01**	4.28E+01	**4.19E+01**	**3.92E+01**	**3.39E+01**
F_{sch}	**3.07E+03**	3.12E+03	3.21E+03	**2.88E+03**	**2.08E+03**	**2.99E+03**	**3.05E+03**
F_{sal}	7.16E-01	**7.16E-01**	7.20E-01	7.72E-01	**7.16E-01**	**7.11E-01**	**6.84E-01**
F_{wht}	1.12E+03	**8.93E+02**	1.18E+03	**1.06E+03**	1.14E+03	**1.08E+03**	**1.09E+03**
F_{pn1}	**2.33E-12**	2.49E-03	**5.85E-15**	**8.79E-17**	2.49E-03	2.49E-03	2.49E-03
F_{pn2}	**1.57E-21**	3.51E-16	3.49E-20	8.87E-16	9.16E-20	1.61E-21	6.39E-02
F_1	1.83E-29	**1.41E-29**	2.29E-29	**1.23E-29**	2.68E-29	**1.21E-29**	1.84E-29
F_2	**9.71E+00**	1.03E+01	1.90E+01	9.83E+00	1.08E+01	1.08E+01	1.07E+01
F_3	**5.06E+06**	5.42E+06	5.22E+06	5.25E+06	**4.88E+06**	5.55E+06	5.39E+06
F_4	**1.77E+04**	2.01E+04	1.88E+04	1.95E+04	1.88E+04	1.79E+04	2.00E+04
F_5	**8.93E+03**	9.39E+03	**8.75E+03**	**8.61E+03**	9.35E+03	9.06E+03	9.65E+03
F_6	4.92E+01	**4.55E+01**	**4.35E+01**	**4.64E+01**	**4.51E+01**	**4.40E+01**	**4.07E+01**
F_7	1.92E-03	**1.18E-03**	**7.63E-04**	2.88E-03	**1.03E-03**	2.31E-03	2.57E-03
F_8	2.11E+01	**2.11E+01**	**2.11E+01**	**2.11E+01**	**2.11E+01**	**2.11E+01**	**2.11E+01**
F_9	**6.33E+01**	6.74E+01	6.53E+01	**6.12E+01**	6.61E+01	6.53E+01	7.14E+01
F_{10}	4.29E+02	**4.19E+02**	4.52E+02	**4.18E+02**	**4.06E+02**	**4.19E+02**	4.32E+02
+/=/-		3/17/0	2/17/1	0/20/0	2/18/0	3/17/0	1/17/2

Fot the sake of better seeing the effect of the initialization method on high-dimensional functions, we extended the dimensions of decision space to 100, 300, and 500 respectively. The test function here selects 9 functions in [25]. The results of the experiment can be seen in following three tables.

Table 3. Error(Mean) obtained by different initial methods (D = 100)

Fun	U(0,1)	Be(2.5,2.5)	Rayl(0.4)	Chaos	UD	OBL	QI
F_{sph}	1.54E-31	**1.06E-31**	**1.42E-31**	**1.15E-31**	3.80E-31	**8.51E-32**	**1.42E-31**
F_{ros}	1.57E+02	**1.50E+02**	1.69E+02	**1.53E+02**	1.59E+02	1.59E+02	**1.54E+02**
F_{ack}	2.29E+00	**2.07E+00**	**2.06E+00**	**2.11E+00**	**2.09E+00**	**2.10E+00**	**2.07E+00**
F_{grw}	**3.15E-03**	6.68E-03	4.62E-03	**2.46E-03**	**9.33E-17**	**2.95E-03**	6.10E-03
F_{ras}	8.48E+01	**7.44E+01**	**7.88E+01**	**7.95E+01**	8.92E+01	**7.54E+01**	**7.46E+01**
F_{sch}	**6.93E+03**	7.12E+03	7.48E+03	7.32E+03	**4.31E+03**	**6.79E+03**	7.10E+03
F_{sal}	**1.85E+00**	1.85E+00	**1.84E+00**	**1.77E+00**	1.90E+00	1.90E+00	**1.84E+00**
F_{pn1}	4.98E-03	**2.49E-03**	9.96E-03	1.49E-02	**2.49E-03**	9.95E-03	8.71E-03
F_{pn2}	4.86E-01	**1.28E-01**	1.57E+00	**3.06E-01**	1.22E+00	**5.02E-02**	1.04E+00
+/=/-		3/6/0	1/7/1	1/8/0	2/7/0	0/9/0	1/8/0

From the Table 3 we can see that Be(2.5, 2.5) and UD have advantage compared with U(0,1) and better than U(0,1) in 3 and 2 functions respectively. This shows that other improved initialization methods are basically better than or at least similar to uniform distribution initialization in high-dimensional functions.

The situation is similar in the Table 4 and Table 5. Although there is no significant improvement of F_{grw} and F_{sch} in D = 300 and F_{ras} and F_{grw} in

Table 4. Error(Mean) obtained by different initial methods (D = 300)

Fun	U(0,1)	Be(2.5,2.5)	Rayl(0.4)	Chaos	UD	OBL	QI
F_{sph}	2.28E-28	**1.59E-28**	**2.13E-28**	3.66E-28	2.60E-28	2.69E-28	2.59E-28
F_{ros}	6.51E+02	**6.45E+02**	6.78E+02	6.55E+02	**6.35E+02**	6.64E+02	**6.37E+02**
F_{ack}	4.45E+00	**4.04E+00**	**4.31E+00**	4.45E+00	**4.23E+00**	**3.80E+00**	4.64E+00
F_{grw}	**1.37E-02**	1.86E-02	3.53E-02	5.61E-02	1.94E-02	4.24E-02	1.76E-02
F_{ras}	3.29E+02	**2.95E+02**	**3.19E+02**	3.32E+02	3.89E+02	**2.86E+02**	**3.05E+02**
F_{sch}	**2.85E+04**	2.97E+04	3.09E+04	2.87E+04	**1.56E+04**	2.90E+04	2.86E+04
F_{sal}	8.48E+00	**7.80E+00**	**8.25E+00**	8.54E+00	**7.88E+00**	**7.88E+00**	8.78E+00
F_{pn1}	3.84E-02	**1.12E-02**	**3.76E-02**	4.43E-02	**1.83E-02**	**1.41E-02**	4.98E-02
F_{pn2}	1.00E+02	**4.99E+01**	**9.28E+01**	1.09E+02	**9.81E+01**	**3.81E+01**	1.15E+02
+/=/-		5/3/1	0/8/1	0/9/0	2/6/1	4/5/0	0/9/0

Table 5. Error(Mean) obtained by different initial methods (D = 500)

Fun	U(0,1)	Be(2.5,2.5)	Rayl(0.4)	Chaos	UD	OBL	QI
F_{sph}	4.08E-25	**3.18E-25**	**2.14E-25**	**3.84E-25**	**2.28E-25**	**3.57E-25**	**2.36E-25**
F_{ros}	**1.12E+03**	1.15E+03	1.18E+03	1.17E+03	1.19E+03	1.20E+03	1.20E+03
F_{ack}	5.61E+00	**4.71E+00**	**5.21E+00**	5.65E+00	**4.92E+00**	**4.75E+00**	**5.50E+00**
F_{grw}	**1.36E-02**	1.06E-01	5.15E-02	2.29E-02	3.05E-02	4.78E-02	1.12E-01
F_{ras}	6.13E+02	**5.67E+02**	**6.01E+02**	**5.80E+02**	1.00E+03	**5.34E+02**	**5.85E+02**
F_{sch}	**5.94E+04**	6.15E+04	6.23E+04	**5.92E+04**	**2.51E+04**	5.97E+04	5.97E+04
F_{sal}	1.51E+01	**1.36E+01**	**1.49E+01**	1.52E+01	**1.42E+01**	**1.37E+01**	1.53E+01
F_{pn1}	6.00E+02	**1.15E-02**	**3.30E-02**	**4.12E-02**	1.39E+03	**1.77E-02**	**1.69E+02**
F_{pn2}	1.94E+02	**7.02E+01**	2.28E+02	2.27E+02	**1.57E+02**	**8.26E+01**	2.17E+02
+/=/-		5/2/2	1/7/1	0/9/0	3/4/2	4/5/0	0/8/1

D = 500, the advanced initial methods can boost CS in remaining functions(i.e., F_{ack}, F_{sal}, F_{pn1} and F_{pn2}.)

In addition, with the help of Friedman test from Table 6, we can see that when D = 100, Be(2.5, 2.5) ranks first, followed by OBL, QI, Chaos, UD, U(0,1) and Rayl(0.4). When D = 300 and D = 500, the performance of Be(2.5, 2.5) and OBL also has a certain advantage, which rank two among all initialization method.

Taken together, the initialization method based on OBL and Be(2.5, 2.5) perform well in high-dimensional functions.

We can see that the use of the initial method can enhance the quality of the solution for most high-dimensional functions. This is because in high-dimensional functions, the the solution produced by uniform initial method is more likely to be unevenly distributed, so it may be unable to explore part of the space in the subsequent iteration process, and would be more likely to get to the local optimum. What's more, the population in search space would be more spare as the dimension of functions increased. And it is nature to consider that increase population size in high dimensional functions may improve the performance of algorithm.

Table 6. Average ranking of different initial methods by the Friedman test (D = 100, 300, 500)

Algorithm	U(0,1)	Be(2.5,2.5)	Rayl(0.4)	Chaos	UD	OBL	QI
D=100	4.50	**3.20**	4.80	3.90	4.40	3.50	3.70
D=300	3.90	**2.50**	4.40	5.70	3.50	3.40	4.60
D=500	4.30	**2.80**	4.40	4.30	3.80	3.20	5.20

4.2 Impact of Population Size

After comparing the impact of different initial methods on the CS algorithm, we explored the impact of population size under high-dimensional functions of the Be(2.5, 2.5), which is the best in previous stage. Here we set the dimension of the function to D = 300, and other unmentioned parameters are consistent with the experimental settings. we set NP = 50, 100, 150 and 200 respectively and compare with NP = 30, which is seen as excellent in low dimension functions. The comparison is list in Table 7.

Table 7. Error(Mean) obtained by different population size (D = 300)

Fun	NP=30		NP=50		NP=100		NP=150		NP=200
F_{sph}	**1.59E-28**	-	1.88E-21	-	1.92E-09	-	8.58E-05	-	1.83E-02
F_{ros}	**6.45E+02**	=	6.79E+02	=	6.96E+02	-	7.47E+02	-	8.29E+02
F_{ack}	4.04E+00	+	**1.95E+00**	+	**1.86E+00**	+	**1.76E+00**	+	**1.31E+00**
F_{grw}	1.86E-02	+	**1.38E-03**	=	**1.31E-09**	=	**7.70E-05**	=	**5.78E-03**
F_{ras}	**2.95E+02**	-	3.57E+02	-	7.83E+02	-	1.04E+03	-	1.23E+03
F_{sch}	**2.97E+04**	-	4.21E+04	-	5.71E+04	-	6.13E+04	-	6.43E+04
F_{sal}	7.80E+00	+	**4.95E+00**	+	**4.04E+00**	+	**4.45E+00**	+	**5.02E+00**
F_{pn1}	1.12E-02	=	**1.24E-03**	=	**2.44E-05**	=	**8.89E-03**	-	1.73E-01
F_{pn2}	4.99E+01	+	**2.39E+01**	+	**1.32E+00**	+	**4.67E+00**	+	**1.42E+01**
+/=/-		4/2/3		3/3/3		3/2/4		3/1/5	

From Table 7 we can see that when NP = 50, Be(2.5, 2.5) algorithm performances better on F_{ack}, F_{grw}, F_{sal} and F_{pn2}. When NP = 100, it performs better on F_{ack}, F_{sal} and F_{pn2}, which is the same with NP = 150 and NP = 200. And CS performs better on F_{sph}, F_{ras} and F_{sch} at NP = 30. It can be seen that when the NP is larger, it performs better for the multi-mode function. When the NP is smaller, the performance is better on the single-mode function. Besides, according to Wilcoxon test, "4/2/3" represents that NP = 50 is superior to NP = 30 in 4 functions, equal in 2 function, and inferior in 3 functions. And NP = 100, NP = 150, NP = 200 both dominate 3 functions compared with NP = 30.

In addition, in order to make it easier to see the difference in performance under different population sizes, we conducted a Friedman test on these results. The comparison is list in the Table 8, from which we can see that NP = 100 ranks first, followed by NP = 50, NP = 30, NP = 150 and NP = 200 respectively. This shows that the population size would influence the performance of the algorithm. And as the dimension increases, we need to increase the population size appropriately to guarantee the performance of the algorithm. This is because the density of the population in the decision space is relatively diluted as the dimension increases, . So under the high-dimensional function, we should also increase the size of the population appropriately.

Table 8. Average ranking of different population size by the Friedman test (D = 300)

Algorithm	NP=30	NP=50	NP=100	NP=150	NP=200
D=300	3.10	2.70	**2.20**	3.00	4.00

5 Conclusion

The initialization method has a significant impact for some algorithms. In this paper, some classic initialization methods (chaotic map, uniform design, opposition-based learning, quadratic interpolation) and some promising statistic distribution based methods are studied for the CS algorithm. In addition, considering the impact of 20 test functions in different dimensions and the impact of population size in high dimension, we can get the following conclusions according to the experimental results:

- The initialization method in the low-dimensional space has little effect on the algorithm, but the initialization method in the high-dimensional space (greater than or equal to 50) can deeply influence the performance of the algorithm, especially for multi-mode problems. This may be because population is diluted as dimension increased and advanced initialization method can scatter solution better in search space. So it performs best in high dimensions (greater than or equal to 100). And the initialization effect based on opposition-based learning and beta distribution are perform well.
- We need increase population size in CS algorithm properly as dimension size extended and 100 population size is appropriate in high dimension(below 1000 dimension) as we test.

Some of the above studies can draw some preliminary conclusions, but there are still fewer functions used for high-dimensional spaces, so follow-up studies may consider the impact of more test functions. And we may study initialization in deep neural network, where many of weights need to be initialized.

Acknowledgement. The author would like to express sincere thanks to all the editors and review experts who helped the paper to refine. This work is supported by the Nature Science Foundation of Fujian Province of P. R. China (No. 2019J01401, No.2021J01127), and the Special Fund for Scientific and Technological Innovation of Fujian Agriculture and Forestry University (No. CXZX2019117S, No. CXZX2020148C, No. CXZX2020150C), and Educational Research Project for Young and Middle-aged Teachers of Fujian Provincial Department of Education(JA170179), and in part by the Open Project Program of Digital Fujian Tourism Big Data Institute under Grant DFJTBDRI2020103.

References

1. Yang, X.S., Deb, S.: Cuckoo search via lévy flights. In: 2009 World Congress on Nature & Biologically Inspired Computing (NaBIC), pp. 210–214. IEEE (2009)
2. Mareli, M., Twala, B.: An adaptive cuckoo search algorithm for optimisation. Appl. Comput. Inf. **14**(2), 107–115 (2018)
3. Wang, L., Yin, Y., Zhong, Y.: Cuckoo search with varied scaling factor. Front. Comput. Sci. **9**, 623–635 (2015). https://doi.org/10.1007/s11704-015-4178-y
4. Wang, G.G., Deb, S., Gandomi, A.H., Zhang, Z., Alavi, A.H.: Chaotic cuckoo search. Soft Comput. **20**(9), 3349–3362 (2016)
5. Wang, L., Zhuang, Y., Chen, Z., Zhong, Y.: An enhanced cuckoo search using dimension selection. Int. J. Innovative Comput. Appl. **10**(1), 3 (2019)
6. Peng, H., Zeng, Z., Deng, C., Wu, Z.: Multi-strategy serial cuckoo search algorithm for global optimization. Knowl-Based Syst. **214**, 106729 (2021)
7. Salgotra, R., Singh, U., Saha, S., Gandomi, A.H.: Self adaptive cuckoo search: analysis and experimentation. Swarm Evol. Comput. **60**, 100751 (2021)
8. Wang, L., Zhong, Y., Yin, Y.: A hybrid cooperative cuckoo search algorithm with particle swarm optimisation. Int. J. Comput. Sci. Math. **6**(1), 18–29 (2015)
9. Hu, X., Yin, Y.: Cooperative co-evolutionary cuckoo search algorithm for continuous function optimization problems. Pattern Recognit. Artif. Intell. **26**(11), 1041–1049 (2013)
10. Long, W., Cai, S., Jiao, J., Xu, M., Wu, T.: A new hybrid algorithm based on grey wolf optimizer and cuckoo search for parameter extraction of solar photovoltaic models. Energy Convers. Manage. **203**, 112243 (2020)
11. Yang, X.S., Deb, S.: Multiobjective cuckoo search for design optimization. Comput. Oper. Res. **40**(6), 1616–1624 (2013)
12. Cai, X., Niu, Y., Geng, S., Zhang, J., Cui, Z., Li, J., Chen, J.: An under-sampled software defect prediction method based on hybrid multi-objective cuckoo search. Concurrency Comput. Pract. Experience **32**(5), e5478 (2020)
13. Zhang, M., Wang, H., Cui, Z., Chen, J.: Hybrid multi-objective cuckoo search with dynamical local search. Memetic Comput. **10**(2), 199–208 (2018)
14. Cui, Z., Zhang, M., Wang, H., Cai, X., Zhang, W.: A hybrid many-objective cuckoo search algorithm. Soft Comput. **23**(21), 10681–10697 (2019). https://doi.org/10.1007/s00500-019-04004-4
15. Chen, L., et al.: Solving multi-objective optimization problem using cuckoo search algorithm based on decomposition. Appl. Intell. **51**(1), 143–160 (2021)
16. Li, Q., Liu, S.Y., Yang, X.S.: Influence of initialization on the performance of metaheuristic optimizers. Appl. Soft Comput. **91**, 106193 (2020)
17. Ali, M., Pant, M., Abraham, A.: Unconventional initialization methods for differential evolution. Appl. Math. Comput. **219**(9), 4474–4494 (2013)
18. Rahnamayan, S., Tizhoosh, H.R., Salama, M.M.: Opposition-based differential evolution. IEEE Trans. Evol. Comput. **12**(1), 64–79 (2008)
19. Peng, L., Wang, Y., Dai, G.: Ude: differential evolution with uniform design. In: 2010 3rd International Symposium on Parallel Architectures, Algorithms and Programming, pp. 239–246. IEEE (2010)
20. Pant, M., Ali, M., Singh, V.P.: Differential evolution using quadratic interpolation for initializing the population. In: 2009 IEEE International Advance Computing Conference, pp. 375–380. IEEE (2009)
21. Gao, W.f., Liu, S.Y.: A modified artificial bee colony algorithm. Comput. Oper. Res. **39**(3), 687–697 (2012)

22. Devaney, R.: An Introduction to Chaotic Dynamical Systems. CRC Press, Boca Raton (2018)
23. Gao, W.f., Liu, S.Y., Huang, L.I.: Particle swarm optimization with chaotic opposition-based population initialization and stochastic search technique. Commun. Nonlinear Sci. Numer. Simul. **17**(11), 4316–4327 (2012)
24. Zhang, J., Feng, J., Chen, G., Yang, X.: Artificial bee colony algorithm combined with uniform design. In: Pan, J.-S., Li, J., Tsai, P.-W., Jain, L.C. (eds.) Advances in Intelligent Information Hiding and Multimedia Signal Processing. SIST, vol. 157, pp. 47–56. Springer, Singapore (2020). https://doi.org/10.1007/978-981-13-9710-3_5
25. Suganthan, P.N., et al.: Problem definitions and evaluation criteria for the CEC 2005 special session on real-parameter optimization. KanGAL Rep. **2005005**(2005), 2005 (2005)

Tumbleweed Algorithm and Its Application for Solving Location Problem of Logistics Distribution Center

Qing-yong Yang[1], Shu-Chuan Chu[1], Anhui Liang[2], and Jeng-Shyang Pan[1(✉)]

[1] College of Computer Science and Engineering, Shandong University of Science and Technology, Qingdao 266590, China
yang_qy@sdust.edu.cn

[2] College of Electronic and Information Engineering, Shandong University of Science and Technology, Qingdao 266590, China
liangah8@sdust.edu.cn

Abstract. In this paper, a new swarm intelligence optimization algorithm named Tumbleweed Algorithm (TA) is proposed. The TA algorithm simulates the two processes of tumbleweed from seedling to adulthood and the propagation of tumbleweed seeds after adulthood. And by introducing the concept of growth cycle, the two stages are combined. In order to verify the effectiveness of the new algorithm proposed to solve the problems, this paper uses the CEC2013 function set to test, and compares the 10D, 30D and 50D dimensions with six swarm intelligence optimization algorithms. By comparing the experimental results under different dimensions, the TA algorithm proposed in this paper is generally superior to other intelligent optimization algorithms compared, and has strong optimization ability and competitiveness. Finally, the TA algorithm is applied to the location problem of logistics distribution center to verify the practicability of the algorithm. In solving this problem, the TA algorithm can also obtain better optimization results.

Keywords: Tumbleweed algorithm · Swarm intelligence · CEC2013 · Location of logistics distribution center

1 Introduction

At present, with the increasing development of science and technology and the continuous improvement of productive capacity, the problems that need to be optimized have become more and more complicated [1–4]. It is difficult for traditional optimization algorithms to obtain satisfactory results in reasonable time. The emergence of intelligent optimization algorithms provides a certain solution direction for solving large-scale and complex optimization problems [5–7], and they are widely applied in various fields [8–11].

The inspiration of the intelligent optimization algorithm comes from the observation of various biological habits, physical and chemical phenomena in

S.-C. Chu et al. (Eds.): ICGEC 2021, LNEE 833, pp. 641–652, 2022.
https://doi.org/10.1007/978-981-16-8430-2_58

nature. It can be roughly divided into two categories, one is evolutionary algorithms; the other is swarm intelligence optimization algorithms [12]. Evolutionary algorithms mainly include Genetic Algorithm (GA) [13–16], Differential Evolution (DE) algorithm [17], Simulated Annealing (SA) algorithm [18] and so on. The swarm intelligence optimization algorithms mainly include Particle Swarm Optimization (PSO) algorithm [19,20], Gray Wolf Optimization (GWO) algorithm [21], Flower Pollination Algorithm (FPA) [22,23], QUasi Affine TRansformation Evolutionary (QUATRE) algorithm [24,25], Artificial Bee Colony (ABC) algorithm [26–28], etc.

The new swarm intelligence optimization algorithm proposed in this paper is inspired by tumbleweed. Tumbleweed, also known as Prickly Russian thistle, is a desert plant with strong vitality, which usually lives in Gobi and desert [29]. The growth process of tumbleweed at the seedling stage is similar to that of other plants. Unlike other plants, tumbleweed is not static. When drought comes, the whole adult tumbleweed will pick up its roots, moving randomly follow the wind, and spreading its seeds in the process of moving. After finding a suitable environment, the roots will continue to penetrate deep into the soil to absorb water. The algorithm in this paper simulates the two processes of tumbleweed seedling growth and seed propagation. According to the different characteristics of the two stages, different evolution formulas are used to update individuals. In order to verify the effectiveness of the TA algorithm, this paper uses CEC2013 function set [30] to simulate and test. We selects six representative swarm intelligence optimization algorithms for comparison: the PSO algorithm [19], the SCA algorithm [31], the WOA algorithm [32], the CSO algorithm [33], the BA algorithm [34] and the BOA algorithm [35]. The final test results show that, compared with the same type of algorithms, the Tumbleweed Algorithm (TA) proposed in this paper can show stronger optimization ability and competitiveness.

Finally, in order to prove the effectiveness of the algorithm, we select the location problem of logistics distribution center [36] as the application object of the TA algorithm. It is an effective way to improve the efficiency of logistics distribution to select the address of distribution center reasonably and effectively from multiple distribution addresses. This paper selects a case of selecting logistics distribution centers from 31 cities to solve. The test results show that the TA algorithm can also solve this problem well.

The remaining chapters of this paper are arranged as follows. The Sect. 2 describes the implementation principle and process of Tumbleweed Algorithm. The experimental results and analysis are stated in the Sect. 3 and the application of Tumbleweed Algorithm in logistics location problem is shown in Sect. 4. The Sect. 5 is the summary of this paper.

2 Tumbleweed Algorithm

According to the previous introduction, the TA algorithm simulates the seedling growth and seed propagation of tumbleweed. To be convenient for the following description, firstly, the related symbols and structure of the algorithm are defined and assumed.

Assume that the current space is Ω, there are N tumbleweed seedlings scattered in different areas in the space. Each seedling is represented by X_i (i = 1,2,...,N), and its fitness is represented by $fit(X_i)$. Among the N seedings, there is one seedling that grows most vigorously, and it is represented by the symbol X_{gbest}. This seedling can eventually grow into a complete tumbleweed, and then spread the seeds. The following is an introduction to the two stages of the algorithm.

2.1 Seedling Growth Stage

Like other plants, the growth of tumbleweed seedlings is also affected by the surrounding environment. In the Tumbleweed Algorithm (TA), we assume that the environmental impact mainly comes from two parts. One part is the impact of strong seedlings on surrounding seedlings, represented by $Factor1$. The other part is the influence of factors such as light, soil, moisture, etc., represented by $Factor2$. The expression equations of the two influencing factors $Factor1$ and $Factor2$ are shown in Eq. 1 and Eq. 2.

$$Factor1 = r_1 * (X_{gbest} - X^i_{old}) \tag{1}$$

$$Factor2 = \frac{r_2}{\frac{fit(X^i_{old})}{sum(fit(X_{old}))} + 1} \tag{2}$$

where the parameter r_1 is the influence coefficient of strong seedling, the value decreases linearly from the preset value t to 0 with the iterative process of the algorithm, as shown in Eq. 3. The parameter r_2 represents the environmental impact coefficient, which is a random number between 0 and 0.1.

$$r_1 = t * (1 - \frac{gen}{Max\_gen}) \tag{3}$$

Finally, the state update equation of the first stage is shown in Eq. 4.

$$X^i_{new} = X^i_{old} + Factor1 + Factor2 \tag{4}$$

2.2 Seed Propagation Stage

According to the literature [37], the propagation of seeds can be expressed by the following equation.

$$x = \frac{\mu H}{F} \tag{5}$$

where x is the seed propagation distance, H represents the seed release height, and F is the seed sedimentation rate. For a specific plant, both H and F are constants.

In the Tumbleweed Algorithm, we integrate the above Eq. 5. The parameters H and F are corresponding to the current iteration number gen and the maximum iteration number Max_gen respectively. The equation of seed propagation stage is as follows.

$$X^i_{new} = X_{gbest} + V_i * \frac{gen}{Max\_gen} \tag{6}$$

where V_i is a random number uniformly distributed in the interval $[lb, ub]$.

In the process of implementation, we introduce the concept of growth cycle. For example, if every 50 generations is set as a growth cycle, seedling growth and seed propagation will each take up half of the growth cycle. Algorithm 1 is the overall flow chart of the Tumbleweed Algorithm.

Algorithm 1. Tumbleweed Algorithm

Input: $f(x)$: objective funtion; ps: population size; dim: problem dimension; gc: growth cycle; Max_gen: maximum number of iterations.
Output: optimal X_{best} and optimal value $f^*(X_{best})$.

Initialize population X;
repeat
    Calculate the fitness of each individual;
    Update X_{best} and $f^*(X_{best})$;
    if $mod(gen, gc) < gc/2$ **then**
        compute $Factor1$ using Eq. 1;
        compute $Factor2$ using Eq. 2;
        update r_1 using Eq. 3;
        compute X_new using Eq. 4;
    else
        compute X_new using Eq. 6;
    end if
    Boundary detection;
until ($gen < Max\_gen$ or Satisfy convergence constraints)

---

3 Experimental Results and Analysis

In order to verify the convergence and accuracy of the proposed algorithm, this paper selects the single objective constraint function set CEC2013 to test. CEC2013 contains 28 test functions, including three types: unimodal ($F1-F5$), basic multimodal ($F6-F20$) and composition ($F21-F28$). For details, please refer to the reference [30]. At present, there are many swarm intelligence optimization algorithms. This paper selects six representative swarm intelligence optimization algorithms for comparison. In the experiment, the maximum number of tests for each function of the optimization algorithm is 30, the maximum number of particles is 30, and the maximum number of iterations is set to 1000. All experiments are carried out on the same machine. All algorithms are programmed by MATLAB, and the MATLAB version is 2019a. The result is recorded as the average value of the difference between the result obtained by the optimization algorithm and the optimal value of the function. The smaller the average value, the stronger the optimization ability of the algorithm.

Table 1, 2 and 3 records the experimental results of the algorithms in different dimensions of 10D, 30D and 50D. In the r column of the table, + means that TA algorithm obtains better results than comparison algorithm, − means that TA algorithm is worse than comparison algorithm, and = means that two algorithms obtain the same results. In the 10D comparison result, the number of + obtained by the TA algorithm and the PSO algorithm is relatively close. In comparison with PSO, the TA algorithm is poorly optimized on the 10D multimodal functions. In functions $F7-F14$, the TA algorithm only wins on $F10$. However, with the increase of dimensionality, this situation has improved, and the number of + obtained by the TA algorithm is more than that of the PSO algorithm. Although TA algorithm is better than BA algorithm in the comparison of $10D$, with the increase of dimension, the number of better results obtained by the two algorithms is similar. Compared with the other four algorithms, TA algorithm obtains more + as a whole. However, in comparison with the CSO algorithm, the TA algorithm is inferior in solving composite functions of different dimensions.

Figure 1 shows the convergence curves of all test algorithms on some 30D test functions. It can be seen from the figure that the TA algorithm also has better convergence performance. However, compared with the other six intelligent optimization algorithms, the early convergence curve of the TA algorithm is slower. The reason is that the TA algorithm introduces the concept of growth cycle. So every time the growth cycle is completed, it is the reinitialization of the population. But as the iteration progresses, all individuals will gradually gather to the optimal individual to achieve the convergence of the algorithm. In summary, in the comparison of the results of 10D, 30D and 50D, the TA algorithm is generally better than the other six swarm intelligence optimization algorithms.

4 Application in Solving Location Problem of Logistics Distribution Center

With the rise of e-commerce and the rapid development of the logistics industry, how to efficiently and reasonably select the location of the logistics distribution center becomes particularly important [38]. The location model of logistics distribution center is generally a nonlinear model with complex constraints, which is a typical NP-Hard problem [36,39]. The location problem of logistics distribution center can be described as: Some locations are selected from a limited number of city locations as logistics centers, and the selected centers provide goods for the locations in need of services. The total cost is the lowest under the condition of meeting the requirements of each supply location. In this article, the goal is to minimize the sum of the product of the distance between the demand of each location and the nearest distribution center, and thus the following model can be established [40].

$$C^*_{min} = \sum_{i \in N} \sum_{j \in M_i} h_i d_{i,j} z_{i,j} \tag{7}$$

$$s.t. \quad \sum_{j \in M_i} z_{i,j} = 1, i \in N \tag{8}$$

Table 1. Comparison of 10D test results.

Function name	TA	PSO	r	SCA	r	WOA	r	CSO	r	BA	r	BOA	r
F1	0.032628	378.4484	+	789.2704	+	5.976163	+	619.2113	+	0.537397	+	9952.396	+
F2	2472373	4764942	+	6746082	+	5004603	+	4616068	+	193148.2	−	37753961	+
F3	2.47E+08	5.76E+09	+	1.07E+09	+	3.58E+09	+	1.72E+09	+	54863750	−	2.01E+11	+
F4	10777.28	19314.28	+	10099.91	−	40524.69	+	7050.739	−	29360.01	+	16324.77	+
F5	13.75237	295.4153	+	180.65	+	52.89088	+	100.2654	+	0.393429	−	4433.1	+
F6	31.53817	53.43113	+	63.9599	+	52.0278	+	65.83961	+	3.455319	−	698.8836	+
F7	78.48597	76.56838	−	47.31335	−	112.7963	+	59.02465	−	321.9186	+	3847.18	+
F8	20.45672	20.42433	−	20.42848	−	20.47214	+	20.26786	−	20.48609	+	20.41425	−
F9	7.042353	6.287766	−	8.853693	+	8.454759	+	8.236776	+	9.627562	+	8.751761	+
F10	2.681012	103.1626	+	107.7925	+	21.49251	+	84.26221	+	0.949901	−	1120.49	+
F11	55.82679	40.71239	−	61.77174	+	71.16828	+	61.07384	+	118.7734	+	139.7753	+
F12	50.97548	44.21038	−	64.04998	+	88.02857	+	64.20604	+	123.171	+	132.6911	+
F13	50.99125	46.477	−	62.7536	+	87.30749	+	60.01905	+	146.9234	+	130.6789	+
F14	1388.636	1187.941	−	1542.283	+	1257.086	−	1381.7	−	1459.143	+	1829.555	+
F15	1165.821	1356.077	+	1659.923	+	1205.51	+	965.5233	−	1376.119	+	1654.246	+
F16	0.682473	1.345912	+	1.357963	+	0.898703	+	0.932697	+	1.06337	+	1.501161	+
F17	61.91293	59.9013	−	69.30154	+	83.57515	+	64.14496	+	215.5543	+	124.1281	+
F18	61.32802	59.50354	−	72.0703	+	90.68317	+	67.14618	+	193.6911	+	121.204	+
F19	3.328938	573.4498	+	15.88423	+	8.429177	+	9.89102	+	8.609695	+	13134.13	+
F20	3.876737	3.643549	−	3.729763	−	4.074657	+	3.60842	−	4.552248	+	4.81214	+
F21	344.1697	386.0444	+	415.4718	+	384.3934	+	411.5635	+	374.6267	+	633.973	+
F22	1565.965	1260.551	−	1788.296	+	1650.336	+	1780.072	+	1678.597	+	2115.323	+
F23	1626.926	1373.871	−	1855.512	+	1649.515	+	1442.778	−	1711.723	+	2058.679	+
F24	221.7403	224.058	+	225.6209	+	224.2452	+	223.0385	−	229.6022	+	196.849	−
F25	222.7663	223.8866	+	226.4556	+	225.6368	+	221.4498	−	222.6429	−	200.9843	−
F26	233.4764	223.1607	−	196.2135	−	211.3994	−	201.6189	−	253.2656	+	189.7804	−
F27	592.639	597.0026	+	627.4505	+	633.0358	+	534.2285	−	657.9603	+	717.5031	+
F28	415.2295	425.6188	+	691.5294	+	793.372	+	768.8465	+	1158.616	+	986.5541	+
Summary		+	15	+	23	+	26	+	17	+	22	+	24
		−	13	−	5	−	2	−	11	−	6	−	4
		=	0	=	0	=	0	=	0	=	0	=	0

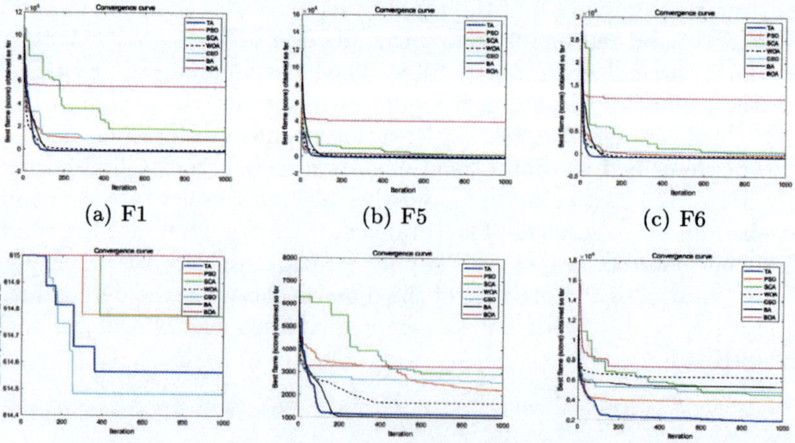

(a) F1 (b) F5 (c) F6

(d) F20 (e) F21 (f) F28

Fig. 1. Convergence curves of 30D benchmark functions: F1, F5, F6, F20, F21, F28.

Table 2. Comparison of 30D test results.

Function name	TA	PSO	r	SCA	r	WOA	r	CSO	r	BA	r	BOA	r
F1	29.2247	7318.213	+	17847.87	+	877.6761	+	12432.9	+	6.265458	−	56667.62	+
F2	43573527	1.26E+08	+	2.43E+08	+	1.17E+08	+	2.06E+08	+	6022044	−	1.29E+09	+
F3	2.49E+10	2.8E+12	+	3.04E+12	+	5.08E+10	+	9.51E+10	+	1.15E+09	−	3.31E+21	+
F4	69072.54	93158.57	+	59211.23	−	110703.1	+	43783.7	−	103356.5	+	67290.54	−
F5	220.9407	4207.388	+	3913.956	+	1010.437	+	1721.461	+	2.185295	−	36562.64	+
F6	123.6164	549.0521	+	1271.45	+	314.9942	+	1415.031	+	50.57627	−	14887.36	+
F7	230.9028	364.1388	+	281.1557	+	4812.176	+	300.0442	+	1176469	+	12860350	+
F8	21.02555	21.04086	+	21.05669	+	21.05284	+	20.92895	−	21.07999	+	21.05096	+
F9	35.51634	32.00123	−	41.70362	+	38.33048	+	38.48434	+	37.7165	+	40.05535	+
F10	112.4554	1603.667	+	2434.114	+	580.6878	+	1824.119	+	2.802315	−	10195.36	+
F11	384.3829	365.7541	−	442.8471	+	599.1057	+	390.2361	+	908.3909	+	871.3567	+
F12	405.2565	333.5895	−	452.1945	+	625.2455	+	428.1716	+	847.5644	+	858.7017	+
F13	389.6997	371.4334	−	459.4434	+	572.8587	+	448.132	+	971.7841	+	828.3325	+
F14	5987.001	6362.343	+	7678.172	+	6090.86	+	7723.347	+	5206.891	−	8209.245	+
F15	5699.647	7421.565	+	8029.813	+	6642.501	+	6940.369	+	5147.015	−	8026.656	+
F16	1.985266	2.992785	+	3.157483	+	2.054132	+	2.295565	+	2.686389	+	3.148998	+
F17	445.734	473.5535	+	613.358	+	695.5921	+	504.4385	+	1511.386	+	855.7605	+
F18	453.6951	437.2396	−	578.629	+	682.5849	+	513.1983	+	1514.442	+	858.6954	+
F19	24.86877	95455.8	+	13524.83	+	273.302	+	6312.148	+	43.56355	+	751182.1	+
F20	14.85368	13.58901	−	14.65786	−	14.98338	+	14.34153	−	14.99948	+	15	+
F21	386.4567	864.089	+	2106.374	+	1079.329	+	1998.936	+	365.4429	−	2569.219	+
F22	7134.69	7028.339	−	8302.984	+	7556.611	+	7801.311	+	6537.506	−	8782.825	+
F23	6563.612	7539.58	+	8464.068	+	7332.481	+	8079.602	+	6525.869	−	8875.445	+
F24	312.3388	305.0246	−	327.3417	+	321.1875	+	304.2838	−	379.3456	+	427.9473	+
F25	320.9114	317.0136	−	335.5742	+	330.0076	+	300.4989	−	294.7885	−	377.5548	+
F26	381.8224	365.0752	−	236.8795	−	377.939	−	218.8767	−	371.8253	−	336.3657	−
F27	1292.298	1209.488	−	1430.534	+	1378.16	+	1317.505	+	1501.837	+	1906.425	+
F28	1285.302	2836.372	+	3199.198	+	4798.8	+	3341.502	+	6273.719	+	6106.754	+
Summary		+	17	+	25	+	27	+	22	+	15	+	26
		−	11	−	3	−	1	−	6	−	13	−	2
		=	0	=	0	=	0	=	0	=	0	=	0

$$z_{i,j} \leq e_j, i \in N, j \in M_i \tag{9}$$

$$\sum_{j \in M_i} e_j = p \tag{10}$$

$$d_{i,j} \leq s \tag{11}$$

where N is the set of serial numbers of all demand points, and M_i is the set of alternative distribution centers whose distance to demand point i is less than the upper limit of distance s. h_i represents the demand at demand point i, and $d_{i,j}$ is the distance from demand point i to distribution center j. Both $z_{i,j}$ and e_j are 0–1 variables. When $z_{i,j}$ is 1, it means that the demand at demand point i is supplied by distribution center j. When e_j is 1, it means that point j is selected as the distribution center.

Equation 7 is the objective function. Equation 8 and Eq. 9 ensure that each demand point and its demand can only be supplied by one distribution center.

Table 3. Comparison of 50D test results.

Function name	TA	PSO	r	SCA	r	WOA	r	CSO	r	BA	r	BOA	r
F1	546.9425	16014.29	+	39418.52	+	4541.048	+	30184.75	+	18.83502	−	79628.69	+
F2	1.23E+08	3.39E+08	+	7.94E+08	+	2.05E+08	+	5.77E+08	+	12058930	−	3.62E+09	+
F3	8.25E+10	1.05E+12	+	4.49E+11	+	1.92E+11	+	2.13E+11	+	2.33E+09	−	1.39E+21	+
F4	149787.9	154285	+	92700.78	−	99703.54	−	66525.7	−	163716.1	+	101092.5	−
F5	857.2595	8433.313	+	5228.166	+	1518.451	+	3038.507	+	6.6791	−	27682.1	+
F6	266.9345	1072.856	+	3121.95	+	673.0121	+	2339.731	+	83.72754	−	11127.52	+
F7	273.2948	540.1117	+	299.755	+	2202.326	+	253.3634	−	45856.53	+	96545.07	+
F8	21.20888	21.21487	+	21.21139	+	21.21104	+	21.12149	−	21.25256	+	21.2176	+
F9	65.88419	61.98078	−	76.27432	+	73.10061	+	69.78575	+	69.67203	+	73.8103	+
F10	653.7293	2751.905	+	5251.518	+	1502.445	+	4611.834	+	8.73604	−	14919.88	+
F11	915.2612	724.9416	−	822.6217	−	897.6709	−	775.9852	−	1354.363	+	1136.899	+
F12	1010.875	735.3048	−	878.6523	−	1032.527	+	811.804	−	1275.558	+	1243.682	+
F13	817.7027	752.8725	−	859.2192	+	1094.325	+	825.9204	+	1491.259	+	1240.107	+
F14	11524.96	12821.21	+	14246.2	+	11401.4	−	14733.12	+	9243.296	−	14938.15	+
F15	11281.04	14399.63	+	15145.74	+	12952.03	+	14342.79	+	9849.34	−	15417.93	+
F16	2.703507	4.054107	+	4.06891	+	3.033938	+	3.292927	+	3.566042	+	4.29673	+
F17	963.3281	1060.532	+	1192.749	+	1256.798	+	961.0977	−	2928.36	+	1371.947	+
F18	995.5679	1002.687	+	1155.819	+	1269.093	+	1005.31	+	2831.762	+	1367.64	+
F19	116.1827	455982.2	+	213400.6	+	1981.643	+	54285.22	+	82.83917	−	1408906	+
F20	24.80622	23.71676	−	24.49121	−	24.85274	+	24.38139	−	24.99457	+	24.94717	+
F21	1182.313	2495.865	+	4180.524	+	2914.916	+	3928.895	+	797.324	−	4598.763	+
F22	12814.8	13042.45	+	15327.9	+	13544.08	+	15018.61	+	12381.4	−	16307.78	+
F23	13219.73	14270.68	+	15876.35	+	14290.1	+	15228.64	+	11963.1	−	16662.25	+
F24	413.1954	407.235	−	440.9169	+	422.8146	+	394.2294	−	552.1235	+	944.3542	+
F25	427.9264	417.9576	−	459.6503	+	447.0926	+	384.9042	−	380.124	−	519.276	+
F26	440.2492	454.7012	+	490.7425	+	469.9969	+	252.5915	−	462.2346	+	452.8166	+
F27	2177.392	2064.106	−	2411.243	+	2361.449	+	2205.287	+	2546.098	+	3673.665	+
F28	2726.142	4504.284	+	6244.913	+	8183.815	+	6613.181	+	10738.99	+	10609.73	+
Summary		+	20	+	24	+	25	+	18	+	15	+	27
		−	8	−	4	−	3	−	10	−	13	−	1
		=	0	=	0	=	0	=	0	=	0	=	0

Equation 10 represents the generation of p logistics distribution centers. Equation 11 ensures that the demand point should be within the distribution range of its corresponding distribution center.

In order to solve this problem efficiently and prove the effectiveness of the proposed new algorithm in solving practical problems, this paper selects 6 distribution centers from 31 cities as an example to solve. Table 4 records the experimental results of the four swarm intelligence optimization algorithms, and the convergence curves of the four algorithms are recorded in Fig. 2. In the experiment, the number of particles of each algorithm is 50, and the maximum number of iterations is 100. The results of 10 consecutive tests are recorded. It can be seen from the table that in the comparison of the four algorithms, the TA algorithm achieves better optimal value and mean value. At the same time, by comparing the convergence curves of the four algorithms, the TA algorithm can converge faster and better when solving the location problem.

Table 4. The results of four algorithms for the location of 31 cities.

Algorithm	Best	Mean	Std	Location
TA	**549648.3**	**568407.1**	15052.1	9-5-17-12-20-27
PSO	554125.2	573141.3	19159.03	25-27-18-12-5-9
SCA	564954.8	588242.2	13807.29	18-28-5-12-9-25
WOA	568052	583001.8	14518.61	27-18-25-14-5-9

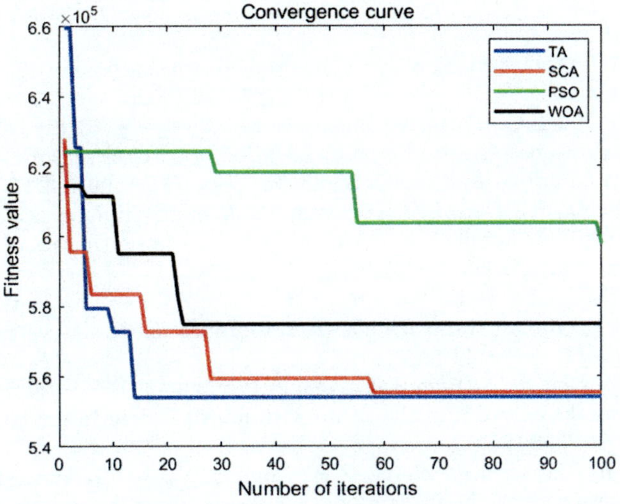

Fig. 2. Convergence curves of TA, SCA, PSO and WOA.

5 Conclusion

Inspired by the living habits of plant tumbleweed, this paper proposes a new swarm intelligence optimization algorithm called Tumbleweed algorithm (TA). The TA algorithm simulated the seedling growth and seed propagation of tumbleweed. By introducing the concept of growth cycle, the combination of the two stages is realized. By comparing the experimental results of different dimensions of CEC2013 test function, the TA algorithm can obtain better optimization results, which is better than the six swarm intelligence optimization algorithms. Finally, in the process of solving the practical application problem of logistics distribution center location, the TA algorithm can also effectively solve it. In the future, the TA algorithm will continue to be improved to enhance its optimization performance. And the application fields of the TA algorithm to solve practical problems will be gradually expanded.

References

1. Pan, J.S., Meng, Z., Chu, S.C., Xu, H.R.: Monkey king evolution: an enhanced ebb-tide-fish algorithm for global optimization and its application in vehicle navigation under wireless sensor network environment. Telecommun. Syst. **65**(3), 351–364 (2017)
2. Xue, X., Zhang, J.: Matching large-scale biomedical ontologies with central concept based partitioning algorithm and adaptive compact evolutionary algorithm. Appl. Soft Comput. **106**, 107343 (2021)
3. Wang, X.D., Chen, R.C., Yan, F.: High-dimensional data clustering using k-means subspace feature selection. J. Network Intell. **4**(3), 80–87 (2019)
4. Tu, T.N.: A fuzzy approach of large size remote sensing image clustering. J. Inf. Hiding Multimedia Signal Process. **11**(4), 187–198 (2020)
5. Chai, Q.W., Chu, S.C., Pan, J.S., Zheng, W.M.: Applying adaptive and self assessment fish migration optimization on localization of wireless sensor network on 3-D Terrain. J. Inf. Hiding Multimedia Signal Process. **11**(2), 90–102 (2020)
6. Pan, J.S., Wang, X., Chu, S.C., Nguyen, T.: A multi-group grasshopper optimisation algorithm for application in capacitated vehicle routing problem. Data Sci. Patt. Recogn. **4**(1), 41–56 (2020)
7. Xu, X.W., Pan, T.S., Song, P.C., Hu, C.C., Chu, S.C.: Multi-cluster based equilibrium optimizer algorithm with compact approach for power system network. J. Netw. Intell. **6**(1), 117–142 (2021)
8. Huang, H.C., Chu, S.C., Pan, J.S., Huang, C.Y., Liao, B.Y.: Tabu search based multi-watermarks embedding algorithm with multiple description coding. Inf. Sci. **181**(16), 3379–3396 (2011)
9. Gao, M., Pan, J.S., Li, J.P., Zhang, Z.P., Chai, Q.W.: 3-D Terrains deployment of wireless sensors network by utilizing parallel gases Brownian motion optimization. J. Internet Technol. **22**(1), 13–29 (2021)
10. Pan, J.S., Chu, S.C., Roddick, J., Dao, T.K., et al.: Optimization localization in wireless sensor network based on multi-objective firefly algorithm. J. Netw. Intell. **1**(4), 130–138 (2016)
11. Kou, X., Feng, J.: Matching ontologies through compact monarch butterfly algorithm. J. Netw. Intell. **5**(4), 191–197 (2020)
12. Chu, S.C., Huang, H.C., Roddick, J.F., Pan, J.S.: Overview of algorithms for swarm intelligence. In: International Conference on Computational Collective Intelligence, pp. 28–41. Springer (2011). https://doi.org/10.1007/978-3-642-23935-9_3
13. Holland, J.H.: Genetic algorithms. Sci. Am. **267**(1), 66–73 (1992)
14. Pan, J.S., Kong, L., Sung, T.W., Tsai, P.W., Snášel, V.: A clustering scheme for wireless sensor networks based on genetic algorithm and dominating set. J. Internet Technol. **19**(4), 1111–1118 (2018)
15. Xue, X., Wu, X., Chen, J.: Optimizing ontology alignment through an interactive compact genetic algorithm. ACM Trans. Manage. Inf. Syst. (TMIS) **12**(2), 1–17 (2021)
16. Wang, J., Pan, B., Wang, Q.R., Ding, Q.: A chaotic key expansion algorithm based on genetic algorithm. J. Inf. Hiding Multimedia Signal Process. **10**(2), 289–299 (2019)
17. Qin, A.K., Huang, V.L., Suganthan, P.N.: Differential evolution algorithm with strategy adaptation for global numerical optimization. IEEE Trans. Evolut. Comput. **13**(2), 398–417 (2008)

18. Kirkpatrick, S., Gelatt, C.D., Vecchi, M.P.: Optimization by simulated annealing. Science **220**(4598), 671–680 (1983)
19. Kennedy, J., Eberhart, R.: Particle swarm optimization. In: Proceedings of ICNN 1995-International Conference on Neural Networks, vol. 4, pp. 1942–1948. IEEE (1995)
20. Wu, J., Xu, M., Liu, F.F., Huang, M., Ma, L., Lu, Z.M.: Solar wireless sensor network routing algorithm based on multi-objective particle swarm optimization. J. Inf. Hiding Multimedia Signal Process. **12**(1), 1–11 (2021)
21. Mirjalili, S., Mirjalili, S.M., Lewis, A.: Grey wolf optimizer. Adv. Eng. Software **69**, 46–61 (2014)
22. Yang, X.S., Karamanoglu, M., He, X.: Flower pollination algorithm: a novel approach for multiobjective optimization. Eng. Optimization **46**(9), 1222–1237 (2014)
23. Pan, J.S., Zhuang, J., Luo, H., Chu, S.C.: Multi-group flower pollination algorithm based on novel communication strategies. J. Internet Technol. **22**(2), 257–269 (2021)
24. Meng, Z., Pan, J.S.: QUasi-affine transformation evolution with external archive (QUATRE-EAR): an enhanced structure for differential evolution. Knowl.-Based Syst. **155**, 35–53 (2018)
25. Liu, N., Pan, J.S., Chu, S.C.: A competitive learning quasi affine transformation evolutionary for global optimization and its application in CVRP. J. Internet Technol. **21**(7), 1863–1883 (2020)
26. Karaboga, D., Basturk, B.: On the performance of artificial bee colony (ABC) algorithm. Appl. Soft Comput. **8**(1), 687–697 (2008)
27. Xiao, S., Wang, H., Wang, W., Huang, Z., Zhou, X., Xu, M.: Artificial bee colony algorithm based on adaptive neighborhood search and Gaussian perturbation. Appl. Soft Comput. **100**, 106955 (2021)
28. Wang, H., Wang, W., Xiao, S., Cui, Z., Xu, M., Zhou, X.: Improving artificial bee colony algorithm using a new neighborhood selection mechanism. Inf. Sci. **527**, 227–240 (2020)
29. Pammel, L.: Botany of Russian thistle. Bulletin **3**(26), 3 (2017)
30. Liang, J., Qu, B., Suganthan, P., Hernández-Díaz, A.G.: Problem definitions and evaluation criteria for the CEC 2013 special session on real-parameter optimization. Computational Intelligence Laboratory, Zhengzhou University, Zhengzhou, China and Nanyang Technological University, Singapore, Technical Report, vol. 201212(34), pp. 281–295 (2013)
31. Mirjalili, S.: SCA: a sine cosine algorithm for solving optimization problems. Knowl.-Based Syst. **96**, 120–133 (2016)
32. Mirjalili, S., Lewis, A.: The whale optimization algorithm. Adv. Eng. Software **95**, 51–67 (2016)
33. Chu, S.C., Tsai, P.W., Pan, J.S.: Cat swarm optimization. In: Pacific Rim international Conference on Artificial Intelligence, pp. 854–858. Springer (2006). https://doi.org/10.1007/978-3-540-36668-3_94
34. Yang, X.S., He, X.: Bat algorithm: literature review and applications. Int. J. Bio-inspired Comput. **5**(3), 141–149 (2013)
35. Arora, S., Singh, S.: Butterfly optimization algorithm: a novel approach for global optimization. Soft Comput. **23**(3), 715–734 (2019)
36. Kayikci, Y.: A conceptual model for intermodal freight logistics centre location decisions. Procedia-Soc. Behav. Sci. **2**(3), 6297–6311 (2010)
37. Nathan, R., et al.: Mechanisms of long-distance dispersal of seeds by wind. Nature **418**(6896), 409–413 (2002)

38. Bensassi, S., Márquez-Ramos, L., Martínez-Zarzoso, I., Suárez-Burguet, C.: Relationship between logistics infrastructure and trade: evidence from Spanish regional exports. Transp. Res. Part A: Policy Pract. **72**, 47–61 (2015)
39. Gammelgaard, B., et al.: The emergence of city logistics: the case of Copenhagen's Citylogistik-kbh. Int. J. Phys. Distrib. Logist. Manage. **45**(4), 333–351 (2015). https://doi.org/10.1108/IJPDLM-12-2014-0291
40. Hu, W.: An improved flower pollination algorithm for optimization of intelligent logistics distribution center. Adv. Prod. Eng. Manage. **14**(2), 177–188 (2019)

Research on Evasion Strategy of Aircraft Based on Deep Reinforcement Learning

Ye Li[1], Lin Zhang[2], Kaixuan Wang[1], Fuquan Zhang[3](✉), and Yi Fan[1]

[1] China Academy of Launch Vehicle Technology, Beijing 100076, People's Republic of China
[2] Capital Aerospace Machinery Company, Beijing 100076, People's Republic of China
[3] Fujian Provincial Key Laboratory of Information Processing and Intelligent Control, Minjiang University, Fuzhou 350121, People's Republic of China

Abstract. In order to improve the autonomous penetration capability of aircraft, a method for generating aircraft maneuver avoidance strategies based on deep reinforcement learning is proposed. In order to solve the sparse reward problem of penetration reinforcement learning, the initialization process and sampling method of deep Q Network are improved, and the convergence efficiency of the learning algorithm is improved. Firstly, a penetration threat scene and an offensive and defensive confrontation motion model are established, and the relative motion state of the interceptor and the typical maneuver mode of the balanced flight section are selected as the state space and the action space. In the learning process, the aircraft is used as an agent to perform a reinforcement learning exploration process through confrontation simulations to train the maneuver value network so that it can finally make autonomous decisions based on the threat situation. Finally, the simulation experiment results show that the algorithm proposed in this paper can effectively generate maneuver penetration strategies and support the intelligent game of aircraft.

Keywords: Deep reinforcement learning · Aircraft penetration · Offensive and defensive confrontation · Guidance

1 Introduction

At present, offensive aircraft have the advantages of large airspace, strong maneuverability, and long-range precision strikes. Correspondingly, aircraft defense technology is also developing in a targeted manner. The defense systems are also continuously improved and upgraded, and their interception capabilities have posed a great threat to aircraft penetration.

Attacking aircraft using maneuver evasion to achieve penetration can give full play to the capabilities of the aircraft platform without adding other equipment. Therefore, maneuver penetration has become the focus of aircraft research. During mission execution, the aircraft needs to deal with interceptors with a certain randomness in the number, direction, and speed of the attack, and at the same time, after successfully evading, it needs to be accurately guided to the specified position and meet the energy constraints

S.-C. Chu et al. (Eds.): ICGEC 2021, LNEE 833, pp. 653–665, 2022.
https://doi.org/10.1007/978-981-16-8430-2_59

required for effective damage. Through the above analysis, it can be seen that maneuver avoidance is a highly dynamic, multi-constrained, and highly uncertain decision-making process.

Aiming at the problem of maneuver penetration, earlier studies focused on the design of procedural maneuver penetration methods [1], mainly discussing the penetration effects of fixed forms of maneuvering methods such as C-type and S-type. In subsequent research, relevant scholars analyzed the problem from the perspective of optimal control considering constraints, and solved the problem by solving methods such as pseudospectral method [2]. Furthermore, relevant research considers the interceptor interception law in the penetration problem, and uses matrix game, differential game and other methods to calculate the penetration strategy [3]. However, the above methods are mainly suitable for fixed constraints and single confrontation scenarios when solving maneuvering decision problems, and are not suitable for intercepting complex and random changes in threats. In order to solve similar complex scene problems, reinforcement learning methods have been introduced in the areas of unmanned system path planning and autonomous drone combat decisions [4, 5], so that the corresponding decision-making problems can be effectively solved.

Aiming at the dynamics and unknowns of the maneuvering confrontation process, this paper is based on the theory of Deep Reinforcement Learning (DRL), and uses the confrontation and interaction between the aircraft and the mission scene to solve the problem. Based on the establishment of a mathematical model of the confrontation situation, the scene return relationship is established by analyzing the flight mission profile, and the deep neural network is used to fit the maneuvering strategy to form an intelligent maneuvering autonomous decision-making method for the aircraft to enhance its ability to effectively complete the strike mission.

2 Maneuvering Confrontation Kinematic Model

2.1 Description of Confrontation Situation

This article chooses a single attacking aircraft to attack the target of the aircraft carrier as the problem scenario. The aircraft carrier formation randomly arranges several frigates in the direction of the incoming attacking aircraft. Each frigate is within 50 km from the center aircraft carrier. Each frigate uses a single-wave interception against the aircraft. Each frigate launches a medium- and short-range interceptor missile for each aircraft in each wave, and the interception target will not be changed during the interception. The schematic diagram is shown in Fig. 1.

Under the setting of this article, the offensive and defensive confrontation problem of multiple bombs to multiple bombs is simplified to a one-to-one problem. In the confrontation process, it is assumed that both parties can obtain the other's information in real time, but the information of both parties will contain random measurement noises of different magnitudes.

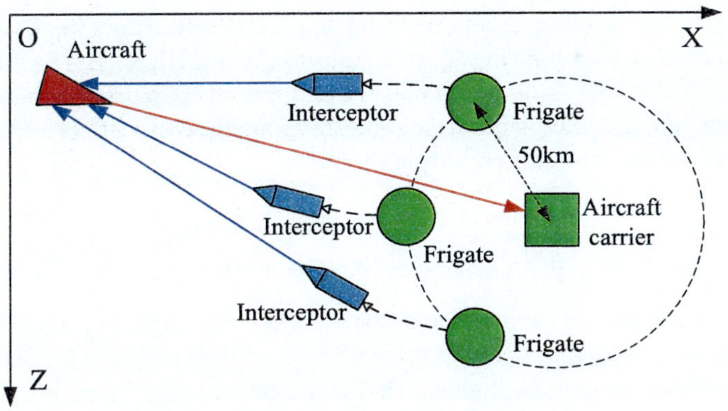

Fig. 1. Confrontation schematic diagram

2.2 Confrontation Kinematic Model

Due to the short time of the confrontation process and the low flight altitude, ignoring the rotation and curvature of the earth, the centroid vector dynamics model of the offensive aircraft and interceptor is established as:

$$\frac{d^2\vec{r}}{dt^2} = \vec{g} + \frac{\vec{R}}{m} + \frac{\vec{P}}{m} \tag{1}$$

where, $\vec{r}$ is the position vector of the missile, $\vec{g}$ is the gravitational acceleration, $\vec{R}$ is the aerodynamic force, $\vec{P}$ is the engine thrust, m is the aircraft quality.

The offensive aircraft is in the equilibrium flight stage in the situation set in this article, then the engine thrust $\vec{P} = 0$, relying only on aerodynamic force $\vec{R}$ for maneuvering flight. Meanwhile, with tilt-turn maneuver the aircraft mode, sideslip angle $\beta \equiv 0$, Using the assumption of instantaneous balance and ignoring the dynamic process of the missile body rotation, the dynamic rotation equation is established as:

$$\begin{cases} mV\frac{d\theta}{dt} = R_Y \cos\gamma_V - mg\cos\theta \\ -mV\frac{d\psi_V}{dt} = R_Y \sin\gamma_V \end{cases} \tag{2}$$

where, θ and ψ_V is the ballistic trajectory inclination and declination, α is the angle of attack, γ_V is the Speed dip angle, R_Y is the Aerodynamic lift.

The longitudinal guidance method of the balanced flight section without maneuvering adopts a jump-free descent based on the balanced gliding condition, as shown in formula (3), the lateral guidance uses the proportional guidance method to guide the target point according to the target line of sight information, see formula (4).

$$R_Y \cos\gamma_V + m\left(V\dot{\theta} - g\right)\cos\theta \le 0 \tag{3}$$

$$R_Y \sin\gamma_V = mV\left(k_d \cdot \dot{\eta} + \frac{k_p}{T_g}\eta\right) \tag{4}$$

In the above formula η and $\dot{\eta}$ is the deflection angle and deflection rate, k_d and k_p is the guidance parameters. Combine the above equations (Where lift R_Y is a function of angle of attack α), the amount of guidance control α_{need} and γ_{need} can be obtained.

The interceptor uses side-slip turning maneuver method, set its speed and tilt angle as $\gamma_V \equiv 0$, the dynamic rotation equation is:

$$\begin{cases} mV\frac{d\theta}{dt} = P\sin\alpha + R_Y - mg\cos\theta \\ mV\frac{d\psi_V}{dt} = P\sin\alpha\sin\beta - R_Z \end{cases} \tag{5}$$

During the forward flight of the interceptor, it points laterally to the predicted hit direction during launch. After entering the final stage, the interceptor seeker is turned on, and the interceptor conducts proportional guidance directly towards the interception target. The guidance method is described as follows:

$$\begin{aligned} n_y &= K_1|\dot{r}_{MT}|\dot{\varepsilon} + \cos\theta \\ n_z &= -K_2|\dot{r}_{MT}|\dot{\eta} \end{aligned} \tag{6}$$

where, n_y and n_z is the normal and lateral overload instructions, $\dot{\varepsilon}$ and $\dot{\eta}$ is the rate of change of the inclination and declination of the line of sight, $\dot{r}_{MT}$ is the change rate of distance between aircraft and target.

3 Maneuver Training Model Based on Reinforcement Learning

In this paper, a reinforcement learning framework is used to solve the maneuver avoidance problem. First, the problem needs to be described as a Markov Decision Process (MDP). MDP is a theoretical framework for achieving goals through interactive learning [6], It can be represented by a tuple $\langle \mathbf{S}, \mathbf{A}, \mathbf{P}, \mathbf{R} \rangle$ (state, action, transition probability, return). Maneuver evasion of offensive aircraft, **S** is the collection of threats facing the aircraft, **A** is the collection of aircraft maneuvering strategy, **P** is state transition probability matrix, In the Markov model, the probability of the current state $\boldsymbol{s}_t$ only depends on the previous state $\boldsymbol{s}_{t-1}$ and action $\boldsymbol{a}_{t-1}$, **R** is the environment's benefits for taking different actions in different states.

3.1 State Space

The state space describes the collection of useful information that the aircraft agent can perceive in the learning environment, and is the input for the agent to make decisions. The influencing factors of the offensive aircraft selection strategy mainly consider two aspects. One is to judge from the energy plan how much energy is used to evade the interceptor in different situations and how much energy is used to reach the target. The second is to increase the amount of interception misses through different maneuvers based on different interception directions and interception time from the game of athletic ability. Therefore, the state space should contain the following information, and the related definitions are shown in Fig. 2:

A) Attacking aircraft speed V (representing the energy of the aircraft);

B) The remaining voyage from the target L;
C) The line of sight angle ε, η relative to the interceptor and the distance r_{MT} relative to the interceptor.

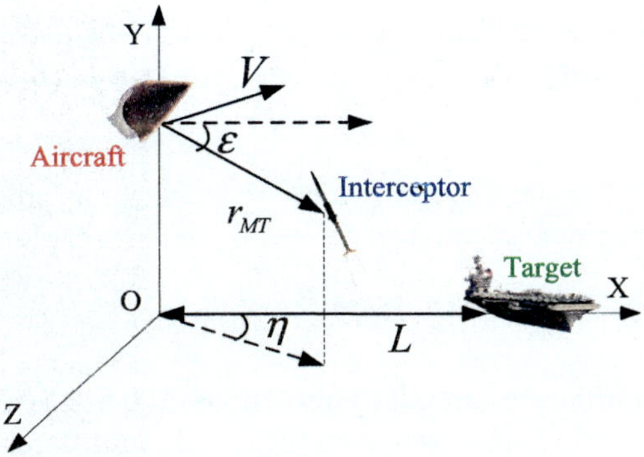

Fig. 2. State definition schematic diagram

The state space is $s = \{V, L, \varepsilon_1, \eta_1, r_{MT1}, \varepsilon_2, \eta_2, r_{MT2} \cdots \varepsilon_n, \eta_n, r_{MTn}\}$.

State space dimension is $3n + 2$, n is the largest number of frigates participating in the interception.

3.2 Gain Signals

In the reinforcement learning framework, the maximum total benefit of obtaining environmental feedback is the learning goal of the agent, so a reasonable setting of the benefit value is the key to solving the goal problem. In the question in this article, it is assumed that the maximum reward benefit can be obtained when the aircraft completes the flight mission, and the corresponding penalty benefit will be given when the aircraft fails to be intercepted, exhausted, fell to the ground, or flees backwards.

The above has completed the conversion of the maneuver evasion problem to the deep learning model. It can be seen that the learning problem has a large state space and can only obtain positive benefits when it reaches the final specific state. It is a sparse reward problem and will appear when using traditional reinforcement learning methods. The problem of lack of positive reinforcement signal, slow learning or even failure to converge.

This article first improves the sparse reward problem from the perspective of improving revenue design and increasing reward intensity. For the final flight strategy, it should conform to the mode of guiding toward the target without maneuvering avoidance. According to this prior information, it is set to give a small reward benefit when the aircraft is in the guidance state of the target in each decision period T_d, to improve the learning efficiency of the decision model. Suppose the reward is a unit return value of dR. Due to the cumulative effect of the reward, the negative return value of other failure situations must offset the positive cumulative value in order to play a punitive effect. Considering the above situation comprehensively, each return parameter is set as follows:

A) Get $2dR \cdot \mathrm{E}[t_g]/T_d$ when the flight mission is completed;
B) Get $-0.8dR \cdot \mathrm{E}[t_g]/T_d$ when intercepted;
C) Get $-dR \cdot \mathrm{E}[t_g]/T_d$ When the energy is exhausted;
D) Get $-2dR \cdot \mathrm{E}[t_g]/T_d$ when falling to the ground or flying in the opposite direction;
E) Get $dR + 0.02dR \cdot \mathrm{E}[t_g]/T_d$ When interceptors is failed and the aircraft is guided to the target.

Among them, $\mathrm{E}[t_g]$ is the expected value of a deduction process time, which is estimated from the sample mean value generated by random exploration.

3.3 Action Space

There are two types of offensive aircraft flying actions. The first is to guide to the target, and the second is to use maximum overload maneuvers in different directions. Define aircraft action space A as $A = \{0, 1, 2, 3, 4, 5, 6, 7, 8\}$.

The meaning of each action is shown in Table 1.

Table 1. Maneuver meaning

Action	Action meaning	Angle of Attack	Inclination angle
0	Guide to the target	$\alpha_{cmd} = \alpha_{need}$	$\gamma_{cmd} = \gamma_{need}$
1	Maneuver straight up	$\alpha_{cmd} = \alpha_{\max}$	$\gamma_{cmd} = 0°$
2	Maneuver to the upper right		$\gamma_{cmd} = 45°$
3	Maneuver to the right		$\gamma_{cmd} = 90°$
4	Maneuver to the lower right		$\gamma_{cmd} = 135°$
5	Maneuver down		$\gamma_{cmd} = 180°$
6	Maneuver to the lower left		$\gamma_{cmd} = -135°$
7	Maneuver to the left		$\gamma_{cmd} = -90°$
8	Maneuver to the upper left		$\gamma_{cmd} = -45°$

4 Strategy Learning Algorithm Based on Deep Q Network

This paper is based on the Deep Q-Learning Network (DQN) [7] method to solve the problem of state space continuity, and uses a deep neural network to fit the value of maneuvers under different threat situations.

The maneuver value network Q_{net} is composed of an input layer, a hidden layer and an output layer. It adopts a fully connected neural network structure, and the activation function adopts a rectified linear unit.

4.1 Initialize the Empirical Sample Pool with a Priori Assumptions

In order to achieve random sampling of historical samples, the training process needs to build an empirical sample pool to store sample sequences, and each group of samples records the implementation of the current strategy $\langle s_t, a_t, r_t, s_{t+1}\rangle$. Set the maximum capacity $N_{\max}$ for the empirical sample pool. When the maximum capacity is exceeded, the latest sample data is used to replace the oldest sample data to reduce the influence of old samples on network training. In order to enhance the diversity of samples, this article assumes that no action value network training is performed when the experience sample pool is not full, and this stage is defined as the experience sample pool initialization stage.

As analyzed in the previous section, the learning process of this article needs to overcome the sparse reward problem, which can be improved by guiding the increase of positive samples in the initialization phase of the sample pool. This paper introduces the following prior information, and believes that the following strategies used by offensive aircraft in flight missions can increase the probability of obtaining positive samples: (1) The closer the interceptor is, the greater the probability of performing maneuver evasive actions; (2) When there is no interceptor threat, the offensive aircraft should always be in a state of flying towards the target.

Compared with the traditional DQN method, in the initialization phase, use a random strategy based on a priori hypothesis is used for exploration, which effectively improves the probability of obtaining the final reward during the exploration process and increases the sample pool In the positive direction, improve training efficiency. Based on the above experience, the random exploration strategy can be expressed as:

$$P\{a=0\}=\begin{cases}1-e^{K(r_{\min}/D)}\ ,\ 0<r_{\min}\le D\\ 1\ ,\ r_{\min}=0\ \text{ or }\ r_{\min}>D\end{cases} \tag{7}$$

The P represents the probability distribution of the guided action to the target in the process of random exploration; where D is the threshold distance. When the threat distance is less than the threshold, the strategy will have a certain probability to select maneuvering actions, and K is the parameter of the probability model, $r_{\min}$ is the distance of the closest interceptor, expressed as $r_{\min}=\min\{r_{MT1},\cdots,r_{MTn}\}$.

4.2 Priority Replay Based on Revenue Experience

The initial goal of experience priority replay [8] is to solve the problem of low convergence efficiency of traditional Q-Learning by improving the utilization of high-value samples. In the literature [8], the difference between the target value and the predicted value of each round is used as the evaluation criterion of the sample value, which can be understood as the sample with the larger deviation from the prediction result of the current decision-making model, the more available information it contains, and the sample used The probability should be greater. In this paper, referring to the above-mentioned ideas, the sample priority is marked with the total return of a complete exploration, and the probability of high-priority samples being used for learning is increased to improve the sparse reward problem.

The samples are stored once after each complete exploration. When the capacity of the sample pool is exceeded, the oldest sample is discarded, and its sample index is assigned to the latest sample, and the index storage pool is updated at the same time. When sampling, first randomly sample a index from each index storage pool, which is the total number of samples sampled each time; after sampling the index, extract a batch of sample values from the empirical sample pool according to the index. The storage and sampling process diagram is shown in the Fig. 3.

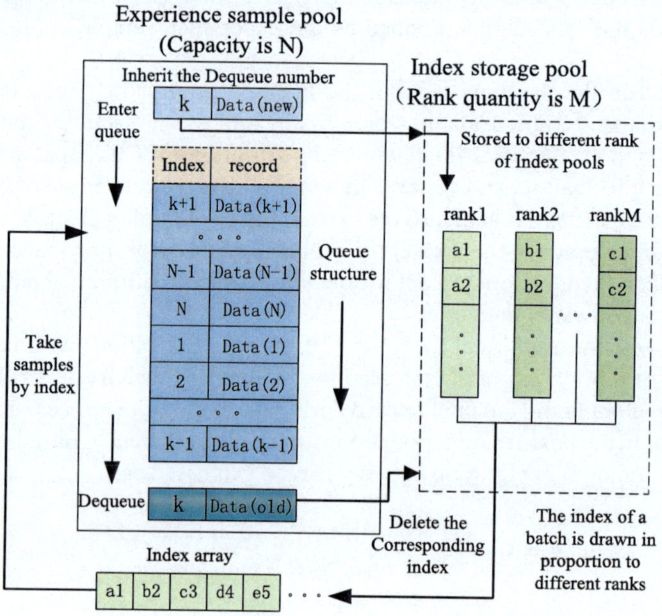

Fig. 3. Prioritized experience replay process

Among them, the priority is divided into M level according to the value range of the total income, which is $rank1$, $rank2$ to $rankM$ according to the income value from large to small. In order to simplify the computational complexity of the experience priority playback process, the experience sample pool is set up with M index storage pools to store the index of samples of different levels. When sampling, the sampling probability of samples of different levels is equivalent to the sampling ratio of different index storage pools, and the sampling ratio of samples of $ranki$ levels is set to P_{rki}, and the following formula is satisfied:

$$\sum_{i}^{M} P_{rki} = 1, \quad P_{rk1} > P_{rk2} > \cdots > P_{rkM} \tag{8}$$

4.3 Maneuver Evasion Strategy Deep Reinforcement Learning Algorithm Flow

In this paper, based on the DQN method, the problem of solving maneuver avoidance strategies is transformed into the problem of training maneuver value network Q_{net}. The goal of training for maneuver Q_{net} is to make the target network estimate the action value as accurately as possible. The pseudo code of the algorithm flow is as follows:

Initialize the experience sample pool, Set the capacity to $N_{\max}$;
Initialize $Q(s,a)$, Assign a random initial value to its network parameter θ.
for episode = 1,*M* **do**
  Initialize the simulation environment to obtain the initial situation s_t.
  for episode = 1,*T* **do**
    When the sample pool is not full, explore randomly based on a priori assumptions, skip the following steps;
    When the sample pool is full, randomly choose a flight action a_t with probability ε, otherwise choose $a_t = \max_a Q(s_t, a \mid \boldsymbol{\theta})$;
    Carry out simulation deduction with action a_t to get the corresponding income r_t and the threat situation s_{t+1} at the next moment;
    Randomly draw mini-batch samples from the sample pool (s_t, a_t, r_t, s_{t+1});
    make $y_i = \begin{cases} r_i & , s_{i+1} \text{ is terminal state} \\ r_i + \gamma \max_{a'} Q(s_{i+1}, a' \mid \boldsymbol{\theta}), & s_{i+1} \text{ is not terminal state} \end{cases}$;
    Calculate the loss function: $\text{loss}_i = (y_i - Q(s_t, a \mid \boldsymbol{\theta}))^2$;
    Perform gradient descent according to loss_i to update parameters θ.
  end for
  According to the rank of income, a complete trajectory sample sequence is deposited into the sample pool at one time.
end for

In the above training process, since the network parameters are updated every round, the labels used for training frequently change, and the network convergence is difficult to guarantee. In response to this problem, this article uses two neural networks in the actual training process, evaluate network $Q(s, a|\boldsymbol{\theta})$ and target network $Q(s, a|\boldsymbol{\theta}^-)$. The evaluation network updates the network parameters every round according to the above algorithm flow. The target value y_i in the loss function is calculated by the evaluation network; while the target network temporarily freezes the network in C cycles, and updates once in each cycle. The current value in the loss function is $Q(s_t, a)$ determined by The target network is calculated, and the training process is shown in Fig. 4.

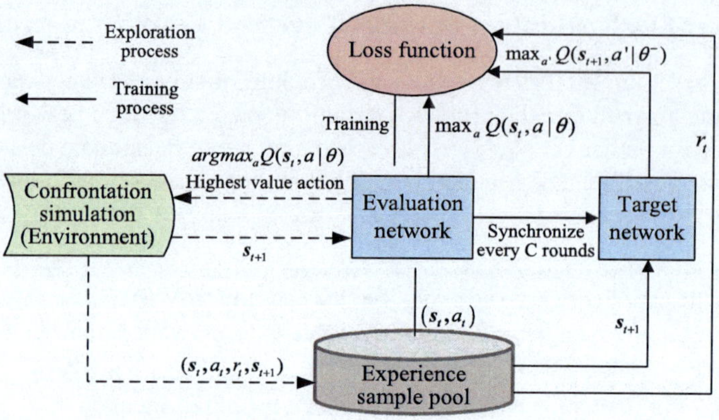

Fig. 4. Maneuver strategy training

5 Simulation

The attacking aircraft is 400 km away from the target and the decision period is 2 s. the maximum number of frigates is 3. When the aircraft arrives at a distance of 30 km from the target and has a speed of 2 Mach or more in the direction of the target, it is determined that the mission is successful.

For offensive aircraft, there is no deviation in the binding target point, the perception error of the interceptor is 10 m, and the perception delay is 200 ms. The interception system radar has a measurement error of 10 m and a perception delay of 100 ms; the interceptor seeker's starting distance is 40 km, the seeker's line-of-sight angular rate measurement noise is 0.2/s, and the seeker's measurement delay is 50 ms.

Reinforcement learning discount factor $\gamma = 0.9$, strategy network learning rate $\alpha_{\ln} = 1 \times 10^{-4}$, experience sample pool capacity is 10000, and the target network is updated every 50 cycles. According to the above scenarios and parameter values, the agent conducts 12000 rounds of intensive learning training and records the cumulative return each time. The results are shown in Fig. 5.

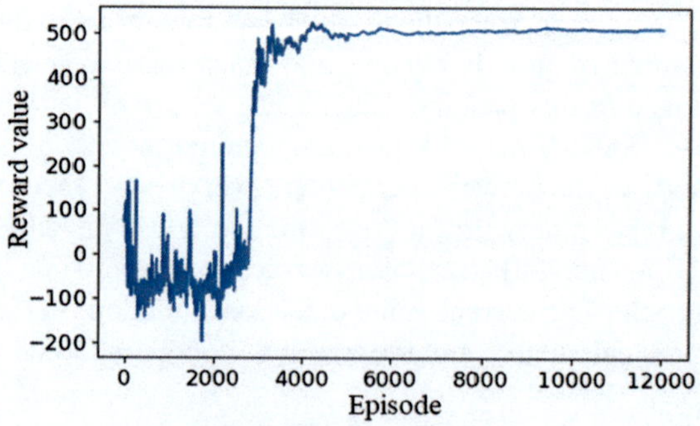

Fig. 5. Training process

At the beginning of training, it can be seen that due to the method of initializing the empirical sample pool with a priori assumptions, the reward value is maintained at a high level, which provides sufficient positive samples for subsequent training. The algorithm starts to converge in about 3000 rounds of training, and the reward value stabilizes above 500 after 6000 rounds. The characterization strategy network can effectively guide the completion of the flight mission.

In order to verify the generalization ability of the strategic network in this paper, after expanding the range of the frigate's random distribution, 100 evaluation tests were carried out, and the ballistic curves of the offensive and defensive sides generated by the test are shown in Fig. 6.

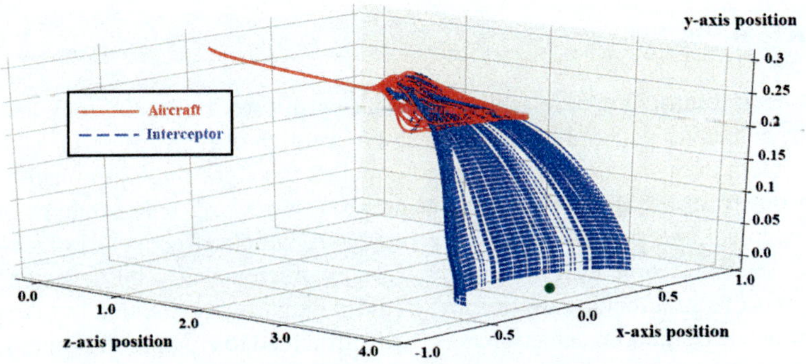

Fig. 6. 3D curve of test trajectory

Observe the figure above, we can see that the strategy trained by this method adopts the approach of approaching and getting rid of for evasive maneuvers. After judging that there is no threat, the speed is quickly pulled back to the direction of the shooting surface. Since the training process will experience confrontation simulations in various incoming directions, and random noise is considered in the simulation, the $\max_a Q(s_t, a|\theta)$ value of the strategy network after accumulating a certain sample will eventually converge to the maneuver that minimizes the probability of interception. The 100 offensive projectile trajectories in Fig. 6 are distributed in a cone shape. It can be seen that the strategy network after the final training selects different maneuvering directions according to the different incoming states of the interceptor.

The result statistics are shown in Fig. 7. In 100 test simulations, the interceptor's minimum misses were all greater than 70 m, and the rewards were all above the threshold that satisfies the flight success conditions. This proves that the method proposed in this paper can enable the aircraft to maneuver and avoid autonomously during the completion of the flight mission.

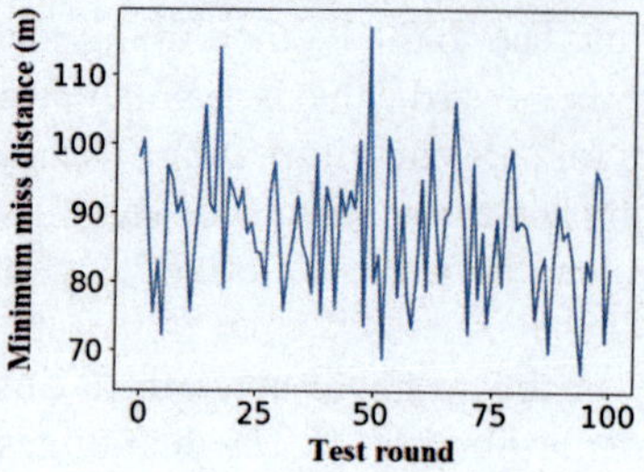

a) Minimum miss statistics

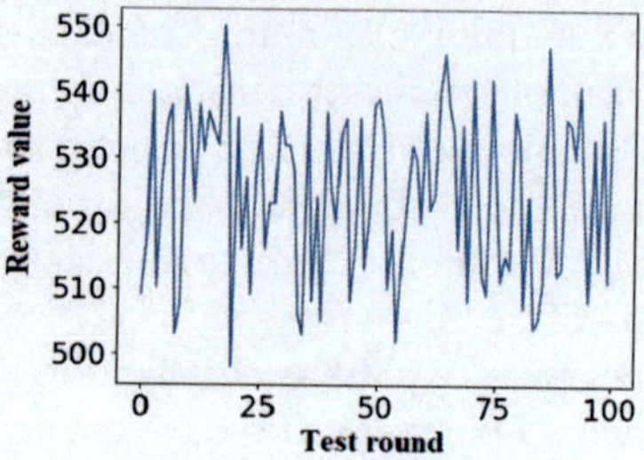

b) Reward value statistics

Fig. 7. Test result

6 Conclusions

Based on deep reinforcement learning, this paper proposes a decision-making method for aircraft maneuver penetration. Aiming at the problem of penetration into the sparse reward problem of the reinforcement learning model, the algorithm optimization design is carried out from the three levels of design process reward, enhancement of initial positive samples, and priority experience playback, which effectively overcomes the weak positive reinforcement signal and low learning efficiency. The bottleneck problem. In the process of strengthening the training of the Q network, the random threat situation and the influence of random detection noise are introduced to enhance the generalization ability of the strategy network. The final decision-making model can generate reasonable maneuvering strategies for interceptors in different directions and different numbers of attacks, and control their own energy to complete the final flight mission. The research results of this paper have important theoretical value and practical significance for the intelligent autonomous penetration of aircraft.

References

1. Jiang, Y., Cui, J.: Effectiveness of weaving maneuver strategy of a missile. J. Beijing Univ. Aeronaut. Astronaut. **28**, 133–136 (2002)
2. Li, J., Chen, W., Min, C.: Terminal hypersonic trajectory modeling and optimization for maneuvering penetration and precision strike. J. Beijing Univ. Aeronaut. Astronaut. **44**, 556–567 (2018)
3. Wang, Y., Ning, G., Wang, X., Hao, M., Wang, J.: Penetration strategy of near space vehicle based on differential game. J. Aeronaut. **41**, 69–78 (2020)
4. Zhang, H., Huang, C., Xuan, Y., Tang, S.: Maneuver decision of autonomous air combat of unmanned combat aerial vehicle based on deep neural neural network. Acta Armamentarii **41**, 1613–1622 (2020)
5. Zhang, K., Li, K., Shi, H., Zhang, Z., Liu, Z.: Autonomous guidance maneuver control and decision-making algorithm based on deep reinforcement learning UAV route. Syst. Eng. Electron. **42**, 1567–1574 (2020)

6. Sutton, R.S., Barto, A.G.: Reinforcement Learning: An Introduction. Massachusessts Institute of Technology (2018)
7. Mnih, V., et al.: Human-level control through deep reinforcement learning. Nature **518**(7540), 529–533 (2015)
8. Schaul, T., Quan, J., Antonoglou, I., et al.: Prioritized experience replay. Computer Science (2015)

Wireless Sensor Networks and Internet of Things

Research on NPE Algorithm of Ring Main Unit Fault Detection Based on Internet of Things Technology

Liu Baojun(✉)

Electrical and Information Engineering College, JiLin Agricultural Science and Technology University, Jilin, China

Abstract. With the increasing of equipment and types in the power network, it will inevitably affect the safety and reliability of the power supply system. Therefore, the real-time fault detection of the ring main unit is particularly important. This paper combines neighborhood preserving embedding (NPE) algorithm and Internet of things technology with a focus on real-time data acquisition, analysis and treatment. Real-time data information such as loop cabinet measurement variables and environment variables is obtained at the sensing extension layer of the Internet of Things, and information is put into the edge computing layer by means of communication to extract effective content and save it to the cloud platform. Secondly, proofread and transmit the extracted information. The Internet of Things (IOT) network transport layer is used to proofread effective information to ensure accurate and safe transmission of information to the IOT platform application layer. In the platform application layer, NPE algorithm is used to extract local features of effective information, and an on-line fault monitoring model based on data features is constructed. Finally, through analysis, the application of Internet of Things technology can further enhance the stable, economic, normative and friendly effect of power supply system. At the same time, the innovation of fault detection technology in the application layer of the platform improves the effect of fault detection and the ability of safe and stable operation of the loop cabinet.

Keywords: Internet of Things technology · NPE algorithm · Ring main unit · Fault detection · On-line monitoring

1 Introduction

The ring main unit is an important part and control node in the urban power supply network, and its operation state is directly related to the safety and reliability of the power grid. As the external environment is complex and changeable, it will inevitably affect the operation of the ring main unit, so it is particularly important to monitor the operation status of the relevant equipment in the ring main unit in real-time. The application of the Internet of Things technology in the monitoring of the equipment in the ring main unit can improve the level of intelligent monitoring and enhance the ability of information management. As a new design principle and implementation method, this

S.-C. Chu et al. (Eds.): ICGEC 2021, LNEE 833, pp. 669–678, 2022.
https://doi.org/10.1007/978-981-16-8430-2_60

technology provides effective technical support for the fault detection of the ring main unit and provides a guarantee for strengthening the construction of the power supply system [1].

At present, many scholars have done a lot of research on ring main unit data acquisition and intelligent fault diagnosis. In reference [2], a detection device for the outdoor ring main unit is designed, which uses cable partial discharge, SF_6 leakage concentration, and temperature information to judge fault information. And realize the synchronous monitoring of background and worksite. In reference [3], an online ozone detection device was designed to detect the early discharge phenomenon inside the ring main unit and offer an early warning. Compared with conventional devices, its rate of fault diagnosis and isolation is greatly improved, and the reliability of the equipment has been enhanced, meanwhile, it greatly reduces the labor intensity of the operators. In reference [4–6], an online intelligent monitoring system for power cable is designed, which can predict whether the cable joint is abnormal only by monitoring the temperature of the cable joint and its environment. In terms of fault diagnosis, references [7] and [8] respectively proposed a big data-driven interactive fault diagnosis method for distribution switchgear and the KPCA (Kernel Principal Component Analysis) model solving method. The research shows that these methods can effectively reduce the fault false alarm rate and improve the fault detection effect of the ring main unit. However, with the development of intelligence and information technology, there are higher requirements for comprehensive fault detection of ring main unit.

The above research puts forward effective solutions for online monitoring and fault diagnosis of ring main unit, but with the intellectualization of ring main unit, the actual operational requirements are more complicated, and the problem of risk vulnerability is gradually prominent. This research combines the core technology of the Internet of Things and NPE detection method to realize the intelligent information acquisition, transmission, processing, and other steps in the ring main unit, enhances the operation reliability and safety of the ring main unit, and improves the comprehensive efficiency of fault handling of ring main unit. In the following sections, I first introduce the application of Internet of things technology and NPE algorithm in the fault detection of ring main unit. Then, the second part is a simulation with real data. And the last part analyzes the progress and limitations of this study.

2 Research on Fault Detection of Ring Main Unit

2.1 Internet of Things Technology for Fault Detection

The Internet of Things technology is to use widely existing perception points to collect data information and external environment information through the relevant equipment of power enterprises to realize the comprehensive perception of the intelligent shared power system [9].

Firstly, aiming at the problem of data information collection in the ring main unit, since the power Internet of Things is built on the basis of perception extension layer, the measurement information of ring main unit can be collected by this layer and transmitted to the Internet of Things. Thus, the environment information collection system of the ring main unit is established, as shown in Fig. 1. The system consists of a central control host, which is connected with the monitoring subsystem through the network switch in the way of communication, and the subsystem is connected with each sensor through Ethernet and uses the sensor to search and process relevant information. In this system, the main environmental information collected is as follows: temperature and humidity in the cabinet, SF6 concentration, O3 concentration, water level, and smoke. At the same time, the system can detect the partial discharge information and transmit the discharge parameters and position information to the remote service system through wireless or wire. The environmental information collection system of the ring main unit designed on this layer integrates the Internet of Things, which makes this layer not only responsible for instruction scheduling but also receive the operation instructions from the cloud platform, so as to realize the autonomous response of secondary equipment information of the ring main unit. In the smart gateway of the Internet of Things, the data acquired by the perception layer is connected with the edge computing layer through the port with distributed computing capabilities, to realize the comprehensive processing of information and transmit the effective information to the cloud platform for unified disposal.

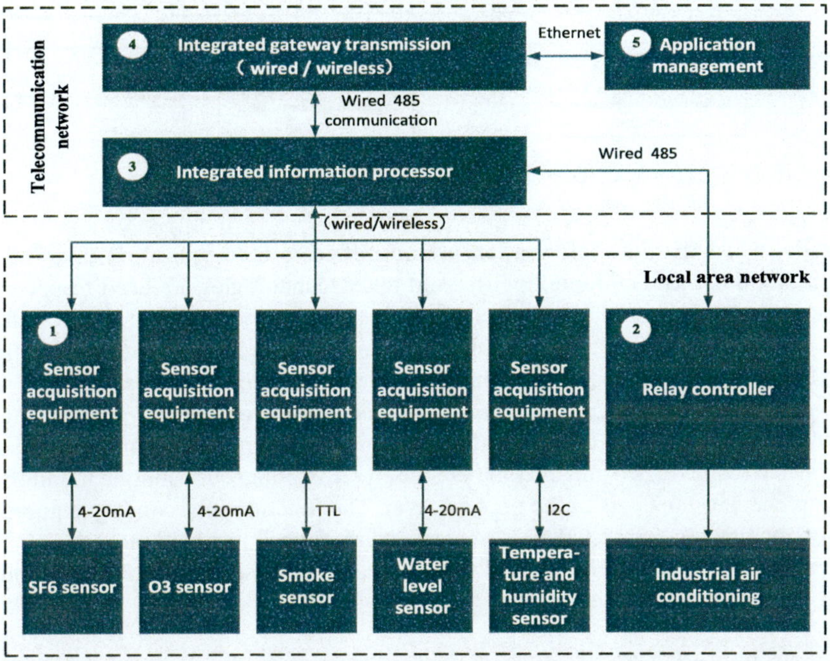

Fig. 1. Block diagram of an environmental information acquisition system for ring main unit.

To ensure the accuracy and availability of data transmission, the transport layer is set as the channel to transmit data information from the perception extension layer to the platform application layer. This layer mainly through the local LAN and remote communication network to ensure the smooth, efficient, and low-cost operation between the perception extension layer and the edge computing nodes to provide high-quality, effective, and barrier-free communication technical support between cloud platform and ring main unit. Therefore, it is necessary to realize the online checking of data in this layer. As shown in Fig. 2, upload the collected data to the cyclic redundancy check code and compare it with the standard redundancy check code of the whole station system. If there is any difference, an alarm will be issued.

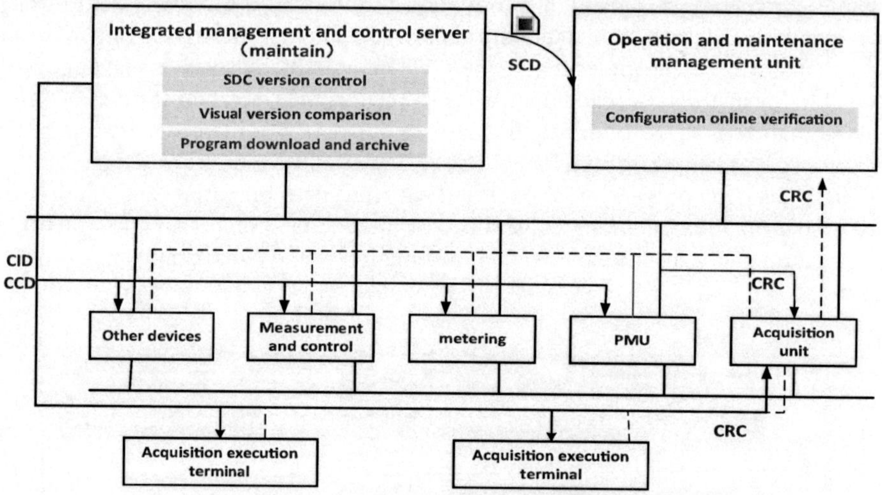

Fig. 2. Data files are verified online.

Finally, when the data information is transmitted to the application layer of the platform, big data, artificial intelligence, and other technologies are used to perceive, process, analyze and control the received data, to establish a management cloud platform with comprehensive application capability, promote the high-quality development of energy distribution and electricity management, and realize the data interaction and sharing among power plant, management department of the power grid and consumers in the Internet of Things. In the application layer, it is based on the acquisition unit and operation and maintenance unit to complete the effective connection, online monitoring, and comprehensive analysis of the process layer. The function of secondary equipment condition monitoring and visualization is realized through algorithm innovation. In this study, the NPE algorithm will be used to realize online monitoring and fault alarm of the ring main unit in this layer.

2.2 NPE Algorithm for Fault Detection

NPE algorithm uses the local features of data to obtain the overall manifold structure features. In the algorithm, the objective function with reconstruction weight and weight vector is constructed between each sample point and its neighborhood points, and the projection matrix is constructed. The reconstructed weight vector is used to find the low dimensional manifold information hidden in the original variable space, and the dimension reduction of the data is realized by keeping the weight of each neighborhood in the low dimensional space, to obtain the irrelevant feature information of the original data.

The algorithm in this study belongs to a multivariable statistical analysis algorithm. Because the abnormality of the ring main unit will be reflected in multiple variables, the detection based on abnormal related variables is of great significance for fault prediction of the ring main unit. The nonlinear complex models are usually composed of abnormal variables. NPE algorithm uses the local linear characteristics of abnormal variables to explore the data mapping relationship among abnormal variables, to obtain the data characteristic information. This algorithm is combined with traditional research methods, such as principal component analysis, independent component analysis, Fisher linear discriminant analysis, etc., to establish the system monitoring model, and then realize the real-time online fault detection of the ring main unit.

Algorithm Model. Firstly, it collects operating data under normal conditions of the ring main unit, such as information obtained from sensors of the cabinet temperature, humidity, cable joint, and core temperature, CO gas volume, cable current, etc., as well as information of outdoor environmental conditions and customer power load. This information is constructed into data set matrix $X(x_1, \cdots, x_n) \in R^{D\times n}$, D is the number of process variables. Storing this data in the platform application layer data monitoring center for backup, pre-processing, and component new data training set $\overline{X} \in R^{n\times D}$. Next, the K-NN method is used to determine the nearest neighbor of each sample point from near to far $\{x_{ij}\}_{j=1}^{k}$. The reconstruction coefficient matrix W for each sample point is obtained by linear reconstruction of the nearest neighbor points, and the reconstruction error is obtained by minimizing Eq. (1).

$$\varphi(W) = \sum\nolimits_i \left\| x_i - \sum_{j=1}^{n} W_{ij} x_j \right\|^2 \tag{1}$$

Equation (1): W_{ij} is the number of weight of data points x_j in the k neighborhoods of B when reconstructing x_i.

Based on the objective function of Eq. (1), the optimal weight matrix $W \in R^{n\times n}$ is constructed for the training data $\overline{X}$, as in Eq. (2).

$$W_j^i = \frac{\sum\limits_{m=1}^{k} (Q^i)_{jm}^{-1}}{\sum\limits_{p=1}^{k} \sum\limits_{q=1}^{k} (Q^i)_{pq}^{-1}} \tag{2}$$

Equation (2): $Q_{jm}^i = (x_i - x_{ij})^T (x_i - x_{im})$, x_{ij} (j = 1, 2, ..., k) are k nearest neighbors of x_i; W_j^i is the value of the weight between x_i and x_{ij}, which satisfies $\sum\limits_{i=1}^{k} W_j^i = 1$.

The low-dimensional projection $Y(y_1, \cdots, y_d) \in R^{n\times d} (d \leq D)$ of the $\overline{X}$ data is obtained according to the local weight reconstruction matrix W and the dimensionality reduction target of the neighborhood preserving embedding algorithm. The data relationships are as follows:

$$\begin{aligned}\overline{X} &= \hat{X} + X = YB^T + E = \overline{X}B(B^TB)^{-1}B^T + E \\ &= \overline{X}B(B^TB)^{-1}B^T + (\overline{X} - YB^T)\end{aligned} \tag{3}$$

Equation (3): $A = \overline{X}B(B^TB)^{-1} \in R^{D\times d}$, d is the dimensionality reduction; X is the residual space of the data; E is the residual matrix.

Through the above steps, the original data is divided into model space and residual space, and the projection matrix of the residual space is: $\overline{A} = [a_{d+1}, a_{d+2}, \cdots, a_D]$. For the real-time measurement data x_{new} of the ring main unit, the data is pre-processed firstly and then projected into the model space and residual space by using the projection matrix.

$$\begin{aligned}x_{new} &= \hat{x}_{new} + \tilde{x}new = y_{new}B^T + (x_{new} - y_{new}B^T) \\ &= y_{new}B^T + x_{new}[I - B(B^TB)^{-2}B^T]\end{aligned} \tag{4}$$

Equation (4): $y_{new} = x_{new}B(B^TB)^{-1} = x_{new}A$, $\hat{x}_{new} = y_{new}B^T$ are the reconstructed vectors of x_{new}.

Hotelling T^2 and square predicted error (SPE) are constructed in the above two spaces to measure the fluctuation of the two spaces respectively and to realize online real-time monitoring and fault alarm of the ring main unit. The statistics and statistical limits associated with the data point y are as follows:

$$T^2 = y\Lambda^{-1}y \le T^2_{\text{lim}} = \frac{d(n^2-1)}{n(n-d)}F(d, n-d; \alpha) \tag{5}$$

$$\begin{aligned}SPE &= \|x_{new} - \hat{x}_{new}\|2 \\ &\left\|x_{new}(I - B(B^TB)^{-1}B^T)\right\|^2 \le g \times x^2_{h,\alpha}\end{aligned} \tag{6}$$

Equation (5): $\Lambda^{-1} = [Y^TY/(n-1)]^{-1}$; $F(d, n-d; \alpha)$ is the F-Distribution with significance level α and degrees of freedom d and $n-d$; g and h are the parameters of the χ^2-Distribution satisfying the conditions $g = {}^{v}/_{2m}$, $h = {}^{2m}/_{v}$, m and v are the mean value and variance of the SPE statistics estimated by training samples, respectively.

For each real-time data monitoring statistic comprehensive calculation, and determine whether the statistic exceeds the modeling statistic limit corresponding to formula (5 and 6), by which we can determine whether the data is normal at this time. If there is a fault warning, then the relevant fault elimination processing operation should be carried out immediately. If the data statistics are below the safe statistical limit, it means that the data is normal at that moment. The flow chart of the online monitoring model of the ring main unit based on NPE algorithm is shown in Fig. 3.

In Fig. 3, when monitoring statistics of real-time data, the size of the statistical limit is compared with that of modeling, and the data state is screened through this judgment. If the data statistics are lower than the modeling statistical limit, it means that the data is normal. Otherwise, the fault is indicated, and corresponding measures should be taken immediately to avoid the expansion of the fault range.

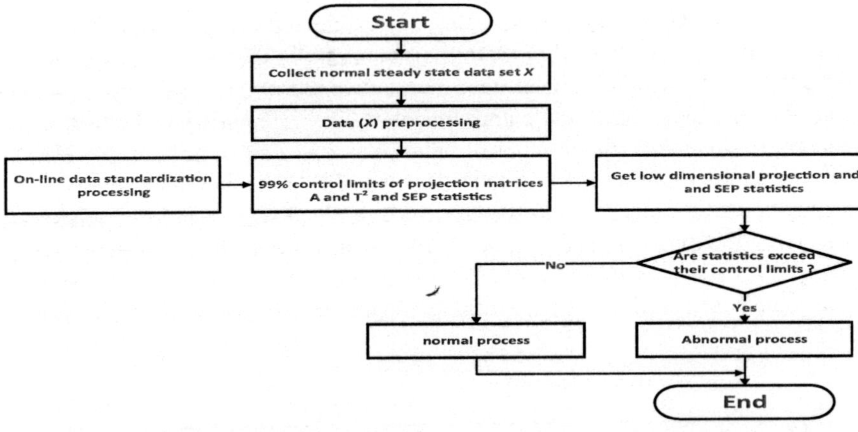

Fig. 3. Flow chart of on-line monitoring of ring main unit.

Data collection, preprocessing, and acquisition of projection matrix and statistical limited process data in Fig. 3 are all off-line modeling. The standardization of online data processing, the judgment of online data security performance index, the calculation of statistical limit and whether the online data statistics exceed the standard, and the judgment of whether the process is faulty are all in the scope of online monitoring.

Model Analysis. The NPE algorithm model proposed in this study is applied to the platform application layer of the Internet of Things. In this network layer, the data sampling values and operational information such as ring main unit are accepted, and after precise analysis and parsing, they are passed to the operation and maintenance control department, and then the valid information is transferred to the inside of the station control layer switch, and the actual operational messages and dynamic data monitoring of the ring main unit are realized using the neighborhood hold embedding algorithm. The method mines the intrinsic structural characteristics of the ring main unit through the information of actual data features. In terms of model construction, it not only relies on the precise process mathematical models with simple and clear model construction mechanism, but also uses the process data to obtain real-time status and increase model accuracy. Therefore, for complex processes where data models and comprehensive process knowledge cannot be accurately established, online fault detection of ring main units through the use of data-driven in the Internet of Things technology has greater advantages.

3 Example Analysis

In this study, we use the actual data provided by a company for a ring main unit, in which there are two compartments. Each partition has A group of A, B, C three-phase cable, measuring 6 groups of wire core temperature; At the same time, the ambient temperature and humidity in the two partitions were measured respectively, namely, 4 groups of variables, and the sampling frequency was set as 2 times/min.

First of all, 20,000 data training samples were collected. To detect the fault detection performance of NPE algorithm, 3500 normal samples were taken as the test set. According to the characteristics of actual process data, 2000 fault data were simulated and tested. The number of neighbors and dimensionality reduction should be specified in the model. In this paper, the number of neighbors k was set as 15, and the variance contribution rate method was used for the dimensionality reduction method. According to the principle of this method, in order to make the data interpretation degree reach more than 90%, the dimensionality reduction $d = 4$, and the confidence level of statistics are 99%. As shown in Figs. 4 and 5, the normal data and fault data monitoring results obtained based on the NPE algorithm at the platform application layer are shown respectively.

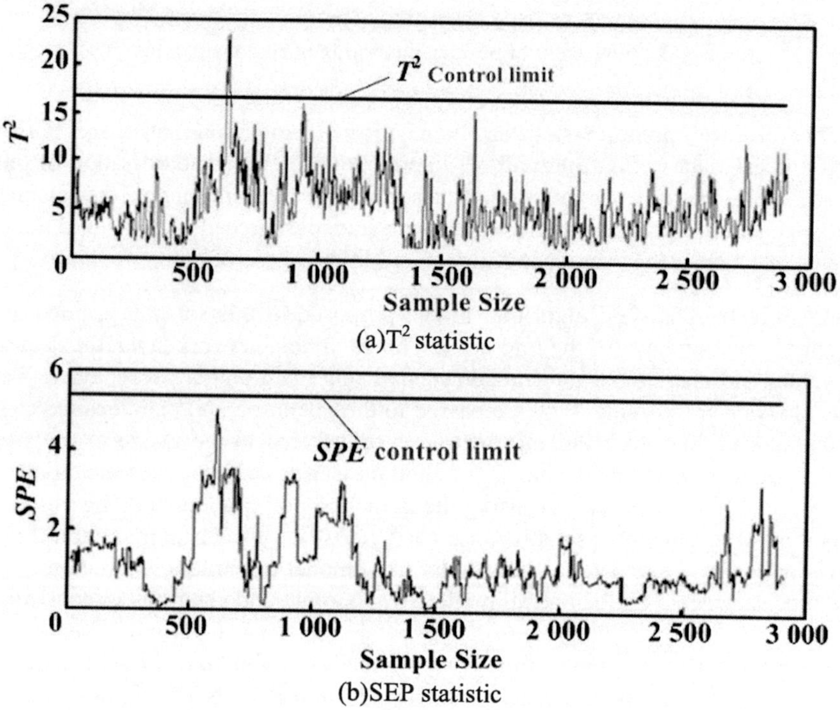

Fig. 4. The normal data monitoring result chart of ring main unit.

The change curve and straight line in Figs. 4 and 5 are the real-time T^2 and SPE statistics and limits, respectively. If the statistics are all located below the statistical limits, it indicates the process is normal. It can be seen from Fig. 4 that most of the normal data lie below the statistical limit, indicating that the NPE algorithm model can improve the accuracy of normal data detection and reduce the false alarm rate. It can be seen from Fig. 5 that T^2 statistics are all above the limit value, which is sensitive to fault detection and has a high detection rate. SPE statistics part of the data is located below the

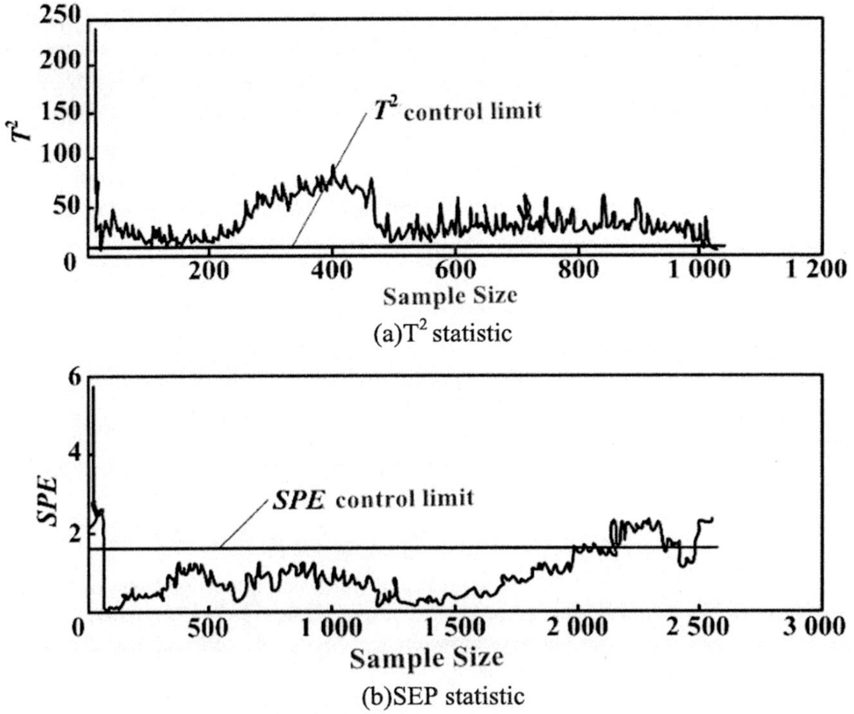

Fig. 5. The fault data monitoring result chart of ring main unit.

statistical limit, this is because the fault data is based on the distribution characteristics of the process, so the residual space of the data is not reflected in the SPE statistics.

4 Concluding Remarks

The emergence of the Internet of Things is a major breakthrough in information technology that has driven rapid development in various fields. In this study, the Internet of Things technology is fully applied in the intelligent monitoring system of the ring main unit, and the embedded algorithm is maintained in the application layer combination field of the Internet of Things platform. By using multivariable and local relationship information, the change characteristics of the ring main unit data are obtained effectively, and the innovation of the online fault detection method is realized. The experimental results prove the feasibility and effectiveness of the neighborhood preserve embedding algorithm in real-time online fault detection. At the same time, the application of the Internet of Things technology makes the power loop network cabinet system develop in the direction of information, network, and intelligence, so as to make the power network run safely and reliably.

The neighborhood preserving embedding algorithm proposed in this study is a typical unsupervised manifold learning method, which has achieved remarkable results in exploring nonlinear dimensionality reduction of data. However, this method can only best describe the data characteristics, but can not achieve the best classification, unable to identify the classification information of data samples. Will study in follow-up studies, therefore, supervision of manifold learning method, embedded in neighborhood algorithm based on the introduction of discriminant neighbor embedding algorithm or the concept of attractor, after making the dimension reduction of similar data samples is more compact, heterogeneous data samples more distant, finally to keep the Internet of things platform application layer neighborhood embedding algorithm has better performance.

References

1. Liu, J., Li, X., et al.: Internet of Things and Smart Grid. Electronic Industry Press, Beijing (2012)
2. Jin, Y., Zhang, R.: Study and development of multi-parameter detection system for SF6 outdoor ring main unit. Process Autom. Instrum. **35**(2), 81–84 (2014)
3. Jin, Q., Li, F., Gu, L., et al.: Design and realization of new intelligent switchgear. East Chin. Electr. Power **42**(7), 1373–1376 (2014)
4. Kang, J., Zong, B., Zhang, D., et al.: Design of the online monitoring system for a network cable branch box. Process Autom. Instrum. **38**(1), 93–96 (2017)
5. Zhang, Y., Wu, J., Li, J.: Research on remote temperature monitoring system of the power cable connector based on Zigbee. Electr. Meas. Instrum. **51**(16), 103–107 (2014)
6. You, X.C., Lei, L.I.: Patrol inspection system of underground cable joint fault based on wireless communication. Mach. Tool Hydraul. **42**(16), 125–128 (2014)
7. Chen, G., Li, J., Chen, Y., et al.: Data-driven and interactive fault diagnosis of distribution switches. Adv. Technol. Electr. Eng. Energy **38**(3), 10–17 (2019)
8. Li, X., Zhang, Q., Fan, W., et al.: Fault detection method for ring main unit based on KPCA algorithm. Ind. Saf. Environ. Prot. **46**(8), 22–26 (2020)
9. Zhang, Y., Yang, T., Meng, G.: Review and prospect of ubiquitous power Internet of Things in smart distribution system. Electr. Power Constr. **40**(6), 1–12 (2019)

Mechanical Fault Diagnosis of Circuit Breaker Based on Improved Grey Relational Analysis

Xiaohuan Wu[1,2(✉)], Pengcheng Wang[1,2], Liqun Wang[2], Yinfei Xu[2], and Zhongwei Zhao[3]

[1] State Grid Hangzhou Power Supply Company, Hangzhou 310000, Zhejiang, China
[2] Hangzhou Power Equipment Manufacturing Co., Ltd., Hangzhou 310000 Zhejiang, China
[3] School of Information Technology, Zhejiang Gongshang University, Hangzhou 310000, Zhejiang, China

Abstract. To improve the extraction efficiency of the mechanical fault features of the circuit breaker and the recognition of the fault type. First, extract the 20-dimensional time-domain characteristics of the mechanical vibration signal waveform of the circuit breaker to form a fault feature set, and form a reference matrix of the standard pattern of 4 types of mechanical faults of the circuit breaker, including the normal state. Then, the entropy weight method is used to obtain the objective weight of the time domain feature, and it is multiplied by the gray correlation coefficient of the corresponding time-domain feature. The correlation degree is obtained and sorted by the summation, and the identification result of the mechanical fault type of the circuit breaker is obtained. Finally, the actual measurement data experiment proves the method's effectiveness in improving the feature extraction efficiency and the accuracy of circuit breaker mechanical fault type identification.

Keywords: Circuit Breaker · Mechanical Fault Diagnosis · Time Domain Characteristics · Improved Gray Correlation Analysis · Gray Correlation Coefficient

1 Introduction

High-voltage circuit breakers (HVCB) are related to the reliability of modern power networks [1]. Mechanical failure is factor affecting the reliability of the circuit breaker. By analyzing the vibration signal generated by the HVCB action, various mechanical failures such as insufficient spring energy storage and screw loosening can be found [2].

Traditional signal processing methods carry out time-domain and frequency-domain analysis on the original signal and extract time-frequency domain information. It mainly includes Empirical Mode Decomposition (EMD) [3], wavelet packet analysis method [4], dynamic time warping method [5], short-term energy analysis method [6], zero-phase filtering method [7] Wait. Although some results have been achieved, they all have certain shortcomings. For example, EMD is prone to problems such as over-envelope, under-envelope, and modal aliasing; The dynamic time warping method and the short-term

S.-C. Chu et al. (Eds.): ICGEC 2021, LNEE 833, pp. 679–689, 2022.
https://doi.org/10.1007/978-981-16-8430-2_61

energy analysis method are only sensitive to changes in the signal over time; Traditional time-frequency analysis methods such as wavelet packet are not adaptive, and there are difficulties in the selection of basis functions, and the time and frequency resolution cannot be balanced. At the same time, the processing process of the above method is relatively complicated, and the time complexity is high, which is not suitable for the actual situation of modern power network.

The time domain features can be extracted directly from the original vibration signal to analyze the HVCB fault status. Standard methods of fault diagnosis include Support Vector Machine (SVM) [7, 8], Neural Networks (NNs) [9], and extreme learning machines [10, 11]. Neural networks and support vector machines can identify known multiple types of states with high accuracy. However, due to the many types of mechanical failures of HVCB, the few actions, and the high cost of obtaining failure samples in experiments, it is difficult to get training samples that include all types of failures [12].

They combine the above factors with improving the feature extraction efficiency and the recognition effect of the fault type. First, the 20-dimensional features of the mechanical vibration signal waveform of the circuit breaker are extracted to form a feature set, and the correlation between the reference sequence and the comparison sequence is calculated using the improved gray correlation analysis model and sorted. Then, a reference matrix of standard patterns of 4 types of mechanical failures of circuit breakers, including normal conditions, is formed. A comparison sequence of test samples of mechanical failures of circuit breakers is created. Finally, the actual measurement data is used to verify the new method's effectiveness in the mechanical fault diagnosis of circuit breakers.

2 HVCB Vibration Signal Feature Processing Method Based on Time Domain Features

2.1 Actual Measurement Signal Acquisition System Based on Virtual Instrument

The LW9-72.5 type SF6HVCB using a spring operating mechanism is the object of analysis. Using a piezoelectric accelerometer and NI 9234 data acquisition card produced by National Instruments (NI) to build a vibration signal acquisition system, the system is shown in Fig. 1. The signal sampling rate is 25.6 kS/s. The coordinate zero points of the sampled signal are when the circuit breaker is about to act (the trigger card sends a signal acquisition instruction to the signal acquisition card). Among them, the looseness of the base screw simulates the degree of failure by setting the tightness of the base screw, iron core jamming, and insufficient lubrication, and setting different degrees of resistance. Three kinds of failure samples contain data of varying failure degrees [13].

Part of the measured signal is shown in Fig. 2.

The mechanical fault types and corresponding labels of the circuit breaker considered by this method are shown in Table 1, including normal state, iron core jammed state, screw loose state, and poor lubrication state.

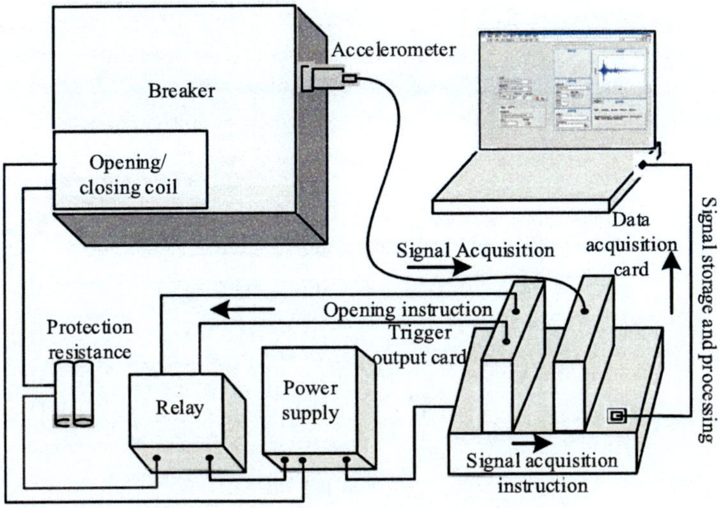

Fig. 1. Vibration signal acquisition system

2.2 HVCB Fault Feature Extraction Based on Time-Domain Features of Original Signals

The key to the recognition effect of the mechanical fault type of the circuit breaker lies in the feature extraction of the mechanical vibration signal of the circuit breaker. Compared with time-frequency analysis and other methods, the feature extraction method of time-domain analysis can reflect the change of signal amplitude, the degree of oscillation attenuation and the time complexity is significantly reduced, meeting the requirements of low-cost terminal edge computing. For example, it can be seen from Fig. 2 that compared with the standard signal; it can be seen from the original signal that the action of the core jamming fault is delayed; the amplitude of the base screw loosening fault is too small, and the attenuation process is slow; the amplitude of the insufficient lubrication of the arm The value is relatively small. Therefore, the differences can be characterized by time-domain characteristics, and the type of mechanical fault of the circuit breaker can be identified. For this reason, the 20-dimensional time-domain characteristics of the mechanical vibration signal waveform of the circuit breaker constitute a feature set. Table 2 lists each feature category and its feature calculation formula [14].

3 Identification of Mechanical Fault Types of Circuit Breakers Based on Improved Grey Relational Analysis

3.1 Traditional Grey Relational Analysis

Regarding identifying the mechanical fault type of the circuit breaker as a gray system, use the calculated correlation degree to determine the automatic fault type of the circuit breaker. Suppose the circuit breaker mechanical failure test sample has the most significant correlation with a specific circuit breaker failure type. In that case, the circuit breaker failure test sample is identified as the circuit breaker mechanical failure type.

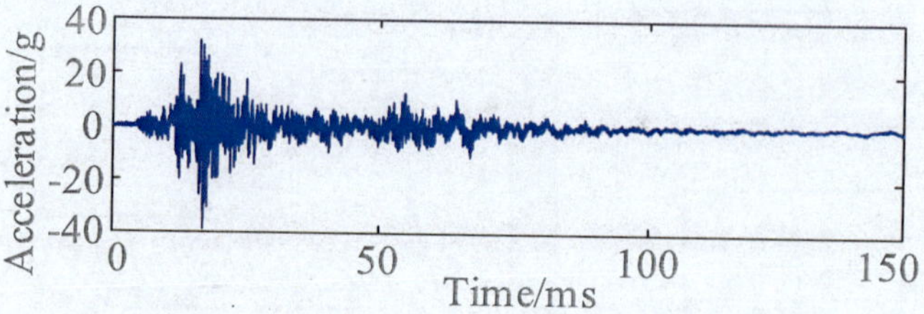

(a) The time domain segmentation of normal signal

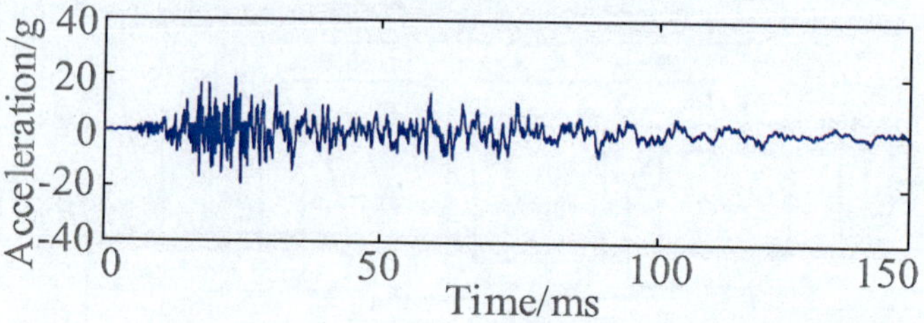

(b) The time domain segmentation of iron core stagnation

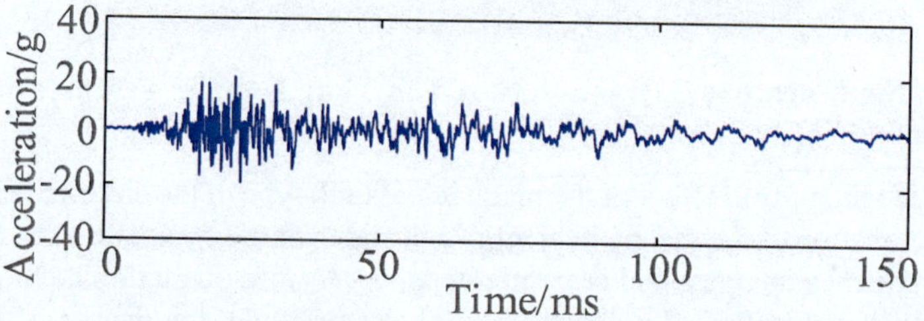

(c) The time domain segmentation of screw loosing

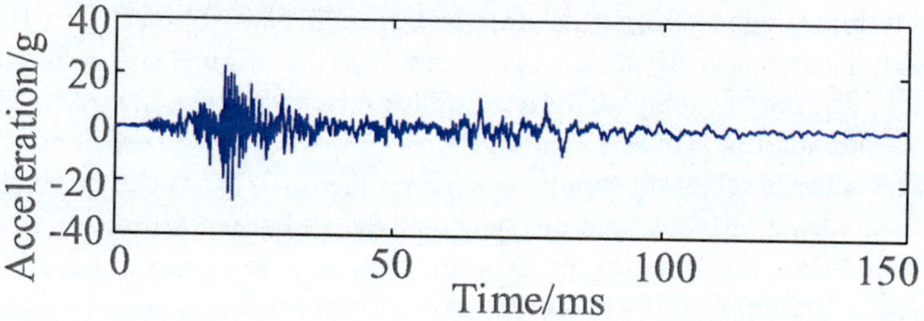

(d) The time domain segmentation of poor lubrication

Fig. 2. The time domain segmentation of vibration

The traditional grey relational analysis calculation process can be divided into the following five steps:

(1) Extract the n-dimensional time-domain features $x_i(j)$ from the mechanical vibration signal samples of known types of circuit breakers to form the reference sequence X_i, and then form the reference matrix R of the standard mode of the mechanical fault type of the circuit breaker. The formed reference sequence is a row vector

Table 1. Breaker mechanical fault types and corresponding labels

Type of mechanical failure	Identification label
Normal status	M1
Core jam state	M2
Screw loose state	M3
Insufficient lubrication state	M4

Table 2. Feature calculation formula

Feature	Formula	Feature	公式
Harmonic mean	$F_{th} = \frac{N}{\sum_{n=1}^{N} \frac{1}{x(n)}}$	Mean	$F_{mv} = \frac{1}{N}\sum_{n=1}^{N} x(n)$
variance	$F_{tv} = \frac{1}{N}\sum_{n=1}^{N}(x(n) - F_{mv})^2$	Root mean square	$F_{rms} = \left(\frac{1}{N}\sum_{n=1}^{N} x(n)^2\right)^{1/2}$
Standard deviation	$F_{std} = \sqrt{\frac{1}{N}\sum_{n=1}^{N}(x(n) - F_{mv})^2}$	Max	$F_{pv} = \max(\lvert x(n)\rvert)$
Average difference	$F_{av} = \frac{1}{N}\sum_{n=1}^{N}\lvert x(n)\rvert$	Minimum	$F_{iv} = \min(\lvert x(n)\rvert)$
Skewness	$F_{sv} = \frac{1}{N}\sum_{n=1}^{N}\left(\frac{x(n) - F_{mv}}{F_{std}}\right)^3$	Difference between maximum and minimum	$F_{ppv} = \max(x(n)) - \min(x(n))$
Kurtosis	$F_{kv} = \frac{1}{N}\sum_{n=1}^{N}\left(\frac{x(n) - F_{mv}}{F_{std}}\right)^4$	Sum of maximum and minimum	$F_{ppv} = \max(x(n)) + \min(x(n))$
entropy	$F_{sh} = -\sum_{n=1}^{N} p_n \log p_n$	energy	$F_{en} = \sum_{k=0}^{N-1} x^2(n)$
Shannon entropy	$F_{se} = -K\sum_{n=1}^{N} p_n \log p_n$	Normal-ized amplitude factor	$F_{fa} = \frac{\max(x(n)) + \min(x(n)) - 1}{2}$
Renyi entropy	$F_{re} = \frac{1}{1-\alpha}\log\sum_{n=1}^{N} p_n{}^{\alpha}$	1/4 cycle energy drop amplitude	$F_{de} = \frac{\min[R_m]}{R_0}$
Tsallis entropy	$F_{te} = -\frac{1}{\alpha - 1}\log(1 - \sum_{n=1}^{N} p_n{}^{\alpha})$	1/4 cycle energy rise	$F_{ss} = \frac{\max[R_m]}{R_0}$

containing 20 time-domain features.

$$R = \begin{bmatrix} X_1 \\ \cdots \\ X_i \\ \cdots \\ X_k \end{bmatrix} = \begin{bmatrix} x_1(1) \cdots x_1(j) \cdots x_1(n) \\ \cdots\ \cdots\ \cdots\ \cdots\ \cdots \\ x_i(1) \cdots x_i(j) \cdots x_i(n) \\ \cdots\ \cdots\ \cdots\ \cdots\ \cdots \\ x_k(1) \cdots x_k(j) \cdots x_k(n) \end{bmatrix} \tag{1}$$

In the formula: i is the number of mechanical faults of the circuit breaker, taking the value 1, 2,…, k ($k = 4$); j is the dimension of the time domain feature, taking the value 1, 2,…, n ($n = 20$).

(2) Extract the n-dimensional time domain features from the mechanical failure test samples of the circuit breaker to form a comparison sequence X_0.

$$X_0 = [x_0(1)\ x_0(2)\ ...\ x_0(j)] \tag{2}$$

(3) Calculate the gray correlation coefficient between the reference sequence X_i and the comparison sequence X_0.
Define the difference between the corresponding values of X_i and X_0 as:

$$\Delta_i(j) = |x_0(j) - x_i(j)| \tag{3}$$

Then the gray correlation coefficient $\zeta_i(j)$ is:

$$\zeta_i(j) = \frac{\min\limits_i \min\limits_j \Delta_i(j) + \rho \max\limits_i \max\limits_j \Delta_i(j)}{\Delta_i(j) + \rho \max\limits_i \max\limits_j \Delta_i(j)} \tag{4}$$

In the formula: $\rho \in [0,1]$ is the resolution coefficient, generally the default value is 0.5.

(4) Calculate the degree of relevance.
By calculating the average value of $\zeta_i(j)$, the correlation degree r_i is obtained.

$$r_i = \frac{1}{n}\sum_{j=1}^{n} \zeta_i(j) \tag{5}$$

(5) Sort the correlation degree r_i, and the one with the largest r_i is the corresponding mechanical fault type of the circuit breaker.

3.2 Improve Grey Relational Analysis

It can be seen from Sect. 3.1 that the correlation degree r_i of the traditional gray correlation analysis model is obtained by weighted average of the gray correlation coefficient $\zeta_i(j)$. That is, the importance of each time domain feature in the feature set is considered to be the same. However, for different types of mechanical faults of circuit breakers, the importance of the same time domain feature may be different. By assigning different weight coefficients to each time domain feature, the accuracy of identifying the mechanical fault type of the circuit breaker is improved.

The entropy method is a method to determine the weight coefficient corresponding to each time-domain feature according to the discrete degree of each time-domain feature of the mechanical vibration signal of the circuit breaker. Compared with human subjective empowerment, it has more credibility, objectivity, and accuracy. The smaller the entropy value of a specific time-domain feature in the feature set, the greater the degree of dispersion of the time-domain feature, the greater the impact on the accuracy of the circuit breaker's mechanical fault type identification, and the greater the weight coefficient should be given. Using the entropy method to revise the formula (5) in the traditional grey relational analysis model, change the calculation method for calculating the relational degree ri, and improve the grey relational analysis model.

The calculation process of the entropy method can be divided into the following four steps:

(1) Data normalization processing.

The original matrix $X = (x_{ij})_{m \times n}$ (m = 20, n = 20) is composed of m samples of mechanical vibration signals of circuit breakers of known types and n-dimensional time-domain features. Normalize $X = (x_{ij})_{m \times n}$ to get $X' = (x'_{ij})_{m \times n}$, and its calculation formula is:

$$x'_{ij} = \frac{x_{ij} - \min(x_j)}{\max(x_j) - \min(x_j)} \tag{6}$$

Where: $\max(x_j)$ and $\min(x_j)$ are the maximum value of the j-th dimension time domain feature and the minimum value of the j-th dimension time domain feature, respectively.

(2) Calculate the proportion of the jth dimension time-domain characteristics of the ith circuit breaker mechanical vibration signal sample.

$$y_{ij} = \frac{x'_{ij}}{\sum_{i=1}^{m} x'_{ij}} \tag{7}$$

(3) The information entropy of the jth dimension time domain feature is calculated.

$$E_j = -\frac{1}{\ln n} \sum_{j=1}^{n} y_{ij} \ln y_{ij} \tag{8}$$

If $y_{ij} = 0$, then $\lim_{y_{ij} \to 0} y_{ij} \ln y_{ij} = 0$.

(4) Calculate the weight coefficient of each time domain feature.

The expression for obtaining the weight coefficient of the jth dimension time-domain feature is:

$$w_j = (1 - E_j) / \sum_{j=1}^{n} (1 - E_j) \tag{9}$$

Therefore, substituting formula (9) into formula (5) in the traditional grey relational analysis model can obtain an improved grey relational analysis, and obtain the degree of relation r_i:

$$r_i = \sum_{j=1}^{n} w_j \zeta_i(j) \tag{10}$$

When calculating the degree of association, the reference sequence should correspond to the weight vector. For example, when calculating the correlation between the comparison sequence and the reference sequence in the normal state, the weight vector $W_1 = (w_j)_{1\times j}$ A corresponding to the normal state should be used. In the same way, the weight vectors $W_i = (w_j)_{1\times j}$ (i = 2, 3, 4) of the other three types of circuit breaker mechanical faults can be determined.

4 Experiment Analysis

In this experiment, 100 circuit breaker mechanical failure sample signals of 4 types of circuit breaker mechanical failure, including normal state, are obtained. In this subject, Gaussian white noise with a signal-to-noise ratio of 20 dB–40 dB is added to the measured signals of the original circuit breaker mechanical failure test samples. Among the four mechanical vibration signal samples of circuit breakers, 20 samples were randomly selected as known samples. The weight vector $W_i = (w_j)_{1\times j}$ corresponding to the mechanical vibration signal samples of various types of circuit breakers is obtained. The standard mode reference matrix R is formed simultaneously. The rest of the data is used as the circuit breaker mechanical test samples. The new method is used to identify the type of mechanical fault of the circuit breaker, and the results are shown in Table 3.

The HVCB vibration signal is a nonlinear and non-stationary signal. The three signal processing methods of EMD, Ensemble Empirical Mode Decomposition (EEMD) [14] and Local Mean Decomposition (LMD) [15], can Its effective decomposition. These methods have been widely used in the field of mechanical fault diagnosis.

To compare the classification capabilities of the new method with the traditional methods based on signal processing to extract time-domain features, three signal processing methods, including EMD, EEMD, and LMD, are used to process vibration

Table 3. Recognition results of mechanical fault types of circuit breakers

Breaker mechanical fault type	Accuracy%			
	No noise	40 dB	30 dB	20 dB
M0	100	99.8	98.8	96.5
M1	98.4	97.6	96.5	93.7
M2	99.7	99.1	97.9	95.4
M3	100	99.5	98.4	95.9
M4	98.4	97.1	95.5	92.9

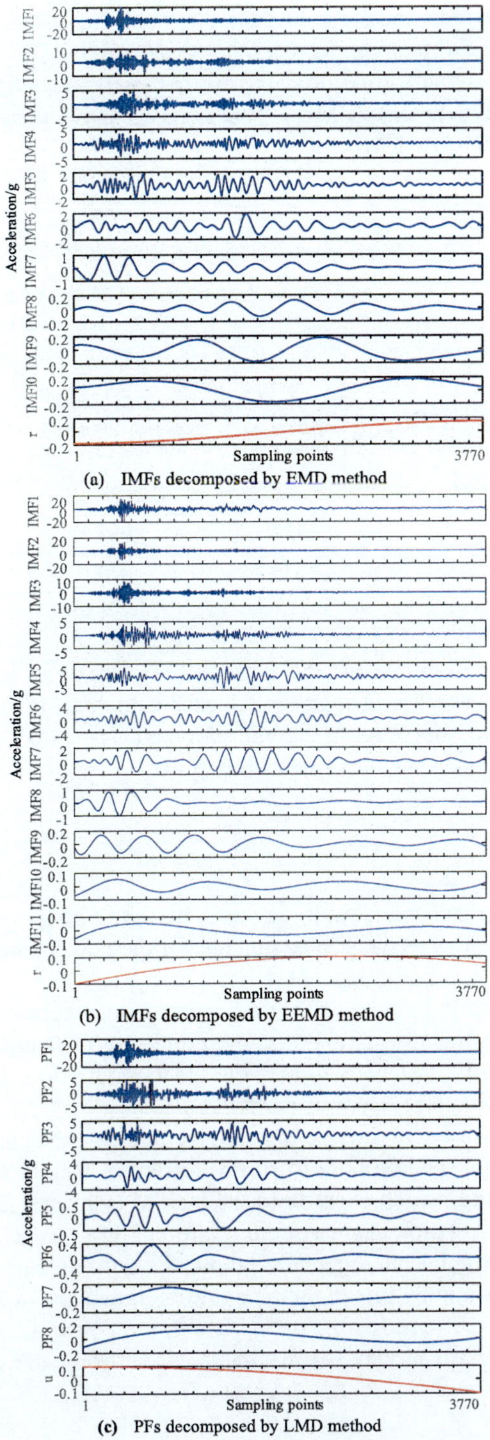

Fig. 3. The time domain segmentation of EMD, EEMD and LMD

signals and extract features, and carry out fault diagnosis experiments. Note: Extract 20-dimensional time-domain features for each layer separately. The time-domain feature dimension of EMD is 220, the time-domain feature dimension of EEMD is 240, and the time domain feature dimension of EMD is 180. Figure 3 shows the signal decomposition results of EMD, EEMD, and LMD.

Table 4 shows the time statistics of different methods for feature extraction and the results of state recognition.

Table 4. The time of feature extraction and diagnosis result

Approach	Average feature extraction time/s	Fault recognition accuracy rate/%
Time domain feature extraction	0.104	99.65
EMD	14.030	86.67
EEMD	17.352	83.33
LMD	2.535	70

It can be seen from Table 4 that the average time to extract features in the direct time domain is 0.204 s, and the average time to extract features in the time domain after EMD, EEMD, and LMD decompose the vibration signal is 14.03 s, 17.352 s, 2.535 s, respectively, and has good classification accuracy.

5 Conclusion

This paper proposes a mechanical fault diagnosis method for circuit breakers based on improved grey relational analysis. The advantages of the new method are:

(1) The feature extraction process of the new method does not require complex time-frequency domain transformation, and the efficiency of feature extraction is significantly improved. Effectively reduce the space required for information storage and reduce the storage pressure of hardware devices. It better meets the needs of real-time analysis of circuit breaker mechanical fault type identification.

(2) The entropy weight method is used to give different weight coefficients to each time-domain feature, which overcomes the lack of equal weight in the process of calculating the correlation degree in the traditional gray correlation model. Compared with the artificial intelligence method, the new method does not have the problems of long training time of the recognition model and difficulty in determining the parameters. The measured data prove that the new method can realize the high-precision identification of the mechanical fault of the circuit breaker.

References

1. Ma, S., Chen, M., Wu, J., et al.: High-voltage circuit breaker fault diagnosis using a hybrid feature transformation approach based on random forest and stacked autoencoder. IEEE Trans. Industr. Electron. **66**(12), 9777–9788 (2018)
2. Yang, Q., Ruan, J., Zhuang, Z., et al.: A new vibration analysis approach for detecting mechanical anomalies on power circuit breakers. IEEE Access **7**, 14070–14080 (2019)
3. Zheng, J., Pan, H.: Mean-optimized mode decomposition: an improved EMD approach for non-stationary signal processing. ISA Trans. **106**, 392–401 (2020)
4. Zhou, H, Wang, H.: A Study on fault diagnosis of rotating machinery combined wavelet transform with VMD. J. Phys. Conf. Ser. **1626**(1), 012136 (2020)
5. Landry, M., Léonard, F., Landry, C., et al.: An improved vibration analysis algorithm as a diagnostic tool for detecting mechanical anomalies on power circuit breakers. IEEE Trans. Power Delivery **23**(4), 1986–1994 (2008)
6. Ma, Q, Rong, M.Z.H., Jia, S.H.L.: Study of switching synchronization of high voltage breakers based on the wavelet packets extraction algorithm and short time analysis method. Proc. CSEE **25**(13), 149–154 (2005)
7. Chang, G., Wang, Y., Wang, W.: Mechanical fault diagnosis of high voltage circuit breakers utilizing zero-phase filter time-frequency entropy of vibration signal. Proc. CSEE **33**(3), 155–162 (2013)
8. Sun, Y.H., Wu, J.W., Lian, S.H.J., et al.: Extraction of vibration signal feature vector of circuit breaker based on empirical mode decomposition amount of energy. Trans. China Electrometrical Soc. **29**(3), 228–236 (2014)
9. You, Z.Y., Wang, N., Li, M., et al.: Method of fan fault diagnosis of gearbox based on EEMD and BP neural network. J. Northeast. Electr. Power Univ. **35**(1), 64–72 (2015)
10. Liu, H., Han, M.: A fault diagnosis method based on local mean decomposition and multi-scale entropy for roller bearings. Mech. Mach. Theory **75**(5), 67–78 (2014)
11. Miao, X.R., Lin, R.C.: Lightning over-voltage fault identification of high-voltage transmission line based on correlation dimension and extreme learning machine. High Voltage Eng. **42**(5), 1519–1526 (2016)
12. Xing, H., Hongli, W., Jinghui, L.: Analog circuit fault diagnosis method based on preferred wavelet packet and ELM. Chin. J. Sci. Instrum. **34**(11), 2614–2619 (2013)
13. Qi, J., Gao, X., Huang, N.: Mechanical fault diagnosis of a high voltage circuit breaker based on high-efficiency time-domain feature extraction with entropy features. Entropy **22**(4), 478 (2020)
14. Gao, W., Qiao, S.P., Wai, R.J., et al.: A newly designed diagnostic method for mechanical faults of high-voltage circuit breakers via SSAE and IELM. IEEE Trans. Instrum. Meas. **70**, 1–13 (2020)
15. Wang, H., Wu, F., Zhang, L.: Fault diagnosis of rolling bearings based on improved empirical mode decomposition and fuzzy c-means algorithm. Traitement du Signal **38**(2), 395–400 (2021)

A Matrix-Analytic Solution to Three-Level Multi-Server Queueing Model with a Shared Finite Buffer

Qigen Zhao[1], Chia-Hung Wang[1,2](✉), Zhenyu Dong[1], Shumeng Chen[1], Qipeng Yang[1], and Yi Wei[1]

[1] College of Computer Science and Mathematics, Fujian University of Technology, Fuzhou, China

[2] Fujian Provincial Key Laboratory of Big Data Mining and Applications, Fujian University of Technology, Fuzhou, China

Abstract. A computing method is presented in this paper for three-level multi-server queues with a shared finite buffer. We derive a matrix-analytic method to determine the steady-state probability for the studied queueing model, which is formulated as a state-dependent Quasi-Birth-and-Death (QBD) process. In addition, a solving algorithm is also developed to calculate the expected waiting time and average queue size. In our numerical experiments, it illustrates the relationship between the system performance and model parameters of interest. The findings in our work could reveal the managerial insights into the stochastic process of a class of queueing systems with three-type heterogeneous servers and a common buffer.

Keywords: Queueing system · Finite capacity · Matrix-analytic method · Markovian process · Routing probability · Embedded Markov chain

1 Introduction

Queueing systems have been commonly applied in many fields, for example, at computer systems [1,2], in telecommunication networks [3,4], in security checkpoints [5–8], etc. For a higher comprehension of the dynamic behavior of the concerned queueing processes, we may need to handle the formulation of mathematical models to give clear explanation and concise description of the stochastic service of randomly arriving requests [9–11]. In queueing theory, the matrix-analytic method is one of the most frequently used approaches to analyze such queueing systems [12,13]. Specifically, it depends on identifying the initial/boundary part and the repetitive part within the structure of the underlying Markov chain [14,15].

The aim of this paper is to derive a matrix-analytic method for the performance evaluation of a queueing system with three-type heterogeneous servers and a common buffer. In the current work, we are going to present an analytic

S.-C. Chu et al. (Eds.): ICGEC 2021, LNEE 833, pp. 690–699, 2022.
https://doi.org/10.1007/978-981-16-8430-2_62

method to evaluate a three-level multi-server queueing system with a shared finite buffer. Besides, we will derive the steady-state probability which is a common basis of evaluating various measures of performance. The main contribution of our paper is to present a matrix-analytic solution to the stochastic process of the concerned queueing system.

The rest of this paper is structured as follows. In the next section, we will introduce the definitions and system description of the studied queueing model. In Sect. 3, we are going to derive a queueing analysis based on a matrix-analytic method. In Sect. 4, we will conduct a series of numerical experiments to evaluate the system performance when varying the model parameters. Lastly, our concluding remarks will be summarized in Sect. 5.

2 A Three-Level Multi-server Queueing Model with Routing Probability and Finite Buffer

The queueing system studied in the present work is shown in Fig. 1, where there are three-type heterogeneous servers and a shared finite buffer. It is assumed that customers arrive to the queueing system according to a stationary Poisson process with average rate λ. Besides, we assume that the incoming customer would be lost if the finite buffer is full, and the capacity of finite buffer is denoted as N. That is, an incoming arrival would be blocked when all N waiting rooms are occupied. Otherwise, when a customer arrives to the buffer where there is an available waiting space, the customer can be admitted immediately and join a queue waiting for service.

For each k-th level of service, we represent the number of servers as c_k, for $k = 1, 2, 3$. In other words, at most c_k customers can be served for the k-th level of service at the same time, for each $k = 1, 2, 3$. An incoming customer would be assigned to the k-th level for service with a routing probability p_k, for each $k = 1, 2, 3$. According to the law of total probability [16], it is easy to obtain

$$p_1 + p_2 + p_3 = 1. \tag{1}$$

For each server in the k-th level, $k = 1, 2, 3$, we consider customers' service times are exponentially distributed with parameter μ_k. Finally, it is given that the service discipline is First-In-First-Out. In Table 1, we summarize all model parameters for the proposed three-level multi-server queueing system with a shared finite buffer.

3 A Matrix Analytic Method

3.1 A Queueing Analysis

Based on the model assumptions, the arrival process follows Poisson distribution, and the service times are exponential distributed. Thus, the stochastic process $\{(q_3(t), q_2(t), q_1(t)) \mid t \geq 0\}$ can be derived as a continuous time Markov

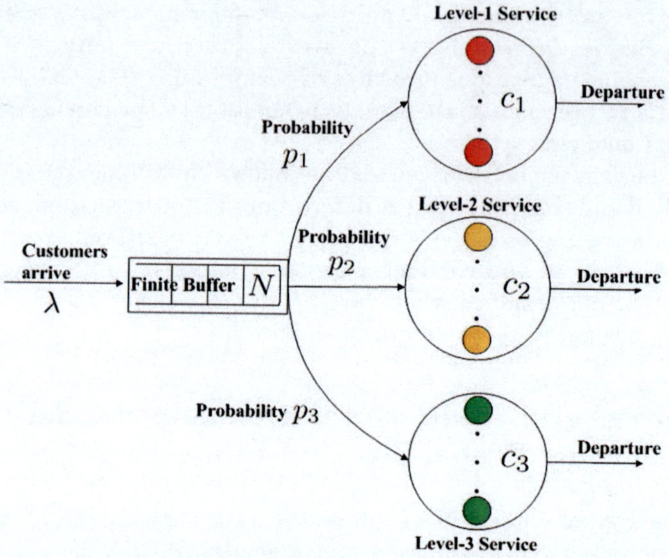

Fig. 1. A three-level multi-server queueing model with routing probability and finite buffer.

chain. Besides, we also know that the queue length $\{q_3(t), q_2(t), q_1(t)\}$ is limited, because

$$q_1(t) + q_2(t) + q_3(t) \leq N + c_1 + c_2 + c_3.$$

Moreover, the state space Ω is also limited. Next, we can construct the infinitesimal generator of the studied Markov chain as follows:

$$\mathbf{Q}_N^* = \begin{pmatrix} \mathbf{A}^*_{(0,0)} & \mathbf{A}^*_{(0,1)} & & & \\ \mathbf{A}^*_{(1,0)} & \mathbf{A}^*_{(1,1)} & \mathbf{A}^*_{(1,2)} & & \\ & \ddots & \ddots & \mathbf{A}^*_{(N+c_3-1,N+c_3)} \\ & & \mathbf{A}^*_{(N+c_3,N+c_3-1)} & \mathbf{A}^*_{(N+c_3,N+c_3)} \end{pmatrix}, \tag{2}$$

where those submatrices $\mathbf{A}^*_{(i,i+1)}$, $\mathbf{A}^*_{(i,i-1)}$ and $\mathbf{A}^*_{(i,i)}$ indicate that the number of customers $q_3(t)$ increases by 1, $q_3(t)$ decreases by 1, and the queue length remains unchanged, respectively.

It is not easy to find those submatrices in $\mathbf{Q}_N^*$. The main challenge to determine them results from the problem that the number of states would be different for different levels. Luckily, it can be observed that the number of states in each level depends on the capacity of finite buffer in other level service. Based on this observation, we can derive a recursive method to construct those submatrices blocks in the infinitesimal generator $\mathbf{Q}_N^*$.

Let $\boldsymbol{\pi} = (\boldsymbol{\pi}_0, \boldsymbol{\pi}_1, \ldots, \boldsymbol{\pi}_{N+c_3})$ denote the stationary distribution of Markov chain $\{(q_3(t), q_2(t), q_1(t)) \mid t \geq 0\}$, where the row vector $\boldsymbol{\pi}_i$ represents as the steady-state probability, for $0 \leq i \leq c_3 + N$. Because the studied Markov chain

Table 1. Notations for the three-level multi-server queueing model

Symbol	Definition of model parameters
N	The capacity of finite buffer
λ	The average arrival rate of customers to the queueing system
p_k	The routing probability that an incoming customer is sent to the k-th level queue for service, for $k = 1, 2, 3$
c_k	The number of servers for the k-th level service, for $k = 1, 2, 3$
μ_k	The average service rate for each k-th level server, for $k = 1, 2, 3$
$q_k(t)$	The number of customers in service or waiting for the k-th level service at time t, for $k = 1, 2, 3$
$E[q_k]$	The average number of customers waiting in the k-th level service queue, for $k = 1, 2, 3$
$E[w_k]$	The average waiting time spent by customers at the k-th level service, for $k = 1, 2, 3$

is irreducible, there exists only non-negative solution, $\boldsymbol{\pi}$, to solve the following simultaneous equations:

$$\boldsymbol{\pi} \cdot \mathbf{Q}_N^* = \mathbf{0} \quad \text{and} \quad \boldsymbol{\pi} \cdot \mathbf{e} = 1, \tag{3}$$

where the column vector $\mathbf{e}$ consists of elements all equal to 1.

3.2 Performance Measure

Here, we are going to obtain the matrix-geometric solution by taking advantage of a tri-diagonal block structure in the infinitesimal generator $\mathbf{Q}_N^*$. First of all, for $N + c_3$ and $N + c_3 - 1$, we determine

$$\boldsymbol{\pi}_{N+c_3} = \boldsymbol{\pi}_{N+c_3-1} \cdot \mathbf{R}_{N+c_3}, \tag{4}$$

where it is given that

$$\mathbf{R}_{N+c_3} = -\mathbf{A}^{(*)}_{(N+c_3,N+c_3)} \cdot (\mathbf{A}^{(*)}_{(N+c_3,N+c_3)})^{-1}. \tag{5}$$

Next, for each $i = 1, 2, \ldots, N + c_3 - 1$, we can determine

$$\boldsymbol{\pi}_i = \boldsymbol{\pi}_{i-1} \cdot \mathbf{R}_i, \tag{6}$$

where it holds that

$$\mathbf{R}_i = -\mathbf{A}^{(*)}_{(i-1,i)} \cdot (\mathbf{A}^{(*)}_{(i,i)} + \mathbf{R}_{i+1} \cdot \mathbf{A}^{(*)}_{(i+1,i)})^{-1}. \tag{7}$$

By making use of (6), we can determine $\boldsymbol{\pi}$ if we find the zero-level $\boldsymbol{\pi}_0$. In order to find $\boldsymbol{\pi}_0$, we need to solve the following system consists of balance equations and normalization equation:

$$\begin{cases} \boldsymbol{\pi}_0 \cdot (\mathbf{A}^{(*)}_{(0,0)} + \mathbf{R}_1 \cdot \mathbf{A}^{(*)}_{(1,0)}) = \mathbf{0} \\ \boldsymbol{\pi}_0 \cdot (\mathbf{e} + \mathbf{R}_1 \cdot \mathbf{e} + \mathbf{R}_1 \cdot \mathbf{R}_2 \cdot \mathbf{e} + \mathbf{R}_1 \cdot \ldots \cdot \mathbf{R}_{N+c_3} \cdot \mathbf{e}) = \mathbf{1} \end{cases}. \tag{8}$$

Many performance indicators can be derived directly by the probability $\boldsymbol{\pi}$. First of all, we focus on the parameters of the third-level queue. Next, the performance indicators of other queues can also be obtained by means of $\boldsymbol{\pi}$.

Definition 1. *The* ***average number of customers in the third-level queue*** *is determined as*

$$E\left[q_3\right] = \sum_{i=0}^{N+c_3} i \cdot \boldsymbol{\pi}_i \cdot \mathbf{e}, \tag{9}$$

where $\mathbf{e}$ *is the column vector of ones.*

Definition 2. *The* ***loss probability*** *is determined as*

$$P_{L_k} = \sum_{(i_1,i_2,\ldots,i_k)\in\Omega:\max\{0,i_j-c_j\}=N(j=1,2,\ldots,k)} \pi_{i_1,i_2,\ldots,i_k}. \tag{10}$$

Definition 3. *The* ***probability that there are*** i ***customers in the*** k***-th service*** *is calculated as*

$$\eta_i(k) = \frac{1}{(1 - P_{L_k})} \sum_{i\in\Omega:\max\{0,i-c_k\}} \boldsymbol{\pi}_i (0 \le i \le c_k + N - 1), \tag{11}$$

for each $k = 1, 2, 3$. *It reflects the effective arrivals to the* k*-th level service after considering the loss probability.*

Definition 4. *The* ***average waiting time*** *is determined as*

$$E\left[w_k\right] = \sum_{i=1}^{N} \frac{i \cdot \eta_{c_k-1+i}(k)}{c_k \cdot \mu_k}, \forall k = 1, 2, 3, \tag{12}$$

where $\eta_{c_k-1+i}(k)$ *can be determined from Eq. (11).*

In Formula (12), the numerator stands for the average number of customers waiting in the k-th queue per unit time, and the denominator indicates the expected number of customer to be served at the k-th queue per unit time.

4 Sensitivity Analysis

In this section, we conduct a series of numerical experiments to figure out the applicability of the proposed queueing model. Our numerical results are calculated with Matlab on the computer composed of 16 GB memory and the processor Intel (R) Core (TM) i7-6500U CPU@2.50 GHz dual-core.

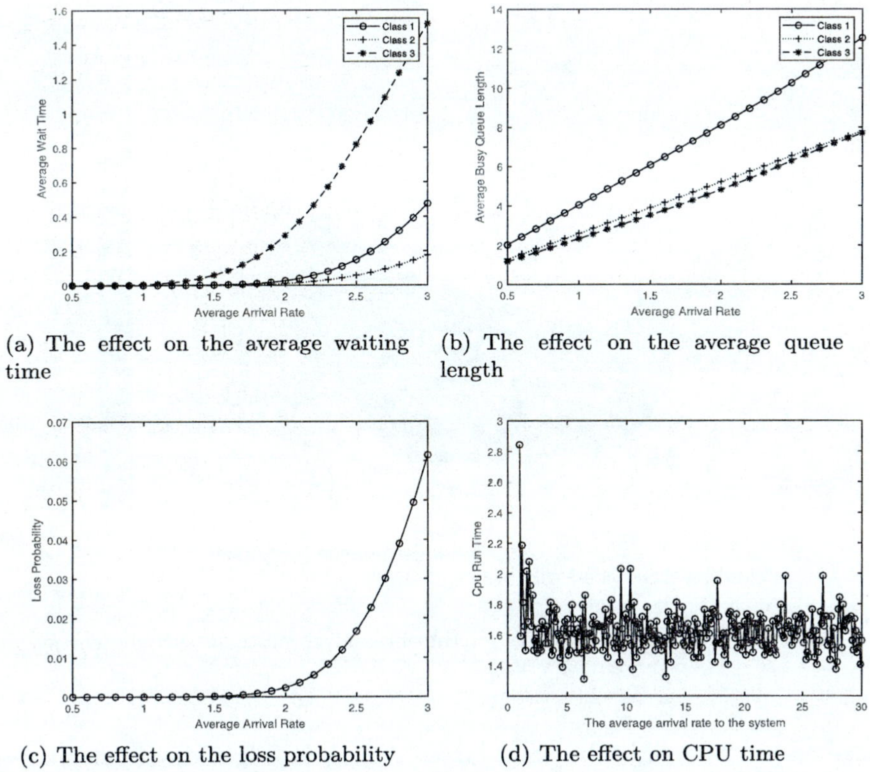

(a) The effect on the average waiting time

(b) The effect on the average queue length

(c) The effect on the loss probability

(d) The effect on CPU time

Fig. 2. Performance evaluation of the queueing system when varying the average arrival rate λ.

4.1 Case Study 1

In the first case, we present an sensitivity analysis of the proposed three-level multi-server queueing system by means of varying the average arrival rate. Here, the average arrival rate λ is going to be varied from 0 to 3 customers per hour. The model parameters used in this numerical experiment are given as follows: $c_1 = 12$, $c_2 = 15$, $c_3 = 8$, $\mu_1 = \mu_2 = \mu_3 = 1/9$, $N = 6$, $p_1 = 0.45$, $p_2 = 0.29$, and $p_3 = 0.26$.

The numerical results of this case study are depicted in Fig. 2. From Fig. 2(a), it is observed that the average waiting time gradually increases when increasing the average arrival rate. The increase in the third-level service is more significant, because the capacity of the third-level service is the smallest. Next, it shows in Fig. 2(b) that the average queue length of the three level services gradually increases when we increase the average arrival rate. Besides, the average queue length of the third-level service is higher than that of the other two level services, which is due to the smallest capacity for the third-level service. In Fig. 2(c), when we increases the average arrival rate from 0 to 2, it is found that the loss probability first increases slowly. Nevertheless, it increases significantly if the

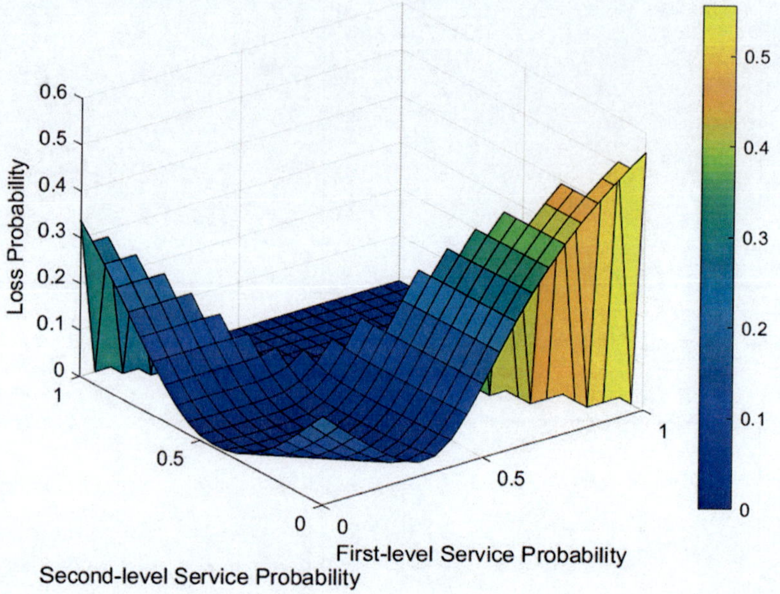

Fig. 3. Effect on total loss probability for the three-level multi-server queueing system

average arrival rate varies becomes larger than 2. In Fig. 2(d), it can be seen that the CPU time is almost the same even if we increase the average arrival rate. The running time for each experiment in this cast study is around 24.11 s.

4.2 Case Study 2

Next, in this numerical experiment, the average arrival rate is fixed as 3 (customers/hour), and vary the values of p_1 and p_2 from 0 to 1. In this numerical experiment, we set up the following parameters: $c_1 = 8$, $c_2 = 12$, $c_3 = 15$, $\lambda = 3$, $\mu_1 = \mu_2 = \mu_3 = 1/6$, $N = 6$, and $p_3 = 1 - p_1 - p_2$.

We show the numerical results in Fig. 3 and Fig. 4, where the running time is around 129.03 s. In Fig. 3, we observe that the loss probability is significantly higher than that of other level services when the probability of being admitted to the first level service increases. The reason is that the number of servers for the first level service is small. Next, it can also be figured out that the average number of customers waiting in the concerned queueing system is also increasing if we increase the routing probability of entering the level service. Those phenomena can also be observed in Figs. 4(a), 4(b), and 4(c), individually. From Figs. 4(d), 4(e), and 4(f), we find that the average waiting time would increase when we increase the routing probability of being admitted to a specific level service.

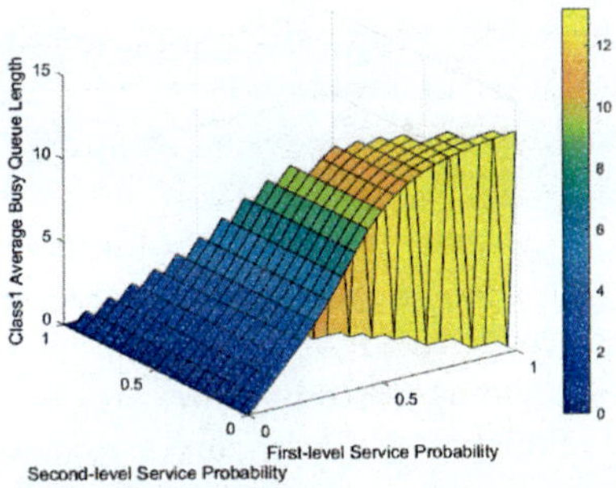

(a) The effect on the average queue length in the first level service

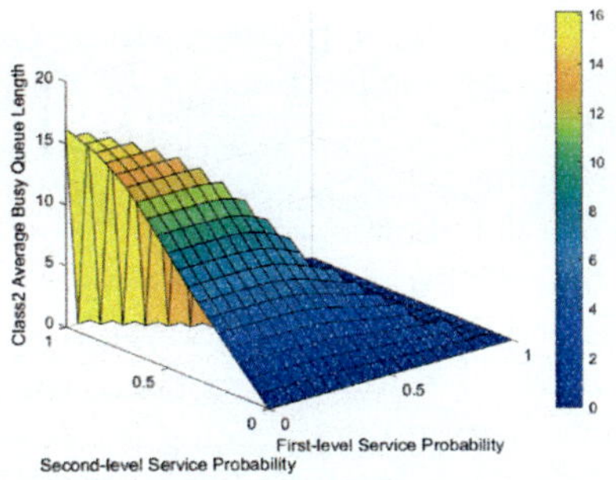

(b) The effect on the average queue length in the second level service

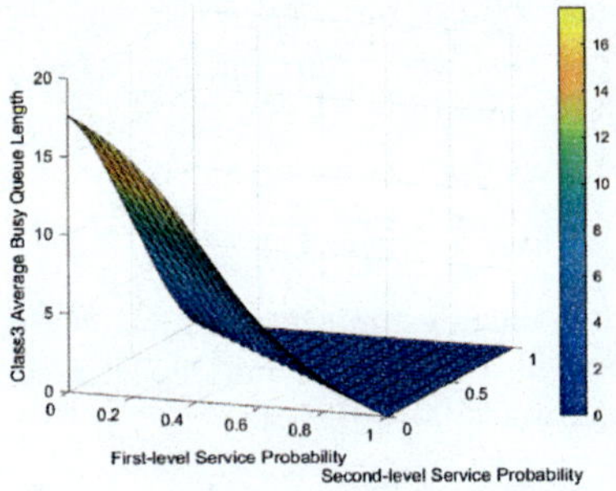

(c) The effect on the average queue length in the third level service

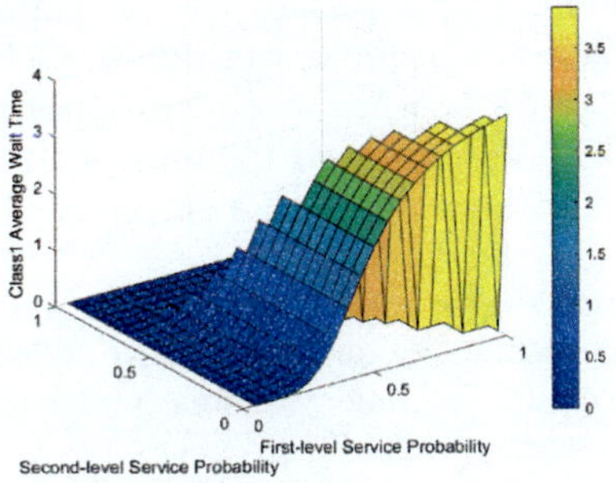

(d) The effect on the average waiting time in the first level service

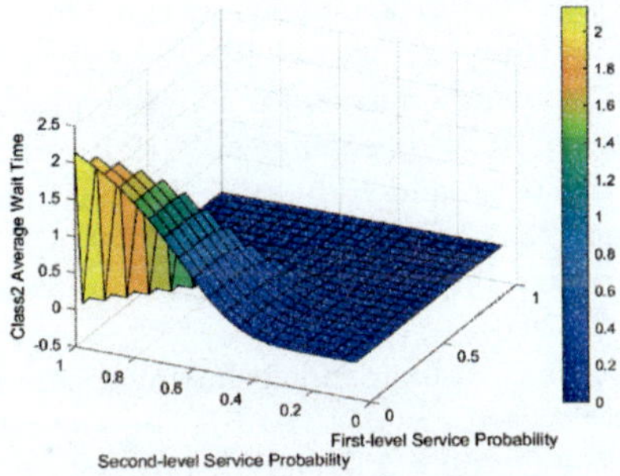

(e) The effect on the average waiting time in the second level service

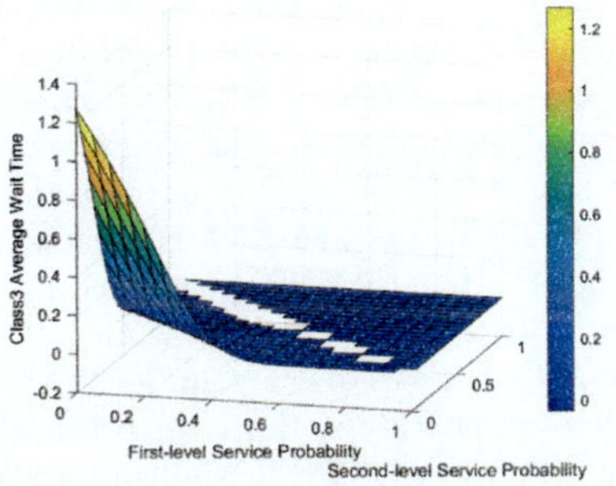

(f) The effect on the average waiting time in the third level service

Fig. 4. Performance evaluation of the three-level multi-server queueing system

5 Conclusions

In summary, we investigated a three-level multi-server queueing model with a shared finite buffer. For the concerned queueing model, we derived the steady-state probability by formulating a state-dependent QBD process. In this paper, we conducted the queueing analysis based on a matrix-analytic method for determining the expected system performance. Moreover, several characteristics revealing the relationship between the model parameters and system performance of interest were also figured out by making use of a series of numerical experiments.

The contribution of our work is to deliver a computing method to determine the steady-state probability and expected system performance of a three-level multi-server queueing model. Besides, it also contributes a more clear comprehension of the operational process of the studied queueing system. Thus, our works could be helpful in improving the operational efficiency of multi-server queueing systems with three-type heterogeneous service and a shared finite buffer.

Finally, more efforts could be devoted in the future study to improve the algorithmic efficiency of the computing method presented in this paper. For a potential application, the presented approach could be applied in handling the high-dimensional and non-negative matrices resulted from industrial applications. As another extended application, if we consider all rejected customers could be re-addressed for service, it would be possible to extend our queueing model with an additional retrial pool having an infinite waiting buffer.

Acknowledgment. This work is supported by the research fund from Fujian University of Technology under Grant No. GY-Z18148, and Fujian Provincial Department of Science and Technology, China, under the grant number 2021J011070.

References

1. Liang, C.-C., Wang, C.-H., Luh, H., Hsu, P.-Y.: Disaster avoidance mechanism for content-delivering service. Comput. Oper. Res. **36**(1), 27–39 (2009)
2. Wang, C.-H., Chen, Y.-T., Hwang, C.-C.: A queueing analysis for job assignment on two-type heterogeneous supercomputer system. In: Proceedings of 2015 International Conference on Computer Information Systems and Industrial Applications (CISIA2015), pp. 375–378, Bangkok, Thailand, 28–29 June 2015
3. Wang, C.-H., Luh, H.P.: Estimating the loss probability under heavy traffic conditions. Comput. Math. Appl. **64**(5), 1352–1363 (2012)
4. Wang, C.-H.,Wu, M.-E., Chung, W.-H.: Perspectives of bandwidth sharing schemes in communication systems with blocking. In: Proceedings of the ASE Big Data & Social Informatics 2015, Kaohsiung, Taiwan, 7–9 October 2015. http://dx.doi.org/10.1145/2818869.2818899
5. Nie, X., Parab, G., Batta, R., Lin, L.: Simulation-based selectee lane queuing design for passenger checkpoint screening. Eur. J. Oper. Res. **219**, 146–155 (2012)
6. de Lange, R., Samoilovich, I., van der Rhee, B.: Virtual queuing at airport security lanes. Eur. J. Oper. Res. **225**, 153–165 (2013)
7. Wang, C.-H.: A three-level health inspection queue based on risk screening management mechanism for post-COVID global economic recovery. IEEE Access **8**, 177604–177614 (2020). https://doi.org/10.1109/ACCESS.2020.3026786
8. Wang, C.-H., Chen, Y.-T., Wu, X.J.: A multi-tier inspection queueing system with finite capacity for differentiated border control measures. IEEE Access **9**, 60489–60502 (2021). https://doi.org/10.1109/ACCESS.2021.3073470

9. Wang, C.-H.: A modelling framework for managing risk-based checkpoint screening systems with two-type inspection queues. In: Proceedings of the Third International Conference on Robot, Vision and Signal Processing (RVSP 2015), pp. 220–223, Kaohsiung, Taiwan, 18–20 November 2015
10. Wang, C.-H.: Arena simulation for aviation passenger security-check systems. In: Pan, J., et al. (eds.) Advances in Intelligent Systems and Computing, vol. 536, Chapter 12, pp. 95–102 (2016). https://doi.org/10.1007/978-3-319-48490-712
11. Wang, C.-H. and Wu, X.J.: Performance analysis of a security-check system with four types of inspection channels for high-speed rail stations in China. In: H. Yang et al. (eds.), Smart Service Systems, Operations Management, and Analytics. 2019 INFORMS International Conference on Service Science (ICSS2019), 27–29 June 2019, Nanjing, China. Springer Proceedings in Business and Economics, pp. 7–16 (2020). https://doi.org/10.1007/978-3-030-30967-1_2
12. Neuts, M.F.: Matrix-Geometric Solutions in Stochastic Models. The John Hopkins University Press (1981)
13. Latouche, G., Ramaswami, V.: Introduction to Matrix Analytic Methods in Stochastic Modelling. ASA & SIAM, Philadelphia, USA (1999)
14. Wang, C.-H.: A matrix-analytic method for evaluating performance of security-check system with two-type inspection queues. In: Proceedings of the 2019 International Conference on Industrial Engineering and Systems Management (IESM 2019), IEEE Xplore, Shanghai, China, 25–27 September 2019
15. Lin, Z., Liu, C., Wang, C.-H.: Performance evaluation of job queue management for a computer center with two heterogeneous clusters. J. Network Intell. **6**(4), 873–885 (2021)
16. Gross, D., Harris, C.M.: Fundamentals of Queueing Theory, 3rd edn. John Wiley & Sons, Inc., New York (1998)

Check for updates

A Novel Learning-Based Approach for Large Outdoor Power Transmission Equipment Detection

Fatana Jafari[1], Rui Xu[1], Gilbert Shyirambere[1], Jean Bosco Ndungutse[1], and Shi-Jian Liu[1,2](✉)

[1] School of Computer Science and Mathematics, Fujian University of Technology, Fuzhou, China

[2] Key Laboratory of Big Data Mining and Applications of Fujian Province, Fujian University of Technology, Fuzhou, China

Abstract. Automatic inspection of large outdoor power transmission equipment via Unmanned Arial Vehicle (UAV) is a significant way to be sure to promote electrical reliability and safety. As a fundamental work of automated inspection, detection of power towers and insulators from aerial images are studied in this research using a deep learning method. Firstly, the aerial images of power towers and insulators with complex backgrounds are collected and well prepared. Secondly, a convolutional neural network based on the latest version of YOLO is trained using the prepared dataset. As the result, the trained model can be used to fulfil the automatic detection. Experiments proved that the proposed network can precisely identify towers and insulators simultaneously and has better performance than existed method.

Keywords: Power tower · Insulator · Deep learning · Object detection · YOLO

1 Introduction

Large outdoor power transmission and repartition network equipment such as power towers and insulators must be reviewed periodically or even continuously to assure the persistent source of electrical power and the great execution of equipment [1] and to prohibit power deficiency and serious disasters. For instance, detection of errors of power equipment at a timely stage can be avoided severe and expensive destruction, even we able to anticipate upcoming variations. Mostly, the crucial troubles in power transmission equipment because of numerous issues such as insistent process underneath electrical and mechanical pressure, tough ecological circumstance, and natural disasters can be wrecked wires, impairment to power tower, insulators, conductor rust.

In the past, such supervision and monitoring are majorly carried out by humans via approaches like manually inspecting and helicopter-assisted inspection, however, both of them have their difficulties, which are extremely endangering the human life, time-consuming, overpriced, and could be failed occasionally. With these problems in mind, manufacturing is enthusiastically looking for resolutions to computerize diverse parts of

S.-C. Chu et al. (Eds.): ICGEC 2021, LNEE 833, pp. 700–711, 2022.
https://doi.org/10.1007/978-981-16-8430-2_63

power transmission equipment to accomplish the inspection automatically without the intervention of a human. To overcome these problems, Unmanned Aerial Vehicles are coming up with alternative resolutions.

In recent years, Unmanned Aerial Vehicles (UAVs) [2] and robots [3] have been paid great attention by the power transmission sectors. which are popularly used to liberate human resources from surveillant data acquisition. For example, the applications of using UAVs for power line and Pylon monitoring are both reported in the work of Zhang et al. [4] and Fang et al. [5]. Advantages of UAV-based inspection include lower price, high efficiency, less risk, etc. However, the automatic inspection is still challenging.

To achieve the autonomous and intellectual identification of power transmission equipment abnormal, a fundamental work is to precisely detect the equipment from the surveillance videos or images. While, this research proposed an innovative approach to detect the large outdoor power transmission equipment, particularly the power towers and insulators captured in aerial images.

The remainder of the paper is organized as follows. Section 2 introduces related work in the field of large outdoor power transmission equipment detection. In Sect. 3, we will present our proposed approaches in detail. Section 4 illustrates the experimental processes and results. Lastly, the conclusion and future work are shown in Sect. 5.

2 Related Work

In this section, we first introduce existed methods for power tower detection and insulator detection respectively. Since our method is based on YOLO (i.e., You Only Look Once), descriptions of the YOLO family will be introduced lastly.

2.1 Power Tower Detection

Before the wide usage of learning-based methods, lots of traditional methods have been proposed for power tower detection. For example, a Graph-cut approach is used by Cheng et al. [6] to do the detection and segmentation of power towers from pictures, which firstly traces the coarse area of the power pole to acquire two constrained areas. Then, a graph-cut framework is utilized in the constrained areas to progress the precision of segmentation. The Canny edge detector and Hough transformation are used in another research to detect and track the transmission towers in video sequences [7]. The towers are detected based on extracted straight lines that are classified as parts of the towers.

Comparing to traditional methods, the detection accuracy improved with the application of the learning-based method. A supervised learning method is proposed by Sampedro et al. [8] for tower detection and classification, based on the HOG [9] features two multi-layer perceptrons (MLP) neural networks are trained as the classifier. Back-ground-foreground segmentation is achieved with the first classifier, whereas classification within four different types of electric towers is achieved with the second one.

Instead of the usage of man-made features described above, Convolutional Neural Networks (CNNs) could be used to find more efficient features to boost the performance. To locate transmission towers and insulators, the YOLOv3 [10] algorithm is applied in

[11]. However, the detection of these two kinds of equipment is done separately. Wang et al. [12], applied two deep learning techniques to differentiate the algorithm's performance for high-voltage power transmission tower detection, and found out that Faster Region-based CNN (R-CNN) [13] has a higher recognition accuracy, while YOLOv3 has a better performance in speed. By contrast, we fulfilled a real-time detection of power towers and insulators simultaneously with high accuracy.

2.2 Insulator and Other Power Equipment Detection

Many studies also have been done for insulator detection with purposes such as defect identification. For instance, a detecting approach based on edge detection of non-sub-sampled contourlet transform is proposed by Zhao et al. [14]. However, the method rests in threshold situations in various states, which confines practical application. Since the texture characteristics contain information richer than saturation element and edge characteristics, a method for detecting insulators through matching the texture features among the insulator and the background is offered by Li et al. [15]. But similar texture features can commonly be found in background and insulators corresponding to aerial images, which make the texture-based method unreliable.

Wu et al. [16] applied the Faster R-CNN algorithm to inspect insulators. Specifically, the Zeiler Fergus Network [17] is used to extract the features of the insulator and the K-mean clustering technique is used to find out the anchor point. In addition, a non-linear penalty element is adopted to boost the non-maximum suppression (NMS) algorithm. The Faster R-CNN model is also employed by Liao et al. [18] to identify the insulator fault. The ResNet-101 model [19] is implemented to extract the features from pictures, and soft-none-maximum suppression is used to increase the performance of the overlaid insulator detection.

Besides Faster R-CNN, a YOLOv3 algorithm is involved by Mohamed et al. [20] to recognize the faulted insulators from the normal ones. Similarly, YOLOv3-dense-network is used by Liu et al. [21] to detect and diagnose insulators with complicated backgrounds.

Detection of other power transmission equipment can be found as well. For example, a deep learning method was proposed by Rabab [22] to detect and segment power lines along with power towers. Chen et al. [23] proposed a method for distribution line pole enumeration, and YOLOv3 is adopted for both upright pole and fall pole detection. YOLO is involved in our method as well considering its nice performance.

2.3 YOLO Family

The YOLO family is famous for its speed and accuracy regarding object detection. The reasons include that YOLO is a single-stage model, which treats the object detection mission as a single regression problem and mitigates the computational complication. The fundamental notion is to exploit the whole images as the input to the system and concurrently predict the bounding boxes and class probabilities in the output layer.

YOLO family contains multiple versions. The first one was launched in 2015 by Redmon et al. [24]. Then other versions came up successively with improvements in terms of accuracy and speed. YOLOv2 [25] utilized Darknet-19 as a feature extractor and came with progress such as batch normalization, great resolution classifier, convolution with anchor boxes, dimension clustering, and fine-grained features. YOLOv3 [10] has raised with application development on the origin of YOLOv2, containing multi-label classification and multi-scale detection with enhanced feature extraction network (i.e., Darknet-53), which has a better sense of balance in terms of detection accuracy and speed performance than previous versions. The Darknet-53, incorporated the concepts of other CNNs [26], has 53 convolution layers and is generally formulated with multiple convolutional layers, and has been proved to be more effective than state-of-the-art architectures including ResNet [19].

In 2020, YOLOv4 [27] algorithm was released with some enhancement over YOLOv3, it improves YOLOv3's accuracy and frame rate by 10% and 12% respectively. Moreover, some important advances in the backbone and have achieved inconceivable outcomes for object detection and classification. The backbone is set up to CSPDarknet53 which comprises 29 convolutions 725×725 receptive fields with 27.6 million parameters. This work is based on the latest version of YOLO due to better accuracy and high speed and great performance has been achieved.

3 Proposed Approach

Here we will cover the whole process of our proposed approach. Firstly, we'll discuss dataset collection and annotation. Then the architecture and implementation of the proposed method will be given.

3.1 Data Preparation

We collected 6000 pictures of power transmission equipment from the Internet. 80% of the pictures are chosen as training data and the rest 20% are utilized for testing purposes. The details of the prepared dataset are shown in Table 1, and Fig. 1 indicates some samples of our dataset.

Table 1. Summary state of the dataset in the paper

Datasets	Category label/class	Sample number
Power transmission equipment	Insulator, Tower	6000

Labelling is an essential stage before training, and the annotation software we used is "LabelImg", which will create an XML file that includes more details about the object such as coordinates of the bounding box, height, and width of the images. And PASCAL VOC [28] protocol is adopted to prepare our object detection dataset.

To prevent the detrimental properties and overfitting during training, data augmentation methods, such as blurring and rotation, are employed in this work as well.

Fig. 1. Indication of some of the dataset samples.

3.2 Structure of Proposed Approach

A real-time object recognition algorithm is proposed in this research which utilizes a one-stage detector to analyze the images and straightens out the location coordinates and classification of objects, while there are also two-stage detectors, such as the R-CNN family.

Figure 2 [27] shows the differences between a one-stage and two-stage detector in network structure. Normally, the structure of the one-stage method takes four main elements into account, namely the input, backbone, neck, and head sequentially. Input is the place where to put the data, features extraction networks like VGG16, Darknet, and ResNet are recognized as the backbone. In our method, CSPDarknet53 is employed as the backbone for it was found to be an optimal model. The neck is used to construct and enhance a decent combination of extracted features, and the Path Aggregation Network (PANet) and Spatial Pyramid Pooling (SPP) are used in our method. Head is used to identify and classify the location of the bounding box, which is the same as used in YOLOv3 in the proposed method.

The method also employs several tricks like the bag of freebies (BoF) and the bag of specials (BoS). The BoF provides changes in the training strategy to improve accuracy. The modified training strategies are Mosaic data augmentation, DropBlock regularization, and Class label smoothing. While the BoS are plugin modules and post-processing methods that only add a fraction to the inference cost, however dramatically improve object detection. Modules are generally used to enhance certain characteristics of a model, such as enlarging receptive fields, adding attention mechanisms, or enhancing feature integration, etc. In order to enhance the receptive field, the following modules are typically used SPP, ASPP, PANet and RFB.

Post-processing is the process of screening model predictions such as NMS which can be used to filter out BBoxes that display poor accuracy, and only keep those with a high response.

In summary, the structural design of the proposed approach is the combination of CSPDarknet, SPP, PANet, and the head as used in YOLOv3.

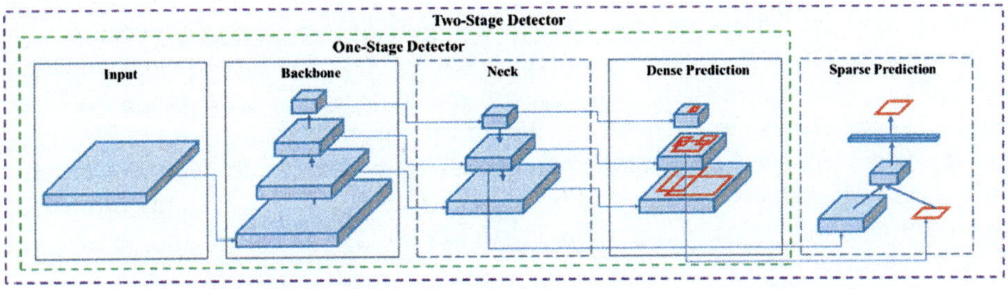

Fig. 2. The general structure of a one-stage and two-stage detector.

3.3 Procedures of Proposed Method

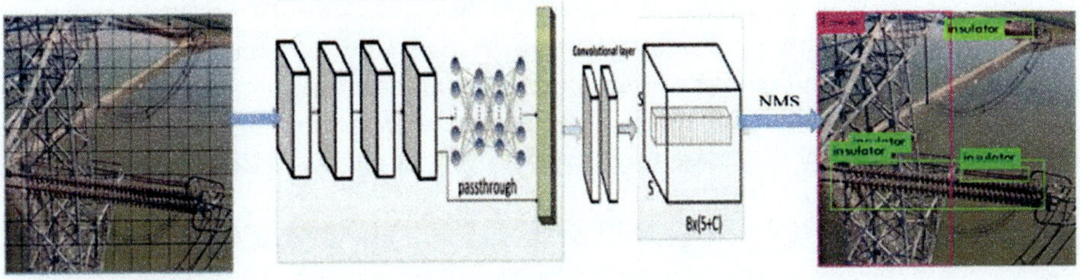

Fig. 3. Procedures of the proposed method.

As illustrated in Fig. 3, taking images as input, the proposed method outputs both the bounding box and class name of the targets. After been fed into the system, the image is firstly divided into multiple gird cells. Next, a CNN is applied to extract features for each image. After that, multiple bounding boxes are created during the prediction process. In order to select a suitable bounding box as a detected target, the NMS algorithm is used to ensure that the appropriate bounding box will be selected according to higher score probabilities and corresponding classes. Details of the procedures are described as follows.

Dataset Arrangement. After the collection of the experiment data, the power transmission equipment particularly the towers and insulators were labelled manually to prepare the dataset, which is then divided into the training set and testing set.

Feeding Images into the System. The inputs are passed through the network and altered to a fixed size of 416 × 416. Next, the proposed method divided the images into multiple M × M grids, if the centre of the bounding box (BB) of an object lying in a grid, then the grid is in charge to detect that object.

Feature Extraction Network. The proposed method using the CSPDarknet53 system to take out all relevant features that can be used for identifying power towers and insulators.

Predication of the Bounding Box. Each grids predicate M bounding boxes outputs in terms of $(i_x, i_y, i_w, i_h, C_i)$, where$i_x$, i_y point to the top-left coordinates of the bounding box. The i_w,i_h shows the bounding box's height and width, and C_i is the confidence score (see Eq. 1) and shows the existence of an object in the gird; if the object exists in the gird, the confidence score is equal to 1, otherwise 0. Regarding the detected object, every grid cell as well predicts C conditional class probabilities (C = 1, 2, …, n), ultimately the prediction of the proposed algorithm is M×M×B×(5+C).

$$C_i = P_r(\text{Object}) * IOU_{pred}^{truth} \tag{1}$$

where C_i indicates the confidence score and P_r (Object) indicates the probability of the object in the cell. IOU_{pred}^{truth} Indicates a calculation of overlap between predicted bounding boxes and their ground truths as described in Eq. 2.

$$\text{IOU} = {}^{B_p \cap B_{gt}}/_{B_p \cup B_{gt}} \tag{2}$$

Confidence score and IoU both are exploited as a measure to define if detection is a True Positive or a False Positive. Using the algorithm described above, every grid cell predicts class C conditional probability of the object, while generating the bounding box and the particular probability class for every grid cell, as shown in Eq. 3.

$$P_r(\text{Class}|\text{Object}) * P_r(\text{Object}) * IOU_{pred}^{truth} = P_r(\text{Object}) * IOU_{pred}^{truth} \tag{3}$$

In addition, to successfully predict correct bounding boxes, the model has to estimate the correct class for these boxes using the Loss Aggregation function, which is the best significant parameter for evaluating the network's performance.

Ultimate Detection Outcome. The proposed algorithm generates multiple bounding boxes to identify and detect the power transmission equipment especially towers and insulators. Preferably for every object in the image, we need only one single bounding box. To opt for the greatest bounding box among the numerous predicted bounding boxes, the proposed approach utilizes the NMS technique to overturn the unwanted Bounding box and preserve the superlative one to detect the target (i.e., the tower or insulator).

4 Experiment and Result

4.1 Environment Setting

As described above, we collect 6000 public images of power transmission equipment from the Internet, then the images are transformed to 1063 × 560 pixels (77 pixels per inch) resolution. The hardware and software we used are indicated in Tables 2 and 3 respectively.

Table 2. The hardware used for the experiment.

Item	Properties
CPU	Intel I Core I i5-10210U, 2.11 GHz
GPU	NVIDIA Tesla Tp100
RAM	8 GB

Table 3. The software used for the experiment.

Item	Properties
Operating System	Windows 10, 64-bit
DL Framework	Darknet
Programing Language	C/Python

4.2 Training and Evaluation of Experimental Results

As proof of the effectiveness, the proposed method has been tested with diverse images, from illuminated scenarios to diverse backgrounds. The results indicate that our method does a satisfactory job of detecting power towers and insulators as shown in Fig. 4. Additionally, detecting a correct sequence depends on how much data the training set contains. As more data become available in the training set, outstanding results will be obtained. Figure 4 clearly illustrates that quite decent results have been achieved through the model we used. In this experiment, the network parameters are set up as follows during the training process: image size 416 × 416 × 3, the momentum is 0.949, the learning rate is 0.001, the batch size is 64, saturation is 1.5, the number of the classes is two, which include tower and insulator.

4.3 Assessment Metrics

On our dataset, the proposed method was successfully trained for a total of 5000 iterations, the moment has come to determine the model's performance, whether it is reasonable. To evaluate the performance, precision (Eq. 4), recall (Eq. 5), and F1 score (Eq. 6)

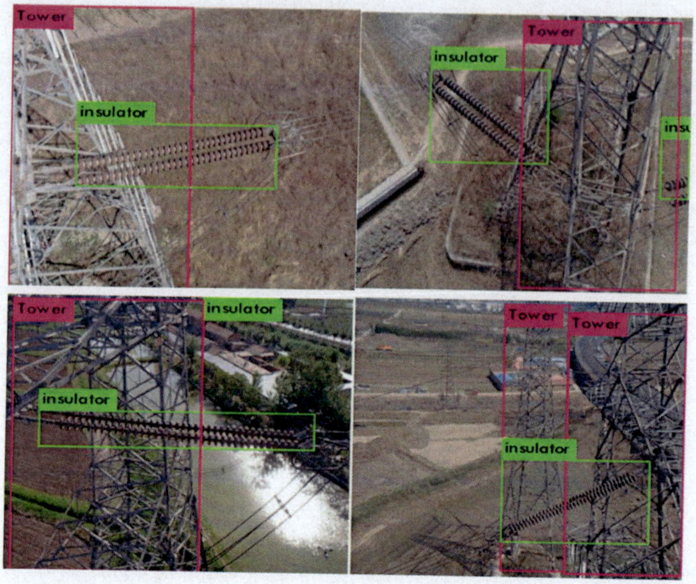

Fig. 4. Demonstration of four typic detection results using the proposed method.

are utilized as the assessment metrics.

$$Precision = \frac{TP}{TP + FP} \tag{4}$$

$$Recall = \frac{TP}{TP + FN} \tag{5}$$

$$F1\ score = 2 \times \frac{Precision \times Recall}{Precision + Recall} \tag{6}$$

where TP (i.e., True Positive) indicates the result of the trained model which predicted the positive class appropriately, FP (i.e., False Positive) points to that the trained model predicates the positive class incorrectly, FN (i.e., False Negative) displays the outcome of the trained model that incorrectly predicts the negative class. Another metric we used is the mAP. To compute the mAP for all classes, first, we should calculate the AP (Eq. 7) for an individual class, then the mAP could be computed according to Eq. 8.

$$AP = \int_0^1 p(r)dr \tag{7}$$

$$mAP = \frac{1}{F}\sum\nolimits_{k=1}^{k=F} AP_k \tag{8}$$

where AP_k shows the number of k classes and F indicates the number of classes. For the confusion matrix of our model, the prediction performance of our system will be assessed through four parameters which are shown in Table 4.

4.4 Comparisons and Analysis

The experimentation section analyses the corresponding precision, recall, mAP, and F1 score for power transmission equipment between the YOLOv3 based method and ours. As shown in Table 4, we also test our method with different image resolutions. The best results are highlighted in the whole table. Table 4 shows that our method with resolution 384 × 384 achieves the best performance. In comparison, Ours (576 × 576) has similar precision, recall, and F1-score, except the mAP is a little bit lower. In summary, our method outperforms the YOLOv3 based method in terms of accuracy.

Table 4. The result of the experiment of the proposed method.

Model	Precision	Recall	mAP	F1 score
YOLOv3 (416 × 416)	**90**	92	90.45	91
Ours (352 × 352)	88	94	95	91
Ours (384 × 384)	**90**	**95**	**96.59**	**92**
Ours (416 × 416)	88	**95**	95.61	91
Ours (512 × 512)	86	**95**	95.77	90
Ours (544 × 544)	88	**95**	96.28	91
Ours (576 × 576)	**90**	**95**	96.35	**92**
Ours (608 × 608)	87	**95**	96.06	91

5 Conclusions

In this paper, a novel deep learning method is proposed for detecting large outdoor power transmission equipment, such as power towers and insulators, based on improved YOLOv4. Initially, after the large outdoor power towers and insulators in numerous circumstances have been captured as images through UAVs, the data were collected, manually labelled, and augmented to produce a dataset. Next, a CNN model is trained using the prepared dataset for power towers and insulators detection. According to the experimental results, our method is more effective in locating and detecting large outdoor power equipment than the previous YOLO versions. In the future, we would try to cover more power equipment detection and diagnose their faults.

Acknowledgement. This work is supported by the National Natural Science Foundations of China (No. 61872085), the Scientific Research Project of Fujian Education Department (JK2017029, JAT190069), and the Scientific Research and Development Foundation of Fujian University of Technology (GY-Z18181, GY-Z20068, XF-X19017).

References

1. Tower Crane Interest Group: CPA Best Practice Guide Inspection, Thorough Examination CPA and Maintenance of Tower Cranes, vol. 9901, no. 1. Tower Crane Interest Group (TCIG) (1999)
2. Fahmani, L., Garfaf, J., Boukhdir, K., Benhadoul, S., Medromi, H.: Unmanned aerial vehicles inspection for overhead high voltage transmission lines. In: 1st International Conference on Innovative Research in Applied Science, Engineering and Technology (IRASET), Meknes, Morocco, pp. 1–7. IEEE (2020)
3. Disyadej, T., Promjan, J., Poochinapan, K., Mouktonglang, T., Grzybowski, S., Muneesawang, P.: High voltage power line maintenance inspection by using smart robotics. In:Innovative Smart Grid Technologies (ISGT), Washington, DC, USA, pp. 1–4. IEEE (2019)
4. Zhang, H., Yang, W., et al.: Detecting power lines in UAV images with convolutional features and structured constraints. https://www.mdpi.com/2072-4292/11/11/1342,last. Accessed 15 Jul 2021
5. Fang, S., Haiyang, C., et al.: A framework of power pylon detection for UAV-based power line inspection. In: IEEE Information Technology and Mechatronics Engineering Conference (ITOEC), Harbin, China, pp. 1–8. IEEE (2020)
6. Cheng, W., Song, Z.: Power pole detection based on graph cut. In: 1st International Congress on Image and Signal Processing (CISP), Sanya, China, vol. 3, pp. 720–724. IEEE (2008)
7. Tragulnuch, P., Chanvimaluang, T., Kasetkasem, T., Ingprasert, S., Isshiki, T.: High voltage transmission tower detection and tracking in aerial video sequence using object-based image classification. In: International Conference of (IC-ICTES), Khon Kaen, Thailand, pp. 1–4. IEEE (2018)
8. Sampedro, C., Martinez, C., Chauhan, A., Campoy, P.: A supervised approach to electric tower detection and classification for power line inspection. In: International Joint Conference on Neural Networks, Beijing, China, pp. 1970–1977 (2014)
9. Using Histogram of Oriented Gradients (HOG) for Object Detection. https://iq.opengenus.org/object-detection-with-histogram-of-oriented-gradients-hog/. Accessed 27 May 2021
10. Redmon, J., Farhadi, A.: YOLOv3: an incremental improvement, April 2018. http://arxiv.org/abs/1804.02767. Accessed 19 May 2021
11. Ohta, H., Sato, Y., Mori, T.: Image acquisition of power line transmission towers using UAV and deep learning technique for insulators localization and recognition. In: 23rd International Conference on System Theory, Control and Computing, Sinaia, Romania, pp. 125–130. IEEE (2019)
12. Wang, H., Yang, G., Li, E., Tian, Y., Zhao, M., Liang, Z.: High-voltage power transmission tower detection based on faster R-CNN and YOLO-V3. In: Chinese Control Conference, CCC, Guangzhou, China, July, vol. 2019, pp. 8750–8755. IEEE (2019)
13. Ren, S., He, K., Girshick, R., Sun, J.: Faster R-CNN: towards real-time object detection with region proposal networks. IEEE Trans. Pattern Anal. Mach. Intell **39**(6), 1137–1149 (2017)
14. Zhao, Z., Liu, N.: The recognition and localization of insulators adopting SURF and IFS based on correlation coefficient. Optik Int. J. Light Electron Opt. **125**(20), 6049–6052 (2014)
15. Li, W., Ye, G., Huang, F., Wang, S., Chang, W.: Recognition of insulator based on developed MPEG-7 texture feature. In: 3rd International Congress on Image and Signal Processing (CISP), Yantai, China, vol. 1, pp. 265–268. IEEE (2010)
16. Wu, J., Tang, S., Li, X.: An insulator inspection method based on deep learning applicable to multi-scale and occlusion conditions. J. Eng. **2021**(4), 216–225 (2021)
17. Detailed depth of ZF network architecture-Programmer Sought. https://www.programmersought.com/article/7240840903/. Accessed 28 May 2021

18. Liao, G.P., Yang, G.J., Tong, W.T., Gao, W., Lv, F.L., Gao, D.: Study on power line insulator defect detection via improved faster region-based convolutional neural network. In: 7th International Conference on Computer Science and Network Technology, Dalian, China, pp. 262–266. IEEE (2019)
19. He, K., Zhang, X., Ren, S., Sun, J.: Deep residual learning for image recognition. In: Computer Society Conference on Computer Vision and Pattern Recognition, Las Vegas, NV, USA, vol. 2016, pp. 770–778. IEEE (2016)
20. Adou, M.W., Xu, H., Chen, G.: Insulator faults detection based on deep learning. In: International Conference on Anti-counterfeiting, Security and Identification, Xiamen, China, vol. 2019, pp. 173–177. IEEE (2019)
21. Liu, C., Wu, Y., Liu, J., Sun, Z.: Improved YOLOV3 network for insulator detection in aerial images with diverse background interference. Electron **10**(7), 1–20 (2021)
22. Abdelfattah, R., Wang, X., Wang, S.: TTPLA: an aerial-image dataset for detection and segmentation of transmission towers and power lines. In: Ishikawa, H., Liu, C.-L., Pajdla, T., Shi, J. (eds.) ACCV 2020. LNCS, vol. 12627, pp. 601–618. Springer, Cham (2021). https://doi.org/10.1007/978-3-030-69544-6_36
23. Chen, B., Miao, X.: Distribution line pole detection and counting based on YOLO using UAV inspection line video. J. Electr. Eng. Technol. **15**(1), 441–448 (2020)
24. Redmon, J., Divvala, S., Girshick, R., Farhadi, A.: You only look once: unified, real-time object detection. In: Computer Society Conference on Computer Vision and Pattern Recognition, Las Vegas, NV, USA, vol. 2016, pp. 779–788. IEEE (2016)
25. Redmon, J., Farhadi, A.: YOLO9000: better, faster, stronger. In: 30th Conference on Computer Vision and Pattern Recognition, Honolulu, HI, USA, vol. 2017, pp. 6517–6525. IEEE (2017)
26. Szegedy, C., Liu, W., et al.: Going deeper with convolutions. In: Computer Society Conference on Computer Vision and Pattern Recognition, Boston, MA, USA, 07–12 Jun 2015, pp. 1–9. IEEE (2015)
27. Bochkovskiy, A., Wang, C.Y., Liao, H.Y.M.: YOLOv4: optimal speed and accuracy of object detection. http://arxiv.org/abs/2004.10934. Accessed 19 May 2021
28. Everingham, M., Eslami, S.M.A., Van Gool, L., Williams, C.K.I., Winn, J., Zisserman, A.: The Pascal visual object classes challenge: a retrospective. Int. J. Comput. Vis. **111**(1), 98–136 (2015). https://doi.org/10.1007/s11263-014-0733-5

Fusion Structure-based Fault State Prediction of Wind Turbine Spindle

Baohua Qiang[1(✉)], Wenqiang Yin[1], Xiaoling Tao[1], Longge Li[1], Baolian Li[2], Bo Peng[2], and YuFeng Wang[2]

[1] Guangxi Key Laboratory of Image and Graphic Intelligent Processing, (Guilin University of Electronic Technology), Guilin 541004, Guangxi, China
qiangbh@guet.edu.cn

[2] The 54Th Research Institute of China Electronics Technology Group Corporation, Shijiazhuang 050002, China

Abstract. Aiming at the problem that there are single characteristics and non-exquisite granularity for the existing key component state prediction methods of wind turbines, a wind turbine spindle fault state prediction model based on fusion structure is proposed. The model based on integration can more fully mine the information representation of wind turbine status. The Factorization Machine (FM) module is used to fine-tune the parameters of first-order and second-order features of the wind turbine. The Cross Network is used to perform a finite-order feature cross-product. The Deep Network module is combined to explore the advantages of high-order wind turbine feature information. The fusion structure model effectively fuses the features of the three modules. The influence of the high and low order features are fully considered to improve the fault state prediction effect of the model. In this paper, Area Under the Curve (AUC) and Log-likelihood Loss (Logloss) are used as evaluation indexes to verify the model's prediction performance in the real wind turbine data set of the SCADA system. The experimental results show that the spindle fault state prediction model based on the fusion structure has better performance than the traditional prediction model and the latest deep learning prediction model.

Keywords: Wind turbine spindle · Fusion structure · Neural network · Fault prediction · High-order features

1 Overview

At present, China is vigorously developing clean energy led by wind power. However, in areas rich in wind energy resources, the environment is often harsh, so it is inevitable that the wind turbine failure. The wind turbine spindle is one of the core components of the wind power transmission system. The failure of this component directly affects the power generation of the wind turbine. Therefore, the real-time prediction of the spindle failure status is an important part of the wind power operation system. The operation cost can be effectively saved, and the faults can be reduced by timely detecting the

S.-C. Chu et al. (Eds.): ICGEC 2021, LNEE 833, pp. 712–723, 2022.
https://doi.org/10.1007/978-981-16-8430-2_64

faults of wind turbine components. The prediction of spindle failure status is a typical classification problem. The literature on the prediction of wind turbine components is based on traditional machine learning algorithms or signal processing methods. The Logistic Regression (LR) model is the most widely used to deal with this kind of problem in the industry [1]. The advantage of the LR algorithm is simple and easy to realize the parallelization of large-scale data. However, it can only deal with the linear separable problem. It is not easy to mine the information of nonlinear high-order features, which restricts the prediction performance of the LR model. Juan et al. [2] apply the Poly2 model to classification prediction scenarios. The Poly2 model considers both the first-order feature and the second-order combined feature information. However, when the input features of the model are sparse feature vectors, the computational complexity of the second-order combined features will increase, which will affect the learning efficiency of the model. At present, using feature engineering to complete the manual feature combination and capture the information contained in high-order features is more prevalent in the industry. He et al. [3] uses Gradient Boosting Decision Tree (GBDT) [4] to perform feature screening combined with LR to improve the prediction effect of the LR model. Xue et al. [5] proposed a problem-specific Compact Evolutionary Tabu Search algorithm (CETS) to efficiently determine the sensor ontology alignment. Feature engineering is a critical link in the problem of wind turbine spindle fault prediction. In wind power, a single wind turbine belongs to multiple systems, and different components have different index systems. Feature engineering for all essential components of wind turbines is unrealistic. In the large-scale data space, these type of prediction methods is challenging to meet the actual requirements. Therefore, an efficient machine learning method suitable for large-scale data space must process data from SCADA systems.

In recent years, nonlinear classification models have gradually gained attention in the field of fault prediction. The Factorization Machine (FM) proposed by Rendle et al. learns the information of second-order feature crossover using the inner product of vectors, which reduces the time complexity of feature crossover calculation [6]. Due to the achievements of deep learning in computer vision, natural language processing, and other fields, deep learning is gradually applied in classification and prediction, such as the DNN model, which can explore implicit high-order features to express information. The Wide&Deep model combines deep learning with a linear structure model to achieve a good classification effect [7]. The DeepFM model is proposed by replacing Wide with a factorization machine (FM) based on Wide&Deep [8]. PNN and FNN models make full use of the advantages of the DNN model for mining high-order feature information. However, PNN and FNN models lack the influence of low-order features on the prediction results. After the experimental demonstration, low-order features are also critical to the performance of wind turbine fault prediction. Therefore, the neural network model applied in wind turbine fault prediction should also consider the role of low-order features. Xue et al. [9] proposed to use both MatchFmeasure, a rough evaluation metric on no reference alignment to approximate f-measure, and Unanimous Improvement Ratio (UIR), a measure that complements MatchFmeasure, in the process of optimizing the ontology alignments by Memetic Algorithm (MA). Wang et al. [10] extreme gradient boosting (XGBoost) algorithm is used for spindle fault prediction. However, the exploration of wind turbine features is single, and it is difficult to mine higher-order feature representation. Jiang et al. [11] applied the multi-scale convolutional neural network to the fault diagnosis of the wind turbine gearbox. This method limited the high-order features of the wind turbine. Luo et al. [12] proposed a new wind turbine gearbox fault prediction based on Pair-Copula structure and BP neural network. Wang [13] firstly proposes a variable selection algorithm based on principal component analysis (PCA) with multiple selection criteria. Su et al. [14] based on the vibration signal, the early fault diagnosis of rolling bearings is achieved by using EMD noise reduction and spectral kurtosis method. The references [15, 16] used this method to analyze the spindle vibration signal based on the kurtosis criterion. They used indicators such as divergence to explore the modal function, including the failure state, and then describe the failure degree of the spindle. The disadvantage of this method is that EMD decomposition is greatly affected by noise. An et al. [17] first decomposes the vibration signal to extract the characteristic components and finally realizes the diagnosis of the spindle failure state. However, its decomposition process easily results in the loss of the characteristic information. Zhao et al. [18] first use the wavelet denoising and take full account of the impact of environmental noise. The original vibration signal is decomposed, and the corresponding feature information is extracted to achieve the diagnosis of the spindle failure state. These methods are only suitable for experiments in limited sample space and are difficult to be applied in large-scale data space. Fault prediction models suitable for wind power need to meet the real-time and accuracy of prediction under large-scale data, thus providing correct guidance for wind turbine operation.

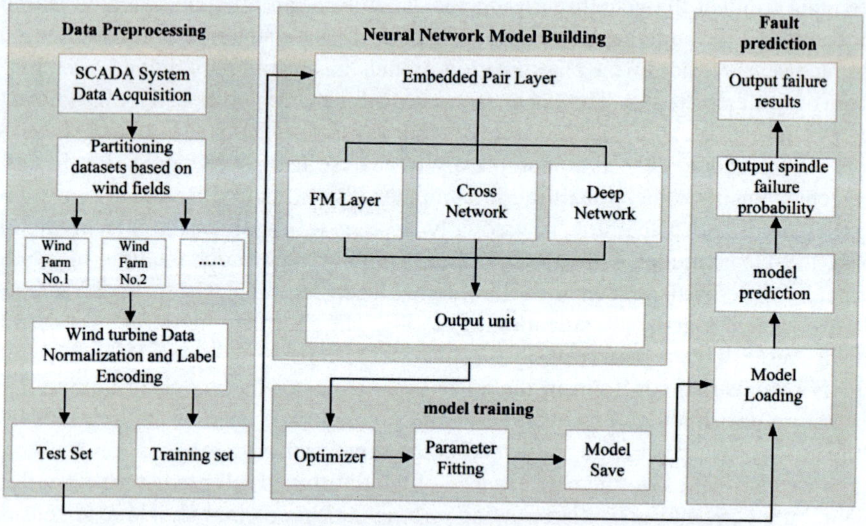

Fig. 1. Overall Flow of Wind Turbine Spindle Failure State Prediction

In order to meet the above requirements, this paper combines the method of Click-Through-Rate (CTR) prediction with the professional knowledge of classification prediction in this field, designs a spindle fault prediction model based on the fusion structure of the deep neural network, which is called DCFM.

2 Fault State Prediction Model for Spindle Based on Fusion Structure

Figure 1 shows the overall process of predicting the failure status of the spindle of a wind turbine using a DCFM fusion structure. Firstly, it is necessary to preprocess the SCADA wind turbine data. The preprocessing mainly aims at the normalization, embedding coding and data set segmentation of the existing wind turbine features. Then, the neural network model is built and the model training is carried out. Finally, the saved DCFM model is loaded to predict the spindle failure status. Several modules have been designed to provide a feasible solution for the rapid diagnosis of the state of wind turbine parts with a large data scale.

2.1 Design Ideas

This paper mainly considers how to construct and extract the wind turbine's high and low-order feature information when designing the DCFM model. On the one hand, the high-order and low-order characteristics of the wind turbine will affect the prediction results of the model, and different wind turbine target parts have different relationships between their characteristics, the high-order features and feature construction methods designed by the feature engineering will also be different. On the other hand, complex feature combinations for different target components in traditional feature engineering carry a huge labor cost, so wind power failure prediction models need to have the ability to automatically explore the high and low order features of wind turbines. To explore low-order features, traditional shallow models such as FM and LR are often introduced in the industrial field. The LR algorithm can make a linear combination of wind turbine characteristics to consider the influence of all first-order wind turbine characteristics fully. However, the LR algorithm only learns first-order feature weight coefficients and does not consider the relationship between wind turbine characteristics. In the actual wind turbine failure prediction scenario, there is a certain relationship between wind turbine characteristics, such as wind turbine power and wind speed, room temperature and wind turbine speed, etc. The DCFM model improves the model's universality by introducing FM to construct the second-order wind turbine characteristics and using more detailed learning of the second-order feature parameters to make the prediction results take into account the role of the second-order wind turbine characteristics. The FM is a shallow machine learning algorithm. Its advantage is that it can not only take into account the first-order wind turbine characteristics and the second-order cross-fan characteristics but also has a good parameter learning ability in the face of sparse data scenes.

Traditional machine learning algorithms, such as FM and LR, can only explore the wind turbine information at a shallow level. These models involve lower order of features, mainly the first and second order, and lack the ability to mine the deep wind turbine information. Compared with such shallow models, DNN has great potential to explore higher-order features. In recent years, it has become a trend to use DNN for classification prediction. The fusion of DNN can make the DCFM model extract the implicit high-order feature representation of the wind turbine. Cross Network module is introduced to process the information which is difficult to capture by implicit high-order features. This module can make DCFM model explore more comprehensive wind turbine status information through the high-order explicit cross of wind turbine characteristics, which can complement DNN and FM well. The information learned by Cross Network module is higher order than FM. Finally, the results of Deep Network, Cross Network and FM are fused and input into the standard sigmoid function to get the failure probability, to realize the fault prediction of the wind turbine spindle. At the same time, it also provides a feasible scheme for improving the data utilization of the SCADA system and realizing the big data diagnosis of wind turbine components.

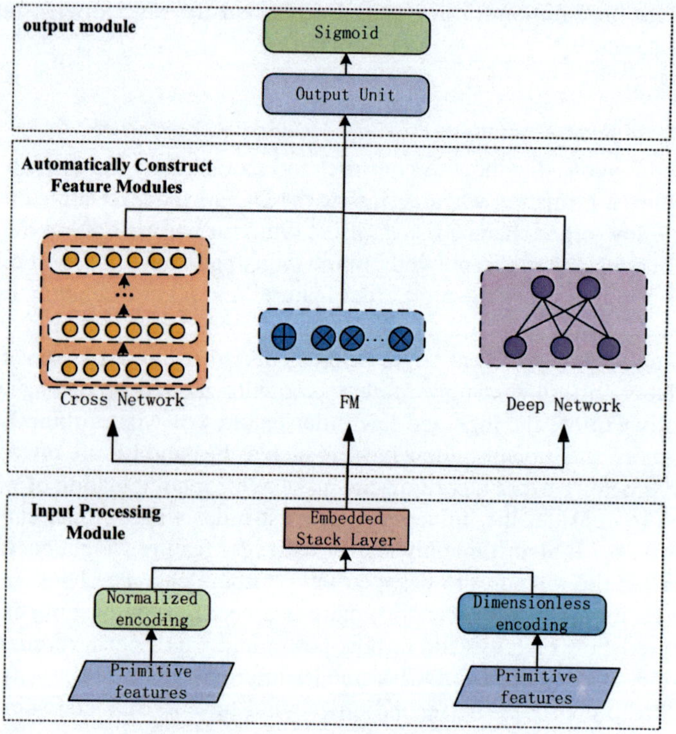

Fig. 2. Fusion Structure Neural Network Model

Figure 2 shows the structure of the DCFM neural network model, including the input processing module, wind turbine feature construction module, and spindle state output module. There are mainly two kinds of wind turbine original features in the input processing module: numerical and discrete. Except for the spindle status label, the rest are numerical features, and the type is float. The measured data collected by sensors are numerical characteristics and need to be normalized. The embedded stack layer consists of numerical features and discrete features, which are embedded and coded. The wind turbine feature construction module starts with the embedded stacking layer of the input processing module and connects three parallel structures to construct the wind turbine feature. The three structures are FM, Deep Network and Cross Network, which can extract the shallow and deep information of the wind turbine and accurately predict the fault state of the spindle. The wind turbine feature construction module starts with the embedded stacking layer of the input processing module and connects three parallel structures to construct the wind turbine feature. This module is the core of the DCFM network and the key to improving spindle fault state prediction performance. The output module connects the parallel Deep Network, Cross Network and FM layers. The connection vector is input into the probability prediction function to obtain the spindle fault state.

2.2 Factorization Machine Learns First-Order and Second-Order Features

The expression of FM is shown in Eq. (1). Based on the conventional linear regression expression, FM adds the weight of the feature cross term and the corresponding term to increase the learning ability of the algorithm for low-order cross terms.

$$y = w_0 + \sum_{i=1}^{n} w_i x_i + \sum_{i=1}^{n-1} \sum_{j=j+1}^{n} w_{ij} x_i x_j \tag{1}$$

In Eq. (1), it can be found that the output value y is affected by x_i and $x_i x_j$ at the same time. n is the number of features of the sample. x_i and $x_i x_j$ represent the first-order wind turbine state features and the intersection of two wind turbine features after normalization coding. After embedding FM, the predictive value y fully considers the role of the first-order wind turbine features and the second-order intersection features through the learning of x_i, $x_i x_j$ weight coefficients w_i and w_{ij} by the model.

The number of weight coefficients corresponding to the cross features is $n(n-1)/2$. In the sparse wind turbine data scene, it is rare to meet the condition that the cross features are not 0, which will make it difficult to learn the weight corresponding to the cross features. Therefore, by introducing the auxiliary vectors v_i and v_j, Eq. (1) can be transformed into:

$$y = w_0 + \sum_{i=1}^{n} w_i x_i + \sum_{i=1}^{n-1} \sum_{j=j+1}^{n} < v_i, v_j > x_i x_j \tag{2}$$

The weight coefficient of the second-order cross feature changes from w_{ij} to the dot product of vector v_i, v_j, which can effectively solve the problem that it is difficult to learn the weight coefficient in a sparse data scene. At the same time, the FM module can learn the second-order feature more finely, and the computational complexity is linear.

2.3 Deep Network Learns Higher-Order Features

Deep Network is an ordinary feedforward neural network in deep learning, which mainly consists of input, output and hidden layer. The hidden layer contains several hidden layers set by super parameters, mainly used to learn the interaction information between high-order nonlinear wind turbine features with the first-order wind turbine features as the model input. In the Deep Network module, all computing nodes in the embedded layer are fully connected with each node in the first hidden layer. The output expression of computing nodes in the first hidden layer adopts Eq. (3), where subscript 1 represents the first hidden layer, $h_1 \in R^{n_1}$ represents the output vector, n_1 represents the number of nodes in the first hidden layer, w_0 represents the weight parameter between the first hidden layer and the embedded layer, $w_0 \in R^{n_1 \times n_0}$, n_0 is the number of nodes in the embedded layer, $x_0 \in R^{n_1}$ represents the model output of the embedded layer in the form of vector, b_0 in Eq. (3) represents the bias term of the first hidden layer, $b_0 \in R^{n_1}$, the calculation node of the hidden layer uses ReLU as the activation function to activate the node.

$$h_1 = f(w_0 x_0 + b_0) \tag{3}$$

In Deep Network, the number of nodes and the number of layers in each hidden layer are set as super parameter indexes. The calculation nodes in the hidden layer are all connected. Equation (4) shows the output term of the $l + 1$ node in the hidden layer. w_1 represents the weight parameters of the nodes of l and $l + 1$, $w_1 \in R^{n_{l+1} \times n_l}$, n_l and n_{l+1} are the number of nodes in the l hidden layer and $l + 1$ hidden layer, $h_l \in R^{n_1}$ represents the output layer l hidden computer node, b_l represents the layer $l + 1$ offset, $b_1 \in R^{n_{l+1}}$, and all computing nodes in the hidden layer are activated by using ReLU as the activation function. In the last layer, the output of the hidden layer can be transformed into a conventional classification or regression problem.

$$h_{l+1} = f(w_l x_l + b_l) \tag{4}$$

In the DCFM network, Deep Network is used to extract implicit high-order wind turbine feature information and share embedded layer input with FM. After the wind turbine features are input into the Deep Network module, the wind turbine feature vector can deal with the nonlinear problems well through the nonlinear transformation of multi-layer calculation nodes. In theory, DeepFM can fit any continuous function, and the final output contains the potential high-order nonlinear linear information of the wind turbine to achieve deep feature extraction.

2.4 Cross Network Learns Cross Features

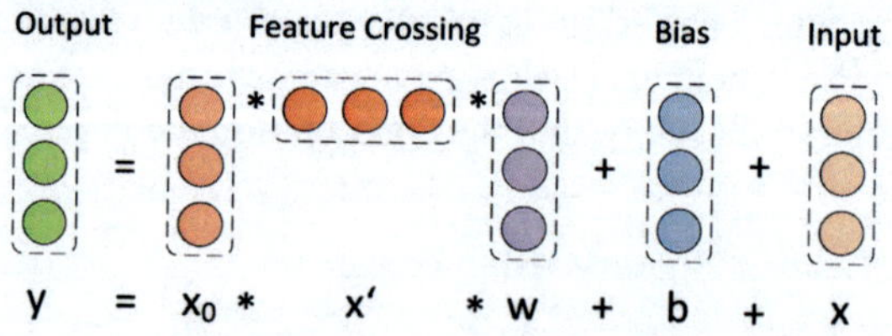

Fig. 3. Characteristic Cross Plot

Cross Network is the core of displaying high-order characteristic structure. The main idea of this module is to exhaust the cross features combination of wind turbines with finite order. Its mathematical expression is shown in Eq. (5):

$$X_{l+1} = X_0 X_l^T W_l + b_l + X_l = f(X_1, W_1, b_1) + X_1 \tag{5}$$

In Eq. (5), X_l represents the output of Cross Network layer l, and x_{l+1} represents the $l+1$ layer output of Cross Network. W_l is the weight parameter to be learned, and b_l is the offset term. In Fig. 3, the cross combination of all wind turbine features can be obtained by using x_0 and $x^{'}$ vectors to do the outer product operation. Combination characteristics of any bounded order can be obtained after cascade. In addition to the cross of different features, Cross Network also increases the cross of features and the order is higher, but the computational complexity is lower. The final output of this module mainly considers the influence of high-order cross features of wind turbines on the state results of wind turbines. The fusion of feature information extracted from Cross Network can make the prediction results more accurate.

2.5 Prediction Process

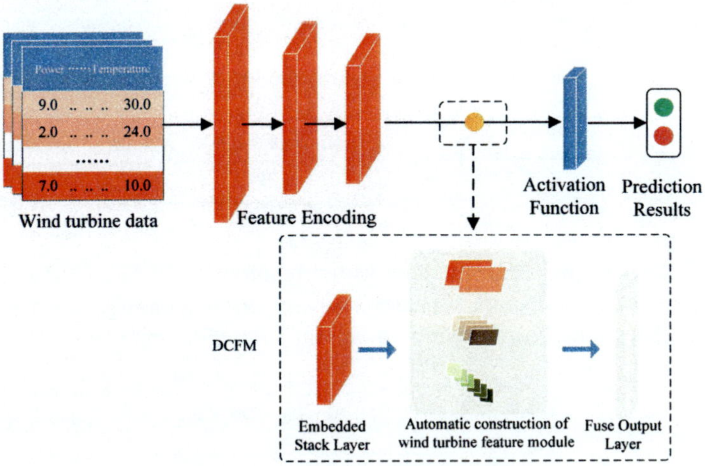

Fig. 4. DCFM Model Application

The overall application of the DCFM model is shown in Fig. 4. The model input is a batch of wind turbine state data samples. After input processing, automatic construction of wind turbine characteristics and output module, the prediction results of the wind turbine spindle fault state will be obtained finally.

The prediction process of the DCFM model is as follows:

Load the state data of wind turbine, define the super parameters of Deep Network, Cross Network and other modules.

The numerical values of wind turbine continuity are normalized, discrete values are embedded coded, and stitched into embedded stacks as model input.

Construct DCFM network structure to start connecting parallel FM, Deep Network and Cross Network with embedded stack layer for automatic construction of high and low order features of the wind turbine.

Fusion output of different types of wind turbine features constructed by three sub-modules.

Predict the probability of the failure state of the main shaft of the wind turbine through the Sigmoid function.

Define thresholds for failure state prediction results.

3 Experiments and Result Analysis

The data set used in this experiment is from the No. 16 wind turbine of a wind farm in Yunnan. In Table 1, there are 32,3788 wind turbine data in normal operation and 125,433 wind turbine data in failure operation. There is a certain balance between normal data and failure data in the data set, which hardly affects the learning of the model. In the statistics of spindle failure variable, the label of spindle failure accounted for 27.9%, the label of normal condition accounted for 72.1%, and the average failure rate of the spindle was about 27.9%. In this experiment, layered sampling and cross-slicing stratify the normal operation data to ensure that the training set and test set have the same proportion as the original dataset category samples.

Table 1. Raw Data Distribution Information

Wind Turbine Number	Normal Operation Data	Failure Run Data
16	323788	125466

In Table 2, the original data set contains ten features, including total shaft failure, power, gearbox inlet oil temperature, gearbox oil temperature, engine room temperature, wind speed impeller speed, yaw position hydraulic oil temperature, etc.

Table 2. Description of Characteristics

Feature Name	Feature Meaning
AI_IPR_RealPower	Power
AI_GBX_InletOilTemp	Inlet Oil Temperature of Gearbox
AI_GBX_OilSumpTemp	Gearbox Oil Temperature
AI_NAC_AirTemp	Cabin Temperature
AI_NAC_OutAirTemp	Extracabin Temperature
AI_NAC_AveWindSpeed10s	Wind Speed
AI_TBN_RotorSpeed	Impeller Speed
AI_NAC_Position	Yaw Position
AI_HYD_OilTemp	Hydraulic Oil Temperature
DI_TBN_ErrBearGlobal	Spindle Failure Status

The experimental environment is Python 3.7, the operating system environment is Ubuntu 16.04, memory 32 GB, Intel Core i7-10700F CPU 3.30 GHz, NVIDIA 1660s

6G, and TensorFlow builds DCFM network in collaboration with DeepCTR. The main details of the network implementation are as follows:

1. Model input: Experiments need to normalize the continuity features, embed the discrete features into coding operations, and stitch the two types of coded wind turbine features into an embedded stack layer for model input.

Optimize: Use Adam optimizer to randomly optimize DCFM model parameters using batch. The size is set to 1024, the epoch is 100, the cross num order of the Cross Network is set to 3, the learning rate is set to 0.0011, and the other parameters are the default parameters for each sub-module.

Evaluation Index: The experiment uses AUC and LogLoss as the main evaluation index, AUC is the area under the ROC curve to judge the performance of the main axis failure prediction model, LogLoss is the cross-entropy loss used to observe the convergence of the model.

Five different types of models, LR, FM, DeepFM, DCN and DNN, were trained in the experiment. The evaluation indexes of the five models were recorded in Fig. 5.

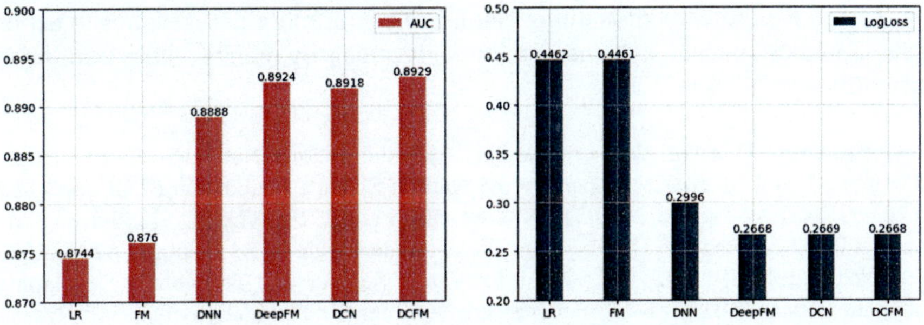

Fig. 5. Performance Evaluation of Multiple Models

In Fig. 5, it can be seen that the AUC of the LR algorithm is 0.8744. The algorithm only uses first-order features to classify but achieves good classification results. The second-order crossover feature is added to the FM model based on the first-order feature, which improves the AUC of the model by 0.0016. This shows that the second-order feature has a positive effect on the improvement of the classification effect.

By exploring the characteristics of implicit high-order wind turbines, DNN models ultimately outperform LR and FM models in performance, which also verifies the influence of high-order fan characteristics on fault prediction results. DeepFM and DCN models have good performance because they both consider the characteristics of high-order wind turbines. From the two indicators of the DCFM model, the model's performance is better than other single structure and shallow models, even over DeepFM and DCN networks. Cross Network can learn the explicit higher-order cross-feature parameters more fully. However, the learning of the lower-order feature parameters is coarser. In contrast, the FM module can learn the lower-order features well based on implicit vectors. The DNN and Cross Network modules can make the model extract the features more fully after combining the lower-order features of FM. The model's classification performance is further improved by fusing the output layer with the wind turbine status information extracted from FM, DNN and Cross Network, which also verifies the validity of the model in predicting the status of the wind turbine parts.

4 Concluding Remarks

In order to solve the problem that it is impossible to explore the data characteristics further when using the original wind turbine characteristics for fault prediction, this paper presents a fault prediction method for the wind turbine spindle based on a fusion structure, which embeds FM, DNN, and cross-network modules. The neural network can learn both the high-order representation, the finite-order intersection of wind turbine characteristics and the first- and second-order feature representation of FM to refine the granularity of data mining. DCFM simplifies the tedious process of feature engineering while guaranteeing the accuracy of classification prediction. At the same time, the FM and crossover network modules of this structure are of linear time complexity and can be used in the field of wind turbine fault prediction with complex working conditions. The next step is to use the reinforcement learning method for reference and apply it in wind power. Reinforcement learning can make use of more data features to achieve better fault prediction results. At the same time, it can easily realize online learning and simplify offline training releases.

Acknowledgments. This work was supported in part by the National Natural Science Foundation of China under Grant 61762025; in part by the Natural Science Foundation of Guangxi under Grants 2017GXNSFAA198226, 2019GXNSFDA185007, and 2019GXNSFDA185006; in part by the Guangxi Key Research and Development Project under Grants AB18126053, AB18126063, AD18281002, and AD19110137; in part by the Guilin Science and Technology Development Program under Grants 20190211–17; in part by the Development Foundation of the 54th Research Institute of China Electronics Technology Group Corporation under Grant SXX18138X017; in part by the Guangxi Key Science and Technology Planning Project under Grants AA18118031 and AA18242028;in part by the Innovation Project of GUET Graduate Education under Grants 2020YCXS052; in part by the Guangxi Colleges and Universities Key Laboratory of Intelligent Processing of Computer Image and Graphics under Grants GIIP201603 and GIIP1806; in part by the Guangxi Key Laboratory of Cryptography and Information Security under Grant GCIS201906.

References

1. Chapalle O, Manavoglu E, Rosales R.Simple and scalable response prediction for display advertising[J].Journal of ACM Transactions on Intelligent Systems and Technology, 2015, 5: 1–34.doi: https://doi.org/10.1145/2532128
2. Juan Yuchin, Zhuang Yong, Chin Wei-Sheng, Lin Chih-Jen. Field-aware factorization machines for CTR prediction//Proceedings of the Association for Computing Machinery Conference on Recommender. Boston, USA, 2016: 43–50. doi: https://doi.org/10.1145/2959100.2959134

3. He Xinran, Pan Junfeng, Jin Ou, et al. Practical lessons from predicting clicks on ads at Facebook//Proceedings of the ACM SIGKDD Conference on Knowledge Discovery and Data Mining. New York, USA, 2014: 1–9.doi: https://doi.org/10.1145/2648584.2648589
4. Friedman, J.H.: Greedy function approximation: a gradient boos-ting machine[J]. The Annals of Statistics **29**(5), 1189–1232 (2001). https://doi.org/10.1214/aos/1013203451
5. Xue, X., Chen, J.: Using Compact Evolutionary Tabu Search Algorithm for Matching Sensor Ontologies. Swarm and Evolutionary Computation **48**, 25–30 (2019). https://doi.org/10.1016/j.swevo.2019.03.007
6. Rendle S. Factorization machines[C]//Proceedings of 13th International Conference on Data Mining, 2010: 995–1000. doi: https://doi.org/10.1109/ICDM.2010.127
7. Cheng H-T, Koc L, Harmsen J, et al. Wide & deep learningfor recommender systems// Proceedings of the Workshop on Deep Learning for Recommender Systems. Boston, USA, 2016: 1–4.doi: https://doi.org/10.1145/2988450.2988454
8. Guo Huifeng, Tang Ruiming, Ye Yunming, et al. DeepFM:Afactorization-machine based neural network for ctr prediction//Proceedings of the International Joint Conference on Artificial Intelligence. Melbourne, Australia, 2017: 1–7.
9. Xue, X., Wang, Y.: Optimizing ontology alignments through a Memetic Algorithm using both MatchFmeasure and Unanimous Improvement Ratio. Artif. Intell. **223**, 65–81 (2015). https://doi.org/10.1016/j.artint.2015.03.001
10. Wang Guilan, Zgao Hongshan, Mi Zengqiang. Application of XGBoost algorithm in prediction of wind motor main bearing fault. Electric Power Automation Equipment[J], 2019, 39 (1): 73–83.doi: https://doi.org/10.16081/j.issn.1006-6047.2019.01.011
11. Jiang, Guoqian, Haibo, et al. Multiscale Convolutional Neural Networks for Fault Diagnosis of Wind Turbine Gearbox[J]. IEEE Transactions on Industrial Electronics, 2019.doi: https://doi.org/10.1109/TIE.2018.2844805
12. Luo, Z., Liu, C., Liu, S.: A Novel Fault Prediction Method of Wind Turbine Gearbox Based on Pair-Copula Construction and BP Neural Network[J]. IEEE Access **8**, 91924–91939 (2020). https://doi.org/10.1109/ACCESS.2020.2994077
13. Wang Y, Ma X, Qian P. Wind turbine fault detection and identification through PCA-based optimal variable selection[J]. IEEE Transactions on Sustainable Energy, 2018:1–1.doi: https://doi.org/10.1109/TSTE.2018.2801625
14. Wensheng, S., Fengtao, W., Zhixin, Z., et al.: Application of EMD denoising and spectral kurtosis in early fault diagnosis of rolling element bearings[J]. Journal of Vibration and Shock **29**(3), 18–21 (2010). https://doi.org/10.3969/j.issn.1000-3835.2010.03.005
15. Jiang Qing, Li Ting, et al. Study of Rolling Bearing SVM Pattern Recognition Based on Correlation Dimension of IMF[C]. 2012 Second International Conference on Intelligent System Design and Engineering Application, 1132–1135 doi:https://doi.org/10.1109/ISdea.2012.665
16. Cheng Junsheng, Y., Dejie, Y.Y.: A Fault Diagnosis Approach for Rolling Bearings Based on EMD Method and AR Model[J]. Mechanical systems and Signal Processing **20**(2), 350–362 (2006)
17. Xueli, A., Dongxiang, J., Jie, C., et al.: Bearing fault diagnosis based on ITD and LS-SVM for wind turbine[J]. Electric Power Automation Equipment **31**(9), 10–13 (2011). https://doi.org/10.3969/j.issn.1006-6047.2011.09.002
18. Hongshan, S.L., Shasha, L.: Method for extracting fault feature of wind turbine main bearing based on maximum signal to noise ratio[J]. Renewable Energy Resources **33**(3), 410–415 (2015). https://doi.org/10.13941/j.cnki.21-1469/tk.2015.03.013

TDOA Location Based on Pigeon-Inspired Optimization Algorithm in WSNs

Jeng-Shyang Pan[1], Ai-Qing Tian[1], Shu-Chuan Chu[1,2](✉), Tsu-Yang Wu[1], and Jin Wang[3]

[1] College of Computer Science and Engineering, Shandong University of Science and Technology, Qingdao 266590, China
jspan@cc.kuas.edu.tw

[2] College of Science and Engineering, Flinders University Sturt Road, Bedford Park, SA 5042, South Australia

[3] Changsha University of Science and Technology, Changsha 410004, China
jinwang@csust.edu.cn

Abstract. The pigeon-inspired optimization (PIO) algorithm belongs to a novel type of meta-heuristic algorithm, which has manifested in many aspects. Wireless sensor networks (WSNs) are a type of information acquisition technology and have been widely used as a new platform for information acquisition and processing. Due to the limited environment and hardware conditions for dealing with a large number of data sets to be processed, this paper uses the PIO algorithm to optimize the positioning method of time difference of arrival (TDOA) which a positioning method in the WSNs. Experimental results show that PIO can improve TDOA higher positioning accuracy and better convergence.

Keywords: Meta-heuristic algorithm · Pigeon-inspired optimization · Time difference of arrival · WSNs

1 Introduction

With the advent of the information age, the amount of data has changed qualitatively. When we are processing these large amounts of data, some original methods are not applicable. Those original methods are likely to cause problems such as low accuracy and long algorithm execution time. Some methods are the most direct way to solve the above problem.

The PIO algorithm is a meta-heuristic algorithm that has appeared in recent years proposed by Duan and Qiao in 2014 [1–3]. The PIO algorithm is to model the behavior of the pigeons returning home. The update and iterative process of the pigeons is an operator that considers two stages. In daily life, pigeons are considered to be a intelligent creatures, and the research on the behavior of individual pigeons has achieved fruitful results. For example, pigeons have the function of consciousness.

S.-C. Chu et al. (Eds.): ICGEC 2021, LNEE 833, pp. 724–733, 2022.
https://doi.org/10.1007/978-981-16-8430-2_65

WSNs are an important method of information transmission [4–7]. TDOA is a positioning technology in WSNs. It belongs to obtaining the distance between an unknown node with different known nodes. When TDOA locates WSNs, using the original algorithm for positioning will cause poor positioning accuracy and sometimes the algorithm cannot get acceptable convergence. PIO is used for estimation in location. The experiment results show that PIO algorithm are used in this paper and it have high accuracy and a small error in positioning of TDOA for WSNs positioning.

The rest of this paper is arranged as follows, Sect. 2 introduces the detailed information for the TDOA and PIO, additionally, the model of TDOA descriptions are given. Section 3 shows how to use the improved model to improve the accuracy of TDOA locations. Section 4 presents the final conclusions and the outlooks for future work based on the results of this paper.

2 Related Work

2.1 Time Difference of Arrive (TDOA)

The TDOA algorithm calculates the corresponding distance according to the time difference between the two signals with different speeds to the unknown node [8–12]. First, node A transmits ultrasonic waves and radio frequency signals together. Due to the different transmission speeds of these two signals, their time to reach node B is also different [13–17]. When node B receives these two signals, it records the arrival time of both, then takes the time difference, and calculates the distance connection between the mobile station and base station according to the transmission speed of the two signals. Suppose the radio frequency signal rate is represented by v_1, the ultrasonic rate is represented by v_2, the arrival times are T_1 and T_2, and the distance between nodes A and B is d, as shown in the Fig. 1.

$$d = \left| \frac{T_2 - T_1(v_1 * v_2))}{v_1 - v_2} \right|, \tag{1}$$

The TDOA algorithm has high positioning accuracy and small error [18–23]. However, it can be seen from the calculation process of the algorithm that TDOA has the same high time requirements. The propagation of wireless signals may also receive noise interference. Changes in the environment will also affect the accuracy of the algorithm. There are two types of devices for transmitting signals. When the anchor node is in the working state, the energy consumption of the node is also relatively large.

$$\begin{aligned} R_{i,1} &= cd_{i,1} = R^0_{i,1} + cn_{i,1} = R_i - R_1 + cn_{i,1} \\ i &= 2, 3, ..., M, \end{aligned} \tag{2}$$

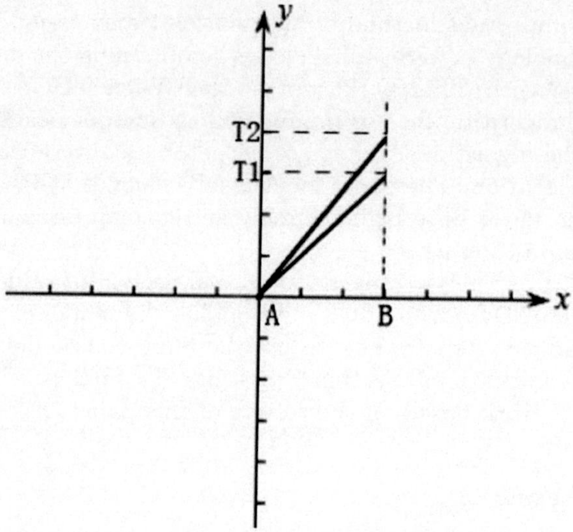

Fig. 1. TDOA positioning distance between A and B

where, c is transmission rate, $d_{i,1}$ is TDOA measure distance values, $cn_{i,1}$ is noise added when TDOA is measured, for convenience, the noise is considered to be white Gaussian noise satisfying the variance of α^2 of the independent and identical distribution.

Suppose there are M anchor nodes randomly placed in a two-dimensional plane. The estimated coordinates of the unknown nodes after successful positioning are represented by (x, y), this right coordinates of the i-th anchor node are represented by (X_i, Y_i), R_i represents the anchor. The distance between unknown position i and this anchor node is calculated using real azimuth coordinates. $R_{i,1}^0$ represents the actual distance difference between any labeled anchor node $i(i \neq 1)$ which is calculated using azimuth coordinates, $R_{i,1}$ is a measure of distance and c is the radio wave transmission speed, as shown in the Fig. 2.

$$\begin{aligned} R_i &= \sqrt{(X_i - x)^2 + (Y_i - y)^2)} \\ R_i^2 &= (X_i - x)^2 + (Y_i - y)^2 \\ &= k_i - 2x_i x - 2y_i y + x^2 + y^2, \end{aligned} \tag{3}$$

$$\begin{aligned} R_{i,1} &= c\tau_{i,1} \\ &= R_i - R_1 \\ &= \sqrt{(X_i - x)^2 + (Y_i - y)^2)} - \sqrt{(X_1 - x)^2 + (Y_1 - y)^2)}, \end{aligned} \tag{4}$$

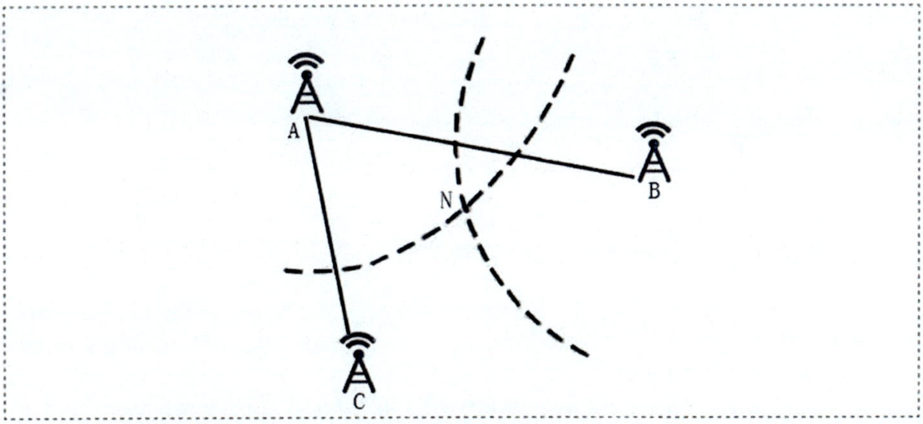

Fig. 2. TDOA positioning principle

In the Eq. (3) and Eq. (4), $K_i = x_i^2 + y_i^2$. And $R_{i,1}$ represents a distance difference, $\tau_{i,1}$ is value of TDOA measurement. From Eq. (3) and Eq.(4), we can get the following equation:

$$R_i^2 = (R_{i,1} + R_1)^2, \tag{5}$$

Taking Eq. (5) back to Eq. (3) and it can gives below equation.

$$\begin{aligned} R_{i,1}^2 &= 2R_{i,1}R_1 + R_1^2 \\ &= K_i - 2x_i x - 2y_i y + x^2 + y^2, \end{aligned} \tag{6}$$

Next, combining Eq. (5) and Eq. (6), we can gives the following formula:

$$R_{i,1}^2 + 2R_{i,1}R_1 = K_i - 2x_{i,1}x - 2y_{i,1}y - K_1, \tag{7}$$

In Eq. (7), $x_{i,1} = x_i - x_1$, $y_{i,1} = y_i - y_1$, x, y, K_1 are unknown variables, in the next formula, the noise effect is added.

$$\begin{aligned} R_{i,1} &= \sqrt{(X_i - x)^2 + (Y_i - y)^2)} - \sqrt{(X_1 - x)^2 + (Y_1 - y)^2)} + cn_{i,1} \\ i &= 2, 3, ..., M, \end{aligned} \tag{8}$$

$$\begin{aligned} \Delta R &= [R_{2,1}, R_{3,1}, ..., R_{m,1}]^T_{(M-1)*1} \\ R_{i,1} &= [R_2, R_3, ..., R_M]^T_{(M-1)*1} \\ R_1 &= [R_1, R_1, ..., R_1]^T_{(M-1)*1} \\ N &= [n_{2,1}, n_{3,1}, ..., n_{M,1}]^T_{(M-1)*1}, \end{aligned} \tag{9}$$

Combining Eq. (8) and Eq. (9) gives the following formula:

$$\Delta R = R_i - R_1 + cN$$
$$= \begin{bmatrix} \sqrt{(X_2-x)^2+(Y_2-y)^2} - \sqrt{(X_1-x)^2+(Y_1-y)^2} \\ \sqrt{(X_3-x)^2+(Y_3-y)^2} - \sqrt{(X_1-x)^2+(Y_1-y)^2} \\ \dots \\ \sqrt{(X_M-x)^2+(Y_M-y)^2} - \sqrt{(X_1-x)^2+(Y_1-y)^2} \end{bmatrix} + cN \quad (10)$$

In this article, set $M > 3$, the use of maximum likelihood estimation method to estimate the unknown node coordinate is the maximum, because $R_{i,1}$ obeys the Gaussian distribution with $R_{i,1}$ mean of $(R_i - R_1)$ and α^2 variance, and each separate value is measured, and therefore the likelihood function is obtained:

$$\prod_{i=1}^{M-1} [\frac{exp^{\frac{(\Delta R - R_i + R_1)^2}{-2\sigma^2}}}{\sqrt{2\pi\sigma}}] == [[\frac{1}{\sqrt{2\pi\sigma}}]^{M-1} exp[\frac{(\Delta R - R_i + R_1)^2}{-2\sigma^2}]], \quad (11)$$

Solving the likelihood function is maximum frame of reference value, which is equivalent to solving
$(x, y) = arg[max[exp[\frac{(\Delta R - R_i + R_1)^2}{-2\sigma^2}]]]$.

$$C = R_i - R_1 = \begin{vmatrix} a_1 \\ a_2 \\ \dots \\ a_{M-1}, \end{vmatrix} \quad (12)$$

In the process of constructing the hyperbolic model, since the difference between the distance from the point on the hyperbola to the two focal points is a fixed value, this fixed value is set to Eq. (12) in this article.

$$(x, y) = \underset{(x,y)\in S}{arg} [min(\Delta R - C)^T (\Delta R - C)], \quad (13)$$

Solving the minimum value of the nonlinear function of Eq. (13) can get the estimated value of the unknown node, S is expressed as a range space of cN floating up and down the points on the hyperbola. It is very difficult to use the analytical method. In this paper, the method of PIO coordinate is used to solve this nonlinear function, and the estimated value of unknown nodes is determined.

2.2 Pigeon-Inspired Optimization

In the target search space, the PIO algorithm first generates a population of N particles representing potential problem solutions, and then the particles in the population follow the best particles in each generation to move continuously in the target space. Individuals in the population gradually seek out the optimal solution on the target space by continuously iterating, in a population the i-th pigeon is represented as a vector in the D-dimensional inter-space. Then in this paper uses Eq. (14) and Eq. (15) on the behalf of position and velocity of the i-th pigeon in the population, and $gBest$ represents the best particle in the population that has been searched so far [24–27].

$$Pos_i = [pos_{i,1}, pos_{i,2}, ..., pos_{i,D}], \tag{14}$$

$$Vel_i = [vel_{i,1}, vel_{i,2}, ..., vel_{i,D}], \tag{15}$$

During the iterative process of the PIO, it consisted of two iterative stages, including the map and compass operator, and landmark operator. The first operators are the way finding methods used by pigeon groups far away from home. The pigeon uses magnetic objects to sense the magnetic field on the earth, and then it can form a map in the pigeon's brain, and then use the sun's height as a compass to adjust the pigeon's flight direction. As the pigeon approached its destination, it became less dependent on magnetic objects and the sun, and instead used landmark operators for navigation. The second stage of the pigeon is to use the landmark operator for navigation. When the pigeon is near the destination, the dependence on the landmark near the destination becomes greater. If the pigeon knows the landmark of the destination, it will fly immediately to the destination, or else just follow close behind the pigeons in the population who know their destination [28–31].

In the PIO, the iterative criteria for the flock in the first phase of the first operator of PIO are shown in Eq. (16) and Eq. (17). The iterative criteria for the pigeons in the second phase of PIO is shown in Eq. (18)–Eq. (21).

Map and compass operator: where R is the first phase factors, and ϕ_1, ϕ_2, ϕ_3 is a stochastic number among 0 and 1. t is the iteration number of the PIO.

$$vel_i^{t+1} = e^{-R \times t} \times vel_i^t + \phi_1 \times (pos_{gBest} - pos_i), \tag{16}$$

$$pos_i^{t+1} = \phi_2 \times pos_i^t + \phi_3 \times vel_i^{t+1}, \tag{17}$$

Landmark operator: pos_{center}^{t-1} is the landmark operator. The pigeons with the ability to know the way in the entire population. Other pigeons without the ability to learn the way follow the pigeon with the ability to learn the way. ϕ_4 is a stochastic number among 0 and 1.

$$No^{t-1} = \frac{No^{t-2}}{2}, \tag{18}$$

$$pos_{center}^{t-1} = \frac{\sum_{i=1}^{No^{t-1}} pos_i^{t-1} \times F(pos_i^{t-1})}{No^{t-1} \times \sum_{i=1}^{No^{t-1}} F(pos_i^{t-1})}, \tag{19}$$

$$pos_i^t = pos_i^{t-1} + \phi_4 \times (pos_{center}^{t-1} - pos_i^{t-1}), \tag{20}$$

$$F(pos_i^{t+1}) = \begin{cases} \frac{1}{fitness(pos_i^{t+1})+\theta} \\ fitness(pos_i^{t+1}), \end{cases} \quad (21)$$

3 Experiments for TDOA in WSNs Problem

The TDOA positioning algorithm determines the location of a mobile station (MS) by measuring the propagation time difference of radio waves traveling from an MS to multiple base stations (BS). The system of hyperbolic equations composed of multiple TDOA measurements is nonlinear and difficult to solve. Many meta-heuristic algorithms have been proposed to solve the above-mentioned nonlinear and difficult problem. The PIO algorithm determines the coordinate location of the MS based on the sector in which the MS is located, and then uses the inverse of the likelihood function as the individual fitness value to select the good individual, determines the crossover rate and the mutation rate, and uses the binary number to perform the chromosome analysis. The code searches the mobile station located within the coordinate range determined by the service desk. Initialize the positioning of the mobile station, compare the final positioning result with TDOA [32–35].

The following problems may occur during the positioning process:

1. Signal impairment: the distance measurement information for indoor positioning is the distance measured under the condition of line-of-sight. If it is not the line-of-sight, such as an obstacle in the middle or it reaches by reflection, it will lead to a longer reception time. The measured distance becomes larger.
2. Base station coordinates error: the measured coordinates are relative to the anchor coordinates of the BS. If the measured coordinates values are wrong, then our positioning data is meaningless.
3. Clock synchronization error: each base station's clock will have a slight gap, but if the gap is 1 ns, there will be a 30 cm error, so if we can synchronize the time of all base stations in the system, we can further improve the positioning accuracy.

Operating environment windows10 i5-8500 3.00 GHz experiment is running on the CPU and 24 GB of memory. The parameters of the PIO algorithm are set to R as 0.2, and the maximum number of evaluations is 10000. The coordinates of the Mobile station are set to (0,0), and the coordinates of the ten Base Stations are set to (50,10) (10,100) (−70,0) (4,−15) (100,−15) (38,54) (21,−32) (15,97) (12,32) (−12,−21).

It can be seen from the Fig. 3, with the continuous increase of noise, there will be an obvious jumping point in TDOA using the least-squares method to solve the problem at 0.4–0.6. When using the PIO and PSO algorithm to solve, although the mean variance will increase slowly, it is very satisfactory compared to using the least-squares method. The PIO performs better than PSO algorithm.

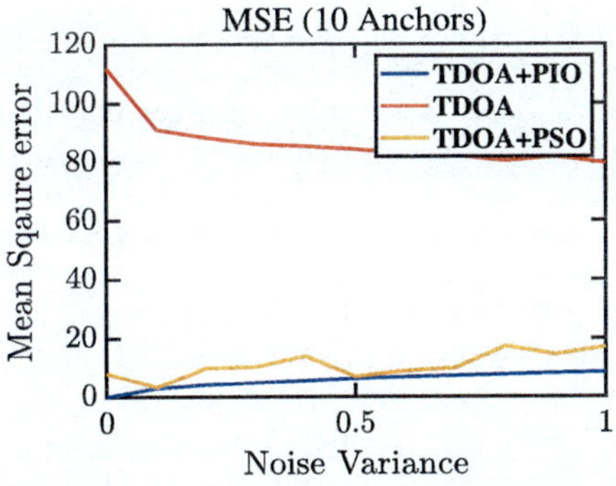

Fig. 3. TDOA positioning principle

4 Conclusion

This paper uses the PIO algorithm to solve the positioning problem of TDOA in WSNs. TDOA is a highly non-linear and realistic problem. The passive positioning method PIO algorithm based on TODA can effectively avoid the stability and accuracy of the positioning results. The realization result proves that the algorithm proposed in this paper can get closer to the optimal solution faster than the traditional solution, which is suitable for solving the positioning problem of wireless sensor networks. The final result proves that it will also perform well in comparison. The proposed method may further improve the efficiency by adopting the fast search algorithms [36–38].

References

1. Tian, A.Q., Chu, S.C., Pan, J.S., Cui, H., Zheng, W.M.: A compact pigeon-inspired optimization for maximum short-term generation mode in cascade hydroelectric power station. Sustainability **12**(3), 767 (2020)
2. Tian, A.Q., Chu, S.C., Pan, J.S., Liang, Y.: A novel pigeon-inspired optimization based MPPT technique for PV systems. Processes **8**(3), 356 (2020)
3. Pan, J.S., Tian, A.Q., Chu, S.C., Li, J.B.: Improved binary pigeon-inspired optimization and its application for feature selection. Appl. Intell. 1–19 (2021). https://doi.org/10.1007/s10489-021-02302-9
4. Chu, S.C., Du, Z.G., Pan, J.S.: Discrete fish migration optimization for traveling salesman problem. Data Sci. Patt. Recogn. **4**(2), 1–18 (2020)
5. Nguyen, T.T., Lin, W.W., Vo, Q.S., Shieh, C.S.: Delay aware routing based on queuing theory for wireless sensor networks. Data Sci. Patt. Recogn. **5**(2), 1–10 (2021)
6. Kuo, C.M., Hsieh, C.H., Tseng, S.C., Wu, J.Y.: Stacked deep convolution network for image super-resolution. Data Sci. Patt. Recogn. **5**(2), 11–27 (2021)
7. Song, W., Huang, C.: Mining high average-utility itemsets based on particle swarm optimization. Data Sci. Patt. Recogn. **4**(2), 19–32 (2020)

8. Meana-Llorián, D., García, C.G., García-Díaz, V., G-Bustelo, B.C.P., Garcia-Fernandez, N.: SenseQ: replying questions of social networks users by using a wireless sensor network based on sensor relationships. Data Sci. Patt. Recogn. **1**(1), 1–12 (2017)
9. Nguyen, T.T., Pan, J.S., Lin, J.C.W., Dao, T.K., Nguyen, T.X.H.: An optimal node coverage in wireless sensor network based on whale optimization algorithm. Data Sci. Patt. Recogn. **2**(2), 11–21 (2018)
10. Wu, J., Xu, M., Liu, F.F., Huang, M., Ma, L.H., Lu, Z.M.: Solar wireless sensor network routing algorithm based on multi-objective particle swarm optimization. J. Inf. Hiding Multimedia Signal Process. **12**(1), 1–11 (2021)
11. Chai, Q.W., Chu, S.C., Pan, J.S., Zheng, W.M.: Applying adaptive and self assessment fish migration optimization on localization of wireless sensor network on 3-D Terrain. J. Inf. Hiding Multimedia Signal Process. **11**(2), 90–101 (2020)
12. Wu, T.Y., Meng, C., Wang, K.H., Chen, C.M., Pan, J.S.: Comments on recent proposed Cui et al'.s Kase and Lu et al'.s Dibeks schemes. J. Inf. Hiding Multimedia Signal Process. **9**(1), 162–169 (2018)
13. Pan, J.S., Kong, L., Sung, T.W., Chu, S.C., Roddick, J.F.: Hierarchical routing strategy for wireless sensor network. J. Inf. Hiding Multimedia Signal Process. **9**(1), 256–264 (2018)
14. Ma, H., Xu, X., Wang, Y., Qin, D., Zhuang, W.: Research on relative coordinate localization of nodes based on topology control. J. Inf. Hiding Multimedia Signal Process. **9**(2), 432–441 (2018)
15. Liao, Y., Liu, B.: Exploration on DOA algorithms in LTE-A HetNet interference suppression. J. Inf. Hiding Multimedia Signal Process. **9**(3), 539–547 (2018)
16. Li, J., Hou, X., Wang, Y., Huo, J.: Quantum genetic uneven clustering algorithm for wireless sensor networks. J. Inf. Hiding Multimedia Signal Processi. **9**(3), 596–605 (2018)
17. Chen, H.: Parameter optimization design for automatic cotton blending based on improved adaptive genetic algorithm. J. Inf. Hiding Multimedia Signal Process. **9**(3), 723–732 (2018)
18. Wang, W.G., Zhou, R., Lv, L.L.: Parameter optimization design for automatic cotton blending based on improved adaptive genetic algorithm. J. Inf. Hiding Multimedia Signal Process. **9**(3), 743–750 (2018)
19. Huang, X., Chen, W.: An improved calculation method of high-frequency winding losses for gapped inductors. J. Inf. Hiding Multimedia Signal Process. **9**(3), 751–759 (2018)
20. Wu, C.M., Yang, X., Gong, H.Q.: Node tacking location algorithm for wireless sensor networks based on improved particle filter. J. Inf. Hiding Multimedia Signal Process. **9**(4), 785–795 (2018)
21. Chen, J.N., Zhou, Y.P., Huang, Z.J., Wu, T.Y., Zou, F.M., Tso, R.: An efficient aggregate signature scheme for healthcare wireless sensor networks. J. Network Intell. **6**(1), 1–15 (2021)
22. Xu, X.W., Pan, T.S., Song, P.C., Hu, C.C., Chu, S.C.: Multi-cluster based equilibrium optimizer algorithm with compact approach for power system network. J. Network Intell. **6**(1), 117–142 (2021)
23. Zhuang, J., Luo, H., Pan, T.S., Pan, J.S.: Improved flower pollination algorithm for the capacitated vehicle routing problem. J. Network Intell. **5**(3), 141–156 (2020)
24. Kou, X., Feng, J.: Matching ontologies through compact monarch butterfly algorithm. J. Netw. Intell. **5**(4), 191–197 (2020)

25. Nguyen, T.T., Pan, J.S., Wu, T.Y., Dao, T.K., Nguyen, T.D.: Node coverage optimization strategy based on ions motion optimization. J. Netw. Intell. **4**(1), 1–9 (2019)
26. Jiang, B.Q., Pan, J.S.: A parallel quasi-affine transformation evolution algorithm for global optimization. J. Netw. Intell. **4**(2), 30–46 (2019)
27. Zhang, S.M., Su, X., Jiang, X.H., Chen, M.L., Wu, T.Y.: A traffic prediction method of bicycle-sharing based on long and short term memory network. J. Network Intell. **4**(2), 17–29 (2019)
28. Huang, Y.F., Hsu, C.H.: Energy efficiency of dynamically distributed clustering routing for naturally scattering wireless sensor networks. J. Netw. Intell. **3**(1), 50–57 (2018)
29. Horng, G.J., Chang, T.Y., Wu, H.T.: The adaptive node-selection mechanism scheme in solar-powered wireless sensor networks. J. Netw. Intell. **3**(1), 58–73 (2018)
30. Chen, Y.J., Horng, G.J., Cheng, S.T.: A distributed cross-layer compromise detection mechanism for wireless sensor networks. J. Netw. Intell. **2**(1), 147–161 (2017)
31. Pan, J.S., Hu, P., Chu, S.C.: Binary fish migration optimization for solving unit commitment. Energy **226**, 120329 (2021)
32. Song, P.C., Chu, S.C., Pan, J.S., Yang, H.: Simplified phasmatodea population evolution algorithm for optimization. Complex Intell. Syst. 1–19 (2021). https://doi.org/10.1007/s40747-021-00402-0
33. Pan, J.S., Liu, N., Chu, S.C., Lai, T.: An efficient surrogate-assisted hybrid optimization algorithm for expensive optimization problems. Inf. Sci. **561**, 304–325 (2021)
34. Fan, F., Chu, S.C., Pan, J.S., Lin, C., Zhao, H.: An optimized machine learning technology scheme and its application in fault detection in wireless sensor networks. J. Appl. Stat. 1–18 (2021)
35. Li, J., Gao, M., Pan, J.S., Chu, S.C.: A parallel compact cat swarm optimization and its application in DV-Hop node localization for wireless sensor network. Wireless Netw. **27**(3), 2081–2101 (2021)
36. Chu, S.C., Roddick, J.F., Pan, J.S.: An efficient k-medoids-based algorithm using previous medoid index, triangular inequality elimination criteria, and partial distance search. In: International Conference on Data Warehousing and Knowledge Discovery, pp. 63–72. Springer (2002). https://doi.org/10.1007/3-540-46145-0_7
37. Pan, J.S., McInnes, F.R., Jack, M.A.: Fast clustering algorithms for vector quantization. Pattern Recogn. **29**(3), 511–518 (1996)
38. Wang, K., Liu, Y., Liu, D., Shi, Q.: Improved particle swarm optimization algorithm based on gaussian-grid search method. J. Inf. Hiding Multim. Signal Process. **9**(4), 1031–1037 (2018)

Author Index

S.-C. Chu et al. (Eds.): ICGEC 2021, LNEE 833, pp. 735–737, 2022.
https://doi.org/10.1007/978-981-16-8430-2